Theoretical Models of Chemical Bonding
Part 3

Molecular Spectroscopy, Electronic Structure and Intramolecular Interactions

Editor: Z. B. Maksić

With contributions by
G. Alagona, F. Bernardi, J. E. Boggs, R. Bonaccorsi,
R. Cammi, E. R. Davidson, M. Eckert-Maksić,
D. Feller, H. P. Figeys, P. Geerlings, C. Ghio,
Y. Harada, E. Heilbronner, E. Honegger,
C. J. Jameson, K. Jug, L. Klasinc, M. Klessinger,
J. Kowalewski, A. Laaksonen, K. T. Leung,
Z. B. Maksić, S. P. McGlynn, K. Ohno, M. Olivucci,
T. Pötter, M. A. Robb, J. Tomasi, K. Wittel

With 172 Figures and 126 Tables

Springer-Verlag Berlin
Heidelberg GmbH

Professor Dr. Zvonimir B. Maksić

Theoretical Chemistry Group
The "Rudjer Bošković" Institute
41001 Zagreb, Bijenička 54, Croatia/Yugoslavia
and
Faculty of Natural Sciences and Mathematics
University of Zagreb
41000 Zagreb, Marulićev trg 19, Croatia/Yugoslavia

ISBN 978-3-642-63493-2   ISBN 978-3-642-58179-3 (eBook)
DOI 10.1007/978-3-642-58179-3

© Springer-Verlag Berlin Heidelberg 1991
Originally published by Springer-Verlag Berlin Heidelberg New York in 1991
Softcover reprint of the hardcover 1st edition 1991

VA51/3020-543210 – Printed on acid-free paper

To the memory of my parents
Olivera and Branko Maksić

Preface

The renowned theoretical physicist Victor F. Weisskopf rightly pointed out that a real understanding of natural phenomena implies a clear distinction between the essential and the peripheral. Only when we reach such an understanding — that is to say — when we are able to separate the relevant from the irrelevant, will the phenomena no longer appear complex, but intelectually transparent. This statement, which is generally valid, reflects the very essence of modelling in the quantum theory of matter, on the molecular level in particular. Indeed, without theoretical models one would be swamped by too many details embodied in intricate accurate molecular wavefunctions. Further, physically justified simplifications enable studies of the otherwise intractable systems and/or phenomena. Finally, a lack of appropriate models would leave myriads of raw experimental data totally unrelated and incomprehensible.

The present series of books dwells on the most important models of chemical bonding and on the variety of its manifestations. In this volume the electronic structure and properties of molecules are considered in depth. Particular attention is focused on the nature of intramolecular interactions which in turn are revealed by various types of molecular spectroscopy. Emphasis is put on the conceptual and interpretive aspects of the theory in line with the general philosophy adopted in the series.

The book commences with the theoretical treatments of vibrations of nuclei, calculations of the force constants and infrared intensities. Then the theoretical basis of photoelectron spectroscopy is laid down followed by the interpretation of the PE spectra provided by the bond orbital model. Considerable attention is paid to the important concepts of through-space and through-bond interactions between more or less separated molecular fragments as mirrored in PE spectroscopy. Penning ionization technique yielding information on the shape of the periphery of molecules is discussed next. The fundamental but elusive concept of atomic charge in molecular environments deserved a separate chapter. It serves as an introduction to the article on ESCA spectroscopy which probably gives the most direct insight into the atomic point-charge distributions in molecules. The next chapter shows what can be learned from studies of molecules in momentum-space. A wealth of information on the fine details of the electronic charge distributions in molecules is provided by the experimental data and the theoretical parameters of NMR and ESR spectroscopies which are considered in the following two articles. The importance of a detailed knowledge of the molecular potential energy surfaces is stressed in discussions of the influence of rovibrations on the average properties and in a description of some salient features of molecules exhibited in excited states. This is followed by an extensive and illuminative survey of the semiclassical methods designed for understanding and interpretation of intramolecular interactions. Finally, the last chapter offers a brief presentation of the diabatic model in analysing potential energy surfaces and its use in treating chemical reactions. It represents a transition to the fourth volume of the series, where intermolecular interactions and reactivity will be elaborated in much more detail.

Ceterum censeo, it is perfectly clear by now that the best description of molecular systems is obtained by combined use of experimental techniques and theoretical methods. The whole series is constructed in a way to build bridges between these two traditionally separated approaches. If we have succeeded, at least partly, in achieving this goal, our efforts will be greatly rewarded.

I would like to express by sincere gratitude to all the authors of this book for their valuable contributions. A part of the Editorial work has been performed at the Organisch-chemisches Institut der Universität Heidelberg and I would like to thank the Alexander von Humboldt-Stiftung for financial support and Professor R. Gleiter for his hospitally.

                                                                        Z. B. Maksić

Table of Contents

Nuclear Vibrations and Force Constants

James E. Boggs

Department of Chemistry, The University of Texas, Austin, Texas 78712, U.S.A.

While valuable information on chemical bonding can be obtained from studies of the static structure of a bound system, even more understanding can be obtained from the internal restoring forces which control molecular vibrational motions and the mechanical response of molecules to external forces. This chapter deals with the description of such motions in terms of molecular harmonic and anharmonic vibrational force constants. Methods are described for the experimental evaluation of these parameters from spectroscopy, but the primary emphasis is on their computational determination since this approach can give more complete and accurate information for medium-sized polyatomic molecules. Consideration is given to techniques which can provide the highest possible accuracy within the harmonic oscillator approximation for polyatomic molecules and also the progress that has been made in fully anharmonic evaluation of the molecular vibrational potential energy hypersurfaces and the anharmonic energy levels on these surfaces.

1 Introduction

Chapter 7 of Volume 1 of this series discusses theoretical methods for determining the equilibrium structure of polyatomic molecules. The result of such studies is information on the static, minimum-energy molecular geometry, not the time-dependent structure of the twisting, stretching, distorting, real molecule with all of the vibrational motions it has in even its ground vibrational state.

In Chapter 6 volume 1, the common experimental approaches for determining structure are surveyed. These do not give direct information on the equilibrium state of the molecule, but rather provide a structure which is some sort of a time average over all the vibrational motions. The type of average produced depends on the experimental method used, and may be taken over only the molecular motion in some particular quantum state or it may be over all of the states occupied by an assembly of molecules at the temperature of the experiment. In any case, even if the experiment does involve a time average, the average is again a static quantity.

Deeper insight into molecular bonding must involve more than an understanding of the bond distances and angles that would be present if the system were in some frozen conformation. Consider how a molecule would be examined if it were of macroscopic size so that it could be picked up and handled. Aside from noting its average shape and size, we would try to analyze its spontaneous squirming motions, deduce characteristic frequencies and amplitudes of motion, squeeze it in different ways to see how it resisted such pressure and how it sprang back after being deformed, and generally poke it and see how it responded. All of this information, as well as static averages, can help in drawing the generalizations about structure and bonding that are at the heart of chemistry.

2 Vibrational Force Fields and Energy Levels

Within the Born-Oppenheimer approximation, the structure and internal dynamics of a molecule are described by the properties of a system of electrons and nuclei moving subject to a potential energy surface which is a function of the position coordinates of the nuclei. This multi-dimensional energy surface can be calculated by the methods of quantum chemistry with varying degrees of accuracy depending on its dimensionality. The positions of local energy minima give equilibrium structures of the system while energy saddle points provide reaction transition states.

While theory can provide information regarding the energy surface with relative ease, it is considerably more difficult to evaluate the steady-state vibrational eigenfunctions of the system and a great deal more difficult to determine the time-dependent solutions which give the reaction dynamics of the system. Experiment, on the other hand, provides direct information on the difference between vibrational eigenvalues, as observed in the transition frequencies seen in infrared absorption or Raman scattering. For reactions, experiment reveals full information about the reactants and products and some newer state-selected experiments give important clues regarding the details of the reaction processes. The current problem is to bring together the insights that are provided by the bulk observations of experiment which average out most of the details and the excessively fine detail which theory provides.

2.1 Potential Energy Surfaces of Nuclear Motions

The classic works on the vibrational behavior of polyatomic molecules include the books by Herzberg [1] and by Wilson, Decius, and Cross [2]. Both of these works continue to be very valuable sources of information. Many textbooks deal with the subject from the viewpoint of spectroscopy, including the volumes by Woodward [3], Graybeal [4] and numerous others. A particularly thorough and readable recent book is the one by Califano [5]. Early and important reviews which emphasize computational methods for determining molecular force fields and vibrational behavior are those of Pulay [6, 7], Fogarasi and Pulay [8, 9], and a more extensive monograph by Hess, Schaad, Carsky, and Zahradnik [10].

2.1.1 Diatomic Molecules

For a diatomic molecule, the molecular energy can be expanded as a function of internuclear distance in the form

$$E = E_{ref} + gq + 1/2\ Fq^2 + 1/6\ F'q^3 + \dots \tag{1}$$

The first term on the right represents the constant absolute energy of the reference configuration around which the expansion is performed. In the second term, g $(= \partial E/\partial q$ evaluated at the reference geometry) represents the force acting on the nuclei along the internuclear coordinate q. If the equilibrium geometry is chosen as the reference, this force is zero and the second term vanishes. In the third term, F is the Hooke's Law force constant, $(= \partial^2 E/\partial q^2$ evaluated at the reference geometry). If the expansion is truncated after this term, as is often done, Eq. (1) describes harmonic oscillator behavior. The fourth term contains the cubic force constant, $F' = \partial^3 E/\partial q^3$ and subsequent terms would involve anharmonicity constants of higher order.

There are, of course, numerous other ways to express the vibrational potential energy of a diatomic molecule, including the Morse equation,

$$E(r) = D[1 - e^{-\alpha(r - r_e)}]^2, \tag{2}$$

where D is the dissociation energy measured from the bottom of the well and r_e is the equilibrium internuclear distance. This approach, as opposed to the use of a power series expansion around the energy minimum, gives a better fit with fewer parameters over the entire coordinate range out to molecular dissociation. There are also more accurate representations that have been used when extensive data are available to be fitted. These potential functions, however, are not useful for general extension to polyatomic molecules since they are designed to describe a stretching vibration that is limited by atomic contact on one side and unlimited on the other.

Hence, they are not appropriate for most bending or torsional motions. It is a regrettable fact, however, that numerous cases can be found in the literature where the Morse potential is used in such inappropriate places.

2.1.2 Polyatomic Molecules

The most common approach to the description of the potential energy surface for polyatomic molecules is a generalization of Eq. (1):

$$E = E_{ref} + \sum_i g_i q_i + 1/2 \sum_i \sum_j F_{ij} q_i q_j + 1/6 \sum_i \sum_j \sum_k F_{ijk} q_i q_j q_k \cdot + \dots \quad (3)$$

where q_i, q_j, q_k, ... represent vibrational coordinates, g_i is the force acting along coordinate i (all such forces being zero if the reference geometry is the equilibrium geometry), F_{ij} is a matrix of harmonic force constants, and F_{ijk} and similar higher order terms represent the anharmonic force and corresponding interaction constants.

In Eq. (3), the harmonic force constant matrix F_{ij} has the form

$$
\begin{array}{llll}
F_{11} & & & \\
F_{21} & F_{22} & & \\
F_{31} & F_{32} & F_{33} & \\
F_{41} & F_{42} & F_{43} & F_{44} \\
\text{etc.} & & &
\end{array}
\quad (4)
$$

where the diagonal elements are harmonic oscillator force constants along the various vibrational coordinates and the off-diagonal elements are coupling constants between pairs of vibrational motions. Thus F_{43} describes the way in which a motion along coordinate 4 is affected by simultaneous motion along coordinate 3. The off-diagonal elements contain a wealth of often ignored information about mechanical coupling within a bonded system of atoms.

Specification of $3n$ coordinates is required to identify the position of n atoms in space. Three of these locate the center of mass of the system and three (or two in the case of a linear molecule) describe the rotational motion of the principal axis coordinate system. Thus, $3n - 6$ coordinates (or $3n - 5$ for a linear molecule) remain as descriptors of internal vibrational motion. It follows that the potential surface to be described is of high dimensionality for anything but the very smallest molecules. For example, benzene has 30 vibrational coordinates, so the vibrational expansion in equation 3 spans a 30-dimensional space. There are 30 values of i, 30 values of j, etc. The number of matrix elements becomes very large for molecules of moderate size — 465 in the case of benzene. Of course, the number of cubic terms, F_{ijk}, quartic terms, F_{ijkl}, etc, is very much larger.

2.1.3 Vibrational Coordinate Systems

Any complete, non-redundant set of vibrational coordinates can be used in Eq. (3). Cartesian coordinates of all of the atoms, with translational and rotational motions subsequently removed, would be a straightforward but not particularly useful choice.

Internal coordinates are often used and there are a number of such systems which may be employed. Consider, for example, the non-linear triatomic molecule SO_2. A suitable choice of the $3n - 6 = 3$ internal coordinates might be the two bond distances and the angle between them. For larger molecules, care must be taken that the internal coordinates chosen, particularly those involving bond and torsional angles, are both independent and form a complete basis for all possible molecular motions. A set of suggested rules for forming such sets has been given [11]. The matrix of force constants expressed in this coordinate system is more nearly diagonal than for some possible alternative choices of internal coordinates.

A special type of internal coordinates known as normal coordinates is often used, although it must be remembered that these are meaningful only within the harmonic oscillator approximation, i.e., when Eq. 3 is simplified to

$$E = E_{ref} + 1/2 \sum_i \sum_j F_{ij} q_i q_j . \tag{5}$$

In a normal mode of vibration all of the atoms in the molecule move at the same frequency and in the same phase but with different amplitudes. Thus, in one of the normal modes of the linear molecule $O=C=O$, the two oxygen atoms move to the right while the carbon atom moves to the left, the motions then being reversed in the other phase of the vibration. A second normal mode of $O=C=O$ involves the in-phase motion of both carbon atoms toward and away from the central carbon. The last two normal modes are bending modes in which the carbon atom moves up while both oxygen atoms move down, and the same process with the plane of the vibration rotated by $90°$ around the molecular axis.

Normal coordinates are used to simplify solution of the vibrational eigenvalue problem with the potential of Eq. (5). In these coordinates, both the potential energy operator and the kinetic energy operator are diagonal matrices, and the Sayvetz conditions [12] permitting the maximum possible separation of vibrational, rotational, and translational energy are satisfied. The physical consequences of this choice of coordinate system are graphically illustrated by Barrow [13] for a simple ball and spring model. The construction of normal coordinates is discussed in detail in Wilson, Decius, and Cross [2] and also in the more recent volume by Califano [5].

Determination of the allowed energy levels subject to the potential of Eq. (3) requires diagonalization of a large ($3n - 6$ by $3n - 6$) matrix (see below). If the molecule under study has elements of symmetry, the solution can be simplified by transforming the coordinates into a new basis which carries a completely reduced unitary representation of the molecular group. The development of symmetry coordinates, together with the necessary group theoretic background, are thoroughly covered by Califano [5]. In symmetry coordinates, the matrix to be diagonalized can be factored into non-interacting symmetry blocks, thereby reducing the dimensions of the problem. Such a treatment leads to groups of non-interacting vibrational energy levels and rigorous selection rules for transitions between such levels. Again, it must be remembered that these treatments are meaningful only within the harmonic oscillator approximation and that for real molecules interaction between all levels is possible, although they may be small, and a "symmetry-forbidden" transition is only one which can be expected to be of weak intensity.

2.2 Harmonic Oscillator Vibrational Energy Levels

Simple treatments of molecular vibrations normally accept the Born-Oppenheimer [14] separation of the vibrational, rotational, and electronic wavefunctions. This is valid in most cases, but it ignores the possibility of Renner and Jahn-Teller effects, which result from vibrational-electronic coupling, and of vibrational-rotational effects such as Coriolis interactions [15]. These effects will not be considered further here, but there are situations in which they cause major distortions in observable phenomena.

The quantum mechanical determination of the energy levels and eigenfunctions for a one-dimensional harmonic oscillator comes from solution of the corresponding vibrational Schrödinger equation, a problem which is used as an introductory example in nearly all quantum mechanics textbooks. Extended to polyatomic molecules, the corresponding solution for the allowed vibrational levels is

$$E_v = \sum_{i=1}^{3n-6} (v_i + 1/2)\, hcv_i \tag{3}$$

where h is Planck's constant, c is the velocity of light, and v_i is the characteristic frequency of the i-th mode. Coupled with the rigid harmonic oscillator selection rule of $\Delta v_i = +1$ for infrared absorption, Eq. (3) leads to a very simple fundamental spectrum of, at most, $3n - 6$ frequencies. The spectrum actually observed experimentally is vastly more complex for a number of reasons. First, each transition is not a single frequency, but takes the form of a band with a set of rotational transitions superimposed on the vibrational transition. These bands spread over an appreciable spectral region and there is a great deal of overlapping of bands in the spectrum of even a moderately complex polyatomic molecule. Next, there is a wide range of absorption probabilities, with corresponding variation in band intensities, so that some of the bands may be very weak or unobservable even though they correspond to allowed transitions within the harmonic oscillator approximation. On the other hand, many weak bands are commonly observed for transitions that violate the harmonic oscillator selection rules (combination bands, difference bands, overtone bands — see Fig. 1 for an illustration of these terms) and these, plus hot bands which are allowed, may be more intense than many of the fundamental bands. Furthermore, there are frequently strong Fermi interactions between bands that perturb their expected frequencies and sometimes drastically change their expected intensities.

For the reasons discussed above and still others, the process of analyzing the absorption spectrum of any but very small polyatomic molecules is a difficult task and can be successfully accomplished only by application of all possible sources of information, normally including absorption spectra from the sample in several phases, Raman spectra with information on polarization effects, and the study of a variety of different isotopic species. Even after all this experimental labor, there are often serious disagreements about the identity of some of the observed bands and the frequency of some of the expected fundamentals. The purpose in dwelling on the experimental difficulties involved in the analysis of a vibrational spectrum is to suggest that this field offers a prime opportunity for assistance from computational results.

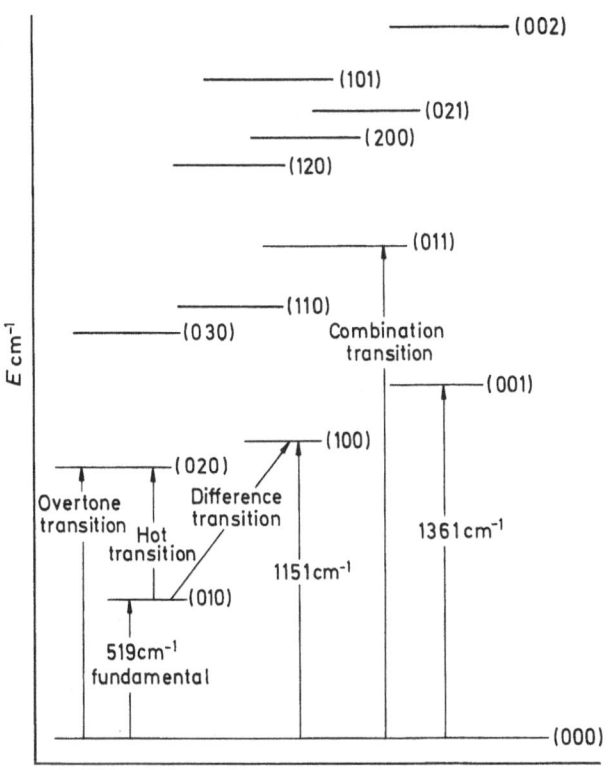

Fig. 1. Schematic representation of the energy levels of SO_2 and the types of transitions that may be observed between them. Taken with permission from Ref. 5

The above discussion of the vibrational spectrum has not yet tackled the problem of relating the characteristic molecular vibrational frequencies, v_i, to quantities pertaining to the absorbing molecule, specifically its vibrational force constants. The problem is normally set up by matrix methods, using the so-called **GF** matrix method of Wilson [16]. Full details cannot be repeated here (see Refs. [2] or [5]), but the secular equation for the vibrational problem in the harmonic oscillator approximation takes the form

$$\mathbf{GFL} = \mathbf{L}\varLambda. \tag{4}$$

In this equation, $\mathbf{F}$ is the force constant matrix in internal coordinates and $\mathbf{G}$ is the inverse of the kinetic energy matrix in internal coordinates. $\mathbf{L}$ is the matrix of the eigenvectors of

$$\mathbf{H} = \mathbf{GF} \tag{5}$$

and $\varLambda$ is the eigenvalue matrix. The required numerical calculations are lengthy for a polyatomic molecule, mostly because of tedious coordinate transformations. Efficient computer programs for solution of the problem are now common, however.

3 Theoretical Determination of the Potential Surface

For a diatomic molecule, the potential energy contribution to the Hamiltonian for the nuclear motion can be computed readily by assuming the Born-Oppenheimer separation of energies, and solving the electronic Schrödinger equation for the system energy at a series of different internuclear distances. (See chap 7 of vol I of this series.) Since only a relatively small number of such points are needed for the one-dimensional surface and the system consists of only two atoms, the computation is comparatively easy and the solution of the Schrödinger equation can be obtained at a high level of approximation and corresponding accuracy.

The situation is entirely different for a polyatomic molecule. For example, each solution of the electronic Schrödinger equation for a molecule containing 10 atoms (assuming it contains 5 times as many electrons as the diatomic molecule) is more time-consuming by a factor ranging from perhaps 500 to 70,000 depending on the accuracy of the method chosen. Coupled with this problem is the fact that the nuclei in a 10-atom molecule without symmetry move on a 24-dimensional potential surface as opposed to the one-dimensional surface of the diatomic molecule. Not only energy changes along each of these 24 dimensions, but also all possible interactions between them, must be evaluated. Considering these difficulties, it seems nearly miraculous that theoretical chemists have made the impressive steps that have been made toward accurate and reliable determination of the vibrational potential surface for polyatomic molecules.

3.1 Computation of Energy Derivatives

A critically important break-through in the determination of energy surfaces for polyatomic molecules was the development of the gradient method, first formulated by Péter Pulay [17] in 1969. With each solution of the electronic Schrödinger equation, not only the energy can be calculated, but also the first derivative of the total energy with respect to every internal coordinate of the molecule. Thus one calculation yields information about the force inhibiting displacement along each direction of the multi-dimensional potential surface. The technique was first implemented for semi-empirical calculations and in the now-obsolete ab initio program MOLPRO, and then in the TEXAS program [18, 19]. It is currently a common feature of all ab initio molecular orbital programs. Further extensive development of the gradient method has involved the direct analytic calculation of higher energy derivatives which are the molecular hamonic and anharmonic force constants and implementation of the concepts into programs utilizing a variety of types of ab initio methods.

3.1.1 Energy Gradients

A clear development of the theory for calculating the gradient, i.e. the first derivative of energy with respect to a nuclear coordinate, from an SCF wavefunction is given by Pulay in Ref. 6. The wavefunction may be written

$$\Psi_i = \sum_k A_k \Phi_k ,$$ (6)

with

$$\langle \Phi_k \mid \Phi_l \rangle = \delta_{kl} ; \qquad \sum_k A_k^2 - 1 = 0$$ (7)

The configurations Φ_k are linear combinations of Slater determinants made from n orthogonal orbitals, φ_i, which are linear combinations of m basis functions χ_r,

$$\varphi_i = \sum_r^m C_{ri} \chi_r$$ (8)

The expectation value for the energy E is

$$\langle \Psi \mid H \mid \Psi \rangle = E(X_a, A_k, C_{ri}, p_i)$$ (9)

where the p_i are parameters giving the dependence of the basis functions on X_a, the nuclear coordinates,

The gradients, i.e. the first derivatives of the energy with respect to the nuclear coordinates, are found as the sum of four terms corresponding to, respectively, the dependence of E on X_a, directly, and through A_k, C_{ri}, and p_i. The first of these terms is the negative Hellmann-Feynman force [20, 21]

$$-f_a^{HF} = \langle \Psi \mid H^a \mid \Psi \rangle$$ (10)

which some earlier papers attempted to use alone as the gradient since it is easily evaluated. The second term involving the derivative with respect to A_k vanishes. The next term, giving the variation with respect to the orbital coefficients, is called the density force. It was shown [17] that this term can be evaluated from easily calculable matrices without need for explicit evaluation of the derivatives $\partial E / \partial C_{ri}$. This discovery was the key step that has made possible the practical calculation of energy gradients. Thus, the remaining term involving $\partial E / \partial p_i$, called the integral force, requires most of the work in computing the gradient.

While the computation of gradients was first developed for closed shell SCF calculations, it has now been implemented for all commonly used wavefunction types at both the Hartree-Fock level and those that include electron correlation. It is a common feature of all ab initio programs that are widely available for general use.

3.1.2 Higher Derivatives

Explicit formulas were given long ago [22, 23] for obtaining second derivatives of molecular energy with respect to nuclear displacement for Hartree-Fock wavefunctions. The 21-year delay between 1958, the date of the earliest such formulation [22] and 1979, the date of the first successful implementation of the procedure [24] illustrates the gap that often exists in quantum chemistry between understanding the basic mathematics of a problem and developing the computer implementation that makes a solution feasible. The difficulty stemmed from the fact that evaluation of the first derivatives of variational parameters cannot be avoided in evaluation of energy second derivatives, while they are not needed for the first derivatives. The problem is described in detail in Ref. [9], which also reviews work through 1984 on the computation of second derivatives for various sorts of wavefunction.

Current work in this field is so active that it is almost pointless to list the current limitations. Energy derivatives through fourth order can be computed analytically and the strongest effort is now being directed toward increasing the variety of types of wavefunctions for which higher derivatives can be obtained. An especially difficult and important task is the development of higher derivative programs for the various types of correlated wavefunctions.

3.2 Accuracy and Source of Errors

It is useful to consider the accuracy of computed force fields separately for two regions of the potential surface: those areas near local minima and the regions well removed from such minima. The situation is reasonably clear in areas closely neighboring a local minimum that corresponds to a stable nuclear conformation for which a single configuration solution of the Schrödinger equation is satisfactory. Fortunately, these are also the regions of interest for studies of equilibrium molecular geometries and properties dependent on small deviations from the equilibrium state. Such properties include the energies of the lower-lying vibrational transitions and the first few overtones and combination bands, the determination of centrifugal distortion constants, Coriolis constants, Fermi interaction magnitudes, mean amplitudes of vibration for assistance in the analysis of electron diffraction experiments, and many other common properties for which theory can be of assistance to experiment.

The best test of accuracy in a computation is comparison with experiment, but a serious problem arises here for polyatomic molecules. Using gas-phase or matrix spectra, it is straightforward to measure infrared band centers to less than 1 cm^{-1} accuracy and modern techniques permit precision up to 1,000 times greater in favorable cases. However, attempts to determine vibrational force fields, a molecular property, from the transition frequencies, a spectral observation, are hampered by the insufficiency of experimental data. Even though the spectrum may be too complex to permit easy and total assignment, the number of independent observations that can be made for all but the simplest molecules is inadequate to permit a complete determination of the vibrational force field even in the harmonic oscillator approximation. Except for molecules containing fewer than half a dozen atoms, experimentally determined force fields are often restricted to the diagonal harmonic elements, and some of these must sometimes be assumed to be the same as in related molecules. Accurate comparison of computed force fields with experimental ones is thus limited to fairly small molecules and it can only be assumed that conclusions drawn there apply also to larger systems.

An alternative approach is to compute some direct observables, such as transition frequencies, from the calculated force constants and compare these with the raw experimental data. This procedure, however, brings in the added question of the accuracy of computation of vibrational energy levels from the calculated energy surface, a problem which is discussed below.

For near-equilibrium regions of the vibrational potential surface, computational errors are mainly due to inadequacy of the finite basis set used in the Hartree-Fock expansion of the single-determinant wavefunction and the neglect or incomplete treatment of dynamic electron correlation. If it can be assumed that experimental force constant elements can be obtained to 1–2% accuracy for very small polyatomic molecules, this level might also be chosen as a goal for the accuracy of computed constants. Such a goal can be achieved, but only with a very high level of computation. An illustration is shown in Table 1 of early calculations [25] on NH_3 and HCN.

Table 1. Experimental and theoretical force constants for NH_3 and HCN[a]

	Computed [25]	Experimental [26, 28]
NH_3:		
F_{rr}	7.187	7.06 $\pm$ 0.13
F_{ss}	0.111	0.113 $\pm$ 0.004
$F_{\alpha\alpha}$	0.689	0.68 $\pm$ 0.07
$F_{rr'}$	-0.020	0.01 $\pm$ 0.19
F_{rs}	-0.136	$-0.21 \pm$ 0.05
$F_{r\alpha}$	-0.158	$-0.17 \pm$ 0.04
HCN:		
F_{rr}	6.301	6.251 $\pm$ 0.004
F_{RR}	19.422	18.703 $\pm$ 0.012
$F_{\alpha\alpha}$	0.272	0.2596 $\pm$ 0.0004
F_{rR}	-0.197	$-0.200 \pm$ 0.004
F_{rrr}	-35.8	$-35.4 \pm$ 2.4
F_{RRR}	-126.1	$-126.0 \pm$ 5.2
F_{rRR}	0.04	0.4 $\pm$ 2.4
F_{rrR}	0.36	0.04 $\pm$ 0.80
$F_{r\alpha\alpha}$	-0.175	$-0.19 \pm$ 0.50
$F_{R\alpha\alpha}$	-0.682	$-0.65 \pm$ 0.40
F_{rrrr}	184	181
F_{RRRR}	685	580

[a] Force constants in units of $aJ/Å^n$ where n is the number of stretching coordinates involved in the constant. For NH_3, see Ref [25] for definition of coordinates. For HCN, r designates $H-C$ stretch; R the $C\equiv N$ stretch

The results in Table 1 show several features that have been verified in many later studies. For NH_3, the complete harmonic force field, both the diagonal force constants and the coupling constants, agrees within the quoted experimental uncertainty with harmonic constants derived by analysis [26] of the experimental spectra. To obtain this level of agreement required use of a triple-zeta basis set augmented by two sets of polarization functions with electron correlation treated by an all singles and doubles configuration interaction method plus use of the Davidson correction [27] to account for unlinked cluster contributions. This already high level of calculation is not the most powerful now available, but it is seen to be adequate to give results within experimental accuracy. Unfortunately, even this computational level is too time-consuming to be used for molecules with as many as a dozen atoms.

The results shown in Table 1 for HCN, calculated at the same quantum chemical level, provide a less encouraging illustration. The reported uncertainties of the harmonic force constants obtained from experiment [28] are much smaller than in the example of NH_3, so that there is never agreement within experimental error. However, it may be mentioned that later experimental studies [29, 30] also fall outside the uncertainty range quoted here. In any case, the computed $H-C$ harmonic stretching constant disagrees by less than 1%. The $C\equiv H$ stretching constant is in error by 3.5%, illustrating a point that has been found to be generally valid. Regions in a molecule having a high electron density, particularly multiple bonds, require a higher level treatment of electron correlation than used here to give a good description of the bond stretching. This is quite reasonable, since the dynamic correlation error arises from incomplete accounting for the tendency for electrons to avoid each other. In regions where the electron density is high, the electron correlation error is large and a more accurate theoretical treatment is needed to reduce it to any desired level.

The results in Table 1 reveal another generalization that can be very useful. Examination of the harmonic coupling constant, F_{rR}, and the cubic constants shows that they are computed within experimental accuracy. While this is largely because of the limited accuracy obtainable in experimental determination of these small correction constants, it suggests that the most reliable present method for determining such constants may be well be ab initio computation in the manner used here.

A more recent and exhaustive study [31] has probed the effect of varying levels of basis set truncation and of electron correlation treatment on the accuracy of computed vibrational force fields, both harmonic and anharmonic. This article confirms the experience of previous workers and serves as a useful reference, although there is good reason to believe that somewhat larger basis sets and higher order electron correlation treatments than recommended in that work are required in order to obtain experimental accuracy, particularly for molecules containing multiple bonds. Further, the highest possible accuracy requires anharmonic constants higher than fourth order so that the analytical derivative methods described in this paper are not fully applicable (see Sect 4.1 below).

With sufficient computational effort, excellent results can thus be obtained for the complete harmonic and anharmonic force constant matrices of very small molecules. While such detailed and accurate calculations have been done at this level for a molecule as large as benzene [32], they are not, in general, practical with existing computers and quantum chemical procedures for molecules containing more than about six to ten atoms, especially if the full higher order anharmonic force constants are required. An attractive alternative approach is to perform the calculation at a lower quantum chemical level and utilize the regularities that are observed in the resulting errors to introduce empirical corrections. This approach is discussed in the following section.

3.3 Empirical Corrections

If harmonic force constants matrices are computed at a modest quantum chemical level, for example with the use of a 6-31G** basis set and complete neglect of electron correlation, it is nearly universally found that the computed force constants are too large by an amount averaging 15–20%. One common approach, then, is to compute the force field at such an easily affordable computational level and multiply at least the diagonal elements of the force matrix by 0.8 or some such figure. Some workers use the same factor for the off-diagonal force constant matrix elements while others choose to leave them uncorrected. Alternatively, if a particular study is interested only in vibrational frequencies, the directly computed frequencies may simply be multiplied by 0.9 since a frequency is related to the square root of a force constant.

While this use of a constant correction factor does not insure that any particular force constant or frequency prediction will be improved, it does appreciably improve the average accuracy. It should be remembered, though, that some predicted force constants deviate from the true value by quite different amounts. It has been shown [33] that the correction factor is as small as 0.45 for the $N-H$ wagging motion in pyrrole, using a level of calculation for which 0.8 would have been a satisfactory average correction factor for all of the other motions. In still other relatively rare cases, the correction factor is larger than unity, i.e., the computed force constant is smaller rather than larger than the true value.

Further improvement by empirical correction of the ab initio predictions relies on the use of chemically meaningful internal coordinates for expression of the force constant matrix rather than Cartesian coordinates or symmetry coordinates. The logic is based on the idea that the electron distribution around a $C-H$ bond, for example, is quite similar from molecule to molecule, and the computational error in predicting a $C-H$ stretching frequency should also be similar. Note that this is not at all the same thing as saying the force constant or vibrational frequency should be identical for all $C-H$ stretching motions. The proposition that the error is similar in different molecular environments represents a second-order approximation, and its accuracy should increase as the error gets smaller.

Even for motions for which the computational error is fairly large, the error does not vary by really large amounts between similar molecules. For example, the factor by which the computed force constant for the $N-H$ wagging motion in pyrrole differs from the true value was quoted above as being 0.45. The factor for the similar motion in imidazole is 0.49 [34] and that in maleimide is 0.51 [35].

3.3.1 Scale Factors

A systematic procedure for exploiting the similarity in the error found in computation of similar motions in related molecules was described by Pulay, Fogarasi, Pang, and Boggs in 1979 [11]. Full details are given in Ref. [11], but the basic idea is to perform similar calculations on a molecule for which the vibrational spectrum is known and on the molecule whose spectrum is to be predicted. Care must be taken to use the best known approximation to the true equilibrium geometry for each molecule as the reference geometry around which the vibrational energy expansion of equation (3) is performed, to use a similar set of internal vibrational coordinates as recommended in Ref. [11], and to use the identical quantum chemical level for the calculations. The procedure was further described in a later paper [36], where the force field obtained was called the "scaled quantum mechanical (SQM) force field".

Use of the SQM force field method can be illustrated by a series of calculations starting with benzene. The harmonic force constant matrix, equation (4), was calculated for benzene using a medium-size (4–21) basis set and no treatment of electron correlation [37]. This matrix was used to predict the fundamental vibrational frequencies of benzene. The internal motions were then divided into six groups of related type and a set of six scale factors was derived to modify the computed force constant matrix to give the best least squares fit of vibrational frequencies computed from the matrix to the experimentally observed frequencies of benzene. The off-diagonal elements of the matrix were taken as the geometric mean of the diagonal constants they connect. The scale factors obtained were in the general range of 0.8, but they varied from 0.739 to 0.911 for the differing types of motion.

The observed vibrational fundamentals of benzene were fit with an average deviation of less than 6 cm^{-1} by this procedure. The more important point, though, is that the derived set of scale factors is expected to be transferable to similar motions in related molecules.

Transferability of the scale factors derived for benzene was first tested by calculations on pyridine [38, 39]. The harmonic force constant matrix was calculated for pyridine taking care to use the same procedures as in the benzene study. The computed pyridine force constant matrix was then scaled using the benzene scale factors, considering the nitrogen in the ring as just another carbon. Next, the fundamental vibrational frequencies of pyridine were calculated from the SQM force field and compared with the experimentally known spectrum. Aside from the C–H stretching frequencies, which were uncertain in the available experimental information, the mean deviation was only 5.7 cm^{-1} for the in-plane motions and 8.5 cm^{-1} for the out-of-plane motions. Including the C–H stretches, the deviations were 9.6 cm^{-1} for the in-plane motions.

The next test involved a similar study with naphthalene [40]. The benzene scale factors were used to scale the computed force constant matrix of naphthalene, and the resulting SQM force field was then used to predict the naphthalene fundamental frequencies. A final comparison with the measured spectrum of naphthalene showed an average difference of only 6.5 cm^{-1}, excluding the C–H stretching frequencies which had not been assigned from experimental evidence.

The same procedure, with scale factors transferred unchanged from benzene, has been used in studies of a variety of related molecules, including aniline [41], toluene [42], fluorobenzene [43], benzonitrile [44], and phenylacetylene [45]. Moreover, the same benzene scale factors have been applied successfully to a number of less closely related 5- and 6-membered heterocyclic rings including even borazine [46] in which the ring consists of alternating boron and nitrogen atoms. These studies served a purpose beyond repeatedly demonstrating the high accuracy that can be obtained by the SQM method. They also clearly pointed out a few misassignments in the analyses of the experimental spectra, and in a number of cases made possible an unequivocal choice between assignments that were previously controversial in the literature.

Even at the low level of calculation used in the studies just described, absolute intensities of the transitions can be calculated with sufficient accuracy for comparison with the designations of very strong, strong, medium, weak, or very weak that are often reported. Closer accuracy requires a much higher level of calculation [31]. The lesser accuracy available from cheaper calculations is often adequate, however, if the purpose is to test experimental assignments.

There is a point that should be remembered about the use of scaled quantum mechanical force fields. Even though they start with a computed harmonic force field, they are scaled to predict real frequencies, which are anharmonic. The scale factors, therefore, correct not only for deficiencies in the basis set and the treatment of electron correlation, but also for the neglect of vibrational anharmonicity. The success which has been demonstrated in the transferability of scale factors shows that there is also considerable similarity in the degree of anharmonicity of similar bond motions in different environments. In some cases, the correction for anharmonicity may be the largest contributor to the scale factors.

3.3.2 Semi-Empirical Methods

Some of the earliest calculations which applied scaling procedures to correct computed vibrational force fields and frequencies were done with semi-empirical quantum chemical methods [47, 48]. A typical, more recent study [49] explores the ability of scaled CNDO/2 force constants to fit the observed spectra of maleimide and of uracil and, more importantly, the degree to which the scale factors for one molecule can be transferred to the other. The scale factors obtained by fitting the spectrum of uracil alone varied from 0.318 to 1.45. This wide variation indicates the futility of trying to make use of the computed force constants without scaling or to use a common scale factor for all modes of vibration. Furthermore, the large deviation of the scale factors from unity means that a large correction is being made (a scale factor of 0.333 is equivalent to correcting a computed value that is too large by a factor of 3), so that transferability of the error from one molecule to another would be expected to be less accurate than with the use of medium- or high-level ab initio methods. As an illustration of this, the scale factor for the $C-C$ stretching modes obtained by a fitting maleimide alone was 0.319 while that obtained by fitting uracil alone was 0.372.

Scaling of vibrational force fields obtained by semi-empirical quantum chemical calculations can lead to reasonably good agreement with the observed spectrum of the molecule from which the scale factors are derived. This can provide a means to check questionable mode assignments or identify unknown bands in spectra where most of the bands have been reliably assigned from experiment. Although the applicability of ab initio methods is making rapid strides toward larger and larger molecules, there will always be a region just beyond the borderline where semi-empirical methods offer the only possibility of obtaining information. In such situations, the procedure should prove to be a valuable adjunct to experimental infrared and Raman spectroscopy. On the other hand, the lower reliability of transferring scale factors between related molecules is a serious drawback of the approach.

3.4 Recommendations

The use of scale factors, as described in Sect. 3.3.1 is only possible if the molecule under investigation has chemical relatives for which the vibrational spectra have been reliably assigned. This is true for most organic compounds, but unfortunately not for many very interesting inorganic and metal-organic species. The difficulty is also present with species which may not be stable minima on the potential surface but are the important transition states for chemical reaction.

When the SQM method is applicable, it is the author's opinion that it is clearly the method of choice. Use of a single scale factor for all force constants does not take full advantage of the information content that is available from the computational effort expended. It takes very little more human work to use the transferable sets of scale factors, and the increase in accuracy of the agreement with experiment fully justifies the effort. The SQM method cannot provide vibrational frequencies of experimental accuracy, but it gives predictions with a high enough certainty that they can be used to verify or disqualify observed band centers as being vibrational fundamentals. Unfortunately, many common ab initio programs including the very popular GAUSSIAN series concentrate so heavily on the calculation of vibrational frequencies rather than force constant matrices that they make it unneccessarily difficult to use a set of internal coordinates that permits ready transfer of scale factors between different molecules.

The choice of computational approach is more difficult when the SQM method is not applicable. The highest possible level of calculation considering the size of the molecule under investigation may still not be high enough to give accuracy better than 10% or so. A single scale factor can be used if the experimenter believes that the rule of thumb developed for organic molecules still applies to whatever exotic species may be under investigation. This must be more a matter of faith than of experience.

While the accuracy obtainable for medium-sized molecules for which the SQM method cannot be used may not be as high as for more routine studies of organic molecules, it may still be high enough to answer the chemical question that led to the computational investigation. For example, an infrared band suspected of arising from a stretching motion of an unusual bond between two inorganic main group elements may be identifiable even though the prediction has an error as large as 10%.

In many cases, there may be little gain in going to the most elaborate calculation possible. A calculation at the 6-31G** level with an MP2 treatment of electron correlation may give agreement with experiment as good as or better than a more extensive calculation if account is not taken of vibrational anharmonicity. There is often a certain amount of cancellation of errors with medium-level computational schemes. Whether such chance cancellation of errors is a situation to be sought is a question which the investigator must decide.

4 Determination of Anharmonic Vibrational Energy Levels

The harmonic oscillator approximation is a wonderfully useful tool, leading to simple relations for vibrational energy levels and transition frequencies, symmetry factorization of the matrices into independent blocks, spectra without the complications of overtone, combination, or difference bands, no Fermi interactions, etc. The only difficulty is that real molecules are not harmonic oscillators and the observed spectra contain many significant features that cannot be explained within the harmonic oscillator assumption.

Advance beyond the treatment of molecules as simple springs obeying Hooke's law brings considerable complexity, but also much better agreement with observed reality. The additional problems may be divided into two parts: (1) determination of the anharmonic potential surface on which the motion occurs, and (2) determination of the molecular eigenfunctions and eigenvalues for nuclear motion on that surface. Both of these tasks bring in additional computational difficulties, but considerable progress has been made in recent years toward their solution.

4.1 Anharmonic Vibrational Potential Energy Surfaces

For many purposes it is sufficient to know the geometry of the potential energy hypersurface in a region relatively near to a local minimum so that energy levels bound to that molecular configuration can be obtained and the spectral properties can be calculated for that conformation of the molecular species under study. In such cases it is often adequate to use a single configuration basis for the solution of the electronic Schrödinger equation, although quite a high level quantum chemical calculation must be used to obtain useful accuracy. Other cases, particularly those of transition states of reactions which involve bond breaking or formation, often should be treated with a multiconfiguration procedure. The present discussion will consider only situations near potential minima and will assume that a single configuration basis suffices for the calculation, in addition to making the common assumption that the Born-Oppenheimer separation of electronic and vibrational motions is adequate.

Near a potential minimum, the molecular energy can be expanded in a set of nuclear coordinates as described in Sect. 2.1.3. The energy surface is then described by the harmonic and anharmonic force matrices, F_{ij}, F_{ijk}, etc of Eq. (3), and the problem is the quantum-chemical determination of these molecular constants. This may be done numerically, analytically, or by some combination of the two approaches. In the numerical method, the molecular energy is obtained by solving the Schrödinger equation for various sets of nuclear coordinates judiciously chosen near the potential minimum in question. For a polyatomic molecule with many vibrational degrees of freedom, a very large number of such points must be calculated, especially if higher order anharmonic constants are desired so that many points must be taken for each mode of vibrational displacement and each combination of such modes.

Alternatively, it is possible to calculate at least some of the required constants analytically from the molecular wavefunction obtained at the potential minimum, as described in Sect. 3.1.1 and 3.1.2. While this adds significantly to the computational time for a single run, it limits the number of runs that are required and results in a great overall saving of time, a saving which increases dramatically with increase of molecular size. Unfortunately, for very high precision work, at least some of the diagonal force constants are needed through sixth order and algorithms for such calculations have not yet been devised. Similarly, analytical higher derivatives are not yet available for many types of wavefunction that may be needed for a particular problem, i.e., various sorts of correlated wavefunctions, open shell molecules, etc. Progress in this direction is proceeding rapidly and it is likely that there will be a much more flexible choice of approaches possible within a few years. In any event, it is useful to use analytical derivatives of the highest order available, since this reduces the number of separate calculations that must be done at displaced geometries.

A word should be added about the computer requirements and practical molecular size for calculations of anharmonic potential surfaces. There would seem to be no point in doing calculations of vibrational anharmonicity unless high accuracy is achieved since the effect of anharmonicity being observed, while it may be decisively important in some applications, is generally small. For such high accuracy, the electronic energy calculation should be done with the equivalent of at least a triple zeta basis set augmented with polarization functions and with a treatment of electron correlation at least of the MP4 level. Simandiras et al. [50] have recently shown the importance of f functions in the basis set of carbon to obtain accuracy at the 1% level in the bending frequencies of acetylene. Such computational levels are possible on the largest existing computers only for molecules no larger than, as a rough example, benzene, and for benzene the calculation is an extremely expensive and time-consuming calculation. However, computers are being improved rapidly and quantum chemical methodology is improving even more rapidly, so that this limit should be a transitory one.

4.2 Anharmonic Vibrational Energy Levels and Spectra

Once the potential energy surface on which the molecular vibration occurs has been determined, the task remains of computing the vibrational eigenfunctions and eigenvalues which are related to observable molecular properties, spectroscopic and otherwise. The classic method for diatomic molecules is by use of perturbation theory, and many investigators now extend this procedure to polyatomic molecules. Especially noteworthy is the work of the group with Professor N. C. Handy at Cambridge University which has pioneered the analytical calculation of higher derivatives and has made extensive use of the perturbation method for obtaining anharmonic energy levels.

An excellent illustration of the use of perturbation theory to obtain anharmonic energy levels is a paper by Lee et al. [51] on the prediction of the vibrational spectrum of the cyclopropenyl cation, $C_3H_3^+$ and its deuterated isotopomers. A more extensive set of applications to smaller linear molecules has been described by Allen et al. [31], who analyze in detail the effect of various computational approximations on the accuracy obtained. In the opinion of the present author, a major limitation of perturbation theory is that resonances occur with increasing frequency as the molecular size increases, leading to divergence and perturbation estimates which are too large.

A different approach has been developed making use of a variational calculation of the vibrational energy levels. Such techniques date back at least 15 years [52], but early applications were made only to molecules with three modes or to the stretching frequencies of linear molecules. The early history is reviewed briefly in Ref. [53]. The major advantages of the variational method arise, however, in applications to the more complex spectra of larger molecules with many modes. A new, general program system for variational solution of the vibrational problem has recently been developed and applied to HCN [53], CH_3F and CD_2HF [54], CD_3H [55]. $C_3H_3^+$ and its isotopomers [56], and $C_3F_3^+$ [57]. In the latter two papers, the full 12-mode system is treated and all fundamental and first overtone vibrational frequencies are calculated. Comparison of the predicted spectra with measured spectra obtained in low-temperature matrices gives agreement within the range of uncertainty that can be attributed to minor matrix interactions. The two studies of the cyclopropenyl cation, $C_3H_3^+$, which were unknowingly carried out independently and simultaneously in different laboratories, furnish an excellent comparison of the variational [56] and perturbation [51] approaches.

One advantage of the variational approach is that all interactions between modes are automatically evaluated. The Fermi resonances in $C_3H_3^+$, for example, are reproduced with good accuracy. A limitation is the great density of states that occurs in the higher frequency overtone region of larger molecules, making it difficult to separate out and identify a particular desired eigenvalue in the final matrix diagonalization.

The variational procedure is described in detail in the original references [53–57] and also in the program description and input instructions which are available [19]. Basically, the calculation is a vibrational configuration interaction scheme, with a three-level configuration selection procedure analogous to one used in electronic structure calculations. The explicit formation of the Hamiltonian matrix is avoided, as in the direct CI method of electron correlation theory, the desired eigenvalues being determined by the Davidson matrix diagonalization method [58].

4.3 Utility of Anharmonic Calculations

If the purpose of an investigation is to determine the fundamental vibrational frequencies of the ground state of a molecule which is related chemically to other known molecules, there is little reason to go to the considerable computational effort required to do an accurate anharmonic calculation. The scaling procedures based on calculations at a relatively low quantum chemical level as described in Sect. 3.3 and 3.4 can provide predictions within 10 cm^{-1} or so, which is adequate for most practical purposes. Furthermore, with the present level of computational ability, it is not feasible to attempt anharmonic treatments of molecules much beyond the 6-atom, 12-mode systems described above. Such a situation can change rapidly, of course, but it is likely that the vibrations of very much larger molecules will always be capable of analysis within the harmonic oscillator approximation than with a full anharmonic treatment.

On the other hand, there are many interesting problems that do not have the characteristics postulated in the preceding paragraph. There are small inorganic molecules that do not have close chemical relatives from which scaling factors can be transferred. Reaction intermediates and other highly reactive species may again have exotic forms of bonding that defy prediction of the extent to which computational errors and anharmonicity corrections can be assumed similar to those in known species. Molecules such as these may show up in chemical studies of flames and rocket exhausts, high temperature chemical processing, species from the upper atmosphere of the earth or other planetary atmospheres, or interstellar molecular clouds − all subjects of intense chemical interest which would warrant the computational effort to do accurate vibrational calculations with a high quantum chemical level and an accurate treatment of the vibrational anharmonicity.

There are, unfortunately, many interesting species for which such a treatment is still impossible. Among the most tantalizing of these are metalorganics and other compounds of elements from the lower regions of the periodic chart where relativistic and spin effects greatly complicate a highly accurate quantum chemical treatment. There is still much work to be done.

5 Applications

Applications of quantum chemical evaluation of vibrational force fields and spectra are now so commonplace that any issue of any of a dozen journals can provide illustrations. It is true that some of these still do little more than report the calculation of the spectrum of fundamental frequencies of a molecule, compare the prediction with the known spectrum, and express varying degrees of satisfaction with the comparison. Even studies such as these can serve a useful chemical purpose, however, in that the "known" spectrum of a medium-sized polyatomic molecule as derived from experiment nearly always involves missing transitions, transitions the assignment of which is in dispute, or even complete misassignments of certain vibrational bands. In defense of the experimentalists, it must be remembered that the real spectrum as observed in nature generally has complexities such as overtone, combination, and hot bands, some very weak bands that are difficult to identify, and varying degrees of Fermi resonance. Cases are known where an overtone transition may be as intense as the strongest fundamental in the spectrum. Such experimental difficulties have been found even for very common molecules, and can usually be cleared up by straightforward calculations of the type described in this chapter.

Here it may be more interesting to describe applications that differ in some manner from those common studies that treat a molecule that has already been studied experimentally in order to complete and improve the existing experimental assignment of its vibrational spectrum.

One study which illustrates the marked invariance of computational errors and anharmonicity corrections for a given mode in related molecules involved acrylic acid. Although acrylic acid forms strongly bound dimers, it was possible to detect the monomer spectrum from a sample frozen in a nitrogen matrix [59]. Too many bands were present to be accounted for by one species, and it was assumed that *cis*- and *trans*-forms (see Fig. 2) were both present, as is true in the gas phase. Irradiation with ultraviolet light made it possible to distinguish two sets of bands, one increasing in intensity and one decreasing after irradiation. It was apparent that Set 1 and Set 2 corresponded in some order to the *cis*- and *trans*-conformations, but there was no experimental method to determine which was which. The frequency differences between corresponding bands were small, and attempts to distinguish them by a normal coordinate analysis failed. A resolution of the problem was achieved by use of SCF-level ab initio calculation of the spectra of the two forms [60]. The directly computed force fields were not highly accurate, but they were then scaled using scaling factors previously derived [36] to fit observed spectra of ethylene, formaldehyde, glyoxal, acrolein, and butadiene. A comparison of the observed differences between the corresponding frequencies of Set 1 and Set 2 and of the computed differences between the *cis*- and *trans*-conformers is shown in Table 2. While all of the frequency shifts accompanying the conformational change are small, some very small, the pattern of correspondence leaves no doubt of the assignment of Set 1 as arising from the *cis*-form of the molecule. Most important for the present discussion, the faithfulness in reproducing the very small spectral changes illustrates the precision with which the errors due to neglect of electron correlation and use of a small basis set and the neglect of vibrational anharmonicity are similar in corresponding modes of these closely related molecules.

A different sort of application of computed force fields is illustrated by studies of the vibrational spectra of 4H-pyran-4-one [61] and 4H-pyran-4-thione [62]. Contrary to the common practice of calculating the spectrum of a molecule after the measurement has been made, these studies made use of a scaled prediction to analyze the spectra as they were obtained. Identification and assignment of the observed transitions were greatly facilitated by having predicted frequencies accurate to 20 cm^{-1} at the worst, together with approximate band intensities. It was possible to get complete and, hopefully, unequivocal assignments on the basis of observations on the single, most common isotopic species. The experiment served to confirm the theoretical predictions and to improve the accuracy of the frequencies by approximately a factor of 10. This method is so powerful that it is difficult to see why it should not be used in all future efforts to assign the infrared and Raman spectra of moderately complex molecules. The calculation at this level is now quite inexpensive and there is a saving of much experimental labor.

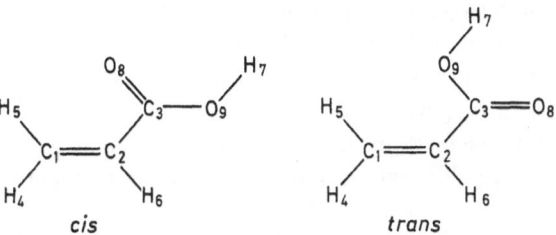

Fig. 2. *Cis*- and *trans*-conformers of acrylic acid

Table 2. Frequency differences between *cis-* and *trans*-acrylic acids[a]

Approximate description	Calculated trans-cis	Observed Set 1–Set 2
C=O deformation	+39	+30
C–C–O bend	−45	−53
C–C stretch	− 4	− 4
CH$_2$ rock	−48	−40
C–O stretch	+ 9	+ 5
CH$_2$ scissors	0	0
C=C stretch	−10	−12
C=O stretch	+12	+12
C–C torsion	− 4	−10
C=O wag	+ 1	0
C–O torsion	−35	−41
C–H & C–O wag	+ 3	− 7
CH$_2$ twist	+ 2	0
CH$_2$ wag	0	− 6

[a] Ref. [60]. Frequency differences in cm^{-1}. Set 1 increased intensity on ultraviolet irradiation relative to Set 2

Vibrational frequencies and intensities are not the only useful quantities that can be derived from computed force fields. In a recent study [63], the quartic centrifugal distortion constants of benzonitrile were derived by analysis of a very large number of submillimeter rotational transitions. The same study presented the values of the same constants computed from a separately determined [44] scaled quantum mechanical force field of benzonitrile. Except for one constant which was not well determined by the data, the agreement was within a few percent. Most significantly, when the computed constants were used to predict the observed submillimeter frequencies, the average deviation was only 0.342 MHz. This result means that vibrational force fields computed at a relatively cheap level have sufficient accuracy to make highly accurate predictions of the centrifugal distortion effects in submillimeter spectra. The technique should be very useful in radioastronomy and other scientific fields where precise prediction of rotational transition frequencies is needed.

Another area of potential application of computed force fields is in the interconversion of different kinds of experimentally determined geometrical structures. As discussed in detail in Part 1, chap 6 of this series, every experimental method for determining the structure of molecules produces some sort of vibrationally averaged structure, and the type of average is different for each method. The differences between the various experimental definitions of structure are significantly larger than the experimental errors in good quality experiments, so that comparison of the results obtained by one method with those arising from another is not straightforward. Also, as explained in Part 1, chap. 7, all of the kinds of structural parameters coming from experiment are different from the equilibrium structure obtained from quantum chemical computation.

Conversion from one type of vibrationally averaged structure to another is possible if the harmonic and anharmonic force fields of the molecule are known. It is conventional to look to vibrational spectroscopic results or to force constants transferred from other, related molecules to obtain this information, and the accuracy obtained in this way is generally adequate to bring the "definition error" in the structure down below the level of the experimental error. There are, however, interesting systems for which sufficient vibrational spectroscopic information is not available and where there are no well-studied related molecules to furnish a good guess at the force constants. High accuracy in the force field is not needed to produce structural conversions, so the quadratic and cubic force fields obtained from relatively low level ab initio calculations should readily furnish the needed data.

In a similar manner, the data obtained from a gas-phase electron diffraction experiment is usually insufficient to determine all of the molecular interatomic distances and vibrational amplitudes. The latter are often fixed at values calculated from simple normal coordinate analysis schemes. When unusual bonding patterns occur in a molecule, however, the results obtained from ab initio calculation of the force constants should lead to improved accuracy.

6 Acknowledgement

The major portion of the work described here that was done at The University of Texas was supported by a series of grants from The Robert A. Welch Foundation. Support is also acknowledged by grants from Cray Research, Inc and from the Texas Advanced Technology Program.

7 References

1. Herzberg G (1945) Molecular spectra and molecular structure. Vol. II. Infrared and raman spectra of polyatomic molecules, Van Nostrand Reinhold, New York
2. Wilson EB, Decius JC, Cross PC (1955) Molecular vibrations, McGraw-Hill, New York
3. Woodward LA (1972) Introduction to the theory of molecular vibrations and vibrational spectroscopy, Oxford University Press, Oxford
4. Graybeal JD (1988) Molecular spectroscopy, McGraw-Hill, New York
5. Califano S (1976) Vibrational states, John Wiley, New York
6. Pulay P (1977), Direct use of the gradient for investigating molecular energy surfaces, chap 3. In: Schaefer HF III Modern theoretical chemistry. vol 4. Applications of electronic structure theory, Plenum, New York
7. Pulay P (1981) In: Deb BM (ed.), The force concept in chemistry, Van Nostrand Reinhold, New York, p 449
8. Fogarasi G, Pulay P (1984) In: Durig JR (ed.) Vibrational spectra and structure, Elsevier, Amsterdam p 125
9. Fogarasi G, Pulay P (1984) Ann rev phys chem 35: 191
10. Hess A Jr., Schaad LJ, Carsky P, Zahradnik R (1986) Chem rev 86: 709
11. Pulay P, Fogarasi G, Pang F, Boggs JE (1979) J am chem soc 101: 2550
12. Sayvetz A (1939) J chem phys 6: 383
13. Barrow GA (1962), McGraw-Hill, New York, pp 116 et seq
14. Born M, Oppenheimer R (1927) Ann physik 84: 457

15. Herzberg G (1966) Molecular spectra and molecular structure, vol III. Electronic spectra, Van Nostrand, p 23
16. Wilson EB (1941) J chem phys 9: 76
17. Pulay P (1969) Mol phys 17: 197
18. Pulay P (1979) Theor chim acta 50: 299
19. The current version of the TEXAS program system for ab initio molecular orbital calculations and calculations of harmonic and anharmonic vibrational behavior may be obtained from Dr. James E. Boggs, Department of Chemistry, the University of Texas, Austin, Texas 78712, U.S.A. The program is presently implemented for operation on Cray computers with UNICOS operating systems
20. Hellmann J (1937) Einführung in die Quantenchemie, Deuticke & Co., Leipzig
21. Feynman RP (1939) Phys rev 56: 340
22. Bratoz S (1958) Colloq intl CNRS 82: 287
23. Gerratt J, Mills IM (1968) J chem phys 49: 1719
24. Pople JA, Krishnan R, Schlegel HB, Binkley JS (1979) Int j quantum chem symp 13: 225
25. Pulay P, Lee, J-G, Boggs JE (1983) J chem phys 79: 3382
26. Duncan JL, Mills IM (1964) Spectrochim acta, Part A, 20: 523
27. Langhogg SR, Davidson ER (1974) Int j quantum chem 8: 61
28. Strey J, Mills IM (1973) Mol phys 26: 129
29. Murrell JN, Carter S, Halonen LO (1982) J mol spectrosc 93: 307
30. Suzuki I, Pariseau MA, Overend LO (1966) J chem phys 44: 3561
31. Allen WD, Yamaguchi Y, Császár AG, Clabo DA, Jr., Remington RB, Schaefer HF III (in press)
32. Saebø S, Pulay P, Boggs JE, work to be published
33. Xie Y, Fan K, Boggs JE (1986) Mol phys 58: 401
34. Fan K, Xie Y, Boggs JE (1986) J mol struct (theochem)) 136: 339
35. Császár P, Császár AG, Harsányi L, Boggs JE (1986) J mol struct (theochem) 136: 323
36. Pulay P, Fogarasi G, Pongor G, Boggs JE, Vargha A (1983) J am chem soc 105: 7037
37. Pulay P, Fogarasi G, Boggs JE (1981) J chem phys 74: 3999
38. Pongor G, Pulay P, Fogarasi G, Boggs JE (1984) J am chem soc 106: 2765
39. Pongor G, Fogarasi G, Boggs JE, Pulay P (1985) J mol spectrosc 114: 455
40. Sellers H, Pulay P, Boggs JE (1985) J am chem soc 107: 6487
41. Niu Z, Dunn K, Boggs JE (1985) Mol phys 55: 421
42. Xie Y, Boggs JE (1986) J computat chem 7: 158
43. Fogarasi G, Császár AG (1988) Spectrochim acta, Part A, 44: 1067
44. Császár AG, Fogarasi G (1988) Spectrochim acta, Part A, 45: 845
45. Császár AG, Fogarasi G (1980) J phys chem 93: 7644
46. Lemert RF, Dunn K (1990) Thirteenth Austin Symposium on Molecular Structure, Austin, Texas, March 12–14, 1990
47. Pulay P, Török F (1973) Mol phys 25: 1153
48. Török F, Hegedüs A, Pulay P (1973) Theor chim acta 32: 145
49. Harsányi L, Császár P (1983) Acta chim hung 113: 257
50. Simandiras ED, Rice JE, Lee TJ, Amos RD, Handy NC, in press
51. Lee TJ, Willetts A, Gaw JF, Handy NC (1989) J chem phys 90: 4330
52. Whitehead JR, Handy NC (1975) J mol spectrosc 55: 356
53. Dunn KM, Boggs JE, Pulay P (1986) J chem phys 85: 5838
54. Dunn KM, Boggs JE, Pulay P (1987) J chem phys 86: 5088
55. Dunn KM (1987) Chem phys lett 139: 165
56. Xie Y, Boggs JE (1989) J chem phys 90: 4320
57. Xie Y, Boggs JE (1989) J chem phys 91: 1066
58. Davidson ER (1975) J comp phys 17: 87
59. Charles SW, Cullen FC, Owen NL, Williams GA (1987) J mol struct 157: 17
60. Fan K, Boggs JE (1987) J mol struct 157: 31
61. Császár P, Császár A, Somogyi A, Dinya Z, Holly S, Gál M, Boggs JE (1986) Spectrochim acta 42A: 473
62. Somogyi A, Jalsovszky G, Fülöp C, Stark J, Boggs JE (1989) Spectrochim acta A 45: 679
63. Wlodarczak G, Burie J, Demaison J, Vormann K, Császár AG (1989) J mol spectrosc 134: 297

Some Aspects of the Quantumchemical Interpretation of Integrated Intensities of Infrared Absorption Bands

H. P. Figeys[1], P. Geerlings[2]

[1] Department of Organic and Physical Organic Chemistry Faculty of Sciences Free University of Brussels 50, Av. F. D. Roosevelt 1050 Brussels (Belgium)
[2] Eenheid Algemene Chemie (ALGC) Fakulteit Wetenschappen Vrije Universiteit Brussel Pleinlaan 2 1050 Brussels (Belgium)

Integrated Intensities of Infrared Absorption Bands corresponding with fundamental transitions, calculated in the Double Harmonic Approximation and with single determinantal wavefunctions can be analyzed in various ways in order to obtain an insight into the various electronic and vibrational factors determining the intensities. The computational strategy for obtaining dipole moment derivatives, with respect to the normal coordinates governing these intensities, is discussed together with the decomposition in non-local and LMO contributions.

Examples from work on aliphatic nitriles and pyramidal AX_3 molecules illustrate the influence of the quality of the wavefunction, the force field and the localization procedure on the results. The interplay between vibrational and electronic factors is analyzed in some typical cases. Decomposition of dipole moment derivatives in non-local contributions shows the possible failure of a classical point charge model in interpreting integrated intensities. The LMO decomposition is used in order to investigate the contribution of the various structural entities to the calculated intensity (evolutions), to discuss in a non-empirical way the validity of the bond moment hypothesis, and to investigate incomplete orbital following effects in stretching and bending modes. Finally, this procedure is used to obtain an insight into the OH stretching intensity increase in formic acid upon dimerization, a typical example of the order-of-magnitude increase of a AH stretching intensity upon H-bond formation.

1 Introduction

For several decades the frequency and intensity characteristics of an infrared spectrum have provided important experimental information for revealing the molecular structure and for gaining an insight into the electron distribution in a molecule at equilibrium geometry and upon distortion during molecular vibrations [1]. It is well known that the integrated intensity of an IR absorption band is much more sensitive than the frequency to intra- and intermolecular perturbations, e.g. changes in substitution pattern or changes in aggregation state and solvent. The experimental data for the frequency and intensity of the CN stretching vibration in the chloro-acetonitriles $CH_{3-n}Cl_nCN$ (n = 0, 1, 2, 3) are a beautiful illustration of this phenomenon [2] (Table 1).

For a given aggregation state or solvent, the intensity may vary by a factor 10 upon changing the number of Cl-atoms and for a given molecule the intensity may change up to a factor 6 when passing from one medium to another. The frequency range however is only 18 cm^{-1} which amounts to only 0.8 % of the average value.

A second well-known example is the evolution of the vibrational characteristics of the protondonor group A—H upon interaction with a protonacceptor B upon hydrogen bonding. Whereas the frequency of the A—H stretching vibration decreases typically by 10%, the integrated intensity is drastically enhanced (up to a factor 10). This combination of phenomena has already been used a long time ago for the detection of hydrogen bonding interaction [3].

This type of experimental data was often interpreted in terms of the electronic structure of the molecule and its response to out-of-equilibrium displacements of the nuclei during molecular vibrations, taking into account that frequencies are mainly related to the second derivatives of the electronic energy with respect to the normal coordinates Q_i, whereas intensities are, in a first approximation, proportional to the square of the first derivatives of the molecular dipole moment μ with respect to these normal coordinates, $(\partial\mu/\partial Q_i)_0$ [4, 5].

Physicochemical concepts such as electronegativity, hybridization, inductive and mesomeric effects, successfully used to interpret structure and reactivity of inorganic and organic compounds, mostly account for the frequency changes but often fail for the interpretation of this experimental intensity sequence. As an

Table 1. The chloroacetonitriles ($CH_{3-n}Cl_nCN$): experimental frequencies (v in cm^{-1}) and integrated intensities (A in km mol^{-1}) for the CN stretching vibration

	Gasphase		CCl$_4$ solution		CHCl$_3$ solution	
	v	A	v	A	v	A
n = 0	2267	1.4	2255	4.2	2255	8.6
n = 1	2266	0.78	2260	0.76	2261	0.90
n = 2	2261	6.2	2259	2.2	2260	1.1
n = 3	2256	11	2249	6.4	2250	4.8

example we give in Table 2 the CO stretching frequency and intensity (gas phase) of X_2CO molecules (X=H, Cl, F). When passing from H to F via Cl we expect a destabilization of the dipolar resonance structure (2) due to an increase in electronegativity of X. A CO bond strengthening along the series H, Cl, F is then expected in agreement with the experimentally found decrease in the CO bond distances [10–12] and the corresponding increase in force constants [13–15]. Neglecting mass effects and coupling phenomena, an increase in the CO stretching frequency is then foreseen in agreement with experiment. If we now try to interpret the CO intensity evolution, as often done in a more qualitative context [16], by relating it to the **static** charge distribution at equilibrium, as reflected in the total dipole moment, we then expect an intensity lowering along the H, Cl, F series as the contribution of the dipolar form (2) decreases. Experimentally, however, a strong intensity increase is found. This increase again is larger than the frequency shift (a factor 6.5 between H and F as compared to a 10% frequency increase).

Relatively simple models also fail when one tries, starting from experimental intensity data, to obtain the "polar properties of valence bonds" such as bond moments and their derivatives. The failure of simple bond-moment theory [17] leading to internally inconsistent results, is a striking example [17, 18]. In this model the total molecular dipole moment is written as a sum of bond moments μ_i and in its zero-order form [19] the hypothesis is put forward that, if a single bond is elongated, the moments associated to all other bonds are unchanged. Upon refinement of the model (first order theory [20]) by introducing cross terms of the type $(\partial \mu_i / \partial R_j)_0$ (i ≠ j) the number of (so-called) electro-optical parameters to be determined from experimental data rapidly becomes too large [21]. Orville-Thomas and co-workers tried to cope with the deficiency of the model by the introduction of theoretically calculated correction terms [22].

Among other interpretational techniques (for a detailled account see [23]) presented to reduce **experimental** integrated intensities we mention the interpretation

Table 2. X_2CO (X = H, Cl, F) molecules: experimental frequencies, v (cm^{-1}) and integrated intensities A (km mol^{-1}) (gasphase) for the CO stretching vibration

X	v	A
H	1746 [6]	53 [8] [9]
Cl	1827 [7]	246 [7]
F	1928 [7]	383 [7]

via the derivatives of the total molecular dipole moment with respect to internal (R_i) or symmetry (S_i) coordinates. The corresponding derivatives $(\partial\mu/\partial R_i)_0$ and $(\partial\mu/\partial S_i)$ can be obtained from the experimental $(\partial\mu/\partial Q_i)_0$ values if the L_R^{-1} and L_S^{-1} matrices, connecting internal or symmetry coordinates with the normal coordinates, are known [4]. This approach has been followed, among others, in order to resolve the sign ambiguity in the $(\partial\mu/\partial Q_i)_0$ components as only the square of $(\partial\mu/\partial Q_i)_0$ is experimentally available. A different method was put forward by Morcillo et al. [24] in their atomic polar tensor formalism, which was thereafter reformulated by Person [25]. The quantities of interest in this theory and in the related atomic effective charge theory developed by King [26] are the partial derivatives of the total molecular dipole moment with respect to the cartesian displacement coordinates x_α, y_α, z_α of the various nuclei α.

From this very concise description of interpretational techniques it follows that none of these offers a **non-empirical** description of the actual reorganization of the molecular charge distribution in the various **bonds** of the molecule upon vibration. This quantity would yield some "chemical" insight into vibrational intensities. These considerations formed the starting point for the interpretational method the present authors presented some years ago [27] and which was applied to a variety of cases later on. Starting from quantum-chemically calculated $(\partial\mu/\partial Q_i)_0$ values, a decomposition of the total dipole moment into a sum of bond dipole and lone pair moments μ_j, via the localized molecular orbital technique, both for the equilibrium geometry and for a displacement along Q_i, yields $(\partial\mu_j/\partial Q_i)_0$ values which reflect the electronic reorganization in the various structural entities of the molecule upon vibration. In this approach we also decomposed the dipole moment derivatives in non-local terms (point charge, hybridization and homopolar terms) in order to obtain a parameter free estimate of the importance of e.g. the "classical" point-charge term. Before starting the LMO-analysis, we usually performed a decomposition of $(\partial\mu/\partial Q_i)_0$ into $(\partial\mu/\partial S_j)_0$ in order to obtain information on the interplay between electronic and vibrational factors in determining the intensity. This procedure was (eventually partly) applied to study the stretching vibrations in cyanoacetylene [27], the $C\equiv N$ (and partially $C-C$) vibrations in the chloroacetonitriles and HCN [28] [29] [30], the stretching and bending vibrations in pyramidal AX_3 type molecules $(A = N, P; X = H, F)$ [31] [32] [33] and some selected vibrations in the formic acid monomer and dimer [34] [35].

The combined approach sketched above was not found elsewhere in the literature, neither at the start of our work in this field, nor recently. An LMO approach was also followed by Bruns [36], shortly after the communication of our first results [37], however restricted to the analysis of CNDO-type $(\partial\mu/\partial S_i)_0$ values in NH_3. NDO-type functions at that time were considered by us to be insufficient for these purposes, however [27]. The same remark also pertains to the more recent work on $(\partial\mu/\partial R_i)_0$ derivatives by Shinoda and Miyazaki [38, 39]. The dipole moment decomposition in non-local contributions is also present in the work of Orville-Thomas and co-workers [22, 40] but as mentioned above used in a different context. Consequently, in this account, we will mainly concentrate on our own contributions to this field. After a brief description of the theoretical background in Sect. 2, we will illustrate in Sect. 3 (Results and Discussion) the main conclusions of our work using some selected examples.

2 Theoretical Background

2.1 The Basic Formula for the Integrated Intensity in the Double Harmonic Approximation

Within the Born-Oppenheimer approximation scheme [41] the wavefunction describing the nuclear vibrations is derived from an eigenvalue problem in which the electronic energy for fixed nuclei arrangements occurs as the potential. The eigenvalues of the vibrational Schrödinger equation can be identified as the vibrational energy levels and their differences as the various transition energies observed in infrared spectroscopy. Knowing the eigenfunctions Ξ resulting from the vibrational Schrödinger equation one can calculate the transition dipole matrix elements:

$$\int \Xi_p^* \mu_{op} \Xi_{p'} \, d\tau$$

governing the integrated intensity of the vibrational transition between the states with labels p and p'.

$$\int \Xi_p^* \mu_{op} \Xi_{p'} \, d\tau$$

If the potential is approximated by a series expansion, around the equilibrium geometry, involving only quadratic terms (mechanical harmonic approximation) Ξ_p can be factorized into a product of one-dimensional Harmonic Oscillator functions $\xi_{v_i}(Q_i)$ where v_i denotes the vibrational quantum number of the i-th oscillator (i = 1, 2 ... 3m-6 where m is the number of atoms; (3m-5) in the case of a linear molecule) with associated normal coordinate Q_i. Expanding the dipole moment μ as a series in the Q_i and retaining only the constant and linear terms (electrical harmonic approximation):

$$\mu = \mu_{eq} + \sum_i \left(\frac{\partial \mu}{\partial Q_i} \right)_0 Q_i \tag{1}$$

it can easily be shown that the only non-necessarily vanishing matrix elements:

$$\int \Xi_p^* \mu \, \Xi_{p'} \, d\tau$$

are those in which only one of the ξ_{v_i} functions changes its quantum number by unity [4, 5]. These transitions are the only ones allowed in the so-called **double harmonic approximation** (DHA), their intensity being obviously proportional to the square of the corresponding dipole moment derivative after summation over "hot transitions". This DHA proved to be adequate in the large majority of cases for the fundamentals [42→45]. The final expression for the integrated intensity A_i of the fundamental associated to the i-th mode of vibration is given by:

$$A_i = \frac{N\pi}{3c^2} d_i \left| \left(\frac{\partial \mu}{\partial Q_i} \right)_0 \right|^2 \tag{2}$$

where N represents Avogadro's constant, c the velocity of light, and d_i the degeneracy of the mode [5].

Obviously, a quantum-chemical calculation of A_i relies upon the knowledge of the normal coordinate Q_i and the variation of the total dipole moment as a function of Q_i.

2.2 Computational Strategy for $(\partial\mu/\partial Q_i)_0$ Derivatives

The normal coordinates are obtained via the classical Wilson **GF**-matrix procedure [4], using Schachtschneider's programs [46]. Experimental geometries and force fields were used throughout. The dipole moment derivative $(\partial\mu/\partial Q_i)_0$ is calculated in a finite difference approach as:

$$\langle\mu(Q_i = 0.01 \text{ amu}^{1/2} \text{ Å})\rangle - \langle\mu_{eq}\rangle/0.01 \tag{3}$$

i.e. by evaluating the molecular dipole moment at the equilibrium geometry and for a geometry distorted from equilibrium along Q_i. The nuclear cartesian coordinates needed for the latter calculation, gathered in a column vector **X**, are given by:

$$\mathbf{X} = \mathbf{X}_{eq} + \mathbf{TQ} \tag{4a}$$

where $\mathbf{X}_{eq}$ denotes the corresponding matrix of the equilibrium cartesian coordinates. **Q** is the column matrix of the normal coordinates and the matrix **T** is given by:

$$\mathbf{T} = \mathbf{M}^{-1}\mathbf{B}^t\mathbf{G}^{-1}\mathbf{L} \tag{4b}$$

where **M** denotes the (diagonal) nuclear mass matrix. **B** connects the internal (**R**) and cartesian displacement coordinates and **L** relates internal to normal coordinates:

$$\mathbf{R} = \mathbf{BX} \qquad \mathbf{R} = \mathbf{LQ} \tag{5}$$

In order to calculate the cartesian coordinates (in Å) for $Q_i = 0.01$ amu$^{1/2}$ Å^{-1}, Eq. (4a) indicates that one simply multiplies the elements of the i-th column of the **T** matrix by 0.01 and adds the results to the elements of $\mathbf{X}_{eq}$ (in Å).

For the dipole moment calculations we always worked at the single determinantal level for the electronic wavefunction, considering only closed shell systems. Semi-empirical (INDO [47]) and a deorthogonalized [48] version) and ab initio methods were used, the latter with basis sets varying from minimal to near Hartree-Fock (in some selected cases). Use was made of the CNINDO program [49] (with deorthogonalization [50]) for the semi-empirical calculations and the GAUSSIAN 70 and 76 ab initio programs [51, 52]. In order to test the validity of the finite difference

---

[1] Conversion to SI units: 1 amu$^{1/2}$ Å $= 4.07496\times10^{-24}$ kg$^{1/2}$ m. Conversion to SI units of other units used in the present work: 1 Å $= 10^{-10}$ m; 1 D $= 3.33564\times10^{-30}$ Cm; 1 D Å$^{-1} = 3.33564$ $\times 10^{-20}$ C; 1 D amu$^{-1/2}$ Å$^{-1} = 8.18570\times10^{-7}$ kg$^{-1/2}$ C.

approach, the $\mu = \mu(Q)$ curve was calculated for the CN vibration in HCCCN both with semi-empirical methods and with the minimal basis set (STO-3G) ab initio [53] method. An almost perfect linear relationship between μ and Q_{CN} (taken to vary between 0 and 0.10 amu$^{1/2}$ Å) was obtained in all cases [27]. Note that the maximal Q_{CN} value considered corresponds to a CN distance increase of 0.0288 Å which is of the order of the calculated root mean square amplitude of vibration [54]. Similar conclusions were recently obtained for polarizability derivatives [55], governing Raman vibrational intensities [56], the polarizability tensor components being obtained via an improved virtual orbital [57, 58] technique with symmetry adapted basis [59, 60]. We finally mention that in this account a positive Q_i value should always be interpreted as a displacement corresponding to an increase in length or angle of the internal coordinate which lends its name to the normal coordinate considered.

2.3 Dipole Moment Decomposition in Non-local Terms

We consider a 2N-electron closed shell molecule whose electronic wavefunction is approximated by a Single Slater determinant Φ, built up from a set of N doubly occupied molecular orbitals $\{\varphi_i\}$. The molecular dipole moment can then be written as the expectation value of the corresponding operator μ_{op}:

$$\langle \mu \rangle = \sum_A Z_A R_A - 2 \sum_{i=1}^N \langle r_i \rangle \tag{6}$$

where R_A denotes the position vector of nucleus A, with charge $Z_{A'} \langle r_i \rangle$ is defined as:

$$\langle r_i \rangle = \int \varphi_i^*(r) \, r \varphi_i(r) \, dr \tag{7}$$

Using an expansion of φ_i in a set of real basis functions χ_λ:

$$\phi_i = \sum_\lambda C_{\lambda i} \chi_\lambda \tag{8}$$

we can decompose $\langle \mu \rangle$ into a sum of five terms:

$$\langle \mu \rangle = \sum_A \left(Z_A - \sum_\lambda^A P_{\lambda\lambda} \right) R_A - \sum_A R_A \sum_{\lambda \neq \sigma}^A P_{\lambda\sigma} S_{\lambda\sigma} - \sum_A \sum_{\lambda \neq \sigma}^A P_{\lambda\sigma} \varrho_{\lambda_A \sigma_A}$$

$$\qquad\qquad (1) \qquad\qquad\qquad (2) \qquad\qquad\qquad (3)$$

$$- \sum_{A \neq B} \frac{R_A + R_B}{2} \sum_\lambda^A \sum_\sigma^B P_{\lambda\sigma} S_{\lambda\sigma} - \sum_{A \neq B} \sum_\lambda^A \sum_\sigma^B P_{\lambda\sigma} \varrho_{\lambda_A \sigma_B} \tag{9}$$

$$\qquad\qquad\qquad (4) \qquad\qquad\qquad\qquad (5)$$

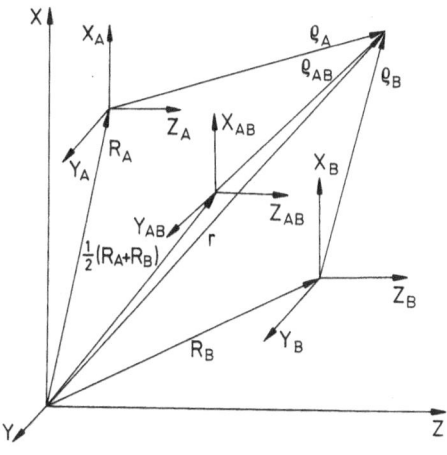

Fig. 1. Orientation of the various systems of axes in the non-local dipole moment decomposition scheme. From Figeys HP, Geerlings P, Van Alsenoy C, J. Chem. Soc., Faraday Transactions II, 75: 528 (1979). Reprinted by permission of the Royal Society of Chemistry

where $S_{\lambda\sigma}$ denotes the overlap integral between orbitals χ_λ and χ_σ and $P_{\lambda\sigma}$ is an element of the density matrix $\mathbf{P}$:

$$P_{\lambda\sigma} = 2 \sum_i^N C_{\lambda i} C_{\sigma i} \tag{10}$$

$\varrho_{\lambda_A\sigma_A}$ and $\varrho_{\lambda_A\sigma_B}$ denote matrix elements of the vectors ϱ_A and ϱ_{AB}:

$$\varrho_{\lambda_A\sigma_A} = \int \chi_{\lambda_A}(\varrho_A)\, \varrho_A\, \chi_{\sigma_A}(\varrho_A)\, d\varrho_A$$

$$\varrho_{\lambda_A\sigma_B} = \int \chi_{\lambda_A}(\varrho_{AB})\, \varrho_{AB}\, \chi_{\sigma_B}(\varrho_{AB})\, d\varrho_{AB} \tag{11}$$

defined in Fig. 1: the origin of ϱ_A is at nucleus A, the local system of axes X_A, Y_A, Z_A taken parallel to the (X, Y, Z) system of axes. The origin of ϱ_{AB} is at the midpoint of the line joining A and B. The local system of axes X_{AB}, Y_{AB}, Z_{AB} is again taken parallel to the (X, Y, Z) system of axes.

The summations over A and B run over all nuclei of the molecule. If wavefunctions of the NDO-type are used, term (**2**) vanishes due to the orthonormality of the valence basisfunctions on a single atom whereas term (**4**) may not be retained due to translational invariance requirements [50]. In this case term (**2**) (the hybridization term) is usually retained [61] whereas the last term (**5**) (the homopolar term [40, 62]) is dropped. If the NDO wavefunctions are deorthogonalized, the diatomic overlap term (**4**) should be retained due to translational invariance requirements; for reasons of internal consistency, the homopolar term should then also be included (although not necessary on the basis of translational invariance requirements). The monoatomic overlap term (**2**) still vanishes. This is no longer the case in ab initio methods where this term should be taken into account. Grouping terms in Eq. (9) we can write $\langle\mu\rangle$

as the sum of the contributions which individually obey translational (and also rotational) invariance requirements [27]:

$$\langle \mu \rangle = \sum_A \left[Z_A - \sum_\lambda^A \left(P_{\lambda\lambda} - \sum_{\substack{\sigma \\ \neq \lambda}}^A P_{\lambda\sigma} S_{\lambda\sigma} - \sum_{\substack{B \\ \neq A}}^B \sum_\sigma P_{\lambda\sigma} S_{\lambda\sigma} \right) \right] \mathbf{R}_A$$

(I)

$$- \sum_A^A \sum_{\substack{\lambda\neq\sigma}}^A P_{\lambda\sigma} \varrho_{\lambda^A \sigma^A} - \sum_{A \neq B}^A \sum_\lambda \sum_\sigma^B P_{\lambda\sigma} \varrho_{\lambda^A \sigma^B} \tag{12}$$

(II)   (III)

On the basis of the Mulliken population analysis [63] the first term (**I**) can be identified as the point charge term:

$$\sum_A Q_A \mathbf{R}_A$$

where Q_A represents the charge on atom A. The second term is an extension of the hybridization term in Eq. (10) and includes hybridization effects between orbitals differering in their principal quantum numbers, this concept loosing its meaning in the case of large extended basis sets. The final term (**III**) is the homopolar term. The decomposition represented in Eq. (12) can be used to decompose dipole moment derivatives via Eq. (3).

2.4 LMO Decomposition of the Dipole Moment

2.4.1 Localization Procedures

If use is made of a single determinantal wavefunction, the molecular orbitals $\{\varphi_i\}$ obtained via the Hartee-Fock-Roothaan equations [64], which are usually delo-calized over the entire molecular system, can be localized via a unitary transformation under which the total wavefunction is invariant (expect for a phase factor) [65]. Using a proper transformation matrix, the classical electronic building blocks of a molecule (two center bonds, lone pairs, inner shells . . .) are thus retrieved. Various localization procedures for the obtention of this transformation matrix have been proposed in the literature [66]. In our work two localization procedures have been used. For the NDO-type wavefunctions and (in an initial phase) the STO-NG wavefunctions we used an INDO approximation to Von Niessen's charge density localization procedure [67], previously developed by us [68], which proved to be successful in the interpretation of directly bonded CH and CC NMR coupling constants [69, 70].

It is based on the extremalization of the localization function:

$$L = \sum_i \iint \phi_i^2(\mathbf{r}_1) \, \delta(\mathbf{r}_1 - \mathbf{r}_2) \, \phi_i^2(\mathbf{r}_2) \, d\mathbf{r}_1 \, d\mathbf{r}_2 \tag{13}$$

where δ stands for the Dirac delta function. This method is based on the same principles as the Edmiston-Rüdenberg procedure [71], namely the extremalization of the intra orbitalar Coulomb repulsion:

$$\sum_i \iint \phi_i^2(\mathbf{r}_1) \, r_{12}^{-1} \phi_i^2(\mathbf{r}_2) \, d\mathbf{r}_1 \, d\mathbf{r}_2 \tag{14}$$

the r_{12}^{-1} interaction being replaced by a delta function. In this way, the computation of two electron integrals is replaced by their much simpler one electron counterparts. Nevertheless, upon expanding the MOs as a linear combination of AOs the number of one-electron integrals to be calculated is still very large. Therefore, an INDO approximation [47] was introduced, yielding a localization function:

$$L = \sum_i \left\{ \sum_A^A \sum_\lambda C_{\lambda i}^4 [\lambda^2 \| \lambda^2] + 3 \sum_{\lambda \neq \sigma} \sum C_{\lambda i}^2 C_{\sigma i}^2 [\lambda^2 \| \sigma^2] \right.$$

$$\left. + \sum_{A \neq B} \sum_\lambda^A \sum_\sigma^B C_{\lambda i}^2 C_{\sigma i}^2 \zeta_{AB}^{n_\lambda l_\lambda, \, n_\sigma l_\sigma}(s, s) \right\} \tag{15}$$

Here, $[\lambda^2 \| \sigma^2]$ stands for the one center charge density overlap integral between the orbitals χ_λ and χ_σ:

$$[\lambda^2 \| \sigma^2] = \iint \chi_\lambda^2(\mathbf{r}_1) \, \delta(\mathbf{r}_1 - \mathbf{r}_2) \, \chi_\sigma^2(\mathbf{r}_2) \, d\mathbf{r}_1 \, d\mathbf{r}_2 = \int \chi_\lambda^2(\mathbf{r}) \, \chi_\sigma^2(\mathbf{r}) \, d\mathbf{r} \tag{16}$$

For the two center integrals an average value $\zeta_{AB}^{n_\lambda l_\lambda, \, n_\sigma l_\sigma}(s, s)$ is computed over s-functions in order to ensure rotational invariance. Moreover, it can be shown that, if for third row atoms an average value is taken for integrals of the type $[2p^2 \| 3p'^2]$, rotational invariance is completely accounted for [68, 72]. A program LOCALIS was written to localize NDO or ab initio STO-NG functions by this procedure [73].

For the localization of more accurate wavefunctions, involving extended bases, we used the well-known Boys' localization procedure [74], based upon the maximalization of the distances between the centroids of charge of the different orbitals. A program BOYLOC was written for this purpose [75] and coupled to the GAUSSIAN 76 package.

2.4.2 Bond Moment (Derivative) Calculation

Having obtained the localized molecular orbitals for a given nuclear configuration, the bond moments μ_i can easily be calculated as:

$$\langle \mu_i \rangle = \sum_A Z_A^{(i)} \mathbf{R}_A - 2\langle \mathbf{r}_i \rangle \tag{17}$$

with

$$\langle \mu \rangle = \sum_i \langle \mu_i \rangle \tag{18}$$

where the summation extends over all doubly occupied LMOs φ_i.

The nuclear charge partitioning is as follows (cfr. [76]): $Z_A^{(i)} = +1$ if φ_i represents a two center bond LMO involving atom A; $Z_A^{(i)} = +2$ if φ_i represents a lone pair or inner shell centered on A; $Z_A^{(i)} = 0$ in all other cases. Obviously:

$$\sum_i Z_A^{(i)} = Z_A \qquad \sum_A Z_A^{(i)} = +2$$

Combining now Eqs. (3), (17), and (18) we obtain:

$$\left(\frac{\partial \langle \mu \rangle}{\partial Q_i}\right)_0 \approx \frac{\langle \mu(Q_i = 0.01\,\text{amu})\rangle - \langle \mu_{eq}\rangle}{0.01}$$

$$= \sum_j \frac{\langle \mu_j(Q_i = 0.01\,\text{amu})\rangle - \langle \mu_{j,\,eq}\rangle}{0.01} \approx \sum_j \left(\frac{\partial \langle \mu_j \rangle}{\partial Q_i}\right)_0 \tag{19}$$

The working formula:

$$\left(\frac{\partial \langle \mu \rangle}{\partial Q_i}\right)_0 \approx \sum_j \frac{\mu_j(Q_i = 0.01\,\text{amu}) - \langle \mu_{j,\,eq}\rangle}{0.01} \tag{20}$$

is justified by the linearity of $\langle \mu \rangle$ as a function of Q_i in the vicinity of the equilibrium geometry (cfr. Sect. 2.2).

3 Results and Discussion

3.1 Factors Influencing Calculated (Bond) Dipole Derivatives

3.1.1 Influence of the Quality of the Wavefunction

The study of the stretching vibrations of cycanoacetylene and its deuterated isotopomer (for which the gas-phase intensities of all eight stretching vibrations are known [77]) was performed both with semi-empirical functions (INDO and INDO/D) and ab initio minimal basis set (STO-3G) functions. From the ratio between calculated and experimental $|(\partial \mu / \partial Q_i)_0|$ values, proportional to the square root of the intensity (Table 3), it was clear that the INDO method yields insufficient accuracy. An average deviation factor of 4.3 between theoretical and experimental values was found. The use of deorthogonalized functions, which mostly yield better equilibrium charge distributions [50, 80] did not improve the result (average deviation factor 4.3).

This indicates that NDO-type wavefunctions are not suited for our purposes and that care should be taken when using these functions to resolve the sign ambiguity as often done in the literature [81]. They have not been used any longer in our later studies. A considerable improvement occurs when passing to the STO-3G functions where an average deviation factor of 2.1 is found (the results for STO-6G deviate only slightly from STO-3G). Moreover, an examination of all possible sign combinations for the experimental $(\partial \mu^{(x)} / \partial Q_i)_0$ values (the x-axis being the mole-

Table 3. Ratios of calculated and experimental $|(\partial\mu/\partial Q_i)_0|$ values in HCCCN and DCCCN. Calculations performed with Cyvin's force field [78] and Costain's r_s geometry [79]

Wavefunctions →	INDO		INDO/D		STO-3G	
Isotope →	H	D	H	D	H	D
Normal coordinate						
$Q_{C-H(D)}$	0.12	0.26	0.34	0.09	1.88	1.52
$Q_{C\equiv N}$	2.35	2.24	2.50	2.60	1.37	1.15
$Q_{C\equiv C}$	1.90	0.14	2.42	0.19	4.61	2.59
Q_{C-C}	0.24	0.23	2.71	4.95	1.31	0.38

cular axis) reveals that the most acceptable choice for relative signs is the one resulting from the STO-3G calculations.

Obviously, STO-3G performs rather well in this case. This is not always so (vide infra) and in some cases (e.g. the NH_3 molecule), the excellent agreement obtained may be qualified as fortuitous as emerged from our studies on AX_3-type molecules (Table 4). In a study on the stretchings and bendings in NH_3, PH_3, NF_3, and PF_3 an average deviation factor theory/experiment of 1.66 was found for STO-3G [32], increasing to 2.03 for Double Zeta [83] calculations [32, 33], indicating that a basis set extension of this type is not advisable.

The results with the "3.1" basis [84] (a 4-31G basis [85] to which is added a small-exponent polarizing p-GTO on hydrogen) are, however, better and are well distributed among stretching and bending modes (1.49 for the stretching modes, 1.56 for the deformation modes). The results obtained with this basis are close to those obtained for NH_3 with the much more elaborate DTZPD basis (Double Zeta on N, Triple Zeta on H, including polarization and diffuse functions) (average deviation factor 1.44). Note that throughout the complete series the signs of the components of $(\partial\mu/\partial Q_i)_0$ are unchanged (expect for the Q_3 (antisymmetric stretching mode) of NH_3 for which the sign instability is well known [86]) and correspond to the "experimental" ones proposed in the literature [33]. The same 3.1 basis performs very well in our calculations on some stretching intensities in the formic acid monomer and dimer (average deviation factors 1.2 and 1.3, respectively [34, 35]). On the contrary, in the monomer case unsatisfactory STO-3G results are obtained for the OH(D) and CH(D) stretching vibrations.

Concluding this section we may say that the influence of the quality of the wavefunction (even within the single determinantal framework) is considerable and that before passing to an interpretation of a calculated intensity (sequence) the results obtained should be dritically examined [87]. NDO-type wavefunctions are not suited and minimal basis set theories perform well in some cases but not in others. The "3.1" basis seems to represent a nice compromise between costs and efficiency. In order to have deviation factors smaller than say 1.2 to 1.3, correlation effects should be included [88], hampering, however, some of the interpretational techniques we used.

Table 4. Theoretical $(\partial\mu^{(\alpha)}/\partial Q_i)_0$ values [32] [33] (in D amu$^{-1/2}$ Å^{-1}) for the pyramidal AX$_3$ series (A = N, P; X = H, F) calculated with the STO-3G, DZ and 3.1 basis [82]. The non-vanishing component α is the z-component for Q_1 and Q_2 and the y-component for Q_{3a} and Q_{4a} the latter transforming as the y component of the doubly degenerate representation. The extended basis (DTZDP) results for ammonia are also included. Experimental intensity data are given for comparative purposes

	Q_1 (symm. stretching)	Q_2 (symm. bending)	Q_{3a} (asymm. stretching)	Q_{4a} (asymm. stretching)
NH$_3$				
STO-3G	−0.373	−1.713	−0.198	0.260
DZ	−0.271	−2.530	−0.027	0.797
3.1	−0.181	−2.029	0.111	0.662
DTZPD	−0.183	−2.165	0.257	0.743
Exp.	±0.40(±0.42)	±1.8(±1.8)	±0.19(±0.21)	±0.54(±0.61)
PH$_3$,				
STO-3G	−0.926	1.247	−1.015	−0.948
DZ	−1.168	0.633	−1.731	−0.745
3.1	−0.936	0.442	−1.441	−0.577
Exp.	±1.2	±0.66	±0.83	±0.52
NF$_3$				
STO-3G	−0.453	0.339	−0.789	−0.128
DZ	−1.229	0.423	−2.145	−0.049
3.1	−1.176	0.431	−1.967	−0.067
Exp.	±0.84	±0.19	±2.2	±0.13
PF$_3$				
STO-3G	−0.570	1.001	−1.088	−0.439
DZ	−1.835	1.349	−2.100	−0.579
3.1	−·1.719	1.373	−1.966	−0.597
Exp.	±1.6	±0.76	±2.2	±0.32

Orientation of the molecules

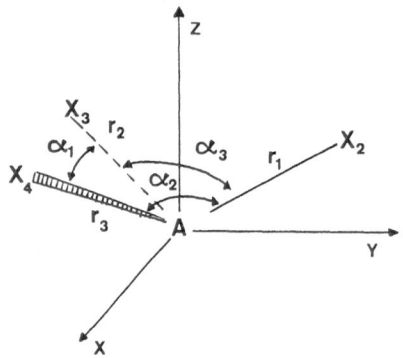

3.1.2 Influence of the Quality of the Force Field

In order to judge the relative influence of the quality of the force field on the calculated intensities, all calculations on the stretching modes in H—C≡C−·C≡N and D—C≡C—C≡N were performed using normal coordinates obtained from two strongly different experimental force fields: the above-mentioned general harmonic force field developed by Cyvin [78], obtained by introducing small interaction constants in Turrell's simple valence force field [89] and the force field published by

Table 5. Influence of the force field on the STO-3G calculated $(\partial\mu^{(x)}/\partial Q_i)_0$ values for HCCCN and DCCCN (in D amu$^{-1/2}$ Å^{-1})

Normal mode		Cyvin and Klaeboe [78]	Uyemura-Maeda [77]
$Q_{C-H(D)}$	HCCCN	+2.248	+2.240
	DCCCN	+1.055	+1.070
$Q_{C\equiv N}$	HCCCN	+0.666	+0.503
	DCCCN	+0.588	+0.442
$Q_{C\equiv C}$	HCCCN	+0.986	+1.086
	DCCCN	+1.453	+1.495
Q_{C-C}	HCCCN	−0.051	−0.131
	DCCCN	−−0.008	−0.090

Orientation of the molecule

$N \equiv C - C \equiv C - H(D) \xrightarrow{x}$

Uyemura and Maeda containing only two interaction constants [77]. The STO-3G results for $(\partial\mu/\partial Q_i)_0$ components obtained with both force fields [27] are given in Table 5. It is clear that, as compared to the influence of the quality of the wavefunction, the influence of the force field is relatively small, except for Q_{C-C} which is, however, an extremely low-intensity mode. Obviously, differences may become larger if, e.g., ab initio calculated (and unscaled) force fields are compared with force fields obtained by fitting procedures using experimental frequencies. The same features were also encountered in a study on NH_3 where two quadratic force fields, both published by Duncan and Mills [82e], were used. These force fields were obtained by fitting procedures on either harmonic or anharmonic frequencies and the differences obtained are again much smaller than those obtained between $(\partial\mu/\partial Q_i)_0$ values calculated with different wavefunctions (STO-3G, 3.1, DZ, DTZPD ...) for a given force field (cf. Sect. 3.1.1) [31].

3.1.3 Influence of the Localization Procedure

The interpretation of the STO-NG results for cyanoacetylene [27], the chloroacetonitriles [28] and (initially) NH_3 [31] was performed with our INDO-approximate charge density localization procedure [68]. The results obtained later on for the complete series of AX_3 type molecules [32, 33] and formic acid monomer and dimer [34, 35] were obtained via the Boys' localization method [74]. In Table 6 we present a comparison between the LMO partitioning of the STO-3G $(\partial\mu^{(z)}/\partial Q_1)_0$ value for NH_3 obtained with both localization procedures. It is seen that the small inner shell contribution in the INDO approximate method is further reduced when passing to Boys' localization, yielding a larger absolute value for both the NH-bond and N lone pair contributions. The global trends in the values are however identical indicating that the INDO-charge density results which will be reported later on can be looked upon with confidence as far as their qualitative aspects are concerned (an exception is the N lone pair contribution in the Q_2 vibration in NH_3 which is strongly underestimated in the INDO-charge density localization; LMO interpretation of this intensity will only be reported with Boys' localization).

Table 6. Influence of the localization procedure (INDO-Charge Density [68] vs Boys [74]) on the LMO partitioning of the STO-3G calculated $(\partial\mu^{(z)}/\partial Q_1)_0$ value for NH_3** (values in D amu$^{-1/2}$ Å^{-1})

LMO contribution	INDO-Charge Density	Boys
N inner shell	+0.052	+0.005
NH bonds*	−0.598	−0.726
N lone pair	+0.172	+0.347
Total	−0.373	−0.373

 * Sum of three equivalent contributions
** Orientation of the molecule: see Table 4. Force field and equilibrium
    geometry: see [82].

A word of caution should be written here: for very large bases such as the DTZPD basis in NH_3, the LMO interpretational method reaches its limits as the radial maximum of the most diffuse basis-function on N is situated at 1.146 Å from the N-nucleus, which is "beyond" the H nuclei.

3.2 Combined Influence of Electronical and Vibrational Factors on Calculated Dipole Moment Derivatives

3.2.1 The C≡N Stretching Intensity in the Chloroacetonitriles

The C≡N stretching gas-phase intensity in the chloroacetonitriles in Table 1 shows a remarkable behaviour as a function of the number of chlorine atoms. When going from acetonitrile to monochloroacetonitrile, the intensity slightly decreases from 1.4 to 0.78 km mol^{-1}, but when more chlorine atoms are introduced, an intensity increase to 6.2 (n = 2) and 11 (n = 3) km mol^{-1} is observed. Various explanations for this sequence have been offered [2, 90–93]. One of them, proposed by Roshchupkin and Popov [91], suggests that vibrational impurities of the CN normal coordinate are at the origin of the intensity variation: large differences in (bond) dipole moment derivatives with respect to the R_{CC} internal coordinate would be responsible for the large increase between n = 0 and n = 3. In the framework of our approach we can investigate whether vibrational factors might indeed contribute to this sequence.

A normal coordinate analysis for Q_{CN} shows that, whatever the force field used [94], the CN stretching normal coordinate is almost unchanged throughout the whole series: it is essentially a CN stretching movement (the $L^{-1}_{Q_{CN}, R_{CN}}$ values being: +2.274 (n = 0), +2.303 (n = 1), +2.284 (n = 2), +2.257 (n = 3)) with a non-negligible but practically constant CC shortening contribution (the $L^{-1}_{Q_{CN}, R_{CC}}$ values are: −0.447 (n = 0), −0.403 (n = 1), −0.431 (n = 2), −0.466 (n = 3)).

If we look at the calculated dipole moment derivatives (minimal basis STO-6G) in Table 7, we see that $(\partial\mu/\partial Q_{CN})_0$ increases almost linearly with the number of chlorine atoms and that no minimum at n = 1 is found. The ratio between theoretical and experimental values for n = 0, 2, and 3 lies between 2 and 2.5 (cfr. Sect. 3.1.1), whereas for n = 1 a totally unacceptable value of 5.1 is found. Also for the C−C

Table 7. STO-6G calculated $(\partial\mu^{(\alpha)}/\partial Q_{CN})_0$ and experimental dipole moment derivatives for the choroacetonitriles $CH_{3-n}Cl_nCN$ (n = 0, 1, 2, 3) (in D amu$^{-1/2}$ Å^{-1})

	n = 0	1	2	3
$\alpha = x$	−0.389	−0.688	−0.957	−1.230
$\alpha = y$	0	−0.070	+0.065	0
$|(\partial\mu/\partial Q_{CN})_0|$	0.389	0.692	0.959	1.230
Exp.	0.185	0.136	0.384	0.502
Ratio Theory/Exp.	2.103	5.088	2.497	2.450

Orientation of the molecules

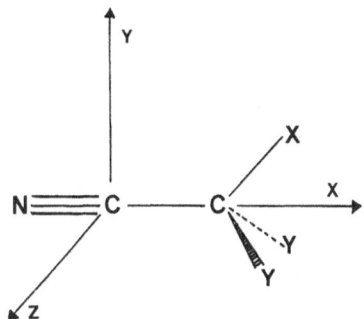

$$
\begin{array}{ll}
n = 0 & X = Y = H \\
n = 1 & X = H \quad W = Cl \\
n = 2 & X = Cl \quad W = H \\
n = 3 & X = Y = Cl
\end{array}
$$

Force field and geometries used: see [94]

intensities ratios of the order of 2 are found [28], so we suggest that a remeasurement of the CN intensity for n = 1 might be rewarding. Combining our theoretical results with the normal coordinate analysis we can say that the calculated intensity *evolution* cannot be ascribed to a vibrational effect, i.e. an important change of the CN normal coordinate upon successive chlorination. However, the *magnitude* of the intensity is strongly influenced by the vibrational impurity of the CN normal coordinate. Decomposing $(\partial\mu/\partial Q_{CN})_0$ in its most important contributions:

$$(\partial\mu^{(x)}/\partial Q_{CN})_0 = (\partial\mu^{(x)}/\partial R_{CN})_0 \, L_{R_{CN}, Q_{CN}} + (\partial\mu^{(x)}/\partial R_{CC})_0 \, L_{R_{CC}, Q_{CN}} + \cdots$$

it turns out that the values for these leading terms are −0.059 and −0.368 for CH_3CN and −0.646 and −0.442 for CCl_3CN indicating that the R_{CC} contribution to the CN stretching intensity is very important. This factor is often neglected in simplified treatments [92, 93]. The interplay between electronic and vibrational factors thus strongly influences the total intensity in this case but not the intensity sequence, which is essentially electronic in origin and shows up in large variations in $(\partial\mu^{(x)}/\partial R_{CN})_0$ values (e.g. −0.152 DÅ^{-1} for CH_3CN and −1.665 DÅ^{-1} for CCl_3CN).

3.2.2 The Symmetric Stretching and Bending Intensities in Pyramidal AX_3 Molecules: NH_3, PH_3, NF_3, and PF_3

As stated in Sect. 3.1.1, the "3.1" basis performed well in a series of calculations on both the stretching and bending (symmetric and asymmetric) intensities in pyramidal molecules of the AX_3 type (NH_3, PH_3, NF_3, PF_3): all experimental trends turned out

Table 8. Integrated intensities (theoretical "3.1"-basis vs experiment) for the symmetric stretching (v_1) and bending (v_2) vibrations in NH_3, PH_3, NF_3 and PF_3 (in km mol^{-1})

		Theory	Exp. [82h–82l]
NH_3	A_1	1.38	7.6
	A_2	174	138
PH_3	A_1	37.0	58.4
	A_2	8.3	18.5
NF_3	A_1	58.4	29.7
	A_2	7.81	1.54
PF_3	A_1	125	110
	A_2	80.0	24.1

to be correctly reproduced [33]. Concentrating for the sake of interpretational simplicity on the symmetric vibrations, we see from Table 8 that for v_1, the experimental sequence $NH_3 < NF_3 < PF_3$ is also found in the theoretical values (in view of the v_1–v_3 separation problem in PH_3 this molecule is nor considered) and that, in addition, the large range of experimental values is reproduced quite well. For v_2 the experimental sequence $NH_3 > PF_3 > PH_3 > NF_3$ is correctly reproduced. Although the intensity for the fluorinated compounds is overestimated, the increase found experimentally (one order of magnitude) when passing from NF_3 to PF_3 is correctly reproduced. For each molecule the experimental sequence for A_1 and A_2 is also found in the theoretical results. For NH_3 and PH_3 A_2 is calculated to be one order of magnitude larger than A_1 whereas for the fluorinated compounds the inverse sequence is found. It's interesting to analyse how these values result from a combination of vibrational and electronic factors.

Decomposing for this purpose the $(\partial\mu/\partial Q_i)$ (i = 1, 2) values as:

$$(\partial\mu/\partial Q_i)_0 = \sum_j L_{ij}^\tau (\partial\mu/\partial S_j)_0 \tag{21}$$

where the summation over j extends over the symmetric stretching (j = 1) and bending (j = 2) symmetry coordinates defined as (cf. Table 4):

$$S_1 = \frac{1}{\sqrt{3}}(r_1 + r_2 + r_3); \qquad S_2 = \frac{1}{\sqrt{3}}(\alpha_1 + \alpha_2 + \alpha_3)$$

we easily see that:

$$L_{ij}^\tau |(\partial\mu/\partial S_j)_0| / |(\partial\mu/\partial Q_i)_0|$$

can be interpreted as a kind of fractional contribution of the j-th internal symmetry coordinate to the square root of the intensity of the i-th normal mode (note that their sum is not necessarily equal to 1 in view of different sign combinations). The nominator in this expression is equal to a product of a purely vibrational factor (L_{ij}^τ) related to the force field, equilibrium geometry and nuclide masses and a purely

Table 9. $(\partial\mu^{(z)}/\partial S_1)_0$ and $(\partial\mu^{(z)}/\partial S_2)_0$ values (in D Å^{-1}) and fractional contributions of elongation (S_1) and deformation (S_2) displacements to the $(\partial\mu^{(z)}/\partial Q_i)_0$ values, symbolized by Q_1 and Q_2

	$(\partial\mu^{(z)}/\partial S_j)_0$	Q_1	Q_2
NH$_3$			
Elongation	−0.045	0.25	0.00
Deformation	−1.742	0.75	1.00
PH$_3$			
Elongation	−0.850	0.91	−0.21
Deformation	0.543	0.09	1.21
NF$_3$			
Elongation	−1.106	0.27	−0.29
Deformation	2.442	0.73	1.29
PF$_3$			
Elongation	−3.365	0.54	−0.04
Deformation	5.484	0.46	1.04

electronic term $(\partial\mu/\partial S_j)_0$ reflecting the molecular charge redistribution upon molecular distortions along internal symmetry coordinates. Calculating these $(\partial\mu^{(z)}/\partial S_j)_0$ values (the only non-vanishing component) according to the inverse of Eq. (21):

$$(\partial\mu/\partial S_i)_0 = \sum_j L_{ij}^{-1}(\partial\mu/\partial Q_j)_0 \tag{22}$$

and recombining them with the $\mathbf{L}^\tau$ matrix elements we obtain the results in Table 9.

For NH$_3$ and NF$_3$, the $(\partial\mu^{(z)}/\partial S_2)_0$ derivatives are found to be roughly of the same order of magnitude; they do however have a different sign. Their absolute value is appreciably larger than the corresponding $(\partial\mu^{(z)}/\partial S_1)_0$ values. For both molecules the main contribution not only to $(\partial\mu^{(z)}/\partial Q_2)$ but also to $(\partial\mu^{(z)}/\partial Q_1)$ is that of the deformation displacement along S_2. Both stretching and bending intensities are essentially due to the bending term in Eq. (21). The intensity increase in v_2 when passing from NH$_3$ to NF$_3$ $((\partial\mu^{(z)}/\partial Q_2)$ varying from 0.432 to −2.029) cannot be ascribed to a corresponding fivefold increase in $(\partial\mu^{(z)}/\partial S_2)_0$ but essentially to the fact that the L_{22}^τ in front of this derivative in Eq. (21) is about five times larger for NH$_3$ than for NF$_3$ (1.68 vs 0.23) due to the much larger mass of the fluorine atoms as compared to the hydrogen atoms. For the stretching vibration Table 9 reveals that in both molecules the calculated intensity finds its origin in the bending movement for about 75%. However in this case NF$_3$ has a much larger $|(\partial\mu^{(z)}/\partial Q_1)_0|$ values than NH$_3$ (1.176 vs 0.181) due to the much larger L_{12}^τ term (−0.353 vs 0.078) indicating a much lower vibrational "purity" of the stretching mode in the fluorinated compound. The combined effects lead to an A$_2$/A$_1$ ratio much larger than 1 in NF$_3$ and much smaller than 1 in NH$_3$.

A comparable situation is encountered for Q_2 in NF$_3$ and PF$_3$ but in these cases the elongation contribution to $(\partial\mu^{(z)}/\partial Q_1)_0$ is larger than 50%. It is the larger $(\partial\mu^{(z)}/\partial S_2)_0$ value for PF$_3$, as compared to NF$_3$, which is the reason for the observed and calculated A$_2$ intensity sequence PF$_3$ > NF$_3$ as the L_{22} weight factors are almost the same in both cases (0.260 and 0.228). We conclude that the A$_2$ sequence

($PF_3 < NF_3$) is electronic in origin, whereas the A_1 sequence ($NH_3 < NF_3$) is essentially due to vibrational factors. These results indicate that care should be taken when interpreting experimental integrated intensities solely in terms of electronic rearrangements occurring upon distortion of the molecule along the symmetry coordinate lending its name to the mode under consideration.

3.2.3 The Influence of Deuteration on the CH and OH Stretching Intensities in Formic Acid

In the monomer of formic acid, an at-first-sight unexpected evolution of the CH stretching intensity upon deuteration is experimentally found in the gas phase [95]: the ratio A_{CD}/A_{CH} turns out to be of the order of 1.5. In the case of perfectly localized vibrations with associated normal coordinates:

$$Q_{CH(D)} = \sqrt{m_{CH(D)}}\, R_{CH(D)}$$

where $m_{CH(D)}$ stands for the reduced mass of the oscillator, an intensity ratio of the order of 0.5 is expected:

$$A_{CD}/A_{CH} = |(\partial\mu/\partial Q_{CD})_0|^2/|(\partial\mu/\partial Q_{CH})_0|^2 = \frac{1}{m_{CD}}\bigg/\frac{1}{m_{CH}} = \frac{m_{CH}}{m_{CD}} = \frac{7}{13} = 0.54$$

$$(23)$$

Table 10 shows the experimental and calculated values, the latter being obtained with the "3.1" and the DTZPD bases discussed in Sect. 3.1.1 [34]. The experimental ratios A_{CD}/A_{CH} turn out to be correctly reproduced with both bases: the values for the "3.1" basis, for example, are: HCOOH/DCOOH: 1.62; HCOOD/DCOOD: 1.43, the experimental values being 1.51 and 1.47, respectively. The "anomalous" ratio of the order of 1.5 is due to the non-perfectly localized nature of the CH and CD vibrators. This phenomenon shows up in the $\mathbf{L}^{-1}$ matrix. Table 11 indicates that,

Table 10. Theoretical ("3.1" and DTZDP basis) and experimental gasphase OH(D) and CH(D) stretching intensities for formic acid monomer and its deuterated isotopomer (values in km mol^{-1}). Normal coordinates obtained [34] using Bellet's r_s geometry [96] and Redington's harmonic force field [97]

	Frequency	Mode Description	Integrated Intensity		
	(calc.)		"3.1"	DTZDP	Exp. [95]
HCOOH	3591	OH stretching	100	122	87
	2940	CH stretching	45	39	51
DCOOH	3591	OH stretching	99	122	87
	2216	CD stretching	73	80	77
HCOOD	2940	CH stretching	46	39	47
	2619	OD stretching	61	73	45
DCOOD	2619	OD stretching	65	77	47
	2214	CD stretching	66	71	69

Table 11. L^{-1} matrix elements relating normal (Q_i) and internal (R_j) coordinates for the OH(D) and CH(D) vibrations

Q_i		$R_{OH(D)}$	$R_{CH(D)}$	$R_{C=O}$	R_{C-O}	δ_{OCO}
HCOOH						
	OH	0.972	0.004	0.006	−0.018	0.009
	CH	−0.009	0.948	−0.160	−0.047	0.024
DCOOH						
	OH	0.972	0.002	0.005	−0.018	0.009
	CD	−0.014	1.185	−0.625	−0.105	0.056
HCOOD						
	OD	1.329	−0.025	0.027	−0.054	0.024
	CH	0.022	0.948	−0.157	−0.048	0.024
DCOOD						
	OD	1.329	0.023	0.013	−0.055	0.026
	CD	0.050	1.195	−0.609	−0.104	0.054

as opposed to the highly localized nature of the OH(D) vibration (vide infra), the CH(D) vibration contains important impurities, mainly from the C=O stretching coordinate, whose contribution is moreover increased by a factor 4 when passing from CH to CD. These effects obviously also show up in the L^r matrix involved in the decomposition of $(\partial\mu/\partial Q_{CH(D)})_0$, in terms of $(\partial\mu/\partial R_i)_0$ contributions according to Eq. (21). Table 12A shows that important contributions from "off-diagonal" terms ($R_{C=O}$ and δ_{CO_2}) are present in both vibrations. Their importance, however,

Table 12.

A

	$(\partial\mu/\partial Q_{CH(D)})_0$	$\theta^{(0)}$	$R_{C-H(D)}$	$R_{C=O}$	δ_{CO_2}*	Σ
HCOOH ($\approx$HCOOD)	1.036	1.8	0.524	0.156	0.386	1.066
DCOOH ($\approx$DCOOD)	1.312	2.5	0.315	0.613	0.396	1.324

Decomposition of $(\partial\mu/\partial Q_{CH(D)})_0$ values (in D amu$^{-1/2}$ Å^{-1}) in contributions of internal coordinates [Eq. (21)]. Only the component parallel to the $(\partial\mu/\partial Q_{CH(D)})_0$ vector is given. Their sum is indicated as Σ. θ denotes the angle between the $(\partial\mu/\partial Q_{CH(D)})_0$ vector and the CH(D) internuclear axis

* $\delta = \dfrac{1}{\sqrt{6}} (2\alpha - \beta - \gamma)$; α, β, and γ denoting the O$\hat{C}$O, H$\hat{C}$—O and H—$\hat{C}$=O angles, respectively

B

	$(\partial\mu/\partial Q_{OH(D)})_0$	$\theta^{(0)}$	$R_{O-H(D)}$	δ_{CO_2}	Σ
HCOOH ($\approx$DCOOH)	1.536	14	1.397	0.163	1.560
HCOOD ($\approx$DCOOD)	1.197	19	0.988	0.237	1.225

Decomposition $(\partial\mu/\partial Q_{OH(D)})_0$ values (in D amu$^{-1/2}$ Å^{-1}) in contributions of internal coordinates [Eq. (21)]. Only the component parallel to the $(\partial\mu/\partial Q_{OH(D)})_0$ vector is given.
Their sum is indicates as Σ. θ denotes the angle between the $(\partial\mu/\partial Q_{OH(D)})_0$ vector and the OH(D) internuclear axis

still increases when passing from CH to CD and in the deuterated compound the $R_{C=O}$ contribution is even twice as large as the R_{CD} contribution. This effect is also related to the much larger value of $(\partial\mu/\partial R_{C=O})_0$ as compared to $(\partial\mu/\partial R_{CH(D)})_0$ which differ by one order of magnitude (5.707 DÅ^{-1} vs 0.517 DÅ^{-1}) [34]. The ratio of the $R_{CH(D)}$ contributions (0.36) is of the order of the values expected in the case of a perfectly localized harmonic oscillator: the far from perfectly localized nature of the CH vibrator (a vibrational factor), coupled with the order of magnitude difference of the $(\partial\mu/\partial R_{C=O})_0$ and $(\partial\mu/\partial R_{CH})_0$ values (an electronic factor) results in the observed and calculated ratio of 1.5.

Turning now to the OH(D) stretching intensity, Table 10 shows a calculated ratio A_{OD}/A_{OH} of the order of 0.5 as expected on the basis of a perfectly localized oscillator. The calculated values for the "3.1" basis are 0.61 (HCOOD/HCOOH) and 0.65 (DCOOD/DCOOH) as compared to the experimental values 0.52 and 0.54. The highly localized nature of the OH(D) vibration is clearly seen in the L^{-1} matrix elements of Table 11. The decomposition of the $(\partial\mu/\partial Q_{OH(D)})_0$ values in Table 12B shows that, despite the fact that $(\partial\mu/\partial R_{OH})_0$ (1.339 DÅ^{-1}) is small as compared to, e.g., $(\partial\mu/\partial R_{C=O})_0$, the OH stretching vibration can be said to be "pure" from the "intensity point of view", the most important impurity being associated with the bending movement. The effect of deuteration on the OH intensity is thus purely vibrational in origin.

3.3 Decomposition of Dipole Moment Derivatives in Non-local Terms: Failure of the Classical Point Charge Model

The decomposition of the dipole moment derivatives in HCCCN clearly illustrates the possible failure of a classical point charge model in the interpretation of experimental and theoretical intensity values. Table 13 gives, as an example, the results of the decomposition of the dipole moment derivatives with respect to the CH and CN normal coordinates, together with the equilibrium dipole moment decomposition [27]. The direction of the point charge term in the latter case is in agreement with the polarity of the CN triple bond. The important hybridization term is mainly due to the N lone pair, whereas the CH bond is the main contributor to the homopolar term in view of the difference in size of the C and H atoms [62]. Both terms do have an opposite sign so that the direction of the total equilibrium dipole moment is the same as that of the point charge term.

Table 13. STO-6G partitioning of the equilibrium dipole moment (in D) and its derivatives (in D amu$^{-1/2}$ Å^{-1}), with respect to the CH and CN stretching normal coordinates, in non-local contributions [Eq. (12)] for HCCCN. Orientation of the molecule: see Table 5

Term [Eq. (12)]	$\mu^{(x)}$	$(\partial\mu^{(x)}/\partial Q_{CH})_0$	$(\partial\mu^{(x)}/\partial Q_{CN})_0$
Point charge	+2.470	+2.036	+0.180
Hybridization	+1.955	−0.795	+0.396
Homopolar	−1.298	+1.056	+0.079
Total	+3.127	+2.327	+0.655

Table 14. STO-6G partitioning of the equilibrium dipole moment (in D) and its derivatives (in D amu$^{-1/2}$ Å^{-1}) with respect to the CN stretching normal coordinate, in non-local contributions [Eq. (2)] for the chloroacetonitriles $CH_{3-n}Cl_nCN$ (n = 0, 1, 2, 3). Orientation of the molecules: see Table 7. Only the x component of the dipole moment and its derivative is considered

	Term	n = 0	n = 1	n = 2	n = 3
$\mu^{(x)}$	Point Charge	+2.519	+1.567	+0.928	+0.423
	Hybridization	+1.748	+1.204	+0.805	+0.261
	Homopolar	−1.159	−0.725	−0.401	+0.012
	Total	+3.108	+2.046	+1.332	+0.696
$(\partial\mu^{(x)}/\partial Q_{CN})_0$	Point Charge	−0.927	−1.197	−1.430	−1.681
	Hybridization	+0.634	+0.593	+0.557	+0.537
	Homopolar	−0.096	−0.084	−0.084	−0.086
	Total	−0.389	−0.688	−0.957	−1.230

Turning now to the dipole moment derivatives, we see that for the CN stretching vibration, involving the highly polar CN bond, the hybridization term is much more important than the point charge term. Moreover, the LMO analysis reveals that the latter term is due to changes in almost all bonds of the molecule. In the CH stretching vibration, however, involving the less polar CH vibrator, the point charge term now turns out to be dominant. The contributions from hybridization and homopolar terms, mainly localized in the CH bond, are not negligible but their sign is different so that a partial cancellation occurs.

As a second example of this decomposition we give in Table 14 the results for the equilibrium dipole moment and the derivative with respect to the CN normal coordinate in the chloroacetonitriles [28].

The equilibrium dipole moment decomposition clearly reflects the electron withdrawing effect of the chlorine atoms: the point charge term appreciably decreases on successive introduction of chlorine atoms in the molecule, without changing its sign, however. The hybridization term is also strongly influenced by the introduction of chlorine atoms, the CCl bonds partly compensating the large value in the +x direction of the N lone pair. The absolute value of the homopolar term follows the same evolution: starting from a large contribution in the −x direction for CH_3CN, due to the large homopolar moment associated to each of the CH bonds, the magnitude of this term diminishes upon successive chlorination. The C—Cl bonds indeed do have homopolar moments with opposite sign to that of the CH bond (C^-Cl^+, C^+H^-) in agreement with the fact that the term is essentially determined by the relative magnitude of the atoms involved in the bonds.

Turning to the derivatives we see that, as opposed to the CN stretching in HCCCN, the point charge term is now the largest term in all four cases. However, an almost linear increase is calculated for the contribution of this term with increasing n, as opposed to the global polarity of the molecules (and also the local polarity in the CN-bond (vide infra)) reflected by the equilibrium dipole moment. The changes in the hybridization and homopolar terms are one and two orders of magnitude smaller and do not contribute to the calculated intensity evolution. Note, however, the important contribution of the hybridization term to the intensity for a given molecule.

Concluding this section we point out the danger associated with an interpretation of infrared intensities via a "classical" point charge model in which a relationship is supposed to exist between the dipole moment derivatives and the static charge distribution.

3.4 LMO Decomposition of Dipole Moment Derivatives $(\partial\mu/\partial Q_i)_0$

3.4.1 Contribution of the Various Structural Entities to the Calculated Intensity (Evolution)

3.4.1.1 The CN Stretching Intensity in the Chloroacetonitriles

As a first example we consider the LMO decomposition of the dipole moment derivative with respect to the CN stretching normal coordinate in the chloro-acetonitriles [29]. The equilibrium dipole moment decomposition is also given in Table 15. It is seen that the N lone pair yields the largest but almost constant contribution to the equilibrium dipole moment. The CN bond contribution gradually diminishes from $+1.482$ D in CH_3CN to $+0.682$ D in CCl_3CN, in agreement with the electron withdrawing effect of the chlorine atoms. It is important to note that no sign reversal in μ_{CN} is obtained. The hypothesis formulated by Besnainou [92] and

Table 15. LMO decomposition of STO-6G calculated equilibrium dipole moment (in D) and its derivative with respect to the CN stretching normal coordinate $(\partial\mu^{(x)}/\partial Q_{CN})_0$ in (D amu$^{-1/2}$ Å^{-1}) for the chloroacetonitriles $CH_{3-n}Cl_nCN$ (n = 0, 1, 2, 3). Orientation of the molecules: see Table 7. Only the x-component is considered

	LMO	n = 0	n = 1	n = 2	n = 3
$\mu^{(x)}$	Inner Shells[a]	+0.018	+0.111	+0.185	+0.256
	C—C Bond	+0.213	+0.153	+0.094	+0.024
	C—H Bond(s)[b]	−1.203	−0.944	−0.580	0
	C—Cl Bonds(s)[c]	0	−0.139	−0.353	−0.634
	C≡N Triple Bond[d]	+1.482	+1.186	+0.907	+0.682
	N Lone Pair	+2.598	+2.600	+2.606	+2.605
	Cl Lone Pairs[e]	0	−0.921	−1.527	−2.237
	Total	+3.108	+2.046	+1.332	+0.696
$(\partial\mu^{(x)}/\partial Q_{CN})_0$	Inner Shells[a]	+0.049	+0.077	+0.097	+0.123
	C—C Bond	−0.143	−0.080	−0.044	−0.033
	C—H Bond(s)[b]	+0.110	+0.019	−0.022	0
	C--Cl Bond(s)[c]	0	+0.042	+0.099	−0.022
	C≡N Triple Bond[d]	+0.773	+0.423	+0.031	−0.339
	N Lone Pair	−1.178	−1.183	−1.153	−1.127
	Cl Lone Pairs[e]	0	+0.014	+0.035	+0.168
	Total	−0.389	−0.688	−0.957	−1.230

[a] Sum of the contributions of all the inner shells of the molecule
[b] Sum of the contributions of the (3—n) CH bonds
[c] Sum of the contributions of the CCl bonds
[d] Contributions of the C≡N triple bond (C—N bent bonds)
[e] Sum of the contributions of the three lone pairs on each of the n Cl atoms

Orville-Thomas [93], in which the CN intensity evolution is ascribed to a sign reversal in μ_{CN} (assuming a proportionality between the magnitude of the equilibrium dipole moment and its derivative) could be rejected on this basis.

Turning to the dipole moment derivative we see that the largest contribution to the calculated $(\partial\mu/\partial Q_{CN})_0$ values comes again from the N lone pair yielding an almost identical contribution in all cases. However, the factor which governs the *evolution* of the calculated intensity in the chloroacetonitriles comes from the CN triple bond whose contribution steadily varies from $+0.773$ in CH_3CN to -0.339 in CCl_3CN. The combination of this term with the constant and negative N lone pair contribution yields a total CN bond contribution sequence (-0.405, -0.780, -1.122–1.466 D amu$^{1/2}$ Å) which parallels the calculated intensity values. Further analysis shows that the essential difference between the four molecules is the way in which the polarity of the CN bond changes during vibration. In CH_3CN the CN bond becomes more polar upon elongation of the CN bond along the CN normal coordinate, while in CCl_3CN it becomes less polar. We now return to the CN intensity interpretation given by Roshchupkin and Popov [91] already mentioned in Sect. 3.2.1. According to these authors, the combination of electro-optical parameters:

$$(\partial\mu^{(x)}_{CC}/\partial R_{CC})_0 + (\partial\mu^{(x)}_{CN}/\partial R_{CC})_0 + (\partial\mu^{(x)}_{CCl}/\partial R_{CC})_0$$

should strongly differ between CH_3CN and CCl_3CN (including a sign seversal), whereas:

$$(\partial\mu^{(x)}_{CN}/\partial R_{CN})_0 + (\partial\mu^{(x)}_{CC}/\partial R_{CN})_0$$

should not differ too much. We performed an LMO analysis on the $(\partial\mu^{(x)}/\partial R_{CC})_0$ and $(\partial\mu^{(x)}/\partial R_{CN})_0$ quantities (Table 16). The data indicate that the former combination is almost unchanged between both molecules (and by interpolation throughout the whole series of the chloroacetonitriles): the values obtained are: $+1.618$ DÅ^{-1} in CH_3CN and $+1.512$ DÅ^{-1} in CCl_3CN. The latter combination, however, changes drastically from $+2.776$ DÅ^{-1} in CH_3CN to $+0.558$ DÅ^{-1} in CCl_3CN. The LMO analysis clearly offers a non-empirical basis for not retaining Roshchupkin and Popov's explanation for the CN intensity behaviour.

Table 16. LMO analysis of STO-6G calculated $(\partial\mu^{(x)}/\partial R_i)_0$ values in CH_3CN and CCl_3CN (in D Å^{-1})

	R_{CN}		R_{CC}	
	CH_3CN	CCl_3CN	CH_3CN	CCl_3CN
Inner Shells	$+0.100$	$+0.175$	-0.056	-0.138
C—C Bond	$+0.278$	-0.094	$+0.926$	-0.098
C—H Bonds	$+0.330$	0	$+0.052$	0
C—Cl Bonds	0	-0.156	0	-0.566
C≡N Triple Bond	$+2.498$	$+0.652$	$+0.640$	$+2.176$
N Lone Pair	-3.358	-3.244	-0.374	-0.532
Cl Lone Pair	0	$+0.690$	0	$+0.820$
Total	-0.152	-1.665	$+1.188$	$+1.662$

The LMO analysis can also be used to discuss a hypothesis which is often used when interpreting IR intensities of stretching vibrations, namely that the more polar a given bond "i" is (i.e. the higher its bond moment $|\mu_i|$) the higher should be the corresponding bond moment derivative $|(\partial\mu_i/\partial R_i)_0|$ and the higher the intensity A_i of the associated stretching vibration (related to $|(\partial\mu_i/\partial Q_i)_0|$) [92, 93]. For CH_3CN and CCl_3CN, the results in Tables 15 and 16 indicate that only the first trend is observed for the CN bond. As far as the C—C bond is concerned, its equilibrium bond moment is much smaller than that of the $C\equiv N$ bond (a factor 7 in CH_3CN and 28 is CCl_3CN). However, the $|(\partial\mu/\partial Q_{CC})_0|$ derivative (calculated value in CH_3CN: $+0.333$ D amu$^{-1/2}$ Å^{-1}) is of the same order of magnitude as its CN counterpart in CH_3CN whereas in CCl_3CN it has even a larger value ($+1.430$ D amu$^{-1/2}$ Å^{-1}) than $|(\partial\mu/\partial Q_{CN})_0|$. Again (cf. Sect. 3.3), the possible failure of IR intensity explanations based on relationships between the static charge distribution and the dipole moment derivatives shows up.

3.4.1.2 The Symmetric Bending Vibration in Pyramidal AX_3 Molecules: NH_3, PH_3, NF_3, and PF_3

From Sect. 3.2 it emerges that the symmetric bending intensity A_2 in these molecules can be interpreted almost exclusively in terms of variations in $(\partial\mu/\partial S_2)_0$ values multiplied by a weight factor L_{22} involving the atomic masses. In Table 17 we give the LMO decomposition of the $(\partial\mu^{(z)}/\partial Q_2)_0$ values, for the 3.1 basis [33], in order to gain an insight into the electronic factors governing this intensity. It is seen that the A lone pair substantially contributes to this derivative in the two non-fluorinated compounds, the sign being, however, different between NH_3 and PH_3. Its contribution in PF_3 is smaller than in PH_3 by a factor 3, whereas in NF_3 it is almost reduced to zero. Its contribution is more important than that of the AX bonds in PH_3 and PF_3. The discussion of the AX bond contributions will be given in Sect. 3.4 treating incomplete orbital following. In the fluorinated compounds, the F lone pair contribution is the most important. Its positive sign can be interpreted on the basis of geometrical rearrangements upon increase of the $X\hat{A}X$ angle. It is interesting to confront the results for the A lone pair contributions with the McKean and Schatz model [82h] based upon classical hybridization-geometry arguments. These authors derived

Table 17. LMO Analysis of "3.1" $(\partial\mu^{(z)}/\partial Q_2)_0$ values (in D amu$^{-1/2}$ Å^{-1}) in pyramidal AX_3 type molecules

	NH_3	PH_3	NF_3	PF_3
Inner Shells[a]	0.000	0.004	−0.001	−0.003
AX Bonds[b]	−1.546	0.061	−0.594	−0.085
A Lone Pair	−0.482	0.377	0.003	0.134
F Lone Pairs[c]	—	—	1.002	1.327
Total	−2.029	0.442	0.431	1.373

[a] Sum off all non-valence LMO contributions
[b] Sum of the three equivalent AX-bond LMO contributions
[c] Sum of the contributions of the nine F lone pairs

the following relation between the lone pair moment of atom A ($\mu_z(A_{l.p})$) and the XÂX angle α, both being directly related to the hybridization of the central atom A:

$$\mu_z(A_{l.p}) = C\,M_{sp}\,\frac{(-\cos\alpha - 2\cos^2\alpha)^{1/2}}{1 - \cos\alpha} \tag{24}$$

M_{sp} denotes the integral $\langle 2s_A | z | 2p_{zA} \rangle$ (for N) or $\langle 3s_A | z | 3p_{zA} \rangle$ (for P) where 2s and $2p_{zA}$ denote the valence AOs on atom A; C is a constant. Differentiation of μ_z with respect to α yields a maximum for $\alpha = 101.5°$.

In order to see in how far the results from our parameter free LMO analysis are compatible with this more qualitative semi-empirical picture we first calculated the values of M_{sp} for N and P by inserting into Eq. (24) the equilibrium LMO N and P lone pair moments (3.238 (NH_3), 3.817 (NF_3), 5.200 (PH_3) and 5.622 (PF_3) (all values in D)) and the experimental equilibrium XÂX angles (NH_3: 106.7°, NF_3: 102.4°, PH_3: 93.3° and PF_3: 96.9° [82a–d]). The value for CM_{sp} for N is then obtained by taking the average of the results obtained for NH_3 and NF_3 (11.9 and 13.2 D), namely 12.6 (D). For P, the values for PH_3 (24.4 D) and PF_3 (20.8 D) yield an average of 22.6 D. Using these results the $\mu_z = \mu_z(\alpha)$ curves were drawn for N and P. Figure 2 shows that, as expected on the basis of the larger "size" of the atom, the curve for P lies above the curve for N. When calculated LMO lone pair moments are inserted in the diagram, we see that the McKean and Schatz model accounts for the relative order of

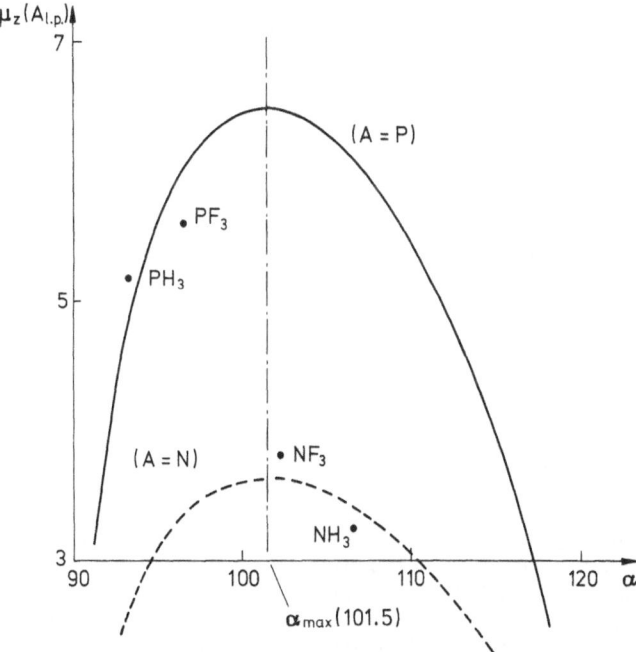

Fig. 2. Curves representing $\mu_z A_{(l.p)}$ [Eq. (24)] as a function of α for A = P (full line) and N (dashed line). LMO calculated A lone pair moments are inserted at the corresponding values at equilibrium geometry. α_{max} denotes the value for which α attains a maximum. From Berckmans D, Figeys HP, Geerlings P, J. Mol. Struct. (Theochem.), 149: 243 (1987). Reprinted by permission of Elsevier Science Publishers BV

magnitude of the equilibrium lone pair moments. Moreover, we see that this model also accounts for the variation in sign of the calculated $(\partial\mu^{(z)}_{Al.p.}/\partial Q_2)_0$ values.

Indeed, on the basis of Fig. 2 we expect a lone pair moment increase upon opening the AX_3 umbrella (as occurring in the symmetric bending), which is effectively the case for PH_3 and PF_3 for which positive $(\partial\mu^{(z)}_{A.l.p}/\partial Q_2)_0$ values were found. For NH_3 α is larger than $101.5°$ and a positive sign of the derivative is expected in agreement with the results of Table 17. The small difference between the NF_3 α equilibrium value and α_{max} suggests an almost vanishing derivative as again indeed resulting from our LMO calculations.

3.4.2 Testing the Bond Moment Hypothesis

In order to test the validity of the bond moment hypothesis [17] described in the Introduction, we investigated whether an elongation of a given internal stretching coordinate R_i induces variations in the bond moments μ_j associated with the other internal coordinates R_j. Therefore, we performed an LMO analysis of the dipole moment derivatives with respect to stretching internal coordinates $(\partial\mu^{(x)}/\partial R_i)_0$ in cyanoacetylene. They were obtained in a finite difference approach as $[\mu^{(x)}(R_i = R_{i,eq} + 0.005) - \mu^{(x)}(R_{i,eq})]/0.005$ where $R_{i,eq}$ denotes the equilibrium distance (in Å) between the two atoms involved in the bond(s) to which the i-th internal coordinate is associated. The results are given in Table 18 together with the equilibrium dipole moment decomposition [27].

It is seen that in the case of the CH stretching coordinate, the largest bond moment variation occurs in the CH bond itself; the contribution of the adjacent C_2C_3 triple bond, however, amounts to 58% of that of the CH bond. The $C\equiv N$ bond which is separated from the CH bond by two bonds yields a contribution which still constitutes 21% of the CH bond contribution. For the $C\equiv N$ stretching, the N lone pair contribution dominates. The $C\equiv C$ triple bond contribution is however very important (47% of that of the N lone pair) and is even larger (by a factor three) than that of the $C\equiv N$ bond itself. For the $C\equiv C$ stretching, the $C\equiv C$ bond contribution is the most important one, but the other three bonds again contribute in a non-

Table 18. LMO analysis of STO-6G results for the equilibrium dipole moment $\mu^{(x)}$ (in D) and the dipole moment derivatives with respect to internal stretching coordinates $(\partial\mu^{(x)}/\partial R_i)_0$ (in D Å^{-1}) for HCCCN. Orientation of the molecule: see Table 5

	$\mu^{(x)}$	$\left(\dfrac{\partial\mu^{(x)}}{\partial R_{CH}}\right)_0$	$\left(\dfrac{\partial\mu^{(x)}}{\partial R_{C\equiv N}}\right)_0$	$\left(\dfrac{\partial\mu^{(x)}}{\partial R_{C\equiv C}}\right)_0$	$\left(\dfrac{\partial\mu^{(x)}}{\partial R_{C-C}}\right)_0$
Inner Shells[a]	+0.030	−0.006	+0.112	−0.026	+0.002
$C_1\equiv N$ triple bond[b]	+0.594	+0.306	+0.504	−0.948	+2.808
C_1C_2 bond	+0.032	−0.024	+0.110	−0.716	+0.092
$C_2\equiv C_3$ triple bond[b]	+1.385	+0.828	+1.500	+2.952	−3.240
C_3−H bond	−1.542	+1.434	−0.024	+0.564	+0.046
N Lone Pair	+2.628	−0.022	−3.182	+0.078	−0.434
Total	+3.127	+2.516	−0.980	+1.748	−0.726

[a] Sum of the C_1, C_2, C_3 and N inner shell contributions
[b] Sum of the contributions of the three equivalent bonds

negligible way: nore that here the N lone pair contribution amounts to only 2% of that of the lone pair, a situation comparable to the case of the CH stretching: the farther the lone pair is remote from the bond which is elongated, the less it contributes to the corresponding $(\partial\mu/\partial R)_0$ value. The most remarkable result is obtained for an elongation of the C—C bond: the change in its own dipole moment is now completely negligible as compared to those calculated for the adjacent C≡C and C≡N bonds. Even the N lone pair contribution is about five times as important as that of the C—C bond itself.

The following conclusions can be drawn: when a given bond is elongated, the variation in its own bond moment has the tendency to be dominant. If a lone pair is present on one of the atoms constituting the bond, a very important change occurs in its bond moment. The variation of the "other" bond moments, certainly those adjacent to the bond considered, are often of the same order of magnitude as the variation of the moment of the bond which is elongated. It is even possible that these bonds contribute to a (much) larger extent as was the case for the C—C bond elongation. All this offers a non-empirical quantum-chemical interpretation of the failure of the zero-order bond moment hypothesis in which all $(\partial\mu_i/\partial R_j)_0$ (i ≠ j) terms are put equal to zero. A large number of "off-diagonal" terms $(\partial\mu_i/\partial R_j)_0$, (i ≠ j) should be included in order to get reliable results leading to a rapid increase in the number of electro-optical parameters which should be determined from a very limited number of experimental data [21].

3.4.3 Incomplete Orbital Following Effects

The concept of "incomplete orbital following" is based on the idea that the electron cloud does noet always completely follows the nuclei when they are displaced from their equilibrium positions as occurring upon molecular vibrations. It was introduced by Linnett and co-workers in several force field studies [98, 99]. Later on Coulson et al. performed pioneering calculations on its influence on the intensity of bending vibrations [100, 101, 102]. The LMO analysis enables us to perform a quantitative, parameter-free study of this effect.

We start with a discussion of this phenomenon in *bending vibrations*, based on the results for the symmetric bending in pyramidal AX_3 molecules [31, 32, 33]. In this account we concentrate on the "3.1" calculations coupled with Boys' localization technique, reported in [33]. Starting from the results in Table 17 we discuss the contri-

Table 19. LMO calculated δ and β angles, characterizing the degree of orbital following upon distortion along the Q_2 normal coordinate (see text)

	$\delta_{eq}{}^a$	$\delta\,(Q_2 = 0.01)$	$\Delta\beta(X)$	$\Delta\beta(\langle r\rangle)$	% following
NH_3	9.53	10.08	0.354	0.242	68
NF_3	4.84	4.92	0.062	0.038	61
PH_3	−3.19	− 2.98	0.238	0.158	66
PF_3	1.03	1.04	0.134	0.089	66

a: A positive δ value indicates that $\langle r\rangle$ lies inside the AX_3 pyramid; $\langle r\rangle$ denotes the position of the centroid of negative charges in the AX bond LMO.

bution of the AX bonds to the integrated intensity of the symmetric bending and their relation to the effect of (in)complete orbital following. We first note that the contribution of the AX bonds is much larger in the nitrogen compounds than in PH_3 and PF_3. A measure of the degree of orbital following is the variation of the angle δ between the AX bond moment vector and the AX internuclear axis. In Table 19 the values obtained on the basis of our LMO results are given, both for the equilibrium geometry and for a deformation of 0.01 amu$^{1/2}$ Å along the corresponding normal coordinate Q_2.

The δ values at equilibrium point out that the AX bonds are slightly bent, as already remarked in early LMO studies on NH_3 [103]. Distortion along Q_2 results in all cases in an incomplete orbital following. For NH_3, NF_3, and PF_3, δ becomes more positive for positive Q_2 values (i.e. upon "opening" the AX_3 umbrella) indicating that the electron cloud in the AX bond orbital does not completely follow the nuclear displacements. For PH_3, δ becomes less negative indicating that the internuclear axis moves closer to the bond moment vector. We defined the percentage of orbital following as the ratio of the variation of the angle between AC and the z-axis, C denoting the position of the centroid of the negative charges of the AX bond LMO ($\Delta\beta\langle r\rangle$) and the corresponding angle variation for A–X, $\Delta\beta(X)$, upon vibration. If $\Delta\beta(\langle r\rangle)$ equals $\Delta\beta(X)$ the orbital following is 100%. If $\Delta\beta(\langle r\rangle) = 0$, no angular displacement of the electron cloud occurs and a situation of 0% orbital following is encountered. Table 19 shows that the degree of orbital following is of the same order of magnitude ($\approx 65\%$) in the four molecules, indicating a considerable lack of orbital following in all cases. This phenomenon exemplifies Nakatsuji's findings [104], based upon the Hellman-Feynman theorem [105, 106] and showing that upon a topological deformation of a stable nuclear configuration, an incomplete orbital following of the centroid of charge of the electronic cloud localized near the moving nuclei takes place. This effect can be invoked to explain the sign of the AX bond contributions to $(\partial\mu^{(z)}/\partial Q_i)_0$. The equilibrium AX bond moments all have negative z components, the calculated values (in D) being: -1.526 (NH_3), -4.655 (PH_3), -2.181 (NF_3) and -5.798 (PF_3). In NH_3 and PH_3 the negative sign can be explained by the difference in size of the A and X atoms, yielding a homopolar term pointing in the $-z$ dirdction, and by the absence of hybridization possibilities in H. In NF_3 and PF_3 the main factor accounting for the minus sign is the difference in electronegativity between N and P on one hand and F on the other hand. The effect is however reinforced by the difference in size between phosphorus and fluorine. When the molecule is distorted along Q_2 the results in Table 19 indicate that, if δ_{eq} is positive, δ will still increase corresponding to a more negative $\mu_{AX}^{(z)}$ value (the z-component of the vector joining the centroid of negative charges in the AX bond and the midpoint of the nuclear axis will increase in absolute value). All this results in a negative $(\partial\mu_{AX}^{(z)}/\partial Q_2)_0$ value in agreement with the results in Table 17. In PH_3 the reverse situation occurs: δ becomes less negative, provoking a less negative $\mu_{AX}^{(z)}$ value and a positive $(\partial\mu_{AX}^{(z)}/\partial Q_2)_0$ value. When relating the extent of orbital following to the magnitude of $(\partial\mu_{AX}^{(z)}/\partial Q_2)_0$ it should be realized that these values are not only influenced by the angular factor just mentioned but also by a radial factor, namely the displacement of the centroid of charge along the internuclear axis. Anyway, as the variation in δ for NH_3 is by far the most important, a large negative $(\partial\mu_{AX}^{(z)}/\partial Q_2)_0$ for this molecule may be expected. As also the A lone pair contribution is negative

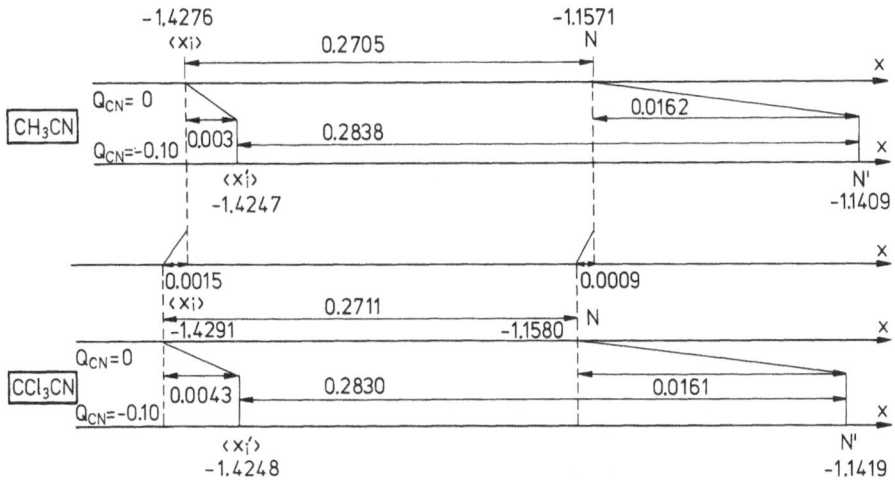

Fig. 3. Incomplete orbital following in stretching vibrations: analysis of the position of the N nucleus and the centroid of negative charges of the N lone pair LMO $\langle x_i \rangle$ in CH_3CN and CCl_3CN for the equilibrium geometry ($Q_{CN} = 0$ amu$^{1/2}$ Å) and for a Q_{CN} value of -0.10 amu$^{1/2}$ Å (primed symbols). Orientation of the molecules: see Table 7. All coordinates and displacements are given in Å. From Figeys HP, Geerlings P, Van Alsenoy C, J. Chem. Soc., Faraday Transactions II, 75: 524 (1979). Reprinted by permission of the Royal Society of Chemistry

(cf. Section 3.4.1.2) a high symmetric bending intensity for this molecule is expected as also experimentally found. For NF_3, Table 17 reveals that the F lone pairs yield the dominating contribution. As its sign is opposite to that of the NF bonds, and the N lone pair does not contribute significantly, the incomplete orbital following effect leads in this case to a decrease in intensity. For the P-containing molecules the situation is less clear: the small contribution of the AX LMOs to the calculated intensity probably results from internal compensation of the various electronic effects.

Incomplete orbital following effects can also be present in the case of *stretching vibrations*. An example emerges from our study on the CN stretching intensity in the chloroacetonitriles [28]. In Sect. 3.4.1.1 it was shown that a large negative and almost constant N lone pair contribution to the CN intensity was found in this series. This contribution arises (see Fig. 3, the CH_3CN case) from an incomplete following of the N lone pair electrons upon displacement of the N nucleus upon vibration. For a relativity large Q_{CN} value of -0.10 amu$^{1/2}$ Å, corresponding to a CN distance decrease of 0.0388 Å, and a displacement of the N nucleus of 0.0162 Å in the $+x$ direction, a displacement of only 0.0030 Å of the centroid of negative charges of the N lone pair $\langle x_i \rangle$ was found, i.e. only 18% of the N-nucleus displacement. This phenomenon increases the $N - \langle x_i \rangle$ distance and yields a higher N lone pair moment upon CN bond compression, i.e. a (large) negative $(\partial \mu^{(x)}_{Nl.p.}/\partial Q_{CN})_0$ value. The situation for CCl_3CN is very similar: the displacement of the centroid of negative charges amounts to 26% of that of the N nucleus, resulting in a similar $|(\partial \mu^{(x)}_{Nl.p.}/\partial Q_{CN})_0|$ value.

3.4.4 Complexes: Influence of Intermolecular Interactions

As mentioned in the Introduction, hydrogen bonding between a proton donor group A—H and a proton acceptor induces at least an order of magnitude increase in the integrated intensity of the AH vibration. In the early seventies, the first ab inito calculations on the influence of hydrogen bonding on integrated intensities appeared, focusing on the evolution of $(\partial\mu/\partial R_{AH})_0$ values upon H bond formation [107→111]. In the following decade, relatively few calculations of this type were published, most of them, however, concentrating on the evaluation of $(\partial\mu/\partial Q_{AH})_0$ [29, 112→115]. As a whole it can be stated that in these studies the important intensification of the A—H vibration is recovered. Little attention was devoted to the interpretation of these results. In those cases where it occurred [29, 112] it was exclusively performed in terms of atomic charges obtained via a Mulliken population analysis which on the basis of our experience (Sect. 3.3) yields only partial, and in some cases misleading, information. The same trend is seen in calculations which appeared in recent years [116→125] showing, except for the Zilles-Person work [118], little interest in a rationalization of the intensity increase of the AH vibration upon Hydrogen Bonding. In a study on the dimer of H_2O these authors applied a charge charge flux overlap (CCFO) analysis to the calculated atomic polar tensors [24, 25] and ascribed the intensity increase upon dimerisation to dynamical charge transfer and polarization effects. In this section we give a brief account of the results obtained when applying our LMO approach to this kind of problems. The system chosen is formic acid and its dimer [34, 35].

It is exceptional in this sense that the experimental gasphase intensities for various stretching modes (C=O, O—H, CH) are known both for the monomer and the dimer in which H-bond formation occurs [95]: for the OH vibration the typical behaviour of frequency decrease and large intensity increase, accompanied by band broadening, is observed upon dimerization. A comparison between the monomer and dimer OH stretching intensities for HCOOH and its deuterated isotopomers was performed with the "3.1" basis. In order to judge the reliability of the results, we give in Table 20 the calculated intensity values for the monomer and the dimer. For the monomer the normal coordinates were obtained with Bellet's geometry and Redington's harmonic force field. For the dimer use was made of Almenningen's geometry [126]; the force field used, described in [35], is obtained via a refinement of the one proposed by Kishida [127]. The IR active normal coordinate in this case is the antisymmetric combination of OH stretchings (B_u symmetry) in which an increase in OH bond length in one monomeric unit is accompanied by a decrease in the OH bond length in the other one. Table 20 indicates that the agreement between the "3.1" theoretical and experimental results both for the monomer and the dimer is highly satisfactory and that the theoretical dimer/monomer intensity ratios are close to the experimental ones. In all cases an intensification by a factor of 20 to 30 is calculated, hereby recovering an important diagnostical tool for hydrogen bond formation [3].

Turning now to the LMO analysis, Table 21 reveals that for the monomer the dominating contribution to the intensity arises from the OH bond, the positive sign being an indication of an incomplete orbital following effect. Other but negative contributions (C—O bond, O lone pairs) are present and partly compensate this

Table 20. "3.1" theoretical and experimental OH(D) stretching intensities in formic acid monomer and dimer. The ratio between the intensities in dimer and monomer is also given

		Theory	Experiment [95]
HCOOH	Monomer	100	87
	Dimer	254×10^1	289×10^1
	Ratio	25	33
HCOOD	Monomer	61	45
	Dimer	137×10^1	105×10^1
	Ratio	22	23
DCOOH	Monomer	99	87
	Dimer	254×10^1	328×10^1
	Ratio	26	38
DCOOD	Monomer	65	47
	Dimer	149×10^1	115×10^1
	Ratio	23	25

feature. In Fig. 4a the in-plane contributions of the different LMO's to $(\partial\mu/\partial Q_{OH})_0$ are plotted, their origin being taken at the centroid of negative charges of the corresponding LMO. It is seen from the displacement of the centroid of negative charge of the OH-LMO upon elongation of the OH bond (corresponding to a positive Q value) that this motion induces an increase in polarity of the OH bond (this oxygen atom is denoted as O_3).

For the dimer also the OH bond contribution is dominant and increases by a factor 2.4 as compared to the monomer (Table 21). Note, however, that if only this effect was operating, the intensity increase between monomer and dimer would only be by a factor 5.9, whereas the experimental ratio is 33 and the calculated value 25. As compared to the monomer case, some other contributions (especially the C=O bond and the lone pairs on this oxygen (O_5)) become much more important and, moreover, both of them are in the dimer of the same sign as the OH contribution,

Table 21. LMO decomposition of $(\partial\mu/\partial Q_{OH})_0$ (in D amu$^{-1/2}$ Å^{-1}) in HCOOH and its dimer. Only components parallel to the $(\partial\mu/\partial Q_{OH})_0$ vector are given. For the dimer, the values divided by $\sqrt{2}$ are given in parentheses in order to allow direct comparison with the monomer; identical contributions from the two monomeric units have been summed

	Monomer	Dimer
CH Bond	−0.181	0.411 (0.291)
C=O Bond	0.008	0.593 (0.419)
O_5 Lone Pairs	−0.162	1.255 (0.887)
C—O Bond	−0.190	−0.133 (−0.084)
O_3 Lone Pairs	−0.193	0.128 (0.090)
OH Bond	2.255	5.490 (3.882)
Total	1.536	7.748 (5.479)

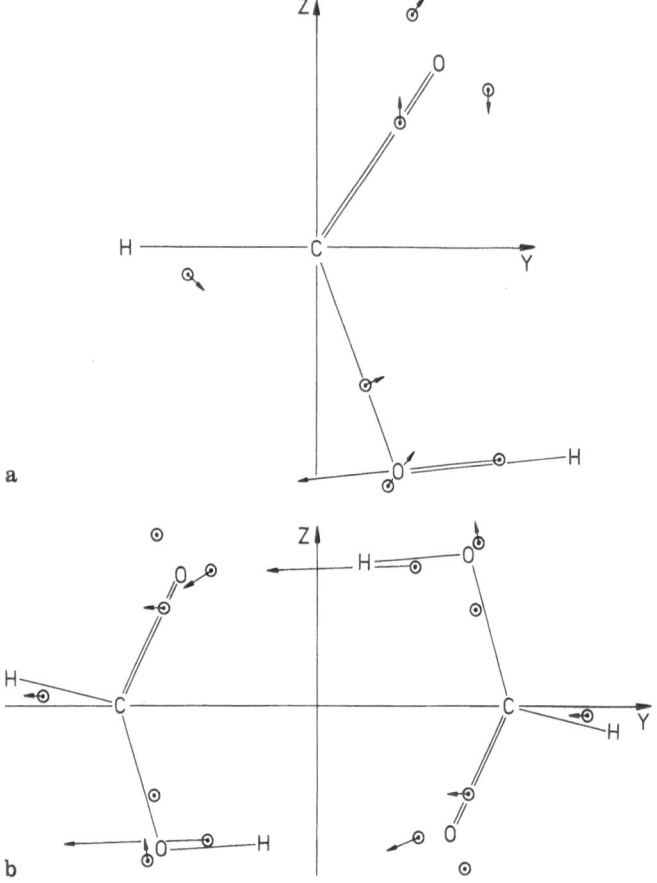

Fig. 4a, b. Contributions $(-\partial\mu_k/\partial Q_{OH})_0$ (k = k th LMO) to $(\partial\mu/\partial Q_{OH})_0$ in HCOOH (**a**) and (HCOOH)$_2$ (**b**) represented as vectors with the centroid of negative charges of the corresponding LMO as the origin. The scales for the internuclear distances at equilibrium and the displacement of the centroids are different. A positive Q_{OH} value corresponds to an elongation of the OH bond (in the left monomeric unit for the dimer).

Only in plane components are given. In order to obtain a better visualization of the relative displacements of the centroid of negative charge associated to the k-th LMO the sign of these components has been inverted.

From Berckmans D, Figeys HP, Geerlings P, J. Phys. Chem., 87: 61 (1988) and Berckmans D, Figeys HP, Maréchal Y, Geerlings P, J. Phys. Chem., 87: 66 (1988). Reprinted by permission of the American Chemical Society

resulting in the much larger dimer/monomer intensity ratio. Figure 4b visualizes these data: the increase in polarity of the OH bond (also seen in the monomer) is now accompanied by an attraction of the lone pairs on the opposite oxygen by the proton and an important bending of the C=O bond occurs. This is schematically shown in Fig. 5; a similar electronic rearrangement occurs in nucleophilic substitution reactions S_N2 [128]. It can easily be seen that these results are in line with the analysis presented by Zilles and Person [118] on the H_2O dimer and with the calculations by Bosi, Zerbi,

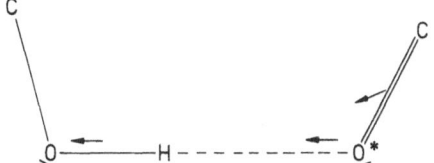

Fig. 5. Electronic rearrangements upon OH bond elongation in the dimer. From Berckmans D, Figeys HP, Maréchal Y, Geerlings P, J. Phys. Chem., 87: 66 (1988). Reprinted by permission of the American Chemical Society

and Clementi who attributed the large $B_u - A_g$ splitting of the $C=O$ stretching vibration frequency to an intermolecular charge flow through the hydrogen bond upon elongation of the $C=O$ bond in the dimer [129].

4 Conclusions

The interpretational techniques discussed in this chapter can be used successfully in the interpretation of quantum-chemically calculated IR integrated intensities of fundamentals, using the double harmonic approximation and single determinantal wavefunctions. It has been shown that vibrational impurities, which are not obvious in oversimplified representations of molecular vibrations, may significantly contribute to the observed intensities due to the interplay between vibrational and electronic factors. The danger of relating intensities to static charge distributions emerges at various places. The failure of the classical bond moment hypothesis is obvious from the LMO analysis. This technique also provides a non-empirical tool for analyzing incomplete orbital following effects in stretching and bending vibrations. On the basis of the formic acid results, the intensification of the AH stretching intensity upon H-bond formation can be interpreted as resulting from a negative charge flow from the proton acceptor, along the H-bond, towards the A-atom, upon AH bond elongation.

5 Acknowledgement

The authors wish to thank Dr. D. Berckmans for his enthusiastic and fruitful collaboration in various parts of their work.

6 References

1. Person WB, Zerbi G (eds) (1982) Vibrational intensities in infrared and Raman spectroscopy. Elsevier, Amsterdam
2. Jesson JP, Thompson HW (1958) Spectrochim. Acta 13: 217
3. Pimentel GC, McClelland AL (1960) The hydrogen bond. WH Freeman, San Francisco

4. Wilson EB Jr, Decius JC, Cross PC (1955) Molecular vibrations. McGraw Hill, New York
5. Overend J (1963) in: Davies M (ed) Infrared spectroscopy and molecular structure, Elsevier, Amsterdam, p 345
6. Duncan JL, Mallinson PD (1972) Chem. Phys. Lett. 23: 597
7. Hopper MJ, Russell JW, Overend J (1968) J. Chem. Phys. 48: 3765
8. Eggers DF, Hisatsune LC (1955) J. Chem. Phys. 23: 487
9. Jalsovsky G (1973) J. Mol. Struct. 19: 783
10. Duncan JL (1970) Mol. Phys. 28: 1177
11. Nakata M, Kohata K, Fukuyama T, Kuchitsu K (1980) J. Mol. Spectr. 83: 105
12. Nakata M, Kohata K, Fukuyama T, Kuchitsu K, Wilkins CJ (1980) J. Mol. Struct. 68: 271
13. McKean DC, Duncan JL (1971) Spectrochim. Acta A27: 1879
14. Schnöckel H (1975) J. Mol. Struct. 29: 123
15. Mallinson PD, McKean DC, Holloway JH, Oxton IA (1975) Spectrochim. Acta A31: 143
16. See for example Brown TL (1958) Chem. Rev. 58: 581
17. Hornig DF, McKean DC (1955) J. Phys. Chem. 59: 1133
18. Steele D (1964) Quart. Rev. 18: 21
19. Davies PR, Orville-Thomas WJ (1969) J. Mol. Struct. 4: 163
20. Jalsovsky GJ, Orville-Thomas WJ (1971) J. Chem. Soc., Faraday Transactions II 67: 1894
21. Gribov LA (1964) Intensity theory for polyatomic molecules. Consultants Bureau, New York
22. For a comprehensive account see: Orville-Thomas WJ, Suzuki S, Riley G (1977) in: Barnes AJ, Orville-Thomas WJ (eds) Vibrational spectroscopy — modern trends. Elsevier, Amsterdam, chapter 13
23. See reference 1, section B
24. Biarge JF, Herranz J, Morcillo J (1961) An. Fis. Quim. A57: 81
25. Person WB, Newton JH (1974) J. Chem. Phys. 61: 1040
26. King WT, Mast GB, Blanchette PP (1972) J. Chem. Phys. 56: 4440
27. Figeys HP, Geerlings P, Van Alsenoy C (1979) J. Chem. Soc., Faraday Transactions II 75: 528
28. Figeys HP, Geerlings P, Van Alsenoy C (1979) J. Chem. Soc., Faraday Transactions II 75: 542
29. Figeys HP, Geerlings P, Berckmans D, Van Alsenoy C (1981) J. Chem. Soc., Faraday Transactions II 77: 721
30. Geerlings P (1981) J. Mol. Struct. 72: 295
31. Figeys HP, Berckmans D, Geerlings P (1979) J. Mol. Struct. 57: 271
32. Figeys HP, Berckmans D, Geerlings P (1981) J. Chem. Soc., Faraday Transactions II 77: 2091
33. Berckmans D, Figeys HP, Geerlings P (1987) J. Mol. Struct. (Theochem.) 149: 243
34. Berckmans D, Figeys HP, Geerlings P (1988) J. Phys. Chem. 87: 61
35. Berckmans D, Figeys HP, Maréchal Y, Geerlings P (1988) J. Phys. Chem. 87: 66
36. Bruns RE, Bassi ABMS, Kuznesof P (1976) J. Am. Chem. Soc. 98: 3432
37. Figeys HP, Geerlings P, Van Alsenoy C, Raeymaekers P (1976) in: "Computation of IR integrated intensities and NMR coupling constants in large organic molecules and their interpretation in terms of localized orbitals" Second International Congress of Quantum Chemistry, New Orleans, April 1976
38. Shinoda H, Miyazaki T (1983) Bull. Chem. Soc. Jap. 56: 1924
39. Shinoda H (1987) Bull. Chem. Soc. Jap. 60: 2355
40. Riley G, Suzuki S, Orville-Thomas WJ, Galabov B (1978) J. Chem. Soc. 74: 1947
41. Born M, Oppenheimer JR (1927) Ann. Physik 84: 457
    Born M, Huang K (1954) Dynamical theory of crystal lattices. Oxford, chapter IV and appendix VII and VIII
42. Geerlings P, Berckmans D, Figeys HP (1979) J. Mol. Struct. 57: 283
43. Botschwina P (1982) Mol. Phys. 47: 241
44. Botschwina P (1983) Chem. Phys. 81: 73
45. Berckmans D, Figeys HP, Geerlings P (1986) J. Mol. Struct. 148: 81
46. Schachtschneider JH (1962) Vibrational analysis of polyatomic molecules: FORTRAN IV programs for solving the vibrational secular equation and for the least squares refinement of force constants. Shell Development Company, Emeryville, California
47. Pople JA, Beveridge DL, Dobosh PA (1967) J. Chem. Phys. 47: 2026
48. Giesner-Prettre C, Pullmann A (1968) Theor. Chem. Acta 11: 159
49. Dobosh P "CNINDO" (QCPE n° 141). Deutsches Rechenzentrum, Darmstadt

50. Figeys HP, Geerlings P, Van Alsenoy C (1975) J. Chem. Soc., Faraday Transactions II 71: 1375
51. Hehre WJ, Lathan WA, Ditchfield R, Newton MD, Pople JA GAUSSIAN 70, QCPE program n° 236, QCPE. Indiana University, Bloomington
52. Binkley JS, Whitehead RA, Hariharan PC, Seeger R, Pople JA, Hehre WJ, Newton MD GAUSSIAN 76, QCPE program n° 368, QCPE. Indiana University, Bloomington
53. Hehre WJ, Ditchfield R, Stewart RF, Pople JA (1970) J. Chem. Phys. 52: 2769
54. Cyvin SJ (1968) Molecular vibrations and mean square amplitudes. Elsevier, Amsterdam
55. Raeymaekers P, Figeys HP, Geerlings P (1988) J. Mol Struct. (Theochem L. Pauling Volume) 169: 509
56. Long DA (1977) Raman spectroscopy. McGraw-Hill, New York
57. Kelly HP (1964) Phys. Rev. B136: 896
58. Huzinaga S, Arnau C (1970) Phys. Rev. A1: 1285
59. Raeymaekers P, Figeys HP, Geerlings P (1988) Mol. Phys., 65: 925
60. Raeymaekers P, Figeys HP, Geerlings P (1988) Mol. Phys., 65: 945
61. Pople JA, Beveridge DL (1970) Approximate molecular orbital theory. McGraw-Hill, New York
62. Mulliken RS (1935) J. Chem. Phys. 3: 375
63. Mulliken RS (1955) J. Chem. Phys. 23: 1833
64. Roothaan CCJ (1951) Rev. Mod. Phys. 23: 69
65. Fock V (1930) Z. Physik 61: 126
66. For a review see for example: England W, Salmon LS, Rüdenberg K (1971) Fortsch. Chem. Forsch. 23: 31
    See also chapter 10 in Part 2 of this series: Edmiston C, Interpretation of molecular behaviour by localized MO's
67. Von Niessen W (1970) J. Chem. Phys. 56: 4290
68. Figeys HP, Geerlings P, Raeymaekers P, Van Alsenoy C (1975) Theor. Chim. Acta 40: 253
69. Figeys HP, Geerlings P, Raeymaekers P, Van Lommen G, Defay N (1975) Tetrahedron 31: 1731
70. Van Alsenoy C, Figeys HP, Geerlings P (1980) Theor. Chim. Acta 55: 87
71. Edmiston C, Rüdenberg K (1963) Rev. Mod. Phys. 35: 457
72. Figeys HP, Geerlings P, Van Alsenoy C (1977) Int. J. Quantum Chem. XI 705
73. Geerlings P, Berckmans D, unpublished
74. Foster M, Boys SF (1960) Rev. Mod. Phys. 32: 300
75. Berckmans D (1984) Ph. D. Thesis, Free University of Brussels
76. England W, Gordon MS (1972) J. Am. Chem. Soc. 93: 4649
77. Uyemura M, Maeda S (1974) Bull. Chem. Soc. Japan 47: 2930
78. Cyvin SJ, Klaeboe P (1965) Acta Chem. Scand. 19: 697
79. Costain CC (1958) J. Chem. Phys. 29: 864
80. Van Catledge FA (1974) J. Phys. Chem. 78: 763
81. Person WB, Steele D (1974) in: Barrow RF, Long DA, Millen DJ (eds) Molecular spectroscopy (Spec. Period. Rep., vol II). The Chemical Society, London, p 357
82. Experimental geometries used
82a. Benedict WS, Plyler EK (1957) Can. J. Phys. 35: 1235 (NH$_3$)
82b. Helms DA, Gordy W (1977) J. Mol. Spectr. 65: 206 (PH$_3$)
82c. Otake M, Matsamura C, Morino Y (1968) J. Mol. Spectr. 28: 316 (NF$_3$)
82d. Hirota E, Morino Y (1970) J. Mol. Spectr. 33: 460 (PF$_3$)
    Force field used
82e. Duncan JL, Mills IM (1964) Spectrochim. Acta 20: 523 (NH$_3$, PH$_3$)
82f. Allan A, Duncan JL, Holloway JH, McKean DC (1969) J. Mol. Spectr. 31: 368 (NF$_3$)
82g. Hirota E, Morino Y (1970) J. Mol. Spectr. 33: 460 (PF$_3$)
    Experimental intensities
82h. McKean DC, Schatz PN (1955) J. Chem. Phys. 24: 316
82i. Smit WMA, Van Dam T (1980) J. Chem. Phys. 72: 3658
82j. Koops TH, Visser T, Smit WMA (1983) J. Mol. Struct. 96: 203 (NH$_3$, PH$_3$)
82k. Schatz PN, Levin IW (1971) J. Chem. Phys. 29: 475 (NF$_3$)
82l. Levin IW, Adams OW (1971) J. Mol. Spectr. 39: 380 (PF$_3$)
83. Dunning TH (1970) J. Chem. Phys. 53: 2823
84. Maroulis G, Sana M, Leroy G (1981) Int. J. Quantum Chem. 19: 431
85. Hehre WJ, Ditchfield R, Stewart RF, Pople JA (1970) J. Chem. Phys. 52: 2769

86. Pulay P, Meyer W (1972) J. Chem. Phys. 57: 3337
87. For a recent compilation of ab initio IR frequency and intensity results, see Hess BA, Schaad LJ, Carsky P, Zahradnik K (1986) Chem. Rev. 86: 709
88. Amos RD (1987) in: Lawley KP (ed) Ab initio methods in quantum chemistry II. John Wiley, New York, p. 241 (Adv. Chem. Phys., LXVIII)
89. Turrel GC, Jones WC, Maki A (1957) J. Chem. Phys. 26: 1544
90. Caldow GL, Cunliff-Jones D, Thompson HW (1960) Proc. Roy. Soc. A 254: 17
91. Roshchupkin GP, Popov EM (1963) Optics Spectr. 15: 109
92. Besnainou S, Thomas B, Bratoz S (1966) J. Mol. Spectr. 21: 113
93. Thomas BH, Orville-Thomas WJ (1971) J. Mol. Struct. 7: 123
94. Force fields
94a. Ref. 28 (CH_3CN)
94b. Crowder GA (1972) Mol. Phys. 23: 707 (CH_2ClCN)
94c. Ref. 28 ($CHCl_2CN$)
94d. Johansen W (1965) Z. Phys. Chem. 230: 240 (CCl_3CN)
      Geometries
94e. Ref. 79 (CH_3CN)
94f. Wada K, Kikuchi Y, Matsamura C, Hirota E, Morino Y (1961) Bull. Chem. Soc. Japan 34: 337 (CH_2ClCN)
94g. Ref. 28 ($CHCl_2CN$)
94h. Baker JG, Jenkins DR, Kenney CN, Sugden TM (1957) Trans. Faraday Soc. 53: 1397 (CCl_3CN)
95. Bournay J, Maréchal Y (1975) Spectrochim. Acta 31A: 1351
96. Bellet J, Deldalle A, Samson C, Steenbeckeliers G, Wertheimer R (1971) J. Mol. Struct. 9: 65
97. Redington RL (1977) J. Mol. Spectr. 65: 171
98. Linnett JW, Wheatley PJ (1949) Trans. Faraday Soc. 45: 33
99. Heath DF, Linnett JW, Wheatley PJ (1950) Trans. Faraday Soc. 46: 137
100. Cohan NV, Coulson CA (1956) Trans. Faraday Soc. 52: 1163
101. Coulson CA, Stephen MJ (1957) Trans. Faraday Soc. 53: 272
102. Coulson CA (1959) Spectrochim. Acta 14: 161
103. Kaldor U (1967) J. Chem. Phys. 46: 1981
104. Nakatsuji H (1974) J. Am. Chem. Soc. 96: 24
105. Hellmann H (1937) Einführung in die Quantenchemie. Deuticke, Vienna, p. 285
106. Feynman RP (1939) Phys. Rev. 56: 340
107. Kollman PA, Allen LC (1969) J. Chem. Phys. 51: 3286
108. Kollman PA, Allen LC (1970) J. Am. Chem. Soc. 92: 753
109. Kollman PA, Allen LC (1970) J. Chem. Phys. 52: 5085
110. Dreyfus M, Pullman B (1970) Theoret. Chim. Acta 19: 20
111. Diercksen GHF (1971) Theoret. Chim. Acta 21: 335
112. Curtiss LA, Pople JA (1973) J. Mol. Spectr. 48: 413
113. Janoschek R (1973) Theoret. Chim. Acta 32: 49
114. Pecul K, Janoschek R (1974) Theoret. Chim. Acta 36: 25
115. Bouteiller Y, Allavena M, Leclerq J (1980) J. Chem. Phys. 73: 2851
116. Allavena M, Silvi B, Cipriani J (1982) J. Chem. Phys. 76: 4573
117. Kecki Z, Sadlej J, Sadlej AJ (1982) J. Mol. Struct. 88: 71
118. Zilles B, Person WB (1983) J. Chem. Phys. 79: 65
119. Akagi K, Tanabe Y, Yamabe T (1983) J. Mol. Struct. 102: 103
120. Leclercq JM, Allavena M, Bouteiller Y (1983) J. Chem. Phys. 78: 4806
121. Wojcik MJ, Hirakawa AY, Tsuboi M, Kato S, Morokuma K (1983) Chem. Phys. Lett. 100: 523
122. Hayashi S, Umemura J, Kato S, Morokuma K (1984) J. Phys. Chem. 88: 1330
123. Swanton DJ, Backsay GD, Hush NS (1983) Chem. Phys. 82: 303
124. Datka J, Geerlings P, Mortier WJ, Jacobs PA (1985) J. Phys. Chem. 89: 3488
125. Amos RD (1986) Chem. Phys. 104: 145
126. Almenningen A, Bastiansen O, Motfeldt T (1969) Acta Chem. Scand. 23: 2848
127. Kishida S, Nakamoto K (1964) J. Chem. Phys. 41: 1558
128. Breslow R (1969) Organic reaction mechanisms. WA Benjamin, New York
129. Bosi P, Zerbi P, Clementi E (1977) J. Chem. Phys. 66: 3376

The Orbital Concept as a Foundation
for Photoelectron Spectroscopy

S. P. McGlynn[1], K. Wittel[1] and L. Klasinc[2]

[1] Department of Chemistry, Louisiana State University, Baton Rouge, LA 70803, USA
[2] The Ruđer Bošković Institute, University of Zagreb, Zagreb, Croatia, Yugoslavia

Chemistry assumes that electrons moving in the potential field of nuclei are crucial to the existence of those forces that define the formation, properties and reaction of chemical compounds. A theoretical approach that adopts this view is the orbital (atomic, AO, or molecular, MO) concept. The orbital concept is concerned with the motion of electrons in the average field of a complicated system of particles (including electrons), and it assumes that the results bear on reality. The success of the method is well documented and the difficulties it encounters, particularly when applied to molecular spectroscopy, have been elaborated [1].

Much of the attraction of photoelectron spectroscopy (PES), in its early years at least, was vested in the belief that one of the most basic concepts of modern chemistry, namely the orbital concept, would be interpretively sufficient and, further, that this belief was experimentally vindicated by the existing PES data. Now, after a quarter of a century, it is clear that PES has outgrown this convenient, but naive expectation, that the whole area has become much less transparent and quite multilayered, and that a particular interpretive mode will often be determined by the viewpoint, intention and purpose of the modern investigator. None the less, while the "now" investigator tends to move away centrifugally from the familiar picture of photoejection of electrons from orbitals and photo-creation of holes in these orbital levels, and while he may, for example, present his story as a time-dependent scattering process, he and/or the users of his results will invariably refer the conclusion back to the orbital concept because of the simplicity and utility which this referral confers. Indeed, if the orbital concept were to lose such utility vis-a-vis PES data, it is quite sure that the PES technique would lose its somewhat special status and become no more than another physical method for the investigation of matter.

In our formulation of the orbital concept as a theoretical foundation for PES, it is not our intention to diminish "the higher levels of truth" embedded in the modern theoretical treatments of the photoionization process. It is our intention, however, to reassert the significance of preserving the orbital interpretive model. This we believe, is important for a number of reasons:

(i) It provides chemists and biologists an interpretive mode that is a part of their existing theoretical armamentarium.

(ii) Many of the existing semiempirical many- and one-electron-methods have been specifically parametrized for PES, and they provide a convenient, first-run interpretive approach that can be refined using various correlative techniques.

(iii) It is a technique which, when wedded to experience, is usually pretty much on target.

1 Photoelectron Spectroscopy

Release of electrons from a sample is termed *ionization*. If the energy for the release is provided by light (photons), the ejected electrons are called *photoelectrons*. The method which measures the kinetic energy and counts the number of photoelectrons is *photoelectron spectroscopy* [2]. The sample may be gaseous, liquid, or solid (depending upon the aggregate state of the sample), molecular or bulk (depending on sample

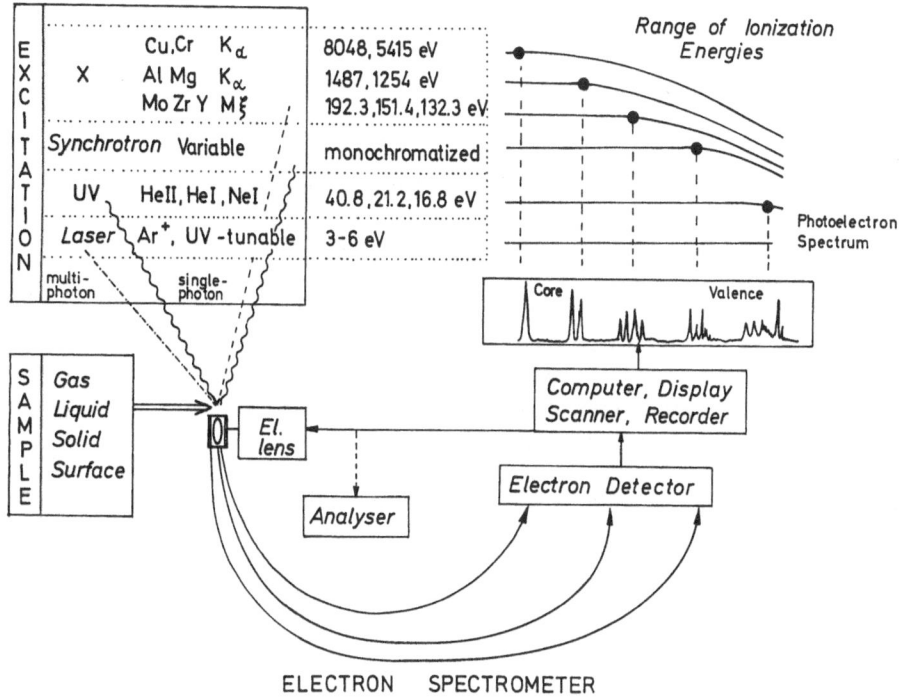

Fig. 1. Various aspects of photoelectron spectroscopy

dispersion) and radiation can be UV, X-ray, synchrotron, or laser, single-photon or multi-photon, monochromatic or polychromatic (See Fig. 1).

The physical basis of photoelectron spectroscopy is contained in the photoelectric effect (Hertz, 1887) and its formulation by Einstein (1905)

$$E_{kin} = h\nu - E_i$$

where E_{kin} is the kinetic energy of the ejected electron, $h\nu$ the energy of the exciting photon, and E_i the system ionization energy. The photon energy is usually maintained constant and the analyzer, an electrostatic deflection, retarding grid or time-of-flight system, is scanned over the appropriate range of electron energies. If the analyzer transmits only zero-energy electrons, the method is termed *threshold PES*.

Technically, PES is but one example of a generic electron spectroscopy, namely all spectroscopies which count electron numbers and measure electron energies (e.g. β-decay and electron transmission, electron bombardment). The particular technique of photoelectron spectroscopy was made possible by the development of high-performance vacuum systems and differential pumping technique that made the use of windowless apparatus possible. These technical developments are now so advanced that even extra-thin jets or fast moving films of liquids may be allowed into the high-vacuum system, and the photoelectrons ejected directly from the liquid phase may be measured [3]. These vacuum techniques also make windowless access to synchrotron radiation [4] feasible and, thereby, make polychromatic PES possible.

2 The Orbital Concept

2.1 Introduction

Chemistry, particularly organic chemistry, is heavily dependent on MO concepts.

The various electronic orbital types are labeled, in self-evident terminology and in the approximate order of decreasing binding energy: *core*, *valence*, and *Rydberg*. The inter-relations of these types of orbitals are rarely specified.

Our aim is to relate orbital properties to state properties in a way which is useful and meaningful for PES. Consequently, we will begin with the general many-electron, many-nuclei problem. While retaining as much rigor as is warranted, we proceed to a discussion of certain trenchant theorems, their derivations and the approximations inherent in their use. In this connection, we will pay considerable attention to Koopmans' theorem and its extension to core ionization events. For the sake of cohesion, we will limit ourselves to molecules with closed-shell ground states and we will restrict ourselves to electronic excitation events which can be induced by photons [5].

2.1.1 Orbital Classification

Orbitals, or electrons, may be categorized as "core", "valence" or "Rydberg". This categorization depends on differences of the binding energies and the spatial extents of the orbitals, and is approximate. It is important to emphasize here that terms such as "valence electron", "Rydberg electron", etc., while very common and convenient, have little theoretical justification. A look at the Slater determinant of p. 9 followed by the question: "Is electron 1 more in orbital a than in orbital b?" is enough to drive this point home. Koopmans' theorem (vide infra) does supply a rationale, at least for electrons of the ground configuration; that is, we can associate an ejected electron with a specific orbital if we understand that the orbital in question is a difference of total electron densities of N and N-1 electron systems. But, while exhibiting some worry about this problem, we blithely proceed to such a categorization in Table 1.

The core-electron binding energies characterize the atom. Thus, when the atom is part of a molecule, the perturbation of the core binding energy by all the other atoms of the molecule rarely exceeds 10%; in fact, it is usually considerably less. In absolute terms, the perturbation lies in the range $0.001 - 0.1 \times 10^6$ cm^{-1}. One could say that the core orbitals are so compressed about their specific atomic centers that all neighboring atomic centres of the molecule, being quite far away, exert only a small perturbative influence.

On the other extreme are the Rydberg orbitals. The binding energy of a Rydberg electron is quite small, usually $<3 \times 10^4$ cm^{-1}. Indeed, the highly-excited Rydberg electron behaves as a quasi-hydrogenic electron whose binding energy is given by $Z^2R/(n^*)^2$ where R is the Rydberg constant, n* is an effective principal quantum number, and Z is the charge "seen" by the Rydberg electron. One could say that Rydberg orbitals are so large spatially that the details of molecular architecture become insignificant and merely exert perturbative effects on an otherwise hydrogenic electron.

Table 1. The various orbital regimes

Orbital regime	Binding Energy (Absolute, cm^{-1})	Major energy contribution (zero-order)	Minor energy contribution (perturbation)	Orbital extent
Core	$\geq 10^6$ [a]	Attraction by one particular nucleus	Neighboring nuclei; valence electrons	Atomic-like; tightly packed around one particular nucleus; localized
Valence				
Bonding	$1-5 \times 10^5$	"Whole molecule" in nature		Delocalized; of same size as molecule
Lone pair	$5-10 \times 10^4$	Attraction by one particular nucleus and its complement of core electrons	Remainder of the molecule	Localized
Antibonding (virtual)	$\leq 4 \times 10^4$	"Whole molecule" in nature		Delocalized; of same size as molecule
Rydberg	$\leq 3 \times 10^4$	Attraction by whole molecule acting as a point charge	Penetration of core; deviation of potential from spherical symmetry	Atomic-like; large; diffuse; hyper-molecular

[a] Except lithium, for which $IE_{1s} = 0.44 \times 10^6 \ cm^{-1}$

The valence electrons, for the most part, are intermediate in energy and spatial extent. They represent a coupling domain which is uniquely molecular. In specific, they are not so tightly bound that they can be considered to be "nearly atomic" (i.e., core) electrons, nor so weakly bound that they can be supposed to be "quasi-hydrogenic" (i.e., Rydberg) electrons. As a result, it is the valence electrons which pose the most difficult theoretical problem. In fact, if a good discussion of the valence electron problem were available, it should be possible, by appropriate extension or shrinkage of the valence orbitals, to extract a pertinent description of the Rydberg electron or core electron extremes, respectively. Consequently, we will begin our discussion with the valence MO's.

There is another reason, however, for starting with the valence orbitals. The valence orbitals are the "uniquely molecular" orbitals. For example, they determine chemical bonding characteristics, molecular structure, and the characteristics of chemical reactions. It is not surprising, then, that very many sub-categorizations of them exist and that discussions of them are heavily laced with empirical and heuristic connotations. The removal of these connotations and/or their replacement with theoretically valid and empirically useful concepts is another one of the major aims of this essay.

A glossary of the notations used here is given in Table 2.

Table 2. A glossary of notations

Wavefunctions

$\Psi = \Psi(\vec{x}, \vec{R})$	the general many-electron wavefunction
$\Psi_{el} = \Psi(\vec{x}; \vec{R})$	the many-electron wavefunction in any "fixed-nucleus" approximation; the dependence on $\vec{R}$ is now understood to be parametric, as is emphasized by the use of a semi-colon
$\psi = \psi(\vec{x}; \vec{R})$	a many-electron Slater determinant wavefunction
$\varphi = \varphi(\vec{r}_1; \vec{R})$	a molecular spin-orbital
$\phi = \phi(\vec{r}_1; \vec{R})$	a molecular (space) orbital
$\alpha(\vec{\omega}_1), \beta(\vec{\omega}_1)$	spinfunctions for $m_s = 1/2$ and $-1/2$, respectively
$\chi = \chi(\vec{x}_{1A})$	an atomic spinorbital
$\Xi = \Xi(\vec{R})$	a nuclear wavefunction, where the dependence on $\vec{R}$ is no longer merely parametric
$\psi_0 = \|\varphi_a \varphi_b \varphi_c \cdots \varphi_l \varphi_m \varphi_n\|$	determinant for the ground configuration
$\Psi_{-m} = \|\varphi_a \varphi_b \varphi_c \cdots \varphi_l \varphi_n\|$	determinant for a Koopmans' configuration
$\Psi_m^v = \|\varphi_a \varphi_b \varphi_c \cdots \varphi_l \varphi_v \varphi_n\|$	determinant for a singly-excited configuration
$\varphi_{lm}^{\mu v} = \|\varphi_a \varphi_b \varphi_c \cdots \varphi_\mu \varphi_v \varphi_n\|$	determinant for a doubly-excited configuration

Coordinates

$\vec{x} = \{\vec{x}_1, \vec{x}_2, \ldots\} = \{1, 2, \ldots\}$	the space and spin coordinate set for electrons 1, 2, ...
$\vec{x}_1 = \{\vec{r}_1, \vec{\omega}_1\}$	the space and spin coordinate set for electron 1.
$\vec{r}_1$	electronic space coordinate set for electron 1
$\vec{\omega}_1$	electronic spin coordinate for electron 1
$\vec{R} = \{\vec{R}_A, \vec{R}_B, \ldots\}$	nuclear space coordinate set

Subscripts and superscripts

a, b, c, ... , k, l, m, n	occupied (spin)orbitals of ground configuration
$\mu, v, \ldots$	unoccupied (or virtual) (spin)orbitals
A, B, C, ...	nuclei or sub-molecular parts
i, j ...	electron numbering

Operators

$\hat{H}$	general Hamiltonian
$\hat{H}_{el}$	electronic Hamiltonian
$\hat{H}^c$	core Hamiltonian; the one-electron part of $\hat{H}_{el}$
$\hat{F}$	Fock operator
$\hat{F}_m$	Fock operator with spinorbital φ_m of ψ_0 deleted
$\hat{T}$	kinetic energy operator
$\hat{P}_{12}$	permutation operator which interchanges coordinates of electrons 1 and 2
$\hat{\Omega}(i)$	general one-electron operator
$\hat{\Omega}(i, j)$	general two-electron operator

Miscellaneous

$\delta_{ij} = 1$ (or zero) for $i = j$ (or $i \neq j$)	Kronecker delta
$\langle \varphi_i \| \varphi_j \rangle = \int \varphi_i^*(1) \varphi_j(1) \, d\vec{x}_1$	
$\langle \varphi_a \varphi_b \| (1 - \hat{P}_{12})/r_{12} \| \varphi_c \varphi_d \rangle =$	$\iint \varphi_a^*(1) \varphi_b^*(2) [\varphi_c(1) \varphi_d(2) - \varphi_c(2) \varphi_d(1)] \dfrac{1}{r_{12}} \, d\vec{x}_1 \, d\vec{x}_2$
N	total number of electrons
Z_A	atomic number of atom A
S	total spin quantum number
M_s	component of S along a particular axis

2.2 The MO Approximation

One of the aims of quantum chemistry is to solve the time-independent Schrödinger-equation

$$\hat{H}(\vec{x}, \vec{R})\, \Psi(\vec{x}, \vec{R}) = E\Psi(\vec{x}, \vec{R}) \tag{1}$$

where $\hat{H}$ is the Hamilton operator. Both $\hat{H}$ and Ψ, as indicated in Table 2, depend on the coordinates of all nuclei and all electrons in the molecule. The goal of this section is to simplify the multi-dimensional problem of Eq. 1: To reduce it, if possible, to one which is dependent on the coordinates of only one electron.

We neglect, for convenience, all relativistic terms in $\hat{H}$ and rewrite it as

$$\hat{H}(\vec{r}, \vec{R}) = \sum_A \hat{T}_A + \sum_{A<B} Z_A Z_B / R_{AB} + \sum_i \hat{T}_i + \sum_{i<j} 1/r_{ij} - \sum_i \sum_A Z_A / r_{iA}. \tag{2}$$

The terms of Eq. 2, in order of appearance, describe the kinetic energy of the nuclei, the electrostatic repulsion of the nuclei, the kinetic energy of the electrons, the electron-electron repulsion, and the nuclei-electron attraction [6].

2.2.1 The Separation of Nuclear and Electronic Motions

The first step is to approximate $\Psi(\vec{x}, \vec{R})$ by a product such that one of the parts is only weakly dependent on $\vec{R}$:

$$\Psi(\vec{x}, \vec{R}) = \Psi(\vec{x}; \vec{R})\, \Xi(\vec{R}). \tag{3}$$

The function $\Psi(\vec{x}; \vec{R})$, which is dependent on $\vec{R}$ only in a parametric fashion, is known as the electronic wavefunction and, accordingly, is often denoted $\Psi_{el}(\vec{x}; \vec{R})$. The function $\Xi(\vec{R})$ is known as the nuclear wavefunction. These functions are determined by

$$\hat{H}_{el}\Psi_{el}(\vec{x}; \vec{R}) = E_{el}(\vec{R})\, \Psi_{el}(\vec{x}; \vec{R}) \tag{4}$$

$$\hat{H}_n \Xi(\vec{R}) = E\Xi(\vec{R}) \tag{5}$$

where the Hamilton operators, to the exclusion of relativistic effects, are given by

$$\hat{H}_{el} = \sum_i \hat{T}_i + \sum_{i<j} 1/r_{ij} - \sum_i \sum_A Z_A / r_{iA} + \sum_{A<B} Z_A Z_B / R_{AB} \tag{6}$$

$$\hat{H}_n = \sum_A \hat{T}_A + E_{el}(\vec{R}). \tag{7}$$

Relative to Eq. 3, Eqs. 4–7 involve approximations [7] such as the neglect of various cross terms (e.g. $\langle \Psi_{el}(\vec{x}; \vec{R}) | \hat{T}_A | \Psi_{el}(\vec{x}; \vec{R}) \rangle$).

This separation into an electronic and a nuclear part is known as the Born-Oppenheimer approximation [8]. It permits the use of several well-known concepts. These are:

— *Electronic state:* The energy of an electronic state is parametrically dependent on $\vec{R}$ and this dependence yields the familiar potential energy well which governs vibrational motion.

— *Equilibrium nuclear geometry:* This geometry is defined as that which exist at the absolute minimum of the (multi-dimensional) potential energy surface $E_{el}(\vec{R})$.

— *Electronic energy and vibrational energy:* The energy separates into distinct electronic and vibrational parts.

All these concepts lose validity when Eq. 3 is not a good approximation to the solutions of Eq. 1. Thus, the Born-Oppenheimer approximation will usually be invalid when the energy differences between electronic states are smaller than the vibrational energy increments associable with any of the electronic states. In particular, when the electronic states are non-accidentally degenerate, Jahn-Teller or Renner instabilities [8] will arise from a coupling of the electronic and vibrational motions.

In a more physical vein, the possibility of a separation into distinct nuclear and electronic motions is based on the comparatively larger mass of the nuclei: The electrons have less inertia and can adjust their motions, more or less instantaneously, to any rearrangement of the nuclear positions.

2.2.2 The Independent Particle Approximation

The electronic wavefunction $\Psi_{el}(\vec{x}; \vec{R})$ is still dependent on the coordinates of all electrons. If a complete set of orthonormal, many-electron wavefunctions $\{\psi_i\}$ is available, $\Psi_{el}(\vec{x}; \vec{R})$ can be expanded as

$$\Psi_{el}(\vec{x}; \vec{R}) = \sum_i \psi_i c_i \tag{8}$$

where

$$\cdot \langle \psi_i | \psi_j \rangle = \delta_{ij} \,. \tag{9}$$

We now investigate such an expansion, the manner of its simplification, and the nature of the functions ψ_i.

2.2.2.1 Determinantal Wavefunctions

The Pauli principle dictates that $\Psi_{el}(\vec{x}; \vec{R})$ must be antisymmetric with respect to any interchange of electron coordinates. Given the abbreviations

$$\Psi_{el}(\vec{x}; \vec{R}) \equiv \Psi_{el}(\vec{x}_1, \vec{x}_2, \dots, \vec{x}_i, \dots; \vec{R}) \equiv \Psi_{el}(1, 2, \dots, i, \dots; \vec{R})$$

the antisymmetry requirement is

$$\Psi_{el}(1, 2, 3, \dots, i, \dots, k, \dots; \vec{R}) = -\Psi_{el}(1, 2, 3, \dots, k, \dots, i, \dots; \vec{R}) \,. \tag{10}$$

Since antisymmetry is one of the more unique characteristics of determinantal wavefunctions, one convenient and complete set of many electron functions is provided by a set of Slater determinants built from a complete set of spinorbitals $\{\varphi_i(j; \vec{R})\}$

$$\psi(1, 2, \ldots, N; \vec{R}) = \frac{1}{(N!)^{1/2}} \begin{vmatrix} \varphi_a(1; \vec{R}) & \varphi_b(1; \vec{R}) & \varphi_c(1; \vec{R}) & \ldots & \varphi_n(1; \vec{R}) \\ \varphi_a(2; \vec{R}) & \varphi_b(2; \vec{R}) & \varphi_c(2; \vec{R}) & \ldots & \varphi_n(2; \vec{R}) \\ \varphi_a(3; \vec{R}) & \varphi_b(3; \vec{R}) & \varphi_c(3; \vec{R}) & \ldots & \varphi_n(3; \vec{R}) \\ \vdots & & & & \vdots \\ \varphi_a(N; \vec{R}) & \varphi_b(N; \vec{R}) & \varphi_c(N; \vec{R}) & \ldots & \varphi_n(N; \vec{R}) \end{vmatrix}$$

(11)

$$= |\varphi_a(1; \vec{R}) \varphi_b(2; \vec{R}) \varphi_c(3; \vec{R}) \ldots \varphi_n(N; \vec{R})|$$

$$= |\varphi_a \varphi_b \varphi_c \ldots \varphi_n| . \tag{12}$$

Since the electron-occupancy number of any spinorbital is either *one* or *zero*, it follows that the number of occupied spinorbitals equals the number of electrons. The functions φ_i depend on the coordinates of one electron only; they are termed "spinorbitals" [9]. If the set of spinorbitals $\{\varphi_i\}$ is complete and, for convenience, orthonormal [10], it may be shown that the set of all possible determinants, $\{\psi_i\}$, which is constructed from these spinorbitals is also complete and orthonormal.

We now subdivide the set of determinantal wavefunctions in a very specific way. The ground state configuration [11]. ψ_0, is obtained by placing the N electrons, one at a time, into the n most tightly-bond spinorbitals. Singly-excited configuration functions, ψ_k^μ, are obtained by promoting an electron from the kth to the μth, most tightly-bound spinorbital. Doubly-excited configuration functions are generated by the promotion of two electrons out of the spinorbitals contained in ψ_0; etc. These classes of determinantal functions are defined as

$$\begin{aligned} \psi_0 &= |\varphi_a \varphi_b \ldots \varphi_k \ldots \varphi_l \ldots \varphi_n| \\ \psi_k^\mu &= |\varphi_a \varphi_b \ldots \varphi_\mu \ldots \varphi_l \ldots \varphi_n| \\ \psi_{kl}^{\mu\nu} &= |\varphi_a \varphi_b \ldots \varphi_\mu \ldots \varphi_\nu \ldots \varphi_n| . \end{aligned} \tag{13}$$

The expansion of Eq. 8 is now rewritten as

$$\Psi_{el} = \psi_0 c_0 + \sum_{k,\mu} \psi_k^\mu c_k^\mu + \sum_{\substack{k,l \\ \mu,\nu}} \psi_{kl}^{\mu\nu} c_{kl}^{\mu\nu} + \ldots \tag{14}$$

The expansion of Eq. 14 is known as the CI (CONFIGURATION INTERACTION) expansion. This expansion does not contain any approximation beyond those of Eqs. 3–5.

2.2.2.2 The Hartree-Fock Operator

It would be very convenient if we could choose the set of spinorbitals, $\{\psi_i\}$, in such a way that, for the ground state, the coefficient c_0 would be approximately unity and others approximately zero. We now inquire into this possibility. In order words, we seek that set of spinorbitals, $\{\varphi_a, \varphi_b, \ldots, \varphi_n\}$, which gives the best approximation to the correct electronic wavefunction of the ground state while simultaneously retaining only the first term of Eq. 14. Equivalently, we wish to approximate the many-electron wavefunction $\psi_{el}(\vec{x}; \vec{R})$ of the ground state by a single Slater determinant.

The choice of the "best" Slater determinant involves the use of the VARIATION principle: If the "best" Slater determinant were available, the electronic energy associated with it would be a minimum with respect to any variations in the spin orbitals [12]. The expectation value of energy for a Slater determinant is given by Slater's rules [13]. These rules are presented in Table 3. These lead to

$$E_{el} = \langle \psi | \hat{H}_{el} | \psi \rangle$$

$$= \sum_{k \in occ} \langle \varphi_k(1) | \hat{H}^c(1) | \varphi_k(1) \rangle + \sum_{A < B} Z_A Z_B / R_{AB}$$

$$+ \sum_{k < l \in occ} \langle \varphi_k(1) \varphi_l(2) | (1 - \hat{P}_{12})/r_{12} | \varphi_k(1) \varphi_l(2) \rangle \tag{15}$$

where the summation indices, k or l, are members of the set of spinorbitals which are occupied in the configurational function of interest. The permutation operator $\hat{P}_{12}$ is defined as

$$\hat{P}_{12} | \varphi_k(1) \varphi_l(2) \rangle = | \varphi_k(2) \varphi_l(1) \rangle \tag{16}$$

and the core Hamiltonian, $\hat{H}^c(1)$, is defined by

$$\hat{H}^c(1) = \hat{T}_1 - \sum_A Z_A / r_{iA} . \tag{17}$$

A variation of ψ_0 is produced by replacing some specific φ_i, where φ_i is a member of the set of occupied spinorbitals of ψ_0, by $\varphi_i + \delta\varphi_i$. The variation $\delta\varphi_i$ must be of such a nature that it does not disrupt any of the existing spinorbital orthonormalities; thus, we require

$$\langle \delta\varphi_i | \varphi_l \rangle = 0 \tag{18}$$

for all φ_l which are occupied. The consequent variation of total energy is given, following the Slater rules of Table 3, as

$$\delta E_{el,i} = \langle \delta\varphi_i(1) | \hat{H}^c(1) | \varphi_i(1) \rangle$$

$$+ \sum_{l \in occ} \langle \delta\varphi_i(1) \varphi_l(2) | (1 - \hat{P}_{12})/r_{12} | \varphi_i(1) \varphi_l(2) \rangle . \tag{19}$$

Table 3. A collection of pertinent Slater rules
(For definition of $\hat{\Omega}(i)$ and $\hat{\Omega}(i, j)$, see Table 2)

Overlap

$$\langle \psi_0 | \psi_0 \rangle = 1$$

$$\langle \psi_0 | \psi_k^\mu \rangle = \langle \psi_0 | \psi_{kl}^{\mu\nu} \rangle = 0$$

One-electron operators

$$\langle \psi_0 | \sum_i \hat{\Omega}(i) | \psi_0 \rangle = \sum_{k \in occ} \langle \varphi_k(1) | \hat{\Omega}(1) | \varphi_k(1) \rangle$$

$$\langle \psi_0 | \sum_i \hat{\Omega}(i) | \psi_k^\mu \rangle = \langle \varphi_k(1) | \hat{\Omega}(1) | \varphi_\mu(1) \rangle$$

$$\langle \psi_0 | \sum_i \hat{\Omega}(i) | \psi_{kl}^{\mu\nu} \rangle = 0$$

Two-electron operators

$$\langle \psi_0 | \sum_{i<j} \hat{\Omega}(i, j) | \psi_0 \rangle = \sum_{k < l \in occ} \{ \langle \varphi_k(1) \varphi_l(2) | \hat{\Omega}(1, 2) | \varphi_k(1) \varphi_l(2) \rangle$$
$$- \langle \varphi_k(1) \varphi_l(2) | \hat{\Omega}(1, 2) | \varphi_k(2) \varphi_l(1) \rangle \}$$

$$\langle \psi_0 | \sum_{i<j} \hat{\Omega}(i, j) | \psi_i^\mu \rangle = \sum_{k \in occ} \{ \langle \varphi_k(1) \varphi_l(2) | \hat{\Omega}(1, 2) | \varphi_k(1) \varphi_\mu(2) \rangle$$
$$- \langle \varphi_k(1) \varphi_l(2) | \hat{\Omega}(1, 2) | \varphi_k(2) \varphi_\mu(1) \rangle \}$$

$$\langle \psi_0 | \sum_{i<j} \hat{\Omega}(i, j) | \psi_{kl}^{\mu\nu} \rangle = \langle \varphi_k(1) \varphi_l(2) | \hat{\Omega}(1, 2) | \varphi_\mu(1) \varphi_\nu(2) \rangle$$
$$- \langle \varphi_k(1) \varphi_l(2) | \hat{\Omega}(1, 2) | \varphi_\mu(2) \varphi_\nu(1) \rangle$$

$$\langle \psi_0 | \sum_{i<j} \hat{\Omega}(i, j) | \psi_{klm}^{\mu\nu\varrho} \rangle = 0$$

The restrictive conditions of Eq. 18 (which constrain the extremal problem of Eq. (19) lead, in Lagrangian undetermined multiplier form, to the conclusion that $E_{el, i}$ is a minimum when

$$\delta E_{el, i} - \sum_{l \in occ} \varepsilon_{il} \langle \delta \varphi_i | \varphi_l \rangle = 0 \tag{20}$$

or, equivalently, when

$$\langle \delta \varphi_i(1) | \hat{H}^c(1) | \varphi_i(1) \rangle + \sum_{l \in occ} \langle \delta \varphi_i(1) \varphi_l(2) | (1 - \hat{P}_{12})/r_{12} | \varphi_i(1) \varphi_l(2) \rangle$$

$$= \sum_{l \in occ} \varepsilon_{il} \langle \delta \varphi_i(1) | \varphi_l(1) \rangle . \tag{21}$$

Since the variations $\delta \varphi_i$ are arbitrary — by virtue of the implicit incorporation of the orthonormality restrictions of Eq. 18 into Eq. 20 — we conclude that we have reached a stationary point if

$$\hat{F}(1) \varphi_i(1) = \sum_{l \in occ} \varepsilon_{il} \varphi_l(1) \tag{22}$$

where $\hat{F}$, which is known as the Fock operator, is defined as

$$\hat{F}(1) \equiv \hat{H}^c(1) + \sum_{l \in occ} \int d\vec{x}_2 \cdot \varphi_l^*(2)[(1 - \hat{P}_{12})/r_{12}] \varphi_l(2) . \tag{23}$$

The set of equations like Eq. 22, one for each φ_i in that set of spinorbitals which is occupied in ψ_0, is known as the Hartree-Fock set of equations.

2.2.2.3 The Self-consistent Field

We have reached a time for pause: The multidimensional problem of Eqs. 1 and 4 has been reduced to the formal one-electron problem of Eq. 22. However, whereas the derivation of the Hartree-Fock equations has been straight-forward, solving this set of coupled equations is a more formidable task. Although formally one-electron, Eq. 22 depends on all other electrons because of the integrals occur in the Fock-operator of Eq. 23. The method for solving these equations is known as the SELF-CONSISTENT FIELD (SCF) method: One starts by guessing a set of spinorbitals $\{\varphi_i^{(0)}\}$, inserts these into Eq. 23 to obtain the Fock operator $\hat{F}^{(0)}$, and solves Eq. 22 for a new set of spinorbitals $\{\varphi_i^{(1)}\}$; this "first-improved" set of spinorbitals, $\{\varphi_i^{(1)}\}$, is reinserted into Eq. 23 to obtain a "first-improved" $\hat{F}^{(1)}$ and, thence, a "second-improved" set $\{\varphi_i^{(2)}\}$; etc. The cyclic iteration is repeated until self-consistency is reached, that is until the difference $\{\varphi_i^{(m+1)}\} - \{\varphi_i^{(m)}\}$, or $|E_{el}^{(m+1)} - E_{el}^{(m)}|$, lies below some initially prescribed limit.

2.2.2.4 Canonical Spinorbitals

Although the energy obtained by the Hartree-Fock SCF procedure is unique, the spinorbitals φ_i are not. This lack of uniqueness follows from the determinantal form of Eq. 11: The determinant of a matrix is invariant to a unitary transformation [14] of the matrix.

Thus, we are free to *choose* a very special set of spinorbitals, namely that which reduced Eq. 22 to the simple form

$$\hat{F}(1) \varphi_i'(1) = \varepsilon_{ii} \varphi_i'(1) = \varepsilon_i \varphi_i'(1) . \tag{24}$$

This set of spinorbitals, which diagonalizes the matrix $\varepsilon = (\varepsilon_{ii})$, is known as the set of "canonical spinorbitals". The ε_i are known as the "spinorbital energies". The canonical set, as will be shown, acquires special significance with regard to electron excitation events.

The physical interpretation of Hartree-Fock equations, in the form of Eq. 24, is especially simple. Equation 24 is a Schrödinger equation for an electron moving simultaneously in the field of the nuclei and in the average field produced by all other electrons. This *average* arises because of the integration performed in Eq. 23. A "self-interaction" is excluded because of the $1 - \hat{P}_{12}$ term. And the exchange term, $-\hat{P}_{12}$, takes care of antisymmetrization requirements.

It is appropriate to belabor a few points. These are:

— The electronic Hamiltonian operator is *not* a sum of Fock-operators. Thus,

$$\hat{H}_{el}(1, 2, 3, \ldots, N) \neq \hat{F}(1) + \hat{F}(2) + \hat{F}(3) + \ldots + \hat{F}(N) . \tag{25}$$

As a result, we find

$$E_{el} \neq \sum_i \varepsilon_i . \tag{26}$$

It must be emphasized that the Hartree-Fock equations are merely the result of a condition which we impose on the spinorbitals in order to obtain the best possible electronic ground-state wavefunction in a single Slater-determinant format.

— The use of orbital concepts such as "canonical orbital" or "orbital energy" does not imply the neglect of electron interactions. These repulsions are explicitly introduced by the integrals which occur in the Fock-operator of Eq. 23.

— The probability of finding the *1st* electron in the vicinity of a given coordinate, $\vec{x}_1$, is *not* given by $\varphi_1^*(\vec{x}_1)\,\varphi_1(\vec{x}_1)\,d\vec{x}_1$. The correct expression is

$$d\vec{x}_1 \iint \ldots \int \psi^*(\vec{x}_1, \vec{x}_2, \ldots, \vec{x}_N)\, \psi(\vec{x}_1, \vec{x}_2, \ldots, \vec{x}_N)\, d\vec{x}_2\, d\vec{x}_3 \ldots d\vec{x}_N$$

which reduces to

$$\sum_{l \in occ} \varphi_l^*(\vec{x}_1)\, \varphi_l(\vec{x}_1)\, d\vec{x}_1 . \tag{27}$$

Since the electrons are indistinguishable, this result should have been expected.

2.2.2.5 LCAO Approximation

Some comment concerning the manner of solving Eq. 24 is pertinent. Rather than solve the Hartree-Fock differential equations, these are first transformed to a matrix problem by introducing the LINEAR COMBINATION OF ATOMIC ORBITALS (LCAO) approximation

$$\varphi_i = \sum_\alpha \chi_\alpha c_{\alpha i} \tag{28}$$

where χ_a is an atomic spinorbital. The task of finding the energies ε_i and the molecular spinorbitals φ_k is thereby reduced to the matrix problem

$$\mathbf{Fc} = \mathbf{Sc\varepsilon} \tag{29}$$

where

$$\mathbf{F} = (F_{\alpha\beta}) = (\langle \chi_\alpha | \hat{F} | \chi_\beta \rangle) \tag{30}$$
$$\mathbf{S} = (S_{\alpha\beta}) = (\langle \chi_\alpha | \chi_\beta \rangle) \tag{31}$$
$$\mathbf{\varepsilon} = (\varepsilon_{ij}) = (\varepsilon_i \delta_{ij}); \quad \mathbf{c} = (c_{\alpha i}) . \tag{32}$$

Solutions are obtained from the secular determinant

$$\det |\mathbf{F} - \varepsilon_i \mathbf{S}| = 0 . \tag{33}$$

The ε_i are obtained by diagonalization. The eigenvector $\mathbf{c}_i$, which is the adjoint of the row vector $(c_{1i}, c_{2i}, c_{3i}, ...)$, is obtained by inserting ε_i into Eq. 29 and solving for the c_{ki}.

This LCAO approach has several advantages:

— It is convenient to program since it makes use of the powerful techniques of matrix algebra.

— The first set of orbitals (i.e., eigenvectors) can sometimes be guessed with some accuracy.

— Since the molecular spinorbitals are expressed in terms of atomic spinorbitals, the concept of an "atom as part of a molecule" becomes viable.

As a result, all kinds of qualitative and semiquantitative notions which relate to our knowledge of atomic systems become feasible.

2.2.2.6 Restricted and Unrestricted Hartree-Fock

The final comment refers to the specific form employed for the one-electron eigenfunctions. So far, in fact, these have been spinorbitals. Since these spinorbitals are either singly-occupied or not occupied at all, their use results in a very simple form for the various equations. However, we now wish to stress the implicit meaning of a *restricted* formalism. The restricted Hartree-Fock formalism, in the context of interest to us, consist of two parts:

— For each molecular spinorbital $\varphi_k(\vec{x}; \vec{R}) = \varphi_k(\vec{r}; \vec{R}) \, \alpha(\vec{\omega}_1)$, where φ is a molecular orbital and α (or β) is the spin function for $m_S = 1/2$ (or $-1/2$) there is another molecular spinorbital $\varphi_{k'}(\vec{x}; \vec{R})$ for which $\varphi_{k'}(\vec{x}; \vec{R}) = \varphi_k(\vec{r}_1; \vec{R}) \, \beta(\vec{\omega}_1)$.

— Since the number of electrons, N, is even and since we have constrained our interests to "closed shells", it is well to be specific about what we mean by the term "closed-shell ground state". We assert that φ_k and $\varphi_{k'}$ are either both occupied (i.e., contained in the determinant which is optimized) or both unoccupied. Alternatively, all space orbitals φ_k in the determinant which is optimized are assumed to be either doubly occupied or not occupied at all. States with either double or zero occupancy of all space orbitals are known as "closed-shell" systems; otherwise, they are known as "open-shell".

Open-shell Hartree-Fock theory [15] is encumbered with several difficulties. We can attack the problem, or attempt to attack it, within the restricted Hartree-Fock formalism. Thus, we write

$$\varphi_k(\vec{x}_1; \vec{R}) = \phi_k(\vec{r}_1; \vec{R}) \, \alpha(\vec{\omega}_1)$$
$$\varphi_{k'}(\vec{x}; \vec{R}) = \phi_k(\vec{r}_1; \vec{R}) \, \beta(\vec{\omega}_1) \tag{34}$$

and, in this manner, permit occupation numbers of one, two, or zero for the space-orbitals. The advantage of the restricted formalism of Eq. 34 is that it ensures that certain states can be represented correctly by a single Slater determinant. These states are those for which the spin quantum numbers S and $|M_S|$ are maximal.

If we now drop the restriction of Eq. 34 and proceed to the unrestricted Hartree-Fock formalism, where $\varphi_{k'} = \phi_k \beta$ with $\phi_{k'} \neq \phi_k$, this last advantage is lost. Specifically, since the unrestricted version allows different space orbitals for different spins, single determinantal wavefunctions are no longer eigenfunctions of the spin operator $\hat{S}^2$, when these represent non-closed shell states. In the restricted version for S and $|M_S|$ not maximal, and in the unrestricted version for *all* S and $|M_S|$, a multi-determinantal form must be used; otherwise, the wavefunction may not fulfil the basic permutation symmetry requirements. This leads immediately to non-integer occupation numbers for spinorbitals and, therefore, to a departure from the basic MO approximation.

Whether or not the restricted Hartree-Fock formalism, with its retention of integer occupation numbers, is embraced by the term "MO approximation" becomes a question to taste. It is for these reasons that we have confined ourselves to closed-shell systems.

2.2.2.7 Beyond the MO Approximation

It is now clear how one proceeds beyond the MO approximation. One merely includes more and more terms in the CI expansion of Eq. 14. The truncation of such an expansion, and the selection of the determinants which most efficiently truncate it, appears to be an art in itself. Finally, the actual computations are both expensive and labor-intensive, particularly when one seeks "really good" wavefunctions.

We provide one example. For this purpose, we use the medium-sized molecule H_2CS. First, however, we recall that the better calculation is the one which yields the lower total energy. Computation[16] for H_2CS gave a total energy of $-1,878.068$ eV, which is very close to the Hartree-Fock limit. Improvement of the wavefunction[17] by a very elaborate CI method decreased the total energy by 7.269 eV to yield $-11,885.337$ eV. This improvement, which amounts to 0.06%, could be considered to be minimal. However, in terms of chemical significance, 7.3 eV is indeed quite a large quantity of energy. In other words, while it is certainly comforting that the MO approximation accounts for 99.94% of the total energy, an error of 7.269 eV is not tolerable because chemistry deals with quantities of just this magnitude. Thus, whether one emphasize the 0.06% or the 7.269 eV becomes a matter of taste.

2.3 A Configuration Description of Electronic Excitation Processes

The approximations inherent in the MO concept are straightforward. Before discussing their appropriateness, we will use the MO concept to describe different electronic states, and transitions among them. This description will reduce some of the more abstract concepts to a familiar form. The major impact of the MO approximation is that it provides a simple and useful way of classifying spectroscopic data. Thus, the idea of an orbital (or orbital energy) as a numerical convenience or as a first approximation to the results of a more elaborate treatment, is merely a minor aspect of its over-all utility.

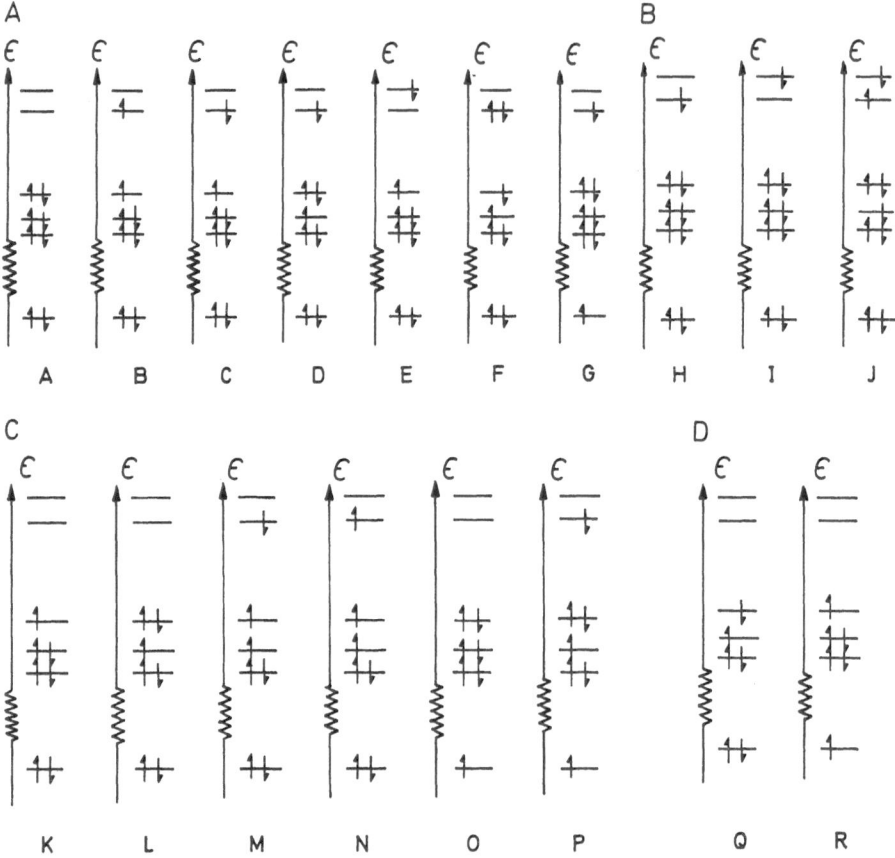

Fig. 2. Electron configuration diagrams for a system containing one core MO, three valence MO's, and two virtual MO's: (**A**) N = 8 electrons, (**B**) N = 9 electrons, (**C**) N = 7 electrons, (**D**) N = 6 electrons. This figure suggests the possibility of a number of different sorts of experiment which have yet to be investigated. Furthermore, of the $18 \times 17 = 306$ possible transitions, only a very few will be discussed. The reader should consult Tables 4 and 5, Fig. 3, and the text

2.3.1 Configuration Excitations

We describe excited states using a set of fixed orbitals and the set of occupation numbers, $\{0, 1, 2\}$. Our description is pictorial: The space orbital ϕ_k is a line on an diagram; the spin function α is represented by an "up-arrow"; and β by a "down-arrow". The MO set consists of one core MO, three valence MO's which are fully occupied in the ground state, and two virtual MO's — either valence or Rydberg — which are unoccupied in the ground state. A set of selected configurations for systems containing N, N+1, N−1 and N−2 electrons is shown in Figures 2A, 2B, 2C and 2D, respectively. Some of the properties of these configurations are collected in Table 4.

Transitions between these configurations may be supposed to correspond to the energy differences measured by various spectroscopic techniques. The correspondence between the different types of configurational excitation and the various spectroscopic

Table 4. The electronic configurations depicted in Fig. 2

Configuration	Number of electrons	M_S	S	Comments
A	N	0	0	Ground state
B	N	1	1	Lowest-excited state; triplet
C	N	0	0, 1	Singly-excited configurations;
D	N	0	0, 1	do not correspond to pure-spin
E	N	0	0, 1	states, but to mixtures of singlet and triplet states
F	N	0	0, 1	Doubly-excited configuration; not a pure spin state
G	N	0	0, 1	Highly excited configuration; not a pure spin state
H	N + 1	$\frac{1}{2}$	$\frac{1}{2}$	Ground configuration of the N + 1 electron system
I	N + 1	$\frac{1}{2}$	$\frac{1}{2}$	Excited configuration of the N + 1 electron system
J	N + 1	$\frac{1}{2}$	$\frac{3}{2}, \frac{1}{2}$	Excited configuration of the N + 1 electron system; does not correspond to pure spin state; enters both quartet and doublet states.
K	N − 1	$\frac{1}{2}$	$\frac{1}{2}$	Ground configuration of the N − 1 electron system; doublet state; Koopmans' configuration
L	N − 1	$\frac{1}{2}$	$\frac{1}{2}$	Excited configuration of the N − 1 electron system; doublet state; Koopmans' configuration
M	N − 1	$\frac{1}{2}$	$\frac{3}{2}, \frac{1}{2}$	"Shake up" configuration; contains quartet and doublet states.
N	N − 1	$\frac{3}{2}$	$\frac{3}{2}$	"Shake up" configuration; pure quartet state.
O	N − 1	$\frac{1}{2}$	$\frac{1}{2}$	"Core-ionized" configuration; doublet state
P	N − 1	$\frac{1}{2}$	$\frac{3}{2}, \frac{1}{2}$	"Shake up" configuration conected with core ionization; enters both quartet and doublet states.
Q	N − 2	0	1, 0	Valence "shake off" configuration; not a pure spin state
R	N − 2	1	1	Core "shake off" configuration; triplet state.

processes is illustrated further in Table 5 and Fig. 3. Indeed, it was the correspondence of Table 5 which dictated the selection of configurations made in Fig. 2. The processes of Table 5 are restricted to techniques which use photons as primary excitation means; however, all of them could be excited by other means [5].

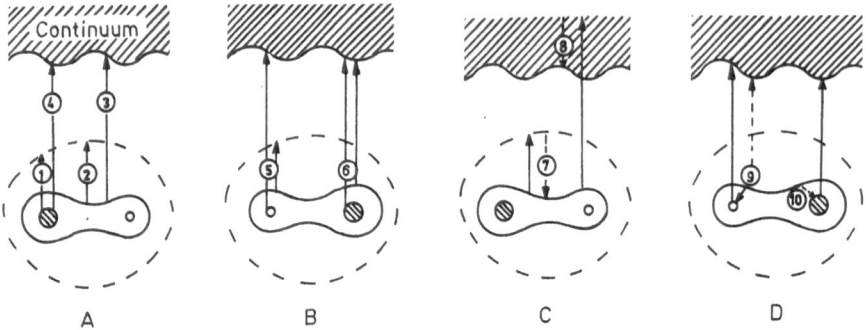

Fig. 3. Pictorial representation of different transitions: (**A**) contains "simple" events; (**B**) contains simultaneous events; (**C**) and (**D**) contain consecutive events

The terminology for the experimental processes (See Table 5 and Fig. 3) is traditional and not necessarily logical. For example, the acronym "PES" stands for Photo Electron Spectroscopy; UPS and XPS for UV and X-ray induced photoelectron spectroscopy, respectively; and ESCA for Electron Spectroscopy for Chemical Analysis. XPS and ESCA are the same techniques. The acronym ESCA is quite misleading in that the technique has a much wider range of applicability than this name implies. The acronyms PES and UPS are often used interchangeably, although PES is usually considered to embrace both UPS and XPS. The term "Electron Spectroscopy" describes a technique in which the kinetic energies and fluxes of electrons are measured; the word "photoelectron" relates to the creation of electrons by photons; and the letters, "U" and "X" refer to the energies (i.e., UV or X-ray) of the exciting photons. Although this terminology may be confusing, its use is quite precise. In the same, somewhat arbitrary fashion we reserve the term "excitation" for bound → bound transitions; the term "ionization" for bound → unbound transitions; and the term "transition" for both excitation and ionization.

The terms "shake up" and "shake off" originated in photoelectron spectroscopy and are quite graphic. Photon excitation of an atom or molecule may cause ionization of one electron and leave the system so disturbed — in the vernacular: so "shock up" — that another electron may suffer excitation (i.e., "shake up") or even ionization (i.e., "shake off"). These "shake up/shake off" processes are concurrent, not consecutive.

One limitation of Fig. 2 is already evident: The configurations C, D, E, F, G, J, M, P and Q do not represent pure spin states. All of these configurations participate in at least two state functions between which $\Delta S = 1$. This difficulty characterizes open-shell systems. A correct spin permutation symmetry requires that more than one determinant be used to describe most open-shell electronic states.

2.3.2 Transition Probabilities

If the orbital set of Fig. 2 is identical for all configurations, certain selection rules will govern transition probabilities. For example, if the transition $\psi_a \rightarrow \psi_b$ is electric

Table 5. Photo-induced electron transitions

Configuration excitation	Name of transition	Name of experimental process	Pictorial process of figure 3
A → B, C, D, E, F	Electronic excitation	VIS/UV absorption spectroscopy	1
A → G	Core excitation	X-ray absorption spectroscopy	2
B → A	Phosphorescence ⎫	Emission spectroscopy	7
^{1}C → A	Fluorescence ⎭		
B → ^{3}E	Electronic excitation	Triplet-triplet absorption	
H → A, B	Ionization of the N + 1 electron system ⎫		
A → K, ... , R	Ionization	⎱ Photoelectron spectroscopy	3, 4
A → M, N, P	"Shake up"	⎰ Photoionization	5
A → Q, R	"Shake off"		6
(A →) O → Q	Auger transition	Auger spectroscopy	9
(A →) E → K	Autoionization		8
(→) O → K	X-ray emission	X-ray emission spectroscopy	10

dipole allowed, the transition moment $\Sigma \, \vec{r}_i$ is a one-electron operator, nonzero values of $\langle \psi_a | \sum_i \vec{r}_i | \psi_b \rangle$ can arise only when the two determinants ψ_a and ψ_b differ by no more than one spin orbital (cf. Table 3). Thus, the optical transitions A → ^{1}C, ^{1}D, ^{1}E, ^{1}G, K, L and O are allowed, in the absence of space-symmetry inhibitions, by the Slater rules. The optical transitions A → ^{1}F, ^{2}M, ^{2}N and ^{2}P, on the other hand, are forbidden by the same rules. If these latter transitions occur at all, their intensities must be "stolen" by configuration interaction. For example, the transition A → ^{1}F could acquire intensity by a mixing of ^{1}F with any or all of ^{1}C, ^{1}D, ^{1}E, or ^{1}G. The transition A → ^{4}M, ^{4}N, Q and R are also forbidden; however, they can obtain intensity by mixing with continuum orbitals or by spin-orbit coupling. The optical transition A → B, ^{3}C, ^{3}D, ^{3}E and ^{3}G are also electric-dipole forbidden but may acquire intensity by spin-orbit coupling.

The experimental counterparts of the forbidden processes described above do occur. Therefore, in an experimental sense, they are "allowed". This allowedness is a consequence of the fact that our description of all states, A to R, by configurations constructed from a single set of orbitals is inadequate. Nonetheless, the theoretical distinction between forbidden and allowed processes is one which holds up well in practice. In other words, even when observed, these "forbidden" events are weak. Thus, the theoretical distinctions are empirically useful.

2.3.3 Less-Familiar Processes

Some comment on the less-familiar processes of Tables 4 and 5 is in order:
1. Distinction between *direct* and *consecutive* processes are feasible.
   — The event (A →)D → L is an auto-ionization and it consists of two consecutive processes. The event A → L is a simple direct ionization event. Although the energies of the emitted electrons, ionized and auto-ionized, are equal in both instances, the

actual band shapes, flux versus energy, may be very different. Direct ionization is characteristic of the geometry of A. Autoionization, however, is characteristic of the geometry of D. In other words, the intermediate state D, having a lifetime of the order of a few vibrational periods, relaxes from a geometry characteristic of A to one more nearly characteristic of its own equilibrium requirements before it releases an electron into the continuum.

— The event $(A \to)O \to Q$ is an Auger transition. It is a consecutive event, in contrast to the "shake off" event $A \to Q$ which is a direct event. Both processes, Auger and "shake off" involve the same two terminal and initial states, Q and A. The "shake off" process (i.e., the direct ejection of two electrons) involves two initial particles, one being a photon, and three final particles. Given that the kinetic energy of the doubly-ionized atom or molecule is small because of the large system inertia, it follows that the sum of energies of the two ejected electrons is $E(1 + 2) = h\nu - (E_Q - E_A)$. Consequently, the energy of either one of these two electrons is continuously variable within the limits $0 \leqq E(1 \text{ or } 2) \leqq h\nu - (E_Q - E_A)$. The energy of an Auger electron, on the other hand, is quite precise, is given by $E_O - E_A$, and is independent of the input photon energy.

2. X-ray and Auger emission processes are competitive. The relative probability of the X-ray process increases with atomic number.

3. Transitions which initiate in a state other than the ground state require a fairly high population of this initial state. Such experiments are difficult to perform. With the advent of laser technology, they have become fairly common. On the other hand, consecutive excitation/de-excitation events are easily observed. Most of the commonly observed processes (e.g. fluorescence, phosphorescence, auto-ionization, Auger events, X-ray emission, etc.) fall in this latter category.

2.3.4 Limitations

It is obvious that the classification schemes of Tables 1, 4 and 5 and Fig. 1, 2 and 3 do not arise solely from the MO approximation. In fact, these classification schemes impose another, distinct approximation: They imply that one set of orbitals suffices for the various states. There are three different aspects to this approximation:

— The first is practical. It asks whether or not the classification schemes work.

— The second deals with numerical accuracy. It enquires into the correspondence between experiment and theory which can be achieved using a single determinantal wavefunction and only one set of orbitals.

— The third is purely theoretical. It is concerned with the conditions under which these approximations can be embedded in or justified by theory.

The first question is easily answered: These classification schemes are exceedingly useful. In fact, in some form or other, they provide the basis for almost all spectroscopic discussions. The answer to the second question is equally clear: Although the numerical agreement between quantities calculated on the MO level and their experimental counterparts may be imperfect, the MO description holds fairly well, especially for the ground state and for "normal" bond distances [18]. The third question is the most difficult and it is the one on which we will concentrate. We will deal mainly, but not exclusively, with the interrelations of the different electronic states of one given moleculer or atom. In fact, we will develop Koopmans' theorem as a central motif because Koopmans' theorem invests the canonical orbitals with a physical significance distinct from that of all other Hartree-Fock orbitals.

3 Koopmans' Considerations

3.1 Valence Orbitals

The distinction between valence orbitals and core orbitals is valid in both experimental and theoretical senses.

On the experimental side, the photon energies required to ionize a core electron are considerably larger than those required to ionize valence electrons (See Table 1). The larger photon energies and the ancillary instrumental demands for better vacuum, better analysis and better detection technologies, have caused many instruments for core studies to deviate markedly from that for valence studies.

On the theoretical side, the removal of a core electron is a considerably larger and very different perturbation than that caused by removing a valence electron. For example, the removal of a C_{1s} electron of CH_4 yields an excited CH_4^+ ion which is essentially identical to a ground state NH_4^+ ion: In specific, removal of a C_{1s} electron is equivalent to increasing the effective nuclear charge of carbon by one unit. It is difficult, on the other hand, to imagine any valence ionization process capable of generating a CH_4^+ entity which is remotely similar to any state of an ammonium ion. The point or relevance, of course, is that many of the concepts which are useful in UPS studies may well be useless in XPS studies, and vice versa.

3.1.1 Valence Ionization

The UPS study of valence ionizations has generated a wealth of evidence in favor of the MO approximation [19–21]. Indeed, the evidentiary tilt is so overwhelming that one tends to forget that the MO approximation is, in fact, an approximation at all. For this reason, we intend to derive Koopmans' theorem [22], outline its approximational content, and discuss it in relation to photoelectron spectroscopy.

3.1.1.1 Koopmans' Theorem: The First Part

We describe the ground state of an N-electron system in the MO approximation by

$$\Psi_0^N = \psi_0^N = |\varphi_a(1)\, \varphi_b(2) \dots \varphi_n(N)| . \tag{35}$$

We also approximate the cationic state obtained in a particular ionization process, Ψ_{-n}^{N-1}, by a single Slater determinant. Finally, we impose the restriction that Ψ^N and Ψ_{-b}^{N-1} be fabricated from an identical set of spinorbitals. Consequently, we write

$$\Psi_{-b}^{N-1} = \psi_{-b}^{N-1} = |\varphi_a(1)\, \varphi_b(2) \dots \varphi_m(N-1)| . \tag{36}$$

Under these conditions, the ionization energy equals the negative of the energy of the missing spinorbital

$$\text{IE}_n = -\varepsilon_{nn} \tag{37}$$

This statement constitutes the first part of Koopmans' theorem. Its proof follows directly from Eqs. 15 and 22.

Proof:
The ionization energy is given by

$$\text{IE}_n = E(\psi_{-n}^{N-1}) - E(\psi_0^N) \tag{38}$$

where

$$E(\psi_{-n}^{N-1}) = \langle \psi_{-n}^{N-1} | \hat{H}^{N-1} | \psi_{-n}^{N-1} \rangle \tag{39}$$

$$= \sum_{\substack{k \in occ \\ k \neq n}} \langle \varphi_k(1) | \hat{H}^c(1) | \varphi_k(1)$$

$$+ \sum_{\substack{k < l \\ k, l \neq n}} \{ \langle \varphi_k(1) \varphi_l(2) | 1/r_{12} | \varphi_k(1) \varphi_l(2) \rangle$$

$$- \langle \varphi_k(1) \varphi_l(2) | 1/r_{12} | \varphi_k(2) \varphi_l(1) \rangle \} \tag{40}$$

and

$$E(\psi_0^N) = \langle \psi_0^N | \hat{H}^N | \psi_0^N \rangle \tag{41}$$

$$= \sum_{k \in occ} \langle \varphi_k(1) | \hat{H}^c(1) | \varphi_k(1) \rangle$$

$$+ \sum_{k < l \in occ} \{ \langle \varphi_k(1) \varphi_l(2) | 1/r_{12} | \varphi_k(1) \varphi_l(2) \rangle$$

$$- \langle \varphi_k(1) \varphi_l(2) | 1/r_{12} | \varphi_k(2) \varphi_l(1) \rangle \} . \tag{42}$$

If we now multiply Eq. 22 from the left by ψ_n^* and integrate, we find

$$\langle \varphi_n(1) | \hat{F}(1) | \varphi_n(1) \rangle = \sum_l \varepsilon_{nl} \langle \varphi_n(1) | \varphi_l(1) \rangle \tag{43}$$

$$= \varepsilon_{nn} \tag{44}$$

where

$$\varepsilon_{nn} = \langle \varphi_n(1) | \hat{H}^c(1) | \varphi_n(1) \rangle$$

$$+ \sum_{k \in occ} \{ \langle \varphi_n(1) \varphi_k(2) | 1/r_{12} | \varphi_n(1) \varphi_k(2) \rangle$$

$$- \langle \varphi_n(1) \varphi_k(2) | 1/r_{12} | \varphi_n(2) \varphi_k(1) \rangle \} \tag{45}$$

The terms contained in Eq. 42 but not contained in Eq. 40 are identical to those of Eq. 45 but of opposite sign. Hence, it follows that

$$IE_n = -\varepsilon_{nn} \, . \tag{46}$$

3.1.1.2 Koopmans' Theorem: The Second Part [23]

Provided the set of spinorbitals $\{\varphi_k\}$ is the canonical Hartree-Fock set for the N-electron system, the negative of the orbital energy, $-\varepsilon_n$, is the best ionization energy. If the word "best" means "best in a variational sense", this statement is the second part of Koopmans' theorem.

The $N-1$ electron wavefunction of the cation can be expanded over a set of determinants (i.e., a CI expansion for the cation states)

$$\Psi^{N-1} = \sum_{k \in occ} \psi_{-k} c_k + \sum_{\substack{k, l \in occ \\ \mu \notin occ}} \psi^{\mu}_{-kl} c^{\mu}_{kl} \tag{47}$$

where, for example, ψ^{μ}_{-kl} denotes a determinant from which the spinorbital φ_k has been deleted and in which φ_l has been replaced be φ_{μ}; and where we have dropped the $N-1$ superscript on ψ^{N-1} to avoid crowding.

Simplification of Ψ^{N-1} might consist of truncation to

$$\Psi^{N-1} = \sum_{k \in occ} \psi_{-k} c_k \tag{48}$$

However, what we really want is truncation to

$$\Psi^{N-1} = \psi_{-k} \tag{49}$$

This gross simplification of Eq. 47 to Eq. 49 is equivalent to the demand that we find an orthogonal transformation of the set of Hartree-Fock spinorbitals so that, simultaneously, the cationic state can be represented by one single determinant constituted from this set (i.e., Eq. 49) and the neutral molecule ground state can be represented by one single determinant constituted from the same set, namely by

$$\Psi^N_0 = \psi_0 \tag{50}$$

Koopmans' theorem asserts this possibility and identifies the appropriate spinorbital set as the canonical Hartree-Fock set. The proof of these assertions follows:
Proof:
If we can show that

$$\langle \psi_{-l} | \hat{H}^{N-1} | \psi_{-k} \rangle = 0 , \quad k \neq l \tag{51}$$

it follows that Eq. 49 represents the best variational approximation to Eq. 48. The proof of Eq. 51 proceeds in straightforward fashion. The Lagrangian multipliers ε_{kl} for a canonical set are zero when $k \neq 1$. Consequently

$$0 = \langle \varphi_1(1)| \hat{F}(1) |\varphi_k(1)\rangle$$

$$= \langle \varphi_1(1)| \hat{H}^c(1) |\varphi_k(1)\rangle + \sum_{m \in occ} \{\langle \varphi_1(1) \varphi_m(2)| 1/r_{12} |\varphi_k(1) \varphi_m(2)\rangle$$

$$- \varphi_1(1) \varphi_m(2)| 1/r_{12} |\varphi_k(2) \varphi_m(1)\rangle\} \tag{52}$$

$$= \langle \psi_{-1}| \hat{H}^{N-1} |\psi_{-k}\rangle \tag{53}$$

where the last step (i.e., Eq. 53) follows from Table 3.

3.1.2 Electron Affinities

It may be shown, by a series of very similar approximations, that the "best" electron affinity is also given by

$$EA_\mu = -\varepsilon_\mu \tag{54}$$

Relatively little informations is available for electron affinities [22]. In addition to experimental determination (based e.g. on electron capture detection [23a], correlation of gas-phase EA's with solution ones based on CT-complexes and polarographic half-wave potentials [23b, c, d] photodetachment methods [23e, f] and ion molecule equilibria in the gas phase [23g, h, i] and computational methods, also qualitative rules have been used for estimation of EA's. Thus, Lowe [24a] related molecular affinity values to EA's of corresponding atomic constituents, and Myers [24b] considered EA's of atoms in relation to IP's of their neighbour in the table of elements with Z larger by one and found smooth curvilinear relations for vertical groups in the periodic table. A relation

$$EA = (0.12 \pm 0.07)\ IP$$

for atoms has been found by Parr and Bartolotti [24c] and extended to atoms and molecules by Liegener [24d]. In the following we will concentrate on ionization potentials.

3.1.3 Photoelectron Spectroscopy

The observables of photoelectron spectroscopy consist of the flux and energy of electrons emitted at a specific angle relative to the ionizing photon beam. These observables, after conversion to ionization cross-sections and energies, are the quantities which must be related to theory.

The photoelectron spectrum of a typical molecule, ethylene, is shown on the right of Fig. 4. It consists of a plot of electron flux versus ionization energy. Each band

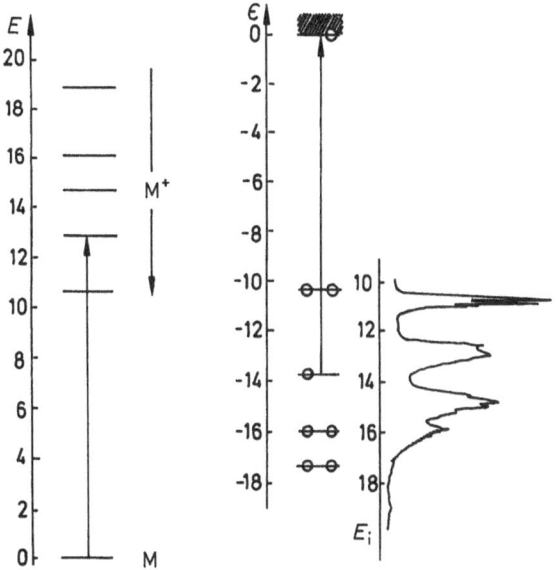

Fig. 4. Energy levels of ethylene, neutral and cationic; orbital energies of neutral ethylene [Brundle CR, Robin MB, Kuebler NA, Basch H (1972) J. Am. Chem. Soc. 94: 1451]; and the photoelectron spectrum of ethylene (Ref. 2c)

corresponds to one electronic level of the ethylene cation. By Koopmans' theorem, the ionization energies can be approximated by negative orbital energies. Consequently, we also show, in the middle part of Fig. 4 the result of an actual MO calculation. Two points are obvious: Firstly, all intense structures can be related to the removal of an electron out of an occupied orbital (i.e., "shake up" processes are not very likely) and, secondly, the numerical agreement is better than 10%.

Koopmans' theorem provides a salient experiment/theory interface. Since we have gone to some trouble to "derive" it, and since its limits of validity are implicit in that derivation, we now outline these.

3.1.3.1 Fixed-Nuclei Approximation

It is the Born-Oppenheimer approximation which allows the notion of a "molecular geometry". Thus, in addition to this approximation, it is also understood that the cationic $N-1$ electron system which is the immediately terminal state of the process

$$N\text{-electron system} + h\nu = (N-1)\text{-electron system} + e^-$$

is identical in all geometric detail to the initial state of the N-electron system. This, of course, is the Franck-Condon approximation. Consequently. Koopmans' theorem is pertinent only to vertical ionization events.

The vertical ionization energy [25] which we denote IE_v, is usually taken to lie at the maximum of the cross-section versus IE plot (i.e., at the "top" of the photoelectron band). However, it is the *1st* moment of this band, particularly if the band

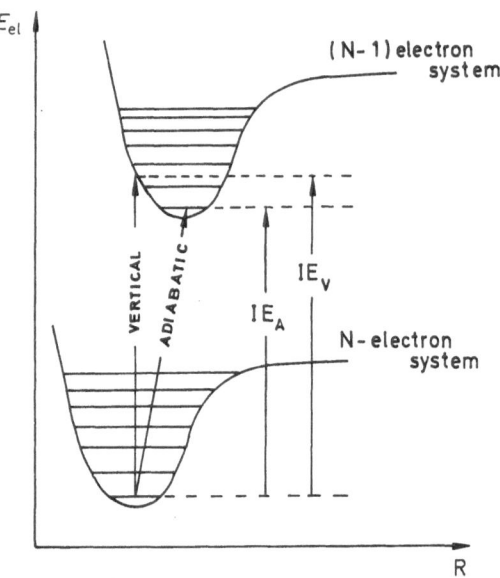

Fig. 5. The vertical and adiabatic processes for the ionization event N-electron system $+ h\nu = (N-1)$-electron system $+ e^-$

is either structured or skewed, which is the correct designation for IE_V [25]. In any event, IE_V must be distinguished from the adiabatic ionization energy, IE_A. The adiabatic ionization energy is the difference in energies of the vibrationally- and rotationally-unexcited states of the N- and (N − 1)-electron system [25 a]. Koopmans' theorem is not valid for IE_A, unless $IE_A = IE_V$. The distinction between IE_A and IE_V is shown in Fig. 5.

3.1.3.2 The Correlation Energy

The neglect of correlation energy is intrinsic to the Hartree-Fock approximation. The correlation energy is attributable to the fact that electrons adjust their motions to the *instantaneous* charge distribution, and not to an *average* charge distribution (as is assumed in the Hartree-Fock equations). In fact, the correlation energy is the difference between the correct energy and the Hartree-Fock energy associated with any given electronic Hamilton operator. If relativistic effects are small, the latter is well known, and the "correct energy" is equivalent to the experimental energy.

Electrons of opposite spin will usually tend to stay considerably further apart (i.e., correlate their motions better) than a single determinantal wavefunction will allow. Thus, the correlation energy can be quite substantial. Nonetheless, while large for any one state, it is only the difference of that between two states, namely that between the initial N- and terminal (N − 1)-electron sates, which is of significance to photoelectron spectroscopy. This difference may well be small, Koopmans' theorem implies that it is zero.

It is important to emphasize that correlation energies for molecules may be quite substantial. Thus, Koopmans' theorem predicts correct numerical ionization energies if, inter alia, the difference in correlation energies were, in fact, zero. If, on the other hand, a not unreasonable estimate of 2 eV per electron were appropriate for correlation energies generally a constant deviation of about 2 eV of Koopmans' and experimental ionization energies should occur. On the other hand, there exist molecules, ones with nearly completely-filled valence shells, such as molecular fluorine, in which the changes of correlation energies that occur for the $N - 1$ electrons of the cation overwhelm the correlation energy of the Nth (i.e. ionized electron prior ionization) and produce a "break-down" of Koopmans' theorem, such that reversals in the order of the experimental ionization events occur relative to predictions based on Koopmans' theorem. Perturbation theory vested in ab initio [26] and semiempirical wavefunctions [27], provides some understanding of such effects.

3.1.3.3 The relaxation Energy

The same set of spinorbitals is used to construct the Slater determinants for the N- and $(N - 1)$-electron systems. This supposition implies that the electrons of the cation do not adjust any way to the reduction of inter-electronic repulsions which must characterize the $(N - 1)$-electron system. This supposition is known as the "frozen-core" or "frozen-orbital" approximation.

3.1.3.4 The Non-Relativistic Approximation

This approximation is not a consequence of the functional nature of the wavefunctions; it is, rather, a defect caused by the omission of relativistic terms from the Hamilton operator. We have omitted these terms solely for convenience. The various relativistic terms — for example, spin-orbit or spin-spin interactions — might have been included in the Fock operator in a way which would not have altered any of our prior conclusions. In fact, in his original paper [22]. Koopmans' included relativistic effects explicitly [28] — and to no ill effects whatsoever in the form of Eq. 37.

3.1.3.5 Restriction to Closed-Shell Systems

Koopmans' theorem is restricted to closed-shell N-electron systems. Thus, at least in the form expressed here, it is specifically inapplicable to non-closed-shell systems (e.g., many transition metal complexes).

Koopmans' theorem is also restricted to certain $N - 1$ electron cationic states. These are referred [29] to as "Koopmans' configurations". For example, the configurations K, L and O of Fig. 2 are Koopmans' configurations, whereas the "shake up" configurations M, N and P are not. Ionizations of the ground state which terminate on a "shake up" configuration are electric-dipole forbidden. Such "shake up" transitions, if observed at all, usually have low cross-sections.

3.1.4 A Critique of Koopmans' Theorem

3.1.4.1 Energies

Koopmans' theorem, when applied to MO calculations, is a means of computing approximate IE_v's. The reasonableness of the approximations involved is a matter of taste. The errors in a MO calculation may amount to $1-3$ eV. Such errors are large relative to the magnitudes of IE_v's as obtained by UPS. Nonetheless, relative to the total electronic energies of a moderately small molecule — which may amount to thousands of electron volts — such an error is quite small.

However, a $1-3$ eV error range is crucial to the assignment of UPS ionization energies. Thus, it is dangerous to attempt the assignment of cationic states which differ by less than 1 eV. As confirmation of this cautionary note, we emphasize three points:
— Several breakdowns of Koopmans' theorem are known. That is, there exist systems where the sequence of cationic states differs from the sequence of canonical orbitals.
— The known breakdowns are few. However, if one considers the information needed to verify such a breakdown, one rapidly concludes that such breakdowns might be more common than not. The information required involves good Hartree-Fock calculations as well as calculations which extend beyond the limitations of the Hartree-Fock approach and/or an experimental assignment of the cationic states. All of these are difficult to obtain. Hence, the widespread conclusion that the sequence of canonical orbitals matches that of the cationic states is more the result of a dearth of contradictory evidence than of any actual confirmatory data. Nonetheless, the existence of certain empirical relations, for example [30]

$$IE_k = -0.92\varepsilon_k \tag{55}$$

implies just such a matching of the two sequences.
— The validity of Koopmans' theorem does not hinge on the use of Hartree-Fock orbitals. However, it makes no sense to discuss the validity of the approximations inherent to Koopmans' theorem within the context of computational schemes which are of lower quality than Hartree-Fock. Indeed, the much parametrized quantum chemical computational schemes (e.g., CNDO/S, AM1, HAM3, MNDO) [31] should be viewed as prescriptions for the calculation of IE's, rather than as MO calculations.

3.1.4.2 Intensities

Koopmans' theorem is also an approximate selection rule for photoelectron spectroscopy. This facet of the theorem has been touched on previously and it probably is its most important content. If Koopmans' theorem were exact, only transitions

$$\Psi_0 \rightarrow \psi_{-k} \tag{56}$$

would be allowed, whereas all "shake up" (removal of one electron and excitation of another) and "shake off" (double ionization) transitions would be forbidden. This is the aspect of Koopmans' theorem which is best supported by experiment and computation and which, in turn, provides the best buttressing for the extensive use of poor-quality, semi-empirical calculations.

"Shake up" transitions, which have zero probability within the Koopmans' context, are often observed, particularly in conjunction with core ionizations. Usually, they are of low cross-section. If they are of high intensity, it is probable that, in this instance,

Cl_2CS

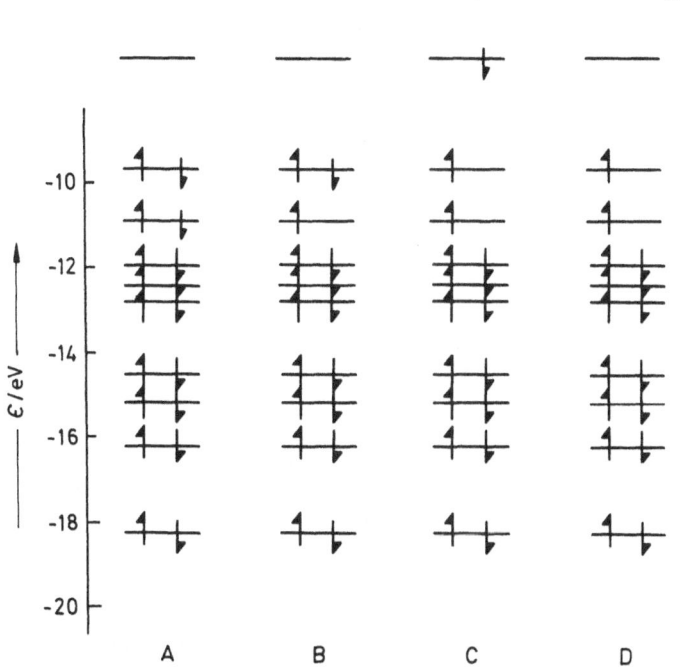

Fig. 6. Electronic configurations of thiophosgene, Cl_2CS: (**A**) the ground state, (**B**) a Koopmans' configuration, (**C**) a shake-up, and (**D**) a shake-off configuration. Note that (**A**) refers to N-, (**B**) and (**C**) to (N − 1)-, and (**D**) to (N − 2)-electron systems. (**A**) is drawn using the experimental ionization energies and Koopmans' theorem; hence, the energies are MO energies

the equality $IE_k = -\varepsilon_k$ is improper. Conversely, when it is known that $IE_k = -\varepsilon_k$ is a poor equality, it might be expected that "shake up" transitions $\psi_0 \rightarrow \psi^\mu_{-kl}$ should also be relatively intense.

Different configurations of thiophosgene, Cl_2CS are shown in Fig. 6. Transitions from the ground configuration (A) to Koopmans' configurations (B) are allowed and account for all the major structures in the photoelectron spectrum of Fig. 7. Transitions to the "shake up" (C) and "shake off" (D) configurations, if they occur at all, contribute only to the weak background beyond 15 eV. A rough estimate of "shake up" transition energies is obtained by adding ionization energies to ultraviolet excitation energies, yielding a lower limit of ~14 eV. By a similar argument, "shake off" energies should exhibit a lower limit of ~21 eV. The one-to-one correspondence between the number of photoelectron bands and the number of orbitals — in other words, the validity of the selection rule aspects of Koopmans' theorem — is not founded on theory, but on the experience gleaned from a great deal of experimental work.

3.1.4.3 Uniqueness of the Canonical Set

Photoelectron spectroscopy, by virtue of its constant referral to Koopmans' theorem, has enhanced the credibility of the MO concept. It has attached a quality of an

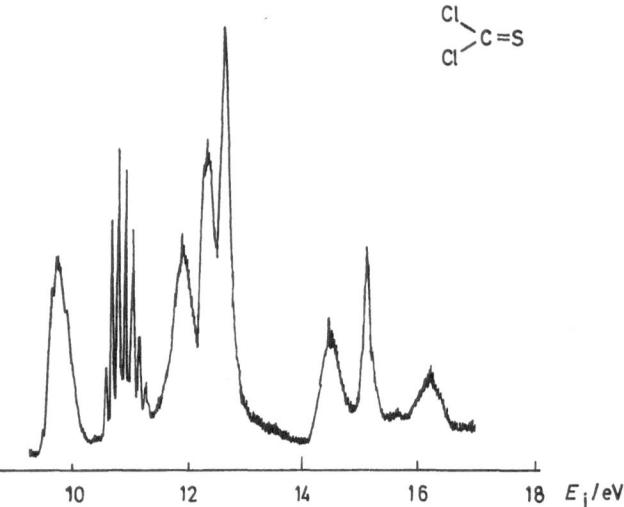

Fig. 7. The photoelectron spectrum of Cl_2CS. The intensity scale is linear

"observable" to the canonical MO energy and, thereby, has intensified the use of MO concepts in all of chemistry. Unfortunately, it is often forgotten that ψ_k and ε_k have direct experimental significances only for ionization events. In fact, the set of canonical MO's, although uniquely defined in a mathematical sense, is totally arbitrary. Localized MO's, for example, are heavily exploited in chemistry because of their transferability from molecule to molecule and because they are more closely connected with traditional chemical ideas concerning homology and reactive groupings. Indeed, any set of MO's which is obtained from the canonical set by a linear transformation of this set is equally as good as the canonical set itself. Thus, the pertinent set of orbitals depends on taste and may be canonical (usually delocalized), localized or otherwise — the only requirement is that the *total* electron distribution remain unaltered. Koopmans' theorem, however, invests the canonical set with a certain uniqueness for ionization events. And it is this investiture, coupled with the accessibility of PES data, which makes Koopmans' theorem so important.

4 The Orbital Perspective

We hope the reader will agree that we have been reasonably successful in describing a variety of spectroscopic results of relevance to photoelectron spectroscopy within the confines of the MO concept. Indeed, it was clear from the outset that success would be ours: That, after all, is what we set out to achieve! Anyway, it seems clear that the MO concepts would not be so popular, if they were not also useful ... And it is not surprising that we should concur in this conclusion.

We hope, however, that our success has been of a slightly different turn. It was not *ab initio* obvious, we think, that we could provide a theoretical justification for the MO approach in all, or even many of the instances in which we have used it. The fact that we have been able to provide such justifications, regardless of how heuristic they might be, encourages us to delve further into the basis for MO success, past and future.

The basic desideratum in any discussion of electronic transitions is to reduce the problem from one which involves two different states, each described by a (multi)-determinantal wavefunction, to one which involves one single set of orbitals. The ability to reduce the problem in such a way hinges on the twin difficulties of electron correlation and relaxation effects. The change of correlation energy between the two states must be small; and the relaxation effects must not be so large that they inhibit a one-to-one correspondence of the orbitals of one electronic state with those of the other.

The use of terms such as "an electron is excited from one orbital to another" implies that the orbitals do not change much during the course of a transition. To the extent that such a statement is valid, it is implied that the orbitals in the two states are either identical or so similar that a knowledge of them in one state is adequate for the identification of them in another. In the same way, a great deal of experience indicates that the similarities between certain electronic states of quite different molecular entities can be translated into a requirement for similarity at the orbital level. Thus, very simple (i.e., one-electron) perturbation schemes can be used to describe the electronic states of one molecule in terms of those of another related molecule.

We now synopsize some of the more pertinent theoretical characteristics of the MO approach. We do so in terms of one-, two-, and many-orbital properties.

One orbital properties * *:* These consist of ionization energies and electron affinities. Both of these events can be described in terms of one single orbital energy. The theoretical justification for such a description is vested in Koopmans' theorem. The second part of Koopmans' theorem states that the canonical MO's of the N-electron system are also optimal (i.e., "good") for the $(N-1)$-electron system. The experimental evidences, based on the near identity of IE_n with $-\varepsilon_n$ and on the general correctness of the selection rule for photoelectron spectroscopy, constitute the best "proof" of the orbital structure of both atoms and molecules. Not surprisingly, this "goodness" holds not only for the familiar valence orbitals, but also for the strongly-coupled core orbitals. For further discussion of OES see ref.

Two orbital properties: These, for the most part, consist of electronic transitions. At best, these cannot be described by less than two orbitals. However, there exists no one set of optimal orbitals which is adequately descriptive of different excited states: Configuration interaction must be considered from the outset. In addition to the non-existence of a single set of orbitals which is "good" for a large number of different states, difficulties arise in connection with the degeneracies due to spatial and spin (i.e., permutational) symmetry. As a result of this unfortunate situation, computational and interpretational problems are rather severe in VIS, UV and VUV spectroscopy. And, as a further consequence, the number of excited state assignments which meet with general accord are relatively few. In view of all these difficulties, we suggest that electronic spectra are best discussed within the "optimum excited state orbital" framework. These OES orbitals should be extremely useful for limited classes of transitions (i.e., all those which arise by excitation from one fixed orbital).

---

* the difference between "one-electron" and "one-orbital" should be noted

Many-orbital properties: These may include one-electron properties such as dipole moments or electron densities. They also include total energies. There is no specific reason to prefer any particular set of orbitals (e.g. canonical) in these instances. However, since it really does not matter [14] which kind of orbitals one uses, the canonical set is just as good as any other Hartree-Fock set. Although there is little or no theoretical reason to expect that a *simple* description of many-orbital properties should obtain, chemical ingenuity, by subsuming different physicochemical effects into one or a few parameters, has achieved a considerable simplicity even in these difficult areas.

We conclude therefore, that canonical MO's are most appropriate for one-orbital properties, that OES MO's may well be those of choice for two-orbital properties, and that any set of HF MO's is just as good as any other for many-orbital properties. We also conclude that the basic MO idea (i.e., the use of a single determinant to describe a many-electron wavefunction) is surprisingly excellent.

5 Acknowledgement

Much of this material is taken from Wittel and McGlynn (1977) Chem. Rev. 77: 745. We have received so many requests for this material, that we thought it pertinent to up-date/re-present it.

This work was supported by the U.S. Department of Energy (OHER). We are grateful for that support.

6 References

1. Wittel K, McGlynn SP (1977) Chem. Rev. 77: 745
2. General reading in the field of photoelectron spectroscopy is available in the five volumes of Electron spectroscopy: theory techniques and applications, vols I–V, Brundle CR, Baker AD (eds) Academic, London, 1977, 1978, 1979, 1981, and 1984 where selected specialists review the general field and their own specific research. Several books with emphasis on specific aspects are also recommended
   (a) Siegbahn K, Nordling C, Fahlman A, Nordberg R, Hamrin K, Hedman J, Johansson G, Bergmark T, Karlsson S-E, Lindgren I, Lindgren B (1967) Electron spectroscopy for chemical analysis-atomic, molecular and solid state structure studies by means of electron spectroscopy, Almqwist and Wiksells, Stockholm;
   (b) Siegbahn K, Nordling C, Johansson G, Hedman J, Heden PE, Hamrin K, Gelius U, Bergmark T, Werme LO, Manne R, Baer Y (1969) ESCA applied to free molecules, North Holland, Amsterdam;
   (c) Turner DW, Baker C, Baker AD, Brundle CR (1970) Molecular photoelectron spectroscopy. Wiley-Interscience, London;
   (d) Baker AD, Betteridge D (1972) Photoelectron spectroscopy, chemical and analytical aspects, vol 53, Pergamon, Oxford;

(e) Eland JHD (1974) Photoelectron Spectroscopy, Butterworths, London;

(f) Carlson TA (1975) Photoelectron and auger spectroscopy, Plenum, New York;

(g) Rabalais JW (1977) Principles of ultraviolet photoelectron spectroscopy, Wiley-Interscience, New York;

(h) Berkowitz J (1979) Photoabsorption, photoionization and photoelectron spectroscopy, Academic, London;

(i) Kimura K, Katsumata, S, Achiba Y, Yamazaki T, Iwata S (1981) Handbook of HeI photoelectron spectra of fundamental organic molecules, Halsted, New York;

(j) Robin MB: Higher excited states of polyatomic molecules, vols 1–3, Academic, Orlando, 1974, 1975 and 1985 respectively;

(k) Hollas JM (1982) High resolution spectroscopy, Butterworths;

(l) Carlson TA (ed) (1978) X-ray photoelectron spectroscopy. In: Kaufman JJ, Koski WS (eds) Benchmark Papers in Physical Chemistry and Chemical Physics/2, Dowden, Hutchinson and Ross, Stroudsburg;

(m) West AR (ed) (1977) Molecular spectroscopy, Heyden and Son, The Institute of Petroleum, London;

(n) Mehlhorn W (ed) (1982) Handbuch der Physik, vol 31, Corpuscles and radiation in matter, Springer, Berlin Heidelberg New York

3. Siegbahn K (1985) J. Electron Spectrosc. 36: 113; Siegbahn H, Lundholm M (1982) J. Electron Spectrosc. 28: 135; Siegbahn H, Asplund L, Kelfve P., Siegbahn K (1975) J. Electron Spectrosc. 7: 411; Siegbahn K (1986) Phil. Trans. R. Soc. Lond., A318: 3

4. Koch EE, Gurtler P (1985) In: McGlynn SP et al. (eds) Photophysics and photochemistry in the vacuum ultraviolet, D. Reidel, Dordrecht. A series of 7 review articles on the use of synchrotron radiation in chemistry appeared in (1986) Chem. Brit. 22: 803; Keller PR, Taylor JW, Carlson TA, Grimm FA (1984) J. Electron Spectrosc. 33: 333 and references therein

5. The other important method of excitation uses electrons: the book by Christophorou (Christophorou LG (1971) Atomic and molecular radiation physics, Wiley-Interscience, New York) provides an excellent introduction to electron excitation methods

6. We use atomic units in all equations

7. For a fuller discussion, see Slater JC (1963) The quantum theory of molecules and solids. I. Electron structure of molecules, McGraw Hill, New York

8. The actual Born-Oppenheimer approximation uses $\psi_{el}(\bar{x}; \vec{R}_0)$ as opposed to the specification given here. We have used the so-called "adiabatic" approximation. The differences become important for various perturbative schemes. See, for example, Englman R (1972) The Jahn-Teller effect in molecules and crystals, Wiley-Interscience, New York

9. An orbital — whether atomic, molecular, canonical, localized or otherwise — is strictly a one-electron function. A spin orbital, in contrast to a space orbital, depends on spin and space coordinates. We prefer the unrestricted Hartree-Fock formalism; hence, we will usually work in a spin orbital context.

10. That is, $\langle \varphi_i | \varphi_j \rangle = \delta_{ij}$

11. A configuration specifies the set of "occupied" orbitals. For closed shell systems, each configuration corresponds to one single determinant

12. This, of course, is our definition of "best". One could equally well seek that determinant which would, in a least square sense, reproduce the electron density as closely as possible. This turns out to be a much more formidable problem

13. See, for example, Watanabe H (1966) Operator methods in ligand field theory, Prentice-Hall, Englewood Cliffs, NJ

14. As an example, consider a two-electron system for which

$$\psi = (1/2)^{1/2} \begin{vmatrix} \varphi_1(1) & \varphi_2(1) \\ \varphi_1(2) & \varphi_2(2) \end{vmatrix}$$

$$= (1/2)^{1/2} [\varphi_1(1)\, \varphi_2(2) - \varphi_1(2)\, \varphi_2(1)]$$

If we define

$$\varphi_+ \equiv (1/2)^{1/2} (\varphi_1 + \varphi_2)$$
$$\varphi_+ \equiv (1/2)^{1/2} (\varphi_1 - \varphi_2)$$

we obtain

$$\psi' = (1/2)^{1/2} \begin{vmatrix} \varphi_+(1) & \varphi_-(1) \\ \varphi_+(2) & \varphi_-(2) \end{vmatrix}$$

$$= (1/2)^{1/2} [\varphi_+(1) \varphi_-(2) - \varphi_+(2) \varphi_-(1)]$$

$$= (1/8)^{1/2} \{[\varphi_1(1) + \varphi_2(1)] [\varphi_2(2) - \varphi_1(2)] - [\varphi_1(2) + \varphi_2(2)] [\varphi_2(1) - \varphi_1(1)]\}$$

$$= (1/8)^{1/2} [\varphi_1(1) \varphi_2(2) - \varphi_2(1) - \varphi_1(2) \varphi_2(1) + \varphi_2(2) \varphi_1(1)]$$

$$= (1/2)^{1/2} [\varphi_1(1) \varphi_2(2) - \varphi_1(2) \varphi_2(1)] = \psi$$

15. Roothaan CCJ (1960) Rev. Mod. Phys. 32: 179
16. Solouki B, Rosmus P, Bock H (1976) J. Am. Chem. Soc. 98: 6054
17. Meyer W (1971) Int. J. Quant. Chem. 55: 341; (1973) J. Chem. Phys. 58: 1017
18. Schaefer HF III (1972) The electronic structure of atoms and molecules. A survey of rigorous quantum mechanical results, Addison-Wesley, Reading, MA
19. Bock H, Ramsey BG (1973) Angew. Chem. 85: 773; (1973) Internat. Edit. 12: 734
20. Koopmans T (1933) Physica 1: 104
21. This second part elevates Koopmans' theorem from the status of a mere algebraic identity to that of a theorem. This development is given in Koopmans' original paper. However, we have chosen to follow the more explicit treatment of Newton MD (1968) J. Chem. Phys. 48: 2825
22. Electron affinities are reviewed by Hotop H, Linneberger WC (1975) J. Phys. Chem. Ref. Data 4: 539; Radzig AA, Smirnow BM (1985) Reference data on atoms, molecules and ions, Springer, Berlin Heidelberg New York; Gutsev GL, Boldyrev AI (1985) Adv. Chem. Phys. 61: 169
23. (a) Chen ECM, Wentworth WE (1983) J. Phys. Chem. 87: 45;
    (b) Briegleb G (1964) Angew. Chem. Int. Ed. Engl. 3: 617;
    (c) Batley M, Lyons LE (1962) Nature (London) 196: 573
    (d) Chen ECM, Wentworth WE (1975) J. Phys. Chem. 63: 3183 and references therein;
    (e) Drzaić PS, Marks J, Brauman JI (1984) Electron photodetachment from gas phase molecular ions. In: Bowers MT (ed) Gas phase ion chemistry, vol 3, Academic, New York;
    (f) Mead RD, Stevens A, Lineberger WC (1984) Photodetachment in negative I on beams, Academic, New York;
    (g) Hogg WM, Kebarle P (1965) J. Chem. Phys. 40: 449; Kebarle P (1977) Annu. Rev. Phys. Chem. 28: 445; Henderson WG, Taagepera D, Holtz D, McIver RT, Beauchamp JL, Taft RW (1972) J. Am. Chem. Soc. 94: 4728
    (h) McIver RT, Fukuda EK (1982) In: Hartman H, Wanzek KP (eds) Lecture Notes in Chemistry, Springer, Berlin Heidelberg New York, p 164;
    (i) Grimsrud EP, Caldwell G, Chowdhury S, Kebarle P (1985) J. Am. Chem. Soc. 107: 4627
24. (a) Lowe JP (1977) J. Am. Chem. Soc. 99: 5557; (1978) Quantum chemistry, Academic, New York;
    (b) Myers RT (1981) J. Inorg. Nucl. Chem. 43: 3083;
    (c) Parr RG, Bartolotti LJ (1982) J. Am. Chem. Soc. 104: 3801;
    (d) Liegener CM (1987) MATCH 22: 215
25. A discussion of this aspect is given by Smith WL (1973) Mol. Phys. 26: 361
    (a) Although a vertical ionization energy is well defined, the concept of a vertical transition, when carried to the extreme, encounters problems connected with the Heisenberg uncertainty principle. See, for example, Schwartz SE (1973) J. Chem. Educ. 50: 608
26. (a) Cederbaum LS (1974) Chem. Phys. Letters 25: 562;
    (b) Cederbaum LS, Hohlneicher G, von Niessen W (1973) Chem. Phys. Letters 18: 503; (1973) Mol. Phys. 26: 1405;
    (c) Pickup BT, Goscinski O (1973) Mol. Phys. 26: 1013;
    (d) Chong DP, Hering FG, McWilliams D (1974) J. Chem. Phys. 61: 74, 958, 3567; (1974) Chem. Phys. Letters 25: 568
27. Wittel K (1974) Thesis, Frankfurt; cf Wittel K (1976) J. Electron Spectrosc. 8: 245
28. For a use of Koopmans' theorem including spin-orbit coupling, see Wittel K (1972) Chem. Phys. Letters 15: 555; Manne R, Wittel K, Mohanty BS (1975) Mol. Phys. 29: 485
29. Jungen M (1972) Theor. Chim. Acta 27: 33
30. Brundle CR, Robin MB, Basch H (1972) J. Chem. Phys. 53: 2196

31. HAM3: Fridh C, Åsbrink L, Lindholm E (1972) Chem. Phys. Letters 15: 282; Lindholm E,
    Åsbrink L (1985) Molecular orbitals and their energies, studied by the semiempirical HAM
    method, Lecture Notes in Chemistry, vol 38, Springer, Berlin Heidelberg New York;
    MNDO: Dewar MJS, Thiel W (1977) J. Am. Chem. Soc. 99: 4899;
    CNDO/S: DelBene J, Jaffe HH (1968); J. Chem. Phys. 48: 1807
    AM1: Dewar MJS, Zoebisch EG, Healy EF, Stewart JJP (1985) J. Am. Chem. Soc. 107: 3902

The Equivalent Bond Orbital Model and the Interpretation of PE Spectra

Evi Honegger[1] **and Edgar Heilbronner**[2]

Institut für Physikalische Chemie, Universität Basel, Klingelbergstr. 80, CH-4056 Basel, Switzerland

---

1 Present address: Digital Equipment Corporation AG, Münchensteinerstraße 43, Postfach, CH-4002 Basel
2 Present address: Grütstraße 10, CH-8704 Herrliberg

1 Introduction

1.1 The Ionization Process

The low-energy part of the He(Iα) or He(IIα) PE spectrum of a hypothetical molecule M (shown in Fig. 1) consists of four bands, ① to ④, which — for simplicity — we

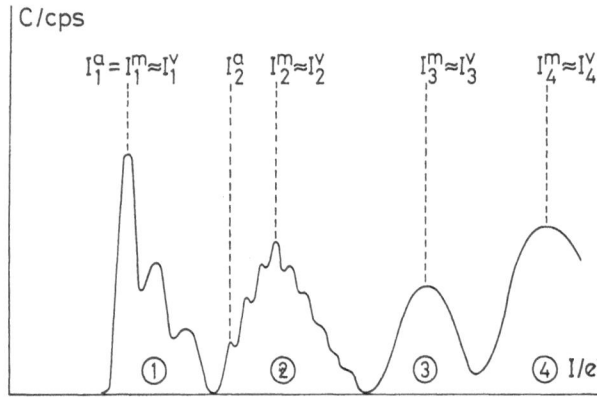

Fig. 1. Low-energy part of the PE spectrum of a hypothetical molecule M

assume not to overlap. The bands ① and ② exhibit resolved vibrational fine structure, whereas the fine structure of ③ and ④ cannot be resolved because of line broadening, of random noise and/or the resolution of the recording, the latter being typically of the order of 20 meV.

The position of a band ⓑ on the abscissa I/eV can be characterized by the ionization energy I_j^m corresponding to the maximum of the Franck-Condon envelope $C = C(I)$. For all practical purposes, this quantity is usually a sufficient approximation of the vertical ionization energy I_j^v. The position of the first vibrational component (disregarding possible hot bands) corresponds to the adiabatic ionization energy I_j^a. Finally, the "centre of gravity" of a band

$$I_j^c = \int_{ⓑ} I\, C(I)\, dI / \int_{ⓑ} C(I)\, dI \tag{1.1}$$

can be used as an alternative approximation for I_j^v, the integrals in Eq. (1.1) being taken over the range of band ⓑ.

In the following we shall be interested in the ionization of a neutral, closed shell molecule M in its electronic singlet ground state, $^1\Psi_0$, which yields according to

$$M(^1\Psi_0; v'') + h\nu \rightarrow M^+(^2\tilde{\Psi}_j; v') + e^-(T_j(v'', v')) \tag{1.2}$$

a radical cation M^+ in its electronic doublet ground state, $^2\tilde{\Psi}_0$ (j = 0) or in an electronically excited doublet state, $^2\tilde{\Psi}_j$ (j > 0). In formula (1.2), v'' and v' symbolize the (set of) vibrational quantum number(s) of M and M^+, respectively. The kinetic energy $T_j(v'', v')$ of the photoelectron e^- is the primary quantity measured in the PE spectrometer. The photon energy $h\nu$ being known, one obtains the ionization energy according to

$$I_j(v'', v') = h\nu - T_j(v'', v').$$ (1.3)

If $v'' = v' = 0$, then $I_j(0,0) \equiv I_j^a$, is the adiabatic ionization energy, and if $v' \leftarrow v''$ corresponds to the most intense vibrational component of band ⑤, then $l_j(v'', v') \equiv I_j^m$. (For further information about PE spectroscopy, see [1]).

1.2 Some Comments on Theory

Within the standard HMO formalism [2], formula (1.2) corresponds to the removal of an electron from a doubly occupied Hückel-MO φ_j^{HMO}. The energy needed is $-\varepsilon_j^{HMO}$ so that

$$I_j^{HMO} = -\varepsilon_j^{HMO} = -(\alpha + x_j\beta).$$ (1.4)

Although Eq. (1.4) looks suspiciously like Koopmans' theorem [3] (see below), it has nothing to do with it. It is simply the consequence of assuming independent electrons. For quantitative purposes the HMO parameters α and β can be calibrated by least squares techniques using the experimental $I_j^m(M_r)$ values of the molecules $M_1, M_2, \dots M_r$ of a calibration set $\{M\}$ as independent variables [4]. The method can be improved by the inclusion of additional calibration parameters (e.g. by taking care of bond-order changes $\Delta p_{\mu\nu}$ resulting from electron ejection) to yield more flexible multilinear regressions [5]. (Concerning the limitations of such regressions see [6]).

The most common procedures used for the systematization and rationalization of PE-spectroscopic results, are semiempirical [7] or ab-initio [8] SCF treatments, usually of the restricted type. The ground state $^1\Psi_0$ of M is approximated by the ground configuration $^1\psi_0 \approx {}^1\Psi_0$, written as a single Slater determinant in terms of the SCF canonical orbitals φ_j:

$$^1\psi_0 = \|\varphi_1\bar{\varphi}_1\varphi_2\bar{\varphi}_2 \cdots \varphi_j\bar{\varphi}_j \cdots \varphi_{HOMO}\bar{\varphi}_{HOMO}\|.$$ (1.5)

Electron ejection from φ_j leads to the doublet configurations $^2\tilde{\psi}_j$ of the radical cation M^+

$$^2\tilde{\psi}_j \begin{cases} \|\varphi_1\bar{\varphi}_1 \cdots \varphi_j \cdots \varphi_{HOMO}\bar{\varphi}_{HOMO}\| \\ \|\varphi_1\bar{\varphi}_1 \cdots \bar{\varphi}_j \cdots \varphi_{HOMO}\bar{\varphi}_{HOMO}\| \end{cases}$$ (1.6)

assumed to be a sufficient approximation of the doublet state $^2\tilde{\Psi}_j$. If it is postulated that the CMOs φ_j in expression (1.6) are exactly the same as those in Eq. (1.5), i.e. if we neglect changes due to electron reorganisation in M^+, then it can be shown that

$$E(^2\tilde{\psi}_j) - E(^1\psi_0) = I_j^v = -\varepsilon_j,$$ (1.7)

where ε_j is the orbital energy of φ_j, defined as

$$\varepsilon_j = h_j + \sum_k (2J_{jk} - K_{jk}).$$ (1.8)

This is known as Koopmans' theorem [3]. The assumption implied in Eq. (1.7) is that M and M^+ have identical geometric structure and that the frozen SCF CMOs φ_j of M can be carried over unchanged for the description of the configurations $^2\tilde{\psi}_j$ of M^+. We shall call such configurations "Koopmans Configurations", KC.

An obvious refinement consists in introducing a configuration interaction (CI) between the KC and excited configurations of M^+ involving virtual orbitals, which allows one to take care of electron reorganization and electron correlation. In this respect the use of the Green-function technique has proved particularly useful [9]. Finally, vibronic mixing between states can be taken into account [10]. Although it has been shown that these improvements lead to an almost perfect match between observed and calculated band positions or Franck-Condon band envelopes, respectively [11], experience has shown that — with rare, but important exceptions [12] — they do not yield changes in the sequence of states (as set down by their irreducible representations), especially in the low-energy region of the PE spectra of larger molecules. In other words, the state sequence predicted by a SCF treatment is more often than not the same as that derived by more sophisticated treatments.

1.3 Towards a Heuristic Model

Mainly due to the work of Dewar [13] and of Woodward and Hoffmann [14], chemists have learned to express themselves in a molecular orbital language. It is therefore not surprising that there should be a pronounced tendency to discuss PE spectra of large (organic) molecules also in those terms.

From this point of view, the HMO model is a rather obvious candidate, being an independent electron treatment which lends itself optimally to different first- and second-order perturbation treatments [15]. These allow a transparent and easy discussion of ionization energy shifts δI, caused by changes in geometry (i.e. changes δR in bond length, $\delta\theta$ in bond angles, $\delta\tau$ in twist angles) or by substituent effects.

In semiempirical or ab-initio SCF calculations the orbital picture is preserved because the Koopmans-configurations $^2\tilde{\psi}_j^{KC}$ are uniquely tied to the CMOs φ_j which have lost the photoelectron. Thus, the usual orbital arguments can still be applied to the KCs postulated for the radical cation M^+, albeit with obvious modifications. The major disadvantage is that the φ_j underlying the KC [expression (1.6)] are expressed as linear combinations of a large number of nonorthogonal basis functions, which renders the application of simple perturbation calculations rather cumbersome. The remedy proposed in this chapter consists in combining the equivalent-bond-orbital (EBO) approach [16] with the concept of localized molecular orbitals (LMOs) [17, 18] in view of expressing the CMOs φ_j in terms of a small number of orthogonal EBOs. This approach preserves (almost) all the ease of manipulation typical of the independent electron HMO model, e.g. the use of perturbation or calibration procedures [19]. Needless to say that the method draws heavily on the pioneering work by Hall [20], Lorquet [21], Brailsford and Ford [22], Herndon [23], Murrell and Schmidt [24], and Gimarc [25].

2 Concepts and Procedures

2.1 SCF Formalism

Neglecting relativistic and electromagnetic effects, the electronic, time-independent Schrödinger equation for the molecule M (Sect. 1.1) consisting of K nuclei and 2N electrons is written as

$$\hat{H}\Psi = E\Psi, \tag{2.1}$$

$$\hat{H} = \sum_{r=1}^{2N} \hat{h}_r + \sum_{r=1}^{2N} \sum_{s>r}^{2N} e^2/R_{rs}, \tag{2.2}$$

$$\hat{h}_r = -(\hbar^2/2m_e)\,\Delta_r - \sum_{A=1}^{K} Z_A e^2/R_{Ar}. \tag{2.3}$$

We solve Eq. (2.1) in the restricted Hartree-Fock SCF-MO approximation [26], the ground-state wave function Ψ_0 being approximated by the Slater determinant [cf. Eq. (1.5)]

$$^1\psi_0 = \|\eta_1\bar{\eta}_1 \ldots \eta_j\bar{\eta}_j \ldots \eta_N\bar{\eta}_N\|, \tag{2.4}$$

where the N orbitals η_j form an orthonormal set, written as a row vector:

$$\eta = (\eta_1, \eta_2, \ldots \eta_j, \ldots \eta_N). \tag{2.5}$$

Minimization of the energy expectation value

$$E(^1\psi_0) = \langle ^1\psi_0| \hat{H} |^1\psi_0\rangle = 2 \sum_{i=1}^{N} h_i + \sum_{i=1}^{N} \sum_{j=1}^{N} (2J_{ij} - K_{ij}), \tag{2.6}$$

in which the individual terms have their usual meaning, i.e.

$$h_i = \langle \eta_i| \hat{h} |\eta_i\rangle; \qquad \hat{h} = \sum_{r=1}^{2N} \hat{h}_r,$$

$$J_{ij} = \langle ij| G |ij\rangle \equiv (ii \mid jj), \tag{2.7}$$

$$K_{ij} = \langle ij| G |ji\rangle \equiv (ij \mid ij),$$

leads to the Hartree-Fock equations for the orbitals η_j,

$$\hat{F}\eta_j = \sum_{i=1}^{N} \eta_i F_{\eta,ij}, \tag{2.8}$$

where $F_{\eta,ij}$ is an element of the matrix

$$\mathbf{F}_\eta = (F_{\eta,ij}) \equiv (\langle \eta_i| \hat{F} |\eta_j\rangle). \tag{2.9}$$

Following Roothaan [27] the η_j are written as linear combinations of appropriate basis functions ϕ_μ (e.g. Slater-type AOs or gaussians):

$$\eta_j = \sum_\mu \phi_\mu c_{\mu j}. \tag{2.10}$$

The solutions of Eq. (2.8) are only determined up to an orthogonal (or − if the η are complex − a unitary) transformation, which means that there is a great ambiguity of orbitals. In the following we are interested in only two types of transformation, the first of which is an orthogonal transformation, defined by an orthogonal matrix $\mathbf{X}$, i.e. $\mathbf{X}^{-1} = \mathbf{X}^T$, such that $\mathbf{XX}^T = \mathbf{X}^T\mathbf{X} = 1$, which transforms $\mathbf{F}_\eta$ into a diagonal matrix $\mathbf{F}_\varphi \equiv \varepsilon$, i.e.

$$\mathbf{X}^T\mathbf{F}_\eta\mathbf{X} = \varepsilon = \mathrm{diag}(\varepsilon_1 \ldots \varepsilon_j \ldots \varepsilon_N). \tag{2.11}$$

As a consequence, the previous set of orbitals $\boldsymbol{\eta} = (\eta_1, \eta_2, \ldots \eta_j, \ldots \eta_N)$ transforms into the so-called canonical molecular orbitals (CMOs)

$$\boldsymbol{\varphi} = (\varphi_1 \ldots \varphi_j \ldots \varphi_N) \tag{2.12}$$

related to the previous row vector according to

$$\boldsymbol{\varphi} = \boldsymbol{\eta}\mathbf{X}. \tag{2.13}$$

The orthonormal CMOs φ_j are those to which the CMO energies ε_j[cf. Eq. (2.11)] referred to in Koopmans' theorem [Eq. (1.7)] correspond. Note that the traditional numerical handling of the Hartree-Fock equations (2.8) yields directly the CMO orbital energies [Eq. (2.11)] and the set of CMOs φ_j [Eq. (2.12)] belonging to irreducible representations of the molecular point group.

2.2 Localized Molecular Orbitals (LMOs)

The second orthogonal (or unitary) transformation we are interested in consists in creating orthonormal "localized" molecular orbitals (LMOs) λ_j

$$\boldsymbol{\lambda} = (\lambda_1 \ldots \lambda_j \ldots \lambda_N) \tag{2.14}$$

from the CMOs φ_j [Eq. (2.13)] according to

$$\boldsymbol{\lambda} = \boldsymbol{\varphi}\mathbf{L}. \tag{2.15}$$

Doubly occupied LMOs λ_j are closely related to the chemist's intuitive idea of a two-centre, two-electron bond. The localization criteria embodied in the localization matrix $\mathbf{L}$ are rather arbitrary, and in fact different criteria have been proposed. Of special interest is the intrinsic localization criterium proposed by Edmiston and

Ruedenberg [28] (cf. [18]), which demands that the sum of the coulombic self-repulsions $J_{jj}(\lambda) = \langle \lambda_j(1) \lambda_j(2)| G|\lambda_j(1) \lambda_j(2)\rangle$ [cf. Eq. (2.7)] become a maximum:

$$D_{E-R}(\lambda) = \sum_{j=1}^{N} J_{jj}(\lambda) = \text{Maximum} . \tag{2.16}$$

This ensures that the electrons are as concentrated by pairs, i.e. as close together within the LMOs λ_j, as is compatible with the SCF treatment. Because of the invariance of the total sum of the coulombic and exchange energies, Eq. (2.16) could also be interpreted as corresponding to a minimization of the sum of the K_{ij} and the J_{ij} with $i \neq j$, i.e. between different LMOs λ_i and λ_j.

Based on the latter point of view, Foster and Boys have proposed an alternative procedure for the evaluation of $D(\lambda)$ [17], namely the maximization of the sum of the squares of the distances between the centres of the LMOs λ_j

$$D_{F-B}(\lambda) = \sum_{i=1}^{N} \sum_{j \neq i}^{N} (\langle \lambda_i| \vec{r} |\lambda_i\rangle - \langle \lambda_j| \vec{r} |\lambda_j\rangle)^2 = \text{Maximum} . \tag{2.17}$$

For practical purposes, the differences between the LMOs λ_j generated according to Eqs. (2.16) or (2.17) is marginal, but Eq. (2.17) is more convenient from a computational point of view.

Independent of the criterion used for defining the localization matrix L one finds for real orbitals that L is an orthogonal matrix $L^{-1} = L^T, LL^T = L^TL = 1$ and that

$$F_\lambda = L^T \varepsilon L \tag{2.18}$$

where $F_\lambda = (F_{\lambda,ij})$ with

$$F_{\lambda,ij} = \langle \lambda_i| \hat{F} |\lambda_j\rangle , \tag{2.19}$$

$\hat{F}$ being the same Fock-operator as in Eq. (2.8).

2.3 Linking CMOs and LMOs

It follows from Eq. (2.18) that the diagonalization of F_λ, i.e. of the Hartree-Fock matrix in localized basis, i.e.

$$F_\lambda L^T = L^T \varepsilon , \tag{2.20}$$

yields the orbital energies ε_j — collected in ε — of the CMOs φ_j, which are now written as

$$\varphi_j = \sum_{i=1}^{N} \lambda_i L_{ji} \tag{2.21}$$

$$\varphi = \lambda \mathbf{L}^T . \tag{2.22}$$

The great advantage of Eqs. (2.21) and (2.22) is that the CMOs φ_j are now expressed as linear combinations of a *small* number of *orthonormal* LMOs λ_j, instead of linear combinations of a large number of non-orthogonal basis functions ϕ_μ according to Eq. (2.10). From a practical point of view this means that it is now very easy to visualize the CMOs φ_j in terms of the λ_j, e.g. to recognize at a glance their nodal and symmetry properties. As an example, Fig. 2 shows the two highest occupied MOs of *trans,trans*-perhydroanthracene $C_{14}H_{24}$ I [29]. Open and full bonds refer to LMOs λ_j of opposite phase, and the numbers given are the absolute values $|L_{ji}|$ of the expansion coefficients in Eq. (2.21), if $|L_{ji}| > 0.1$. The LMOs λ_j affected with $|L_{ji}| < 0.1$ are indicated by thin lines, irrespective of sign. In addition, it is now possible to make use of the types of perturbation treatments to which one has become accustomed in HMO theory, and which demand an orthonormalized basis, now provided by the LMOs λ_j.

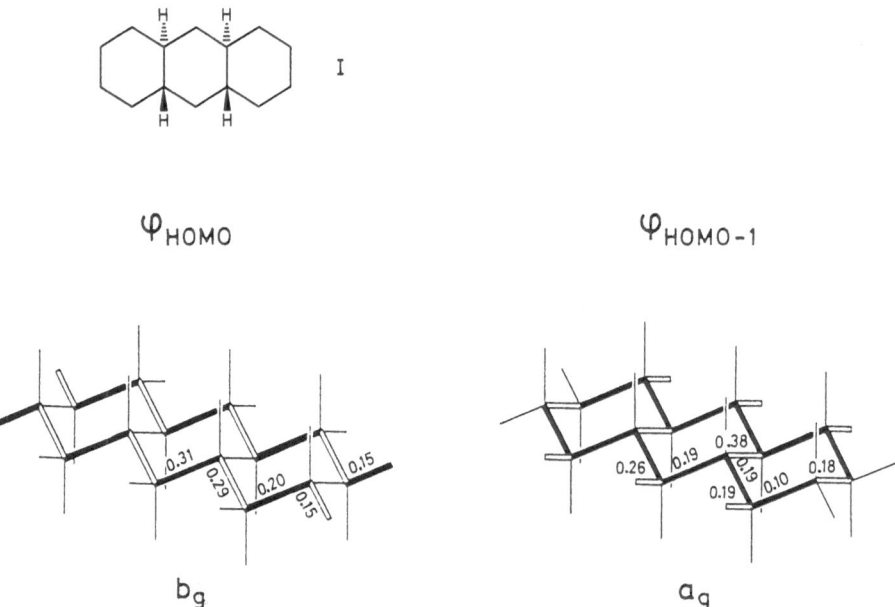

Fig. 2. Orbital diagrams in LMO basis for the two highest occupied CMOs φ_{HOMO} and φ_{HOMO-1} of perhydroanthracene I $(C_{14}H_{24})$

As mentioned before, the CMOs φ_j have the unique distinction that the corresponding Hartree-Fock matrix $\mathbf{F}_\varphi \equiv \varepsilon$ is diagonal, and that the φ_j belong to irreducible representations of the point group of the molecule M. In contrast, the Hartree-Fock matrix $\mathbf{F}_\lambda$ belonging to the LMOs λ_j is now a *full* matrix. However, because the matrix elements $F_{\lambda,ij}$ refer to LMOs λ_j of strongly local character, they exhibit a high degree of transferability from one molecule to another. It is therefore reasonable to designate these transferable $F_{\lambda,ij}$ by special symbols. With reference to the formula

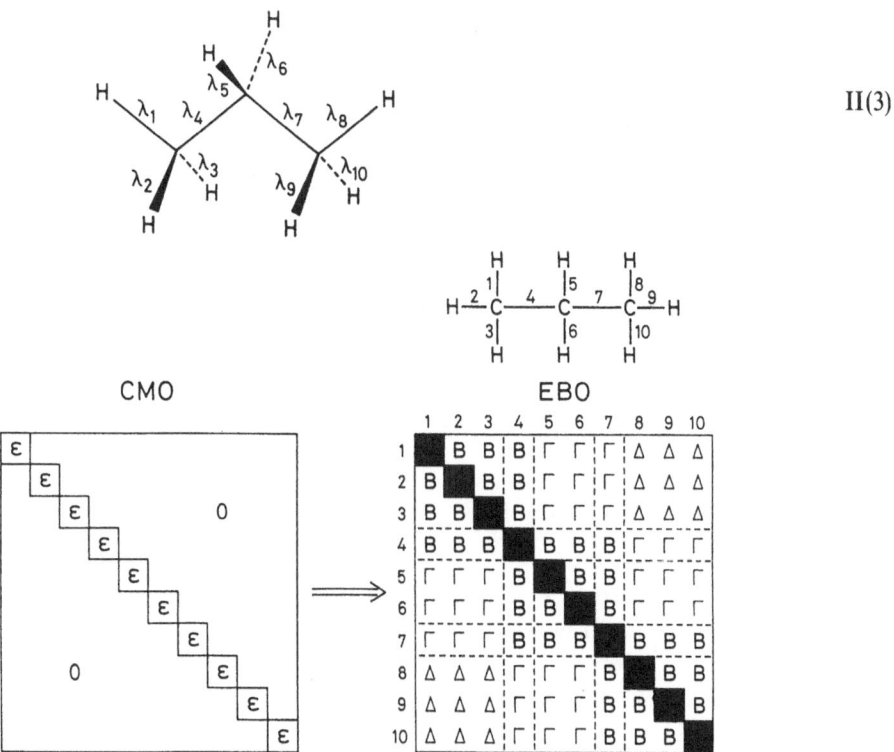

Fig. 3. Schematic representation of the orthogonal transformation of Eq. (2.18) with reference to propane II(3) (C_3H_8)

of propane II(3) as an example and to the schematic representation of transformation (2.18) shown in Fig. 3, we use the following conventions:

a) The *diagonal* matrix elements $F_{\lambda,jj}$ are designated as self-energies $A_{jj} = A_{\text{Bond-type}}$, e.g. $A_{11} = A_{\text{CH}}$, $A_{44} = A_{\text{CC}}$.

b) Interaction cross terms $F_{\lambda,ij}$ between *geminal* LMOs λ_i, λ_j are designated by B_{ij}, i and j referring to the geminal bonds in question, e.g.

$$B_{1,2} = B_{\text{CH,CH}}, \; B_{1,4} = B_{\text{CH,CC}}, B_{4,7} = B_{\text{CC,CC}}.$$

c) In analogy to (b), interaction cross terms $F_{\lambda,ij}$ between *vicinal* LMOs λ_i, λ_j are designated by $\Gamma_{i,j}$, e.g. $\Gamma_{1,5} = \Gamma_{\text{CH,CH}}$, $\Gamma_{1,7} = \Gamma_{\text{CH,CC}}$.

Interaction cross terms $F_{\lambda,ji}$ between two LMOs λ_i, λ_j separated by two or more bonds are usually small and can be neglected in a first approximation, except in a few particular cases (see below).

2.4 Symmetry-Adapted Semilocalized MOs

If the molecule belongs to a symmetry group with irreducible representations $\Xi^{(1)} \ldots \Xi^{(r)} \ldots \Xi^{(t)}$ one can form symmetry-adapted, semilocalized molecular orbitals (SLMOs) ϱ_j according to

$$\varrho = \lambda \mathbf{R} \tag{2.23}$$

where $\varrho = (\varrho_1 \ldots \varrho_j \ldots \varrho_N)$ is the row vector of the SLMOs. Without loss of generality the orthogonal (unitary) transformation matrix $\mathbf{R}$ can always be chosen in such a way that the SLMOs ϱ_j are ordered in ϱ according to the conventional sequence of the irreducible representations $\varXi^{(r)}$ to which they belong:

$$\varrho = (\underbrace{\varrho_1 \ldots \varrho_h}_{\varXi^{(1)}}; \ldots; \underbrace{\varrho_i \ldots \varrho_n}_{\varXi^{(r)}}; \ldots; \underbrace{\varrho_s \ldots \varrho_N}_{\varXi^{(t)}})$$

$$\downarrow \qquad\qquad \downarrow \qquad\qquad \downarrow$$

$$\varrho = (\varrho^{(1)}; \quad \ldots; \quad \varrho^{(r)}; \quad \ldots; \quad \varrho^{(t)}). \tag{2.24}$$

Under these conditions the Hartree-Fock matrix

$$\mathbf{F}_\varrho = (\langle \varrho_i | \hat{F} | \varrho_j \rangle) = \mathbf{R}^T \mathbf{F}_\lambda \mathbf{R} \tag{2.25}$$

is blocked out in submatrices $\mathbf{F}_\varrho(\varXi^{(r)}) \equiv \mathbf{F}_\varrho^{(r)}$ belonging to the different irreducible representations $\varXi^{(r)}$ of the group:

$$\mathbf{F}_\varrho = \mathbf{F}_\varrho^{(1)} \oplus \mathbf{F}_\varrho^{(2)} \oplus \ldots \oplus \mathbf{F}_\varrho^{(r)} \oplus \ldots \oplus \mathbf{F}_\varrho^{(t)}. \tag{2.26}$$

If the molecule has no symmetry (i.e. belongs to C_1), then the SLMOs are identical with the LMOs. However, it may still be convenient to chose two (or more) LMOs, e.g. the π LMOs λ_a, λ_b of a diene, and to transform them into semilocalized molecular orbitals, e.g.

$$\varrho_+ = (\lambda_a + \lambda_b)/\sqrt{2},$$
$$\varrho_- = (\lambda_a - \lambda_b)/\sqrt{2}, \tag{2.27}$$

all other LMOs λ_j ($j \neq a, b$) remaining unchanged ($\varrho_j = \lambda_j$).

2.5 Partial Diagonalization of $\mathbf{F}_\lambda$

In the following we restrict our discussion to the subset $(\varrho_i \ldots \varrho_k \ldots \varrho_n) = \varrho^{(r)}$ of SLMOs belonging to a given irreducible representation $\varXi^{(r)}$ [see Eq. (2.24)]. (The extension to the full set ϱ is trivial). To simplify the discussion we assume that we are interested only in the behavior of a single preselected SLMO ϱ_k of the set $\varrho^{(r)}$ in relation to the remaining $(n - 1)$ SLMOs ϱ_j ($j \neq k$). The most convenient way of achieving this comparison consists in forming orthonormal linear combinations ψ_j from the $(n - 1)$ SLMOs ϱ_j ($j \neq k$) in such a way that all interaction matrix elements $\langle \psi_i | \hat{F} | \psi_j \rangle = 0$ for $i \neq j$ and $i, j \neq k$. We call such orbitals ψ_j "precanonical **molecular** orbitals" (PCMOs). Obviously, the cross terms $\langle \psi_j | \hat{F} | \varrho_k \rangle$ between the PCMOs ψ_j and the SLMOs ϱ_k will differ from zero so that the resulting submatrix $\mathbf{F}_\psi^{(r)}$ is of the form

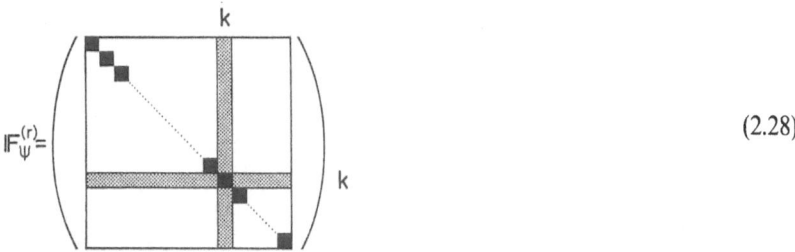

$$\mathbf{F}_\psi^{(r)} = \qquad \qquad \qquad \qquad \qquad \qquad \qquad \qquad (2.28)$$

To simplify the following formulae, we drop the superscript r, it being understood that they refer to the SLMOs and PCMOs belonging to a single given, irreducible representation. To compute $\mathbf{F}_\psi$ we first delete all off-diagonal elements in row k and column k of the matrix $\mathbf{F}_\varrho$ generating $\mathbf{F}_{\varrho,k}$. Diagonalization of $\mathbf{F}_{\varrho,k}$ yields

$$\mathbf{P}_k^T \mathbf{F}_{\varrho,k} \mathbf{P}_k = \mathrm{diag}\, (\varepsilon_1^{PCMO} \ldots \varepsilon_j^{PCMO} \ldots) \qquad (2.28)$$

with $\varepsilon_k^{PCMO} = F_{\varrho,kk}$. The corresponding row vector of the PCMOs,

$$\psi = (\psi_1 \ldots \psi_{k-1}, \varrho_k, \psi_{k+1} \ldots \psi_N), \qquad (2.29)$$

is obtained by the transformation

$$\psi = \varrho \mathbf{P}_k. \qquad (2.30)$$

Finally, we calculate the matrix $\mathbf{F}_\psi$ [shown diagrammatically in Eq. (2.28)] by applying the transformation $\mathbf{P}_k$ to the original matrix $\mathbf{F}_\varrho$ according to

$$\mathbf{F}_\psi = \mathbf{P}_k^T \mathbf{F}_\varrho \mathbf{P}_k. \qquad (2.31)$$

As shown diagrammatically in Eq. (2.28) all off-diagonal elements $F_{\psi,ij}$ are zero, with the exception of the elements $F_{\psi,jk}$ and $F_{\psi,kj}$ in column k and row k. These elements link the SLMO $\varrho_k \equiv \psi_k$ to the set of PCMOs ψ_j (j ≠ k).

3 Saturated Hydrocarbons

3.1 SCF Computation of F_λ in LMO Basis

We begin by calculating the CMOs φ_j of a saturated hydrocarbon C_nH_m by using some chemically unbiased ab-initio procedure. Because our investigations of the PE spectra of organic molecules were begun 20 years ago, the following results refer to the STO-3G model [8, 30]. In view of the subsequent calibration procedure, the choice of a particular ab-initio model is largely irrelevant. Standard geometries [8] have been used wherever possible. The CMOs φ_j have been transformed into the LMOs λ_j, and the corresponding eigenvalue matrix ε into the Hartree-Fock matrix $\mathbf{F}_\lambda$ according to Eqs. (2.15) and (2.18), using the Foster-Boys localization procedure [17]. The matrix elements of $\mathbf{F}_\lambda$ so obtained are characterized by the superscript "0".

a) *Self-Energies* A_{jj}^0. The self-energies $A_{jj}^0 \equiv F_{\lambda,jj}^0$ i.e. the diagonal terms, yield the mean values $A_{CC}^0 = -17.50$ eV and $A_{CH}^0 = -16.95$ eV with negligible scatter. The difference $A_{CC}^0 - A_{CH}^0 \approx 0.5$ eV, albeit significant, is small, so that for some purposes both values can be assumed to be equal, i.e. $A = A_{CC} = A_{CH}$ in a first approximation, e.g. for the rationalization [31] of the C2s-band system of He(IIα) PE spectra of saturated hydrocarbons [32, 33] (see below). The near equality of A_{CC}^0 and A_{CH}^0 is the main reason why the CMOs φ_j extend rather uniformly over a given saturated hydrocarbon. It follows that the positive charge distribution of the radical cation $C_nH_m^+$ in a given Koopmans configuration ${}^2\tilde{\psi}_j$ [cf. Eq. (1.6)] is also rather smooth and uniform.

b) *Geminal Cross Terms* B_{ij}^0. All cross terms $B_{ij}^0 \equiv F_{\lambda,ij}^0$ between geminal LMOs λ_i, λ_j, are the same within small limits of error, independent of the nature of the two LMOs λ_i, λ_j. Therefore, $B_{CC,CC} = B_{CC,CH} = B_{CH,CH} = -2.89$ eV $= B$ is a perfectly acceptable approximation, as long as all bond angles are the same, i.e. close to tetrahedral (109° 27′). As will be discussed later, $|B|$ increases with decreasing bond angle.

c) *Vicinal Cross Terms* Γ_{ij}^0. The matrix elements $\Gamma_{ij}^0 \equiv F_{\lambda,ij}^0$ between two vicinal LMOs λ_i, λ_j which bracket a given CC-bond joining λ_i to λ_j, depend on the local conformation characterized by the twist angle τ_{ij}. For a *syn*-planar conformation, with $\tau_{ij} = 0°$ by definition, the cross term Γ_{ij}^0 is negative, for an *anti*-planar conformation with $\tau_{ij} = 180°$ it is positive. From a series of calculations with τ_{ij} spanning the interval $0° \leq \tau_{ij} \leq 180°$ one finds that

$$\Gamma_{ij}^0 = \Gamma^0 \cos \tau_{ij} \tag{3.1}$$

without a statistically significant constant term. Within the required limits of error the factor Γ^0 of Eq. (3.1) is independent of the nature of the LMOs λ_i, λ_j, so that, for a given value $\tau_{ij} = \tau$, one has $\Gamma_{CC,CC}(\tau) = \Gamma_{CC,CH}(\tau) = \Gamma_{CH,CH}(\tau) = \Gamma(\tau)$.

d) *Higher Interaction Terms*. The 1,4-interaction terms Δ_{ij}^0, which refer to two LMOs λ_i, λ_j separated by two linking bonds b_1, b_2, e.g. the two λ_{CH} LMOs of the two methyl groups in propane II(3), and which depend on two twist angles, have been discussed in [19]. Although small, they are significant, as shown by the values of Table 1.

On the other hand, if two LMOs λ_i, λ_j are separated by three or more bonds, the corresponding interaction term $F_{\lambda,ij}^0$ becomes in general so small that it can safely be neglected.

Table 1. Dependence of the 1,4-interaction term $\Delta^0_{CH,CH}(\tau_i, \tau_j)$ on the twist angles τ_i, and τ_j in propane II(3)

$\Delta^0_{CH,CH}(\tau_i, \tau_j)/eV$

τ_i	τ_j						
	0°	60°	120°	180°	240°	300°	360°
0°	−0.84	−0.33	−0.21	0.33	0.21	−0.33	−0.84
60°	−0.33	0.02	0.15	0.13	0.01	−0.29	−0.33
120°	0.21	0.15	−0.08	−0.18	−0.13	0.01	0.21
180°	0.33	0.13	−0.18	−0.29	−0.18	0.13	0.33
240°	0.21	0.01	−0.13	−0.18	−0.08	0.15	0.21
300°	−0.33	−0.29	0.01	0.13	0.15	0.02	−0.33
360°	−0.84	−0.33	0.21	0.33	0.21	−0.33	−0.84

3.2 Equivalent Bond Orbital Models

We are now in a position to develop improved EBO models, calibrated to yield predictions of ionization energies of saturated hydrocarbons, assuming the validity of Koopmans' theorem [3]. All that has to be done is to take over the significant matrix elements $F^0_{\lambda,ij}$ derived above, and to construct the reduced matrix F_λ for the hydrocarbon of interest. Diagonalization of F_λ yields the desired result. It is sufficient to restrict the off-diagonal elements to the B^0_{ij}, Γ^0_{ij}, and Δ^0_{ij} only, the other elements yielding contributions too small to be relevant for all practical purposes.

We classify our models depending on the type of matrix elements used, as follows:

AB model: includes only A_j and B_{ij},

$A\Gamma$ model: includes A_j, B_{ij}, and Γ_{ij},

$A\Delta$ model: includes A_j, B_{ij}, Γ_{ij}, and Δ_{ij}.

An upper index "0" indicates that the matrix elements used are those obtained directly from the ab-initio SCF procedure without change. Otherwise, they are calibrated on the basis of the ionization energies of a calibration set, as explained with reference to the $A\Delta$ model.

For each of the hydrocarbons C_nH_m of the calibration set, we construct first the matrix F^0_λ of the $A\Delta^0$ model, the diagonalization of which yields the eigenvalues $\varepsilon_j^{A\Delta0}$. These eigenvalues − for the complete calibration set − are then fitted by a least-squares calculation to the observed ionization energies I_k^m to yield the regression

$$I_k^m = p + q(-\varepsilon_k^{A\Delta0}).$$

(3.2)

For example, if the calibration set consists of the molecules methane II(1), ethane II(2), propane II(3), butane II(4), isobutane III(3), and cyclohexane IV(6), and if the STO-3G procedure is used to calculate the matrix elements incorporated in $\mathbf{F}^0_\lambda$, one obtains $p = (3.100 \pm 0.122)$ eV, $q = 0.750 \pm 0.006$, with standard error $s(I^m_j) = 0.320$ eV and a correlation coefficient $r = 0.998$.

$$H-(CH_2)_N-H \qquad\qquad H_{4-N}C(Me)_N \qquad\qquad \bigcirc (CH_2)_N$$

$$\mathrm{II}\,(N) \qquad\qquad\qquad \mathrm{III}\,(N) \qquad\qquad\qquad \mathrm{IV}\,(N)$$

We now calibrate our $A\Delta^0$ model to yield the parametrized $A\Delta$ model. This is achieved by adding to the diagonal terms A^0_j an increment δA and by multiplying the off-diagonal elements B^0_{ij}, Γ^0_{ij}, Δ^0_{ij} with a common factor f. Because of the orthogonality of the N basis LMOs $\lambda = (\lambda_1 \dots \lambda_j \dots \lambda_N)$, the orbital energy $\varepsilon^{A\Delta 0}_k$ belonging to the molecular orbital $\varphi_k = \lambda\mathbf{c}_k$ of a hydrocarbon C_nH_m is given by

$$\varepsilon^{A\Delta 0}_k = \mathbf{c}^T_k\mathbf{F}_\lambda\mathbf{c}_k, \tag{3.3}$$

which yields for the hydrocarbon C_nH_m

$$\varepsilon^{A\Delta 0}_k = (b_{CC}/b)\,A^0_{CC} + (b_{CH}/b)\,A^0_{CH} + B^0_k + \Gamma^0_k + \Delta^0_k, \tag{3.4}$$

where $b_{CC} = 3n - m/2$ and $b_{CH} = m$ are the number of CC and CH bonds, i.e. of λ_{CC} and λ_{CH} LMOs, respectively ($b = b_{CC} + b_{CH} = 3n + m/2$), and where B^0_k, Γ^0_k, and Δ^0_k are defined as

$$B^0_k = \sum_{i=1}^b \sum_{j=1}^b c_{ik}c_{jk}B^0_{ij},$$

$$\Gamma^0_k = \sum_{i=1}^b \sum_{j=1}^b c_{ik}c_{jk}\Gamma^0_{ij}, \tag{3.5}$$

$$\Delta^0_k = \sum_{i=1}^b \sum_{j=1}^b c_{ik}c_{jk}\Delta^0_{ij}.$$

We now shift A^0_{CC} and A^0_{CH} by δA and we multiply B^0_k, Γ^0_k, Δ^0_k by a factor f in such a way that the eigenvalues $\varepsilon^{A\Delta}_k$ of the calibrated matrix $\mathbf{F}_\lambda$ are within the error limits of Eq. (3.2):

$$-I^m_k = \varepsilon^{A\Delta}_k = (b_{CC}/b)\,(A^0_{CC} + \delta A) + (b_{CH}/b)\,(A^0_{CH} + \delta A)$$
$$+ f(B^0_k + \Gamma^0_k + \Delta^0_k). \tag{3.6}$$

From Eqs. (3.2), (3.4), and (3.6) one obtains

$$\delta A = -p + [(q-1)/b]\,(b_{CC}A^0_{CC} + b_{CH}A^0_{CH}),$$
$$f = q. \tag{3.7}$$

Obviously, the result would have been the same if we had used the AB^0 or the $A\Gamma^0$ model as a start. Thus, the calibrated EBO model is now defined as follows:

EBO-*Model for* C_nH_m

---

1) Calculate F^0_λ according to one of the models defined at the beginning of this section.
2) Multiply all off-diagonal terms B^0_{ij}, Γ^0_{ij} and/or Δ^0_{ij} by f, and,
3) add to all diagonal terms A_i the value δA computed according to Eq. (3.7).
4) Diagonalize F_λ so obtained.

Note that according to Eq. (3.7) δA depends on the composition of the hydrocarbon C_nH_m, although not very critically. On the other hand, f is the same for all hydrocarbons. For the calibration set quoted above the resulting linear regression of I^m_j on $-\varepsilon^{AA}_j$ is shown in Fig. 4.

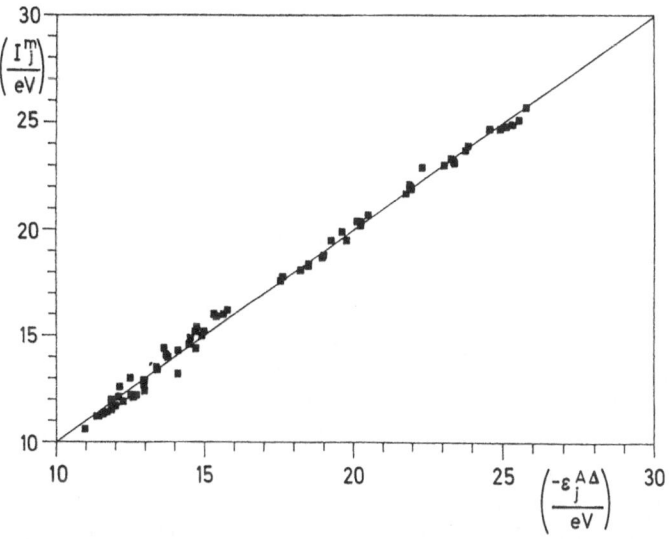

Fig. 4. Linear regression of observed ionization energies I^m_j on $-\varepsilon^{AA}_j$ [i.e. after calibration according to Eq. (3.7)] for the calibration set methane II(1), ethane II(2), propane II(3), butane II(4), isobutane III(3) and cyclohexane IV(6). Here Eq. (3.7) is explicitly given by $\delta A = -3.100\ eV + (0.250/b)\ (b_{CC}\ 17.50\ eV + b_{CH}\ 16.95\ eV)$ and $f = 0.750$

3.3 The C2s Manifold and the AB Model

In this section we consider an ionization process [expression (1.2)] in which the vacated CMO φ_j belongs to the C2s inner valence shell of the hydrocarbon C_nH_m. The corresponding bands in the PE spectrum occupy the region of ionization energies extending typically from 15 to 26 eV. Such spectra have been first studied by Price et al. [32] and by Potts and Streets [33]. In Fig. 5 are reproduced the He(IIα) PE spectra of the series methane II(1) to neopentane III(4), taken from [32]. It is immediately apparent that there is a natural separation of the two groups of bands labelled as C2p type and C2s type, especially prominent among the smaller members of

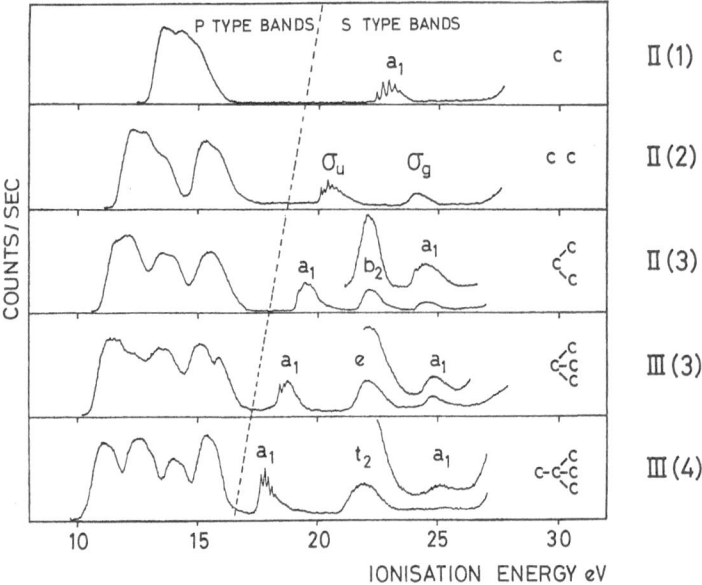

Fig. 5. He(IIα) PE spectra of methane II(1), ethane II(2), propane II(3), isobutane III(3) and neopentane III(4) [32]

the homologous series. Whereas there is profuse overlap of the bands in the C2p-type region, the C2s-type bands are well separated. The correspondence of the number of C2s bands (including degeneracies and/or overlap) to the number n of carbon atoms in C_nH_m is evident. He(IIα) PE spectra of a large number of saturated and unsaturated hydrocarbons can be found in [34] and [35], and the references given therein.

The simplest interpretation of the observed data was provided by Potts and Streets [33], who proposed that the CMOs φ_j of the C2s manifold are well represented by linear combinations of C2s AOs only, i.e. $\phi_\mu \equiv 2s_\mu$ (cf. Eq. (2.10)). Assigning basis energies α_s to the $2s_\mu$ and limiting the interactions β_{ss} to linked 2s AOs, they obtained a Hückel-type treatment, which yielded a rather satisfactory rationalization of the observed C2s band positions I_j^m. This suggested that in the absence of significant long-range interactions, an EBO model, limited to geminal interactions only, i.e. the AB model (including only A_j and B_{ij}), should prove to be a heuristically useful approximation [19, 23].

In the AB^0 version one would use $A_{CC}^0 = -17.50$ eV, $A_{CH}^0 = -16.95$ eV, $B_{CC,CC}^0 = B_{CC,CH}^0 = B_{CH,CH}^0 = -2.89$ eV. However, in view of the drastic simplification of the matrix, the differentiation between A_{CC}^0 and A_{CH}^0 becomes rather irrelevant. Thus, the only adequate approximation for such a model is to assume equal self-energies A for all EBOs and equal cross terms B for all geminal interactions. Thus, the AB model is now defined by the following matrix elements:

AB Model

$$A_{CC} = A_{CH} = A \,,$$

$$B_{CC,CC} = B_{CC,CH} = B_{CH,CH} = B \,, \tag{3.8}$$

All other cross terms (e.g. Γ_{ij}, etc.) $= 0 \,.$

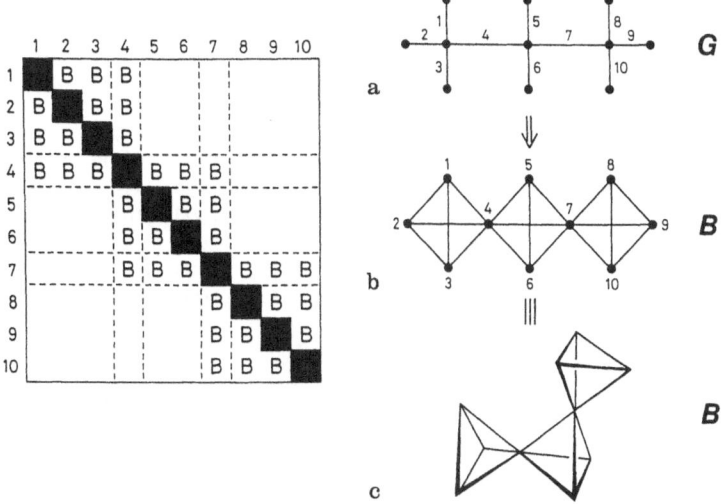

Fig. 6a–c. F_λ matrix of propane ($CH_3CH_2CH_3$) according to the AB model. **a)** Structure graph G of propane II(3); **b)** Line graph B of G; **c)** Three-dimensional representation of B

Under these circumstances the Hartree-Fock matrix F_λ of the AB model degenerates to that of a standard HMO treatment based on the line graph B of the structure graph G as shown for propane II(3) in Fig. 6. Each vertex of B corresponds to a bond in G, and each line in B to a geminal interaction. In other words, if A is the adjacency matrix $A = (a_{ij})$ of B — with $a_{ij} = 1$ if λ_j and λ_j are geminal, and $a_{ij} = 0$ otherwise — then F_λ of the AB model takes the form (cf. Fig. 6)

$$F_\lambda = A1 + BA. \tag{3.9}$$

Diagonalization [Eq. (2.20)] of F_λ yields the CMO eigenvalues ε_j. If these are written as

$$\varepsilon_j = A + x_j B, \tag{3.10}$$

the problem reduces to the search for the "spectrum" $x = \text{diag}(x_1 \ldots x_j \ldots x_N)$, i.e. for the eigenvalues x_j of the adjacency matrix A of B:

$$\det(A - x1) = 0. \tag{3.11}$$

The corresponding CMOs φ_j are given by Eqs. (2.21) and (2.22). As an example we present the results for propane II(3), the numbering of the EBOs λ_j being that given in Figs. 3 and 6a.

Before commenting on the results of Table 2, it is amusing to note that a three-dimensional representation of the graph B (Fig. 6b) corresponds to the classical van't Hoff model of this hydrocarbon, i.e. three tetrahedra, each pair joined by a

Table 2. CMOs φ_j of propane II(3), given as linear combinations of LMOs λ_μ

j	1	2	3	4	5	6	7	8	9	10
x_j	4.529	3.303	1.832	−.303	−.362	−1.000	−1.000	−1.000	−1.000	−1.000
μ 1	.218	.326	.253	.245	.298	.289	.224	.354	.354	.500
2	.266	.425	.390	−.565	−.527	0.000	0.000	0.000	0.000	0.000
3	.218	.326	.253	.245	.298	.289	.224	− .354	−.354	−.500
4	.503	.326	−.180	.245	.121	−.577	−.447	0.000	0.000	0.000
5	.285	0.000	−.433	0.000	−.177	0.000	.447	.500	−.500	0.000
6	.285	0.000	−.433	0.000	−.177	0.000	.447	−.500	.500	0.000
7	.503	−.326	−.180	−.246	.120	.577	−.447	0.000	0.000	0.000
8	.218	−.326	.253	−.246	.297	−.289	.223	.354	.354	−.500
9	.266	−.425	.390	.566	−.526	0.000	0.000	0.000	0.000	0.000
10	.218	−.326	.253	−.246	.297	−.289	.224	−.354	−.354	.500

common vertex as shown in Fig. 6c. In fact, using this analogy is the simplest way of setting up the graph B for more complicated saturated hydrocarbons [31]. Note that all such graphs B are strongly nonalternant and that their characteristic values x_j must occur in the interval $-6 \leqq x_j \leqq 6$, because the highest order of a vertex (i.e. the number of edges terminating at a given vertex) is 6, e.g. the vertices 4 and 7 in the graph B of propane (Fig. 6). With regard to Table 2 the following remarks have to be made:

1) The model clearly separates the CMOs corresponding to the C2s manifold from those of the C2p manifold. For the former we have $x_j > 0$, i.e. $\varepsilon_j < A$ and for the latter $x_j < 0$, i.e. $\varepsilon_j > A$.

2) The model is useless for the prediction (or better: the rationalization) of the I_j^m values of the C2p manifold because of the high, accidental degeneracy of the CMOs with $x_j = -1.000$.

3) The AB model is purely topological and thus independent of the conformation of the molecule.

4) The spectrum $\{x_j\}$ is related to the set of eigenvalues $\{\varepsilon_j\}$ collected in ε by a translation A and a common scale factor B. Thus, the AB model is Hückel type, enjoying all its advantages and disadvantages. In particular, the x_j can be matched by a simple linear regression, based on Eq. (3.10), with the corresponding $I_j^m = -\varepsilon_j$, assuming Koopmans' theorem. This allows a simple calibration of A and B, which yields [31]

$$I_j^m = [(15.83 \pm 0.07) + (2.17 \pm 0.03)\,x_j]\,\text{eV}, \tag{3.12}$$

$$r = 0.9961; \quad \phi = 56.$$

In (3.12), r is the correlation coefficient and ϕ the number of degrees of freedom on which the standard deviations of the slope and of the intercept at $x_j = 0$ depend.

The quality of agreement between the bonding x_j values derived from the AB model and the observed I_j^m values of the C2s manifold of a series of acyclic and cyclic saturated hydrocarbons is demonstrated by the comparison shown in Fig. 7.

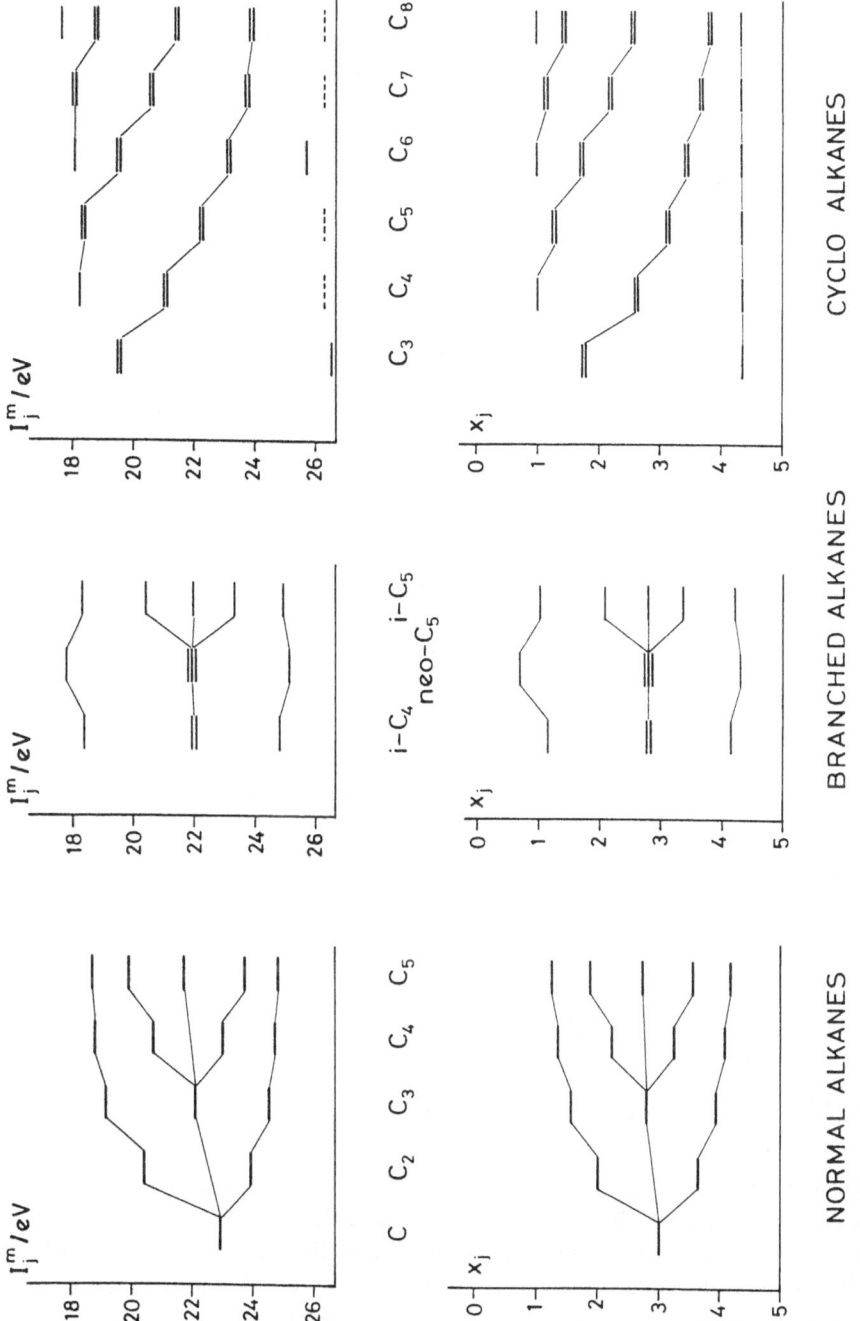

Fig. 7. Comparison of the observed band positions I_j^m of the C2s bands in the He(IIα) PE spectra of saturated hydrocarbons with the eigenvalues x_j of the corresponding graph **A** of the adjacency matrix **A** of the corresponding graph **B**. Normal alkanes $H(CH_2)_N H$ II(N) from methane (N = 1) to pentane (N = 5); branched alkanes isobutane III(3) (i-C$_4$) (i-C$_4$), neopentane III(4) (neo-C$_5$) and isopentane = $CH_3CH_2CH(CH_3)_2$ (i-C$_5$); cyclic alkanes IV(N), $(CH_2)_N$ from cyclopropane (N = 3) to cyclooctane (N = 8) [31]

A diagrammatic representation (3.13) of the lowest three CMOs of propane II(3) (see Table 2) reveals that the LMOs λ_j corresponding to the bonds issuing from a given C atom are either all in phase, or their local linear combination is dominated by the in-phase λ_j:

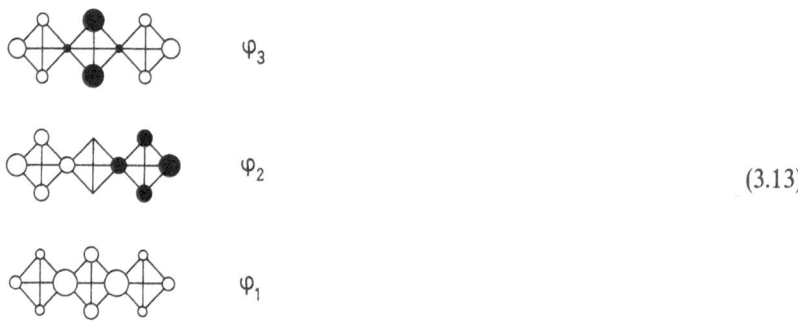

$$\varphi_3$$

$$\varphi_2 \tag{3.13}$$

$$\varphi_1$$

From the point of view of each individual C centre μ, the CMOs (3.13) are therefore rather close to local totally symmetric combinations

$$2s_\mu = (\lambda_r + \lambda_s + \lambda_t + \lambda_u)/2 \tag{3.14}$$

of the LMOs meeting at carbon atom μ, which for symmetry reasons are of 2s type, and in fact dominated by the 2s AO of the C-atom μ. Thus, in a crude approximation, the CMOs (3.13) could be represented as the linear combinations (3.15) of the 2s AOs $2s_\mu$ ($\mu = 1, 2, 3$), with $a^2 + b^2 = 1$:

$$\varphi_3 \approx (b/\sqrt{2})\,(2s_1 + 2s_3) - a2s_2\,,$$

$$\varphi_2 \approx (1/\sqrt{2})\,(2s_1 - 2s_3)\,, \tag{3.15}$$

$$\varphi_3 \approx (a/\sqrt{2})\,(2s_1 + 2s_3) + b2s_2\,.$$

This justifies the naive treatment proposed by Potts and Streets [33], where such orbitals have been obtained from a 2s-AO-based Hückel model.

A special case are the higher homologues of the alkanes II(N) with $N > \approx 13$, which yield PE spectra in which the C2s bands merge to form a double humped, unresolved band system in the interval $15\,\mathrm{eV} < I_j^m < 25\,\mathrm{eV}$ [36] (see Fig. 8). Application to the AB model of such alkanes of a previously proposed orbital treatment for systems consisting of repeated units [37, 38] yields a closed formula which gives the individual C2s-band positions I_j^m ($j = 1$ to N) as a function of N and j with sufficient accuracy. If the calculated I_j^m values are folded with an appropriate gaussian or lorentzian shape function the AB model reproduces the observed Franck-Condon envelopes of the C2s-band system within narrow limits of error as can be seen in Fig. 8.

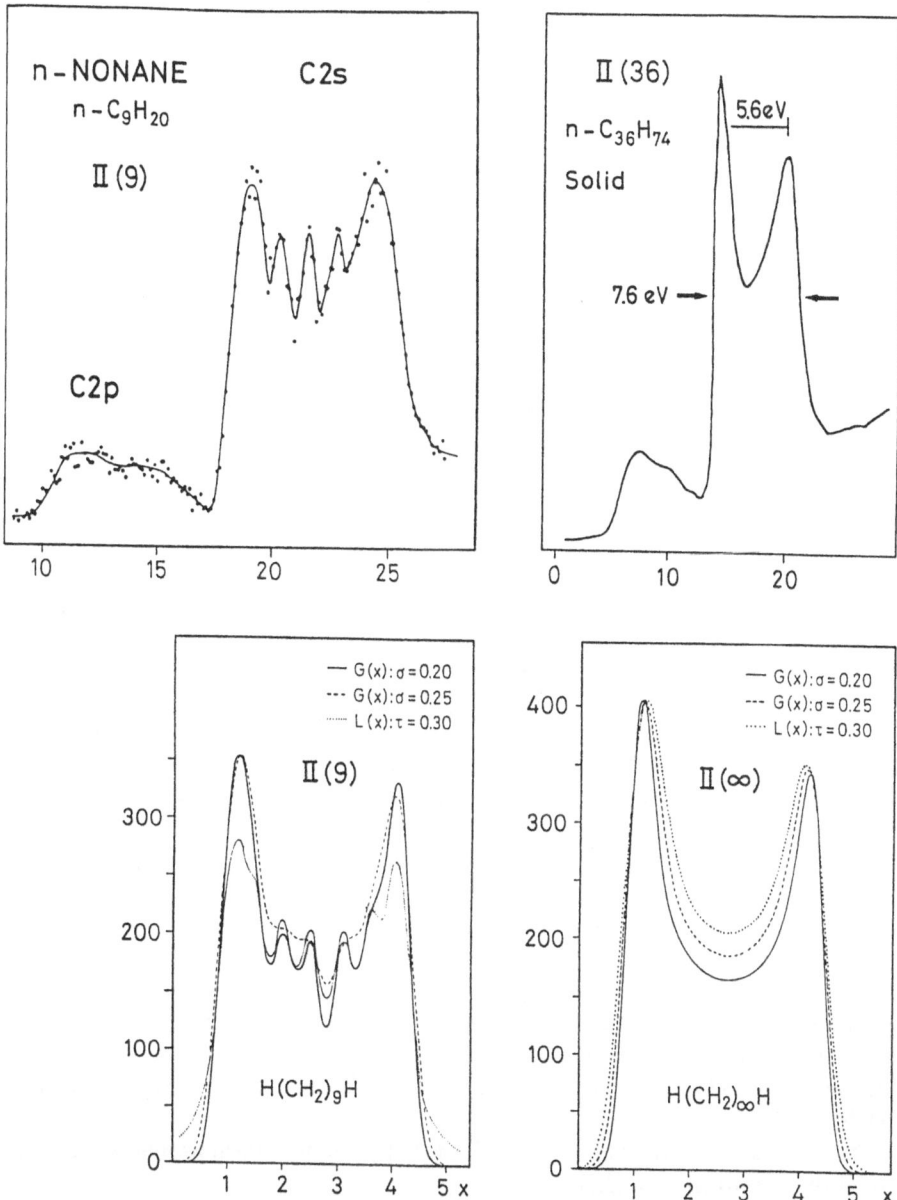

Fig. 8. *Top row*: Experimental PE spectra of nonane II(9) (C_9H_{20}) and hexatridecane II(36) ($C_{36}H_{74}$) taken from [36]. *Bottom row*: Calculated contours of the C2s band systems of nonane and polyethylene II(∞), $H(CH_2)_\infty H$ using the AB model and an empirical line shape function; $G(x)$ = gaussian line shape with full-width-at-half-height = 2σ, $L(x)$ = lorentzian line shape with full-width-at-half-height = 2τ. The values quoted for σ and τ are in units of x

3.4 The C2p Manifold and the *AΓ* Model

As is obvious from the example presented in Table 2, the *AB* model is useless for a rationalization of the observed band positions I_j^m in the C2p band system of hydrocarbons, because of the high accidental degeneracy of the corresponding

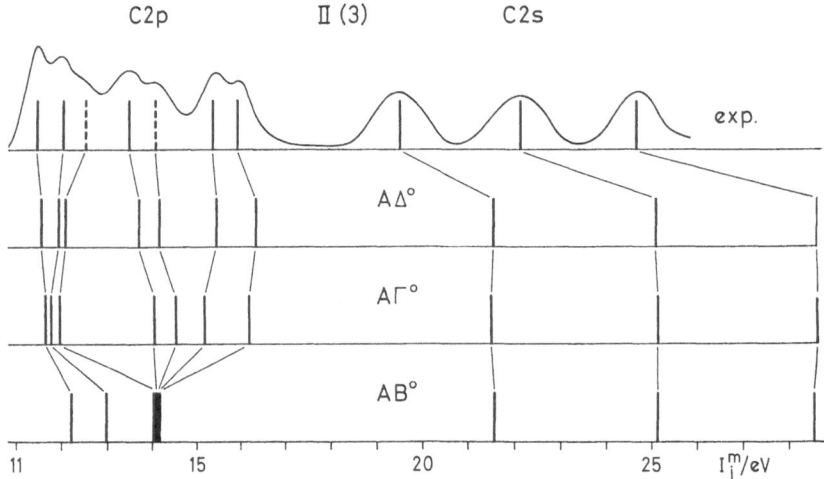

Fig. 9. Comparison of the He(IIα) PE spectrum of propane II(3) (redrawn from [32]) with the orbital energies stemming from an AB^0, $A\Gamma^0$, and $A\Delta^0$-EBO model of propane, assuming the validity of Koopmans' theorem

eigenvalues x_j, e.g. $x_6 = x_7 = x_8 = x_9 = x_{10} = -1.000$ in the case of propane II(3). To lift the degeneracy two ways are open, namely (a) to use different self-energies A_{CC} and A_{CH} for the LMOs λ_{CC} and λ_{CH}, or/and (b) to include higher cross terms Γ_{ij}, Δ_{ij}, etc. (see classification scheme on p. 11). The question is, what are the minimum requirements for a heuristically useful model? The answer is provided by Fig. 9, which shows the influence of the inclusion of Γ_{ij} and Δ_{ij} cross terms on the predictions derived from the $AB^0 \rightarrow A\Gamma^0 \rightarrow A\Delta^0$ model of propane II(3). As can be seen, the inclusion of the interaction matrix elements between vicinal LMOs splits the five degenerate eigenvalues $\varepsilon_j = -14.1$ eV of the AB^0 model into nondegenerate eigenvalues which yield a fair rationalization of the C2p-band system according to Koopmans' theorem. The further improvement resulting from the inclusion of Δ_{ij} cross terms is marginal and hardly worthwhile from a qualitative (or even semi-quantitative) point of view. Note that the orbital sequence predicted by the $A\Gamma^0$ and $A\Delta^0$ models are the same. It must be pointed out, however, that Δ_{ij} cross terms become important in compact, polycyclic hydrocarbons, containing small rings (see below).

The $A\Gamma^0$ model, which mimics STO-3G results, is now defined as follows, making use of the approximation (3.1):

AΓ model

$$A^0_{CC} = -17.50 \text{ eV},$$

$$A^0_{CH} = -16.95 \text{ eV},$$

$$B^0_{CC,CC} = B^0_{CC,CH} = B^0_{CH,CH} = -2.89 \text{ eV}, \qquad (3.16)$$

$$\Gamma^0_{CC,CC} = \Gamma^0_{CC,CH} = \Gamma^0_{CH,CH} = \Gamma^0 \cos \tau_{ij},$$

$$\Gamma^0 = -1.00 \text{ eV}.$$

All other cross terms = 0.

To obtain the $A\Gamma$ model from Eqs. (3.16), we use the calibration procedure outlined in Eqs. (3.2) to (3.7) and summarized in the EBO model for C_nH_m on p. 15, which yields, e.g. on the basis of the calibration set mentioned in the legend to Fig. 4 [cf. Eq. (3.7)]

$$\delta A = -3.10 \text{ eV} - (0.25/b)(b_{CC}A_{CC}^0 + b_{CH}A_{CH}^0),$$
$$f = 0.750. \tag{3.17}$$

Both the $A\Gamma^0$ and the $A\Gamma$ models defined in Eqs. (3.16) and (3.17) assume that the geminal interaction terms B_{ij} between CC and/or CH LMOs have the fixed value $B_{ij}^0 = -2.9 \text{ eV}$, which is only true if all bond angles θ are tetrahedral, i.e. $\theta = 109° 27'$. The same assumptions is implied in the value $\Gamma^0 = -1.0 \text{ eV}$ determining the vicinal cross terms Γ_{ij}^0. If θ deviates from $109° 27'$, e.g. in cyclobutane IV(4) or cyclopropane IV(3), then B_{ij}^0 and Γ_{ij}^0 as well as the self-energies A_j^0 deviate from the standard values given in Eqs. (3.16), as exemplified for A_{CC}^0, $B_{CC,CC}^0$, and $\Gamma_{CC,CC}^0$, of the cycloalkanes IV(N) in Table 3, assuming planar conformations, i.e. $\tau_{CC,CC} = 0°$:

Table 3. Dependence of the self-energies A_{CC}^0 and of the cross terms $B_{CC,CC}^0$ and $\Gamma_{CC,CC}^0$ on the bond angle θ

Cycloalkane	θ	A_{CC}^0/eV	$B_{CC,CC}^0/eV$	$\Gamma_{CC,CC}^0/eV$
IV(3) Cyclopropane	60°	−16.9	−4.8	
IV(4) Cyclobutane	90°	−17.2	−3.5	−2.0
IV(5) Cyclopentane	108°	−17.5	−3.0	−1.0

3.5 Application of the EBO Model to Polycyclic Hydrocarbons

To illustrate the application of the EBO model (presented in Sect. 3.4) for the interpretation of PE band systems due to electron ejection from CMOs of the C2p manifold, we discuss the PE spectra of saturated, polycyclic hydrocarbons. We begin, however, by applying the model to cyclopropane IV(3).

The chemical and physical properties of cyclopropane and of cyclopropyl groups (see below) differ significantly from those of other cycloalkanes or (cyclo)alkyl groups, e.g. by the almost double-bond like behavior of a cyclopropane ring in conjugation with a π system [39], or the reactive behavior of the parent hydrocarbon [40]. Consequently, the description of the electronic structure of cyclopropane IV(3) should allow a rationalization of those properties. In 1947, Walsh proposed the set of three bonding MOs, based on tangential 2p AOs and radial hybrid AOs, which now carry his name [41], whereas Coulson and Moffitt described in the same year the electronic structure of cyclopropane in terms of three bent "banana" bonds [42] in analogy to an earlier proposal by Förster [43].

If the CMOs φ_j of cyclopropane are transformed according to Eq. (2.15) into LMOs λ_j, one obtains, apart from the six LMOs λ_{CH}, three bent "banana" LMOs $\lambda_{CC,1}$,

$\lambda_{CC, 2}$, and $\lambda_{CC, 3}$. These yield, according to Eq. (2.25), the three (symmetry-adapted under D_{3h}) SLMOs ϱ_0, ϱ_s, and ϱ_A, defined as

$$(\varrho_0, \varrho_S, \varrho_A) = (\lambda_{CC, 1}, \lambda_{CC, 2}, \lambda_{CC, 3}) \begin{bmatrix} 1/\sqrt{3} & 2/\sqrt{6} & 0 \\ 1/\sqrt{3} & -1/\sqrt{6} & 1/\sqrt{2} \\ 1/\sqrt{3} & -1/\sqrt{6} & -1/\sqrt{2} \end{bmatrix}. \qquad (3.18)$$

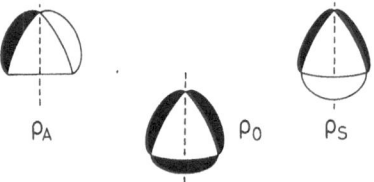

Of these "Förster-Coulson-Moffitt" (FCM) SLMOs, ϱ_0 belongs to the irreducible representation A_1 of D_{3h}, whereas ϱ_S and ϱ_A span the degenerate representation E' (see [44, 45] and references therein).

In contrast to a widespread opinion, the bonding Walsh orbitals are not equivalent to the FCM SLMOs. In fact, there exists no unitary transformation of the doubly occupied CMOs φ_j of cyclopropane, which would project out the traditional Walsh orbitals. In contrast, the FCM SLMOs are accessible via the LMOs as indicated above [45, 46]. The FCM SLMOs are therefore a much better and safer starting set for the qualitative or semiquantitative interpretation of the electronic properties of cyclopropane or, in particular, of its PE spectrum [47].

The highest occupied σ CMOs of benzene are the e_{2g} orbitals (in D_{6h} symmetry) depicted on the left in the notation of Lindholm [48]. These CMOs, which have been called "ribbon orbitals" [49], are not peculiar to benzene but recur in any hydrocarbon with C_6 carbon cycles, e.g. as the e_g CMOs (in D_{3d}) of cyclohexane VI(6), shown in (3.19) or in polycyclic hydrocarbons such as *trans-trans*-perhydroanthracene I (see Fig. 2) or perhydropyrene V (see Fig. 10) [29], where these "ribbon orbitals" are represented schematically as linear combinations of LMOs λ_j, obtained via the $A\Gamma$ model.

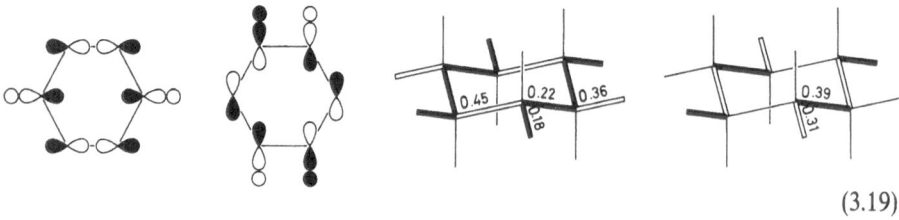

$$(3.19)$$

Calculations by the $A\Gamma$ model suggest that the two frontier "ribbon orbitals" φ_{HOMO}, φ_{HOMO-1} of polycyclic hydrocarbons consisting of all-*trans* connected six-membered rings are well separated in energy from the remaining manifold of σ orbitals by a gap which increases with increasing size of the hydrocarbon. Assuming the

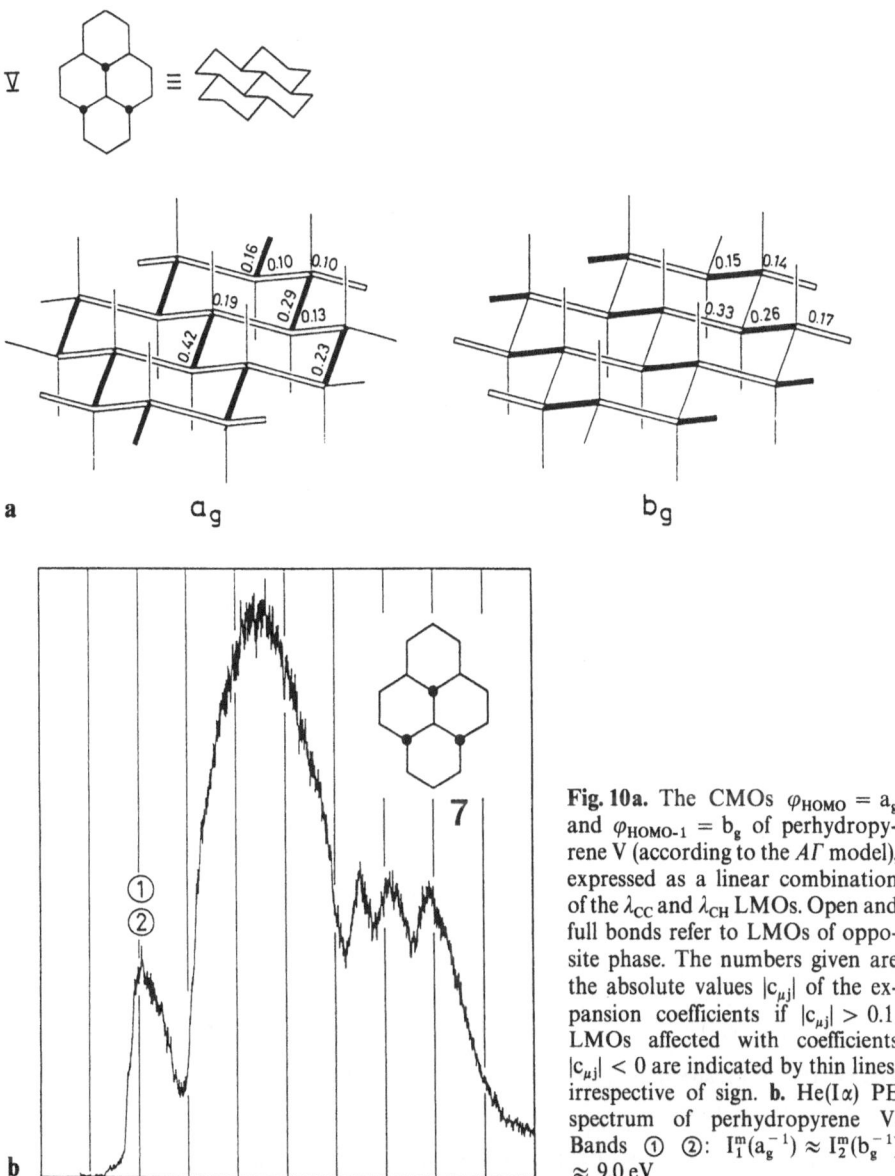

Fig. 10a. The CMOs $\varphi_{HOMO} = a_g$ and $\varphi_{HOMO-1} = b_g$ of perhydropyrene V (according to the $A\Gamma$ model), expressed as a linear combination of the λ_{CC} and λ_{CH} LMOs. Open and full bonds refer to LMOs of opposite phase. The numbers given are the absolute values $|c_{\mu j}|$ of the expansion coefficients if $|c_{\mu j}| > 0.1$. LMOs affected with coefficients $|c_{\mu j}| < 0$ are indicated by thin lines, irrespective of sign. **b.** He($I\alpha$) PE spectrum of perhydropyrene V. Bands ① ②: $I_1^m(a_g^{-1}) \approx I_2^m(b_g^{-1}) \approx 9.0$ eV

validity of Koopmans' theorem, a PE spectroscopic study of such polycyclic hydrocarbons has shown that this is indeed the case [29]. As an example, Fig. 10 shows the PE spectrum of perhydropyrene V in which the double band ①, ②, associated with the φ_{HOMO}^{-1} and φ_{HOMO-1}^{-1} ionization processes is well separated by ≈ 1 eV from the manifold of the other σ bands. According to the $A\Gamma$ model these ribbon orbitals are evenly delocalized over the whole of the molecular frame and they conserve the characteristic phase relationship postulated previously [49] for cyclohexane moieties. Presumably, these high-lying frontier orbitals play an important role in "σ conjugation", as defined by Dewar [50]. In this connection it should be mentioned that the simple pattern of the σ^{-1} ionization energies of aromatic, catacondensed hydrocarbons [51] is also well reproduced by the $A\Gamma$ model [29].

The PE spectra of the polycyclic hydrocarbons containing three- and/or four-membered rings [52], e.g. cubane VI [53], pentaprismane VII [54], and tetrahedrane VIII [55], are of interest, because of the high symmetry of these molecules and because of the small CCC angles which lead to a noticeable increase of the geminal and vicinal interaction terms between their LMOs λ_{CC} (see Table 3).

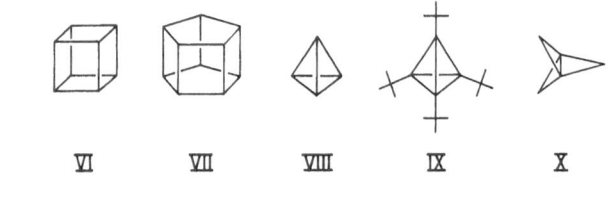

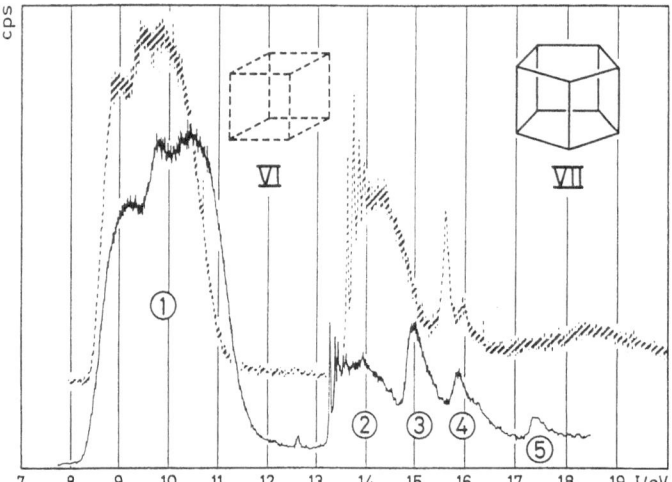

Fig. 11. He(Iα) PE spectra of pentaprismane VII (solid line) and of cubane VI (dashed line)

In Fig. 11 are shown the PE spectra of cubane VI [53] and of pentaprismane VII [54], which differ from those of other hydrocarbons by the extremely large gap of ≈ 3 eV between the first bands corresponding to the degenerate CMOs (t_{2g}, t_{2u} of cubane, e_2'', e_2', e_1'', e_1' of pentaprismane) and the band system extending from ≈ 13.5 eV towards higher ionization energies. This peculiar spacing of the CMO energies can be explained with reference to the very large value of the Γ_{ij} term between the two vicinal LMOs λ_{CC} opposed to each other within a cyclobutane moiety, coupled with the small value of Γ_{ij} if the vicinal λ_{CC} belong to two different rings. As an example, the matrix elements for the Hartree-Fock matrix $\mathbf{F}_\lambda$ of pentaprismane are given in Fig. 12.

Whereas the PE spectrum of the parent hydrocarbon tetrahedrane VIII is still unknown, its tetra-t-butyl derivative IX has been studied in detail [55]. Using the

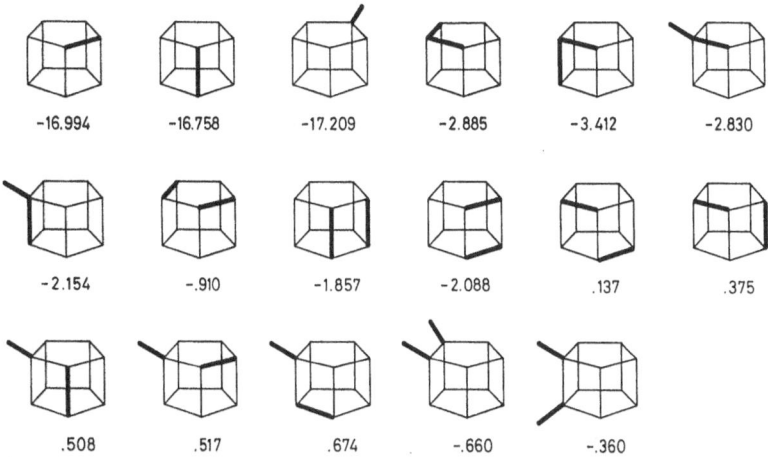

Fig. 12. Matrix elements in eV of the $\mathbf{F}_\lambda$ matrix, according to the STO-3G model of pentaprismane VII, in localized basis. (The values given refer to the self-energy of the localized orbital λ_j (LMO) indicated by a heavy line in the first three formulae, or to the cross term between the two LMOs λ_i, λ_j indicated. All other values are given by symmetry)

symbolic representation of the interaction terms defined in Fig. 12, one obtains from an STO-3G calculation of the parent hydrocarbon the following values:

$$
\begin{array}{cccc}
\text{A} & \boxtimes & \boxtimes & \\
 & -16.8\text{ eV} & -18.2\text{ eV} & \\
\\
\text{B} & \boxtimes & \boxtimes & \\
 & -4.5_5\text{ eV} & -1.9\text{ eV} & \\
\\
\Gamma & \boxtimes & \boxtimes & \boxtimes \\
 & 0.0\text{ eV} & +0.8\text{ eV} & -0.1_5\text{ eV}
\end{array} \tag{3.20}
$$

The remarkable feature is that the vicinal interaction between two LMOs λ_{CC} in the tetrahedrane skeleton vanishes, so that the spacing of the symmetry-adapted linear combinations of the CC LMOs is entirely due to the large geminal terms $B_{CC,CC} = -4.5_5$ eV.

The He(Iα) PE spectrum of [1.1.1]propellane X (Fig. 13) [56] agrees qualitatively and quantitatively with the predicted orbital sequence HOMO = $3a'_1$, $1e''$, $3e'$, $1a'_2$, $2e'$. Because of the extreme deviation from a tetrahedral situation of the bridge-head atoms, ab-initio treatments using an extended polarized basis [57] or treatments of DZ + P quality [58] have to be used. However, the resulting CMOs φ_j can again be expressed in terms of LMOs λ_j, as shown in Fig. 13. The Franck-Condon shape of the first band in the PE spectrum of [1.1.1]propellane X and, in particular, its narrow width at half-height indicate that both the parent molecule X and the radical

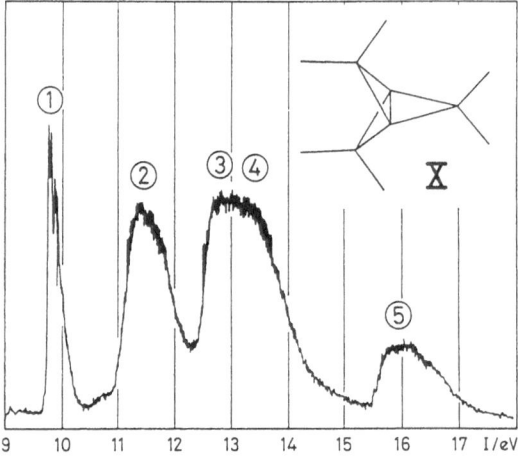

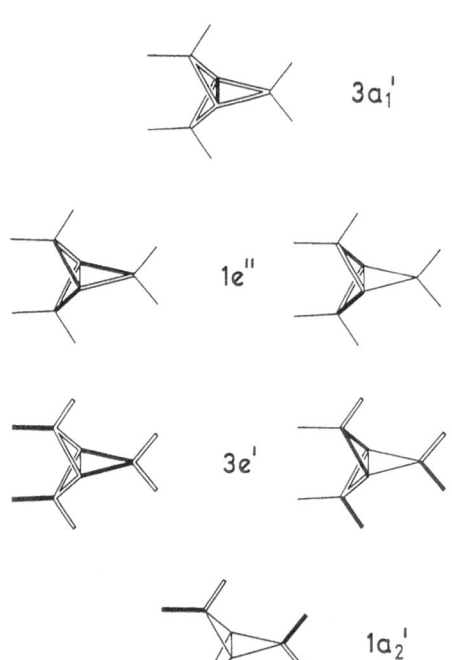

$3a_1'$

$1e''$

$3e'$

Fig. 13. He(Iα) PE spectrum of [1.1.1]propellane X.

Band	I_j^m/eV	CMO
①	9.74	$3a_1'$
②	11.3_5	$1e''$
③	12.6_0	$3e'$
④	(13.4_5)	$1a_2'$
⑤	15.7 (16.1)	$2e'$

The numbering of the symmetry labels within each irreducible representation refers to the valence-shell orbitals only. The diagrams of the CMOs φ_j show qualitatively the phase relationships between the LMOs

$1a_2'$

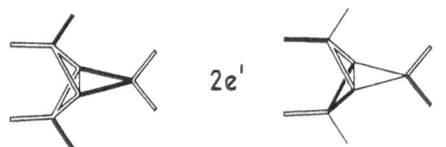

$2e'$

cation X^+ in its electronic ground state $^2A_1'$ differ very little in geometry. This can be interpreted, in qualitative orbital language, as a strong indication that $\varphi_{HOMO} = 3a_1'$ is essentially nonbonding [59].

4 Unsaturated Hydrocarbons

4.1 The LMOs of Ethene and Ethine

If the six valence-shell CMOs φ_j of ethene XI(1) are subjected to a localization transformation (Eq. (2.22)), e.g. by using the Foster-Boys [17] or the Edmiston-Ruedenberg [18, 28] procedure, one obtains two "banana" CC LMOs $\lambda_1, \lambda_2 = \lambda_b$, and four CH LMOs λ_3 to $\lambda_6 = \lambda_{CH}$:

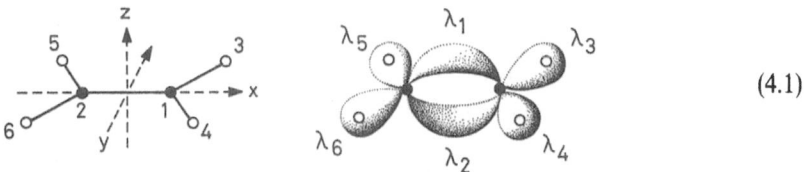

$$(4.1)$$

As shown in Fig. 14, the name "banana" LMO is due to the fact that the line of maximum orbital density is bent away from the line joining the two C centres. The out-of-phase linear combination ϱ_- of λ_1 and λ_2 yields the two-centre π CMO belonging to the irreducible representation B_{1u} of D_{2h} [see also (4.1) and Fig. 14]:

$$\varrho_- \equiv \pi = (\lambda_1 - \lambda_2)/\sqrt{2},$$
$$\varrho_+ \equiv \sigma = (\lambda_1 + \lambda_2)/\sqrt{2}.$$

$$(4.2)$$

The two-centre π orbital $\varrho_- \equiv \pi$ is therefore orthogonal to all other CMOs of ethene XI(1), in particular to the in-phase combination ϱ_+ of λ_1 and λ_2 given in Eqs. (4.2), which represents the two-centre σ orbital of the ethene double bond belonging to A_g of D_{2h}. If, for the reasons given in Sect. 3.1, the STO-3G procedure is used for the calculation of the ethene CMOs φ_j, the following matrix elements $F_{\lambda,ij}$ of the Hartree-Fock matrix F_λ in localized basis are obtained for ethene, the numbering of the LMOs λ_j being defined in diagram (4.1):

$$A_b^0 = A_1^0 = A_2^0 = -15.54 \text{ eV},$$

$$A_{CH}^0 = A_3^0 \text{ to } A_6^0 = -17.47 \text{ eV},$$

$$B_{bb}^0 = B_{1,2}^0 = -6.76 \text{ eV},$$

$$B_{b,CH}^0 = B_{1,4}^0 = B_{1,3}^0 \text{ etc.} = -2.38 \text{ eV},$$

$$(4.3)$$

$$B_{CH,CH}^0 = B_{3,4}^0 = B_{5,6}^0 = -2.83 \text{ eV},$$

$$\Gamma_{CH,CH}^0(Z) = \Gamma_{3,5}^0 = \Gamma_{4,6}^0 = -0.77 \text{ eV},$$

$$\Gamma_{CH,CH}^0(E) = \Gamma_{3,6}^0 = \Gamma_{4,5}^0 = +1.14 \text{ eV}.$$

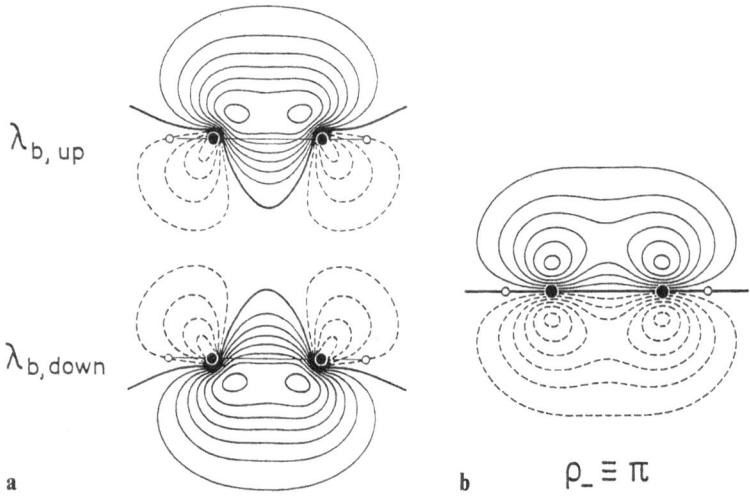

Fig. 14a. The "banana" LMOs $\lambda_{b,up}$ and $\lambda_{b,down}$ of the ethene XI(1) double bond, as obtained from an STO-3G calculation through the Foster-Boys localization procedure. **b.** The out-of-phase combination $\varrho_- = (\lambda_{b,up} - \lambda_{b,down})/\sqrt{2} \equiv \pi$ (cf. Eq. (4.2))

To apply the AB model to ethene XI(1) (and to other unsaturated hydrocarbons containing double bonds) one has to amplify Eqs. (3.8) by taking care of the significant differences between the self-energy A_b of the "banana" LMOs λ_b and the self-energies $A_{CC} \approx A_{CH}$, as well as between the cross term B_{bb} and the remaining ones, i.e. $B_{CC,CC} \approx B_{CC,CH} \approx B_{CH,CH} \approx B_{b,CH} \approx B_{b,CC}$. In a first approximation

---

AB *model* (including double bonds)

$A_{CC} = A_{CH} = A$,

$A_b = A - B$,   (4.4)

$B_{CC,CC} = B_{CC,CH} = B_{CH,CH} = B_{CC,b} = B_{CH,b} = B$,

$B_{bb} = 2B$.

As in the case of the saturated hydrocarbons (see Fig. 6), the AB model defined by Eqs. (4.4) leads to a line-graph B of the structure graph G, a three-dimensional representation of which corresponds again to the van't Hoff model of the molecule, i.e. to two tetrahedra joined by a common edge to represent ethene XI(1) (see Fig. 15). In contrast to the graphs B of the AB model defined in Eqs. (3.8), the graph B of the AB model according to Eqs. (4.4) is now a weighted graph, which means that $-B$ has to be added to those diagonal elements of $\mathbf{F}_\lambda$, as defined in Eqs. (3.19), which refer to a "banana" orbital λ_b, and that the cross term between two such orbitals λ_b spanning the same centres is now $2B$ instead of B. With reference to Eq. (3.11), the adjacency matrix $\mathbf{A} = (a_{ij})$ of the graph B has to be amended by setting $a_{ii} = -1$ for each vertex i corresponding to a "banana" LMO $\lambda_i = \lambda_b$ and $a_{ij} = 2$ for an edge between them.

In Fig. 15 the result of the diagonalization of the amplified adjacency matrix **A** for the *AB* model (4.4) of ethene is presented.

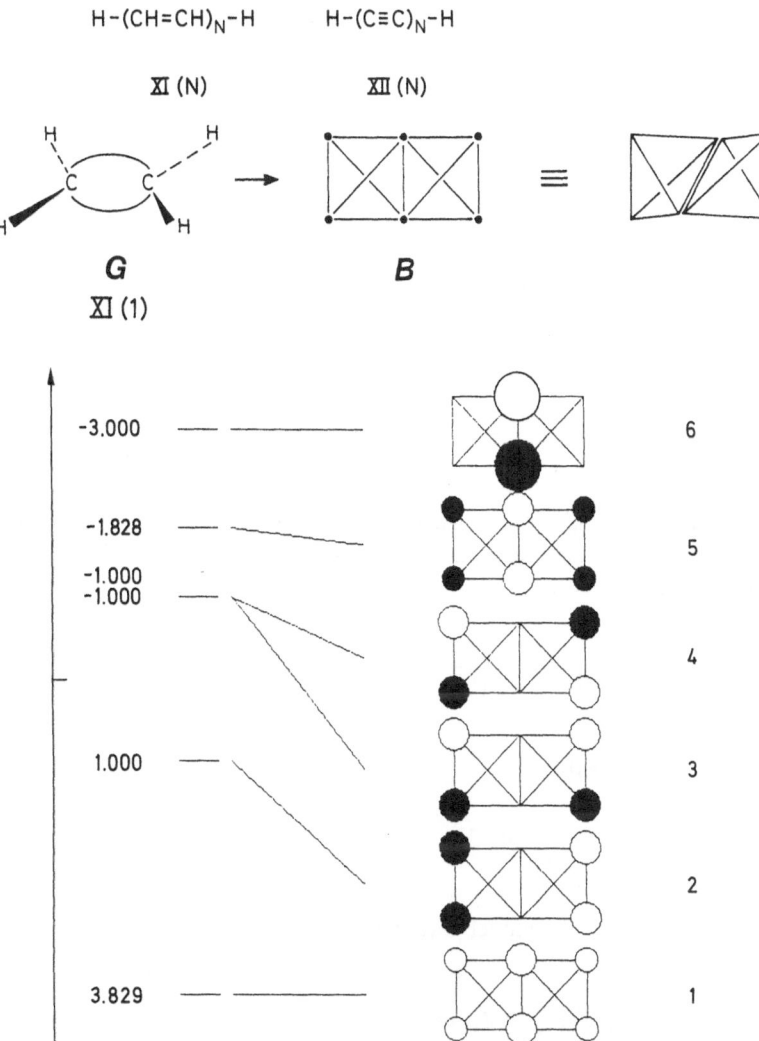

Fig. 15. *Top*: The structure graph *G* and the corresponding line-graph *B* of ethene XI(1). *Bottom*: The eigenvalues x_j and the LMO-based CMO diagrams, obtained from the augmented adjacency matrix **A** of *B* (see text). The experimental I_j^m values of ethene and their assignment are [33, 34, 60]:

I_j^m/eV	CMO	Type
10.5	$1b_{1u}$	π
12.9	$1b_{1g}$	CH_2
14.7	$3a_g$	$\sigma(CC)$
15.9	$1b_{2u}$	CH_2
19.1	$2b_{3u}$	$C2s_-$
23.7	$2a_g$	$C2s_+$

Note that the *AB* model reverses the sequence $1b_{1g}$ above $3a_g$

If the five valence-shell CMOs φ_j of ethine XII(1) are subjected to the same transformation [Eq. (2.22)] as described above for those of ethene, one obtains three CC "banana" LMOs $\lambda_1, \lambda_2, \lambda_3 = \lambda_b$ related to each other by a $C_3(z)$ rotation about the CC z-axis, yielding the structure graph G shown to the left in (4.5). The three-dimensional representation of the corresponding line graph B is the van't Hoff model of ethine, i.e. two tetrahedra sharing a common face:

$$\tag{4.5}$$

$$G \qquad\qquad\qquad B$$

From the LMOs λ_1, λ_2, and λ_3 one constructs the $D_{\infty h}$ symmetry-adapted SLMOs ϱ_1, ϱ_2, ϱ_3, of which two correspond to the π orbitals, and one to the σ orbital of the triple bond. In real representation:

$$\pi_x \equiv \varrho_1 = (2\lambda_1 - \lambda_2 - \lambda_3)/\sqrt{6},$$

$$\pi_y \equiv \varrho_2 = (\lambda_2 - \lambda_3)/\sqrt{2}, \tag{4.6}$$

$$\sigma \equiv \varrho_3 = (\lambda_1 + \lambda_2 + \lambda_3)/\sqrt{3}.$$

In a first approximation the AB model parameters (4.4) may be used.

4.2 Monoenes and Conjugated Polyenes

Monoenes and conjugated polyenes are ideal molecules for PE spectroscopic investigations, because their low-energy π^{-1} ionizations yield well-defined, fine-structured PE bands (cf. Fig. 16). In a first approximation the π CMOs of unsubstituted polyenes can be obtained by using a different type of AB model, namely one based on the linear combination of two-center π orbitals [Eqs. (4.2)], e.g. for the butadiene XI(2) π system:

$$\pi_- = (\pi_a - \pi_b)/\sqrt{2}; \quad b_g,$$
$$\pi_+ = (\pi_a + \pi_b)/\sqrt{2}; \quad a_u. \tag{4.7}$$

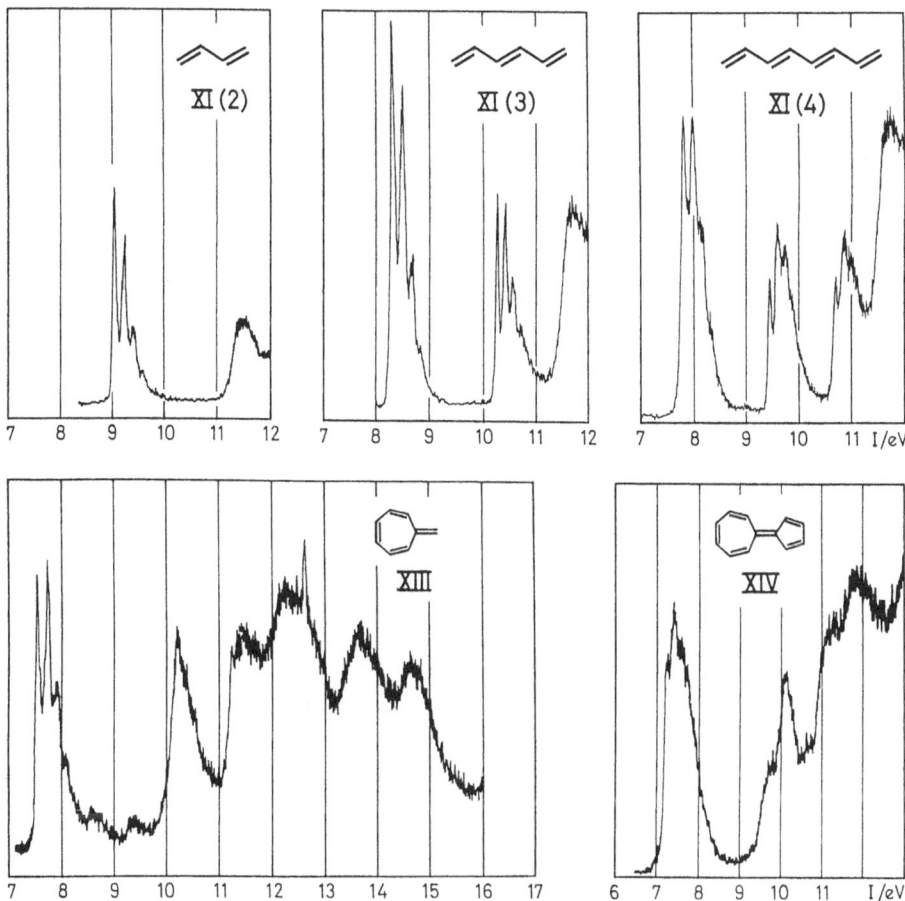

Fig. 16. The He(Iα) PE spectra of 1,3-butadiene XI(2), 1,3,5-hexatriene XI(3), 1,3,5,7-octatetraene XI(4) [65], heptafulvene XIII, and sesquifulvalene XIV

Compd.	I_1/eV	I_2/eV	I_3/eV	I_4/eV
XII(2)	9.03 (1b$_g$)	11.46 (1a$_u$)		
XII(3)	8.29 (2a$_u$)	10.26 (1b$_g$)	11.9 (1a$_u$)	
XII(4)	7.79 (2b$_g$)	9.61 (2a$_u$)	10.89 (1b$_g$)	11.7 (1a$_u$)
XIII	7.69 (3b$_1$)	10.22 (1a$_2$)	11.24 (2b$_1$)	
XIV	7.40 (4b$_1$)	7.6 (2a$_2$)	9.8 (1a$_2$)	10.2 (3b$_1$)

The higher energy states of XIII and XIV involve appreciable configuration interaction with non-Koopmans configuration, which is not taken care of by our EBO model

The corresponding F_λ matrix is again of the form Eq. (3.9), **A** being the adjacency matrix of a graph B in which the vertices now refer to a basis consisting of two centre π LMOs, and the edges to their conjugative interaction cross terms (e.g. for the molecules 1,3-butadiene **XI(2)**, 1,3,5-hexatriene **XI(3)**, 1,3,5,7-octatetraene **XI(4)**, heptafulvene XIII and sesquifulvalene XIV shown in Fig. 16):

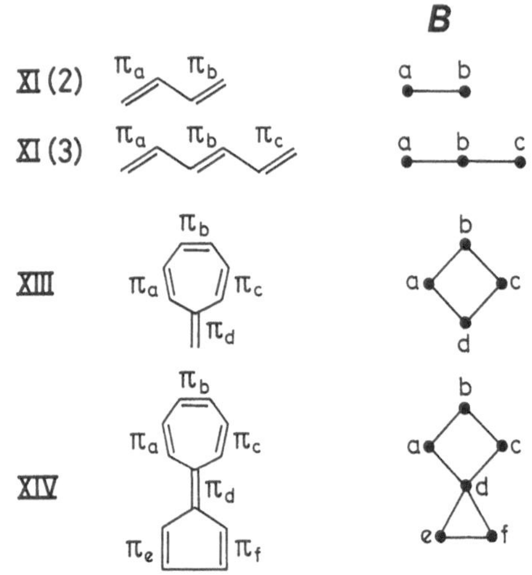

The parameters A_π and $B_{\pi,\pi}$ to be used in Eq. (3.9) are easily obtained from the known PE spectra of such π systems, e.g. ethene XI(1) [60, 61, 62], butadiene XI(2) [60, 63, 64], hexatriene XI(3) [64, 65], fulvene XV [66] and other unsubstituted polyenes. Depending on the calibration set, one obtains roughly $A_\pi \approx -10.2\ \mathrm{eV}$ and $B_{\pi\pi} \approx -1.2\ \mathrm{eV}$.

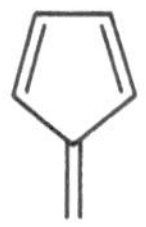

To derive a more refined EBO model which can be used in the general case, it is of advantage to proceed as shown with reference to the following (restricted) calibration set (Me = methyl group):

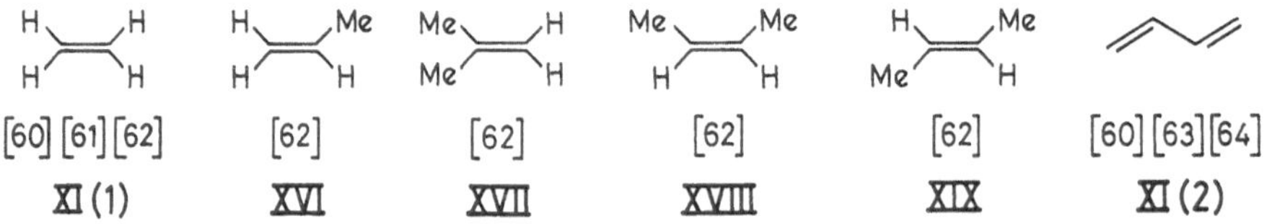

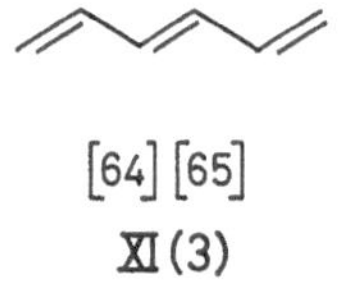

Localization of the CMOs φ_j obtained from an ab-initio treatment (e.g. the STO-3G model) yields the LMOs λ_j of the molecule, with two "banana" LMOs λ_b for each double bond. From the latter we form, in analogy to Eqs. (4.2) the linear combinations $\varrho_- \equiv \pi = (\lambda_{b,up} - \lambda_{b,down})/\sqrt{2}$ and $\varrho_+ \equiv \sigma = (\lambda_{b,up} + \lambda_{b,down})/\sqrt{2}$, the self-energies of which are

$$\varrho_- \equiv \pi: \quad A_\pi = A_b - B_{bb},$$
$$\varrho_+ \equiv \sigma: \quad A_\sigma = A_b + B_{bb}. \tag{4.8}$$

Assuming that the local conformation of the methyl groups is the preferred one, i.e. such that one of the three CH bonds, say bond 3, is eclipsed with the CC double bond, we form the following linear combinations of the three methyl-group LMOs $\lambda_{CH,1}$, $\lambda_{CH,2}$, and $\lambda_{CH,3}$, with respect to the local C_3 symmetry

$$\pi(Me) = (\lambda_{CH,1} - \lambda_{CH,2})/\sqrt{2},$$

$$\sigma_1(Me) = (\lambda_{CH,1} + \lambda_{CH,2} - 2\lambda_{CH,3})/\sqrt{6}, \tag{4.9}$$

$$\sigma_0(Me) = (\lambda_{CH,1} + \lambda_{CH,2} + \lambda_{CH,3})/\sqrt{3}.$$

Of these, $\pi(Me)$ has the proper symmetry to interact (hyperconjugate) with π [cf. Eqs. (4.2) and (4.8)], whereas the other two, $\sigma_1(Me)$ and $\sigma_0(Me)$, will mix with the remaining σ LMOs. Restricting ourselves to only those ionization energies I_j^m of the calibration set which refer to σ^{-1} ionization processes, one finds that the linear regression $(I_j^m/eV) = p + q(-\varepsilon_j^{STO-3G})$ yields $p = 3.36 \pm 0.22$ and $q = 0.750 \pm 0.014$ with standard error $s(I_j^m) = 0.28$ eV and a correlation coefficient $r = 0.992$. (These values are not significantly different from those obtained for saturated hydrocarbons; see Sect. 3.4). If this regression is used for the calibration of an $A\Gamma$ model yielding only the σ CMO energies, the following parameters are found [66]:

$$A_\sigma = -21.3 \text{ eV},$$

$$A_{CC} = -17.0 \text{ eV}, \qquad A_{CH} = -16.4 \text{ eV},$$

$$A_{CC} = -18.1 \text{ eV},$$

cf. (4.9) $A_{Me(\sigma_0)} = -20.4 \text{ eV}, \qquad A_{Me(\sigma_1)} = -14.0 \text{ eV}; \tag{4.10}$

$$B_{\sigma,CC} = B_{\sigma,CH} = -2.5_5 \text{ eV},$$

$$B_{CC,CC}(gem.) = B_{CH,CH}(gem.) = B_{CC,CH}(gem.) = -2.1 \text{ eV},$$

$$\Gamma_{CC,CC} = \Gamma_{CC,CH} = \Gamma_{CH,CH} = -0.5_5 \text{ eV},$$

$$\Gamma_{CC,CC} = \Gamma_{CC,CH} = \Gamma_{CH,CH} = 0.8_5 \text{ eV}.$$

As long as the π system of an unsaturated hydrocarbon is planar, the treatment of the π orbitals within the $A\Gamma$ model is rather similar to the AB model. With respect to the calibration set, the only new parameters are the self-energy $A_\pi(\text{Me})$ of the methyl group and the interaction cross term $B_{\pi,\text{Me}}$ for the hyperconjugative interaction of π [cf. Eqs. (4.2) and (4.8)] with $\pi(\text{Me})$ [cf. Eq. (4.9)]. As $\pi(\text{Me})$ and $\sigma_1(\text{Me})$, defined in Eq. (4.9) are degenerate, it follows from (4.10) that $A_\pi(\text{Me}) = -14.0\,\text{eV}$. The other parameters are calibrated again, using the analogous procedure to the one outlined above for the σ orbitals. However, before we present the results, attention should be drawn to an important special feature of the EBO model, which is quite general, but best illustrated in the context of the particular case of π systems.

Substitution of a π system, e.g. that of ethene XI(1), by methyl groups, lowers the π^{-1} ionization energy $I_1^m(\pi)$ [60, 62] (Table 4).

Table 4. π^{-1} ionization energies of ethene XI(1) and of methyl substituted ethenes

Alkene		$I_1^m(\pi)/\text{eV}$	$\Delta I_1^m(\pi)/\text{eV}$
XI(1)	$H_2C = CH_2$	10.52	
XVI	$H_2C = CHMe$	9.73	-0.79
XVII	$H_2C = CMe_2$	9.23	-1.29
XVIII	$MeHC = CHMe(Z)$	9.13	-1.39
XIX	$MeHC = CHMe(E)$	9.13	-1.39
XX	$MeHC = CMe_2$	8.68	-1.84
XXI	$Me_2C = CMe_2$	8.30	-2.22

According to the classical LFER treatments of organic chemistry [67] (= Linear Free Energy Relationship) the corresponding destabilization $\Delta\varepsilon_\pi = -\Delta I_1^m(\pi)$ of the π orbital ($= \varphi_{\text{HOMO}}$) is explained as being due to the inductive effect of the methyl groups, as characterized, e.g. by the Taft σ^* parameter [67, 68]. Within the EBO treatment, the rationalization of $\Delta I_1^m(\pi)$ is radically different. It is due (a) to the hyperconjugative interaction of π with the $\pi(\text{Me})$ orbitals of the methyl groups, and (b) to a positive shift δA_π of the π-LMO self-energy, depending on the number of attached methyl groups. The origin of the latter shift δA_π should not be confused with the traditional inductive effect (see Sect. 5.2). It is due to the fact that the orthogonality of the LMOs is insured by small "tails" of opposite phase, which a given LMO exhibits in the surrounding space occupied by other (mainly geminal) LMOs. If the latter are changed, e.g. from a λ_{CH} to a λ_{CC} and to the set of methyl-group LMOs, cf. Eqs. (4.9), the requirements of orthogonality inherent in the transformation of Eq. (2.15) will lead to changes in the LMO in question, and thus in its self-energy. It can be shown that this orthogonality-conditioned shift δA_π of the self-energy A_π amounts to roughly

0.1 to 0.2 eV per substituting methyl group. Taking this into account we obtain

---

$AB(= A\Gamma)$ Model of Methyl-Substituted π Systems

$A_\pi = -10.2_5$ eV $+ 0.1_5$ eV $\cdot n_{Me}$,

   (n_{Me} = number of Me groups or sp^3 C centres)

$A_{Me}(\pi) = -14.0$ eV,

$B_{\pi\pi}(\equiv \Gamma_{\pi\pi}) = -1.2_5$ eV,

$B_{\pi, Me}(\equiv \Gamma_{\pi, Me}) = -1.3_5$ eV.

(4.11)

The two partial models (4.10) and (4.11) may be combined into a single, general one, e.g. by calculating the self-energy A_b of a "banana" LMO λ_b and the cross term B_{bb} between two λ_b spanning the same C centres, according to e.g.

$$A_b = (A_\sigma + A_\pi)/2 \approx -15.8 \text{ eV},$$
$$B_{bb} = (A_\sigma - A_\pi)/2 \approx -5.5 \text{ eV},$$

(4.12)

disregarding orthogonality induced shifts. Other individual A_i and B_{ij} values can be obtained in a similar fashion. Applications of such models can be found in the literature (e.g. [69, 70]) in particular in connection with the problem of "through space" vs "through bond" interaction [71] between nonconjugated double bonds [72]. Two special cases are briefly discussed in Sects. 4.4 and 4.5.

4.3 Polyines and Cumulenes

The PE spectra of ethine, XII(1), and of its dideutero derivative DC≡CD, were first recorded by Baker and Turner [73]. Later investigations are summarized in [60] and especially in [74]. The PE spectra of ethine, butadiine XII(2) [60, 73, 74, 75], hexatriine XII(3) [74, 75], and octatetraine XII(4) [74, 75] consist, respectively, of 2, 3, and 4 well-separated π bands in the region from 9 to 13 eV, with well-resolved vibrational fine structure, dominated by the (normal) stretching modes of the radical cation. Because of their inherent simplicity, as evident from the self-explanatory correlation diagram shown in Fig. 17, it is not surprising that this systematic behavior can be easily mimicked by a simple EBO model using linear combinations of the π LMOs π_x and π_y, as defined in Eqs. (4.6).

The influence of the replacement of one or both hydrogen atoms by (an) alkyl group(s) on the ionization energies $I_j^m(\pi)$ of ethine and polyines has been extensively studied for the first time for the special case R = methyl [73, 75]. The most comprehensive study in this field is that of Mouvier et al. [76] who investigated 11 monosubstituted and 25 disubstituted ethines with 16 different kinds of alkyl groups (R) in various combinations. The PE spectra of the symmetrically substituted dialkyl diines XXII(R) with R = methyl, ethyl, n-propyl, and n-butyl are given in [77] and those of the 1,ω-di-t-butylpolyines XXIII(N) with N = 1 to 5 in [78]. Leading

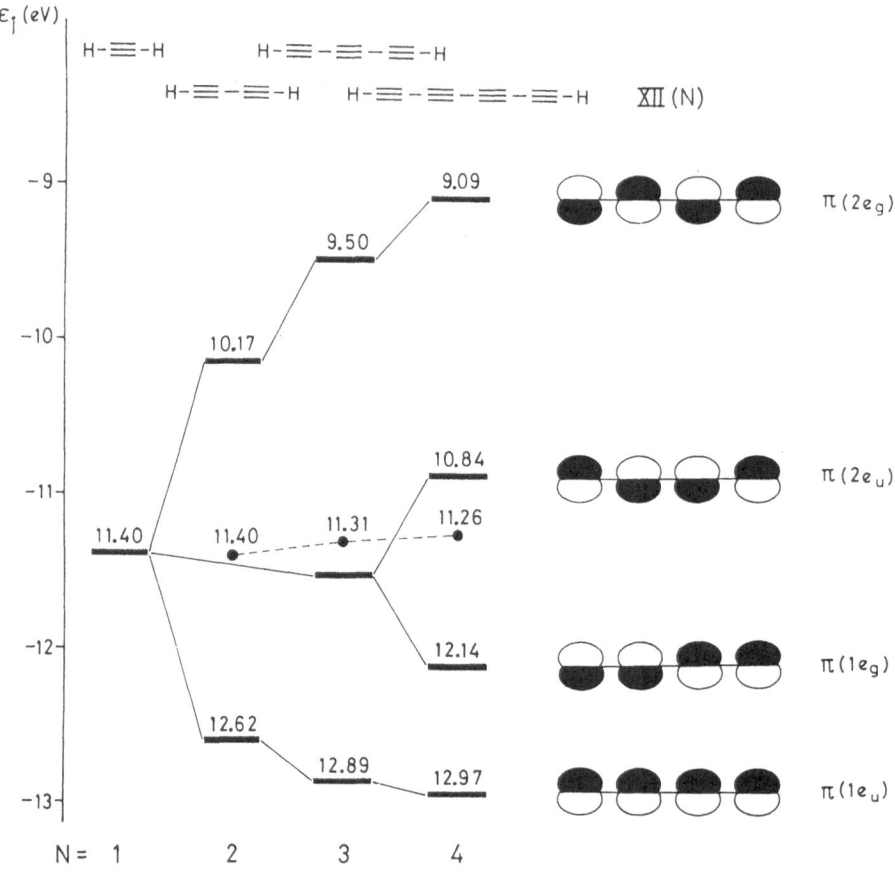

Fig. 17. Correlation diagram of the π orbitals of ethine XII(1) to octatetraine XII(4). The orbital energies ε_j/eV quoted correspond to the experimental energies, i.e. $\varepsilon_j = -I_j^m$. The orbital diagrams of XII(4) show qualitatively the phase relationship of the π LMOs

references for mono- and dihalopolyines XXII(N, X, Y), with X, Y = H, F, Cl, Br, I in various combinations, can be found in [79, 80, 81].

R–C≡C–C≡C–R   tBu–(C≡C)$_N$–tBu   X–(C≡C)$_N$–Y

XXII (R)   XXIII (N)   XXIV (N,X,Y)

As in the case of the polyenes XI(N), an EBO model of the *AB* type is a useful first aproximation for the rationalization of the π^{-1} ionization energies of such molecules, e.g. of the halopolyines XXIV(N, X, Y) [74, 79]. Although the mean self-energy A_π of the basis π-LMOs $\pi_{\lambda,\mu}$ remains practically constant from ethine to octatetraine (XII(N), N = 1 to 4), as is evident from Fig. 17, the spacing of the CMO-orbital energies $\varepsilon_j(\pi)$ is no longer symmetrical with respect to A_π. This means that the basis of the EBO model has to include virtual, antibonding π-LMOs $\pi^*_{\lambda,\mu}$ of self-energy A^*_π. (Depending on the choice of model, the angular momentum quantum number $\lambda = +1$ or -1 has to be replaced by $\lambda = x$ or y, if a real basis is used). The empirical self-energies and cross terms to be used in such an AB model are [74, 79]

$$A_\pi \approx -11.4 \text{ eV},$$

$$A_\pi^* \approx -6.0 \text{ eV},$$
(4.13)

$$B_{\pi\pi} = -B_{\pi^*\pi^*} = B_{\pi\pi^*} = -B_{\pi^*\pi} = -1.2 \text{ eV}.$$

H−C≡C−C≡C−C≡C−C≡C−H   XII (4)

Basis

A_π^*

B

A_π

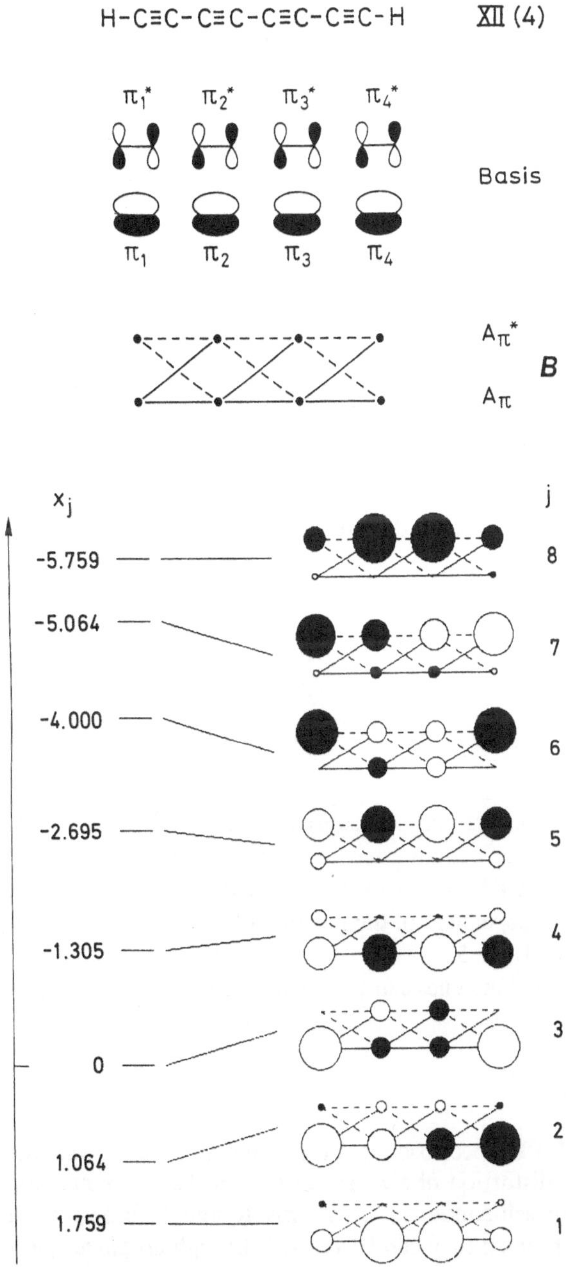

Fig. 18. Calculation of the CMO energies ε_j of octatetraine XII(4) using the model summarized in Eq. (4.13), i.e. including antibonding basis π^* LMOs. *Top row*: Schematic representation of the basis LMOs. *Second row*: The line graph B including vertices representing the π^* LMOs. Full edges correspond to cross terms $B_{\pi\pi}$ and $B_{\pi\pi^*}$ (negative energy value), dotted edges to $B_{\pi^*\pi^*}$ and $B_{\pi^*\pi}$ (positive energy value). The phases in the orbital diagrams $j = 1$ to 8 refer to the basis functions as a whole, which means that the nodes of the π^* orbitals are not shown

Note that the interaction cross term between (a) two consecutive bonding ($B_{\pi\pi}$), (b) two antibonding ($B_{\pi^*\pi^*}$), (c) a bonding and an antibonding ($B_{\pi\pi^*}$) or (d) an antibonding and a bonding ($B_{\pi^*\pi}$) π LMO is assumed to be of the same absolute size, the sign being determined by the relative phases of the basis LMOs. Figure 18 shows the line graph B for octatetraine XII(4), which is now a signed graph, i.e. one with "positive" and "negative" edges. In the same Fig. 18, the corresponding eigenvalues x_j and LMO-based CMO diagrams are presented. As can be seen, inclusion of the antibonding basis functions leads to a relative spacing of the CMO orbital energies ε_j which is in accordance with observation (cf. Fig. 17).

For the halo- and dihalopolyines XXIV(N, X, H), XXIV(N, X, Y), the following parameters have proved to be useful [74, 79, 80]:

$$A_\pi = -11.4 \text{ eV} + \delta A(R_1) + \delta A(R_2),$$

$$A_{\pi^*} = -6.00 \text{ eV}, \tag{4.14}$$

$$B_{\pi\pi} = -B_{\pi^*\pi^*} = B_{\pi\pi^*} = -B_{\pi^*\pi} = -1.2 \text{ eV}.$$

R:	C≡C	F	Cl	Br	I
$\delta A(R)/eV$	0.27	− 0.50	− 0.09	0.02	0.15
A_X/eV		−17.50	−12.96	−11.79	−10.43
$B_{\pi X}/eV$	(−1.23)	− 1.72	− 1.46	− 1.27	− 0.98

The corrections $\delta A(R_1)$ and $\delta A(R_2)$ depend on the left (R_1) and right (R_2) substituents of the particular triple bond.

XXV   XXVI   XVII

Application of the Foster-Boys localization procedure [17] to the STO-3G CMOs φ_j of allene XXV, butatriene XXVI, or pentatetraene XXVII [82] yields again a pair of "banana" LMOs λ_b for each double bond. Their uncalibrated self-energies A_b^0 and cross terms B_{bb}^0 or Γ_{bb}^0 are rather independent of the position of the "banana" LMOs λ_b within the cumulene π system (Table 5).

Application to other, substituted cumulenes can be found in [83] and the references given therein.

4.4 Twisted and Bent π Systems

One of the early problems of PE spectroscopy of organic molecules was the investigation of the effect which distortion of a π system has on the π^{-1} ionization energies. Typical examples were self-contained π systems R and S, linked by a nonessential double bond, about which they can be twisted through an angle τ, e.g.

Table 5. Typical self-energies and cross terms for cumulenes

		XXV	XXVI	XXVII
A_b^0		-16.63	-16.66	-16.83
		-16.63	-17.21	-17.32
		-7.22	-7.07	-7.12
B_{bb}^0		-1.66	-1.64	-1.63
		-7.22	-7.56	-7.43
			-0.95	-0.97
Γ_{bb}^0			1.24	1.27

butadienes XI(2) [84, 64], biphenyls XXVIII [85], or styrenes XXIX [86]. If A_R and A_S are the energies of the semilocalized orbitals π_R and π_S of the partial π systems R, S, and if B_{RS} is their cross term for $\tau = 0$, then the ionization energies $I_1^v = -\varepsilon_1$ and $I_2^v = -\varepsilon_2$ are obtained by solving the eigenvalue problem

XXVIII   XXIX

$$\begin{vmatrix} A_R - \varepsilon & B_{RS}\cos(\tau) \\ B_{RS}\cos(\tau) & A_S - \varepsilon \end{vmatrix} = 0 . \tag{4.15}$$

Unexpectedly, and seemingly in contradiction to the above, it has been observed that even significant distortions of self-contained π systems, e.g. double bonds, triple bonds, or benzene rings, have little, if any influence on their π-ionization energies. Typical examples are provided by strained double or triple bonds in *trans*-cyclooctene XXX [87], *anti*-Bredt hydrocarbons such as XXXI and XXXII [88], cycloalkines of type XXXIII or XXXIV [89], or by the cyclophanes (e.g. [2,2]-paracyclophane XXXV [90]).

This lack of response of π-ionization energies towards distortions of the double or triple bond can be explained in terms of simple rules based on an EBO model. In order to derive the necessary set of parameters for such a treatment, calculations on the prototype system ethene X(1) have been performed [91], subject to various

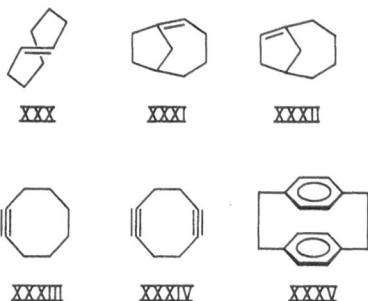

out-of-plane distortions, which can be written as linear combinations of the following normal modes:

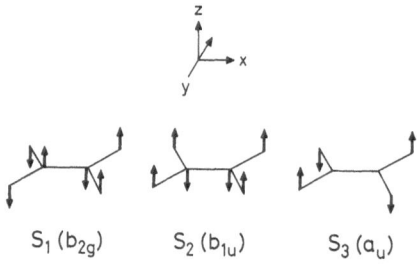

$S_1 (b_{2g})$ $S_2 (b_{1u})$ $S_3 (a_u)$

Actual distortions of an ethene double bond are usually superpositions of S_1, S_2, and S_3, and the occurrence of a single one of these is an exception. The combination of S_2 with some S_3 and a little S_1 is the most likely for the convex surface of a strained alicyclic system. Since the π-orbital energy is a quadratic function of S_1, S_2, and S_3 with coefficients of both signs, the geometry-dependent corrections remain small (except for utterly unrealistic deformations), and, furthermore, partly cancel each other. Expressed in numbers, the effect caused by a distortion with $|S_1|$, $|S_2|$, $|S_3| < 10°$ is below 0.07 eV (in either direction), i.e. less than one-tenth of the typical hyperconjugative shift (≈ 0.8 eV) due to a single alkyl substituent. The main result, is that the $C = C$ double bond is quite insensitive to changes in local geometry, as far as its ionization energy is concerned. Needless to say, other properties, in particular the many-electron properties, such as the total energy or the reactivity, do depend significantly on such deformations. (For a special example see [92]).

5 The Influence of Substituents

5.1 EBO Description of Substituent Effects

Replacement of the H atom of a parent molecule SH (e.g. benzene, S = phenyl) by a substituent R (e.g. R = methyl), to yield SR (e.g. toluene), shifts a particular ionization energy $I_j^m(SH)$ of the parent SH by

$$\delta I_j^m(SR) = I_j^m(SR) - I_j^m(SH) \qquad (5.1)$$

It is only convenient to discuss $\delta I_j^m(SR)$ in terms of traditional "substituent effets", e.g. within a LFER formalism [67, 68], if the "target orbital" from which the electron is ejected, i.e. the CMO $\varphi_j(SH)$ of orbital energy $\varepsilon_j(SH) = -I_j^v(SH) \approx I_j^m(SH)$ carries over — up to second-order changes — to the substituted system SR, which means that $\varphi_j(SR) \approx \varphi_j(SH)$. Under these conditions $\delta\varepsilon_j(SR) \approx -\delta I_j^m(SR)$ can be computed by first- and/or second-order pertubation theory.

The answer to the question which part of a given molecule SR has to be considered as the parent system S, and which as the substituent R is usually obvious, e.g. for SR = 1-methylnaphthalene, where S = 1-naphthyl and R = methyl. However, this is not necessarily always the case. Thus, in iodocyclopropane we might be interested in the shift of the $5p(I)^{-1}$ band(s) of the iodine atom, relative to the corresponding band in HI, or alternatively in the shifts of the Walsh-orbital bands associated with the ϱ_A^{-1} and ϱ_S^{-1} ionization processes [see Eq. (3.18)], relative to their position in the cyclopropane PE spectrum. Once the two moieties S and R have been defined for a particular pair SH and SR, the LMOs are assigned as follows.

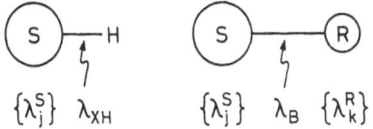

The LMO sets $\{\lambda_j^S\}$ and $\{\lambda_k^R\}$ belong to S and R, respectively, whereas λ_{XH} and λ_B are the LMOs of the σ bonds linking H to S and R to S. (Most of the time one has $\lambda_{XH} \equiv \lambda_{CH}$). The computation of $\delta\varepsilon_j(SR) \approx -\delta I_j^m(SR)$ is carried out as follows [45, 72].

Disregarding λ_{XH} or λ_B one first sets up the Hartree-Fock matrices F_λ^S and F_λ^R of the S and R moieties in localized basis. Their diagonalization yields the PCMOS ψ_j^S and ψ_k^R of S and R, collected in ψ^S and ψ^R [cf. Eq. (2.30)] and their orbital energies ε_j^S and ε_k^R, collected in ε^S and ε^R [cf. Eq. (2.31)]:

$$\begin{aligned} \psi^S &= \lambda^S C_S; & \varepsilon^S &= C_S^T F_\lambda^S C_S, \\ \psi^R &= \lambda^R C_R; & \varepsilon^R &= C_R^T F_\lambda^R C_R. \end{aligned} \qquad (5.2)$$

The complete Hartree-Fock matrices of the molecules SH and SR in localized basis are F_λ^{SH} and F_λ^{SR}. Using the partial matrices C_S and C_R defined in Eqs. (5.2) we construct the two blocked-out matrices

$$C_{SH} = \begin{pmatrix} C_S & 0 \\ 0 & 1 \end{pmatrix},$$

$$C_{SR} = \begin{pmatrix} C_S & 0 & 0 \\ 0 & 1 & 0 \\ 0 & 0 & C_R \end{pmatrix} \qquad (5.3)$$

of the same order as F_λ^{SH} and F_λ^{SR}, respectively. The latter are now subjected to the orthogonal transformations

$$C_{SH}^T F_\lambda^{SH} C_{SH} = F_\psi^{SH},$$
$$C_{SR}^T F_\lambda^{SR} C_{SR} = F_\psi^{SR}. \tag{5.4}$$

The resulting Hartree-Fock matrices F_ψ^{SH} and F_ψ^{SR} refer to the PCMOs of S and R, and to the LMOs λ_{XH} and λ_B which link H to S and R to S in the molecules SH and SR. Taking the molecules SR as an example, the detailed structure of F_ψ^{SR} is depicted in Eq. (5.6), where the abbreviation $\varkappa_{rs} = F_{rs}^{SR}$ has been used for simplicity, and where A_B is the self-energy of the LMO λ_B [cf. Eq. (5.1)].

$$F_\psi^{SR} = \begin{pmatrix} \varepsilon_1^S & & & & & \\ & \varepsilon_j^S & \varkappa_{jB} & & (\varkappa_{jk}) & \\ & \varkappa_{Bj} & A_B & & \varkappa_{Bk} & \\ & & & \varepsilon_1^R & & \\ & (\varkappa_{jk})^T & \varkappa_{kB} & & & \varepsilon_k^R \end{pmatrix} \tag{5.5}$$

Let ψ_j^S be the target PCMO of S, and assume that $\psi_j^S \approx \varphi_j^{SH} \approx \varphi_j^{SR}$. Its orbital energy is ε_j^S. With reference to the matrix in Eq. (5.5), joining S to R via λ_B has the following effects on ε_j^S:

a) In the general case there will be a first-order shift $\delta\varepsilon_j^S(\text{orthog})$ of ε_j^S due to the requirement of orthogonality between the LMOs λ_j, spanned by the target PCMO ψ_j^S, and the new LMO λ_B of the linking bond. (This effect has been discussed in Sect. 4.2.) However, if the target PCMO is a π-type orbital of a planar π system S (e.g. of a phenyl group), and if λ_B is an in-plane σ LMO, then orthogonality is symmetry-conditioned and $\delta\varepsilon_j^S(\text{orthog}) = 0$.

b) If symmetry allows, there will be a second-order interaction between φ_j^S and λ_B (or λ_{XH}) via the cross term $\varkappa_{j,B}$ (or $\varkappa_{j,XH}$). Again this contribution will vanish if φ_j^S and λ_B (or λ_{XH}) are orthogonal. In addition, one finds very often that $\varkappa_{j,B} \approx \varkappa_{j,XH}$ (e.g. if the linking LMO in SH is λ_{CH} and in SR λ_{CC}, because both LMOs have similar self-energies, and similar cross terms with other geminal or vicinal LMOs). Under these conditions the $\delta\varepsilon_j^S$ contributions will be about equal, and thus their difference negligible.

c) Finally, there will be first- or second-order interactions between the PCMOs ψ_j^S and ψ_k^R, depending on their cross terms $\varkappa_{jk}$ and their energy difference $\varepsilon_j^S - \varepsilon_j^k$.

Accordingly, the target orbital energies ε_j^{SH} and ε_j^{SR} are given, in a first approximation by the following formulae, assuming that $|\varepsilon_j^S - A_{XH}|$, $|\varepsilon_j^S - A_B|$ and $|\varepsilon_j^S - \varepsilon_k^R|$ are all large with respect to the relevant coupling cross terms $\varkappa_{rs}$

$$\varepsilon_j^{SH} \approx \delta\varepsilon_j^{SH}(\text{orthog}) + \varkappa_{j,XH}^2/(\varepsilon_j^S - A_{XH}),$$

$$\varepsilon_j^{SR} \approx \delta\varepsilon_j^{SR}(\text{orthog}) + \varkappa_{j,B}^2/(\varepsilon_j^S - A_B) + \sum_k \varkappa_{jk}^2/(\varepsilon_j^S - \varepsilon_k^R).$$
(5.6)

The substituent-induced shift of the PE band associated with the φ_j^{-1} ionization process is therefore

$$\delta I_j^m(SR) \approx -\delta\varepsilon_j^{SR} = \varepsilon_j^{SH} - \varepsilon_j^{SR}.$$
(5.7)

In the particular case of planar or linear π systems, e.g. substituted aromatic hydrocarbons or polyines, Eq. (5.7) reduces to Eq. (5.8) for the reasons quoted above,

$$\delta I_j^m(SR) \approx -\sum_k \varkappa_{jk}^2/(\varepsilon_j^S - \varepsilon_k^R),$$
(5.8)

where the sum includes only terms referring to PCMOs ψ_k^R nonorthogonal to ψ_j^S.

5.2 The Influence of Alkyl Groups, in Particular of the Cyclopropyl Group

The traditional rationalization of the ionization energy-reducing effect of alkyl groups in terms of LFER parameters [67, 68] was already mentioned in Sects. 4.2 and 4.3. This type of parametrization has been reviewed in detail in [93, 94, 95]. In the particular case of $S = \pi$ system, the effects considered are the inductive effect of the alkyl group R and its hyperconjugation with the π system. The main difficulty with this approach is that both effects are — statistically speaking — confounded, as can be shown with reference to Fig. 19. The alkyl group R is attached to the C-centre μ, which carries the 2p-AO p_μ of basis energy α_μ. The LCAO coefficient of p_μ is $c_{\mu j}$ for the target

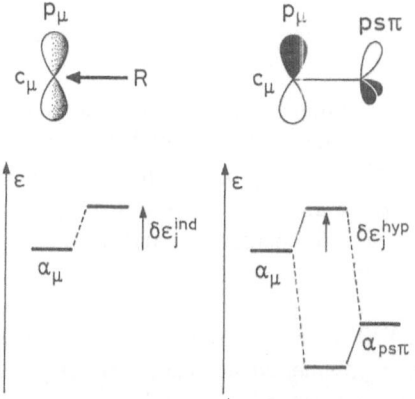

Fig. 19. Inductive and hyperconjugative interaction of a methyl group with a π system

CMO φ_j. The pseudo-π orbital of the alkyl group is designated by psπ and its orbital energy by $\alpha_{ps\pi}$. The inductive effect of R shifts the basis energy α_μ by $\delta\alpha_\mu$, whereas hyperconjugation is described by a resonance integral $\beta_{\mu R}$ between p_μ and psπ. The shifts $\delta\varepsilon_j^{ind}$ and $\delta\varepsilon_j^{hyp}$ induced are therefore in a first, crude approximation

$$\delta\varepsilon_j^{ind} \approx c_{\mu j}^2 \delta\alpha_\mu ,$$
$$\delta\varepsilon_j^{hyp} \approx c_{\mu j}^2 (\beta_{\mu R}^2 / (\alpha_\mu - \alpha_{ps\pi})) . \tag{5.9}$$

(The second equation assumes $\varepsilon_j \approx \alpha_\mu \gg \alpha_{ps\pi}$, which is not too bad an approximation for frontier π orbitals.) Thus, both shifts depend in the last analysis on $c_{\mu j}^2$, i.e. on the same independent variable.

This difficulty does not arise within the EBO model underlying the matrix (5.5), which explains the substituent-induced shifts $\delta\varepsilon_j^{SR}$ only in terms of conjugative interactions. (The very small, orthogonality-induced shifts $\delta\varepsilon_j^{SR}(orthog)$ in Eqs. (5.6) should not be confused with $\delta\varepsilon_j^{ind}$ of Eqs. (5.9)!)

The matrix $\mathbf{F}_\psi^{SR}$ defined in (5.5), allows a straightforward discussion of the interaction between the PCMOs of the S and R moieties in terms of Hoffmann's "through space" and "through bond" interaction scheme [71]. This is shown in (5.10). In the left diagram, the cross term $\varkappa_{jk}$ links the PCMO ψ_j^S directly to the PCMO ψ_k^R, as indicated by the solid line joining ε_j^S to ε_k^R via $\varkappa_{jk}$. This

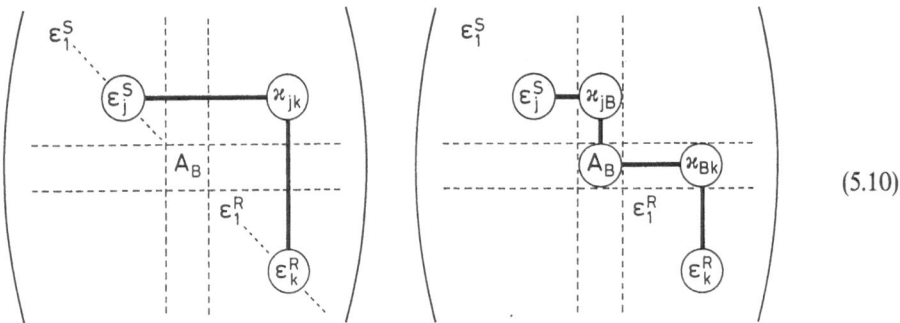

direct link corresponds to pure "throughspace" interaction. On the other hand, ψ_j^S and ψ_k^R interact also via the LMO λ_B as shown by the solid line in the right-hand diagram of (5.10). The size of this interaction, which is "through bond" by definition, is a function of the size of the two cross terms $\varkappa_{jB}$ and $\varkappa_{Bk}$. Obviously, the resultant shifts of the orbital energies ε_j^S and ε_k^R depend both on the size and sign of the above cross terms and on the energy differences $|\varepsilon_j^S - A_B|$ and $|\varepsilon_k^R - A_B|$. Interesting examples for this type of treatment are provided by the cyclophanes, e.g. [2,2]-paracyclophane XXXV [90] and related molecules [92, 96]. In these cases the alkyl groups R are either polymethylene chains bridging two systems S and S', or more complicated moieties, as discussed in the following chapter on "through space" and "through bond" interactions.

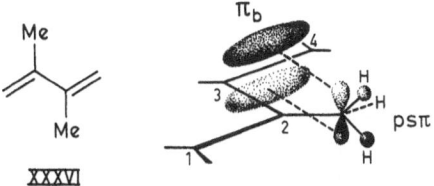

XXXVI

Attention is drawn to the fact that "through space" interaction may occur not only between directly bonded S and R, but also between more distant PCMOs, e.g. between one of the π LMOs of 2,3-dimethyl-1,3-butadiene XXXVI (say the LMO π_b spanning the centres 3 and 4) and the psπ orbital of a methyl group separated from the LMO π_b by two bonds (in this instance the methyl group attached to center 2 [64, 97].

To conclude this section we discuss briefly the interaction between a double bond and a cyclopropyl group, e.g. in vinylcyclopropane XXXVII.

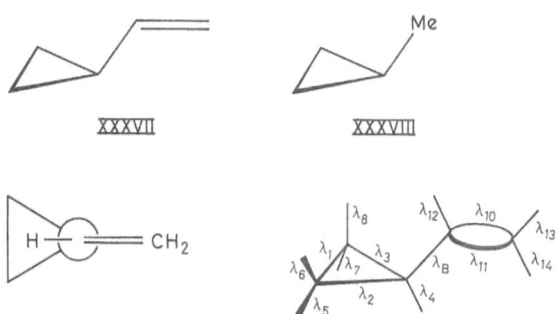

The Förster-Coulson-Moffitt (FCM) SLMOs ϱ_A, ϱ_S, ϱ_0, have been presented in Eq. (3.18). Up to minor corrections (cf. Eqs. 5.6)), which can be neglected, they are identical with the PCMOs ψ_1 to ψ_8, derived from the LMOs λ_1 to λ_8 of the cyclopropyl group of XXXVII, as described in Sect. 5.1. Also, the PCMOs ψ_{10} to ψ_{14} of the vinyl group are similarly obtained from the LMOs λ_{10} to λ_{14} [98].

According to the classical treatment of cyclopropane IV(3) by Walsh [41], the almost double-bond-like propensity of the cyclopropyl group to conjugate with a double bond π orbital, or other p-type orbitals — as contrasted with the less important hyperconjugation of other alkyl groups — is due to a large resonance integral $\beta_{\pi\Delta}$ between the π orbital (π) and the antisymmetric Walsh orbital (Δ) in, e.g., vinylcyclopropane XXXVII. In other words, this model assumes that $|\beta_{\pi\Delta}| \gg |\beta_{\pi Me}|$. In contrast, the EBO model yields the following explanation.

As far as the cross terms are concerned (which, in a Hückel treatment, would correspond to the above-mentioned resonance integrals $\beta_{\pi\Delta}$ and $\beta_{\pi Me}$) there is no difference between the $\varkappa_{ij}$ (cf. (5.10)) linking the pseudo-π orbital psπ of a methyl group to the FCM PCMO $\psi_A \approx \varrho_A$ (Eq. 3.18) in methylcyclopropane XXXVIII, and the π orbital $\pi = (\lambda_{10} - \lambda_{11})/\sqrt{2}$ of the vinyl group to $\psi_A \approx \varrho_A$ in vinylcyclopropane XXXVII, if the latter molecule is in the preferred bisected conformation shown for

XXXVII. The only feature which distinguishes the FCM PCMOs from the pseudo-π orbitals psπ of alkyl groups is their high-lying orbital energy $\varepsilon^{FCM}(\psi_A) \approx \varepsilon^{FCM}(\psi_S)$ ≈ -10.7 eV, which is well above that of a pseudo-π orbital of a methyl group, i.e. $\varepsilon^{Me}(\psi_{ps\pi}) \approx -14$ eV, and close to the orbital energy of a double bond π orbital, i.e. $\varepsilon(\pi) \approx -10.5$ eV. This is the direct consequence of the large size of the geminal interaction terms $B^0_{CC,CC} = -4.8$ eV between the "banana" LMOS λ_b of the cyclopropane ring, as shown in Table 3. In consequence the interaction between the double bond π orbital, π, and the antisymmetric Walsh orbital, $\psi_A \approx \varrho_A$, is practically first order, with a correspondingly large shift $\delta\varepsilon(\pi)$ of the π-orbital energy, whereas hyperconjugation is only a small, second-order effect, because of the large energy gap $|\varepsilon^{Me}(\psi_{ps\pi}) - \varepsilon(\pi)| \approx 4$ eV. For a detailed discussion see [45, 46, 98].

5.3 The Effect of Spin Orbit Coupling

A further advantage of the EBO model is that it allows the incorporation of spin orbit coupling (SOC) in an approximation which is quite sufficient for many practical applications in PE spectroscopy. It is based on a naive Hückel-type treatment of SOC, developed for the discussion of the PE spectra of simple alkyl halides RX [99] (R = alkyl group, X = F, Cl, Br, I) and of the halo- or dihaloethines XXII(N, X, Y) [74, 79].

Localizing the CMOs φ_j of molecules containing halogen atoms X, Y, ... as discussed in Sect. 2.2, e.g. of an alkyl halide RX, one obtains, for each halogen atom X, Y, ..., three lone-pair LMOs $\lambda_1(X)$, $\lambda_2(X)$, $\lambda_3(X)$, which are close to sp^3 hybrids of the valence-shell AOs ns and np of X. If the bond axis R$-$X coincides with a molecular symmetry axis of order 3 or higher (e.g. in Me$-$X), then these LMOs are related to each other by a rotation of 120°. The linear combinations

$$np_+(X) = (\lambda_1(X) + \omega\lambda_2(X) + \omega^2\lambda_3(X))/\sqrt{3},$$
$$np_-(X) = (\lambda_1(X) + \omega^2\lambda_2(X) + \omega\lambda_3(X))/\sqrt{3} \tag{5.11}$$

with $\omega = \exp(2\pi i/3)$, are then eigenfunctions of the angular moment operator with eigenvalues $\hbar$ and $-\hbar$, respectively. The contribution of SOC to the orbital energy $\varepsilon(np(X))$ of an electron in one or the other of the np(X)-type AOs is postulated to be

$$\langle np_+(X)| \hat{H}^{SOC} |np_+(X)\rangle = \zeta(X)/2,$$
$$\langle np_-(X)| \hat{H}^{SOC} |np_-(X)\rangle = -\zeta(X)/2, \tag{5.12}$$

where $\hat{H}^{SOC}$ is the SOC operator defined ad hoc in such a way that Eqs. (5.12) yield the correct sequence of state for the more than half-filled electron shell of X (inversion), i.e. $\varepsilon(^2\Pi_{1/2}) > \varepsilon(^2\Pi_{3/2})$ with $\varepsilon(^2\Pi_{1/2}) - \varepsilon(^2\Pi_{3/2}) = \zeta(X)$. The $\zeta(X)$ are the SOC constants [79, 100]. Figure 20 shows the bromine lone-pair bands in the He(Iα) PE spectrum of methylbromide MeBr, which are split, in accordance with Eqs. (5.12) by $\zeta(Br) = 0.32$ eV.

If the (local) environment of the halogen atom X no longer exhibits rotational symmetry of order 3 or higher, competition between SOC and (hyper)conjugation will affect the two states of $^2\Pi_{1/2}$ and $^2\Pi_{3/2}$ parentage differently. An example is

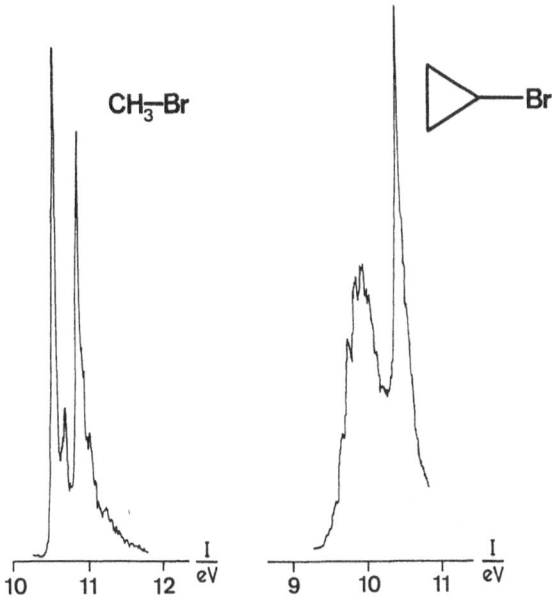

Fig. 20. He(Iα) PE spectra of methylbromide MeBr and of cyclopropylbromide $(CH_2)_2CHBr$

provided by the He(Iα) PE spectrum of cyclopropylbromide, shown in Fig. 20, where the Franck-Condon envelope of the first band is considerably broadened, in contrast to the second one which is still as sharp as in the PE spectrum of methylbromide [101, 99]. Under such conditions it is of advantage to use a real basis for the lone-pair AOs instead of the complex basis (5.11), i.e.:

$$np_x(X) = (2\lambda_1(X) - \lambda_2(X) - \lambda_3(X))/\sqrt{6},$$
$$np_y(X) = (\lambda_2(X) - \lambda_3(X))/\sqrt{2}.$$
$$(5.13)$$

[Note that $ns(X) = (\lambda_1(X) + \lambda_2(X) + \lambda_3(X))/\sqrt{3}$ corresponds to the remaining σ-type lone-pair AO of X.] The self-energies of the SLMOs of Eqs. (5.13) and their cross terms with other basis orbitals are now incorporated into the F_ψ^{SR} matrix (5.5) as before, with the trivial difference that X now takes the place of the substitutent R. The important difference is that we have now SOC cross terms between the SLMOs of Eqs. (5.13), namely

$$\langle np_x(X)|\, \hat{H}^{SOC}\, |np_x(X)\rangle = \langle np_y(X)|\, \hat{H}^{SOC}\, |np_y(X)\rangle = 0,$$

$$\langle np_x(X)|\, \hat{H}^{SOC}\, |np_y(X)\rangle = -i\zeta(X)/2,$$
$$(5.14)$$

$$\langle np_y(X)|\, \hat{H}^{SOC}\, |np_x(X)\rangle = +i\zeta(X)/2.$$

This means that F_ψ^{SR} is now a Hermitian matrix.

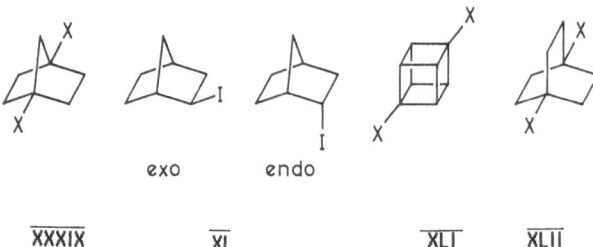

Applications of the EBO model, including SOC, can be found in the literature, e.g.
1,4-dihalonorbornanes XXXIX [102], exo- and endo-2-norbornyl iodides XL [103],
1,4-dihalocubanes XLI [104], and 1,4-dihalobicyclo [2.2.2]octanes XLII [105].

6 Concluding Remarks

For obvious reasons − mainly the lack of space − some important applications of
the EBO model have not been included in this review. These concern, in particular,
the treatment of molecules containing heteroatoms, and the question of the conformation dependence of ionization energies.

The EBO model reviewed in this chapter is not an alternative to more sophisticated
treatments, but rather a convenient and reliable method of providing the necessary
basis for a qualitative or − at best − semiquantitative discussion of experimental
or theoretical results in the framework of the usual orbital language.

It is only fair to draw attention to some of the shortcomings of the EBO model,
although these must be rather obvious. First of all, the models, as described, are
unreliable for very compact, polycyclic systems because of the neglect of the numerous,
albeit small, long-range interaction terms, which can add up to rather sizeable orbital
energy shifts. Secondly, a closer examination of the regression presented in Fig. 4
shows that the points of the 2p manifold, taken by themselves, would lead to a steeper
slope of the partial regression line. Thus, the predicted low-energy ionization energies
I_j^m tend to be on the high side. A third difficulty concerns the higher interaction terms
(Δ_{ij}^0 or higher), which are somewhat too small, relative to B_{ij}^0 and Γ_{ij}^0, if taken from
ab-initio calculations using gaussian functions. Accordingly, long-range "through
space" interactions are sometimes underestimated.

Finally, our EBO model shares major complications with all other orbital models. Due
to the high density of CMOs within the 2p manifold, the orbital sequence must necessarily
be uncertain for close-lying orbitals belonging to different irreducible representations.
However, this may well be irrelevant because the closeness of many states within a small
energy interval is inducive to extensive vibronic mixing, quite apart from the effects of
configuration interaction between Koopmans and non-Koopmans configurations.

7 References

1. Brundle CR, Baker AD (1977) Electron spectroscopy: theory techniques and applications,
   vol 1, Academic Press, London; Eland JHD (1984) Photoelectron spectroscopy, 2 ed,
   Butterworths, London; Siegbahn H, Karlsson L (1982) Photoelectron spectroscopy, Handbuch der Physik, vol XXXI: Corpuscles and radiation in matter I. Mehlhorn W (ed)
   Springer, Berlin Heidelberg New York, p 215

2. Heilbronner E, Bock H (1976) The HMO model and its application, Wiley, London/New York
3. Koopmans T (1934) Physica 1: 104
4. Streitwieser A Jr, Nair PM (1959) Tetrahedron 5: 149; Eland JHD, Danby CJ (1968) Z. Naturforsch. 23a: 355
5. Brogli F, Heilbronner E (1972) Theoret. Chim. Acta (Berl.) 26: 289
6. Heilbronner E, Schmelzer A (1980) Nouv. J. Chim. 4: 23
7 Murrell JN, Harget AJ (1972) Semi-empirical self-consistent molecular-orbital theory of molecules, Wiley-Interscience, London
8. Pople JA, Beveridge DL (1970) Approximate Molecular Orbital Theory, McGraw-Hill, New York
9. Von Niessen W, Schirmer J, Cederbaum LS (1984) Computer Phys. Rep. 1: 57
10. Köppel H, Cederbaum LS, Domcke W, Von Niessen W (1979) Chem. Phys. 37: 303
11. Von Niessen W, Diercksen GHF, Cederbaum LS, Domcke W (1976) Chem. Phys. 18: 469
12. Cederbaum LS, Domcke W, Von Niessen W (1975) Chem. Phys. 10: 459
13. Dewar MJS (1952) J. Am. Chem. Soc. 74: 3341, 3345, 3350, 3353, 3357
14. Woodward RB, Hoffmann R (1970) Die Erhaltung der Orbitalsymmetrie, Verlag Chemie, Weinheim
15. Dewar MJS (1969) The molecular orbital theory of organic chemistry, McGraw-Hill, New York
16. Hall GG (1951) Proc. Roy. Soc. London Ser. A 205: 541; Lennard-Jones J, Hall GG (1952) Trans Faraday Soc. 48: 581
17. Boys SF (1960) Rev. Mod. Phys. 32: 296; Foster JM, Boys SF (1960) ibid 32: 300
18. England W, Salmon LS, Ruedenberg K (1971) Topics Curr. Chem. 23: 31
19. Honegger E, Yang ZZ, Heilbronner E (1984) Croatica Chem. Acta 57: 967
20. Hall GG (1954) Trans. Faraday Soc. 50: 319
21. Lorquet JC (1965) Mol. Phys. 9: 101
22. Brailsford DF, Ford B (1970) Mol. Phys. 18: 621
23. Herndon WC (1971) Chem. Phys. Letters 10: 460; (1979) J. Chem. Educ. 56: 448
24. Murrell JN, Schmidt W (1972) J.C.S. Faraday II 68: 1709
25. Gimarc BM (1973) J. Am. Soc. 95: 1417
26. Hartree R (1928) Proc. Cambridge Phil. Soc. 24: 89; Fock V (1930) Z. Physik 61: 126
27. Roothaan CCJ (1951) Rev. Mod. Phys. 23: 69
28. Edmiston C, Ruedenberg K (1963) Rev. Mod. Phys. 35: 457; (1965) J. Chem. Phys. 43: 597; Ruedenberg K (1965) In: Modern quantum chemistry, vol I: 85, Academic Press, New York
29. Heilbronner E, Honegger E, Zambach W, Schmitt P, Günther H (1984) Helv. Chim. Acta 67: 1681
30. Hehre WJ, Stewart RF, Pople JA (1969) J. chem. Phys. 51: 2657; Hehre WJ, Lathan WA, Ditchfield R, Newton MD, Pople JA, Program No. 236, QCPE, Bloomington, Indiana
31. Bieri G, Dill JD, Heilbronner E, Schmelzer A (1977) Helv. Chim. Acta 60: 2234
32. Price WC, Potts AW, Streets DG (1972) Electron spectroscopy Shirley DA (ed North Holland, Amsterdam, p 187; Potts AW, Williams TA, Price WC (1972) Faraday Disc. Chem. Soc. 54: 104
33. Potts AW, Streets DG (1974) J.C.S. Faraday II 70: 875; Streets DG, Potts AW (1974) ibid. 70: 1505
34. Bieri G, Burger F, Heilbronner E, Maier JP (1977) Helv. Chim. Acta 60: 2213
35. Bieri G, Åsbrink L (1980) J. Electron Spectrosc. Rel. Phenom. 20: 149
36. Pireaux JJ, Svensson S, Basilier E, Malmqvist PA, Gelius U, Caudano R, Siegbahn K (1976) Phys. Rev. A. 14: 2133
37. Heilbronner E (1954) Helv. Chim. Acta 37: 921; Pauncz R (1956) Acta physica Acad. Sci. Hungaricae 6: 15; Hall GG (1987) Synthetic Metals 17: 123
38. Heilbronner E (1977) Helv. Chim. Acta 60: 2234
39. De Meijere A (1979) Angew. Chem. 91: 867; Angew. Chem. Int. Ed. 18: 809; Gleiter R (1979) Topics Curr. Chem. 86: 197
40. De Puy CH (1973) Topics Curr. Chem. 40: 73; Wendisch D (1971) Carbocyclische Dreiring-Verbindungen, in: Methoden der Organischen Chemie, Houben-Weyl-Müller,

Edts., vol VI/3, Thieme Verlag, Stuttgart; Charton M (1970) Olefinic properties of cyclopropanes. In: The chemistry of alkenes, Zabicky J (ed) vol 2, Wiley Interscience, New York

41. Walsh AD (1947) Nature 159: 167, 712; (1949) Trans. Faraday Soc. 45: 179; see also: Sugden TM (1947) Nature 160: 367
42. Coulson CA, Moffitt WE (1947) J. Chem. Phys. 15: 151; Phil. Mag. 40: 1
43. Förster Th (1939) Z. phys. Chem. B 43: 58
44. Honegger E, Heilbronner E, Schmelzer A, Wang JQ (1981) Israel J. Chem. 22: 3
45. Honegger E, Dissertation Universität Basel (1983); Honegger E, Heilbronner E, unpublished results
46. Honegger E, Heilbronner E, Schmelzer A (1982) Nouv. J. Chim. 6: 519
47. Basch H, Robin M, Kuebler NA, Baker C, Turner DW (1969) J. Chem. Phys. 51: 52
48. Jonsson BO, Lindholm E (1969) Arkiv f. Fysik 39: 65
49. Hoffmann R, Mollère PhD, Heilbronner E (1973) J. Am. Chem. Soc. 95: 4860
50. Dewar MJS, McKee ML (1989) Pure & Appl. Chem. 52: 1431; Dewar MJS (1984) J. Am. Chem. Soc. 106: 669
51. Veszprémi T (1982) Chem Phys. Letters 88: 325
52. Gleiter R (1979) Topics Curr. Chem. 86: 199
53. Bischof P, Eaton PE, Gleiter R, Heilbronner E, Jones TB, Musso H, Schmelzer A, Stober R (1978) Helv. Chim. Acta 61: 547
54. Honegger E, Eaton PE, Ravi Shankar BK, Heilbronner E (1982) Helv. Chim. Acta 65: 1982
55. Heilbronner E, Jones TB, Krebs A, Maier G, Malsch KD, Pocklington J, Schmelzer A (1980) J. Am. Chem. Soc. 102: 564; Maier G (1988) Angew. Chem. 100: 317
56. Honegger E, Huber H, Heilbronner E, Dailey WP, Wiberg KB (1985) J. Am. Chem. Soc. 107: 7172
57. Harihan PC, Pople JA (1973) Theor. Chim. Acta 28: 213
58. Huber H (1980) Theor. Chim. Acta 55: 117
59. Newton MD, Schulman JM (1972) J. Am. Chem. Soc. 94: 773; Wilberg KB (1983) ibid. 105: 1227; Jackson JE, Allen LC (1984) ibid. 106: 591; and references given therein
60. Turner DW, Baker C, Baker AD, Brundle CR (1970) molecular photoelectron spectroscopy, Wiley-Interscience, London; Kimura K, Katsumata S, Achiba Y, Yamazaki T, Iwata W (1981) Handbook of HeI photoelectron spectra of fundamental organic molecules, Japan Scientific Societies Press, Tokyo
61. Eland JHD (1969) Int. J. Mass Spectrom. Ion Phys. 2: 4471; Brundle CR, Robin MB, Kuebler NA, Basch H (1972) J. Am. Chem. Soc. 94: 1451; Branton GR, Frost DC, Makita T, McDowell CA, Stenhouse IA (1970) J. Chem. Phys. 52: 802
62. Masclet P, Grosjean D, Mouvier G (1973) J. Electron Spectrosc. Relat. Phenom. 2: 225
63. Whitey RM, Carlson TA, Spears DP (1974) J. Electron Spectrosc. Relat. Phenom. 3: 59; Brundle CR, Robin MB (1970) J. Am. Chem. Soc. 92: 5550
64. Beez M, Bieri G, Bock H, Heilbronner E (1973) Helv. Chim. Acta 56: 1028
65. Allan M, Dannacher J, Maier JP (1980) J. Chem. Phys. 73: 3114; Allan M, Maier JP (1976) Chem. Phys. Letters 43: 94; Jones TB, Maier JP (1979) Int. J. Mass Spectrom. Ion Phys. 31: 287
66. Yang ZZ (1980) Dissertation University Basel; unpublished results
67. Shorter J (1973) Correlation analysis in organic chemistry, an introduction to linear free-energy relationships, Clarendon Press, Oxford; Wells PR (1968) Linear free energy relationships, Academic Press, London
68. Taft RW Jr (1956) Separation of polar, steric, and resonance effects in reactivity. In: Steric effects in organic chemistry, Newman MS (ed) Wiley, London
69. Bickle P, Hopf H, Bloch M, Jones TB (1979) Chem. Ber. 112: 3691
70. Batich C, Heilbronner E, Quinn CB, Wiseman JR (1976) Helv. Chim. Acta 59: 512; Bloch M, Brogli F, Heilbronner E, Jones TB, Prinzbach H, Schweikart O (1978) Helv. Chim. Acta 61: 1388
71. Hoffmann R, Imamura A, Hehre J (1968) J. Am. Chem. Soc. 90: 1499; Hoffmann R (1971) Accounts Chem. Res. 4: 1; Gleiter R (1974) Angew. Chem. 86: 770
72. Heilbronner E (1972) Israel J. Chem. 10: 143; Heilbronner E, Schmelzer A (1975) Helv. Chim. Acta 58: 936

73. Baker C, Turner WD (1967) Chem. Commun. 1967: 797; (1968) Proc. Roy. Soc. A 308: 19
74. Bieri G, Heilbronner E, Jones TB, Kloster-Jensen E, Maier JP (1977) Physica Scripta 16: 202
75. Allan M, Kloster-Jensen E, Maier JP (1976) Chem. Physics 7: 11; Brogli F, Heilbronner E, Hornung V, Kloster-Jensen E (1973) Helv. Chim. Acta 56: 2171; Allan M, Heilbronner E, Kloster-Jensen E, Maier JP (1976) Chem. Phys. Letters 41: 228
76. Carlier P, Dubois JE, Masclet P, Mouvier G (1975) J. Electron Spectrosc. Relat. Phenom. 7: 55
77. Heilbronner E, Jones TB, Maier JP (1977) Helv. Chim. Acta 60: 1967
78. Heilbronner E, Jones TB, Kloster-Jensen R, Maier JP (1978) Helv. Chim. Acta 61: 2040; Haselbach E, Klemm U, Buser U, Gschwind R, Jungen M, Kloster-Jensen E, Maier JP, Christen H, Baertschi P (1981) Helv. Chim. Acta 64: 823
79. Heilbronner E, Hornung V, Maier JP, Kloster-Jensen E (1974) J. Am. Chem. Soc. 96: 4252
80. Bieri G, Heilbronner E, Stadelmann JP, Vogt J, von Niessen W (1977) J. Am. Chem. Soc. 99: 6832
81. Allan M, Kloster-Jensen E, Maier JP (1976) J. Chem. Soc. Faraday Trans. II 73: 1406, 1417
82. Bieri G, Dill JD, Heilbronner E, Maier JP, Ripoll JL (1977) Helv. Chim. Acta 60: 629
83. Basch H, Bieri G, Heilbronner E, Jones TB (1978) Helv. Chim. Acta 61: 46; Kovać B, Heilbronner E, Prinzbach H, Weidmann K (1979) Helv. Chim. Acta 62: 2481
84. Brundle CR, Robin MB (1970) J. Am. Chem. Soc. 92: 5550; Eland JHD (1969) Int. J. Mass Spectrom. Ion Phys. 2: 471; Mollère PD, Houk KN, Bomse DS, Morton TH (1976) J. Am. Chem. Soc. 98: 4732
85. Maier JP, Turner DW (1972) Discuss. Faraday Soc. 54: 149; Daintith J, Maier JP, Sweigart DA, Turner DW (1972) In: Electron spectroscopy. Shirley D (ed) North-Holland, Amsterdam
86. Maier JP, Turner DW (1973) J. Chem. Soc. Faraday Trans. 2 69: 196; Kobayashi T, Yokota K, Nagakura S (1973) J. Electron Spectrosc. Relat. Phenom. 3: 449
87. Klessinger M, Rademacher P (1979) Angew. Chem. 91: 885; Angew. Chem. Int. Ed. 18: 826
88. Robin MB, Taylor GN, Kuebler NA, Bach RD (1973) J. Org. Chem. 38: 1049; Batich C, Ermer O, Heilbronner E, Wiseman JR (1973) Angew. Chem. 85: 302; Angew. Chem. Int. Ed. 12: 312; Batich C, Heilbronner E, Quinn CB, Wiseman JR (1976) Helv. Chim. Acta 59: 512
89. Schmidt H, Schweig A, Krebs A (1974) Tetrahedron Lett. 1471; Bieri G, Heilbronner E, Kloster-Jensen E, Schmelzer A, Wirz J (1974) Helv. Chim. Acta 57: 1265
90. Heilbronner E, Yang Z (1983) Topics Curr. Chem. 115: 1
91. Honegger E, Schmelzer A, Heilbronner E (1982) J. Electron Spectrosc. Rel. Phenom. 28: 79
92. Honegger E, Heilbronner E, Wiberg K (1983) J. Electron Spectrosc. Rel. Phenom. 31: 369
93. Turner DW (1966) Adv. Phys. Org. Chem. 4: 31; van Cauwelaert (1971) Bull. Soc. Chim. Belges 80: 181
94. Cocksey BJ, Eland JHD, Danby CJ (1971) J. Chem. Soc. (B) 790; cf. [1]
95. Heilbronner E, Maier JP (1977) Some aspects of organic photoelectron spectroscopy. In: Electron spectroscopy: theory, techniques, and applications. Brundle CR, Baker AD (eds) Academic Press, London, vol 1
96. Prinzbach H, Sedelmeier G, Krüger C, Goddard R, Martin HD, Gleiter R (1978) Angew. Chem. 90: 297; Fessner WD, Sedelmeier G, Knothe L, Prinzbach H, Rihs G, Yang ZZ, Kovać B, Heilbronner E (1987) Helv. Chim. Acta 70: 1816, and references therein
97. Honegger E, Yang ZZ, Heilbronner E, v. E Doering W, Schmidhauser JC (1984) Helv. Chim. Acta 67: 640
98. Wang JQ, Honegger E, Heilbronner E, Schmelzer A (1982) Scientia Sinica (Series B) 25: 236
99. Brogli F, Heilbronner E (1971) Helv. Chim. Acta 54: 1423
100. Dunn TM (1961) Trans. Faraday Soc. 51: 1441; Cornford JL, Stenhouse IA (1971) J. Chem. Phys. 54: 2651; Potts AW, Price WC (1971) Trans. Faraday Soc. 67: 1242
101. Hashmall JA, Heilbronner E (1970) Angew. Chem. 82: 320; Angew. Chem. Int. Ed. 9: 305
102. Honegger E, Heilbronner E, Hess N, Martin HD (1985) Chem. Ber. 118: 2927
103. Honegger E, Heilbronner E, Dratva A, Grob CA (1984) Helv. Chim. Acta 67: 1691
104. Honegger E, Heilbronner E, Urbanek T, Martin HD (1985) Helv. Chim. Acta 68: 23
105. Honegger E, Heilbronner E, Hess N, Martin HD (1987) Chem. Ber. 120: .187

Through-space and Through-bond Interactions as Mirrored in Photoelectron Spectra

Mirjana Eckert-Maksić

Department of Organic Chemistry and Biochemistry,
Ruđer Bošković Institute, Zagreb, Croatia, Yugoslavia

Useful and transparent orbital models for rationalizing intramolecular interactions between identical functional groups are discussed in some detail. A seminal idea of Hoffmann, Imamura and Hehre to dissect interactions between molecular fragments into through-bond and through-space is found to possess a high interpretive power. A simple perturbational approach employing semilocalized molecular orbitals suffices for most purposes if information at the qualitative level is desired. Quantitative theoretical data are conveniently produced by the method of Heilbronner and Schmelzer. It is gratifying and intellectually pleasing that theoretical results are in good agreement with PES observations. Thus one can say that orbital models represent a useful means for treating and understanding long-range interactions within molecules themselves. They give a theoretical framework to the empirical knowledge which relates properties of molecular systems to the features of their constituent fragments.

1 Introduction

It is a common wisdom that atoms retain their identity in chemical environments. Furthermore, they are frequently clustered in larger molecular subunits called functional groups which give rise to characteristic chemical behaviour [1]. Functional groups are usually well separated from the rest of the molecule and exhibit properties which vary relatively little in a wide variety of compounds. This does not imply that functional groups are completely isolated moieties. On the contrary, they "feel" their chemical environment and a fine variation in their features faithfully reflects differences in chemical bonding in different molecules. Molecules possessing two or more functional groups are particularly interesting because their intramolecular interactions give a direct insight into the nature of the electronic structure of molecules. Hence, their understanding is of paramount importance for rationalizing molecular physical and chemical properties at the electronic level. It is therefore not surprizing that the last two decades have witnessed an enormous increase in a number of papers dealing with the problem of interactions of nonconjugated functional groups. A seminal idea of Hoffmann, Imamura and Hehre [2, 3] to dissect intramolecular interactions into "through-space" (TS) and "through-bond" (TB) terms by using the perturbation theory based on localized and/or semilocalized molecular orbitals, proved extremely useful in this respect. Soon after the introduction of photoelectron spectroscopy (PES) in the 1960s [4] and its extensive application to the investigation of organic molecules [5] it became apparent that consequences of intramolecular orbital interactions are often mirrored in a transparent way in the photoelectron spectra [5, 6]. Interpretation of the latter is enabled by the well-known Koopmans' theorem [7] stating that the vertical ionization energies are given by the negative SCF orbital energies. Hence, the underlying physical picture is that of the independent electrons moving in the average electrostatic field. This approximation is adequate for outer (peripheral) valence electrons and there is indeed a close correspondence between photoelectron events and molecular orbital (MO) energies. It appears that PE spectroscopy provides one of the best experimental probes of the molecular orbital model. We would like to stress again that many-body effects may be important in inner-valence regions thus leading to a break-down of the single-electron picture. Fortunately, localized and semilocalized MOs describing outer valency electrons of the functional groups belong to the realm of orbitals which can be detected and identified by PES technique. Hence, the latter is a method of choice for studying polyfunctional molecules at the orbital level.

There is abundant literature on photoelectron spectroscopic studies of long-range orbital interactions in molecules and several excellent reviews are available [8–13]. Accordingly, a comprehensive survey of this topic is neither intended nor necessary. Instead, we shall briefly present the salient features of a general pattern of the long-range orbital interactions within molecules and describe a simple theoretical model designed for their interpretation. Then some most recent PES results on interactions between π-electron fragments in olefinic, acetylenic and aromatic hydrocarbons are considered in more detail. Finally, interactions of lone pairs is discussed. Two points should be made: (a) bifunctional molecules possessing two symmetry-related equivalent groups or chromophores are examined as a rule with very few exceptions. Splitting of the corresponding MO energy levels is the main indicator of the intramolecular orbital interactions; (b) the most transparent theoretical framework yielding interpretation at the qualitative level is employed in keeping up with the general idea of this series.

2 Theoretical Aspects

2.1 Background

Theoretical framework is provided by the Roothaan-Hartree-Fock (RHF) method where a system of equations:

$$\hat{F}\psi_i = \varepsilon_i\psi_i \qquad (i = 1, \dots n) \tag{1}$$

is solved in an iterative fashion. Here $\hat{F}$ is the customary Fock operator and ψ_i are MOs occupied by two electrons with opposite spins. The resulting total molecular wavefunction $\bar{\Psi}$ is a simple Slater determinant built by MOs ψ_i. The form of the total wavefunction and the definition of the Fock operator imply that electron correlation is not taken into account apart a portion inherent in the antisymmetric requirement imposed on $\bar{\Psi}$. Solutions of the pseudoeigenvalue problem of Eq. (1) ψ_i are called canonical molecular orbitals (CMO) and negative eigenvalues $-\varepsilon_i$ come close to the ionization energies [7]. An implicit assumption made by Koopmans' is that photoionization is a sudden event and all CMOs are correspondingly frozen. This assumption is not quite justified because CMOs undergo some reorganization upon ionization giving rise to relaxation energy. It turns out, however, that for outer valence electrons relaxation and correlation energies practically cancel, thus making the RHF-CMO approach a suitable vehicle for exploring PES ionization energies. For a more detailed discussion the reader is refered to Ref. [14]. One should reiterate a well-known fact that the total wavefunction is determined up to a unitary (orthogonal) transformation which remains unspecified. This is remarkable because a judicious choice of the transformation matrix translates completely delocalized CMOs into equivalent localized and semilocalized MOs, which represent functional groups in a more direct, transparent and intuitively appealing way. This feature enables a quantum-mechanical interpretation of molecular fragments which serve as building blocks (localized bonds, atomic groupings, etc.) of complex systems. Reversing this chain of reasoning one can form by an educated guess local (group) orbitals describing particular fragments and then include subsequently their mutual interactions. Since energies of interactions between chemical bonds in general and functional groups in particular are relatively small in well-localized systems, perturbation theory [15, 16] lends itself as an appropriate tool. This type of modelling is conceptually advantageous because a book-keeping of relevant physical interactions is easily done. Additionally, by using switch-on and -off procedure pertaining to particular modes of interactions one can identify dominant mechanisms. This was exactly the procedure adopted by Hoffmann et al. [2] who put forward through-space and through-bond interactions. The former involves the interaction of two orbitals (AOs or LMOs) through direct overlapping, whereas the latter is mediated by some other relay (bond) orbitals. A slightly different definition, based on an *a posteriori* analysis instead of a direct perturbational treatment was given by Brunck and Weinhold [17] and Imamura et al. [18].

Original formulation of TS and TB interactions was given within the framework of the EHT model [19], which will be employed in the present discussion too for simplicity. Let's denote a TS interaction between two basis orbitals φ_i and φ_j as $B_{ij} = \langle \varphi_i| \hat{H} |\varphi_j\rangle$. As it is usual in semiempirical methods, the resonance integral B_{ij} is expressed by the overlap integral $S_{ij} = \langle \varphi_i | \varphi_j \rangle$:

$$B_{ij} = -kS_{ij} = -k \langle \varphi_i | \varphi_j \rangle , \tag{2}$$

where k is a positive calibration constant. Hence, positive overlapping is energetically favourable and stabilizes a molecular system (vide infra).

The self energies of φ_i and φ_j are:

$$E_i = \langle \varphi_i| \hat{H} |\varphi_i\rangle \quad \text{and} \quad E_j = \langle \varphi_j| \hat{H} |\varphi_j\rangle . \tag{3}$$

Through-space interaction of φ_i and φ_j will lead to the following linear combinations and associated orbital energies:

$$\varphi_+ = c_i\varphi_i + c_j\varphi_j; \quad \varepsilon_+(\varphi_+) = \bar{E} + B_{ij}\left[\left(\frac{\Delta E}{2B_{ij}}\right)^2 + 1\right]^{\frac{1}{2}}, \tag{4a}$$

$$\varphi_- = c_i\varphi_i - c_j\varphi_j; \quad \varepsilon_-(\varphi_-) = \bar{E} - B_{ij}\left[\left(\frac{\Delta E}{2B_{ij}}\right)^2 + 1\right]^{\frac{1}{2}}, \tag{4b}$$

where coefficients c_i and c_j are subject to the normalization conditions; $\bar{E} = (E_i + E_j)/2$ and $\Delta E = E_i - E_j$ are average value and energy difference of the interacting levels, respectively.

This implies that for positive S_{ij} (negative B_{ij}) the orbital sequence will be *natural*, i.e. φ_+ linear combination will be put at lower energy. If φ_i and φ_j are related by symmetry, i.e.:

$$\varphi_+ \equiv \varphi_S = (\varphi_i + \varphi_j)/\sqrt{2} , \tag{5a}$$

$$\varphi_- \equiv \varphi_A = (\varphi_i - \varphi_j)/\sqrt{2} , \tag{5b}$$

where φ_S and φ_A are symmetry-adapted orbitals. Their respective energies are:

$$\varepsilon_+(\varphi_+) \equiv \varepsilon_S(\varphi_+) = E + B_{ij} , \tag{6a}$$

$$\varepsilon_-(\varphi_-) \equiv \varepsilon_A(\varphi_-) = E - B_{ij} . \tag{6b}$$

It was assumed in the derivation of Eqs. (2–6) that φ_i and φ_j are pure atomic orbitals (AOs), but the same formalism *mutatis mutandis* holds for semilocalized (e.g. π-bond) orbitals. Hence, symmetry adapted MOs φ_A and φ_S delocalized over both functional groups have a simple form given by Eq. (5). Their levels are split by $2B_{ij}$ (cf. Eq. (6a) and (6b)) caused by the direct through-space coupling providing that the overlap integral S_{ij} entering the resonance integral B_{ij} is appreciable. However, overlapping decays exponentially and the overlap integrals are negligible for remote functional groups. Nevertheless, there are substantial splittings of the energy levels which would have been degenerate in a hypothetical case of perfectly isolated functional groups, in molecules where two semilocalized fragment orbitals are separated by a very long chain(s) of localized σ-bonds [12, 13]. Obviously, another mechanism is responsible for these long-range interactions. It turns out that information between the two semilocalized orbitals describing corresponding functional groups is transmitted by a relay of σ-orbitals [2, 3, 8, 13]. Let us for the sake of simplicity suppose that there is just one orbital of this kind φ_m, since a generalization poses no problem whatsoever. A mediator orbital φ_m interacts with φ_S and φ_A via resonance integrals:

$$B_{m+} = \langle \varphi_m | \hat{H} | \varphi_+ \rangle = (H_{mi} + H_{mj})/\sqrt{2} = -k(S_{mi} + S_{mj})/\sqrt{2} \qquad (7a)$$

and

$$B_{m-} = \langle \varphi_m | \hat{H} | \varphi_- \rangle = (H_{mi} - H_{mj})/\sqrt{2} = -k(S_{mi} - S_{mj})/\sqrt{2}. \qquad (7b)$$

It is tacitly assumed that a transmitter orbital φ_m has proper symmetry characteristics so that integrals H_{mi} and S_{mi} are different from zero. Hence the resonance integrals B_{m+} and B_{m-} assume different values thus leading to lifting of the φ_S and φ_A degeneracy. Usually a mediating orbital φ_m has much lower energy so that both differences $|\varepsilon_+ - \varepsilon_m|$ and $|\varepsilon_- - \varepsilon_m|$ are considerably larger than the matrix elements B_{m+} and B_{m-}. Then corrections to the energy levels ε_+ and ε_- (Eqs. (6a) and (6b)) are readily obtained by the second-order perturbation theory:

$$\varepsilon'_+ = E + B_{ij} + B^2_{m+}/(\varepsilon_+ - \varepsilon_m) \qquad (8a)$$

and

$$\varepsilon'_- = E - B_{ij} + B^2_{m-}/(\varepsilon_- - \varepsilon_m), \qquad (8b)$$

where it was assumed without any loss in generality that orbitals have real forms. Several comments are in place here. Firstly, φ_m could be a bond orbital σ or excited-state orbital σ^* of appropriate symmetry. However, the latter play a minute role in view of denominators in the last term of Eqs. (8a) and (8b) which become very large. Secondly, both TS and bond-assisted TB interactions are always simultaneously present although the former decays quickly being very small if not negligible in long-range interactions. This is not the case for TB mode of interaction because φ_m is one or more orbitals vicinal to φ_+ and φ_- and the corresponding overlaps are always significantly large. Finally, even at small distances one can find situations with inverted energy level spacing with φ_+ above φ_-. This is a consequence of the fact that $B_{m+}^2/(\varepsilon_+ - \varepsilon_m)$ can be substantially larger than $B_{m-}^2/(\varepsilon_- - \varepsilon_m)$. Therefore the TS splitting can be reinforced, practically cancelled, or overcompensated by the TB interactions [21]. A very simple rule developed independently by Paddon-Row et al. [11, 22] and Verhoeven and Pasman [23] shows that normal or inverted energy level splitting depends on a number of σ-bonds which separate two functional groups, which in turn may be either odd or even. This is the so-called parity rule which will be discussed in the following section. To conclude, TS and TB mechanisms of intramolecular orbital interactions are competitive and the final result depends on the molecule in question. There are some general patterns, however, which will be briefly considered later (vide infra).

2.2 Parity Rule

Dependence of the relative sign of TS and TB terms in nonconjugated bifunctional molecules on the parity of the number (odd-even) of intervening sigma bonds has been clearly recognized by Hoffmann, Imamura and Hehre in their pioneering paper on orbital interactions [2]. It was postulated by using primarily symmetry arguments that the energy level ordering resulting from TB interaction is inverted for an odd number of the relay σ-bonds. In particular, a coupling of two interacting orbitals separated by three sigma bonds was elaborated by using a perturbational MO (PMO) model. Figure 1 depicts the interaction of two lone pairs in 1,2-diaminoethane. To begin with, symmetry-adapted combinations of n_1 and n_2 are formed: $n_+(S) = (n_1 + n_2)/\sqrt{2}$ and $n_-(A) = (n_1 - n_2)/\sqrt{2}$. The symmetry classification (A and S) of delocalized orbitals is determined with respect to the C_2 axis passing through the middle of the C_2–C_3 bond and lying in the molecular plane. The corresponding rotation by 180° interchanges two equivalent lone pairs. The natural level ordering, n_+ below n_-, would follow from the TS interaction, which in turn seems to be rather weak. Symmetry-allowed mixing of n_+ and n_- with bonding and antibonding σ_{23} and σ_{23}^* MOs of the C_2–C_3 bond, respectively, provides at first sight an intuitively tempting description of TB interaction. Since σ_{23} and n_+ possess the same symmetry (S) they repel each other. The σ_{23} being more stable is pulled down whereas the n_+ level is pushed up. The same holds for n_- and σ_{23}^* orbitals of A-symmetry

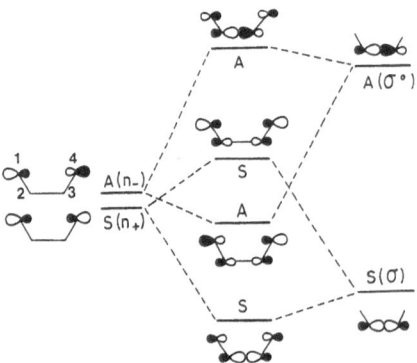

Fig. 1. Interaction diagram illustrating through-bond coupling in 1,2-diaminoethane [8]

which leads to stabilization of the former and destabilization of the latter level. This sort of mixing may easily outweigh a regular TS effect leading to the inverted ordering of the S and A level as shown in Fig. 1. An essential element of the procedure described above is the choice of a unique central bond, which is bisected by the (pseudo)symmetry element that interchanges the interacting p,n, or π-type orbitals. Hoffmann et al. [2, 3] state that "other bonds always enter in symmetry-related pairs" as, e.g. the C_1-C_2 and C_3-C_4 bonds in Fig. 1. Concomitantly, there is a symmetric S combination of $\sigma_{12} + \sigma_{34}$ bond orbitals and that of their antibonding $\sigma_{12}^* + \sigma_{34}^*$ counterparts. The net effect on the n_+ level would roughly cancel out. The same conclusion should hold for mixing of the antisymmetric A and the n_- levels. Thus, other bonds can be left out of the procedure in the first approximation.

Although this type of rationale based on symmetry is plausible and has an intellectual appeal, it is insufficient in several respects. In the first place, it is inadequate in systems containing an even number of sigma bonds. Consider, for example, a case of 2N intervening σ-bonds. There are in total 4N sigma levels consisting of equal number (N) of σ_+, σ_+^*, σ_- and σ_-^* MOs. Hence a p_+ (or n_+) combination of AOs belonging to equivalent functional groups has as many σ_+^* levels that lie above it as there are σ_+ levels below. Accordingly, a net perturbation on the p_+ (or n_+) level would be roughly zero. Analogously, the p_- (or n_-) level would be left intact implying that TB splitting involving an even number of sigma bonds would be negligible. This conjecture is at variance with ab initio calculations [22, 24] and experimental data [22, 24] which conclusively show that TB interactions over four and six sigma bonds are very large. Typical examples are provided by *exo,exo*-1,4:5,8-dimethano-1,4,4a,5,8,8a-hexahydronaphthalene (*1*) and its diepoxy congener 2 in which π,π-splitting energies are 0.87 eV [22] and 0.55 eV [25], respectively.

A more general and satisfactory model for rationalization of the TB coupling was independently put forward by Verhoeven and Pasman [23] and Paddon-Row et al. [11, 22]. They pointed out that mixing of occupied and unoccupied levels is symmetry allowed, but in practice can be safely ignored in view of a very large difference in their energies. Instead, they concentrate on the σ-framework and explicitly consider all bonds on an equal footing by employing a simple C-approximation of Sandorfy and Daudel [26]. In this approach the CC skeleton is described by sp^3 hybrid AOs neglecting all other (e.g. CH) bonds. The procedure requires solution of a Hückel-type secular determinant in which the diagonal elements are Coulombic integrals for sp^3 carbon hybrids (α_c) and the only non-zero off-diagonal elements are resonance integrals for sigma-type bonding between adjacent carbons (β_{cc}) or for two sp^3 hybrids emanating from the same carbon ($m\beta_{cc}$ where $m < 1$). The final MOs resemble those of the MOs resulting from a topologically equivalent array of p-basis orbitals. The most relevant feature for the present discussion is an interesting finding that the splitting and the sequence of MO S and A levels depends only on the parity of the number of sigma bonds, being σ_- above σ_+ for all even values of N (Fig. 2a) and the opposite for all odd values of N* (Fig. 2b) [27].

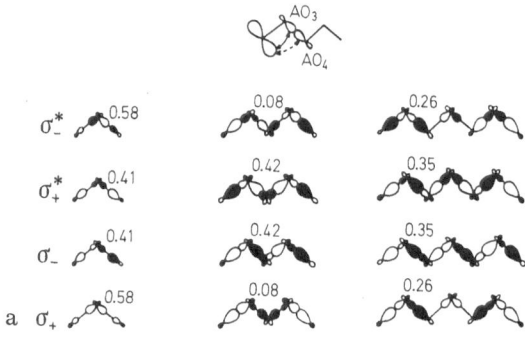

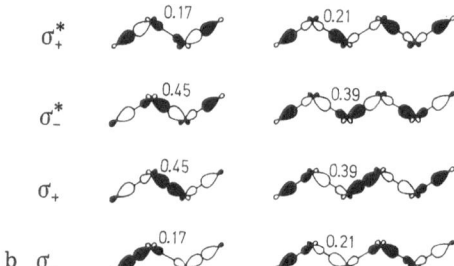

Fig. 2a, b. Sigma orbitals for chains containing (**a**) two, four and six CC sigma bonds, and (**b**) three and five sigma bonds, as obtained from the "C" approximation with m equal to 0.7. The number associated with each MO is the value of the coefficient of AO3 in that MO. Reprinted with permission from Ref. [27]. Copyright 1990 American Chemical Society

Inspection of nodal properties of the highest occupied σ-MO has led Verhoeven and Pasman [23] to conclude that alternation of the sign of through-bond interaction follows directly from the symmetry properties of this particular MO. More precisely, due to the fact that in systems with odd numbers of sigma bonds the terminal coefficients in σ-HOMO have equal sign, TB interaction leads to the destabilization of the symmetric linear combination of the functional group orbitals. Similarly, an opposite sign of the terminal coefficients found in σ-HOMO in systems comprising an even number of σ-bonds implies destabilization of the antisymmetric linear combination of the functional group orbitals. Paddon-Row considers in addition to HOMO the second highest occupied molecular orbital (SHOMO). This approach will be illustrated by diene *1* as a characteristic example [22]. The PE spectrum of *1*, as mentioned above, exhibits two *p*-bands differing in energy by 0.87 eV inspite of the significant separation of π-orbitals which excludes their direct interaction. Furthermore, comparing the $I_{v,j}$'s of *1* [22] with the first ionization energy of the corresponding monoene (*3*) [28] indicates that the stabilization of the π_+ level in *1* (0.5 eV) relative to the π level in the monoene significantly overrides destabilization of the π_- level (0.37 eV). This could be hardly explained by Hoffman's model since it is

---

* σ_+ and σ^- are here symmetric and antisymmetric combinations with respect to the appropriate symmetry element, which is a reflection plane for even values of N and a (pseudo) C_2 symmetry axis for odd values of N.

extremely unlikely that mixing of the π_+ level with a high-energy σ_+^* orbital could be more important than that between the π_- MO and the σ_-.

In order to circumvent this difficulty Paddon-Row suggested a model which includes interaction of both π-levels in *1* with occupied σ-orbitals of the appropriate symmetry [22, 27]. This proposal is illustrated by the interaction diagram in Fig. 3.

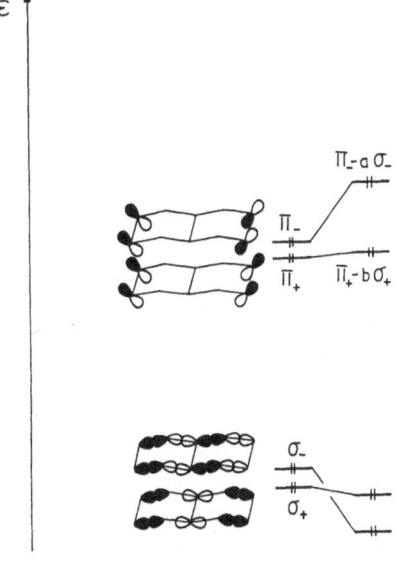

Fig. 3. Interaction diagram depicting through-bond interaction of π-levels in *1*. Reprinted with permission from Ref. [27]. Copyright 1990 American Chemical Society

Its most salient feature is a raise in energy for both symmetry-adapted linear combinations of π levels, with the π_- level being significantly more influenced. This occurs because the hybrid atomic orbital AO3 which is principally responsible for the $\sigma-\pi$ mixing is associated with significantly larger coefficients in σ-HOMO than in σ-SHOMO (see Fig. 2). More specifically the matrix element describing $\sigma-\pi$ mixing $H_{\pi\sigma} = \langle\pi|\hat{H}|\sigma\rangle = \langle\pi|\hat{H}|\sum_m c_m\varphi_m\rangle$, where c_m is the coefficient of the m-th hybrid, is proportional to the sum of the corresponding overlap integrals $-k\sum c_m\langle\pi|\varphi_m\rangle$. Here c_m and $\langle\pi|\varphi_m\rangle$ related to the hybrid AO3 play a decisive role.

Paddon-Row's model is also applicable to a wide variety of molecules incorporating double bonds connected via all even as well as odd σ-coupling units (cf. Ref. [13]). It has also been succesfully applied to the molecules possessing unsaturated groups other than double bonds such as aromatic rings [13, 29]. The model provides a simple explanation of the all-*trans* effect, which is reduced to the extent of vicinal overlapping mentioned above and the size of the corresponding coefficients of AOs predicted by "C"-approximation [11, 13] (cf. Sect. 3.3). Namely, all-*trans* conformation yields always the largest TB interaction. A more detailed discussion of the model and a comprehensive review of its applications can be found in Ref. [13].

2.3 Heilbronner-Schmelzer Approach to the Problem of Intramolecular Orbital Interactions

Hoffmann's qualitative ideas on through-space and through-bond interactions and their development in terms of a local bond model within "C"-approximation provide a conceptual framework for understanding relevant orbital energy shifts in molecules possessing two or more functional groups. It would be useful, however, to have at hand a more rigorous theoretical method providing the same information at the quantitative or at least semiquantitative level. Such a procedure was described by Heilbronner and Schmelzer [30]. It involves several distinct steps which are thoroughly discussed by Honegger and Heilbronner in this book [31]. Therefore only essential features will be concisely mentioned here. In the first stage, canonical HF MOs are transformed into a set of localized molecular orbitals (LMOS):

$$\lambda = \Psi L, \qquad\qquad\qquad (9)$$

where λ and Ψ are row vectors of LMOs and CMOs, respectively, and L is an orthogonal matrix. This can be achieved by a number of prescriptions including those of Foster and Boys [32], Edmiston and Ruedenberg [33], or perhaps the "natural" localization procedure of Weinhold [34]. There are two principal advantages in using LMOs (λ_i) as basis orbitals: they are transferable in a sense that their self-energies $\langle \lambda_i | \hat{F} | \lambda_i \rangle$ and the inter-bond interactions $\langle \lambda_i | \hat{F} | \lambda_j \rangle$ are approximately independent of the surrounding molecular environments [35, 36]. Secondly, the LMOs generated from any of the procedures mentioned above are orthogonal thus lending themselves to a simple Hückel-type treatment which may be illuminative for some purposes. Additionally, usual AO basis sets are not only non-orthogonal but also numerous in most cases, whereas the LMO basis is small and compact. We note in passing that non-orthogonal LMOs are susceptible to criticism [37], but imposed orthogonality has also its imperfections in interpreting TS and TB effects (vide infra).

The magnitude of TS interaction between two localized orbitals λ_i and λ_j is given simply by the corresponding off-diagonal matrix elements $\langle \lambda_i | \hat{F} | \lambda_j \rangle$. Since λ_i and λ_j are two LMOs related by the symmetry operation, their matrix element is always different from zero. Hence, a definition of a threshold value is necessary below which TS interaction is considered as negligible.

Quantitative assessment of TB interaction is much more involved. The dimension of the problem can be diminished by transforming LMOs to the symmetry-adapted localized orbitals (SLMOs) denoted by:

$$\varrho = \lambda R, \qquad\qquad\qquad (10)$$

where the orthogonal matrix R partitions the set ϱ into subsets $\varrho^{(1)}, \varrho^{(2)}, \ldots \varrho^{(r)}$, which belong to different irreducible representations of the molecular symmetry group. Then one can focus attention to the subset $\varrho^{(r)}$ only, which contains by supposition the considered, e.g., $\lambda_+ = (\lambda_i + \lambda_j)/\sqrt{2}$ combination. The TS interaction leading to λ_+ is easily estimated as mentioned above. Its TB mode of interaction is taken into account as follows. The Fockian submatrix $F_{\varrho(r)}$ is formed and all matrix elements involving λ_+ are set equal to zero. Then the submatrix $F_{\varrho(r)}$ is diagonalized thus yielding the so-called pre-canonical molecular orbitals (PCMOs). The matrix elements between λ_+ and PCMOs are different from zero and represent TB interaction contributions which can be conveniently analysed in terms of particular bond or fragment orbitals.

One of the first PE studies where the Heilbronner-Schmelzer approach has been applied was that of a series of 3,6-bridged-1,4-cyclohexa-1,4-dienes $4(1)$–$4(4)$, [30]. Hence this molecular system is chosen for illustrative purposes.

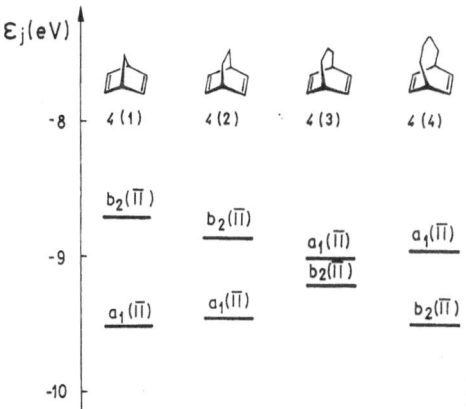

$$4 \ (1) \ n=1$$
$$4 \ (2) \ n=2$$
$$4 \ (3) \ n=3$$
$$4 \ (4) \ n=4$$

A detailed analysis of the PE spectra of $4(1)$–$4(4)$ [38] based on comparison of the measured PE data with that of structurally related compounds (cf. Refs. [21, 39]), has revealed that the sequence of the two highest occupied π-molecular orbitals changes from natural ordering in compounds $4(1)$ and $4(2)$ to inverted in $4(3)$ and $4(4)$ (Fig. 4). This was rationalized in terms of compensating influences of the π MOs interacting both through space and through bond, the former being predominant for $n < 3$ whereas the latter is becoming more important for larger values of n.

$\varepsilon_j(eV)$

-8 $4\,(1)$ $4\,(2)$ $4\,(3)$ $4\,(4)$

$b_2(\bar{\pi})$

$b_2(\bar{\pi})$

$a_1(\bar{\pi})$ $a_1(\bar{\pi})$

-9 $b_2(\bar{\pi})$

$a_1(\bar{\pi})$ $a_1(\bar{\pi})$

$b_2(\bar{\pi})$

-10

Fig. 4. Inversion of the level ordering in the bicyclic dienes $4(n)$

In order to account for the trend in the ionization energies of these hydrocarbons, Heilbronner and Schmelzer have analysed (a) the origin of TS and TB interaction in $1(1)$ and (b) the dependence of the TS and TB interactions on the dihedral angle ω of a hypothetical cyclohexa-1,4-diene system bent along an axis passing through the centres 3 and 6. Calculations were carried out in the framework of the three semiempirical methods available at that time, namely SPINDO [40], CNDO/2 [41], and MINDO/2 [42]. The results of this study are reproduced in Tables 1 and 2, respectively.

A survey of the data in Table 1 reveals that both SLMOs π_+ and π_- are affected by through-bond coupling. In both cases the CH-σ-orbitals of the methylene group and the sp^3–sp^2 CC-orbitals of the six-membered ring are found to be important in

Table 1. Numerical results for the through-space and through-bond interaction between the LMOs $\pi_a \equiv \lambda_a$ and $\pi_b \equiv \lambda_b$ of norbornadiene $4a$ according to the SPINDO, MINDO/2 and CNDO/2 treatments. All values in eV [30]

a	SPINDO	MINDO/2	CNDO/2
$F_{\lambda,ii} = F_{\lambda,jj}$	− 10.44	− 10.70	− 16.41
$F_{\lambda,ij}$	− 0.54	− 0.78	− 2.09
F_{ϱ^+}	− 10.98	− 11.47	− 18.50
F_{ϱ^-}	− 9.90	− 9.92	− 14.32
$A_1 : \tau_+$	0.85	2.07	6.46
$B_2 : \tau_-$	0.34	0.66	1.99
$b_2(\pi)$	− 9.54	− 9.36	− 12.33
$a_1(\pi)$	− 10.13	− 9.40	− 12.05

a Subscripts i and j refer to the LMOs λ_i and λ_j and subscripts $(+)$ and $(-)$ to the SLMOs λ_+ and λ_-, respectively; $F_{\lambda,ii}$, $F_{\lambda,jj}$ refer to self energies of λ_a and λ_b and $F_{\lambda,ij}$ to off-diagonal elements in the corresponding Hartree-Fock matrix F_λ linking λ_i and λ_j; τ_+ and τ_- designate through bond-induced orbital energy shifts and $\varepsilon_{b2(\pi)}$ and $\varepsilon_{a1(\pi)}$ orbital energies of CMOs $a_1(\pi)$ and $b_2(\pi)$, respectively

Table 2. Dependence of the TS and TB contributions to the splitting of the π-orbitals of a hypothetical cyclohexa-1,4-diene with variable dihedral angle ω. All values in eV[a] [30]

Method		ω		
		180°	150°	120°
SPINDO	TS-split	0.33	0.50	1.33
	τ_+	1.04	0.97	0.91
	τ_-	0.00	0.07	0.27
MINDO/2	TS-split	0.66	0.94	2.01
	τ_+	1.86	1.81	1.75
	τ_-	0.00	0.21	0.73
CNDO/2	TS-split	2.50	2.99	4.04
	τ_+	5.80	5.72	6.09
	τ_-	0.00	0.87	4.50

[a] τ_+ and τ_- have the same meaning as indicated in Table 1

this respect. It should be noted, however, that the three semiempirical SCF treatments differ considerably with regard to the size of the individual contributions. In particular the CNDO/2 model predicts a final level ordering in contradiction with the PES results. Perusal of the data displayed in Table 2 shows that the three semiempirical treatments differ considerably with regard to the quantitative aspects of the TS and TB interactions. However, at the qualitative level, all three models agree in predicting a crossing of the π-CMOS $a_1(\pi)$ and $b_2(\pi)$ near $\omega \approx 130°$. As expected the TS interaction between the LMOs λ_i and λ_j decreases with increasing ω, whereas the TB induced shift τ_+ is almost independent of ω. Surprisingly enough, the major reason for the observed crossing of $a_1(\pi)$ and $b_2(\pi)$ MOs is the unexpectedly large increase of τ_- with decreasing dihedral angle ω. The latter contribution has usually been assumed to be negligibly small in qualitative discussions, because the methylene groups lie in the nodal plane of the orbital $b_2(\pi)$ for all values of n and are thus not available for hyperconjugative TB interaction within this orbital. More detailed analysis has revealed that τ_- growth results from the increasing interaction of the PCMOs λ_i and λ_j with the σ-LMO of the two sp^2-sp^3 C–C single bonds on the other side of the six-membered ring. In fact, it is this contribution which is responsible for the observed orbital crossing near $\omega \approx 130°$.

It is important to keep in mind that there are some difficulties in interpreting TS direct coupling by using orthogonalized LMOs. It appears that the latter are not strictly localized and have small tails distributed over all other atomic centres in order to ensure orthogonality. Concomitantly, matrix element $\langle \lambda_i | \hat{H} | \lambda_j \rangle$ includes a contribution of tails which in turn describes in fact TB type of interaction. The latter can considerably obscure a simple TS picture as discussed in some detail by Paddon-Row et al. [43] in the case of cyclohexa-1,4-diene. Employing STO-3G [44a] and 3-21G [44b] basis sets they calculated localized orbitals of π-orbitals by making use of Foster-Boys [32] and Weinhold's natural LMO [34] procedures. It turned out that each of the local π-orbital is "contaminated" by the significant contributions of the pseudo-π AOs of CH_2 groups and tails placed at the vis-a-vis double bond. Calculation of the TS matrix element $\langle \lambda_{\pi 1} | \hat{F} | \lambda_{\pi 2} \rangle$ has shown that intrusion of the CH_2 fragment makes an appreciable contribution which is by no means negligible. Hence, some care has to be exercized in interpreting TS splitting by the Heilbronner-Schmelzer method. It is gratifying that in molecules with more distant functional groups the TS term can be safely abandoned and a conflict is thus automatically resolved [43].

To conclude, the Heilbronner-Schmelzer method is an elegant means for quantitative assessment of TS and TB interactions and their interplay in detemining energy levels in bi(poly)-functional compounds. It should be stressed, however, that optimal basis sets and a localization procedure have yet to be found.

3 PE Spectroscopic Evidence for Interactions of Olefinic π-Orbitals

There is a plethora of measured PE spectra of compounds incorporating double bonds which are conveniently interpreted by the fruitful concept of TS and TB effects. We present in the following sections some characteristic examples which illustrate the utility of the simple orbital model indicating that the latter is one of the most important tools in chemical armamentorium. The employed classification of the intramolecular orbital interactions corresponds to that used by Martin and Mayer in Ref. [12].

3.1 1,3-π,π Interactions

Since the earliest days of photoelectron spectroscopy [4] norbornadiene *4*(1) has played a pivotal role in studying 1,3-π,π orbital interactions [30, 31]. Thus it is appropriate to start this section by considering orbital interactions in structurally related and chemically at least equally attractive system. Sesquinorbornatriene (*5*), which formally incorporates two co-operating norbornadiene subunits, seems to satisfy both conditions [46]. The PE spectra of *syn-* (*5a*) and *anti-* (*5b*) isomers of sesqui-norbornatriene are depicted in Fig. 5 [47].

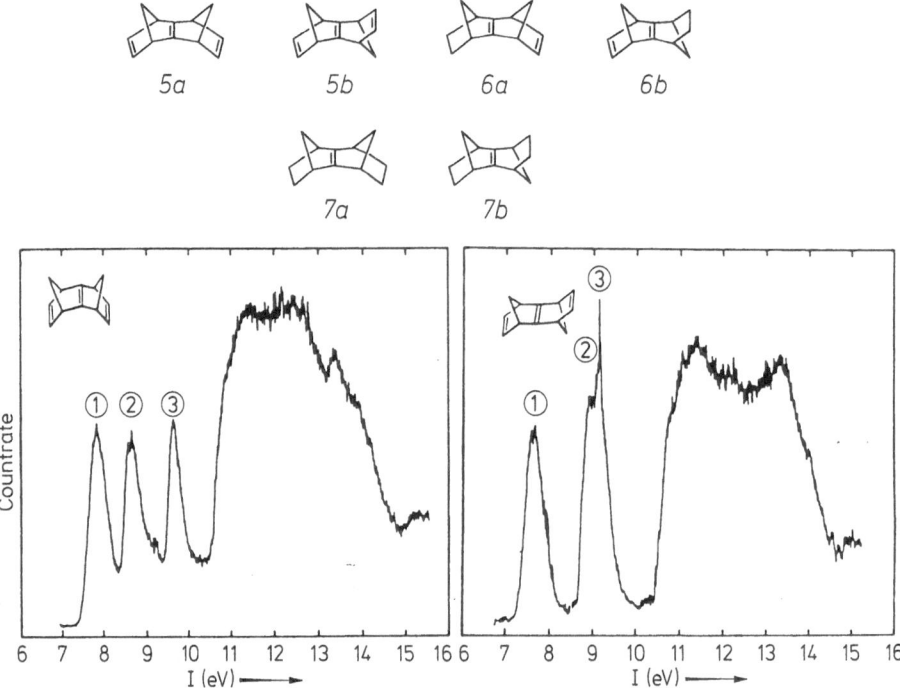

Fig. 5. Photoelectron spectra of *5a* and *5b*. Reprinted with permission from Ref. [47]. Copyright 1990 American Chemical Society

It appears that the PE spectra of two isomers are markedly different. In *syn*-triene (*59*) three well-separated bands are seen below 10 eV, while in the PE-spectrum of *anti*-triene (*5b*) the latter two bands overlap strongly. This striking difference is rationalized in terms of strong through-space interaction (∼0.5 eV) of the peripheral double bonds in the *syn*-isomer. In the *anti*-isomer (*5b*) this type of interaction is, due to the geometrical reasons, virtually, and perhaps, totally absent. This conclusion was corroborated by comparing the PE data of *5a* and *5b* with that of sesquinorborna-dienes (*6a* and *6b*) [47] and the parent sesquinorbornenes (*7a* and *7b*) [48] (Table 3), which showed that ionization energies related to the central and peripheral double bonds in the latter compounds differ by ca. 1 eV. Arguments based on perturbation theory indicate that the interaction between these two levels is negligible. It is worth mentioning that the experimental trends are reproduced with STO-3G calculations (based on MINDO/3 [49] geometry) which predict HOMO of *5a* ($11a_1$) to be mainly localized at the central carbons, while $9b_2$ and $10a_1$ MOs have larger coefficients at

Table 3. Comparison between Vertical Ionization Energies ($I_{v,j}$) and Calculated Orbital Energies ($-\varepsilon_j$) of 5–7.–All values in eV[a]

Compd.	$I_{v,j}$	Orbital Assignment	$-\varepsilon_j$(STO-3G)	$-\varepsilon_j^{corr.}$ [b]
7a	8.12	$11a_1(\pi)$	6.71	8.05
7b	7.90	$10b_u(\pi)$	6.56	7.87
6a	7.87	$20a'(\pi)$	6.48	7.77
	9.09	$19a'(\pi)$	7.70	9.24
6b	7.80	$20a'(\pi)$	6.39	7.67
	9.0	$19a'(\pi)$	7.42	8.90
5a	7.84	$11a_1(\pi)$	6.31	7.57
	8.66	$9b_2(\pi)$	7.24	8.69
	9.63	$10a_1(\pi)$	7.95	9.54
5b	7.65	$10b_u(\pi)$	6.23	7.48
	9.0	$9b_u(\pi)$	7.27	8.72
	9.2	$10a_g(\pi)$	7.62	9.14

[a] Ref. [47]
[b] $-\varepsilon_j^{corr.} = +\varepsilon_j(\text{STO-3G}) + 0.2\varepsilon_j(\text{STO-3G})$

the peripheral double bonds. It is also interesting to note that $9b_2$ MO is influenced, although slightly, with lower lying σ orbital of b_2 symmetry, whereas $11a_1$ and $10a_1$ are unaffected by π/σ interaction (see below).

An entirely different topology for coupling of two and three π-units via 2 sigma bond chains occurs in 2,8-dimethylene- (8) and 2,8,9-trimethylene- (9) [3.3.3]pro-pellanes, respectively [50]. The PE spectra of both compounds exhibit only a minute split (0.2 eV) between π-bands, which is attributed to the rather large distance between the double bonds (~ 3.0 Å). Interaction of the same order of magnitude has been also observed in the structurally related trivinylmethane (10) [51].

An interesting approach for determining the sequence of the highest occupied MOs in 1,3-dimethylenecyclobutane (11) was presented by Martin et al. [52]. It rests on the assumption that a decrease in the dihedral angle (ω) within the cyclobutane moiety should lead to the intensification of trans-annular through-space interaction of the two π-MOs and to the weakening of their interaction mediated through σ-relay. In other words, a decrease in ω should lead to a reduction of $\Delta I_v(\pi)$ if the parent compound exhibits an inverted sequence of the MOs in question and to its increase if their ordering is natural (see Fig. 6). The PE spectrum of 11 was measured by Hemmersbach et al. [53] and its features were rationalized in terms of prevailing through-bond conjugation (S > A). Comparison of the PE spectrum of *11* with that of 5,6-dimethylenebicyclo-[2.1.1]hexane (*12*) [52], which is characterized by significantly smaller ω ($\sim 130°$), demonstrates a large reduction in the energy gap between the bands related to π-ionizations. More specifically, the first two bands in *11* appear at 9.09 and 9.94 eV, while in *12* they are encountered at 9.25 and 9.65 eV, respectively. As indicated above, this finding is compatible only with inverted ordering of the π-MOs (case 1c in Fig. 6), thus confirming the assignment proposed by Hemmersbach et al.

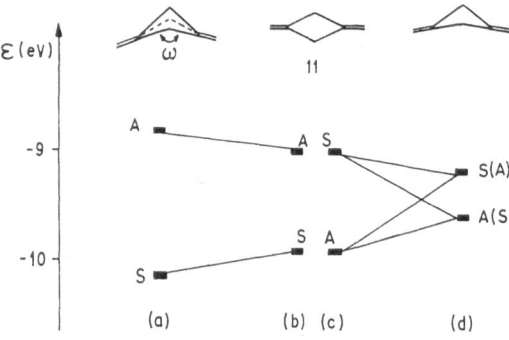

Fig. 6. Orbital correlation diagram illustrating dependence of the assumed ordering of the highest occupied levels in *11* on the change in dihedral angle (ω). If the sequence of levels is natural, then a decrease in ω leads to an increase in the splitting (case (**b**) → (**a**). The opposite is the case if the order of levels is inverted: a decrease in the dihedral angle causes a reduction of the $\Delta\varepsilon$ value (case (**c**) → (**d**)) [52]

3.2 1,4-π,π Interactions

Molecules which incorporate olefinic subunits connected via three σ-bonds comprise a vast majority of compounds studied by PE spectroscopy [11, 12]. A great deal of these studies has been discussed in a review article of Martin and Mayer [12]. Therefore, we shall concentrate on results of the PE studies published after this survey.

Rather interesting candidates for 1,4-π,π interaction are provided by 3,3′-bicyclopropenyl (*13*) and its alkyl derivatives *14–18* [54, 55]. Theoretical considerations including linear combination of bond orbital (LCBO) treatment [31, 56] 4-31G [45] and

semiempirical calculations, and PE measurements indicate indeed that through-bond coupling in these compounds leads to an appreciable splitting of 1.0–1.5 eV [54, 55]. An empirical correlation diagram illustrating its origin in *anti*-conformer of the parent hydrocarbon *13* is shown in Fig. 7. The interaction is viewed as taking place via a single ethanediylidene unit, comprising the central C3–C3′ bond and two attached C–H bonds. The energies of the relevant semilocalized sigma orbitals are derived by using a simple Hückel-type LCBO approach which for the basis orbital energies employs $A_{CC} = A_{CH} = -17.0$ eV and takes into account only geminal interactions $(B_{gem.} = -2.0$ eV) [31, 56]. Through-space interaction between the π-orbitals is assumed to be negligible, whereas the through-bond coupling is included by taking $A_\pi = -11$ eV for cyclopropene double bond and $B_{\pi\sigma}$-value of 1.7 eV. By diagonalization of the corresponding matrix [Eq. (11)]:

$$
\begin{array}{c}
1 \\ 2 \\ 3 \\ 4 \\ 5
\end{array}
\left[
\begin{array}{ccccc}
-17.0 & & & & \\
2.0 & -17.0 & & & \\
2.0 & 0 & -17.0 & & \\
2.7 & 0 & 1.7 & -11.0 & \\
1.7 & 1.7 & 0 & 0 & -11.0
\end{array}
\right], \qquad (11)
$$

the following energies for the highest occupied π-orbitals were obtained $\varepsilon(\pi_+) = -9.3$ eV and $\varepsilon(\pi_-) = -10.6$ eV, where π_+ and π_- denote wave functions with dominant contribution from the π-orbitals. For reasons obvious from the pertinent interaction diagram in Fig. 7, through-bond coupling destabilizes the totally symmetric π_+ combination much more effectively than the π_- combination leading to the appreciable split of 1.3 eV. The LCBO prediction is consistent with the results of nonempirical STO-3G and 4-31G calculations, 1.40 and 1.14 eV [54], indicating that the simple formalism outlined above is basically correct.

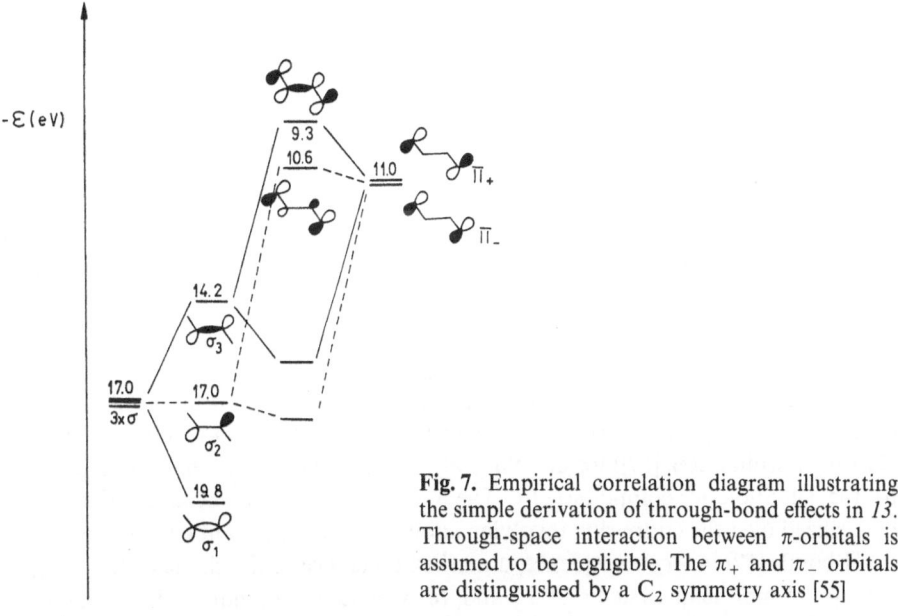

Fig. 7. Empirical correlation diagram illustrating the simple derivation of through-bond effects in *13*. Through-space interaction between π-orbitals is assumed to be negligible. The π_+ and π_- orbitals are distinguished by a C_2 symmetry axis [55]

The essential features of the interaction diagram in Fig. 7 also apply to the alkyl derivatives of 3,3'-bicyclopropenyl *14–18*. This is nicely reflected in their ionization energies displayed in Table 4, which also includes results of MINDO/3 calculations. It is interesting to mention that the rather large difference in $\Delta I_v(\pi)$ for *15* and *16* has its origin in the marked difference in the shape and energy of the highest occupied σ-orbital associated with the cyclopentane and cyclohexane fragment, respectively.

Table 4. Measured vertical ionization energies ($I_{v,j}$) and the calculated orbital energies (ε_j) of *14–18*. All values in eV [55]

Compound	$I_{v,j}$	Orbital assignment	$-\varepsilon_j$(MINDO/3)
14	8.8_4	$11a(\pi_+)$	8.60
	9.5	$10a(W_A^-)$	8.94
	10.0	$10b(W_A^+)$	9.25
	10.3	$9b(\pi_-)$	9.83
	11.5	$9a(W_S^+)$	10.68
	12.2	$8b(W_S^-)$	11.48
15	8.6_3	$12a(\pi_+)$	8.39
	9.1_3	$11a(W_A^-)$	8.58
	9.7_8	$11b(\pi_-)$	9.64
	10.5_1	$10b(W_A^+)$	9.99
	11.4	$10a(W_S^+)$	10.49
	11.7	$9b(W_S^-)$	10.92
		$8b(\sigma)$	11.18
16	8.7_3	$14a(\pi_+)$	8.53
	9.2	$13a(W_A^-)$	8.75
	9.7	$12b(W_A^+)$	9.11
	10.0	$11b(\pi_-)$	9.85
	10.9_2	$12a(W_S^+ - \sigma)$	10.37
	11.6	$11a(\sigma + W_S^+)$	11.16
		$10b(W_S^-)$	11.28
		$10a(\sigma)$	11.31
17	8.6_6	π_+	
	9.2_0	W_A^-	
	9.6_4	π_-	
	10.2	W_A^-	
	10.7		
18	8.6_6	π_+	
	9.1_4	W_A^-	
	9.6	W_A^-	
	10.8	π_-	
	11.2		

Namely, analysis of the highest occupied totally symmetric orbitals of these cycloalkanes indicates that only in the case of *15* such a high-lying σ-orbital is available (12a (σ_{cc})-orbital, Ref. [57]) for coupling with the π_+ linear combination, in contrast to the situation in compound *16*. The corresponding cyclohexane orbitals are delocalized over the entire ring involving additionally a significant contribution from C–H bonds [57]. As a consequence, a marked difference in the shape of the π_+-type orbitals of molecular systems of *15* and *16* is expected as indicated schematically below:

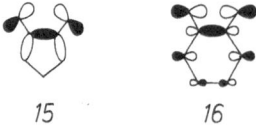

15   *16*

It is apparent that the TB destabilization is relatively inefficient in the case of *16*; moreover, the π-lobes are rotated in a manner so that TS stabilization is increased. In fact, the balance of their contributions is such that a slightly lower binding energy is calculated (and observed) for *15* than for *16* (Table 4). Is should be noted that this trend is opposite of what one might have expected in considering a larger destabilizing inductive efect in the case of *16* and a smaller torsional angle in the case of *15*, favouring TS stabilization.

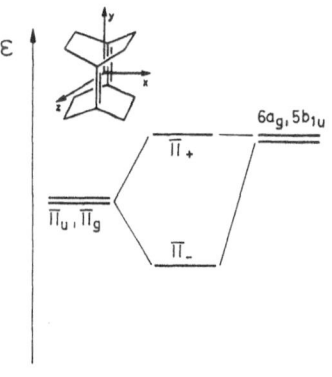

Cyclic polyenes *19–21* exemplify another class of highly interesting hydrocarbons [58, 59] which also incorporate π-subunits connected via 3 sigma bonds. Contrary to the PE spectral features of the 3,3'-bicyclopropenyl derivatives their PE spectra are charac- terized by negligible splitting in the π-manifold [60–62]. More specifically the PE spec- trum of diene *19* exhibits a broad band at 8.3–8.5 eV which is attributed to ionization from both the $b_1(\pi)$ and $a_g(\pi)$ orbitals [60]. The PE spectrum of the triene *20* has peaks at 7.9. 8.1 and 8.3 eV. The peak at 7.9 eV is attributed to ionization from the $a_1'(\pi)$ orbital and the 8.1- and 8.3-eV features to ionization from the $e'(\pi)$ orbitals, with the splitting between the latter arising from a Jahn-Teller distortion [61]. Finally, the PE spectrum of the tetraene *21* is characterized by a broad feature centered near 7.9 eV, which is apparently due to ionization from the $a_{1g}(\pi)$, $e_u(\pi)$, and $b_{1g}(\pi)$ orbitals [62, 63].

Detailed analysis based on application of LCBO approach [60] and its modification based on simple perturbation theory analysis [62] has shown that the origin of the near degeneracy lies in accidental cancelation of the TS and TB interaction terms. The relevant interaction diagram for *19* is shown in Fig. 8.

Fig. 8. Rationalization of the first double band in *19* [60]

In this particular compound it is estimated that the TS interaction splits the symmetric and antisymmetric linear combinations $\pi_+ = (\pi_u + \pi_l)/\sqrt{2}$ and $\pi_- = (\pi_u - \pi_l)/\sqrt{2}$ by about 3.2 and 3.4 eV.* Inspection of the symmetry-adapted linear combinations of the LMOs which belong to the irreducible representations A_g and B_{1u} indicates that

---

* π_u and π_l designate upper and lower pi-bond orbitals, respectively

only the lower-lying linear combination (π_-) can interact appreciably with the bridging groups, causing a shift of $6a_g$ to an energy close to that of $5b_1(\pi_+)$ (Fig. 9). This interpretation was further supported by semiempirical (MINDO/3) and ab initio (STO-3G) calculations. Considerable through-bond interaction between double bonds separated by three bonds and the σ-frame is also found in 1,5-dimethyl-3,7-

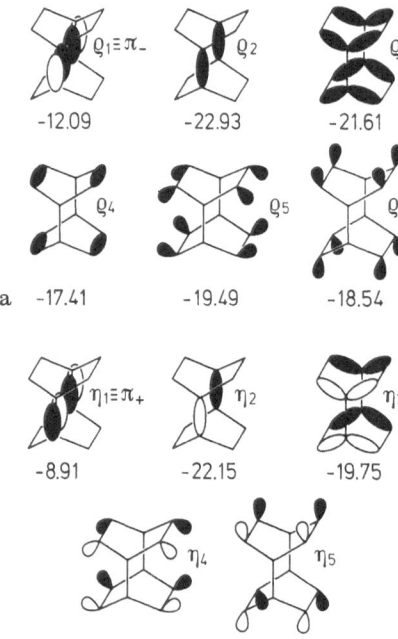

Fig. 9a, b. Symmetry-adapted linear combinations of the π- and σ-LMOs belonging to the irreducible representations $A_g(\mathbf{a})$ and $B_{1u}(\mathbf{b})$, respectively. LMOs are obtained by localization of STO-3G MOs according to the procedure of Foster and Boys (see Ref. [60])

dimethylidene tetracyclo[3.3.0.0²,⁸.0⁴,⁶]octane (22) [64]. The corresponding coupling mechanism is schematically shown in Fig. 10, which additionally includes relevant data for structurally related systems 23 and 24. In the first of them 3-σ bond coupling

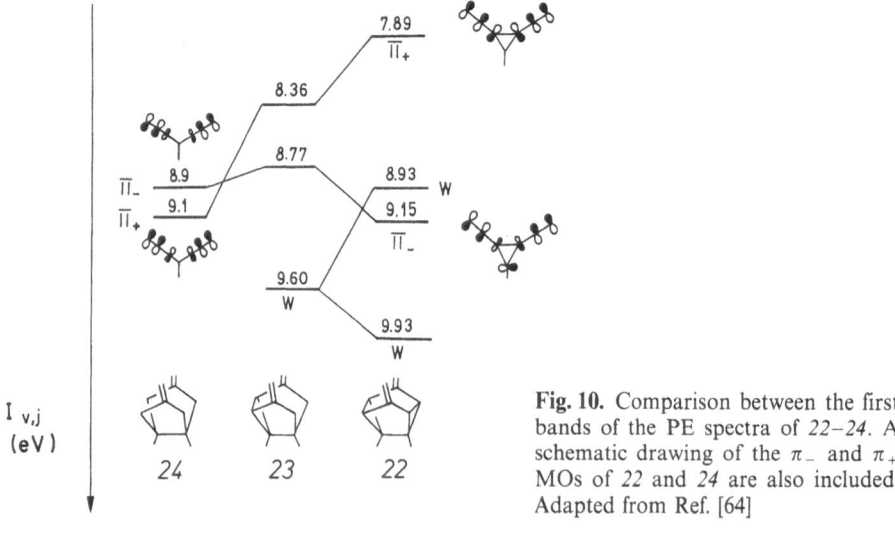

Fig. 10. Comparison between the first bands of the PE spectra of 22–24. A schematic drawing of the π_- and π_+ MOs of 22 and 24 are also included. Adapted from Ref. [64]

units are replaced by 4-σ bond coupling units, while *23* represents an intermediate case. Schematic representations of the wave functions for the two highest occupied MOs of *22* and *23* calculated by MINDO/3 procedure are also displayed. They show marked decrease in the π/σ mixing for the $\pi_+(a_1)$ linear combination in going from *22–24*, whereas the interaction between $\pi_-(b_1)$ and the σ-frame remains essentially constant.

3.3 1,5-π,π Interactions

Within this section we shall consider only the PE spectroscopic results of dime-thanohexahydronaphthalenes *1*, *25* and *26* [22, 65]. They provide namely a most valuable piece of information on the nature of interaction of double bonds through the chain(s) of four sigma bonds and its strong dependence on conformation. The relevant π-ionization energies are summarized in Table 5, together with the energies

Table 5. First vertical Ionization Energies ($I_{v,j}$) of the Compounds *1, 2* and *25–27*. All Values in eV

Compound	$I_{v,j}$	Splitting $\pi \pm \pi$	$-\varepsilon(\text{STO-3G})^a$	$\Delta\varepsilon(\text{STO-3G})^a$
1	8.48$(b_1, \pi_-)^b$ 9.35(a_1, π_+)	0.87	7.46 8.35	0.89
25	8.08$(b_1, \pi_-)^c$ 9.34(a_1, π_+)	1.26	–	1.55
26	8.46$(a', \pi_-)^c$ 8.90(a', π_+)	0.44	7.54 8.06	0.52
27	9.15$(b_1, \pi_-)^d$ 10.0(a_1, π_+)	0.85	–	–
2	9.03$(b_1, \pi_-)^e$ 9.58(a_1, π_+)	0.55	–	–

^{a} STO-3G calculated orbital and splitting energies are included where available; ^{b} Ref. [22]; ^{c} Ref. [65]; ^{d} Ref. [66]; ^{e} Ref. [25]

of the highest occupied π-MOs calculated using the STO-3G procedure. The largest $\Delta I_v(\pi)$ value is observed for *25* (1.23 eV) and is attributed to the cooperative action of through-space and through-bond mechanisms. The relative contributions of TS and TB coupling to the total $\Delta I_v(\pi)$ were estimated to be 0.90 and 0.36 eV, respectively [65b]. Two effects are also involved to explain π,π-splitting in *26*: the TB interactions mediated through 4 σ-bond chains and the hyperconjugative interaction between the double bonds and the CH_2 group at position 9. The former type of interaction predominates. Finally, the splitting energy in *1* (0.87 eV) is, as already mentioned, a consequence of pure through-bond interaction. Parenthetically, this is one of the largest splittings of the π-levels so far reported which arises exclusively from orbital interactions across 4 σ-bonds. Besides this, splitting of the similar order of magnitude has been noted in α-diketone *27*, (Table 5, Ref. [66a]) [4.4.2]propella-3,8,11-triene (*28*) (Ref. [66b]), and diepoxy-hexahydronaphthalene *2* (Table 5, Ref. [25]).

The trend of changes in the splitting energies caused by the through-bond coupling mechanism along the series *1* (0.87 eV) > *26* (0.44 eV) > *25* (0.36 eV) is attributed to the change in conformation of the sigma bond relay from all-*trans* in *1* to *cis-cis* arrangement in *25*. A few comments on the all-*trans* effect are in order here. It is a consequence of the different extent of overlapping between vicinal σ-bonds in question and the homoallylic σ,π-overlap on π_- and π_+ linear combinations in different conformations. It appears that the all-*trans* arrangement of bonds enhances the orbital interactions the most. To illustrate this point let us consider the origin of σ/σ-mixing in the highest occupied π_- MO of *1* and *26*. The relevant components of the σ-skeletons are isolated and schematically shown as fragment structures *29* and *30*, respectively. We shall concisely present arguments put forward originally by

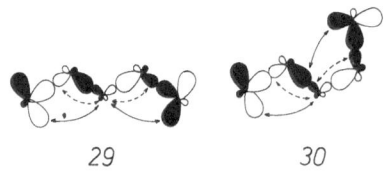

29   *30*

Paddon-Row et al. [22]. The relative orientation of local orbitals in the HOMO π_- MO in mentioned structures indicates that the π_- level in 29 should be of higher energy than that in *30* because one of the vicinal antibonding σ,σ-interactions in the former (the right-hand side dashed line) is converted into a bonding interaction in *30*. In contrast, the corresponding π_+ level is of the lower energy for the all-*trans* conformations of sigma bonds because a vicinal σ,σ interaction in the π_+ MO, which is bonding in the all-*trans* conformations, becomes antibonding in the *cis-trans* arrangement (see Fig. 2). Homoallylic π,σ overlaps of the type indicated by the solid arrows in *29* and *30*, also contribute to the same pattern. Namely, the π_- and π_+ levels in the all-*trans* conformations are raised or lowered, respectively, relative to the corresponding levels in the *cis-trans* form due to the homoallylic interactions. For example, the homoallylic π,σ-overlap, indicated by the right-hand solid arrows are antibonding in *29* but bonding in *30*. Hence, consideration of both vicinal σ,σ

overlaps within π_- and π_+ MOs, together with homoallylic π_-, σ_- and π_+, σ_+ overlaps, leads to the conclusion that TB coupling is maximized for a planar, all-*trans* conformation of σ bonds.

3.4 1,6-π,π Interactions

PE spectroscopic studies of the interactions between π-subunits interconnected via chain(s) involving five sigma-bonds are extremely rare [12, 29, 67]. We shall discuss only two specific cases. The first of them is exemplified with diene *31* and provides evidence in favour of the long-range nature of through-bond interaction. Namely, the π-splitting energy deduced from its PE spectrum amounts to 0.43 eV, which is half the size of the value observed in dimethanonaphthalene system *1*. Its TB origin was confirmed by STO-3G calculations which in full accordance with the parity rule puts π_+ linear combination above π_-. Let us mention here that further extension of the σ-bond relay leads to an even less-pronounced decrease in splitting energy (0.1 eV), as evidenced by *32* (vide infra), indicating that TB interaction between π MOs in

31   *32*

rigid dienes exhibiting all-*trans* arrangement of σ-bonds is of very long range indeed. In fact, by comparing $\Delta I_v(\pi)$'s for a series of structurally related all-*trans* dienes differing only in the number of σ-bonds in the coupling unit, Paddon-Row et al. found that decay of splitting energy follows an approximate exponential form that may be expressed as follows:

$$\ln \Delta I_v(\pi) = -0.46N + 1.56 \,, \tag{12}$$

where N denotes the number of sigma bonds. This implies that a through-bond splitting energy of about 0.003 eV is expected for values of N as large as 16, or, equivalently, for a separation of π-orbitals of 21 Å. Although such a split is negligible from a PES point of view, it is sufficiently large to enhance dramatically the rates of intramolecular electron transfer processes.

A question we would like to address in the second example concerns the extent of (homo)conjugative interaction of the vinylic groups in hydrocarbons *33–35* [67].

33   *34*   *35*

Common to all three hydrocarbons is the arrangement of the vinylic groups in relation to the biphenylene moiety which ensures 1,6-connectivity of the external double bonds. On the other hand, they differ considerably with regard to the spatial disposition of the vinylic groups and thus in the extent of their mutual (homo)conjugative interaction. In compound *33* no conjugation between the termini of vinylic groups is expected, while *34* represents a case with full conjugation between the double bonds. Recently conducted PE spectroscopic investigation of *33–35* [67] sustained with PPP calculations [68] indicated that vinylic groups in the latter compound exhibit sizeable homoconjugation.

3.5 π,π Interactions over C_s- and Longer Chains

By comparing the PE spectrum of "norbornylogues"* *32* and *36* (Table 6) Jørgensen

et al. concluded that $\Delta I_v(\pi)$ of *36* is 1.6 *times larger* than that of *32* [28]. This trend which was also reproduced by PRDO [69] calculations, violates the *trans*-rule, since on going from *32* to *36* the number of *trans* arrangements of vicinal σ bonds decreases. It was suggested that its origin lies in the presence of an additional type of orbital interaction mediated by the presence of a central bridging CH_2 group in *36* which is not present in *32*. Supportive of this view is the observtion that the measured $\Delta I_v(\pi)$ in *37*, which has structural features resembling that of *36* except for the central CH_2 group, is only 0.17 eV [29].

Because of the laticyclic topology [70] in *36*, this type of hyperconjugation was named by Paddon-Row as *laticyclic hyperconjugation* [71].

Table 6. First vertical ionization energies ($I_{v,j}$) of the dienes *32*, and *36–39*. All values in eV

Compound	$I_{v,j}$	$\Delta I_v(\pi)$	$\Delta\varepsilon$(STO-3G)
32	8.58(b_1, π_-)[a] 8.90(a_1, π_+)	0.32	0.31
36	8.24(b_1, π_-)[a] 8.76(a_1, π_+)	0.52	0.51
37	8.63(b_1, π_-)[b] 8.80(a_1, π_+)	0.17	0.18
38	8.31(b_1, π_-)[c] 8.60(a_1, π_+)	0.29	–
39	8.41(π)[d]	0	

[a] Ref. [28]; [b] Ref. [29]; [c] Ref. [73]; [d] Ref. [74]

* The term "norbornylogue" refers to the compounds containing the bicyclo-[2.2.1]heptyl moiety as the repeating unit [74]

Strong support in favour of laticyclic hyperconjugation was gained by model ab initio MO calculations [71, 72]. They also indicated that this type of interaction could be extended over an even larger distance than that found in *36*. This is in accordance with the PE spectrum of the 8-bond diene *38* which exhibits two bands for ionizations related to π-orbitals differing in energy by 0.29 eV (Table 6) [73].

Paddon-Row and co-workers have also developed a simple model of laticyclic hyperconjugation which is summarized in Fig. 11 [71, 72].

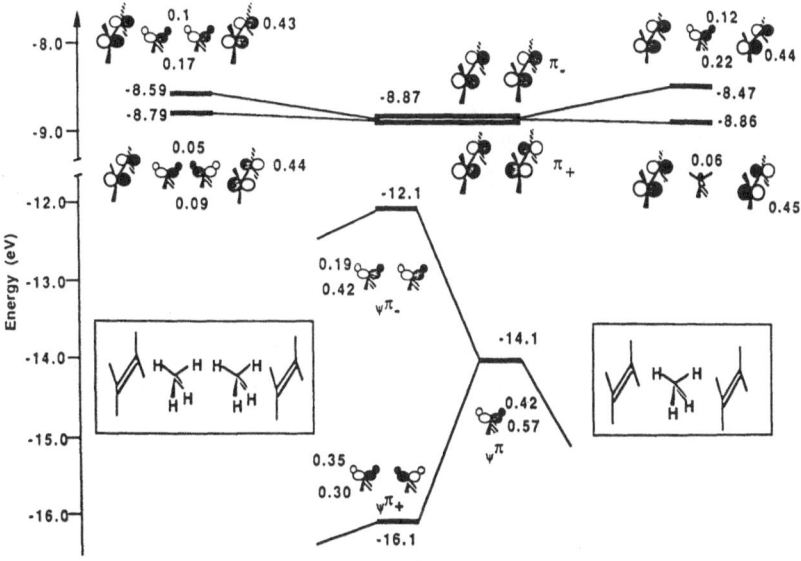

Fig. 11. Energy correlation diagram for model ethene ... $(CH_2)_n$ ... ethene complex for n = 1 and 2. Energies and MO coefficient are from HF/STO-3G calculations*

According to this model laticyclic hyperconjugation in dienes *36* and *38* is modeled by ethene ... CH_4 ... ethene and ethene ... $(CH_4)_2$ ... ethene complexes, respectively. Ethene ... CH_4 and CH_4 ... CH_4 separations are deduced from the crystal structures of the dodecachloro congeners of *36* and *38* [73]. The STO-3G computed energies and MO coefficients for the π and ψ^π MOs are shown in the center of Fig. 11, whereas those for the MOs of monomethane and dimethane complexes are displayed on the right- and left-hand sides, respectively. In the case of the monomethane complex, the level of π_ symmetry-adapted combination of MOs is raised by 0.4 eV through mixing, in an antibonding fashion, with the ψ^π MO of CH_4. The corresponding π_+ MO can not mix with the ψ^π MO for symmetry reasons. It can, however, interact with a lower-lying CH_4 MO of a_1 symmetry (assuming C_2 point group symmetry), but because of the large energy gap this mixing is expected to be energetically negligible. Indeed, the π_+ level is calculated to be raised by only 0.01 eV, and the resulting MO contains only a small contribution from the CH_4 a_1 orbital. The degeneracy of the π levels in the complex is therefore lifted, the calculated energy difference being 0.39 eV.

---

* Reprinted by permission of VCH Publishers, Inc., 220 East 23rd St., New York, N.Y., 1001.0 from Ref. [13]

In dimethano complex, two symmetry-adapted pairs of ψ^π MOs ($\psi^{\pi+}$ and $\psi^{\pi-}$) are formed and each of them is able to mix with one of the π combinations of the correct symmetry. Therefore, both π_+ and π_- levels undergo destabilization, with π_- level being more affected largely because of smaller energy gap between π_- and $\psi^{\pi-}$ basis MOs as compared to the energy difference between π_+ and $\psi^{\pi+}$ basis MOs. As a consequence, the splitting energy for this complex is smaller than that found for ethene ... CH$_4$ ... ethene complex. It is estimated to be 0.20 eV. Let us also mention that the splitting energies in both cases were found to be insensitive to bond angle distortions of the (CH$_4$)$_n$ fragment, as well as on the displacement of the CH$_4$ groups out of the plane defined by the four carbon atoms of the double bonds.

Finally, it is worth mentioning that by applying the same type of model STO-3G calculations to the complexes with more than two CH$_4$ bridges, Paddon-Row et al. came to the conclusion that laticyclic hyperconjugation decays exponentially with respect to the number, N, of intervening methano groups. The corresponding equation reads:

$$\ln \Delta E = -0.69N - 0.25, \tag{13}$$

from which a splitting energy of about 0.03 eV is predicted for N = 8 CH$_4$ units, or equivalently, for an ethene ... ethene separation of about 27 Å! Let us mention, however, that the PE spectrum of 39 [74] in which the ethene subunits are separated

39

by three bridging CH$_2$ groups (12 Å apart) did not show any discernable splitting in the π-manifold (Table 6).

In relation to the model of orbital interactions discussed above it is interesting to mention that PE spectroscopic evidence in favour of laticyclic conjugation between three parallel π-bonds in 40 was reported recently by Hünig et al. [75]. It manifests itself in the significant destabilization of the π-band related to ionization from the norbornene fragment in the PE spectrum of 40 when compared to the PE spectrum

40   41

of structurally related compound 41 where geometry prevents such conjugation. The experimental finding was confirmed by Hückel-type treatment and semiempirical calculation [75].

3.6 Through-bond Interaction of two Mutually Perpendicular π-Systems via Cyclobutane Relay Orbitals

On the basis of model calculations, Bischof, Gleiter and Haider [76] have proposed that the interaction between two perpendicular fragments found in spiro compounds

[77, 78] should be larger when their central atom is replaced by a four-member ring system as shown in Fig. 12. The four-membered ring acts like a sigma relay and replaces the through-space interaction present in the spiro compounds by a through-bond interaction in the resulting tricyclic system. Crucial for the relay effect [76] is the interaction of the a_2 and b_1 Walsh orbitals [79] of the four-membered ring with the corresponding linear combinations of the spiro units (see also Fig. 13).

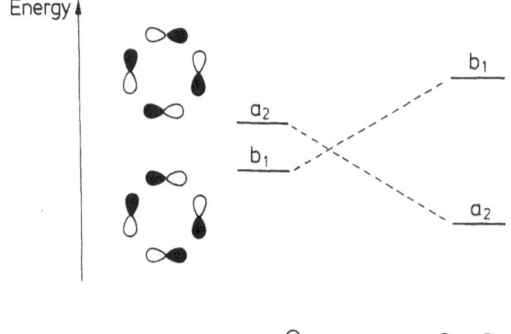

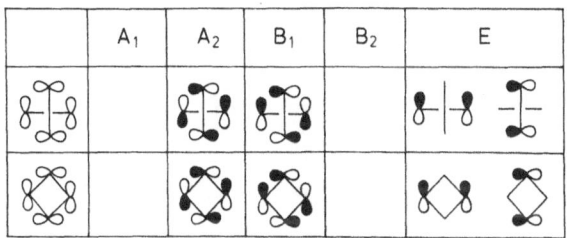

Fig. 12. Orbital sequence for b_1 and a_2 in spiro compounds (*left*) and the corresponding tricyclic compounds (*right*) [76]

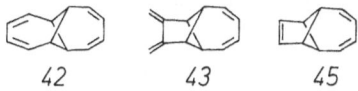

Fig. 13. Comparison of the valence orbitals of a spiro system with the valence (Walsh) orbitals of cyclobutane [76]

The most convincing evidence for strong relay conjugation has been obtained in photoelectron spectroscopic investigation of tricyclo[5.5.0.0^{2,8}]dodecatetraene (*42*) and 9,10-dimethylenetricyclo[5.3.0.0^{2,8}]deca-3.5-diene (43) [80].

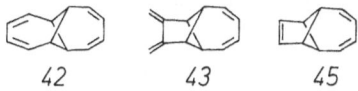

The difference $\Delta I_v(1, 2) = 1.44$ eV between the positions of the first two PE bands of *42* is by 0.2 eV larger from that found in its spiro analog [4.4]nonatetraene (*44*) [81] (Fig. 14). The assignment given in Fig. 14 is substantiated by MO calculations using the fragment molecular approach (FMO) and semiempirical MINDO/3 as well as ab initio STO-3G calculations. All three procedures point to the existence of a strong interaction between $b_1(\pi)$ and $b_1(\sigma)$ in *42*.

The interaction scheme discussed for *42* holds in principle for *43*. Considerably higher energy for the first band of *43* (8.17 eV) as compared to that found in *42* (7.56 eV) and a reduction of $\Delta I_v(\pi)$ from 1.4 eV in *42* to 1.1 eV in *43* is traced back to the different connectivity between the central ring and the butadiene moiety.

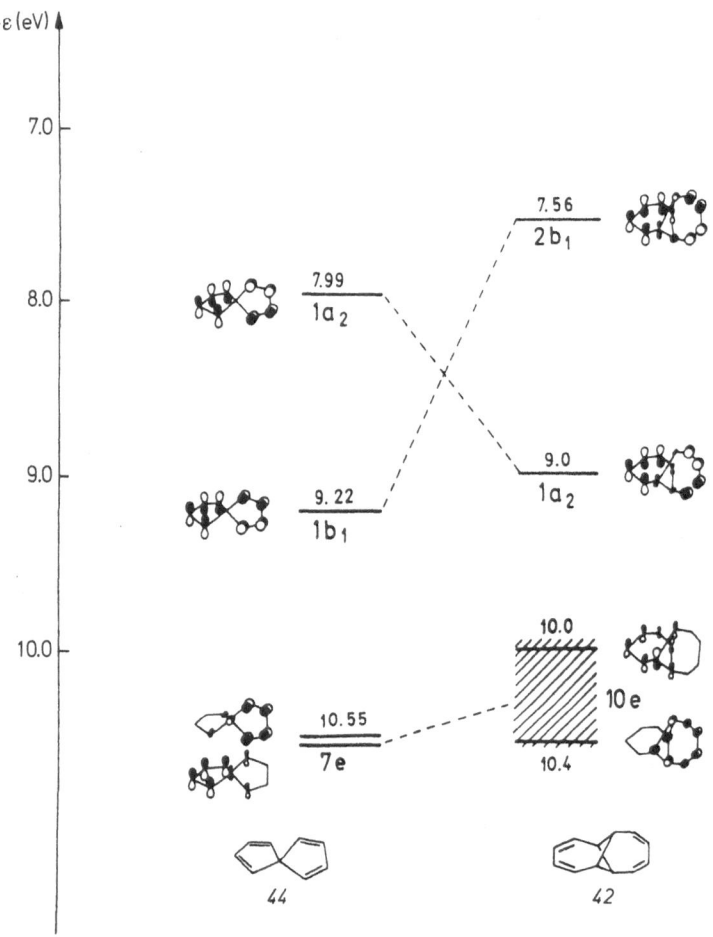

Fig. 14. Comparison between the first bands of the PE spectra of *42* and *44*. Reprinted with permission from Ref. [76]. Copyright 1990 American Chemical Society

Contrary to the systems *42* and *43*, the PE spectrum of *45* reveals the absence of interactions between the two olefinic fragments. An underlying reason is that interaction in the latter compound involves the π^* orbitals of one π-unit and the π orbitals of the other. This qualitative picture is fully confirmed by the results of calculations mentioned above. In addition to a minute $2a_2(\pi) - 3a_2(\pi^*)$ interaction (<0.1%) calculation however suggests a pronounced interaction between the central ring and π units [80].

4 Interactions of Acetylenic π-Orbitals and Their Dependence on Parity of the Intervening σ Bonds

Numerous PE spectroscopic studies of acetylenic compounds which appeared lately [82–91] have been largely focused on two aspects: (a) the effect of lowering or raising the basis orbital energy of the σ-relay and (b) the influence of increasing the length of the σ-coupling unit on the relative contributions of through-space and through-bond interactions. The most interesting results related primarily to the second aspect will be presented within this section. We shall commence discussion by considering the interaction between acetylenic units across two sigma bonds. As representative examples of this class of compounds we shall consider [5]- (*46*) and [6]- (*47*) pericyclyne [82] and 1,1-diethylnylcyclopropane (*48*) [83].

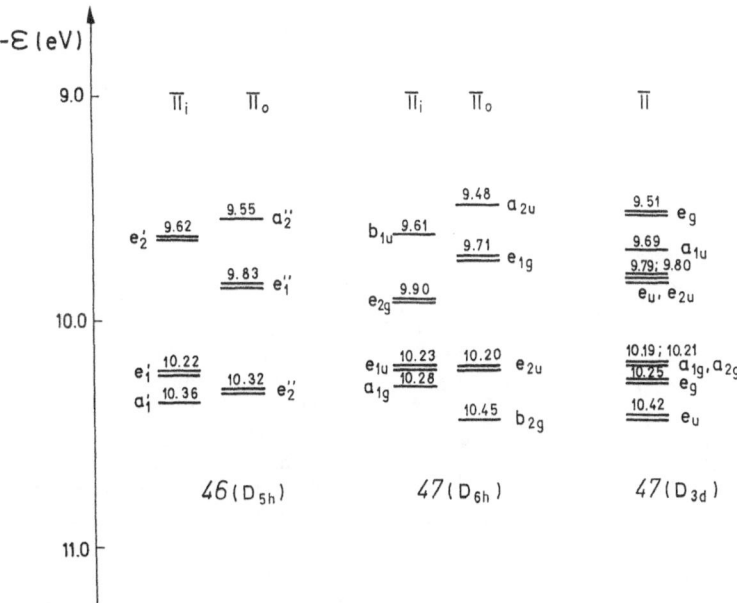

In both families through-space as well as through-bond interactions between the acetylenic units are found to be important, with the former type of interaction dominating in the in-plane (π_i) orbitals and the latter in the out-of-plane (π_o) orbitals.

Fig. 15. Corrected STO-3G [82] orbital energies and symmetries for planar [5]- and planar and chair [6]pericyclyne

STO-3G calculations performed for the planar conformer of the fully demethylated *46*, predict as the highest occupied MOs ten wave functions belonging to A_2'', E_2', E_1'', E_1', E_2'' and A_1' (under assumption of D_{5h} point symmetry group). For the parent [6]-pericyclyne, calculations were done for planar and chair conformers (under the assumption of D_{6h} and D_{3d} point symmetry group, respectively). The calculated ordering of MOs for both conformers together with the results for the [5]-pericyclyne is shown schematically in Fig. 15. Its analysis reveals that all π-orbitals are predicted to be spaced within 1 eV. It is also worth mentioning that HOMO in planar conformers of both compounds corresponds to the out-of-plane acetylenic MO, indicating thus the prevailing influence of TB interaction. In the *chair* conformation of [6]-pericyclyne, a distinction between π_i and π_0 orbitals disappears, and the pattern of mixing becomes more complicated. The PE spectra of *46* and *47* are relatively broad, with only a few characteristic maxima in the region from 8.1 to 10.5 eV and σ-onsets at $\sim$ 11.5 eV. This is in reasonable agreement with the calculated spacing of the π-MOs but precludes any definite conclusion about their exact ordering.

In sharp contrast to that, PE spectroscopic analysis of 1,1-diethynylcyclopropane enables a clear distinction between ionization events from the π_0 and π_i-dominated MOs. More specifically, the first ionization event ascribed to the ionization from the π_0^+ linear combination appears at 9.26 eV, while ionization bands related to the π_i^-, π_i^+ and π_0^- are found between 10.3 to 10.6 eV. The observed trend is in line with the smaller energy gap between the π and σ subunits imposed by replacement of the $C(CH_3)_2$ fragment in *46* (or *47*) with the highly strained cyclopropyl group [83].

The most thoroughly studied representative of cyclic diacetylenes which exhibit interactions over three sigma bonds is cycloocta-1,5-diyne (*49*). Its PE spectrum exhibits three close-lying bands followed by another one separated by $\sim$ 0.5 eV [84]. This is rationalized in terms of strong through-bond interaction of the π_i^+ linear combination with the C–C σ-orbitals involving C3–C4 and C7–C8 bonds. As a consequence, the π_i^+ orbital has an energy similar to the two antisymmetric linear combinations π_i^- and π_0^-, which are not at all affected by TB destabilization, as indicated in Fig. 16.

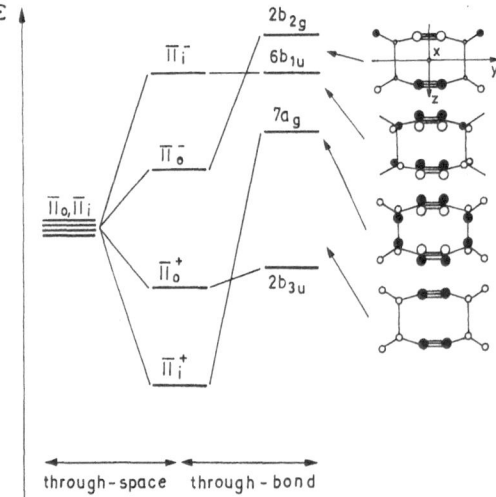

Fig. 16. Qualitative diagram illustrating the interplay of through-space and through-bond interactions of the basis σ-orbitals in 1,5-cyclooctadiene *49* [84]. The symmetry designations refer to D_{2h}. Schematic representation of the π-molecular orbitals of *49* are also shown. Note that in $2b_{2g}$ and $2b_{3u}$ the circles on the methylene groups characterize the pseudo π-orbitals, which are antisymmetric with respect to the y,z-plane, whereas in $6b_{1u}$ and $7a_g$ the 1s-linear combinations of these groups are symmetrical

The influence of TB interactions was found to have an even more pronounced impact on the PE spectrum of 3,3,4,4,7,7,8,8-octamethyl-3,4,7,8-hexasilacycloocta-1,5-diyne (50) [85] where ionization events related to the ionization from π_i^+ MO appear as the first ionization band. This is interpreted by the enhanced π/σ mixing due to the reduction of the energy gap between π and σ orbitals caused by replacing CH_2–CH_2 by Me_2Si–$SiMe_2$ groups and is fully confirmed by semiempirical calculations and MO analysis employing LCBO treatment.

A pronounced change in the interaction pattern of acetylenic fragments is also observed upon replacing ethylenic bridges in 49 with trimethylene chains [86]. To illustrate this point, the first ionization bands of 49 [84] and 51 [86] are compared in Fig. 17.

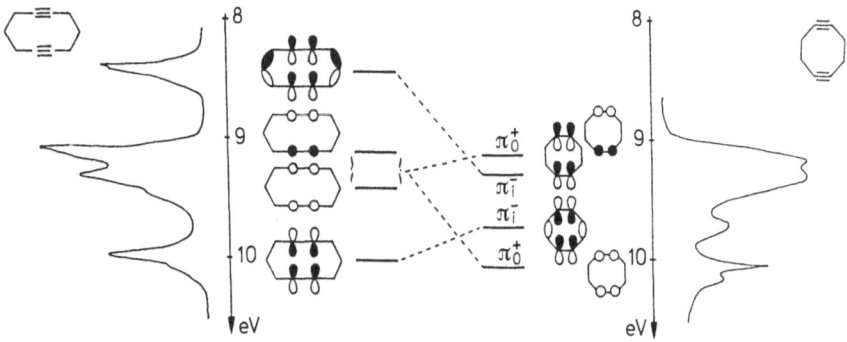

Fig. 17. Comparison of the first ionization bands of the PE spectra of 49 and 51 [86]

The interpretation given in Fig. 17 suggests the change in the nature of HOMO on going from 49 to 51 as a consequence of the overwhelming importance of the TB effect of the two bridging units in the latter compound.

A similar impact of the TS and TB interactions was also noticed in the PE spectra of bridged 1,8-diethynylnaphthalenes 52–54 [87]. The results of this study and the ordering of the highest occupied π MOs as calculated by using MINDO/3 procedure are illustrated in Fig. 18.

Their analysis reveals the existence of a considerable splitting between in-plane MOs only in case of 52. This is attributed to the influence of strong TS interactions due to the close proximity of the acetylenic groups and a destabilization of $12b_1(\pi_i^-)$ due to the pronounced π/σ mixing with the aliphatic bridge*. In 53 and 54 the splitting in energy between π_i levels is significantly smaller than in 52, and in fact identical to that found in 56. This strongly indicates that in determining the energy difference between the molecular orbitals derived from the linear combinations of the acetylenic in-plane π-orbitals in 52–54 through-space interaction has a dominant role.

---

* Contribution from TB interaction involving naphthalene moiety to the overall splitting of π_i orbitals has been judged from the PE spectrum of 55 [88]

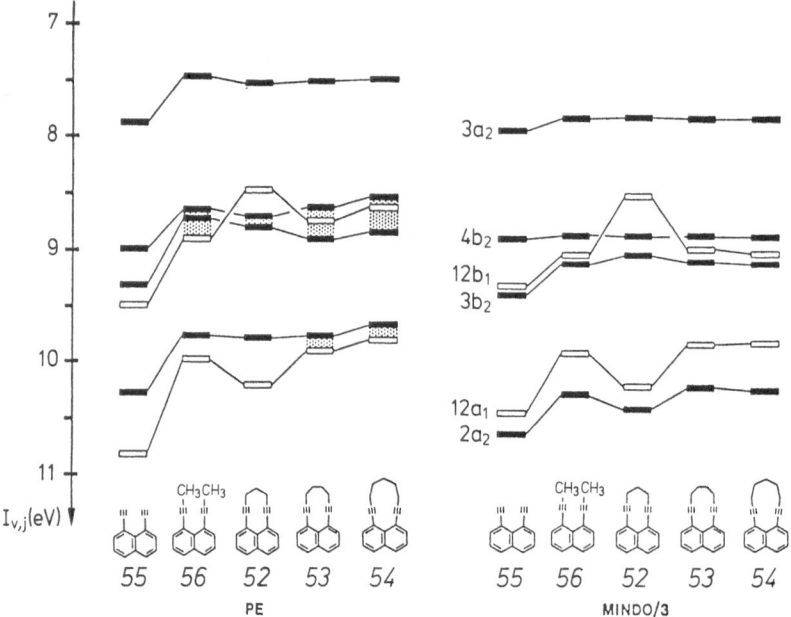

Fig. 18. Correlation diagram between the first PE bands in *52–56* (*left*) and between the highest occupied MOs of *52–56* (right). The irreducible representations refer to the point symmetry groups of prototype molecule *55*. The full bars were used for the out-of-plane π MOs while the open bars denote the in-plane π MOs. Data for *55* are taken from Ref. [88]. Reprinted with permission from Ref. [87]. Copyright 1990 American Chemical society

5 PE Spectroscopic Evidence for Interactions of Aromatic Rings

In discussing interactions between aromatic rings [2.2](1.4) cyclophane (= paracyclophane, *57*) occupies undoubtedly a pivotal place. In this molecule two benzene rings are kept together by two ethano bridges in a strained face-to-face arrangement with

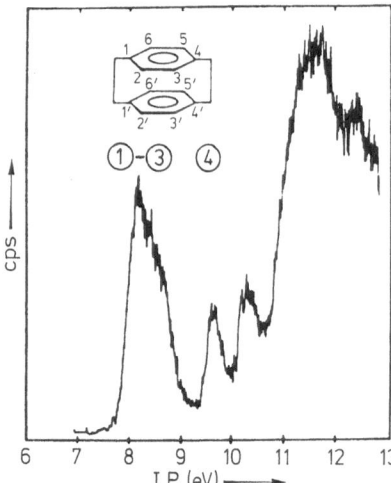

Fig. 19. Photoelectron spectrum of [2.2]paracyclophane [99]

an average distance close to 3 Å, giving rise to exceptional trans-annular interactions and unique chemical and physical properties [92]. The PE spectrum of paracyclophane (Fig. 19) was first recorded by Pignataro et al. who attributed the first ionization feature at ~ 8 eV to two ionization processes [93]. Later work has shown that this feature should be assigned to ionization from three close lying π-orbitals b_{2g}, b_{3g} and b_{3u}, respectively [94, 95]. A schematic drawing of the corresponding wave functions is shown below:

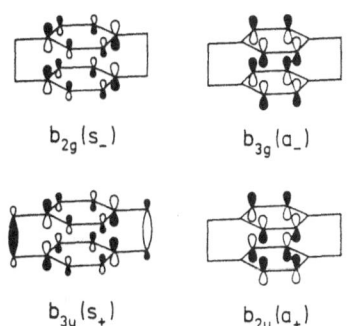

$$b_{2g}(s_-) \qquad b_{3g}(a_-)$$

$$b_{3u}(s_+) \qquad b_{2u}(a_+)$$

The additionally included b_{2u} π-MO is associated with the second ionization band. An examination of the wave functions indicates that the MOs pertaining to irreducible representations b_{3g}, b_{2g} and b_{2u} are pure π MOs, while b_{3u} has a considerable admixture of the C–C σ-bonds connecting the two rings.

The orbital interaction pattern in 57 was first analysed in terms of through-space and through-bond interactions at the extended Hückel level by Gleiter [95]. Since then, considerable discussion has centered around the relative contributions to the splitting of cyclophane π levels arising from direct π overlaps (TS effect) vs. that

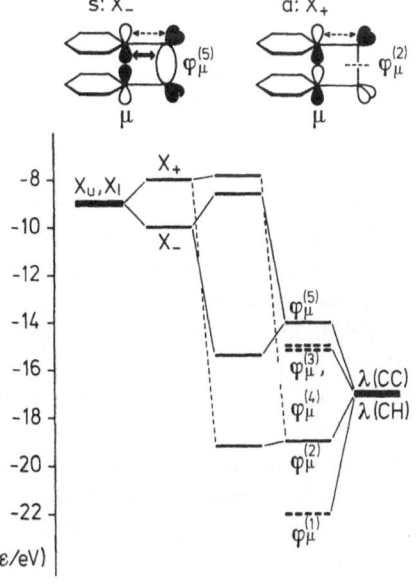

Fig. 20. Orbital diagram for the through-bond interaction between the linear combinations X_+ and X_- of the upper (X_u) and lower (X_l) benzene basis orbitals with the ethano-bridge orbitals $\varphi_\mu^{(j)}$. The parameters implied are $\tau = 1.0$ eV; self energy of $\lambda(CC)$ and $\lambda(CH) = -17$ eV; geminal interaction parameter $B = -2$ eV [96]

arising from π/σ mixing (TB effect). Following the procedure suggested by Gleiter [95], Heilbronner and co-workers [96, 97] have developed a simple molecular orbital model for dissecting through-space and through-bond interactions operating in *57*, according to which the sequence of the lowest ionization events can be interpreted in a straightforward and transparent manner (Fig. 20).

Starting with the e_{1g} orbitals of benzene moieties of the upper (u) and the lower (l) deck (designated as S_u, A_u and S_l, A_l, respectively), being symmetric (S) and antisymmetric (A) with respect to the x,z-plane containing the centres 1,4,1',4' symmetry-adapted linear combinations (D_{2h} point symmetry group) $S_+ (b_{2g})$, $A_+ (b_{3g})$, $S_- (b_{3u})$ and $A_- (b_{2u})$ are formed. Attention is drawn to the fact that the plus-combinations A_+, S_+ are *antisymmetric*, the minus-combinations A_-, S_- *symmetric* relative to the mirror plane passing between the two decks. Hence the degenerate pair A_+, S_+ lies *above* the degenerate pair A_-, S_-. They are split by 2τ as a consequence of pure TS interaction leading to two semilocalized linear combinations designated as X_+ and X_-, respectively, where X_+ stands for A_+ or S_+. The through-bond interaction involving the relay orbitals $\varphi_u^{(j)}$ derived by applying the equivalent bond orbital (EBO) approach [31, 56] is introduced in a second step. From the schematic representation of the bridge orbitals $\varphi_\mu^{(j)}$ and their energies shown below:

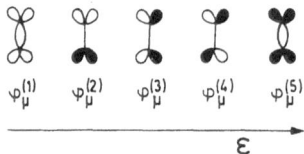

$$\varphi_\mu^{(1)} \quad \varphi_\mu^{(2)} \quad \varphi_\mu^{(3)} \quad \varphi_\mu^{(4)} \quad \varphi_\mu^{(5)}$$

$$\varepsilon \longrightarrow$$

it is obvious that only $\varphi_\mu^{(5)}$ can interact significantly with X_- and only $\varphi_\mu^{(2)}$ with X_+, the latter interaction being negligible. On account of different cross terms of X_- orbitals with $\varphi_u^{(5)}$ ($2\sqrt{3}$ B for S_- and 0.0 for A_-, cf. Ref. [96]), S_- linear combination undergoes a more pronounced destabilization, approaching in energy the X_+ combinations.

The most conclusive confirmation of the sequence of the highest occupied MOs in *57* was provided empirically in the following way. The idea employed for this purpose was to affect the extent of π/σ mixing of the b_{3u} orbital by lowering or raising the energy of the $b_{3u}(\pi)$ orbital which interacts with $b_{3u}(\sigma)$. An increase in energy of $b_{3u}(\sigma)$ will increase π/σ mixing thus lowering the orbital energy $b_{3u}(\pi)$ and vice versa. Heilbronner and Maier [98] have studied a case where the π/σ interaction is lowered by fluorination as e.g. in 1,1,2,2,9,9,10,10-octafluoro[2.2] paracyclophane (*58*). Due to the inductive effect of the fluorine atoms, the ionization from the $b_{3u}(\pi)$ orbital of *57* occurs at considerably higher energy than from $b_{2g}(\pi)$ and $b_{3g}(\pi)$. The opposite effect can be detected in the PE spectrum of [1,2:9,10]dimethano[2.2]paracyclophane (*59*) [99]. In this compound the basis orbital energy of the σ bond of the three-membered rings is higher than the σ bonds in *57*, which leads to an increase of the π/σ interaction thus lowering the energy associated with the ionization from the $b_{3u}(\pi)$ orbital. An even more dramatic effect of raising the energy of the $b_{3u}(\pi)$ orbital is observed by replacing C–C bridges in *57* by Si–Si bonds (to yield *60*) [100]. This is best illustrated by the correlation diagram between the first four bands of the PE spectra of *57–60* which is shown in Fig. 21.

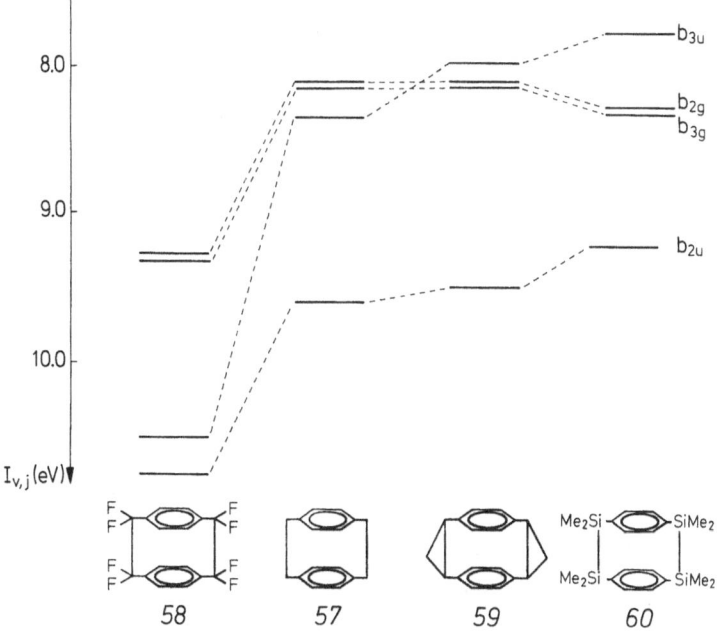

Fig. 21. Comparison of the first four bands of the PE spectra of *57–60*

A more detailed discussion of the PE studies of paracyclophane, as well as a number of higher-order cyclophanes and heterocyclophanes can be found in a review article of Heilbronner and Yang [97].

Face-to-face oriented benzene rings are also present in benzo-annellated hypostrophanes *61* and *62* (see Fig. 22).

Their PE spectra, however, exhibit in the π-energy region highly unresolved bands, which do not allow a definite conclusion to be made about the size of through-space and through-bond interactions. By comparing the measured ionization energies with

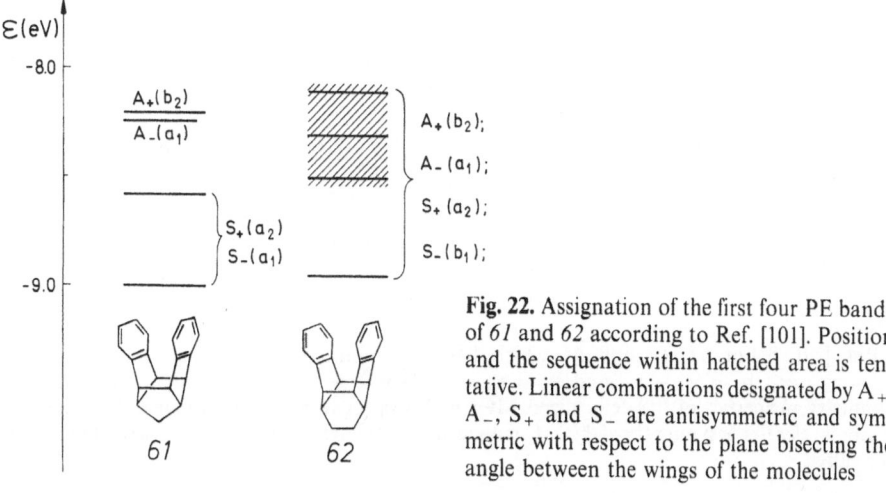

Fig. 22. Assignation of the first four PE bands of *61* and *62* according to Ref. [101]. Position and the sequence within hatched area is tentative. Linear combinations designated by A_+, A_-, S_+ and S_- are antisymmetric and symmetric with respect to the plane bisecting the angle between the wings of the molecules

those of structurally related compounds the assignment of the highest occupied MOs indicated in Fig. 22 was tentatively proposed [101]. In the remainder of this section we shall briefly consider the PE spectra of the dibenzo-annellated norbornylogs 63–65 [29]. Some of them illustrate a long-range character of through-bond interactions (63, 64), whilst 65 is relevant in connection with laticyclic hyperconjugation. The corresponding PE data are given in Table 7 together with some STO-3G splitting energies. Their survey shows that benzene-benzene through-bond interactions, involving the π_S MOs show only a weak dependence on distance, similar to that found for the corresponding dienes. On the contrary, the π_A, π_A splitting energies are extremely small (63) or negligible (64). This simple fact reflects the different magnitudes of the coefficients of the π_S and π_A aromatic MOs at the sites where significant overlap with the sigma framework takes place (0.5 for π_S and 0.29 for π_A, respectively).

63   64   65

The π_S,π_S splitting energy of 0.38 eV for 65 is due mainly to laticyclic hyperconjugation. This is consistent with the lack of any π_A,π_A splitting. Namely, laticyclic hyperconjugative mixing of the π_A orbitals with the array of ψ^π orbitals (cf. Fig. 11) is prohibited for symmetry reasons.

Table 7. First vertical ionization energies ($I_{v,j}$) for 63–65. All values in eV[a]

Compound	$I_{v,j}$	Orbital Assignment[b]	$\Delta I_v(\pi_S)$	$\Delta I_v(\pi_A)$
63	8.10	b_1	0.56	
	8.66	a_1		
	8.80	a_2		0.14
	8.94	b_2		
64	8.23	b_1	0.26	
	8.49	a_1		
	8.81	a_2		0
		b_2		
65	7.98	b_1		
	8.36	a_1	0.38	
	8.81	a_2		0
		b_2		

[a] Ref. [29]; [b] For 63, 64 and 65 which have C_{2v} point group symmetry $a_1 = \pi_S + \pi_S$; $b_1 = \pi_S - \pi_S$; $b_2 = \pi_A + \pi_A$ and $a_2 = \pi_A - \pi_A$. The "+" and "−" symbols refer to the symmetry properties of the delocalized MOs with respect to a reflection plane. "A" and "S" symbols refer to the symmetry of the MO with respect to the symmetry plane that is retained in the derivative

6 PE Spectroscopic Evidence for Interactions of Lone-pair Orbitals

Photoelectron spectroscopy has been widely used for studying the nature of interactions of lone pairs in full analogy with investigations of intramolecular interactions of π-subunits elaborated above. The archetypal case is provided by 1,4-diazobicyclo[2.2.2]octane (DABCO) where the inverted sequence of energy levels due to through-bond interaction have been demonstrated first by theory [2, 3] and then by PES measurements [102]. The PE studies of α, β, and γ-diketones have offered additional valuable pieces of information which shed light on intramolecular interactions over three, four and five σ-bonds [103]. It is interesting to mention in this respect that acyclic dicarbonyl compounds generally obey the parity rule, which does not hold for their cyclic counterparts. Importantly, the first experimental evidence of the exponential decay of through-bond coupling against the increase in a number of intervening σ-bonds was found in molecules possessing lone pairs (diiodopolyace-tylene, [104]). Hence this type of molecular systems deserves a few comments.

Martin and Mayer [12] have reviewed a vast number of PE studies of molecules incorporating lone pairs. Most of them are confined to α-dicarbonyles and polyazo compounds. Since then PE spectroscopic properties of molecules involving functional groups with lone-pair electrons have continued to attract attention of PE spectroscopists [105–113], thus contributing to broadening of our knowledge about these interesting molecules. We shall consider a few representative examples within this section.

Let us commence this consideration with a PE-spectroscopic investigation of pyrazino[2,3-b]pyrazine (66). Its PE spectrum (Table 8) as well as the PE spectra of several of its methyl derivatives are characterized by the appearance of a rather low-lying ionization feature related to ionization from totally symmetric nitrogen "non-bonding" orbital [105]. This is rationalized in terms of strong through-bond coupling of the lone pairs mediated through the intervening C–C σ-bonds. Such an

Table 8. Comparison between vertical ionization energies ($I_{v,j}$) and calculated orbital energies ($-\varepsilon_j$) of 66. All values in eV [a]

$I_{v,j}$	Orbital Assignment	$-\varepsilon_j$		
		HMO	EWMO	HAM/3
9.2	$9a_g(n)$	9.1	9.09	8.84
9.87	$2b_{1u}(\pi)$	9.82	9.59	10.04
10.53	$1a_u(\pi)$	10.74	10.96	10.69
(10.8)	$7b_{3w}(n)$	11.3	10.80	10.08
11.0	$5b_{1g}(n)$	11.7	11.71	10.94
11.7	$6b_{2u}(n)$	11.9	11.90	11.41
12.6	$1b_{3g}(\pi)$	12.40	12.60	12.21
12.7	$1b_{2g}(\pi)$	12.51	12.97	12.36

[a] Ref. [105]

assignment was supported with the results of MINDO/3, HAM [114] and EWMO [115] calculations.

Convincing evidence in favour of a massive contribution of through-bond coupling to the sequence of the highest occupied n-levels for a series of 1,4-disubstituted butanetetrones 67 (a–e) was presented by Gleiter and Dobler [106].

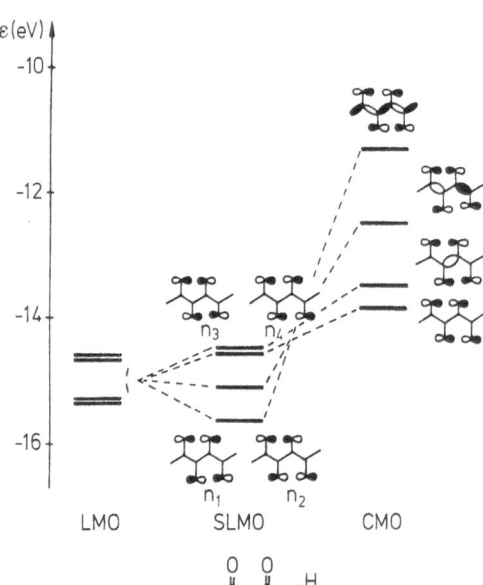

A strong support for this conjecture was offered by analysis based on a LCBO approach which indicated that the extent of interaction between four highest-occupied symmetry-adapted linear combinations of lone-pair orbitals (SLMOs) and precanonical σ-relays' orbitals of appropriate symmetry differ considerably. The most pronounced TB-induced shift, as indicated in Fig. 23, is predicted for the lowest energy SLMO (n_4) [106].

In the following example we shall illustrate the effect of changing the basis orbital

Fig. 23. Orbital correlation diagram illustrating the interplay of through-space and through-bond interactions in butane-tetrones 67 (a–e) (see Ref. [106])

energy of the σ-relay on the extent of through-bond type of interaction in derivatives of cyclohexadiene-1,4-diones *68–73* [107] shown below:

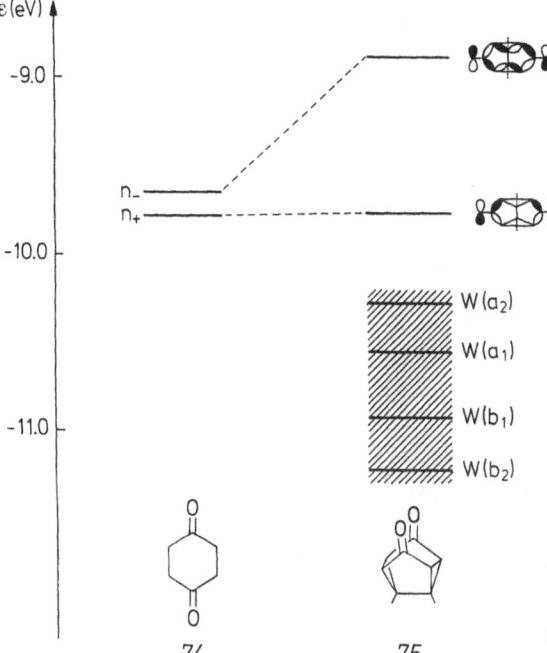

Comparison of the PE spectra of cyclohexa1,4-dione (*74*) [108] and 1,5-dimethyl-tetracyclo[3.3.0.0^{2,8}.0^{4,6}]octane-3,7-dione (*75*) [109] (Fig. 24) demonstrates the value of this concept the best. For the 1,4-dione *74* the splitting between the first two bands (which correspond to ionization events from n_+ and n_- linear combinations) is found to be 0.2 eV [108], whilst for the 1,4-dione 75 an energy difference of 0.9 eV [109] is observed.

Fig. 24. Comparison between the first bands of the PE spectra of *74* and *75*. A schematic drawing of the n_- and n_+ MOs are also included [108, 109]

Analogously, the difference between the first two ionization energies in the PE spectra of 1,4-diones *68–73* increases from 0.2 eV in *68* and *69* to ~ 0.7 eV in *70* and *71*, whereas the corresponding ionization energies in *72* and *73* assume intermediate values. The relevant correlation diagram is shown in Fig. 25.

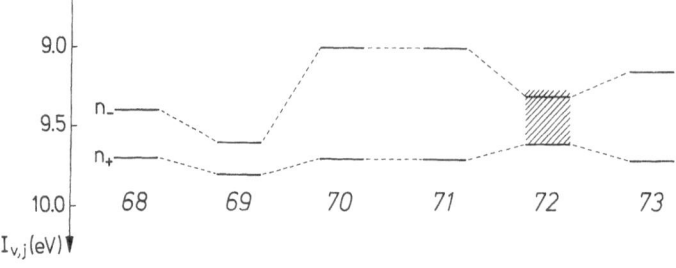

Fig. 25. Correlation of the first two PE bands of *68–73*

The observed increase in splitting energies is interpreted in terms of enhanced interactions between the n_- linear combination, while the n_+/σ interaction remains essentially constant. This is best demonstrated by the schematic representations of the two highest-occupied MOs for *70* and *74* as obtained by MINDO/3 calculations which are shown in Fig. 26.

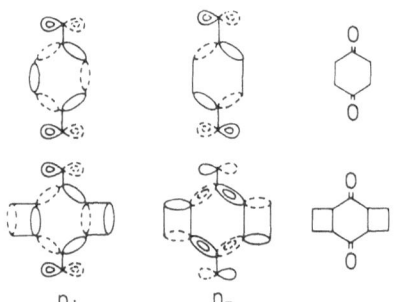

Fig. 26. Schematic representation of the two highest occupied MOs (n_-, n_+) of *70* and *74*. Double contour lines indicate higher density. Reprinted with permission from see Ref. [107]. Copyright 1990 American Chemical Society

An extension of this approach to 1,4-diones with cyclobutene fragments *76–80* has shown, on the contrary, that these moieties do not enhance the n_-/n_+ splitting energy. In accordance with that, MINDO/3 calculations predict a similar increase

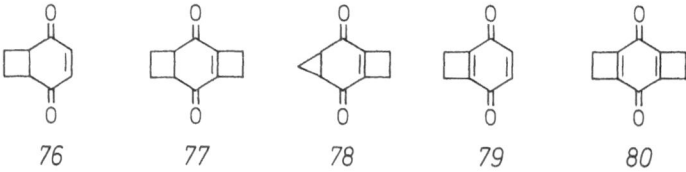

in π/σ participation for both combinations upon cyclobutene attachment to the six-membered ring [106].

Contrary to the molecules discussed above in 2,8-dicarbonyl-(*81*) and 2,8,9-tricarbonyl-(*82*), propellane carbonyl groups are expected to interact primarily via through-space mechanism. Their PE spectra [50], however, exhibit only a small n,n-splitting energies (0.3 eV for *81* and 0.5 eV for *82*, respectively) indicating that

TS interaction is rather weak. This can be understood in terms of rather large separation between the interacting carbonyl units.

Through-space interaction of lone pairs is also found to be operative in 4,5-bis(diemethylamino)fluorene (*83*) [110]. Analysis of its PE spectrum and simple ZDO calculations shows that the split between the two lone pairs of the dimethylamino groups in *83* amounts to 2.2 eV. It is worth mentioning that a slightly smaller value (by 0.2 eV) was observed for 1.8-bis(dimethylamino)naphthalene (*84*) [111], indicating a closer spatial proximity of the two nitrogen centres in the former compound.

A pronounced lone-pair interaction was also detected in the PE spectra of diepoxy compounds *85a* and *85b* [112] and their benzo-annellated analogs *86a* and *86b*. In

85a X-X: CH$_2$-CH$_2$ *85b* X-X: CH$_2$-CH$_2$

86a X-X: ⟨◯⟩   *86b* X-X: ⟨◯⟩

the *syn*-series (*85a* and *86a*, resp.) splitting energy of ∼ 0.9 eV was encountered, while in the case of *anti*-conformers (*85b* and *86b*, resp.) $\Delta I_v(n)$ drops to ∼ 0.6 eV [112].

Finally, let us mention the PE investigation of halogen lone pairs through cubane σ-orbitals in 1,4-dihalocubanes (*87*) reported recently by Honegger et al. [113]. σ-orbital manifold of cubane as suggested by its PE spectrum [116], is divided into two subsystems separated by a gap extended from ∼ 10.5 eV to 13.5 eV. Hence, lone-pair orbitals of chlorine in 1,4-dichloro analog of *87* (*87a*) whose basis energies are falling into this gap, hyperconjugate appreciably with both sets of cubane σ-orbitals. The same holds for 1,4-dibromocubane (*87b*). Interaction with the lower lying set leads to the destabilization ("normal" hyperconjugation), whereas interaction with the set of higher energy levels leads to a stabilization ("inverted" hyperconjugation). As a result, the lone-pair ionization energies of both compounds are found to be significantly larger than expected for alkyl halides [5e, 117] containing the same number of C-atoms as the cubane skeleton. More quantitative rationalization of the PE spectral features of *87a* and *87b* based on application of the EBO model adapted to account for spin-orbit coupling [56] can be found in Ref. [113].

7 Concluding Remarks

Our aim in this chapter has been to illustrate the usefulness of simple orbital model approaches in tackling the complex problem of intramolecular interactions between functional groups within molecules and their manifestations in photoelectron spectra. The treatise focuses on the electronic interactions between identical subunits in a number of representative molecules which display some important types of functional groups and interaction mechanisms.

In the first two sections the most salient features of a general pattern of the long-range orbital interactions and their dissection into through-space and through-bond terms are outlined. Then, simple theoretical models designed for their interpretations are discussed in more detail.

The examples discussed in the third section illustrate the utility of these simple model concepts in rationalization of the PE spectroscopic evidence of the interactions between olefinic subunits in diverse hydrocarbons. Particular attention is paid to the dependence of the relative sign of through-space and through-bond interactions in these molecules on the parity of the number of intervening sigma bonds, as well as to the possible origins of the *trans*-effect and long-range character of through-bond interactions. In most cases the results of analysis based on usage of these simple models are substantiated with reference to the more sophisticated calculational procedures.

Additional examples of long-range orbital interactions between π-units are presented in Sects. 4 and 5, in which PE spectra of hydrocarbons incorporating acetylenic (Sect. 4) and aromatic (Sect. 5) subunits are discussed. Finally, in Sect. 6, intramolecular interactions between functional groups involving lone-pair electrons are considered in the light of the most recent PE spectroscopic results.

The concept described above appears to provide an effective and self-consistent scheme for understanding the nature of intramolecular interactions and their most important PE spectroscopic consequences. Such understanding is of paramount importance for rationalization of the molecular physical and chemical properties at the electronic level, and is likely to provide a basis for designing novel substances exhibiting new and useful chemical and physical features.

8 Acknowledgements

A part of this work has been done at the Organisch-Chemisches Institut der Universität Heidelberg and the financial support of Alexander-von-Humboldt Stiftung is gratefully acknowledged. I am grateful to Professor Rolf Gleiter for his hospitality, continuous encouragement and advice in my PES work over a number of years. Finally, I thank the Council for Scientific Research of Croatia (SIZ) for supporting our ongoing investigations in the area of PE spectroscopy.

9 References

1. March J (1985) Advanced Organic Chemistry. Reactions, Mechanisms and Structure, 3rd ed, Wiley, New York
2. Hoffmann R, Imamura A, Hehre WJ (1968) J Am Chem Soc 90: 1499
3. Hoffmann R (1971) Acc Chem Res 4: 1
4. Al-Joboury MI, Turner DW (1963) J Chem Soc. 5141; Turner DW (1968) Proc Roy Soc A307: 15; Turner DW, Baker C, Baker AD, Brundle CR (1970) Molecular Photoelectron Spectroscopy, Wiley, London

5. (a) Eland JHD (1974) Photoelectron spectroscopy Butterworths, London; (b) Carlson TA (1975) Photoelectron and Auger Spectroscopy, Plenum Press, New York; (c) Rabalais JW (1977) Principles of Ultraviolet Spectroscopy, Wiley, New York; (d) Brundle CR, Baker AD (1977) Electron Spectroscopy: Theory, Techniques and Applications, Vol 1, Academic Press, London; (e) Kimura K, Katsumata S, Achiba Y, Yamazaki T, Iwata S (1981) Handbook of HeI Photoelectron Spectra of Fundamental Organic Molecules, Halsted Press, New York

6. Hoffmann R, Heilbronner E, Gleiter R (1970) J Am Chem Soc 92: 706

7. Koopmans' T (1934) Physica 1: 104

8. Gleiter R (1971) Angew Chem 86: 770; Angew Chem Int Ed. Engl. 13: 696

9. Wittel K, McGlynn SP (1977) Chem Rev 77: 745

10. Verhoeven JW (1980) Rec Trav Chim Pays-Bas 99: 431

11. Paddon-Row MN (1982) Acc Chem Res 15: 245

12. Martin H-D, Mayer B (1983) Angew Chem 95: 281; Angew Chem Int Ed. Engl. 22: 283

13. Paddon-Row MN, Jordan KD (1988) in: Liebmann JF, Greenberg A (eds) Modern Models of Bonding and Delocalization VCH Publ, p 115

14. McGlynn SP, Wittel K, Klasinc L (1991) in: Maksic ZB (ed) Theoretical Models of Chemical Bonding Vol 3, Springer Verlag, Berlin, Heidelberg, New York

15. Dewar MJS, Dougherty RC (1975) The PMO Theory of Organic Chemistry, N. York, London

16. Heilbronner E, Bock H (1978) Das HMO-Modell und seine Anwendung, 2nd ed, Verlag Chemie, Weinheim

17. Brunck TK, Weinhold F (1976) J Am Chem Soc 98: 4392

18. Ohsaku M, Imamura A, Hirao K (1978) Bull Chem Soc Jpn 51: 3443; Imamura A, Ohsaku M (1981) Tetrahedron 37: 2191

19. Hoffmann R (1963) J Chem Phys 39: 1397; J Chem Phys 40: 2745

20. Heilbronner E, Maier JP (1977) Some Aspects of Organic Photoelectron Spectroscopy in Ref. [5d]

21. Heilbronner E (1972) Isr J Chem 10: 155

22. Paddon-Row MN, Patney HK, Brown RS, Houk KN (1981) J Am Chem Soc 103: 5575

23. Verhoeven JW, Pasman P (1981) Tetrahedron 37: 943

24. Balaji V, Jordan KD, Burrow PD, Paddon-Row MN, Patney HK (1982) J Am Chem Soc 104: 6849

25. Eckert-Maksić M, Maksimović Lj (unpublished results)

26. Sandorfy C, Daudel R (1954) R Hebd Seances Acad Sci 238: 93; Sandorfy C (1955) Can J Chem 33: 1337; Herndon WC (1972) in: Prog Phys Org Chem 9: 99; Dewar MJS (1979) Bull Soc Chim Belg 88: 957

27. Balaji V, Ng L, Jordan KD, Paddon-Row MN, Patney HK (1987) J Am Chem Soc 109: 6957

28. Jørgensen FS, Paddon-Row MN, Patney HK (1983) J Chem Soc Chem Commun 573

29. Paddon-Row MN, Patney HK, Peel JB, Willet GD (1984) J Chem Soc Chem Commun 564

30. Heilbronner E, Schmelzer A (1975) Helv Chim Acta 58: 936

31. Honegger E, Heilbronner E (1991) in: Maksić ZB (ed) Theoretical Models of Chemical Bonding Vol 3

32. Foster JM, Boys SF (1960) Rev Mod Phys 32: 300

33. Edmiston C, Ruedenberg K (1963) Rev Mod Phys 35: 457; England W, Salmon LS, Ruedenberg K (1971), Topics Curr Chem 23: 31

34. Foster JM, Weinhold F (1980) J Am Chem Soc 102: 7211; Reed AE, Weinstock RB, Weinhold F (1985) J Chem Phys 83: 735; Reed AE, Weinhold F (1983) ibid 78: 4066; Reed AE, Weinhold F (1986) ibid 108: 3586; Reed AE, Weinhold F (1985) J Chem Phys 83: 1736

35. Carpenter JE, Weinhold F (1988) J Am Chem Soc 110: 368

36. Bieri G, Dill JD, Heilbronner E, Schmelzer A (1977) Helv Chim Acta 60: 2234; Mohraz M, Batich Ch, Heilbronner E, Vogel P, Carrupt P-A (1979) Recl Trav Chim Pays-Bas 98: 1979; Bloch M, Jones TB (1979) Chem Ber 112: 369

37. Weinhold F, Carpenter JE (1988) J Mol Struct (Theochem) 165: 189

38. Batich Ch, Bischof P, Heilbronner E (1973) J Elec Spectr Rel Phenomena 1: 333, Goldstein MJ, Natowsky S, Heilbronner E, Hornung V (1973) Helv Chim Acta 56: 294; Bischof P, Hashmall JA, Heilbronner E, Hornung V (1969) Helv Chim Acta 52: 1745

39. Heilbronner E, Martin H-D (1972) Helv Chim Acta 55: 1490; Hoffmann RW, Schuttler R, Schafer W, Schweig A (1972) Angew Chem 84: 533; Angew Chem Int Ed Engl 11: 511

40. Fridh C, Asbrink L, Lindholm E (1972) Chem Phys Letters 15: 282; Asbrink L, Fridh CF, Lindholm E (1972) J Am Chem Soc 91: 5501

41. Pople JA, Santry DP, Segal GA (1965) J Chem Phys 43: S129; Pople JA, Segal GA (1965) ibid 43: S136

42. Dewar MJS, Haselbach E (1970) J Am Chem Soc 92: 590

43. Paddon-Row MN, Wong SS, Jordan KD (1990) J Chem Soc Perkin Trans 2: 425

44. (a) Hehre WJ, Stewart RF, Pople JA (1969) J Chem Phys 51: 2657; (b) Hehre WJ, Ditchfield R, Stewart RF, Pople JA (1970) J Chem Phys 52: 2769

45. Dietchfield R, Hehre WJ, Pople JA (1971) J Chem Phys 54: 724

46. Paquette LA, Künzer H, Green KE, DeLucchi O, Licini G, Pasquato L, Valle G (1986) J Am Chem Soc 108: 3458 (and references cited therein)

47. Künzer H, Litterst E, Gleiter R, Paquette LA (1987) J Org Chem 52: 4740

48. Brown RS, Buschek JM, Kopecky KR, Miller A (1983) J Org Chem 48: 3692; Gleiter R, Paquette LA (unpublished results)

49. Bingham RC, Dewar MJS, Lo DH (1975) J Am Chem Soc 97: 1285; Bischof P (1976) ibid 98: 6844

50. Gleiter R, Litterst E, Drouin J (1987) Chem Ber 121: 923

51. Gleiter R, Haider R, Bischof P, Lindner HJ (1983) Chem Ber 116: 3736

52. Martin HD, Eckert-Maksić M, Mayer B (1980) Angew Chem 92: 833; Angew Chem Int Ed Engl 19: 807

53. Hemmersbach P, Klessinger M, Bruckmann P (1978) J Am Chem Soc 100: 6344

54. Greenberg A, Liebman JF (1981) J Am Chem Soc 103: 44

55. Spanget-Larsen J, Korswagen de C, Eckert-Maksić M, Gleiter R (1982) Helv Chim Acta 65: 968

56. Honegger E, Yang Z-Z, Heilbronner E (1984) Croat Chem Acta 57: 967

57. Jorgensen WL, Salem L (1970) The Organic Chemist's Book of Orbitals, McGraw-Hill, New York, p 254

58. Wiberg KB, Matturo M, Adams R (1981) J Am Chem Soc 103: 1600; Wiberg KB, Matturro MG, Okarma PJ, Jason ME (1984) J Am Chem Soc 106: 2194

59. Houk KN, Gandour RW, Strozier RW, Rondan NG, Paquette LA (1979) J Am Chem Soc 101: 6797

60. Honegger E, Heilbronner E, Wiberg KB (1983) J Electron Spec Relat Phenom 31: 369

61. McMurry JE, Haley GJ, Matz JR, Clardy JC, VanDuyne G, Gleiter R, Schäfer W (1984) J Am Chem Soc 106: 5018

62. Falcetta MF, Jordan KD, McMurry JE, Paddon-Row MN (1990) J Am Chem Soc 112: 579

63. Gleiter R (unpublished results)

64. Gleiter R, Jähne G, Müller G, Nixdorf M, Irngartinger H (1986) Helv Chim Acta 69: 71

65. (a) Martin H-D, Schwesinger R (1974) Chem Ber 107: 3143; (b) Prinzbach H, Sedelmeyer G, Martin HD (1977) Angew Chem 89: 111; Angew Chem Int Ed Engl 16: 103

66. (a) Bartezko R, Gleiter R, Muthard JL, Paquette LA (1978) J Am Chem Soc 100: 5589; (b) Gleiter R, Heilbronner E, Paquette LA, Thompson GL, Wingard RE (1973) Tetrahedron 29: 565

67. Wilcox CF, Blain DA, Clardy J, Van Duyne G, Gleiter R, Eckert-Maksić M (1986) 108: 7693

68. Pariser R, Parr RG (1953) J Chem Phys 21: 466; Pople JA (1953) Trans Faraday Soc 49: 1375

69. Halgren TA, Kleier DA, Hall Jr. JH, Broen LD, Lipscomb WN (1978) J Am Chem Soc 100: 6595

70. Goldstein MJ, Hoffmann R (1971) J Am Chem Soc 93: 6193

71. Paddon-Row MN (1985) J Chem Soc Perkin 2: 257

72. Craig DC, Paddon-Row MN, Patney HK (1986) Austr J Chem 39: 1587

73. Paddon-Row MN, Engelhardt LM, Skelton BW, White AH, Jørgensen FS, Patney HK (1987) J Chem Soc Perkin Trans 2: 1835

74. Jørgensen FS, Paddon-Row MN (1983) Tetrahedron Lett 5415

75. Hünig S, Martin HD, Mayer B, Peters K, Prokschhy F, Schmitt M, von Schnering HG (1987) Chem Ber 120: 195
76. Bischof P, Gleiter R, Haider R (1978) J Am Chem Soc 100: 1036
77. Hoffmann R, Imamura A, Zeiss GD (1967) J Am Chem Soc 89: 5215
78. Dürr H, Gleiter R (1978) Angew Chem 90: 591; Angew Chem Int Ed Engl 17: 559
79. Hoffmann R, Davidson RB (1971) J Am Chem Soc 93: 5699; Salem L, Wright JS (1969) ibid 91: 5947
80. Gleiter R, Toyota A, Bischof P, Krennrich G, Dressel J, Pansegrau PD, Paquette LA (1988) J Am Chem Soc 110: 5490; Paquette LA, Dressel J (1989) in: de Meijere A, Blechert S (eds) Strain and its Implications in Organic Chemistry, p 77
81. Batich C, Heilbronner E, Rommel E, Semmelhack MF, Foos JS (1974) J Am Chem Soc 96: 7662
82. Houk KN, Scott LT, Rondan NG, Spellmeyer DC, Reinhardt G, Hyun JL, DeCicco GJ, Weiss R, Chen MHM, Bass LS, Clardy J, Jorgensen FS, Eaton TA, Sarkozi V, Petit CM, Ng L, Jordan KD (1985) J Am Chem Soc 107: 6556
83. Eckert-Maksić M, Gleiter R, Zefirov NS, Kozhushkov SI, Kuznetsova TS, Chem Ber (in print)
84. Bieri G, Heilbronner E, Kloster-Jensen E, Schmelzer A, Wirz J (1974) Helv Chim Acta 57: 1265
85. Gleiter R, Schäfer W, Sakurai H (1985) J Am Chem Soc 107: 3046
86. Gleiter R, Karcher M, Jahn R, Irngartinger H (1988) Chem Ber 121: 735
87. Gleiter R, Schäfer W, Flatow A (1984) J Org Chem 49: 372
88. Gleiter R, Schäfer W, Eckert-Maksić M (1981) Chem Ber 114: 2309
89. Honegger E, Heilbronner E, Hess N, Martin HD (1987) Chem Ber 120: 187
90. Gleiter R, Pfeifer KH, Szeimies G, Bunz U (1990) Angew Chem 102: 418; Angew Chem Int Ed 29: 413
91. Gleiter R, Mezger R (1990) Tetrahedron Lett 31: 1845
92. Boekelheide V (1980) Acc Chem Res 13: 65, (1983) Topics Curr Chem 113: 87; Gerson (1983) Topics Curr Chem 115: 57; Kleinschroth J, Hopf H (1982) Angew Chem 94: 485; Angew Chem Int Ed Engl 21: 469
93. Pignataro, S, Mancini V, Ridyard JNA, Lempka HJ (1971) Chem Commun, 142
94. Boschi R, Schmidt W (1973) Angew Chem 83: 408; Angew Chem Int Ed Engl 12: 402
95. Gleiter R (1969) Tetrahedron Lett 4453
96. Kovač B, Mohraz M, Heilbronner E, Boekelheide V, Hopf H (1980) J Am Chem Soc 102: 4314
97. Heilbronner E, Yang ZZ (1983) Topics Curr Chem 115: 1
98. Heilbronner E, Maier JP (1974) Helv Chim Acta 57: 151
99. Gleiter R, Eckert-Maksić M, Schäfer W, Truesdale EA (1982) Chem Ber 115: 2009
100. Gleiter R, Schäfer W, Krennrich G, Sakurai H (1988) J Am Chem Soc 110: 4117
101. Fessner WD, Sedelmeier G, Knothe L, Prinzbach H, Rihs G, Yang ZZ, Kovač B, Heilbronner E (1987) Helv Chim Acta 70: 1816
102. Bischof P, Hashmall JA, Heilbronner E, Hornung V (1969) Tetrahedron Lett 46: 4025
103. (a) Cowan DO, Gleiter R, Hashmall JA, Heilbronner E, Hornung V (1971) Angew Chem 83: 405; Angew Chem Int Ed Engl 10: 401; (b) Meeks JL, Maria HJ, Brint P, McGlynn (1975) Chem Rev 75: 603; (c) Dougherty D, McGlynn SP (1977) J Am Chem Soc 99: 3234; (d) Dougherty D, McGlynn SP (1978) ibid 100: 5597
104. (a) Heilbronner E, Hornung V, Kloster-Jensen E (1970) Helv Chim Acta 53: 331; (b) Heilbronner E, Hornung V, Maier JP, Kloster-Jensen E (1974) J Am Chem Soc 96: 4252; (c) Bieri G, Heilbronner E, Jones TB, Kloster-Jensen E, Maier JP (1977) Phys Scr 16: 202
105. Gleiter R, Spanget-Larsen J, Armarego WLF (1984) J Chem Soc Perkin Trans 2: 1517
106. Gleiter R, Dobler W (1985) Chem Ber 118: 1917
107. Gleiter R, Jähne G, Oda M, Iyoda M (1985) J Org Chem 50: 678
108. Cowan DO, Gleiter R, Hashmall JA, Heilbronner E, Hornung V (1971) Angew Chem 83: 405; Angew Chem Int Ed Engl 10: 401
109. Jähne G, Gleiter R (1983) Angew Chem 95: 500; Angew Chem Int Ed Engl 22: 488; Angew Chem Suppl 661
110. Gleiter R, Schäfer W, Staab HA, Saupe T (1984) J Org Chem 49: 4463

111. Maier JP (1974) Helv Chim Acta 57: 994
112. Eckert-Maksić M, Maksimović L, J Org Chem (in print)
113. Honegger E, Heilbronner E, Urbanek T, Martin HD (1985) Helv Chim Acta 68: 23
114. Lindholm E, Asbrink L (1985) Lecture Notes in Chemistry, Vol 38, Springer-Verlag, Berlin
115. Lindenberg J, Ohrn Y (1973) Propagators in Quantum Chemistry, Academic Press, N
     York, p 82; Lindenberg J, Ohrn Y, Thulstrup PW (1976) in: Quantum Science-Methods
     and Structure, Plenum Press, N York; Spanget-Larsen J, (1974) QCPE11: 246
116. Bischof P, Eaton Ph E, Gleiter R, Heilbronner E, Jones TB, Musso H, Schmelzer A, Stober
     R (1981) Helv Chim Acta 61: 547
117. Brogli F, Heilbronner E (1971) Helv Chim Acta 145: 1423

Penning Ionization — The Outer Shape of Molecules

Koichi Ohno and Yoshiya Harada

Department of Chemistry, The College of Arts and Sciences, The University of Tokyo, Komaba, Meguro-ku, Tokyo 153, Japan

1 Introduction

Important intermolecular interaction processes involving electron transfer, energy transfer, molecular collisions, and chemical reactions take place through a close contact of molecules, since a short-range interaction, such as an exchange-interaction, is responsible for most of the above processes. In order to elucidate such processes, it is interesting and seems almost necessary to study the outer characteristics of molecules by introducing an experimental probe which provides information on the frontier properties of molecules. As shown below in detail, Penning ionization electron spectroscopy (PIES) is a sensitive method for probing electron densities as well as potential energy surfaces at the very frontier of the molecule where the molecule is attacked by the reagent.

2 Penning Ionization Electron Spectroscopy

2.1 Penning Ionization and PIES

In 1927 F. M. Penning suggested an ionization process in which a metastable atom A^* behaves as an energy source for ionization of target system T, when he was studying rare gas discharge [1].

(Penning Ionization; PI) $A^* + T \rightarrow A + T^+ + e^-$ (1)

The Penning ionization process is a kind of chemiionization and related to the Jesse effect found in 1952; W. R. Jesse discovered that a minute amount of contaminants greatly enhances the total ionization of helium gas by alpha particles [2]. The Penning ionization (PI) process has been studied as an elementary process of reaction kinetics and atomic physics by means of flowing afterglow [3] and molecular beam techniques [4].

Table 1. PIES and UPS sources

		PIES		UPS	
		Excitation energy/eV	Life time/s		Energy/eV
He	2^1S	20.616	3.8×10^{-2}	He I	21.218
	2^3S	19.820	4.2×10^3		
Ne	$^3P_0$	16.716	4.3×10^2	Ne I	16.848
	$^3P_2$	16.619	2.4×10^1		16.670
Ar	$^3P_0$	11.723	4.5×10^1	Ar I	11.828
	$^3P_2$	11.548	5.6×10^1		11.623

A kinetic energy analysis of the electrons ejected via the process (1) was first performed by Čermák [5] in 1966 and is now called Penning ionization electron spectroscopy, PIES [6]. This technique is in many respects similar to ultra-violet photoelectron spectroscopy, UPS, in which ultra-violet photons are used as the ionization-energy source in place of the metastable atoms [7, 8].

$$\text{(Photoionization)} \quad h\nu + T \rightarrow T^+ + e^- \tag{2}$$

Table 1 lists the energy sources usually used in these electron spectroscopies. The energy range of ejected electrons are so common for these methods that both spectra are often measured with the same electron energy analyzer. Although the source energy E is generally different, the energy conservation law for a particular ionization potential, IP, is equally expressed as follows;

$$K = E - IP \tag{3}$$

Where K is the kinetic energy of the ejected electron. A kinetic energy analysis of ejected electrons thus provides the energy (IP) required for the reaction process of (1) or (2).

2.2 Experiment

The experimental apparatus for PIES and UPS is composed of vacuum chambers equipped with excitation sources (a metastable source and a UV photon source) as well as an electron analyzer and electronic circuit systems controlling and scanning

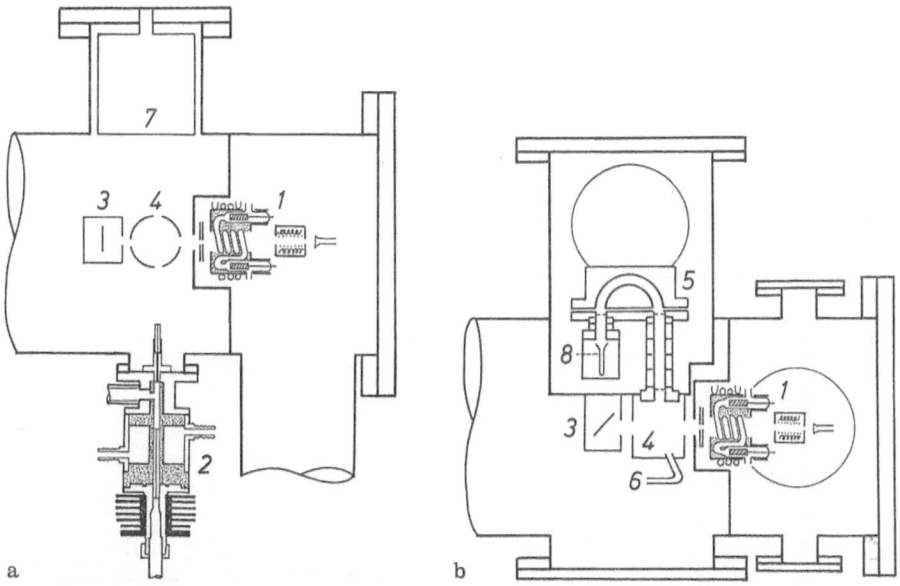

Fig. 1a, b. Electron spectrometer for PIES and UPS. **(a)** Side view, **(b)** Top view. *1*: Metastable atom source; *2*: He discharge lamp; *3*: Faraday cap; *4*: Collision chamber; *5*: Electron energy analyzer; *6*: Sample inlet; *7*: Cold trap; *8*: Electron multiplier

the sampling kinetic energy of electrons as well as detecting and storing (or recording) the number of electron counts. A typical example of the electron spectrometer employed by us for PIES/UPS of gaseous samples [9] is shown in Fig. 1. The electron analyzer is an electrostatic deflection type of hemispheres. The metastable atoms are produced by electron impact or a cold discharge, and UV photons are supplied by a water cooled rare gas discharge in a capillary. The energy dependence of the transmission of the electron spectrometer was determined by a detailed study of the UPS of O_2, CO, CO_2, N_2, C_2H_4, butadiene, and benzene [10].

2.3 Spectral Band Assignment

Figure 2 shows the He*(2^3S) Penning ionization electron spectrum (the spectrum is also denoted by PIES) and the He I ultra-violet photoelectron spectrum (UPS) for N_2 [10]. Abscissas are kinetic energies of ejected electrons (K) which correspond to the ionization potentials (IP) via Eq. (3). The energy differences 1.40 eV between the He*(2^3S) 19.82 eV and the He I resonance line 21.22 eV (see Table 1) is reflected by the relative positions of the abscissas. The spectral band positions in Fig. 2 correspond well to each other between PIES and UPS.

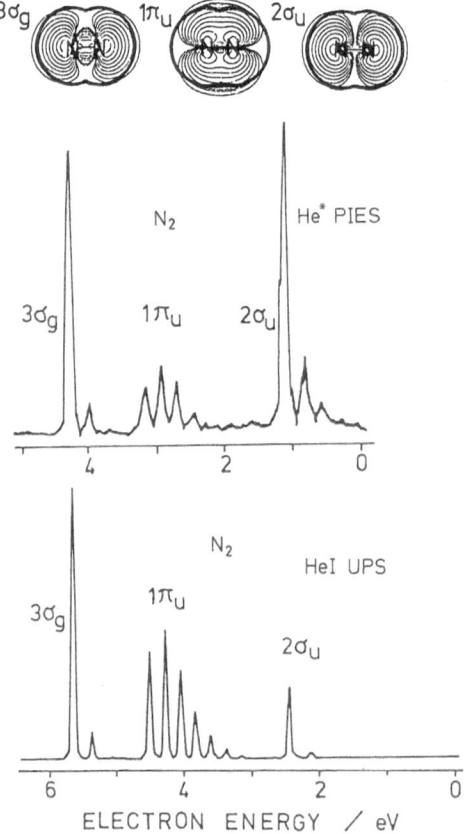

Fig. 2. He*(2^3S) PIES and He I UPS for N_2 and electron density maps for the relevant molecular orbitals. The thick solid curve in the maps indicates the molecular surface

The observed bands are assigned to ionizated states of N_2, $X^2 \Sigma_g^+$, $A^2 \Pi_u$, and $B^2 \Sigma_u^+$ which can be ascribed to an electron removal from $3\sigma_u$, $1\pi_u$, and $2\sigma_u$ molecular orbitals of N_2, respectively. Electron density maps are also shown in Fig. 2 for these orbitals together with the repulsive surface of the molecule. Fine structure in the spectra are due to vibrational progressions which are related to the decrease of the electron energy K associated with the increase of IP due to the vibrational excitation in the ionized states. Slight broading of the PIES bands is caused by modification of electron kinetic energies due to excitation or deexcitation of the relative motion of the interacting particles, A* and T, which can also occur upon ionization. In the case of UPS, the change in the translational energy upon ionization is negligible in comparison with the energy resolution of the apparatus (40 meV).

For most closed shell molecules, observed bands in UPS and those in PIES are assigned to molecular orbitals, as in the case of N_2. It is only exceptional that additional bands appear by electron correlation effects [11].

2.4 Spectral Characteristics and Interpretation

Spectral characteristics of PIES have been noted in comparison with UPS since the pioneering work by Čermák [5, 6].

(i) Peak positions in PIES usually shift to a certain extent (a few meV — several hundreds meV) from those expected for UPS, and band shapes in PIES show somewhat broader features.

(ii) Relative band intensities in PIES are different from those in UPS.

These characteristics have been interpreted on the basis of the following two fundamental models. (a) In 1966, Hermann and Čermák proposed a two-state potential curve model based on the adiabatic approximation for the interacting system [12]. Collision trajectories can be described by the potential curves for A* + T and A + T⁺, and electron ejection process is considered as an autoionization transition from the initial potential curve to a continuum level of the final potential curve. (b) In 1969, Hotop and Niehaus suggested and electron exchange model for the ionization process involving metastable rare gas atoms [13]. In the exchange mechanism, metastable atoms A* are deexcited in such a way that an electron in the outer shell of the metastable atom A* is ejected into a continuum state associated with the transfer of an electron in one of the orbitals of T into the vacant inner shell orbital of A*.

These two models were combined and successfully applied to analyses of many aspects in Penning ionization for simple target systems [14]. The exchange model was confirmed from various studies including mass spectromety [15], measurements of absolute cross sections [16], analyses of angular distributions [17], and theoretical calculations [18, 19]. Physical analyses of PIES have been carried out further especially on the broadened features and the velocity dependence of Penning ionization cross sections, and the above two models have been accepted for Penning ionization involving metastable rare gas atoms and simple molecules [20, 21, 14].

Relative band intensities in PIES or branching ratios (partial cross sections) for Penning ionization reactions yielding various ionic states of target molecules, however, were not understood clearly from theoretical bases until this decade. Many aspects have been observed and various suggestions have been made. Relative enhancement of σ bands in PIES for linear molecules was first pointed out by Hotop and Niehaus in 1970 [22]. Geometrical accessibility of A* to the target orbitals was suggested as being important [3]. Relative diffuseness of π orbitals indicates that they are important for the enhacement of σ bands in N_2 and CO [23]. Dependence on the spatial distribution of electrons in the relevant molecular orbitals was suggested as governing the most probable geometry for Penning ionization of O_2 [24]. A remarkable enhancement of "lone-pair" bands for CN compounds was observed, and this finding was ascribed to large electron density localized on the nitrogen atom in the relevant orbital [25]. Also, similar enhancements of lone-pair bands in PIES were found for various molecules [26–31]. From an analysis of two-electron process yielding $Cd^+(5p)$ ions, it is concluded that probability of the Penning ionization process essentially reflects the overlap of the interacting target electron with the inner shell hole of the metastable rare gas atom [32]. A different type of enhancement in PIES was found for unsaturated hydrocarbon molecules for which π bands in the PIES were considerably enhanced than σ bands in comparison with the UPS, and this finding is suggested to be due to the spatial limitation of σ orbitals and shielding of π orbitals [33]. Subsequent studies of PIES for large π electron systems showed that the higher electron densities of the relevant orbitals in the interacting region are suggested as being most responsible for the observed enhancement [34, 35]. Stimulated by these findings, further studies were made for various systems [36–39].

In a recent study of transmission-corrected PIES for both organic and inorganic molecules, a simple principle, which governs relative intensities of PIES, was established on the basis of comparison of observed band intensities with electron densities of ab initio molecular orbitals; the outer orbital which is exposed outside the molecular surface is active and the inner orbital which is localized inside the molecular surface is inactive [10]. This model yielded a simple concept of the exterior electron density (EED) which was shown to be proportional to the relative PIES intensity [40, 41].

3 A Simple Model for Penning Ionization

3.1 Theoretical Models

Theoretical models for PI were developed on the basis of adiabatic approximation and formulated as an autoionization process [42, 6]. Key factors involved were (V(R) and F(R). V(R) is real potential energy surface for various nuclear configurations. For the entrance and exit channels of PI process, there are at least two potential surfaces, V*(R) and $V^+(R)$ for A* + T and A + T^+, respectively. $\Gamma(R)$ is the autoionization width in energy scale, and $\Gamma/\hbar$ is the autoionization rate which is related to the electronic transition probability for the autoionization process, W(R).

Since the excited (A* + T)-state is resonant with continuum (A + T^+ + e^-)-states, conventional CI methods cannot be used for the determination of V*(R). On account of this difficulty, applications of fundamental theories have been limited to very simple systems, such as hydrogen atoms [18, 43], hydrogen molecules (H_2 and D_2)

[43–47], and rare gas atoms [48, 49] with the aid of several techniques, (1) the stabilization method [50–53], (2) an expansion technique of continuum orbitals in terms of
discrete MO's [43], (3) the use of experimental Li–T potential in place of the He*–T
potential [48], and (4) one-electron model calculations [49]. Further applications of
these techniques cannot be straightforward for more strongly anisotropic systems
because these will require inclusion of higher order terms in the Legendre expansion
of the potential functions.

Further difficulties are involved for the more general systems; crossing, noncrossing,
or near crossing of potential surfaces including $A^* + T, A^+ + T^-, A^- + T^+$,
and some Rydberg states will play important roles in the PI process. Also, trajectory
calculations will become practically impossible for complex anisotropic systems.

In order to calculate branching ratios for the PI reactions or relative intensities in
PIES, $V^+(R)$ and $\Gamma(R)$ must be considered for several ionic states. Thus, it becomes
more difficult to apply rigorous theoretical procedures to PI process involving polyatomic target systems. In the following, some general remarks of $V(R)$ and $\Gamma(R)$ are
summarized in order to find a simple approach to explain relative intensities in PIES.

3.2 Potential Energy Curves

In Fig. 3 two-potential cruves $V^*(R)$ and $V^+(R)$ are schematically shown for an atomic
target T. The curves V^* and V^+ denote the potential curves for the entrance $(A^* + T)$
and the exit $(A + T^+)$ channels, respectively. The excitation energy of a metastable
atom A^*, $E(A^*)$, is assumed to be larger than the ionization potential of an atomic
target T, IP. An electronic transition can occur at all atomic distances larger than that
for the classical turning point, Rc. The separate conservation of the electronic and
nuclear energies in the transition at distance R, leads to:

$$V^*(R) = K(R) + V^+(R) \quad \text{or} \quad K(R) = V^*(R) - V^+(R) \tag{4}$$

and

$$K(\infty) = V^*(R) - V^+(\infty) = E(A^*) - IP(\infty) \tag{5}$$

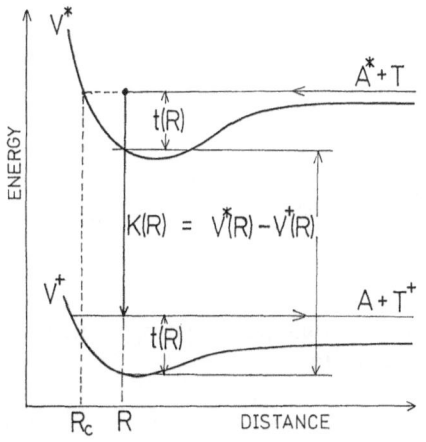

Fig. 3. Penning ionization process for the metastable atom A* and the target sample $T(A^* + T \rightarrow$
$A + T' + e^-)$. The kinetic energy of the ejected
electron $K(R)$ is equal to the potential energy
difference, $V^*(R) - V^+(R)$, provided that the
relative translational energy $t(R)$ is conserved upon
the electronic transition. Rc denotes the classical
turning point

where $K(R)$ is the kinetic energy of Penning electrons. Using the energy shift of the ionization potential with respect to its value at the infinite distance; $\delta(R) = IP(R) - IP(\infty)$, the following equations are obtained:

$$K(R) = E(A^*) - (IP(\infty) + \delta(R)) \tag{6}$$

$$\delta(R) = (V^*(R) - V^+(R)) - (V^*(\infty) - V^+(\infty)). \tag{7}$$

The $IP(\infty)$ is the ionization potential of T at the infinite separation and this quantity is applied to UPS in Eq. (3). $\delta(R)$ is a correction factor of the ionization potential at a finite separation, and its value depends on the difference of the curves, $V^*(R)$ and $V^+(R)$. This leads to peak shifts and band broadening in PIES. Thus, the shift and width of the PIES-band contain information on the interparticle potentials, V^* and V^+.

3.3 Electronic Transition Probabilities

In general, the electronic transition probability $W(R) = \Gamma(R)/\hbar$ is given by

$$W(R) = \Gamma(R)/\hbar = (2\pi/\hbar) \, |T(R)|^2 \, \varrho$$

where ϱ is the density of the final state and $T(R)$ is the transition matrix element between the initial and final states;

$$T(R) = \langle \psi_i | \, \hat{H} - E \, | \psi_f \rangle \tag{9}$$

If one approximates the wavefunctions ψ_i and ψ_f in terms of antisymmetrized products of orthogonal orbital functions, one has the following expressions for the collision between a closed shell molecule (MO; $\phi_1, \phi_2, \dots, \phi_n$) and a metastable singlet or triplet atom (inner shell AO, χ_a) and outer shell AO, χ_b):

$$^{1,3}\psi_i = \frac{1}{\sqrt{2}} \left[|\phi_1 \bar{\phi}_1 \cdots \phi_i \bar{\phi}_i \cdots \phi_n \bar{\phi}_n \chi_a \bar{\chi}_b| \mp |\phi_1 \bar{\phi}_1 \cdots \phi_i \bar{\phi}_i \cdots \phi_n \bar{\phi}_n \bar{\chi}_a \chi_b| \right] \tag{10}$$

and

$$^{1,3}\psi_f = \frac{1}{\sqrt{2}} \left[|\phi_1 \bar{\phi}_1 \cdots \phi_i \bar{\varphi}_e \cdots \phi_n \bar{\phi}_n \chi_a \bar{\chi}_a| \mp |\phi_1 \bar{\phi}_1 \cdots \bar{\phi}_i \varphi_e \cdots \phi_n \bar{\phi}_n \chi_a \bar{\chi}_a| \right] \tag{11}$$

where an electron is removed from MO ϕ_i into a continuum orbital φ_e in the final state. The conservation of the spin multiplicity leads to

$$T(R) = J \pm J - K \tag{12}$$

in which J and K are Coulomb and exchange-type integrals:

$$J = \int\int \phi_i^*(1)\, \varphi_e(1)\, \frac{1}{r_{12}}\, \chi_b^*(2)\, \chi_a(2)\, dv_1\, dv_2 \tag{13}$$

and

$$K = \int\int \phi_i^*(1)\, \chi_a(1)\, \frac{1}{r_{12}}\, \chi_b^*(2)\, \varphi_e(2)\, dv_1\, dv_2 \tag{14}$$

In the above equations, the upper sign is for the singlet and the lower for the triplet states.

For triplet metastable atoms, such as He*(2^3S), Ne*($^3P_0$ and $^3P_2$) and Ar*($^3P_0$, $^3P_2$), the J terms vanish and need not be considered. Even for He*(2^1S) the J terms can be neglected because the value of J is usually very small. This is supported by the following reasons. (i) The radiative transition from 2^1S state to the ground state of the helium atom is a dipole forbidden transition for which an excitation transfer between the particles cannot effectively occur by way of the dipole-dipole mechanism. (ii) The PI cross sections for He*(2^3S) and He*(2^1S) are not much different [13, 15]. (iii) The branching ratios for yielding various ionic states are nearly the same for 2^3S and 2^1S He* atoms [13]. (iv) Both metastable atoms give very similar angular distributions of ejected electrons [17].

Thus, the transition probability $W_i(R)$ for the ejection of an electron from the MO ϕ_i is given by

$$W_i(R) \propto |K|^2 = \left| \int\int \phi_i^*(1)\, \chi_a(1)\, \frac{1}{r_{12}}\, \chi_b^*(2)\, \varphi_e(2)\, dv_1\, dv_2 \right|^2 \tag{15}$$

the value of which depends on the differential overlaps $\phi_i^*(1)\chi_a(1)$ and $\chi_b^*(2)\varphi_e(2)$. It should be noted here that the relative transition probability for an electron removal from a particular MO largely depends on the mutual overlap of ϕ_i and χ_a. Since the mutual overlap generally decays exponentially as a function of the mutual distance R, and approximate form of W(R) has been often assumed to be Aexp($-$aR) [6, 14, 20].

3.4 The Simple Principle for Orbital Activities in PI

In general both V(R) and W(R) are important for describing PI. However, the relative importance depends on what aspects are concerned. If the PI cross section is studied as a function of the relative velocity of A* and T or their scattering angle, the shape of potential surfaces should be known as accurately as possible, although W(R) may be approximated by a simple assymptotic form of Aexp($-$aR) where A and a are considered as parameters [49]. On the other hand for the study of branching ratios, a comparative description becomes important for $W_i(R)$, $W_j(R)$, ... , which are the autoionization transition rates for one-electron removal from MO ϕ_i, ϕ_j, ... [41].

As long as the relative activity of MO in Penning ionization is concerned, V*(R) is common for all MO. Thus, the shape of V* can be treated as parametrized one. The simplest description of V* is the hard sphere model in which A* and T are considered as rigid spheres of certain radii. The sum of these radii corresponds to the classical turning point Rc in Fig. 3. As a crude approximation, these radii can be estimated from geometrical parameters obtained in other experiments. For polyatomic target systems, the single hard sphere model must be replaced by a modified model using a molecular repulsive surface (molecular surface). This molecular surface can be estimated by placing a sphere with an effective radius on each atomic position.

Since the value of W(R) increases exponentially as the metastable atom approaches the target molecule, the most favorable distance for the transition is in the neighborhood of the classical turning point of the collision trajectory, which corresponds to the molecular surface where the interaction potential becomes repulsive. Furthermore, major contributions to the electronic integral K can be attributed to local regions where ϕ_i effectively overlaps with χ_a. Therefore, relative activities of individual orbitals are governed by electron densities of ϕ_i in the exterior region outside the molecular surface [10, 40, 41].

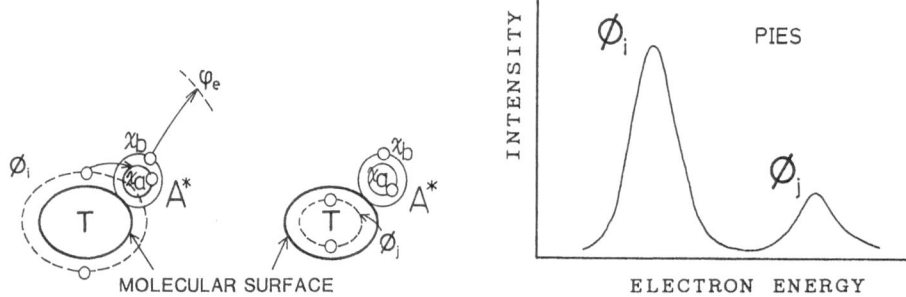

Fig. 4. The principle for orbital activities in Penning ionization and relative PIES intensities. The metastable atom (A^*) can approach the target molecule (T) up to the molecular surface shown by the solid curve. Electron transfer from T to A^* can take place easily for the outer orbital ϕ_i which has a large overlap with the vacant inner-shell orbital of $A^*(\chi_a)$. The Penning ionization process is less likely to occur for the inner orbital ϕ_j because of its small overlap with χ_a even when the metastable atom (A^*) comes up to the minimum distance. Consequently, the ionization from the outer orbital ϕ_i gives rise to a larger intensity in PIES than does the inner orbital ϕ_j

A clear insight is obtained for qualitative understanding of relative activities of target orbitals in PI from Fig. 4 [10]. The solid curve indicates the boundary of the repulsive interaction between the metastable and the molecule (molecular surface). Dashed curves show two extreme cases of orbitals ionized; the orbital ϕ_i is an outer orbital extending out of the molecular surface, and the orbital ϕ_j is an inner orbital localized inside the molecular surface. Since an orbital exposed to the outside more effectively overlap with the inner-shell orbital of A*, the orbital ϕ_i extending outside gives rise to a larger intensity in PIES than does ϕ_j localized inside. This provides a simple principle for orbital activities in PI; the exterior MO which is exposed to the outside is active and the interior MO which is localized inside is inactive.

4 Application of PIES to the Study of the Outer Shape of Molecules

4.1 Stereochemical Properties of Molecular Orbitals and Molecular Surfaces

PIES is considered as a novel technique probing electron densities of individual molecular orbitals in the exterior region outside the molecular surface. Thus, stereochemical properties of molecular orbitals and molecular surfaces can be studied [39, 54]. When molecular surfaces are estimated from some experimental or empirical parameters, exterior electron distributions of molecular orbitals can be probed. If electron distributions of MO are calculated with good accuracy, geometrical characteristics of molecular surfaces can be studied. The metastable atoms can thus be used as probes to detect microscopic properties of molecules, especially their outer shapes reflecting the interplay of MO and the molecular surface.

In the last few years, we have measured PIES for more than 200 molecules, and have found various interesting aspects of stereochemical properties [55, 56]. In the following, typical examples are shown.

4.2 σ and π Orbitals

The relative activities of molecular orbitals can be measured as relative band intensities in PIES. Remarkable correlations were found between orbital activities in PI and orbital types. Typical examples are listed in Table 2. First and foremost, the relative reactivity of σ and π orbitals are discussed here.

As can be seen in Fig. 2, bands are enhanced with respect to π bands in PIES for N_2; the $3\sigma_g$ and $2\sigma_u$ orbitals are more reactive with metastable atoms than the $1\pi_u$ orbitals [22, 10]. Similar examples of $\sigma > \pi$ in the PIES orbital activity were found for linear molecules without hydrogen atoms [10]. The reverse tendency was found for unsaturated hydrocarbon molecules for which π orbitals are more reactive than σ orbitals. Typical example is shown in Fig. 5 [10]. This tendency that $\pi > \sigma$ in the

Table 2. Observed tendency for PIES band intensity[a]

Tendency	Example	Reference
$\sigma > \pi$	N_2, CO, CO_2, N_2O	[22], [10], [62]
$\pi > \sigma_{CH} > \sigma_{CC}$	C_2H_2, C_2H_4, C_6H_6	[33], [10], [41]
	$C_{10}H_8$, $C_{14}H_{10}$	[37], [57]
$n > b$	NH_3, H_2O, H_2S	[10]
	RX (X=NH_2, SH, Cl, Br, I)	[58]
n_x: $5p > 4p > 3p > 2p$	RX (X=Cl, Br, I, F)	[58], [59]
n_x: V > VI > VII	RX (X=NH_2, OH, F, SH, Cl)	[58], [75]
$n_{sp} > \pi$	RCN, RCOR'	[63], [62]
Ligand > Metal–d	$Fe(C_5H_5)_2$, $Fe(CO)_5$	[38], [9]

[a] This tendency shows the relative reactivity of target orbitals upon the electrophilic attack by the metastable rare gas atom

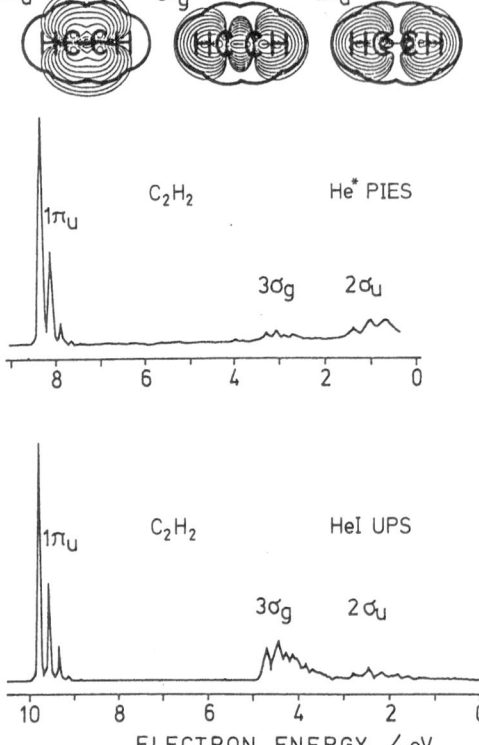

Fig. 5. He*(2³S)PIES and He I UPS for C_2H_2 and electron density maps for the relevant orbitals. The *thick solid curve* in the maps indicates the molecular surface

PIES intensity has been noted as giving valuable supports in the assignments of photoelectron spectra for butadiene, benzene, naphthalene, anthracene, and some substituted derivatives [10, 33, 37, 41, 57].

It is interesting that, contrary to the usual spectroscopy, the symmetry, σ or π, is not decisive in the PIES intensity. In the case of N_2, electron distributions extending outside along the molecular axis are considered to be responsible for the higher activites of σ orbitals (Fig. 2). In the acetylene molecule (Fig. 5) there are two hydrogen atoms in the molecular axis and its σ orbitals have no nonbonding(lone-pair)character. Thus the activity of σ orbitals in acetylene decreases, whereas for π orbitals electron distributions become much more exposed outside because of the lower electronegativity of C atoms in comparison to that of N atoms. These effects explain the reversal of the relative activity of σ and π orbitals.

4.3 Nonbonding and Bonding Orbitals

Strongly bonding orbitals generally concentrate their electron densities in the bond-region inside the molecular surface, and thus their activities are usually low because of their small exterior electron distributions. Nonbonding orbitals with lone-pair characters have relatively large electron distributions in the exterior region and show higher orbital activities. Therefore, metastable atoms tend to make electrophilic

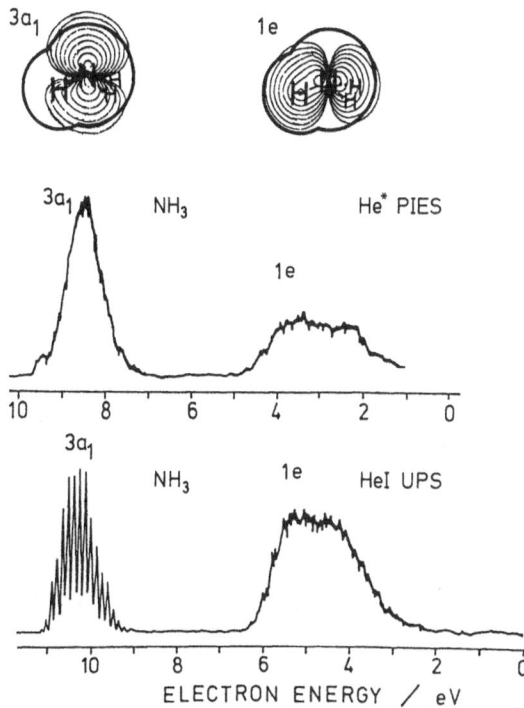

Fig. 6. He*(2^3S) PIES and He I UPS for NH_3 and electron density maps for relevant orbitals. The *thick solid curve* in the maps indicates the molecular surface

attacks upon nonbonding orbitals (n) rather than upon bonding orbitals (b) (Table 2). NH_3 is an example to show $n > b$ in PIES activity (Fig. 6); the $3a_1$ orbital has large exterior electron distributions corresponding to the nitrogen lone-pair, and this orbital gives stronger band in PIES than the $1e$ orbitals of strongly NH bonding character [10]. Very high activities of nonbonding orbitals were found for amines, sulfides and halogenides having lone-pair electrons [58, 59].

4.4 Electronegativity

In general nonbonding orbitals exhibit high activities in PIES. Their relative activities depend on the extent of electron distributions exposing outside of the molecular surface. A remarkable example of the effect can be seen in Fig. 7, in which PIES for C_2H_5X (X = OH, Cl, NH_2, I, SH) are shown [59]. Band 1 is assigned in each case to a nonbonding orbital which has a large population on the heavy-atom contained in the functional group X. When the band intensity I is normalized by the summation for other bands mainly related to the ethyl group, the value of I increases from OH to SH in good accordance with the decrease of the electronegativity. This tendency is also in good agreement with the decreasing order of the orbital exponents from Slater's rule. This fact indicates that the smaller electronegativity of the heavy-atom causes the more extended distribution of the nonbonding orbital to result in the higher orbital activity in PIES [58]. Analogous to this result, studies of monohalogeno benzenes revealed that nonbonding orbitals become much more exposed outside with high activities on going from $2p$ to $5p$ for the character of the nonbonding orbitals having large populations on halogen atoms [59].

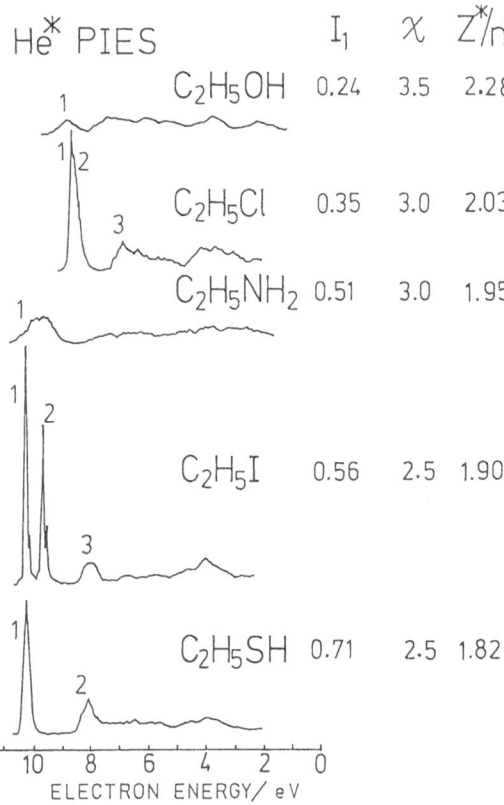

He*(2³S) PIES

	I_1	χ	Z^*/n^*
C_2H_5OH	0.24	3.5	2.28
C_2H_5Cl	0.35	3.0	2.03
$C_2H_5NH_2$	0.51	3.0	1.95
C_2H_5I	0.56	2.5	1.90
C_2H_5SH	0.71	2.5	1.82

ELECTRON ENERGY/ eV

Fig. 7. He*(2³S) PIES for $C_2H_5X(X = $ = OH, Cl, NH_2, I, SH). The relative intensity for the first band (I_1) of the nonbonding character (n_X), the Pauling's electronegativity of the heavy-atom in the functional group (χ), and the Slater's orbital exponent (Z^*/n^*) give good correlation

4.5 Isolobal

Connection of the reactivity with the shape of the relevant orbital has been pointed out by Hoffmann [60], and this connection is called the isolobal analogy by Mingos [61]. Figure 8 shows a typical example of the isolobal analogy in PIES [62]. The PIES for HCOOH contains components corresponding to the spectra of H_2O and HCHO. There are strong correlations in the band intensity and band shape for the bands of similar type orbitals. One of the most remarkable features is the enhancement of the σ_{CO} band in carbonyl compounds; 8a′ in HCOOH and $5a_1$ in HCHO are markedly outstanding in the spectra, and the same tendency has been observed for other carbonyl compounds [62]. Isolobal effects were also found for nitrile compounds [63], transition metal carbonyl complexes [64], and some aromatic compounds [65]. In the case of the C=O group and the C≡N group, the remarkable enhancement should also be related to the attractive nature of the potential well which increases the PIES activity of orbitals which have high electron densities in the well region.

4.6 Through-Space/Bond Interactions

Intramolecular interactions of orbitals have been studied in relation to variation of ionization potentials of related molecules; through-space and through-bond interactions have been discussed [66–71]. Figure 9 shows an example of the through-bond

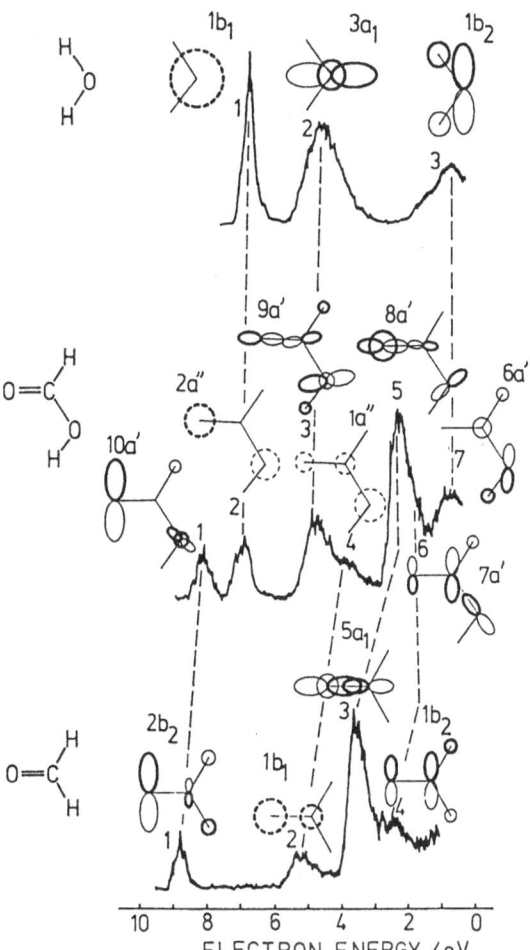

Fig. 8. Correlation of PIES for H_2O, HCHO, and HCOOH

interaction by which the symmetric type nonbonding orbital (n_s) of 1,4-diazabicyclo-(2.2.2)octane(DABCO) has smaller IP than the asymmetric type (n_a) [52]. The PIES of DABCO gave a support that the n_s orbital has smaller exterior electron densities than the n_a orbital; the observed PIES activities were $n_s:n_a = 1.0:1.75$. In the n_a orbital the lone-pair character is weakened by the through-bond interactions at three pairs of C—C bonds, where electron densities are absorbed for increasing the C—C bond characters. Therefore, this result is noted as a typical example of the effect of the through-bond interaction which causes absorption of electron densities into bonding regions and also a decrease of exterior electron densities to result in a considerable reduction of the efficiency in the electrophilic reaction for the relevant orbital [54].

Figure 10 shows an example of the through-space interaction by which the symmetric type combination of π orbitals (π_s) has a larger IP than the asymmetric type (π_a) [54]. The PIES activities for these orbitals were clearly found to be $\pi_s > \pi_a$. This indicates that the exterior electron densities in the endo face is effectively increased for the π_s orbital by through-space interaction which switches on the extraction of electron densities out of the molecular surface into the exterior region to enhance the efficiency of the electrophilic reaction for the relevant orbital [54].

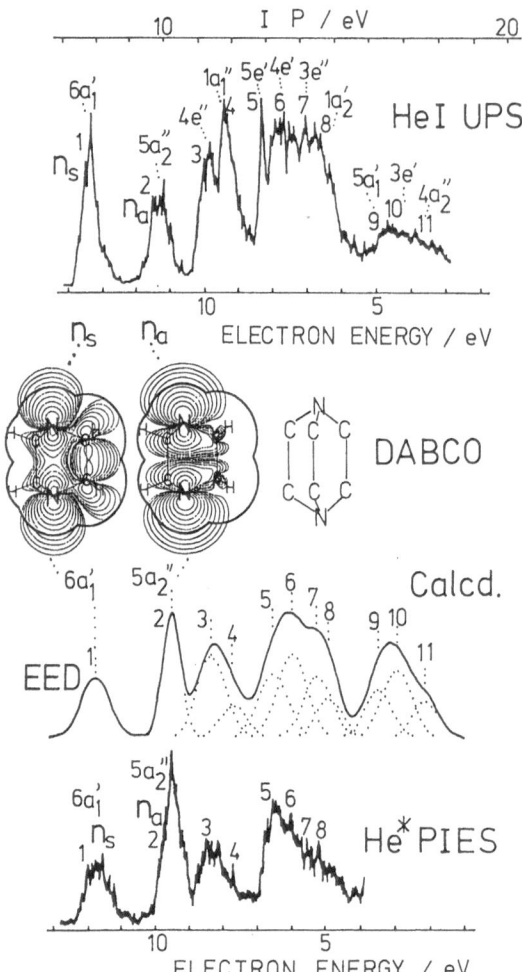

Fig. 9. He*(2^3S) PIES **(bottom)**, theoretical Penning spectrum calculated from EED **(middle)**, and He I UPS **(top)** for 1,4-diazabicyclo(2.2.2)octane (DABCO)

4.7 Inclusion and Sandwich

In general reactive electrons are in the exterior region. However, the situation changes dramatically when the reactive electrons flock to form a new bond; the region where the new bond is built is inside the repulsive surface of the new system.

In transition-metal complexes, metal atoms or ions are usually enveloped by ligands. The geometrical accessibility of incoming species to the metal d orbitals is so reduced in complexes that their PIES activities are very small in comparison with the ligand orbitals which can face the attack by electrophiles. Typical examples are $Fe(CO)_5$ [9], $Cr(CO)_6$ [64], and $Fe(C_5H_5)_2$ [38]. Figure 11 compares orbital activities of the complex molecule $Fe(CO)_5$ and the ligand molecule CO [9]. In CO, the 5σ orbital is most active because of its high exterior electron density on the carbon side. In the complex, the 5σ orbital is used for forming σ bonds between Fe and C atoms, and thus PIES activities for the 5σ orbitals become reduced in $Fe(CO)_5$. On the other hand, the 4σ orbitals having large electron populations on the oxygen side are almost solely out-

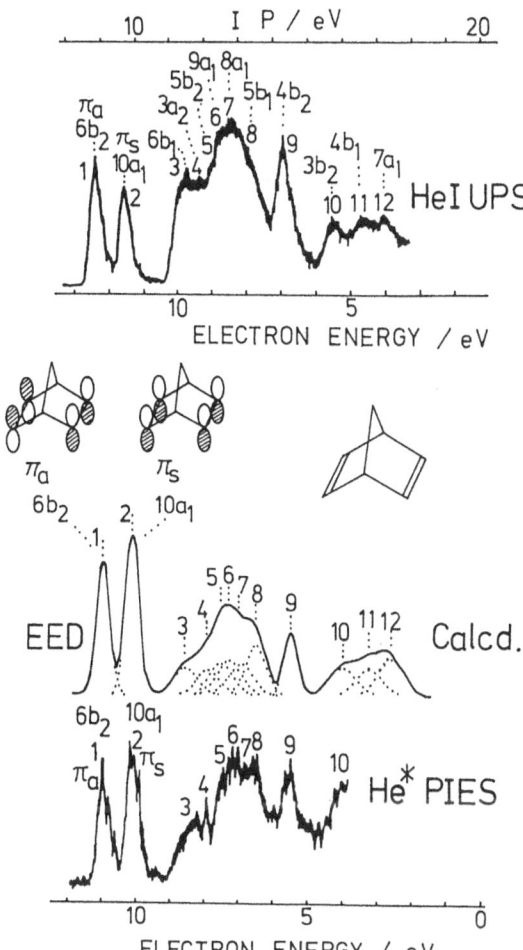

Fig. 10. He*(2^3S) PIES **(bottom)**, theoretical Penning spectrum calculated from EED **(middle)**, and He I UPS **(top)** for norbornadiene

standing in the complex to face with the electrophilic attack by metastable atoms and give very strong PIES activities. Here, it is also noted that Fe $3d$ orbitals, which have most of their electron densities in the center of the complex, show very weak activities in the PIES. This clearly indicates that the Fe $3d$ orbitals are almost completely protected from attacks of incoming reagent (A*) by surrounding five CO groups.

In Fe(C_5H_5)$_2$ iron d orbitals are shielded by a couple of cyclopentadienyl rings. PIES activities for the sandwiched Fe d orbitals were found to be much smaller than those for ring π orbitals [38].

4.8 Substitution

Metastable atoms can be used to probe effects of substitution on the reactivity of orbitals upon the electrophilic attack. Figure 12 shows relative PIES activities of π and n orbitals in a variety of anilines with methylgroups [39]. Relative UPS intensity ratios are also shown in Fig. 12 for a comparison. The ratios for UPS are almost

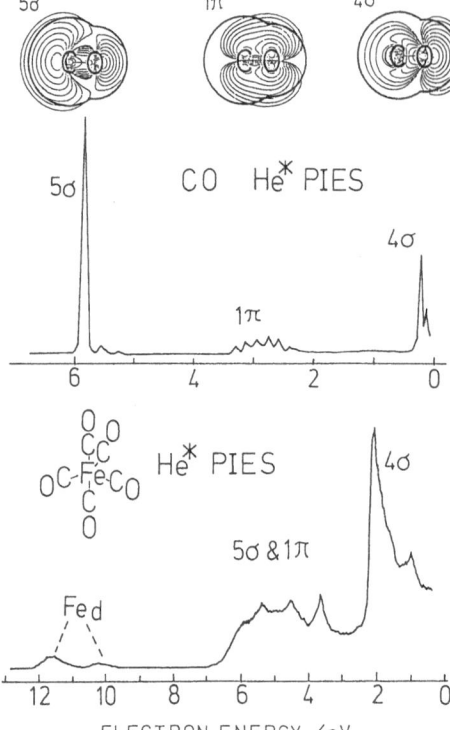

Fig. 11. He*(2³S) PIES for CO **(upper)** and Fe(CO)₅ **(lower)**

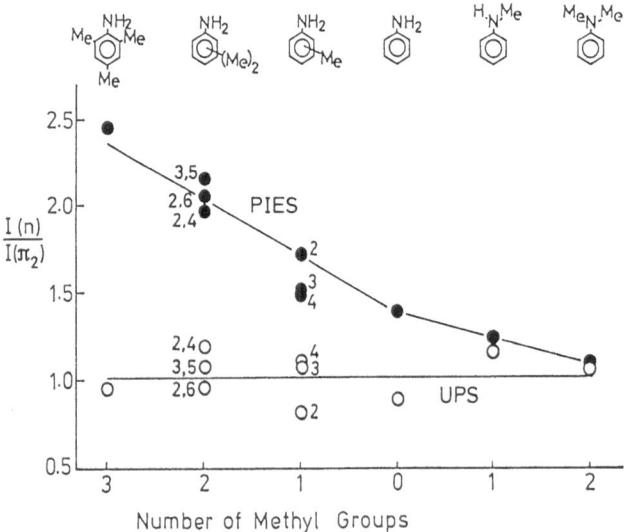

Fig. 12. The integrated intensity ratio for the n and π_2 bands in He*(2³S) PIES and UPS of Methyanilines. *Closed circles* are for PIES and *open circles* are for UPS. The compounds studied are shown at the top of the figure. The positions of methylation at the ring are shown beside the circles for mono- and dimethylanilines

constant and equal to unity within experimental error. This provides a confirmation of the simple propensity rule that photoionization cross sections are nearly equal for MO composed of similar type AO's [72]. Therefore UPS does not show substitution effects in the relative intensities. On the other hand in PIES remarkable substitution effects were found. In aniline, the PIES intensity ratio $I(n)/I(\pi_2)$ is 1.4. This indicates that the nonbonding orbital having a large population on the amino group is more active than the π orbital. When methyl groups are introduced in the amino group, the activity of the n orbital decreases with the number of substitution. Whereas for the ring substitution, the activity of the π orbital decreases with the number of methyl groups. Theses results have been explained from the geometrical accessibility of metastable atoms which decreases when a bulky group is introduced and it acts as a shield or an umbrella against the incoming beams of metastable atoms. Such kinds of steric shielding effects were also found for nitriles [63], halogenobenzenes [59], and trimethyl acetylenes [73].

4.9 Population

Metastable atoms are used to probe variations of electron populations depending on chemical structures. A systematic study of PIES activities in $(CH_3)_4M$ (M = C, Si, Ge, Sn, Pb) has revealed that the relative activity of σ_{MC} orbitals with respect to that of σ_{CH} orbitals decreases with increasing size of the central atom [74]. This can be interpreted by the electron distributions on the methyl group for these orbitals, which diminish on going from $(CH_3)_4C$ to $(CH_3)_4Pb$, as the interaction between the central atom and the methyl groups become weaker associated with the considerable decrease of the IP value for the central atom.

4.10 Intramolecular Hydrogen Bond and Dynamic Change

When a hydrogen bond is formed in a molecule, the PIES activity of the nonbonding orbital, which is relevant to the lone-piar electrons participating the hydrogen bond (H-bond), becomes reduced. In XCH_2CH_2OH molecules, an intramolecular hydrogen bonding of the OH-X type is formed, and the PIES activity of the n_X orbital is found to be much less than that for the corresponding n_X orbital for XCH_2CH_3 molecules [75]. The effect of intramolecular H-bonds on the PIES activity is also interpreted in terms of stereoelectronic characteristics of the PI process; upon the formation of the intramolecular H-bond, lone-pair electrons in the exterior regions are brought into the interior region and also the geometrical accessibility of metastable atoms to the n_X orbital becomes diminished by the steric shielding effect of the OH group.

Dynamic variations of the PIES activity of n_X orbital in the H-bond system have been studied for $H_2NCH_2CH_2OH$ and their derivatives [75]. The observed PIES activity of the n_N orbital is an average of those for the bonded and dissociated forms. An analysis of the temperature dependence has led to some thermodynamic data, H = 1.25 kJ mol^{-1}, S = 29 JK^{-1} mol^{-1} for the dissociation of the intramolecular H-bond in N,N'-dimethylaminoethanol.

5 Exterior Electron Density (EED) Analysis

5.1 The ĖED Model

Since the direct application of the rigorous theoretical approach has been limited to some very simple systems, the following EED model has been introduced for the analysis of the relative PIES activity for polyatomic systems including large molecules [41]. The rate of producing a particular final state i, Z_i, can be expressed as

$$Z_i = \int D(R, P) \, W_i(R, P) \, dRdP \tag{16}$$

where $D(R, P)$ is the statistical probability finding A* at a point (R, P) in the phase space defined with geometrical coordinates of A* with respect to the sample molecule, R, and their conjugate momenta, P. $W_i(R, P)$ is the electronic transition probability at this point (R, P). In the usual experimental condition, sample molecules are randomly oriented in the A* beam, and collision energies are in thermal regions and much smaller than the ionization potentials. Deexcitation probability of A* are approximately of the order of 10^{-1} for a single collision [20]. Since W has an asymptotic character of $A\exp(-aR)$ [20], only the spatial regions just outside the exclusion surface are considered to be important. Various factors together with the above ones lead to the following simplest possible treatment of $D(R, P)$; values of $D(R, P)$ are assumed to be constant for the exterior region outside the exclusion surface and to vanish for the inside. This means that A* cannot penetrate into the molecular surface and that probabilities of finding A* are uniform at any point outside the excluded volume of classical collisions. Neglecting the P-dependence of W and denoting the exterior space outside the exclusion surface as Ω, one obtains the following expression for Z_i:

$$Z_i = A \int_{\Omega} W_i(R) \, dR \tag{17}$$

In this model principal parts of $V^*(R)$ are effectively included in the shape of Ω that may be modified when (i) relative velocities between A* and sample molecules are changed by experimental conditions or (ii) the kind of A* itself is changed.

Further simplification can be made for the treatment of $W_i(R)$ which is essentially dependent upon the mutual overlap of the target orbital, ϕ_i, and the inner-shell orbital of A*, χ_a. The two-electron integral K in Eq. (14) is thus treated as involving a delta function placed approximately in the most effective region of the interaction or the overlap. This δ-function model leads to the following equation:

$$W_i(R) = B|\phi_i(\bar{r})|^2 \tag{18}$$

where $\bar{r}$ is the most effective position of electronic coordinates. This position can be assumed to move with A* and to contact with the molecular surface at the classical collision. Thus one may finally find a simple model for Z_i:

$$Z_i = C(EED)_i \tag{19}$$

and

$$(EED)_i = \int_{EXT} |\phi_i(r)|^2 \, dr \tag{20}$$

where $(EED)_i$ is the exterior electron density (EED) for the relevant target orbital ϕ_i that is the amount of electron densities integrated over the exterior region (EXT) outside the molecular surface. This model leads to the following equation for the relative PIES activity:

$$Z_i/Z_j = (EED)_i/(EED)_j \qquad (21)$$

5.2 EED Calculations

Ab initio MO calculations for obtaining ϕ_i were performed by using a library program of the computer center at the University of Tokyo [76]. EED calculations were made with the use of MO coefficient data by a lattice point method [77, 78]. The repulsive molecular surface was estimated from the van der Waals radii [79]. Conventional basis functions [80], STO-nG, 4-31G, 6-311G, and their modifications together with more extended basis sets were employed for both MO and EED calculations.

5.3 EED Ratios for Simple Molecules

Table 3 compares results of EED calculations for some simple molecules with the observed PIES intensities. The EED ratios obtained by using 4-31G basis sets are in

Table 3. Relative PIES band intensity and EED ratio

		PIES band intensity	EED ratio[a]
NH_3	$3a_1$	1.00	1.00
	$1e$	0.39	0.61
H_2O	$1b_1$	0.61	1.23 (0.68)
	$3a_1$	1.00	1.00 (1.00)
	$1b_2$	0.48	0.72 (0.40)
H_2S	$2b_1$	1.00	1.00
	$5a_1$	0.70	0.84
	$2b_2$	0.44	0.83
N_2	$3\sigma_g$	0.90	1.23 (1.08)
	$1\pi_u$	0.36	1.06 (0.81)
	$2\sigma_u$	1.00	1.00 (1.00)
C_2H_2	$1\pi_u$	1.00	1.00
	$3\sigma_g$	0.51	0.52
	$2\sigma_u$	0.68	0.52

[a] Relative EED values for 4-31'G basis sets are in parentheses

good agreement with the observed relative PIES intensities [10]. As mentioned below, the quality of wavefunction tails depends on the basis-set employed. In the case of N_2 and H_2O, the 4-31'G set, which is produced as a slightly modified set from the 4-31G set [77], gave better agreement with the observation.

5.4 Basis Set Dependence of EED

In order to make further analyses by the EED model, it is necessary to characterize qualities of basis functions. Figure 13 show the basis set dependence of EED values for NH_3, CO, H_2O, and H_2S [40]. SCF energies are shown in a normalized scale; the SCF energy for the STO-6G value was taken to be -1. The EED values are plotted for molecular orbitals for which the sum of EED and interior electron density IED is equal to 1; EED + IED = 1 for each MO. As can be seen from Fig. 13, SCF ener-

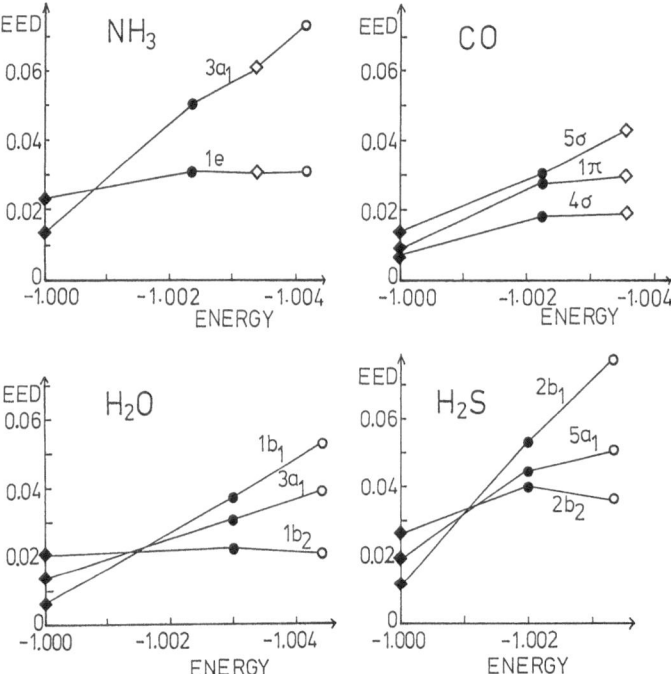

Fig. 13. Basis set dependence of EED values for NH_3, CO, H_2O, and H_2S. Total energies obtained by STO-6G basis sets are normalized to be -1. (◆): STO-6G; (●): 4-31G; (◇): 6-311G; (○): more extended basis sets

gies are refined with the basis size increasing. The amount of refinement, however, is less than 0.5%. EED values generally increase with the basis size increasing, and the increments amount to 200—1000%. Especially for nonbonding orbitals, such as $3a_1$ for NH_3, $1b_1$ for H_2O, and $2b_1$ for H_2S, exterior electron densities tend to be underestimated in the smaller basis sets. This means that the energy variation method for determining wavefunctions is not so effective for estimating exterior electron densities distributed in the remote regions where electron densities do not effectively contribute to the energy.

The agreement of EED ratios with PIES activities is generally refined with the increase of the basis size. A typical example is shown in Table 4 [40, 77]. The experimental intensity ratio is 2.60 for the nonbonding ($3a_1$) and the bonding ($1e$) orbitals

Table 4. Basis set dependence of exterior electron densities for $3a_1$ and $1e$ orbitals of NH_3

Basis set	Energy au	EED($1e$)[a]	FED($1e$)[a]	EED($3a_1$) / EED($1e$)
STO-6G	−55.9882	0.0134	0.0233	0.575
4-31G	−56.1025	0.0504	0.0308	1.64
6-311G	−56.1777	0.0602	0.0304	1.98
56CGTO	−56.221907	0.0736	0.0306	2.41
6-311G*	−56.2012	0.0593	0.0319	1.86
6-311G**	−56.2102	0.0562	0.0319	1.76
6-311G*d'	−56.1861	0.0610	0.0317	1.92
6-311G**$d'p'$	−56.1893	0.0724	0.0323	2.24
HDD2G	−56.1790	0.0736	0.0294	2.50
HDD2G*	−56.2025	0.0720	0.0309	2.33
HDD2G**	−56.2114	0.0688	0.0310	2.22
HDD2G*d'	−56.1846	0.0734	0.0300	2.45
HDD2G**$d'p'$	−56.1853	0.0734	0.0309	2.38
4-31+G	−56.1101	0.0776	0.0312	2.49
6-311+G	−56.1813	0.0786	0.0305	2.58
4-31'G	−56.0506	0.0656		2.15
6-311'G	−56.1517	0.0769	0.0306	2.51
Expt.[b]				2.60

[a] EED value for a moiety of degenerate $1e$ orbitals
[b] Experimental value obtained from Penning ionization electron spectroscopy

of NH_3. This is consistent with the common notion that lone-pair electrons extend much more outside than bonding electrons do. However, the STO-6G set yielded the opposite results. The split valence type of the 4-31G and 6-311G sets gave considerable refinements and qualitatively satisfactory results. The extended basis set employed as a benchmark is the 56CGTO basis which gives a near Hartree-Fock result. This extended set was found to give an excellent values for the EED ratio in good agreement with the observed value.

Although addition of polarization functions, denoted with *, in the basis sets usually yield considerable refinements in energy as well as for various quantities, EED values are underestimated in comparison to the values obtained without polarization functions [77]. This unexpected tendency was found to be the result of an unfavourable absorption of electron densities into the bonding region. When the half value of the standard orbital exponent was used for 6-311G* and 6-311G**, the results (6-311G*d' and 6-311G**$d'p'$) become improved.

Some specially designed basis sets to describe diffuse electronic states, (i) the HDD2G set for Rydberg states, (ii) the 4-31 + G and 6-311 + G sets for negative ionic states, and (iii) the 4-31'G and 6-311'G sets for exterior electron distributions, gave considerably improved EED ratios which are almost completely in agreement with the observed values.

Wavefunction tails are thus found to be very important for describing the reactivity of molecular orbitals. It is of great note that the orbital activity in Penning ionization is governed by a small amount of the exterior electron density which is about 1–10%

for every molecular orbital. This extreme sensitivity of PI to the EED can be used for an experimental verification of wavefunction tails theoretically obtained.

5.5 EED Spectra

A theoretical spectrum for PIES can be synthesized from superposition of gaussian-shaped bands with area proportional to EED values [41]. Since the interest is focussed on the intensity distribution, peak positions and band widths are taken from the observed spectra. The synthesized spectra are denoted as EED spectra.

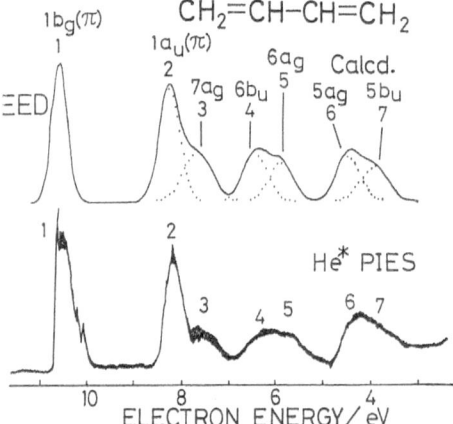

Fig. 14. He*(2³S) PIES **(lower)** and theoretical Penning spectrum calculated from EED **(upper)** for butadiene

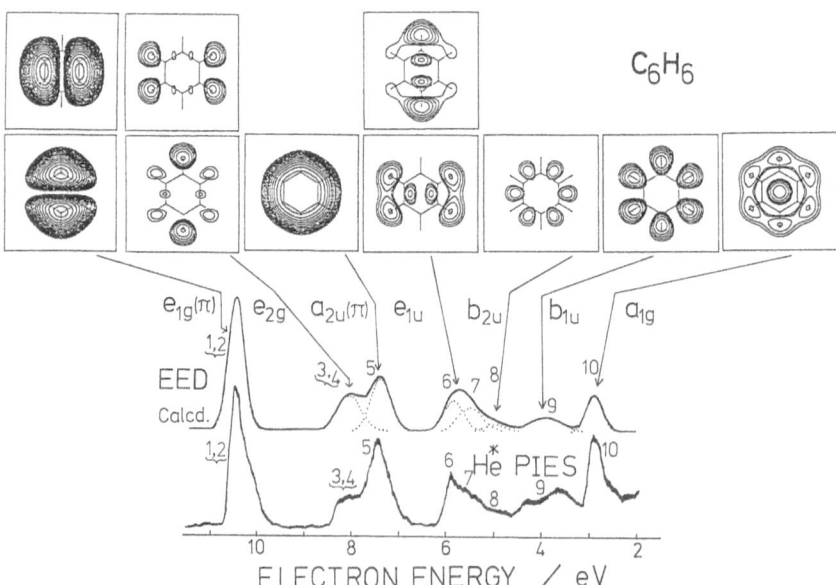

Fig. 15. He*(2³S) PIES **(lower)** and theoretical Penning spectrum calculated from EED **(upper)** for benzene, and electron density maps for the relevant orbitals

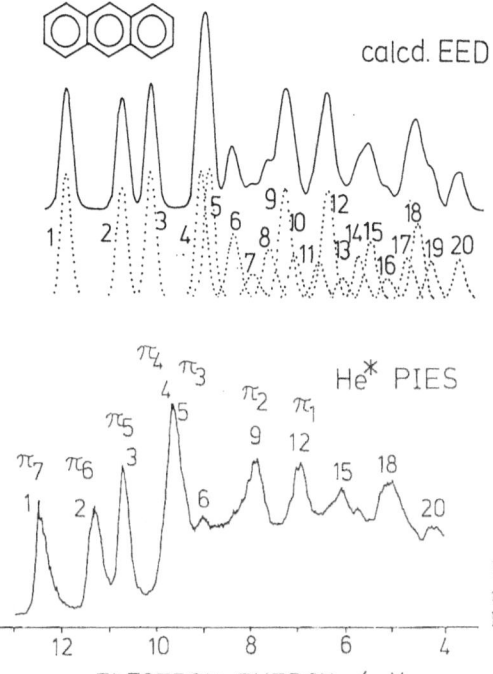

Fig. 16. He*(2³S) PIES **(lower)** and theoretical Penning spectrum calculated from EED **(upper)** for anthracene

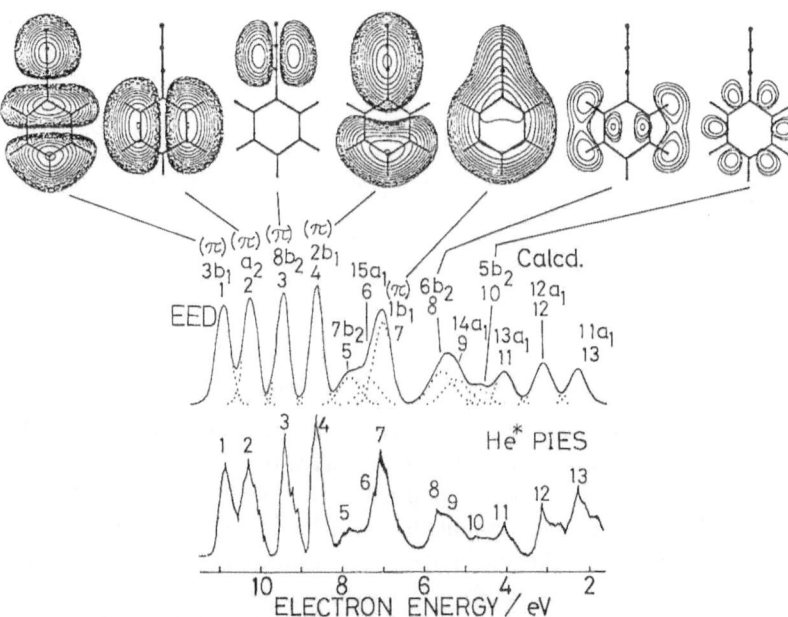

Fig. 17. He*(2³S) PIES **(lower)** and theoretical Penning spectrum calculated from EED **(upper)** for phenylacetylene, and electron density maps for the relevant orbitals

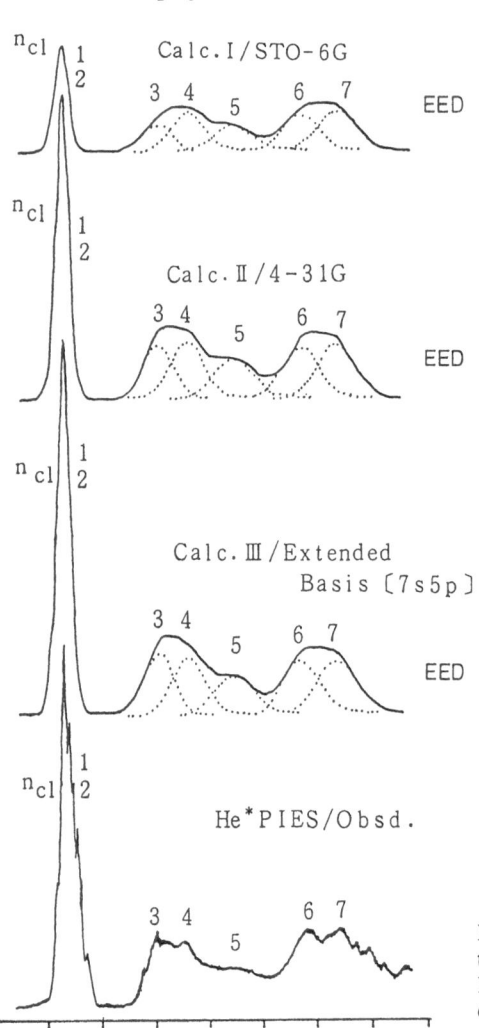

Fig. 18. He*(2^3S) PIES for C_2H_5Cl (lower) and theoretical Penning spectra calculated from EED with various levels of basis sets (*Calc.* I, *Calc.* II, and *Calc.* III)

Figures 14–18 show examples of EED spectra [41, 57]. The observed tendency for unsaturated hydrocarbons, $\pi > \sigma$, is explained in good accuracy, and some subtle differences among σ bands are also reproduced. For example, the b_{2u} orbital of benzene in Fig. 15 is localized on the C—C bonds and has a very weak reactivity, whereas the a_{1g} and e_{1u} orbitals, having large populations on the C–H bonds, show higher activities. The EED spectra for phenylacetylene and anthracene are also very satisfactory.

A careful choice of basis functions was required for systems including lone-pair electrons, as mentioned for NH_3. With the increase of the basis size, EED spectra tend to agree with the observed PIES; a typical example is shown in Fig. 22 [81].

6 Application of PIES to the Study of Solid Surfaces

6.1 Characteristics of Surface PIES

When PIES is employed for the study of a solid surface the following two characteristics are displayed [82, 83[1]]. (1) PIES probes the outermost surface layer of the
solid surface because metastable atoms do not penetrate into the inner layers. This is
not the case in other electron spectroscopies such as photoelectron, Auger, and electron impact spectroscopies, where photons or electrons used for the excitation sources
penetrate into the inner layers. (2) PIES enables us to study the electron distribution
of individual orbitals in the exterior region outside the solid surface. This aspect of
surface PIES can be easily understood on the basis of the gas-phase PIES described
in the preceding sections.

Figure 19 shows the interaction between a metastable atom A* and molecules BC
in the gas and solid phases. The corresponding PIES is shown below in each case.
Since gas-phase molecules are randomly oriented with respect to the direction of the
metastable beam, the relative intensity of the PIES bands reflects the overall spread
of individual orbitals. Thus, an orbital ϕ_2 extending, on the average, more widely than
an orbital ϕ_1 gives a stronger band in PIES (Fig. 19 (a)). On the other hand, on a solid

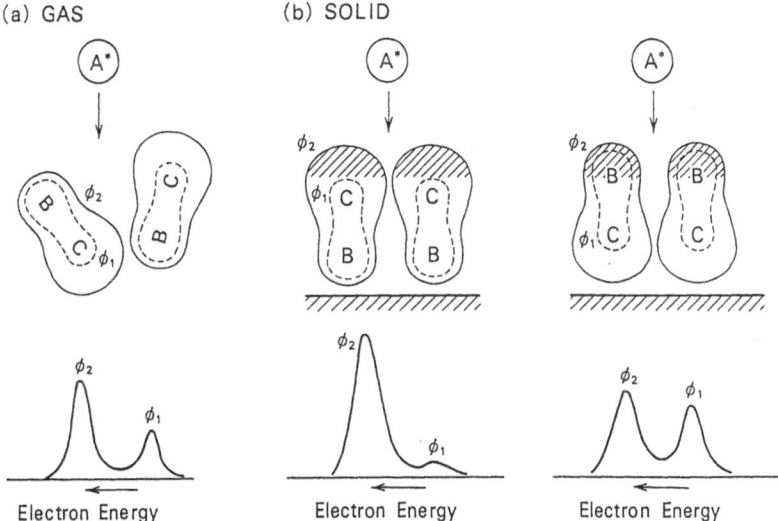

Fig. 19a, b. Interaction between a metastable atom A* and molecules BC in the gas (a) and solid (b)
phases. The corresponding PIES is shown below for each case

---

[1] On a metal surface, an electron in the excited level of a metastable atom tunnels into an empty level
in the metal. The positive ion, thus formed, is then neutralized by an Auger process. Therefore, the
Penning ionization process does not take place on the metal and the electron spectrum is similar to
that obtained by ion neutralization spectroscopy (see, for example, Hagstrum HD (1979) Phys. Rev.
Lett. 43: 1050; Conrad H, Ertl G, Küppers J, Wang SW, Gerard K, Haberland H (1979) Phys. Rev.
Lett., 42: 1082).

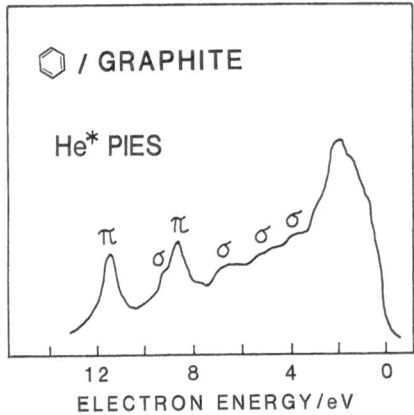

Fig. 20. He*(2^3S) PIES of benzene on a graphite cleavage plane [84]. The coverage of benzene is a few monolayers

surface, a metastable atom interacts with the outermost part (shaded part in the figure) of the regular array of molecules. Hence, if molecules BC are oriented so as to expose the atom C end outside the surface (left of Fig. 19 (b)), orbital ϕ_2, distributed largely on atom C, gives a stronger band; while orbital ϕ_1 distributed predominantly on atom B, gives a weaker band compared to the case of the gas phase spectrum. If the molecular orientation is reversed (right of Fig. 19 (b)), the PIES shows a tendency opposite to the above case. Thus, an analysis of the relative intensity of the PIES bands provides information on the geometrical orientation of molecules at the outermost layer.

Figure 20 shows an example of benzene layer adsorbed on a graphite cleavage plane [84]. The coverage of benzen is a few monolayers. The figure indicates that the relative intensities of the π bands are much stronger and those of the σ bands are much weaker compared to the case of the gasphase PIES in Fig. 15. This observation is accounted for, if we assume that the benzene molecules are oriented flat to the substrate. As can be seen from Fig. 21 (a), if a metastable atom approaches a benzene molecule oriented parallel to the graphite cleavage plane, a σ orbital exposed outside the surface should interact effectively with the metastable atom and gives a strong band in the PIES, while a σ orbital shielded by π orbitals should give a weak band. When

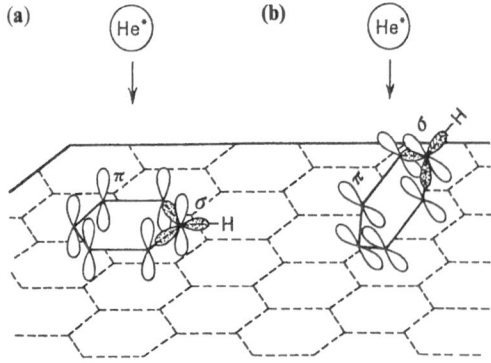

Fig. 21 a, b. Interaction between a metastable helium atom and a benzene molecule. **(a)** The molecule is oriented parallel to a graphite substrate. **(b)** The molecule is tilted to the substrate

surface molecules are tilted on the substrate, the σ bands as well as the π are expected to appear rather strongly in the spectrum, because the σ orbitals are also exposed outside (Fig. 21 (b)). In fact, this effect was observed in thick films (several tens of monolayers) of benzene on graphite [84].

6.2 Molecular Orientation at the Outermost Surface Layer

As stated above, the study of the wave function tail exposed outside the solid surface by PIES enables us to observe the molecular orientation at the outermost surface layer. In this section we will show some examples of pentacene films.

Figure 22 shows the PIES and UPS of pentacene layers deposited by vacuum sublimation onto a graphite substrate (cleavage plane) held at 123 K [85]. The spectra of the clean substrate are also shown. The coverage θ (monolayer unit) was estimated with a quartz-oscillator monitor on the assumption that the pentacene molecules are oriented parallel to the substrate. In the PIES the intensity of the sharp peak (denoted by an arrow) due to conduction bands of graphite is reduced to about one-half at $\theta = 0.5$ and almost zero at $\theta = 1$. This indicates that the PIES provides information on the outermost surface layer selectively. On the other hand, in the UPS involving inner layers as well as the outermost, the sharp peak of graphite appears even at $\theta = 4$.

In Fig. 22 we find that the σ bands are much weaker than the π bands in the PIES for thin layers and their intensities gradually increase with incleasing thickness.

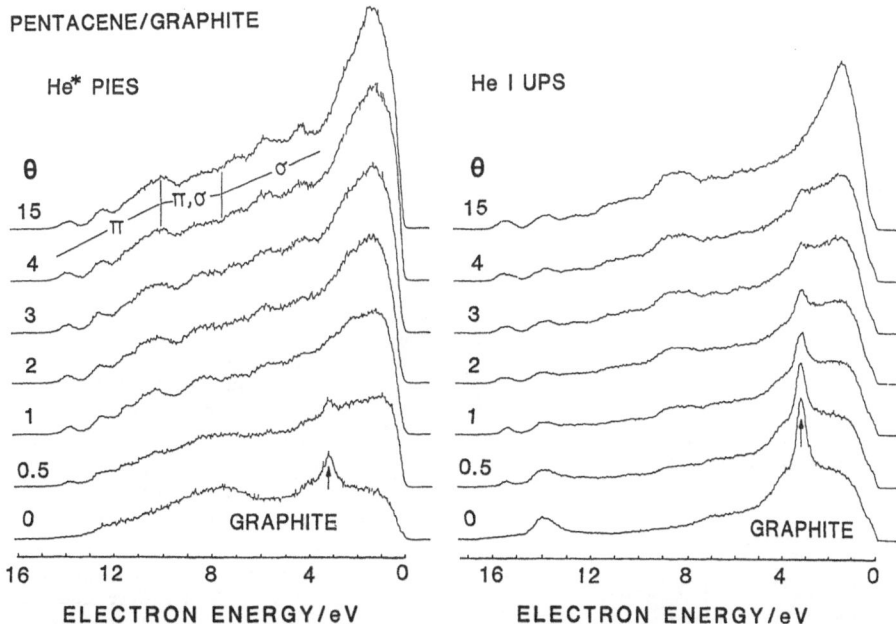

Fig. 22. Change in the He*(2^3S) PIES and He I UPS of a pentacene film on a graphite substrate at 123 K with increasing coverage θ [85]

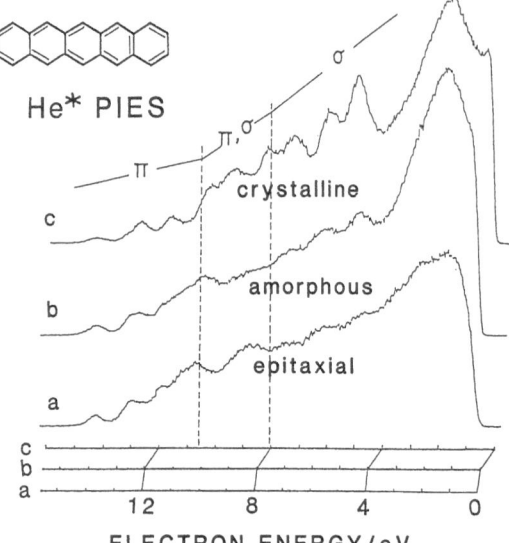

Fig. 23. He*(2^3S) PIES of a crystalline, an amorphous, and an epitaxial pentacene films. The abscissae for the spectra are shifted so that the peak positions of the first bands coincide with one another

As in the case of benzene on graphite (Fig. 20), this indicates that pentacene molecules are deposited flat onto the substrate at low coverage and gradually tilted in subsequent layers with increasing coverage. Thus, in thin layers the molecular orientation is determined mainly by the interaction between the molecules and the graphite

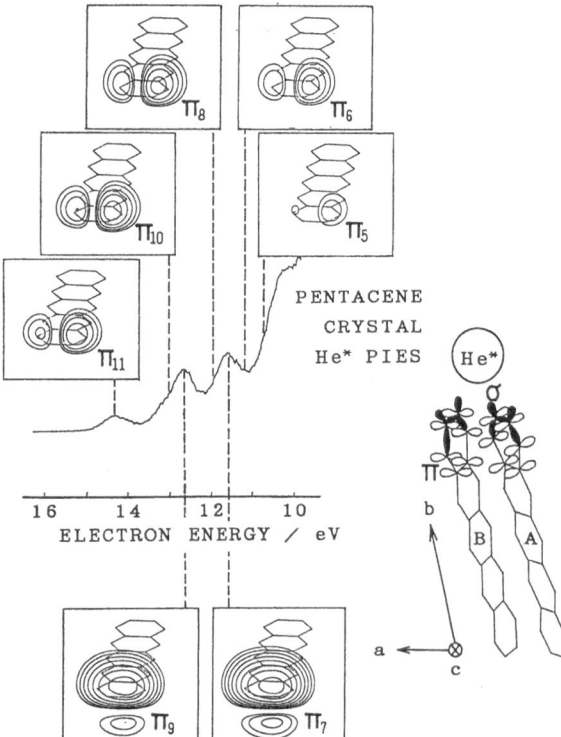

Fig. 24. π bands in He*(2^3S) PIES of a pentacene crystalline film deposited onto a stainless steel substrate at 293 K, together with the electron density maps of the corresponding π orbitals. The arrangement of surface molecules is illustrated on the **right**. The contour lines of the maps were calculated for a surface molecule A, in a plane parallel to the surface at a distance of 1.2 Å from the top hydrogen atom

substrate, whose hexagonal network causes the pentacene molecules to lie on it. With increasing thickness the effect of the substrate-molecule interaction gradually becomes weaker. This example of the pentacene film indicates the usefullness of PIES in probing subtle changes in the electronic state and the molecular orientation of the outermost layer during the epitaxial growth of the film.

The observation of the local electron distribution outside the solid surface also enables us to distinguish between the crystalline and amorphous states. Figure 23 shows the PIES of a crystalline, an amorphous, and an epitaxial pentacene films [86]. The crystalline and amorphous films were prepared on a stainless-steel substrate held at 293 K and 123 K, respectively. The spectrum of the epitaxial film is the same as the one at $\theta = 1$ in Fig. 22. In Fig. 23 the σ bands are much stronger than the π in the spectrum of the crystalline film, because the molecules are oriented with their long axis nearly perpendicular to the substrate (see the insert in Fig. 24) and expose σ orbitals outside the film surface. In the spectrum of the amorphous film the π and σ bands have similar intensity, which indicates that the degree of the molecular tilt is, on the average, much smaller than that in the crystalline film owing to the random molecular orientation. Finally, in the epitaxial film σ bands are much weaker than the π bands, showing the flat molecular arrangement as mentioned before.

Using the characteristics of PIES described in this section, we could detect the change in the molecular arrangement due to the transition, amorphous to crystalline, at the outermost surface layer of a biphenyl film [87]. Further, the structural change, disordered layer → closely packed monolayer → island, could be observed for an ultrathin iron phthalocyanine film [88]. In the case of a naphthacene film, a photochemical reaction that takes place at the outermost surface layer could be sensitively probed [89].

6.3 Spatial Distribution of Surface Molecular Orbitals

As stated before, PIES contain significant information regarding the wave function tails of individual orbitals at a definite molecular part exposed outside. This feature of PIES was first demonstrated for an anthracene crystalline film deposited on a metal substrate [35] and was recently applied for obtaining direct evidence as to which end of the molecule (either head or tail) is exposed outside the surface of LB (Langmuir-Blodgett) films [90, 91]. Next we will show the examples of a pentacene and an iron phthalocyanine films.

Figure 24 shows the PIES due to the highest seven π orbitals of a pentacene crystalline film together with the electron density map of each orbital [86]. The contour lines of the maps are drawn for a surface molecule A in Fig. 24 in a plane parallel to the surface at a distance of 1.2 Å (van der Waals radii of the hydrogen atom) from the top hydrogen atom. As can be seen in Fig. 24, the π_9 and π_7 bands are markedly enhanced relative to the other π bands. This can be explained from the electron density maps; the π_9 and π_7 orbitals without a nodal plane along the long axis of the molecule extend further outside the film surface than the other π orbitals with the nodal plane.

Figure 25 shows the PIES of two kinds of iron phthalocianine (FePc) films [92]. Film A is a monolayer prepared by vacuum sublimation onto a graphite substrate held at 213 K. Film B is a crystalline one deposited on a stainless steel substrate at

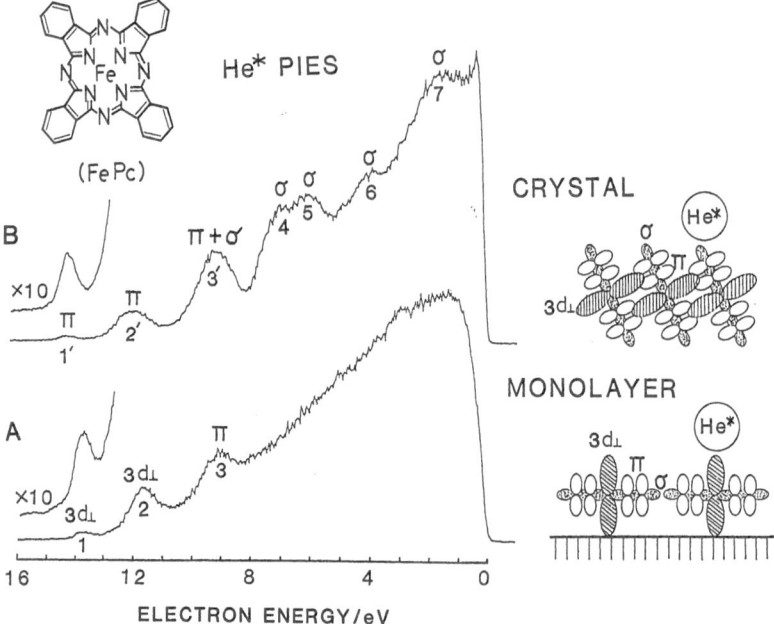

Fig. 25. He*(2^3S) PIES of a monolayer film A and a crystalline film B of iron phthalocyanine (FePc) [92]. Film A was prepared on a graphite substrate held at 213 K, while film B was deposited on a stainless steel substrate at 298 K. The orientation of surface molecules in each film is illustrated on the **right**

room temperature. As illustrated on the right of the figure, molecules are arranged flat to the substrate in film A and are tilted in film B. In the flat molecular orientation, π-type orbitals extending normal to the molecular plane (xy plane) effectively interact with metastables, whereas σ orbitals distributed within the molecular plane and shielded by π-type orbitals scarcely interact with metastables. Among the π-type orbitals, those mainly originated from the iron $3d_{yz}$, $3d_{zx}$, or $3d_{z^2}$ AO ($3d_\perp$-like MO) should interact with metastables more effectively than those derived from the carbon and nitrogen $2p_z$ AO, because the former orbitals protrude outside the molecular surface more prominently than the latter ones. On the other hand, in the tilted molecular orientation, the σ orbitals as well as the π should be effectively attacked by metastables, but the $3d_\perp$-like orbitals with little distribution outside the film surface is expected to be scarcely attacked. As can be seen from Fig. 25, the relative intensity of the PIES bands for films A and B substanciate the above expectations regarding the distributions of the surface orbitals.

Thus, surface PIES reflects the local distribution of individual orbitals at a definite molecular part exposed outside. If we can control the orientation of the surface molecules by means of the selections among various substrates, film preparation methods (vacuum deposition, adsorption, Langmuir-Blodgett technique etc.), deposition conditions (substrate temperature, deposition speed), and film treatments (e.g. annealing), we can probe the distribution of molecular orbitals from various direction and approach their 'shape' or whole picture.

7 Concluding Remarks

We have shown that Penning ionization electron spectroscopy enables us to study outer shapes of molecules. Especially, it provides information on the spatial electron distribution of *individual* molecular orbitals in the frontier region outside the molecular surface. Such information is essential in elucidating various intermolecular processes including chemical reactions, which start with effective overlapping of the tails of molecular wave functions. For solids, Penning spectroscopy probes the distribution of orbitals penetrated outside the outermost layer that essentially determines the properties of the solid surface. This unique feature of Penning spectroscopy should be further explored by its application to various solid surfaces.

8 References

1. Penning FM (1927) Naturwissenschaften 15: 818
2. Jesse WP, Sadauskis J (1952) Phys. Rev. 88: 417
3. Stedman DH, Setser DW (1971) Prog. Reaction Kinetics 6:123
4. Muschlitz EE (1966) Advan. Chem. Phys. 10: 171
5. Čermák V (1966) J. Chem. Phys. 44: 3781
6. Yencha AJ (1984) In: Baker AD, Brundle CR, Baker AD (ed) Electron spectroscopy, theory, technique and applications. Academic Press, New York (vol 5)
7. Turner DW, Baker C, Brundle CR (1970) Molecular photoelectron spectroscopy, Wiley, New York
8. Rablais JW (1977) Principles of ultraviolet photoelectron spectroscopy, Wiley, New York
9. Harada Y, Ohno K, Mutoh H (1983) J. Chem. Phys. 79: 3251
10. Ohno K, Mutoh H, Harada Y (1983) J. Am. Chem. Soc. 105: 4555
11. Masuda S, Ohno K, Harada Y (to be published)
12. Herman Z, Čermák V (1966) Collect. Czech. Chem. Commun. 31: 649
13. Hotop H, Niehaus A (1969) Z. Phys. 228: 68
14. Niehaus A (1981) Advan. Chem. Phys. 45: 399
15. Hotop H, Niehaus A, Schmeltekopf (1969) Z. Phys. 229: 1
16. Hotop H, Niehaus A (1970) Z. Phys. 238: 452
17. Hotop H, Niehaus A (1971) Chem. Phys. Lett. 8: 497
18. Miller WH, Slocomb CA, Schaefer HF III (1972) J. Chem. Phys. 56: 1347
19. Ebding T, Niehaus A (1974) Z. Phys. 270: 43
20. Niehaus A (1973) Ber. Bunsenges. Phys. Chem. 77: 632
21. Hotop H (1974) Radiat. Res. 59: 379
22. Hotop H, Niehaus A (1970) Int. J. Mass Spectrom. Ion Phys. 5: 415
23. Richardson WC, Setser DW (1973) J. Chem. Phys. 58: 1809
24. Berry RS (1974) Radiat. Res. 59: 367
25. Čermák V, Yencha AJ (1976) J. Electron Spectrosc. Relat. Phenom. 8: 109
26. Čermák V, Spirko V, Yencha AJ (1976) J. Electron Spectrosc. Relat. Phenom. 8: 339
27. Yee DSC, Stewart WB, McDowell CA, Brion CE (1975) J. Electron Spectrosc. Relat. Phenom. 7: 377
28. Yee DSC, Hamnett A, Brion CE (1976) J. Electron Spectrosc. Relat. Phenom. 8: 291
29. Yee DSC, Brion CE (1976) J. Electron Spectrosc. Relat. Phenom. 8: 313
30. Yee DSC, Brion CE (1976) J. Electron Spectrosc. Relat. Phenom. 8: 377
31. Brion CE, Yee DSC (1977) J. Electron Spectrosc. Relat. Phenom. 12: 77
32. Gérard K, Hotop H (1976) Chem. Phys. Lett. 43: 175
33. Munakata T, Kuchitsu K, Harada (1979) Chem. Phys. Lett. 64: 409

34. Munakata T, Kuchitsu K, Harada Y (1980) J. Electron Spectrosc. Relat. Phenom. 20: 235
35. Munakata T, Ohno K, Harada Y (1980) J. Chem. Phys. 72: 2880
36. Kubota H, Munakata T, Hirooka T, Kuchitsu K, Harada Y (1980) Chem. Phys. Lett. 74: 409
37. Munakata T, Ohno K, Harada Y, Kuchitsu K (1981) Chem. Phys. Lett. 83: 243
38. Munakata T, Harada Y, Ohno K, Kuchitsu K (1981) Chem. Phys. Lett. 84: 6
39. Ohno K, Fujisawa S, Mutoh H, Harada Y (1982) J. Phys. Chem. 86: 440
40. Ohno K, Matsumoto S, Harada Y (1984) J. Chem. Phys. 81: 2183
41. Ohno K, Matsumoto S, Harada Y (1984) J. Chem. Phys. 81: 4447
42. Miller WH (1970) J. Chem. Phys. 52: 3563
43. Hickman AP, Isaacson AD, Miller WH (1977) J. Chem. Phys. 66: 1483
44. Hickman AP, Isaacson AD, Miller WH (1977) J. Chem. Phys. 66: 1492
45. Cohen JS, Lane NF (1977) J. Chem. Phys. 66: 586
46. Isaacson AD, Hickman AP, Miller WH (1977) J. Chem. Phys. 67: 370
47. Martin DW, Siska PE (1985) J. Chem. Phys. 82: 2630
48. Martin DW, Gregor RW, Jordan RM, Siska PE (1978) J. Chem. Phys. 69: 2833
49. Siska PE (1979) J. Chem. Phys. 71: 3942
50. Taylor HS, Williams JK (1965) J. Chem. Phys. 42: 4063
51. Hazi AU, Taylor HS (1970) Phys. Rev. A1: 1109
52. Fels MF, Hazi AU (1971) Phys. Rev. A4: 662
53. Taylor HS, Thomas LD (1972) Phys. Rev. Lett. 28: 1091
54. Ohno K, Ishida T, Naitoh Y, Izumi Y (1985) J. Am. Chem. Soc. 107: 8082
55. Ohno K (1986) Ionics 133: 11
56. Harada Y, Ohno K (1988) Nippon Kagaku Kaishi (J. Chem. Soc. Japan, Chemistry and Industrial Chemistry) 1988: 1
57. Kajiwara T, Masudo S, Ohno K, Harada Y (1988) J. Chem. Soc. Perkin Trans. II 1988: 507
58. Ohno K, Imai K, Matsumoto S, Harada Y (1983) J. Phys. Chem. 87: 4346
59. Fujisawa S, Ohno K, Masuda S, Harada Y (1986) J. Am. Chem. Soc. 108: 6505
60. Hoffmann R (1982) Angew. Chem. Int. Ed. Engl. 21: 711
61. Mingos DMP (1977) Adv. Organomet. Chem. 15: 1
62. Ohno K, Takano S, Mase K (1986) J. Phys. Chem. 90: 2015
63. Ohno K, Matsumoto S, Imai K, Harada Y (1984) J. Phys. Chem. 88: 206
64. Aoyama M, Masuda S, Ohno K, Harada Y (to be published)
65. Fujisawa S, Masuda S, Ohno K, Harada Y (to be published)
66. Hoffman R, Imamura A, Hehre WJ (1968) J. Am. Chem. Soc. 90: 1499
67. Hoffmann R (1971) Acc. Chem. Res. 4: 1
68. Hoffmann R, Heilbronner E, Gleiter R (1970) J. Am. Chem. Soc. 92: 706
69. Heilbronner E, Muszkat K (1970) J. Am. Chem. Soc. 92: 3818
70. Heilbronner E, Martin HD (1972) Helv. Chim. Acta 55: 1490
71. Broglie F, Heilbronner E, Ipaktschi J (1972) Helv. Chim. Acta 55: 2447
72. Eland JHD (1974) Photoelectron spectroscopy, Butterworths, London
73. Matsumoto H, Akaiwa K, Nagai Y, Ohno K, Imai K, Masuda S, Harada Y (1986) Organometallics 5: 1526
74. Aoyama M, Masuda S, Ohno K, Harada Y, Yew MC, Hua HH, Yong LS (1989) J. Phys. Chem. 93: 1800
75. Ohno K, Imai K, Harada Y (1985) J. Am. Chem. Soc. 107: 8078
76. Kosugi N (1981) Program GSCF2. Program Library, The Computer Center, The University of Tokyo
77. Ohno K, Ishida T (1986) Int. J. Quant. Chem. 29: 677
78. Ishida T, Ohno K (1989) Int. J. Quant. Chem. 35: 257
79. Pauling L (1960) Nature of the chemical bond, 2nd edn, Cornell, Ithaca
80. Hehre WJ, Radom L, Schleyer PvR, Pople JA (1986) Ab initio molecular orbital theory. Wiley, New York
81. Ishida T, Ohno K (unpublished)
82. Harada Y (1985) Surface Sci. 158: 455
83. Harada Y, Ozaki H (1987) Jpn. J. Appl. Phys. 8: 1201
84. Kubota H, Munakata T, Hirooka T, Kondow T, Kuchitsu K, Ohno K, Harada Y (1984) Chem. Phys. 87: 399

85. Harada Y, Ozaki H, Ohno K (1984) Phys. Rev. Lett. 52: 2269
86. Ozaki H, Harada Y (unpublished)
87. Kubota H, Munakata T, Hirooka T, Kuchitsu K, Harada Y (1980) Chem. Phys. Lett. 74: 409
88. Harada Y, Ozaki H, Ohno K, Kajiwara T (1984) Surface Sci. 147: 356
89. Ohno K, Mutoh H, Harada Y (1982) Surface Sci. 115: L128
90. Ozaki H, Harada Y, Nishiyama K, Fujihara M (1987) J. Am. Chem. Soc. 109: 950
91. Mitsuya M, Ozaki H, Harada Y, Seki K, Inokuchi H (1988) Langmuir 4: 569
92. Ozaki H, Harada Y (1987) J. Am. Chem. Soc. 109: 949

The Meaning and Distribution of Atomic Charges in Molecules

Karl Jug[1] and Zvonimir B. Maksić[2]

[1] Theoretische Chemie, Universität Hannover, Am Kleinen Felde 30, 3000 Hannover 1, Federal Republic of Germany

[2] Rudjer Bošković Institute, Bijenička 54, 41001 Zagreb, Croatia, Yugoslavia and Faculty of Natural Sciences, University of Zagreb, Marulićev trg 19, 41000 Zagreb, Croatia, Yugoslavia

Various methods of representing changes in electron densities upon formation of chemical bonds are briefly discussed. Virtues and shortcommings of describing charge (re)distribution in terms of sets of numbers, which can be identified as gross atomic electron populations, are thoroughly considered subsequently. Experimental techniques yielding atomic charges are critically assessed and it is concluded that ESCA spectroscopy gives the most straightforward and transparent insight into distribution of atomic monopoles in molecular systems. Applications of the atomic charge concept in rationalizing a number of physical and chemical properties of molecules are described in some detail. It is concluded that properly defined atomic charges possess a grain of truth and their interpretive power is stressed. Finally, prospects of future developments are briefly sketched.

1 Introduction

The knowledge of the electronic wavefunctions Ψ is the basis for the understanding of the structure and properties of molecules, because in principle, they embody all the information about quantum systems. The square of the absolute value $\Psi^*\Psi$ yields a probability of finding electrons within an infinitesimal volume in three-dimensional space, or more loosely speaking − it gives the electron density distribution ϱ. The latter entity is of utmost importance for several good reasons. It is an observable which can be measured by X-ray diffraction [1] or by a combined use of X-ray and neutron diffraction techniques [2]. The former method gives X-X density maps whereas the latter offers the so called X-N density contours which combine useful characteristics of both methods. An interesting insight into the spin distribution in paramagnetic systems is provided by polarized neutron scattering [3] which is a valuable tool in studying transition metal complexes. Additional experimental information on electron densities is obtained by the electron scattering techniques [4]. On the other hand electron density distribution is theoretically interesting *per se* because of the two fundamental theorems. The first is that of Hohenberg and Kohn [5] which asserts that the total density involves as much information as the corresponding wavefunction and that there exists a functional which gives the total energy straight forwardly from the density ϱ. There are some difficulties in finding this still unknown functional which is definitely not based on the local density alone excluding its derivatives [6, 7], but a lot of the research effort is being put into solving the problem [8]. The second important theorem, set forth independently by Hellman and Feynman [9], relates the electron density to forces (electrostatic) exerted on the nuclei. Its interpretive merits are discussed in detail in several monographs [10].

It is a common knowledge that atoms undergo subtle changes in forming molecules and that their electron density is redistributed in chemical environments. Understanding of charge reorganization in molecular formation is therefore the key to revealing secrets of chemical bonding. A variety of ways have been proposed to describe global and local density distributions. Since there is neither a way of isolating of a particular chemical bond nor a unique way of defining all local properties, a number of models of chemical bonding have evolved. Their relative merit is a matter of debate because they are sometimes contradictory instead of the highly desired complementarity. Broadly speaking, one can classify the methods for analyzing wavefunctions and density distributions into two large categories: (a) differential and (b) integral. In differential methods, plots of density distributions are usually presented in two or three-dimensional forms. To be more specific, the electron distribution could be plotted as a surface above a plane of two molecular coordinates of interest. The plane represents a cut through the molecular volume in the three-dimensional space chosen to describe some critical features. Two-dimensional maps yield contours of equal electron density (isopycnic lines) through the relevant molecular plane, a particular bond etc. They provide useful information about global features. Slices perpendicular to the bond reveal axial symmetry of σ bonds, ellipticity of double bonds and the like. More detailed information is obtained by deformation density maps which clearly show a charge shift into the intramolecular bonded regions, off-center displacement of the lone pairs, bond bending in strained systems etc. However these contours are transparent and helpful in forming our mental pictures of chemical bonding, they frequently give more information than we need. It would be useful to condense and summarize information in a set of numbers which would describe atomic charges (monopoles). A more refined pattern would give higher atomic and bond multipoles. This is the essence of the integral method which will be discussed in the main body of this chapter. Reduction of spatial density distributions into a small set of numbers which reflect all relevant features in a concise manner is by no means a trivial problem. The main difficulty is assignement of charges to atoms in intramolecular environments and there is no unique, or generally accepted way of density partitioning. We will discuss several procedures and try to analyze their relative merits and shortcomings.

2 Spatial Density Distributions

2.1 Total and Deformation Isopycnic Maps

There are several obstacles in determining experimental geometries and charge densities by using X-ray techniques because photons are scattered by electrons implying that atoms with few electrons like hydrogen are less "visible" leading to greater inaccuracies in atomic positions. If the charge distribution is desired, then high-order X-ray data (i.e. at large Bragg's angles) are necessary and careful estimates of the vibrational smearing effect have to be taken into account. Substantial help is provided by additional neutron scattering which locates positions of the nuclei directly. It appears that the total density maps are dominated by inner core electrons leading to high localized peaks which swamp a lot of chemical information as we shall see later.

Theoretically, the total density is given by

$$\varrho(1) = \sum_i n_i \psi_i^2(1),\tag{1}$$

in orbital approximation, where ψ_i stands for orthonormal atomic or molecular orbitals and n_i are their orbital populations. It should be mentioned that real orbitals will be used througout the paper. Electron density distribution is a single particle function because of the identity of electrons. In order to illustrate the spatial distribution of electrons in atoms and molecules, representative planes are selected and the density surface is usually computed for a grid of points in a rectangle which covers the important region. With a number of different planes s, p, d etc. orbitals in atoms and σ, π and δ bonds in molecules can be described. This three-dimensional representation by surfaces can be reduced to two-dimensional contour diagrams. Here cuts of the density with a sequence of carefully selected, usually equidistant planes parallel to the initial basic plane are projected on the latter or given in a perspective drawing [11].

An alternative approach was adopted by Wahl [11]. He produced the first detailed ab initio pictures of density distribution in homonuclear diatomics involving first row atoms. The contour lines were constructed by the condition $\varrho(x, z) = $ const., where z was the symmetry axis of molecules. In this way a systematic trend of changes of bonding from H_2 to F_2 could be visualized. It should be pointed out that a distinct advantage of the theoretical approach is that single orbitals can be carefully examined in contrast to experimental measurements. By the same token one can visualize changes induced by ionization, dilatation or shrinkage of bond distances, determine a shape of unoccupied anti-bonding MOs, describe excited states etc., which is not amenable to experimental observation as a rule. Clearly, if a molecule does not have axial symmetry then a single plane cannot do justice to the distribution of the total density. Wahl for example selected a plane passing through the maxima of the π-electron density in molecules possessing a double bond. This enabled observation of a systematic trend in contraction of density around atoms in the series. A more comprehensive account of both surfaces and density maps was later given by Streitwieser and Owen [13]. In this book diatomics and selected organic polyatomic molecules were treated in a didactive manner appropriate for the practicing chemist in general. Van Wazer and Absar [14] extended the scope of applications to inorganic systems. This was undoubtedly another step forward to present the sequence of density plots of the occupied and important unoccupied orbitals on a single page for easy comparison. Whereas Streitwieser and Owens [12] skipped the representation of total densities, Van Wazer and Absar document the behaviour of total densities in essential planes by surface plots [13]. Thus a comparison along a series of related molecules was possible. It is noteworthy that density maps are used as an interpretive tool in many textbooks [15, 16] and particularly density difference contours, which will be discussed in the next section.

In spite of the pictorial appeal of the presentation of orbital and total density surfaces and contours their usefulness is of a limited value. Small changes introduced by chemical bonding are mostly absorbed in very dense inner cores which remain spherically symmetrical. Much more informative are the difference densities $\Delta\varrho$ between the total molecular density ϱ_{mol} and the sum of all atomic densities ϱ_A placed at the equilibrium positions.

$$\Delta\varrho = \varrho_{mol} - \sum_A \varrho_A . \tag{2}$$

The reference state formed by spherical and noninteractive atoms is called a promolecule. A careful analysis of difference densities[1] (DD) by using a promolecule concept was performed by Bader et al. [17]. The isopycnic $\Delta\varrho$ map for covalent bonding is characterized by a density increase located between and shared by both atoms. On the other hand the $\Delta\varrho$ map for ionic binding exhibits an increase in charge density which is localized in (or very close to) the electronegative atom. There is also a strong polarization away from the bonding region if the participating atom has lone pair(s). Bader et al. also discuss forces exerted on the nuclei in terms of the deformation maps. It turns out that the one-electron spatial density distribution permits a direct physical insight and interpretation of chemical bonding. Bader relates this to intramolecular forces estimated by the Hellmann-Feynman theorem. There are, however, some disturbing features indicating limitations of the notion of the

---

[1] Difference densities are also called deformation densities because they reflect reorganization of the electron density and departure from spherical symmetry.

promolecule as a suitable reference distribution. For example, in O_2 and F_2 molecules very deep troughs were found in the vicinity of the nuclei in the bonding region. There are mild build-ups of charge in the middle of the bonds. The fluorine molecule exhibits pathological increase of density in the π electron region. It was conjectured that these unusual bonding patterns are consequences of the arbitrarily chosen promolecule density distributions [17]. This point is of utmost importance because experimental DD maps are invariably produced by utilizing a promolecule gauge and employing ab initio spherically averaged densities of free atoms. Thus deficits of electron density were found in $O-O$ bonds in hydrogen peroxide [18] and in tetra-aza-tetra-oxatricyclotetradecane [19]. Depletion of charge was obtained in theoretical maps in the F_2 molecule [20] whilst only weak density accumulations were detected for $N-N$, $C-N$ and $C-O$ bonds [21]. This has led to a conclusion that "accumulation of charge (bonding density) in the internuclear region as occurs in the hydrogen molecule may not be characteristic of covalent bonds in general" [19]. It turned out that this statement was not justified since a judicious choice of the reference state reestablishes a full accordance between chemical intuition and $\Delta\varrho$ maps. The point is that atomic charge distributions are not intrinsically spherically symmetrical, noble gas atoms being notable exceptions because they have completed shells. Let us consider the fluorine atom as an example. Its electron configuration is $(1s)^2 (2s)^2 (2p_x)^2 (2p_y)^2 (2p_z)^1$ yielding a ^2P ground state. Population of the $2p_z$ orbital by a single electron is a matter of convention and the electron could be placed on $2p_x$ or $2p_y$ AOs as well, or on any linear combination thereof. Another way of expressing a fact that a real physical space is isotropic is that the orientation of the z-axis is arbitrary. In other words, in a large ensemble of F atoms all orientations are present and atoms behave as if they possess spherical symmetry. This is not the case any more if a particular direction is preferred like in external fields or, what is more important for our discussion, if two F atoms meet and form a molecule. They reorient themselves so that $2p_z$ AOs point to each other thus preparing the corresponding electrons to form electron pair chemical bond in the classical sense of G.N. Lewis. Another important feature is hybridization which is a genuine phenomenon [22]. It enhances accumulation of charge between the nuclei thus contributing to their better screening, shifting at the same time the lone pairs from the bonding region. Concomitantly, nonbonding repulsions are diminished and the Pauli principle stating that electrons with the same spin cannot be described by the same spatial orbital is better satisfied. This physical picture was consistently pursued by Hall in a number of papers [23–25] which stressed the importance of a choice of the reference state in extracting chemically meaningful DD maps. If the valence states of atoms are used, based on the chemically prepared hybrid AOs, then a common picture of covalent bonding emerges indicating strong bumps of electron density along chemical bonds at the expense of peripheral regions. Lone pairs in F_2 are nicely profiled and the "ghost" accumulations of charge in the π electron regions perpendicular to the bond axis appearing both in experimental and some theoretical papers [26] disappear. This result is easy to understand. Neglecting hybridization of the F atoms for a while, the spherical promolecule concept implies uniform population of all three $2p$ AOs by 1.67 electrons. This is too much for the bonding region and too small for the lone pair domains. A more meaningful population is obtained in oriented atoms possessing one electron in the bonding $2p_z$ AO and two electrons in the rest of AOs. Similarly, the unusual feature of the experimental DD maps in tetra-aza-tetra-oxa-tricyclote-tradecane [19] could be ascribed to the arbitrary choice of the promolecule scale [24]. We shall focus attention to N and O atoms. Spherical ground-state of the nitrogen atom is $2s^2 2p^3$ representing the only component of the spectroscopic ^1S state. Hence it seemingly does not pose a problem within a promolecule. This is not true however, because the N atom hybridizes thus moving the doubly occupied lone pair $(2s)^2$ away from the bonding region resulting in singly occupied bonding hybrids and doubly occupied lone pair hybrid expelled from the internuclear domain(s). It follows that the standard DD maps exhibit weak buildups in $C-N$ and $N-N$ bonds because the density of the $(2s)^2$ lone pair is subtracted from the two electron localized MO. Consequently, deep electron density troughs appear in the immediate vicinity of the N atom along the bond direction. On the other hand, lone pair density is exaggerated since the singly occupied $2p$ density is subtracted from resulting localized MO describing lone pair. Hence, substantial increase in density behind N atom in $C-N$ bonds is not caused by the electronegativity difference, but rather by the use of the artificial promolecule reference state. Similarly, the spherical $2s^2 2p^4$ ground state of the oxygen atom leads to 4/3 populations of $2p_x$, $2p_y$ and $2p_z$ AOs. The resulting deformation density in $O-O$ bonds is negative. The use of hybridized valence states clearly shows a flow of charge from the nonbonding regions to the bonding domains and the whole picture is dominated by constructive interference characterizing covalent bonds. This holds true for all $C-N$, $N-N$, $C-O$ and $O-O$ single bonds indicating that a general pattern is valid even for atoms possessing a more than half filled p-subshell. The work of Hall et al. [23–25] underlined once again the importance of hybridization in interpreting chemical bonding. They pointed out that optimal hybridization employed in "prepared-for-bonding" atoms would eventually cancel out lone pair density in DD maps. Incidentally, tetrahedral (but deformed) arrangements of lone pairs are experimentally detected at the N and O atoms [18, 19]. It is noteworthy that the use of hybrids' densities in reference atomic states was found useful in depicting DD contours in complexes as well [27, 28]. Educated use of proper reference density enables separation of hybridization, delocalization and constructive inter-ference contributions to electron DD maps as shown for the first row hybrides [25].

A similar point of view was adopted by Schwarz et al. [29–31]. They determine optimal orientation and population of atomic ground state orbitals in molecular environments which are subsequently used in construction of chemical difference density (CDD) maps. Regrettably, hybridization is not included in optimal AOs and consequently this type of local polarization is not subtracted from DD diagrams. An interesting study of DD maps in first-row diatomics is made by using correlated MOs [29] based on valence configuration space generated from the minimal basis set atomic orbitals of the corresponding atoms in a molecule. These AOs are allowed to deform for optimal adaptation to molecular environments by the MCSCF approach where configuration mixing coefficients are simultaneously obtained. The resulting MOs are called FORS orbitals from Full Optimized Reaction Space calculation. Correlated wavefunctions are important if a variation of bond distances and its influence on DD maps is studied [29]. One of the outcomes of all these studies is that atoms which have quadrupolar density distribution in free state, do maintain this distribution to a large extent when embedded in molecules. This is in line with the conclusion of Hall et al. [23–25]. Schwarz et al. stressed also importance of orbital contraction of

AOs upon molecular formation and discussed in depth the influence of inner-core electrons and a crucial role of orbital overlapping.

Two additional points are worth mentioning. Electron correlation does not change qualitative features of DD maps. It merely diminishes charge accumulation at the bond center [33, 34]. This is in accordance with a general conclusion that the electron density ϱ is correct to the first order at the Hartree-Fock level of approximation [36, 37]. Secondly, it should be stressed that the experimental contours differ from theoretical ones because of thermal smearing. If their direct quantitative comparison is desired, then one should simulate thermal motion theoretically or deduce static experimental maps [38, 39]. Since correction of the static theoretical maps for nuclear motion leave the striking features of the contours unchanged and actually reduces the heights of the peaks and broadens troughs of density, the use of static distributions is fully justified in discussing their qualitative aspects.

The work of Hall et al. and Schwarz et al. yields several important conclusions. There is not a unique way in determining the reference density state in constructing DD maps, but this is more an advantage than a weakness. It is indeed possible to choose various prepared atomic density states which indirectly give a meaningful insight into the changes induced by bonding. Their appropriate selection enables one to delineate various effects like hybridization, orbital contraction, charge transfer thus contributing to our understanding of the chemical bonding process by direct inspection. Judicious choice of the atomic reference state unequivocally shows that there are general charge migration patterns which are common to widely different bonds. In particular, it provides convincing evidence that the conventional view of charge buildup between the bonded nuclei is correct. Earlier notions that the charge deficit in the bond region between electron-rich (electronegative) atoms should be attributed to the Pauli exclusion principle is wrong. It is simply a consequence of the unjustified use of the promolecule reference state where density is obtained by superposition of spherical neutral atoms. On the other hand, deliberate construction of DD maps offers a transparent interpretation of chemical binding in terms of the electrostatic forces.

Finally, we shall briefly comment on the electron density distribution in inorganic compounds. DD maps of complexes involving the central transition metal atom reveal a picture which is expected from the ligand or even crystal field model. A dominant feature is the asphericity of the central atom caused by unequal population of the d orbitals. If the ligands are electronegative atoms or atomic groupings, then the d orbitals avoiding them will be preferred in energy. Consequently, they will be more occupied by electron density than the d AOs directed toward ligands. Since the reference state of the transition metal is spherical with equal occupancy assigned to each d AO, the deformation density exhibits negative troughs in directions of less stable AOs and peaks in positions of the d orbital lobes which avoid ligands. The latter have more than average occupancy. This simple pattern is of course dependent on the number of d electrons and the extent of the ligand field splitting which leads to either high-spin or low-spin complexes. Thus some care must be exercised if the d subshell is more than half filled just like in the p subshell discussed earlier. Assuming that the covalency of the metal-ligand bonds is low one can fit the X-ray or polarized neutron diffraction data by multipole expansion of the "pseudo-atom" densities and then deduce d orbital occupancies [4, 27, 28]. Results are usually in good qualitative accordance with simple ideas as provided by the angular overlap model for example. It is fair to say, however, that there are still many unsolved and puzzling features in the DD maps in some transition metal complexes such as the appearance of some unexpected peaks and some expected density peaks not appearing [44]. Hence, additional and more accurate work, both experimental and theoretical, is necessary here.

2.2 Gradient Vector Field and Laplacian of the Molecular Electron Density

The gradient vector field of the electron density and its Laplacian play a significant role in the rigorous definition of atoms in molecules, chemical bonds and rearrangement of the charge distribution, respectively. In Bader's theory of topological atoms [45–47] immersed in a molecular environment, surfaces defining exclusive atomic domains determined by a condition

$$\nabla\varrho(\vec{r}) \cdot \vec{n}(\vec{r}) = 0,\tag{3}$$

are pivotal. Here $\nabla\varrho$ yields a direction and the amount of the steepest ascent thus being perpendicular to the isopycnic lines and $\vec{n}$ is the unit vector normal to the boundary surface. The vector field $\nabla\varrho$ within an atomic basin produces gradient paths which all terminate at the nucleus as a rule[2]. Hence the electron density maxima are placed in the close vicinity of the nuclei. Two bonded atoms are connected by the bond path — a trajectory of maximum electron density (MED) which has a minimum at $\nabla(\vec{r}_b) = 0$. This is the so called bond critical point which is minimum in the direction of the bond path and maximum in all others.

Topography of the total charge density is important in offering the quantum definition of the chemical bond, but alone is not sufficient. Existence of the MED between two nuclei is a necessary requirement for bonding, but an adequate condition is provided by the local energy density [48, 49]:

$$H(\vec{r}) = G(\vec{r}) + V(\vec{r}),\tag{4}$$

where $\int H(\vec{r}) \, d\vec{r}$ gives the total molecular electronic energy $E = T + V$ with $T = \int G(\vec{r}) \, d\vec{r}$ and $V = \int V(\vec{r}) \, d\vec{r}$. Since $G(\vec{r})$ is always positive, negative values of $H(\vec{r})$ indicate domains where negative potential $V(\vec{r})$ outweighs the contribution of the kinetic energy. Hence, low negative $H(\vec{r})$ values are indicators of strong covalent bonding. On the other hand, zero or small positive $H(\vec{r})$ values are characteristics of repulsive closed shell interactions, e.g. of strong ionic bonding involving full shell anions and cations, and hydrogen or van der Waals bonding [48]. It follows that one can define and specify some aspects of chemical bonding in a fundamental way inspite of the fact that the chemical bond does not correspond to any hermitian quantum mechanical operator.

Bond ellipticity is an important index of the π character of the chemical bond. In a cylindrically symmetrical bond both negative curvatures are equal in magnitude. This is not the case if electronic charge is preferentially accumulated in a given plane

---

[2] There are, however, some exceptions (vide infra).

along the bond path like in π and/or bent bonds. The elliptical profile of the density can be expressed by curvatures at the bond critical point. Many other conventional chemical concepts like delocalization, aromaticity, strain etc. were discussed along the same line [50]. The position of $\vec{r}_b$ does not coincide with the geometric center of the bond if it is formed by heteroatoms differing in electronegativity. The bond critical point is shifted toward the more electropositive atom thus offering a descriptor for the ionic bond character. This property can serve as an aid in defining a topological electronegativity scale [51]. Topological electronegativities can be favorably compared with other scales, notable exceptions being methods employing complete electronegativity equalization.

Topography of the total density is extremely simple. It involves sharp maxima (spikes) of density at the nuclei and not so highly pronounced ridges of electron charge representing chemical bonds. The reason behind missing features lies in the fact that changes caused by chemical bonding are minute. However, they can be magnified by examining Laplacian $-\Delta\varrho(\vec{r})$ which restores the shell structure of spherical atoms and makes the dipolar electron structure of e.g. the C atom in its ground state visible [50]. This is not surprising because the Laplacian of a scalar field determines sites where the field is locally concentrated or depleted [52]. By the same token, one would expect that $-\Delta\varrho$ will reflect lumps of density in the bonding regions and its depletion in antibonding domains. This is indeed the case because the local density $H(\vec{r})$ is related to the Laplacian in the following manner

$$H(\vec{r}) = (\hbar^2/4m)\, \Delta\varrho(\vec{r}) - 2G(\vec{r}). \tag{5}$$

Since the kinetic energy density $G(\vec{r})$ is intrinsically a positive entity, it follows that condition $-\Delta\varrho(\vec{r}) > 0$ leads to a substantially low negative energy density $H(\vec{r})$ [50].

It follows that $-\Delta\varrho > 0$ describes compression whilst $-\Delta\varrho < 0$ means decompression of the electron charge density indeed thus making the Laplacian a useful tool in determining e.g. sites of electrophilic and nucleophilic attacks [50].

To summarize, Bader's definition of the zero-flux atomic basins produces a well defined partition of molecular volume to the mutually exclusive subspaces. But it faces difficulties, e.g. that a zero-flux surface sometimes does not encompass a nucleus as in the Li_2 molecule and in some lithium clusters [61]. In these cases unusual maxima of ϱ are found at the midpoint(s) of $Li-Li$ bond(s) or within triangles in clusters. Lithium atoms are bound to these "attractors" but not within themselves. We shall show later that the atomic charges produced by Bader's concept are not in line with physical and chemical intuition. They are dependent on an atomic reference density and even with this reference the absolute values may not be reliable.

3 Atomic Charges

Three-dimensional molecular density shapes and two-dimensional isopycnic contours give a rough-and-ready insight into electron charge rearrangement upon bond formation. The former 3D representations are useful e.g. in understanding lock-and-key interactions between biological molecules and substrates. The Laplacian of the electron density can detect and rationalize sites of electrophilic and nucleophilic attacks whereas educated choice of the reference state in constructing DD maps enables a dissection of the chemical bonding process into several contributions and stages. However, all this information is qualitative although very transparent and intuitively appealing. For quantitative information one has to resort to an integral description of charge distribution which provides some numbers for relative comparisons. The notion of atomic charges or atomic monopoles is the most important concept in this respect. This is understandable because atoms retain their individuality in molecules being slightly adjusted to a new environment [54]. Formal or effective atomic charges provide a significant descriptor of chemical bonding being close to intuition and the idea of electronegativity [55].

Methods for estimating atomic charges fall into two broad categories: (a) theoretical procedures involving numerous recipes of the mixed charge partitioning and (b) experimental techniques involving X-ray scattering, ESCA and IR spectroscopies. We shall discuss first theoretical approaches which can be based on the orbital approximation population analyses, density matrix partitioning, spatial division of the molecular volume and application of the electronegativity concept. We shall also comment on charges derived from some molecular properties e.g. the electrostatic potential. It should be stressed that atomic charges serve the purpose of summarizing information on electron density distributions in a lapidary way, but a price has to be paid by losing some detail. A better description is obtained by making use of a larger number of point charges which do not all coincide with atomic nuclei and/or employing higher atomic multipoles. These aspects will be briefly considered, too.

3.1 Atomic Overlap Partitioning

Two main methods, the molecular orbital (MO) model and the valence bond (VB) approach, are both based on the use of atomic orbitals (AOs), which is conceptually and computationally advantageous. There are however two fundamental problems in using orbital approximation. The calculated charges are dependent on the quality of the employed basis set. A necessary prerequisite for meaningful atomic charges is the use of sufficiently large and well balanced basis sets. The latter feature cannot be rigorously defined, but essentially it means that the number of basis set functions placed on each atom should be consistent with the contribution of all other atoms. One criterion for appraisal of the bais set balance will be mentioned later. A typical example of an unbalanced and counterintuitive basis set is the one-center expansion of the molecular wavefunction at the site of one (usually the heaviest) atom. In this case all electrons have to be ascribed to the single heavy atom! This is contrary to the generally adopted idea of (modified) atoms in molecules. It is therefore not surprising that the single-center wavefunctions leave much to be desired in many respects. The second fundamental problem is embodied in apportioning the mixed interatomic charge to constituent atoms. Mulliken solved this Gordian knot by cutting the overlap charge in two halves by equipartition [56]. Let us denote an MO by

$$\psi_i = \sum_\mu c_{i\mu} \phi_\mu, \tag{6}$$

where ϕ_μ are atomic orbitals. Normalization condition yields:

$$\langle \psi_i \mid \psi_i \rangle = \sum_\mu c_{i\mu}^2 + \sum_{\mu \neq \nu} c_{i\mu} c_{i\nu} S_{\mu\nu} = 1. \tag{7}$$

Here $S_{\mu\nu} = \langle \phi_\mu \mid \phi_\nu \rangle$ is the overlap integral. Whilst $c_{i\mu}^2$ contributions are well localized corresponding to atomic distributions ϕ_μ^2, the $c_{i\mu} c_{i\nu} S_{\mu\nu}$ term actually belongs to both atoms when ϕ_μ and ϕ_ν are centered on two different nuclei. These two terms can be termed as passive and active charge, respectively. Mulliken apportioned the overlap charge on a 50:50 basis. Hence, the total orbital occupancy $Q_{A\mu}$, the total electron density on atom Q_A and the atomic charge are

$$Q_{A\mu} = \sum_i^{occ} n_i \left[c_{i\mu}^2 + \sum_{\substack{\nu \\ (\nu \neq \mu)}} c_{i\mu} c_{i\nu} S_{\mu\nu} \right], \tag{8a}$$

$$Q_A = \sum_\mu^A q_{A\mu}, \tag{8b}$$

$$q_A = Z_A - Q_A, \tag{8c}$$

where n_i denotes electron population of MO ψ_i. Mulliken named Q_A and q_A as gross and net atomic charges, respectively. This is a practical way of defining partial atomic charges which meets all invariance requirements and the inner need for balance, but it is at the same time arbitrary. Equipartition of the mixed charge is legitimate for homonuclear diatomics but in heteropolar bonds it sometimes leads to unphysical situations. For example, orbital occupancy $Q_{A\mu}$ can exceed 2 thus violating the Pauli principle, or assumes small negative values which does not make sense because it implies that there are less than zero electrons placed in the atomic orbital $\phi_{A\mu}$. This is obviously unacceptable. Some specific examples will be discussed later on. Sometimes negative populations can assume values as low as $-0.7e$ as does sulfur's orbital in SF_6 for a particular basis set. Addition of polarization functions can change the atomic charge by $\approx 0.8e$ without a dramatic change in orbital energies indicating that it is an artefact of the mixed charge equipartition recipe. Mulliken population in general performs rather poorly in highly ionic species and it is not appropriate for extended basis sets which will be discussed later on.

The overlap charge can be partitioned in several different ways. One can introduce weighting factors ω

$$\phi_{A\mu}\phi_{B\nu} = \tfrac{1}{2} [\omega_{A\mu}\phi_{A\mu}^2 + \omega_{B\nu}\phi_{B\nu}^2], \tag{9}$$

where conservation of the overlap charge yields

$$\omega_{A\mu} + \omega_{B\nu} = 2S_{\mu\nu}. \tag{10}$$

Hence, an additional requirement is needed for full characterization of relative weights. Following an earlier suggestion of Löwdin for approximate calculation of multicenter integrals [57], Doggett [58] and Hiller and Wyatt [59] independently proposed a partitioning which preserves the center of density of the overlap charge distribution. Let us consider a diatomic molecule A–B for the sake of simplicity and place the origin at an atom A. We shall also assume that AOs are unhybridized and centered at the corresponding nuclei:

$$\bar{X}_{\mu\nu} = \langle \phi_{A\mu}| \ x \ |\phi_{B\nu}\rangle = \omega_{A\mu}\langle \phi_{A\mu}| \ x \ |\phi_{A\mu}\rangle + \omega_{B\nu}\langle \phi_{B\nu}| \ x \ |\phi_{B\nu}\rangle . \tag{11}$$

Since centers of charge of atomic orbitals coincide with the nuclei, one can easily find that:

$$\omega_{A\mu} = S_{\mu\nu} - \bar{X}_{\mu\nu}/R_{AB}, \qquad \omega_{B\nu} = \bar{X}_{\mu\nu}/R_{AB} . \tag{12}$$

Hence the gross atomic populations read:

$$Q_A = \sum_{\mu}^{A} \left[P_{\mu\mu} + \sum_{\nu}^{B} P_{\mu\nu}(S_{\mu\nu} - \bar{X}_{\mu\nu}/R_{AB}) \right], \tag{13a}$$

$$Q_B = \sum_{\nu}^{B} \left[P_{\nu\nu} + \sum_{\mu}^{A} P_{\mu\nu}\bar{X}_{\mu\nu}/R_{AB} \right]. \tag{13b}$$

This procedure preserves the dipole moment in diatomics provided the hybridization term is properly taken into account by using unmodified $P_{\mu\nu}$ matrix elements. Hillier and Wyatt pointed out that the bond dipole moment conserving population can reverse a sign of charges as compared to classical Mulliken analysis [59]. They also stressed the importance of using realistic atomic charges in the methods of Fenske et al. [60] and in the iterative self-consistent charge (SCC-MO) scheme [61]. Doggett also found some cases where modified atomic charges reverse the sign but in the wrong direction, i.e. the Mulliken population is at least qualitatively correct. An illustrative example is given by the HF molecule. The Mulliken population gives charges for the fluorine atom of -0.153 and -0.230 for Slater and best limited LCAO-MO basis sets, respectively. On the other hand, dipole moment conserving atomic charges obtained by Löwdin partitioning read $+0.105$ and -0.056 respectively for the same sets as mentioned above, indicating that the STO basis set charge distribution cannot be reconciled with the empirical notion of the fluorine atom as a carrier of the negative charge. The best limited set improves the situation regarding sign but not magnitude since extended basis sets with a particular spatial partitioning yield effective charges as much as $+0.27$ (H) and -0.27 (F) (vide infra). It is not surprising therefore that this type of population analysis has not received much attention. An additional problem is posed in trying to describe local atomic dipoles by point charges. Claverie offered a solution by allocating some charges to the nearest neighbours so that local dipoles of each atom are preserved [62]. Thole and van Duijnen [63] preserved local dipoles by distributing partial point charges over all atomic centers in a molecule. An interesting detail is that interatomic charge transfer is kept at minimum. However changes in their weighting function can cause large charge shifts which is an undesirable feature. Additionally, the atomic hybrid moments are not explicitly defined.

A rather general way of accomplishing the same objective, namely charge and dipole moment conservation in a large basis set was suggested by Jug [64]. Starting from a commutation of hermitian or antihermitian operators u, v and w:

$$u = [v, w], \tag{14}$$

and representing it in a matrix form by using a complete set of non-orthogonal functions ϕ_μ one obtains:

$$u_{\mu\nu} = \sum_{\alpha\beta} \{ v_{\mu\alpha}(S^{-1})_{\alpha\beta} w_{\beta\nu} - w_{\mu\alpha}(S^{-1})_{\alpha\beta} v_{\beta\nu} \}, \tag{15}$$

where $u_{\mu\nu} = \langle \phi_\mu | u | \phi_\nu \rangle$ etc. Substituting in Eq. (14) $v = 1$, $w = \vec{r}$ and $v = \vec{r}$, $w = \vec{r}$ the commutator vanishes yielding conditions for preserving charge and dipole moment in partitioning overlap densities. Let us assume that AOs ϕ_μ and ϕ_ν are placed at nuclei A and B, respectively, and that the x-coordinate axis coincides with the interatomic straight line. Then after some algebra one obtains

$$x_{\mu\nu} = \frac{F_{\mu\mu} - F_{\nu\nu}}{(F_{\mu\mu} - F_{\nu\nu})^2} \left\{ \sum_\alpha^A F_{\alpha\nu}x_{\mu\alpha} - \sum_\beta^B F_{\beta\mu}x_{\beta\nu} + \sum_\beta^{B\prime} F_{\beta\nu}x_{\mu\beta} - \sum_\alpha^{A\prime} F_{\alpha\mu}x_{\alpha\nu} \right\}, \tag{16}$$

where $F_{\mu\nu} = \sum_\gamma (S^{-1})_{\mu\gamma} \vec{r}_{\gamma\nu}$. Indices $\beta = \nu$ and $\alpha = \mu$ should be omitted in the last two summations as indicated by primes. It turns out that a two center term $x_{\mu\nu}$ is expressed by single center contributions $x_{\mu\alpha}$ and $x_{\beta\nu}$ of atoms A and B, respectively. The last two terms again involve overlap density contributions thus lending the identity (16) an iterative solution. In a polyatomic molecule one can rotate a molecule-fixed coordinate system to a local one with the x-axis along the A-B bond, perform apportioning of mixed density terms according to Eq. (16), and go back to the original system by making use of appropriate unitary (orthogonal) transformations. It is noteworthy that y and z components of the bond dipole moment disappear in the local coordinate frame. By putting $x = 1$ in Eq. (16) one obtains partitioning of the overlap $S_{\mu\nu}$. This sort of partitioning can be extended to include higher multipole moments than dipole [65] but then the formula corresponding to (16) holds only approximately. Jug population analysis gives results which fall between Mulliken and Löwdin dipole moment partitioning as revealed by a study on selected diatomics [64]. All three methods of determining atomic charges exhibit an undesirable basis set dependence. However, this basis set dependence should gradually decrease in the method described above as the size of balanced basis sets systematically increases. It was suggested that the extent, to which the quadrupole moment is conserved in the expansion like (16) could be used as a criterion for a balanced basis set.

Christoffersen and Baker [66] as well as Politzer and Stout [67] independently devised a partitioning technique which favors the atom with the larger passive charge in the partitioning scheme. Hence, the overlap charge is divided proportionally to coefficients c_μ^2 and c_ν^2. For a diatomic covalent bond A–B one derives:

$$Q_{A\mu} = 2c_\mu^2 + 4c_\mu c_\nu S_{\mu\nu}[c_\mu^2/(c_\mu^2 + c_\nu^2)], \tag{17a}$$

$$Q_{B\nu} = 2c_\nu^2 + 4c_\mu c_\nu S_{\mu\nu}[c_\nu^2(c_\mu^2 + c_\nu^2)]. \tag{17b}$$

One can eliminate overlap integral through the normalization (charge conservation) condition which yields:

$$Q_{A\mu} = 2c_\mu^2/(c_\mu^2 + c_\nu^2), \qquad Q_{B\nu} = 2c_\nu^2/(c_\mu^2 + c_\nu^2). \tag{18}$$

Generalization to polyatomic and many-electron systems is simple, but it turned out that this plausible idea did not perform well. It led to exaggerated charge transfer within molecules, i.e. overpolarization of atoms. Credit to this method should be given, however, for the fact that atomic orbital populations are not smaller than zero and not greater than 2.

An alternative way of weighting atomic ϕ_μ^2 distributions was proposed by Gallup and Norbeck [68]. In their inverse overlap population analysis, occupation number $Q_{A\mu}$ of AOs ϕ_μ is given by

$$Q_{A\mu} = NP_{\mu\mu}/(S^{-1})_{\mu\mu}, \tag{19}$$

where $N = \left[\dfrac{1}{n}\sum_\mu P_{\mu\mu}/(S^{-1})_{\mu\mu}\right]^{-1}$. Here n is a number of electrons in a molecule, $P_{\mu\mu}$ and S^{-1} are conventional diagonal elements of the charge density matrix and inverse of the overlap matrix, respectively. This scheme was applied to the analysis of the VB function in BeH_2. It was concluded that the examined VB function has $\approx 90\%$ of covalent character, $\approx 6\%$ of positive and 4% negative ionic character. The diagonal terms of S^{-1} tend to enhance the importance of the covalent component, which is a desirable feature.

3.2 Atomic Orbital Orthogonalization Approach

The problem of overlap partitioning is avoided altogether if orthogonalized atomic basis sets are employed. In this case the diagonal elements $P_{\mu\mu}$ of the density matrix are equal to the occupation numbers. We shall call this basis λ

$$Q_{A\mu} = P_{\mu\mu}^\lambda. \tag{20}$$

Initial atomic basis sets are genuinely non-orthogonal and the question arises which orthogonalization method should be used as a most appropriate. Löwdin's symmetric orthogonalization [69] seems to be the first plausible choice because it minimally perturbs the original AOs retaining the maximal similarity as measured by the integral least squares' deviations [70]. An additional conceptual advantage is that all AOs are treated in the same way. Orthogonal orbitals are expressed by

$$\lambda_\mu = \sum_\nu (S^{-1/2})_{\mu\nu} \, \phi_\nu \, . \tag{21}$$

The density matrices in these two basis sets are related by the formula

$$P^\lambda = S^{1/2} P S^{1/2} \, . \tag{22}$$

It is found that the symmetric orthogonalization usually results in a reduction of charge transfer as compared to the Mulliken analysis. A useful feature is that gross orbital populations never assume unphysical negative values. On the other hand Löwdin charges suffer from two drawbacks. Although Löwdin AOs are even more localized around the host nucleus, there are still tails distributed over other atoms. Hence, the population of orthogonalized AOs strictly speaking cannot be ascribed to a single atom. Secondly, in symmetric orthogonalization all AOs are uniformly changed implying that this procedure is quite appropriate for atoms which have not a pronounced difference in electronegativity. However, if widely different atoms are involved, some discrimination might be desirable. This goal can be achieved by the weighted Löwdin procedure [71].

The possibility of the weighted symmetric orthogonalization has already been discussed by Carlson and Keller [70]. Recently, Weinhold et al. [72] introduced the so called natural population analysis based on the weighted symmetric orthogonalization which should counteract the effect of excessive charge equalization. The "natural population analysis" represents occupancies of natural atomic orbitals (NAOs) [73]. These natural populations are inherently positive and conserve correctly the total number of electrons. The NAOs are closely related to conventional natural orbitals introduced by Löwdin [74]. The former are orthonormal atomic orbitals possessing maximal occupancy for the given molecular wavefunction. It appears that they converge smoothly toward defined limits as the wavefunction is improved with stable populations. The diagonalization is performed in two steps. The first straightforward operation involves diagonalization of the one-center block of the density matrix which gives the pre-natural AOs. It yields two subsets for each atom: (1) The "minimal" basis set of high occupancy spanning ground state configuration of a free atom and (2) the "Rydberg" set of higher AOs of low populations. In the second step diatomic overlaps are considered. The "minimal" basis set of one atom exhibits strong overlapping with the "Rydberg" set of neighbouring atoms yielding unphysical results for the Mulliken analysis. Hence the "minimal" AOs should not be contaminated by neighbouring "Rydberg" AOs in the orthogonalization procedure. This is minimized indeed by the weighted Löwdin approach. The NAOs are obtained by

$$\lambda_\mu^N = \sum_\nu T_{\mu\nu} \phi_\nu' \, , \tag{23}$$

where ϕ_ν' are pre-natural AOs and

$$T = W(WSW)^{-1/2} \, . \tag{24}$$

Here W denotes a diagonal matrix involving populations of prenatural AOs. The main effect of these transformations is a description of a charge transfer from the electron-poor atoms to the electron-rich atoms, i.e. from the left side to the right side of the same row of elements in the periodic system in a way which is probably too excessive.

Weinhold et al. [72] analysed charge distribution in compounds of type CH_3X and LiX (X = F, OH, NH_2, CH_3, BH_2, BeH, Li, H) by making use of several basis sets. Some characteristic results for Li atom are presented in Table 1 and compared with charges obtained with the Mulliken population analysis (MPA) and density integration (DI) technique [75]. Perusal of the data shows the expected strong and nonuniform dependence of MPA charges on basis sets and their inadequacy in describing ionic bonds. Both DI and NPA population schemes exhibit a large jump in net charge in going from STO-3G to 4-31G basis set. This change is at least one order of magnitude smaller in a step from 4-31G to a more involved 6-31G* set. The NPA results show the largest stability in this respect. It seems also that little physical significance should be attached to STO-3G charges in ab initio methods. The NPA and DI charges are often in qualitative agreement, but should not be based on the STO-3G results.

Weinhold's scheme can be modified in two directions. Large basis sets could be first contracted to minimal basis sets and renormalized. This means contraction to $1s$ AOs for H and He, $1s$, $2s$, $2p$ AOs for the first row atoms from Li to Ne and $1s$, $2s$, $2p$, $3s$, $3p$ and $3d$ AOs for the second row atoms. The d orbitals are necessary in the latter for interpretation of hypervalence. But they have occupancies which are lower than 1. This procedure should reduce the basis set dependence. Secondly, there is nothing sacrosanct about the weighting factors employed by Weinhold et al. [72]. We used the weights $\sqrt{n/2}$ with n = 1, 2 and 3 for Li, Be and B atoms, respectively, and 1 for the rest of the first row elements (C, N, O, F, Ne) [76]. This choice is based on an analytical formula involving bond and lone pairs. It diminishes charge transfer.

An unusual way of combining Löwdin orthogonalization and its series approximation for overlap population was presented by Guerrilot et al. [77] within the framework

Table 1. Comparison of Li atomic charges as computed by several basis sets and different population techniques

Molecule	STO-3G			4-31G			6-31G*		
	MPA[a]	DI[a]	NPA[b]	MPA[a]	DI[a]	NPA[b]	MPA[a]	DI[a]	NPA[b]
LiF	0.226	0.615	0.525	0.715	0.859	0.916	0.639	0.874	0.929
LiOH	0.177	0.588	0.530	0.650	0.814	0.926	0.573	—	0.935
LiNH$_2$	0.202	0.634	0.550	0.558	0.739	0.891	0.486	—	0.901
LiCH$_3$	0.146	0.701	0.442	0.451	0.794	0.793	0.416	0.797	0.805
LiBH$_2$	−0.090	0.595	0.201	0.201	0.688	0.544	0.145	—	0.567
LiBeH	−0.106	0.554	0.008	−0.087	0.593	0.151	−0.117	—·	0.191

[a] MPA and DI denote Mulliken population analysis and density integration method, respectively, Ref. [75].
[b] Natural population analysis results Ref. [72].

of the EHT model. They truncate the series expansion of $(S')^{-1/2}$ after the first term, where $S' = S - 1$ has zero diagonal elements and propose the use of the expression

$$P'_{\mu\nu} = P^{\lambda}_{\mu\nu}S_{\mu\nu}, \tag{25}$$

as a measure of bond population. Here superscript λ refers to Löwdin orthogonal AOs. It seems that this recipe is not theoretically sound. Since the Mulliken overlap population is given by $P^M_{\mu\nu} = P_{\mu\nu}S_{\mu\nu}$, a commutability of S and P and truncation would lead to $P^{\lambda} = PS$. In such a case $P'_{\mu\nu} = (PS)_{\mu\nu}S_{\mu\nu}$ implies that Eq. (25) can not be used for proper counting of electrons in molecules.

Finally, it should be pointed out that an unambiguous atomic population analysis within the adopted FORS framework is furnished by the corresponding minimal basis set AOs optimally adjusted to the molecular environment as discussed in depth by Ruedenberg et al. [32].

3.3 Theoretical Methods Beyond the Independent-Electron Approximation

Although electron correlation is not decisive in calculating atomic charges it is by no means negligible. Therefore, we shall consider methods which can be applied to highly accurate calculated (and/or experimental) electron density distributions in retrieving atomic charges. Some preliminaries are necessary for this purpose. Let us start with a configuration interaction (CI) wavefunction of the form:

$$\Psi = \sum_I A_I \Psi_I, \tag{26}$$

built by Slater determinants which in turn are composed of MOs ψ_i. The latter are expanded in AOs according to Eq. (6). The density operator Γ of the CI wavefunction is a projection operator of the form

$$\Gamma = |\Psi\rangle\langle\Psi| = \sum_{IJ} A_I A_J |\Psi_I\rangle\langle\Psi_J|. \tag{27}$$

The corresponding first-order density operator P is then defined over MOs ψ_i [78]:

$$P = \sum_{IJ} A_I A_J \sum_{ij} \varrho_{IJij} |\psi_i\rangle\langle\psi_j|, \tag{28}$$

where

$$\varrho_{IJij} = \begin{cases} (-1)^{s-t} & \text{for } (I - i_s) = (J - j_t) \\ 0 & \text{otherwise} \end{cases}.$$

The occupation matrix $P_{\mu\nu}$ of the operator P (28) is then obtained by

$$P_{\mu\nu} = \langle\phi_\mu| P |\phi_\nu\rangle. \tag{29}$$

One can express P in the AOs basis set as

$$P = \sum_{\alpha\beta} D_{\alpha\beta} |\phi_\alpha\rangle \langle\phi_\beta|,$$ (30)

with $D_{\alpha\beta} = \sum_{IJ} A_I A_J \sum_{ij} \varrho_{IJij} c_{i\alpha} c_{j\beta}$. We shall refer to $D_{\alpha\beta}$ as the bond order matrix. The relation between $P_{\mu\nu}$ and $D_{\mu\nu}$ can be concisely written as a matrix equation:

$$P = SDS$$ (31)

whereas normalization of the wavefunction is expressed by

$$n = \sum_\alpha (S^{-1}P)_{\alpha\alpha}.$$ (32)

Davidson has tried to fit the accurate charge density $\Gamma(r, r)$ by the sum of all products $\phi_\mu \phi_\nu$ employing the integral least square method [79] where ϕ_μ and ϕ_ν are customary AOs:

$$\varepsilon = \int |\Gamma(r, r') - \sum_{\mu\nu} P^{app}_{\mu\nu} \phi_\mu(r) \phi_\nu(r')|^2 \, dr \, dr' = \min.$$ (33)

In other words, the approximate density is of the form:

$$\Gamma(r, r') \cong \sum_{\mu\nu} P^{app}_{\mu\nu} \phi_\mu(r) \phi_\nu(r').$$ (34)

If the atomic set ϕ_μ has a physical meaning, then ε is small compared to 1. It was found that the sum of $P^{app}_{\mu\mu}$ generally exceeds the number n of electrons in a molecule. This surplus

$$\Delta n = \sum_\mu P_{\mu\mu} - n,$$ (35)

was considered as a measure of electron sharing between atoms. This type of electron population analysis has shown that there is a back-transfer of charge in the π region in BF and CO molecules which acts in the opposite direction to the charge shift in the σ bond. Surprisingly, a larger degree of hybridization is established in more electronegative atoms which contradicts a generally accepted picture [79].

An interesting charge partitioning was suggested by Roby [80]. He argued that if ϕ_μ and ϕ_ν are two overlapping AOs, then an electron placed in ϕ_μ is at the same time partly in ϕ_ν orbitals. Hence, the complete mixed charge should be ascribed to both atoms simultaneously. Thus, it follows that Roby's atomic charges substantially exceed the number of electrons in free atoms. Furthermore, if AOs centered on each atom separately tend to achieve a complete set, then atomic charge of any particular atom approaches the total number of electrons in a molecule. Roby made use of projection operators in population analysis and named the procedure as the "projection-density" approach. Namely, the total density is projected to atomic, bond etc. subspaces by the corresponding projection operators P_A, P_{AB} etc. For a diatomic case the respective operators are defined as

$$P_A = |\phi_A\rangle \langle \phi_A| = \sum_{\mu}^{A} |\phi_{A\mu}\rangle \langle \phi_{A\mu}|, \tag{36a}$$

$$P_{AB} = \sum_{\mu} \sum_{v} |\phi_{A\mu}\rangle (S^{-1})_{\mu v} \langle \phi_{Bv}|, \tag{36b}$$

where both summation indices μ and v run over AOs placed on both atoms A and B. The occupation numbers are:

$$n_A = \sum_i \langle \psi_i| PP_A |\psi_i\rangle \quad \text{and} \quad n_{AB} = \sum_i \langle \psi_i| PP_{AB} |\psi_i\rangle. \tag{37}$$

Since the mixed density is simultaneously attributed to both atoms and counted twice correspondingly, the shared population is defined by subtracting n_{AB} from the sum n_A and n_B:

$$M_{AB} = n_A + n_B - n_{AB}. \tag{38}$$

Roby's method allows definition and calculation of polycentric shared populations which have no counterpart in Mulliken's analysis. For example, the mixed density encompassing three atoms is given by

$$M_{ABC} = n_{ABC} + (n_A + n_B + n_C) - (n_{AB} + n_{BC} + n_{AC}). \tag{39}$$

Roby considered also the most appropriate AOs for description of electron density in molecules. The overlap matrix S is in a pseudodiagonal form with block unit matrices placed on the main diagonal. However, the remaining blocks can also be diagonalized by unitary (orthogonal) transformations resulting in chemically adapted hybrid AOs, which are coupled in pairs and possess optimal overlap [81]. They describe atomic polarization, directional properties of covalent bonding, whereas their populations reflect the intramolecular charge transfer where M stands for the mixed density. Roby's ideas were elaborated and modified by Cruickshank and Avramides [82].They suggest a reduction of atomic occupation numbers by half of the shared charge:

$$Q_A = n_A - (1/2) M_{AB}, \tag{40}$$

which for a diatomic A−B reads

$$Q_A = c_{A\mu}^2 + c_{A\mu}c_{Bv}S_{\mu v} - \tfrac{1}{2} S_{\mu v}^2 (c_{A\mu}^2 - c_{Bv}^2). \tag{41}$$

Hence, in addition to the conventional Mulliken term there is a contribution proportional to the square of the overlap integral. We shall call the charges Eq. (41),

Roby-Cruickshank (RC) populations. Obviously, they are smaller than Mulliken charges because of the last term in Eq. (40). It appears that RC charges are typically between 50% and 70% of the Mulliken charges (Table 2). This is particularly important for highly ionic compounds where Mulliken charges are notoriously exaggerated (cf. CF_4, SiF_4, PF_3 etc.). The projection-density technique is in principle independent of the orbital approximation, but if AOs basis sets are used, then projection operators depend on the choice of AOs. This is probably the reason behind the negative RC charge of carbon atom in CH_4, which most likely is not correct. Roby charges reflect, for a given coordination number, electronegativities of atoms and show to what extent an atom in chemical environment approaches a complete noble gas shell.

Cruickshank and Avramides devised a technique for calculating population of polarization functions which usually represents the largest problem. This approach delineates the unshared polarization contribution which is quite small as a rule. For instance, in NF_3 only 0.14 electrons belong to polarization functions out of 34 electrons. They have also shown that the mixed density M_{AB} is a very good criterion for estimating the strength of the chemical bonds. Mulliken's overlap charge fails in SF_2 and CIF mollecules since it assumes negative -0.126 and -0.130 values, respectively, and yet S–F and Cl–F are "normal" covalent bonds. The RC M_{AB} values are 0.807 and 0.722, respectively, indicating appreciable covalency. Furthermore, M_{AB} is proportional to the overlap integral S_{AB} over a large range of $r = (c_{Bv}/c_{A\mu})$ compatible with a notion of (hybrid) AOs overlapping as a necessary ingredient of covalent bonding. It is noteworthy that the localized molecular orbitals (LMO) in F_2 obtained by Boys method [83] give one σ-bond and three tetrahedral-type lone pairs on each fluorine. The σ-bond LMO contribute 0.708 to the M_{FF} whilst each of the six lone-pairs gives only -0.017. This is another piece of information indicating the importance of hybridization and positive interference of density yielding a shared charge of participating hybrids. It was also found that three banana bonds in N_2 each contribute 0.978 to the M_{NN} in contrast to lone-pair hybrids which yield -0.010 each.

Table 2. Comparison of RC and Mulliken charges in some selected molecules

Molecule	Roby-Cruickshank		Mulliken	
	q_A	q_B	q_A	q_B
CH_4	−0.27	0.07	−0.76	0.19
NH_3	−0.43	0.14	−0.92	0.39
OH_2	−0.46	0.23	−0.78	0.39
FH	−0.32	0.32	−0.47	0.47
CF_4	0.60	−0.15	0.98	−0.25
NF_3	0.40	−0.13	0.54	−0.18
OF_2	0.17	−0.08	0.21	−0.10
SiF_4	1.41	−0.35	2.25	−0.56
PF_3	1.09	−0.36	1.63	−0.54

The seminal ideas of Davidson and Roby gave impetus to Ahlrichs et al. [84, 85] to define modified atomic orbitals (MAO). Davidson [79] used atomic HF orbitals for ϕ_μ in Eq. (33) which has led to a defect measured by ε of $\simeq 0.2$. Ahlrichs [84] employed more flexible MAOs (hybrids) which have been able to minimize ε further to values of ≤ 0.05. They have also shown that the total number of electrons n can be written as a sum of mono- and polycentric (shared) densities

$$n = \sum_A n_A - (1/2!) \sum_{A \neq B} M_{AB} + (1/3!) \sum_{A \neq B \neq C} M_{ABC} - \dots \qquad (42)$$

Since M_{ABC} are appreciable only in molecules possessing three-center bonds like B_2H_6, H_3^+ etc., and densities shared by four and more atoms are small indeed, one obtains

$$n \cong \sum_A n_A - (1/2) \sum_{A \neq B} M_{AB}, \qquad (43)$$

within limitation mentioned above. Hence, two-center mixed density can be equipartitioned between atoms A and B. This is arbitrary just like Mulliken's analysis, but it is compatible with Eq. (40). Comparison of RC charges with MAOs for some small molecules show that they are very close (Table 3). MAOs have substantially smaller charge defect ε. Ahrichs claims that a use of MAOs greatly reduces the basis set dependence of charges. However, the undesirable feature of this procedure is that the total electronic charge is not conserved. On the other hand shared density M_{AB} provides a very good measure of bond energy [85]. A useful index of hypervalency is established too.

Another quite different criterion for charge partitioning was developed in studying experimental densities, but it can be applied to approximate theoretical charge distributions as well. This is the "stockholder" principle put forward by Hirshfeld [86]. He defined atomic fragments of molecular density by dividing the latter at each point in space in proportions determined by the corresponding promolecule contribu-

Table 3. Comparison of RC and EA atomic charges, shared densities and charged defects in some small molecules

Molecule	Atomic charge			M_{AB}		ε	
	Atom	RC[a]	EA[b]	RC	EA	RC	EA
N_2	N	0	0	2.93	2.94	0.106	0.002
CO	C	0.07	0.06	2.59	2.57	0.105	0.002
CO_2	C	0.50	0.63	2.15	2.06	0.191	0.031
H_2O	O	-0.46	-0.36	1.20	1.23	0.077	0.015

[a] Roby-Cruickshank charges;
[b] Ehrhardt-Ahlrichs charges [85]

tion of spherically symmetric neutral atoms. Charge density of a promolecule $\varrho_{pro}(\vec{r}) = \sum_A \varrho_A^{free}(\vec{r})$ is a gauge for defining atomic shares S_A

$$S_A(\vec{r}) = \varrho_A^{free}(\vec{r})/\varrho_{pro}(\vec{r}) . \tag{44}$$

The actual molecular density $\varrho_{mol}(\vec{r})$ is then partitioned by

$$\varrho_A(\vec{r}) = S_A(\vec{r})\varrho_{mol}(\vec{r}), \tag{45}$$

meaning that if an atom invests more of its density in interatomic overlapping regions then it obtains more "profit" by the constructive interference. One immediately observes that the choice of the reference promolecule density is open to a criticism, but charges obtained by the stockholder criterion are by no means unreasonable. Once atomic fragment densities are determined atomic deformation densities are easily produced

$$\delta\varrho_A(\vec{r}) = \varrho_A(\vec{r}) - \varrho_A^{free}(\vec{r}) = S_A(\vec{r})\,\Delta\varrho(\vec{r}), \tag{46}$$

where $\Delta\varrho(\vec{r}) = \varrho_{mol}(\vec{r}) - \varrho_{pro}(\vec{r})$ is the customary difference density DD distribution.
    The total electronic charge of a bound atom is

$$Q_A = \int \varrho_A(\vec{r})\,d\vec{r}, \tag{47}$$

which gives a monopole of the electron density. Atomic charge is concomitantly given by $q_A = Z_A - Q_A$ yielding effective atomic monopole. Since ϱ_A varies too steeply near nuclei it is more convenient to integrate the atomic deformation density providing directly $q_A = -\int \delta\varrho_A(\vec{r})\,d\vec{r}$. Local atomic multipole moments are easily expressed by the $\delta\varrho_A$ density too. We shall be interested in particular in atomic dipole moments

$$\mu_{Ak} = -\int r_k \delta\varrho_A(\vec{r})\,d\vec{r}, \tag{48}$$

where k stands for coordinates x, y and z, respectively. Stockholder atomic charges and local dipoles will be used in later discussions.

3.4 Spatial Partitioning of Molecular Density

Daudel who was first to suggest the use of deformation density in describing rearrangement of electron charge in molecules [87] also introduced the concept of loge — a part of space containing localized electrons [88]. Optimal spatial partitioning into mutually exclusive loges is obtained by information theory. Thus volumes of the real space are deduced involving core electrons, bond and lone pair electron couples corresponding to Lewis classical theory of molecules. For example, the CH_4 molecule has five loges: one spherical involving inner shell pair and four quadrants encompassing C−H bonds. Boundary surfaces are either spheres or planes. It was shown that lone pair loges were more bulky than bond couples. One can say that the best partitioning into loges corresponds to a minimum of the fluctuation of the number of electrons in the loges or to put it in another way − the probability of a fixed number of electrons being separated is greatest. Formalism of the information theory is not easy to implement. A more practical approach is suggested by Bader and Stephens [89]. They have shown that electron localization is optimally achieved if its Fermi correlation hole is totally contained within the same spatial region. Bader and Stephens analysed loge and zero-flux boundary surfaces in some small molecules and found that they are not widely different.

Division of the molecular volume into regions associated with the individual atoms by carefully determined surfaces is conceptually advantageous because atomic charges calculated by integration of $\varrho(\vec{r})$ over the corresponding domains are in principle independent of orbital approximation and transformation of local coordinate systems. In some cases it is natural to use half planes or planes as boundaries. For five-membered heterocycles Hartmann and Jug [90] used half-planes which cut the space in segments like pieces of a cake. Politzer et al. [67, 91] considered linear molecules and determined dissection planes so that integration of the promolecule density over atomic basins gives the number of electrons in neutral free atoms. This is achieved by the electron count function G(z):

$$G(z) = \int\limits_{-\infty}^{z} dz \int\limits_{0}^{\infty} r \, dr \int\limits_{0}^{2\pi} \varrho(r, z, \varphi) \, d\varphi \,, \tag{49}$$

where $\varrho(r, z, \varphi)$ is the electron density distribution expressed in cylindrical coordinates and z-coordinate coincides with the molecular symmetry axis. Politzer's partitioning proved useful in resolving a paradox presented by Kern and Karplus [92] in discussing electron distribution in the H−F molecule. Namely, two large basis set calculations of Clementi (16 functions) and Nesbet (18 functions) gave nearly the same energies (−100.0575 au and −100.0571 au, respectively) and yet Mulliken charges were widely different. They are −0.23 and −0.48 e on the F atom for Nesbet and Clementi wavefunctions, respectively. When Politzer's counting of electrons was applied charges assumed almost the same values: −0.27 and −0.26 for ab initio densities mentioned above. Politzer's result is more consistent with the electroneutrality principle stating that intramolecular charge transfer should be at minimum [55]. This indicates once again an intrinsic weakness of the Mulliken population analysis and importance of realistic criteria of charge (and space) partitioning. Needless to say, Politzer' recipe depends on the definition of the reference free atom density as discussed in Sect. 2.1.

Bader et al. [93] cut the internuclear line in diatomics and linear molecules by a plane at points where charge density reaches its minimum. The resulting populations were further divided into nonbonded and bonded populations for a terminal nucleus and into two bonded populations for a nucleus with two adjacent atoms. These atomic charges and accompanying bonded and nonbonded radii proved useful in classifying and characterization of density distributions in a variety of linear molecules. It was strongly pointed out that molecular dipole moments are poor indicators of the intramolecular charge transfer effect in a point charge approximation.

Maclagan [94] suggested nuclear electrostatic potential and nuclear force criteria for partitioning of geometric space in diatomics. The former yields a surface $(Z_A/r_A) = (Z_B/r_A)$ and the charge density within the volume

$$(Z_A/r_A) > (Z_B/r_B),$$

is ascribed to atom A. Results obtained by this definition were unsatisfactory because too much charge was transfered to the atom with the larger nuclear charge. The force criterion gives all the density within the region satisfying

$$(Z_A/r_A^2) > (Z_B/r_B^2),$$

to atom A. The most realistic charges were obtained when nuclear charges Z_A were substituted by core charges e.g. $Z_A - 2$ for the first row atoms. It is difficult to implement the force criterion in polyatomic molecules and concomitantly this idea did not find wide application.

The topological or virial partitioning of Bader divides the space of a molecular system into non-overlapping fragment building-blocks encompassed by zero-flux surfaces. Electron density is strictly additive as are the other properties, which is highly desirable. However, electron populations suggest an excessive charge transfer effect and dramatically differ from results from all the other methods [95–98]. For instance, oxygen gross atomic populations are typically between 9 and 9.5 electrons suggesting a charge shift of 1.0–1.5 electrons even in molecules such as CO, H_2CO and CH_3OH. This is unrealistic and consequently unsatisfactory, triggering the question why? An answer was found by Maslen and Spackman [99, 100] who found that Bader's scheme yields quite substantial charges even for promolecules where charge transfer is zero by definition. Hence virial partitioning does not yield neutral atoms in passing to the limiting case of noninteracting atoms like e.g. Politzer's procedure. A tempting possibility of defining charge transfer by using promolecule superposition densities as a reference was examined in detail [100] and results are compared with estimates of atomic charges obtained by the stockholder criterion in Table 4 for a large number of diatomics widely differing in bonding features. The molecular wavefunctions are of near Hartree-Fock quality [101, 102], whilst the spherical atom promolecule densities are produced by the analytic HF limit wavefunctions of Clementi and Roetti [103]. Improvement of the results by using the promolecule as a standard is dramatic in all molecules. Some illuminative examples are provided by PN where the charge transfer drops from 1.74 to 0.43 and in BeO where virial partitioning electron population on oxygen is as large as 9.69 being reduced to only 8.54 electrons. The promolecular charges are appreciable as a rule, a notable exception being CH. Modified virial charges are in much better agreement with results offered by the stockholder criterion. It should be pointed out, however, that even modified topological charges sometimes yield unrealistic pictures. A notable example is provided by Li-F, which is a prototype for the ionic bond and yet the predicted charge transfer is only ~ 0.3 electrons (Table 4).

Since lithium compounds have attracted considerable interest lately we note in passing that promolecule corrected Bader charges of Li span a range of values between 0.246 (LiB) and 0.333 (LiF) which is too narrow. On the other hand Bader's virial charges are about three times larger assuming values of 0.761 and 0.940, respectively. This is obviously unrealistically large and contradicts the electroneutrality principle. A reason behind it is the fact that even the $\phi^2_{A\mu}$ density (of an electropositive atom)

Table 4. Atomic charges obtained from the virial and stockholder partitioning schemes*. All Δq values represent electron transfer from A to B in the molecule AB

Mole-cule	Δq^{pro}	$\Delta q^{vir.}$	$\Delta q^{mod.}$	Δq^S	ref.
LiH	0.644	0.911	0.267	0.414	a
BeH	0.750	0.868	0.118	0.193	a
BH	0.720	0.753	0.033	0.075	a
CH	0.019	0.032	0.013	−0.016	a
NH	−0.236	−0.322	−0.086	−0.091	a
OH	−0.428	−0.584	−0.156	−0.164	a
HF	0.585	0.759	0.174	0.228	a
NaH	0.356	0.810	0.454	0.413	a
MgH	0.508	0.796	0.288	0.282	a
AlH	0.613	0.825	0.212	0.228	a
SiH	0.627	0.795	0.168	0.125	a
PH	0.488	0.580	0.092	0.034	a
SH	0.123	0.094	−0.029	−0.050	a
HCl	0.108	0.240	0.132	0.124	a
LiF	0.650	0.937	0.287	0.624	a
LiF	0.607	0.940	0.333	0.619	b
BeF	0.777	0.945	0.168	0.328	a
BF	0.962	0.940	−0.022	0.118	a
CF	0.748	0.781	0.033	0.080	a
NF	0.324	0.439	0.115	0.112	a
NaF	0.425	0.941	0.516	0.677	b
AlF	0.711	0.974	0.263	0.357	b
LiO	0.651	0.932	0.281	0.580	a
BeO	1.151	1.692	0.541	0.647	a
BO	1.216	1.552	0.336	0.376	a
CO	1.220	1.346	0.126	0.139	a
CO	1.241	1.363	0.122	0.138	b
NO	0.380	0.495	0.115	0.086	a
MgO	0.680	1.413	0.733	0.678	b
SiO	1.184	1.633	0.449	0.461	b
LiN	0.666	0.916	0.250	0.542	a
BeN	0.731	1.236	0.505	0.415	a
BN	0.979	0.836	−0.143	−0.001	a
CN	0.810	1.123	0.313	0.198	a
PN	1.310	1.741	0.431	0.289	b
LiC	0.599	0.883	0.284	0.444	a
BeC	0.574	0.853	0.279	0.233	a
LiB	0.515	0.761	0.246	0.257	a
BeB	0.334	0.438	0.104	0.086	a
LiCl	0.624	0.926	0.302	0.553	b
NaCl	0.428	0.915	0.487	0.620	b

* $\Delta q^{mod.}$ refers to modified virial partitioning employing promolecule reference density. Δq^S denotes stockholder charges.[a] Ref. [101], [b] Ref. [102]

has appreciable values at the position of the (electronegative) neighbouring atom B much in agreement with Roby's discussion. Integration of $\phi_{A\mu}^2$ over a domain of the atom B ascribes much of its charge to the neighbouring nucleus B. This feature underlines the importance of the choice of a reference or gauge partitioning. One should always keep in mind that the definition of optimal reference superposition density is an open question and that the influence of atomic reorientation during formation of chemical bonds as well as hybridization (polarization) phenomenon should be explicitly considered and taken into account. Only in this case, can one extract a true charge-transfer effect.

A generalization of the partitioning of space introduced by Politzer [91] was suggested by Streitwieser et al. [75, 104]. He defined a projection function P(x, z):

$$P(x, z) = \int_{-\infty}^{+\infty} \varrho(x, y, z) \, dy, \tag{49a}$$

which in linear systems can be reduced to the electron count function G(z) (49) by additional integration. P(x, z) is a two-dimensional function which projects the total charge on the xz plane by integrating density over all y = const planes. Its volume gives a number of electrons and unlike the density ϱ itself it provides a representation of both σ and π electrons on the same footing. Namely, density $\varrho(x, y, z)$ is not suitable for that purpose because σ and π electrons have different nodal planes. Another distinct advantage of difference maps of the projection function is that it has zero net volume, i.e. charge can be gained in one region only at the expense of another region. Thus, difference maps give a direct measure of the amount of charge redistribution rather than the amount of density redistribution offered by DD maps. Examination of P(x, z) charge topology shows deep minima between bounded atoms thus enabling the determination of atomic charges. Difference of projection functions for two related molecules yields contours which by integration yield a change of particular atomic charge in similar moieties. For instance, carbonyl oxygen in acetaldehyde has more electron density than in formaldehyde [104]. The opposite is the case at the corresponding carbon atom. Integration of the projection function P(x, z) over the x-coordinate projects the total number of electrons on the (bond) z-axis. It should be pointed out that P(x, z) analysis can be applied to any molecule whereas the electron count function G(z) is suitable for linear molecules only. There are however some problems related to integrated projected electron populations (IPP). They are computationally efficient but the definition of IPP charges is somewhat arbitrary as discussed in detail by Glaser [105]. By using the gradient of the projection function $\nabla P(x, z)$ one can define projection critical points $(2, -2)_p$ and $(2, 0)_p$ which correspond to maxima and saddle points of P(x, z), respectively, in full analogy with Bader's critical points of $\nabla \varrho(x, y, z)$. A demarcation line between two atoms originates at the $(2, 0)_p$ point and follows the steepest descent path in both (opposite) directions. Generally, the $(2, 0)_p$ and $(3, 1)_B$ points do not coincide*. Secondly, the rotationally symmetrical zero-flux surface in diatomics is replaced by a vertical curtain. Hence IPP charges differ from Bader's topological estimates and are their approximations. These differences are not pronounced in bonds realized by two atoms largely different in electronegativity. Then values of density $\varrho(r)$ and P(x, z)

---

* See Ref. [46] for definition of critical points.

Table 5. Oxygen charges obtained by the 3-21G* set at optimized geometries estimated by various partitioning methods (in −e)

Compound	MPA		NPA		IPP		IBP	
	Q_0	ΔQ_0	Q_0	ΔQ_0	Q_0	ΔQ_0	Q_0	ΔQ_0
Formaldehyde	8.470	0	8.535	0	9.221	0	9.265	0
Acetaldehyde	8.552	0.082	8.564	0.024	9.229	0.008	9.292	0.027
Acetone	8.502	0.032	8.577	0.042	9.275	0.054	9.263	−0.002
Formic acid	8.597	0.127	8.652	0.117	9.299	0.078	9.346	0.081
Ketene	8.419	0.051	8.499	0.036	9.148	−0.073	9.324	0.059
Acrolein	8.543	0.073	8.558	0.023	9.206	−0.015	9.283	0.017
Carbonmonoxide	8.339	−0.131	8.571	0.036	9.223	0.002	9.386	0.121
Dimethyl ether	9.057	0.587	8.619	0.084	9.165	−0.056	9.323	0.058
Ethylene oxide	8.622	0.152	8.558	0.023	9.100	−0.121	9.177	−0.088

are low in the internuclear region and a precise location of the boundary surface is not crucial. However, if atoms possessing similar electronegativities are in question, then the electron density along the bond is increased and spatial position and shape of the partitioning surface is decisive. Bachrach and Streitwieser [106] examined atomic populations in some oxygenated compounds by using various partitioning techniques. Geometry optimizations was performed by the 3-21G* basis set. Comparison of oxygen atomic electron charges obtained by Mulliken, NPA (natural population analysis), IPP (integrated projection populations) and IBP (integrated Bader or zero-flux populations) is given in Table 5. Perusal of the data shows that the density integration methods IPP and IBP yield very high gross electron charges which are significantly greater than 9. Obviously these numbers have nothing to do with and are not good descriptors of the actual charge transfer (vide supra). Orbitally based gross atomic densities MPA and NPA are closer to empirical feeling and experimental knowledge. More physically meaningful numbers are obtained by taking formaldehyde as a reference molecule. MPA and NPA population differences ΔQ_0 are comparable at the qualitative level with notable exceptions of CO, dimethyl ether and ethylene oxide. Bader's IBP relative charges are in even better accordance with NPA ΔQ_0 populations if discrepancies observed in acetone and ethylene oxide are put aside. On the other hand IPP ΔQ_0 values are at variance with both NPA and IBP populations which casts some doubt on their usefulness. Therefore great care has to be exercised if IPP charges are employed in interpreting molecular properties.

3.5 Electronegativity Estimates of Atomic Charges

A driving force for intramolecular charge transfer is a difference in atomic electronegativities giving rise to an ionic character of chemical bonding. Electronegativity is an intuitively appealing concept defined by Pauling as "the power of an atom in a molecule to attract charge to itself" [55]. Pauling suggested that the bond energy enhancement in A − B molecule relative to the arithmetic mean of bond energies in homonuclear $A - A$ and $B - B$ diatomics can be used as a measure of electronega-tivity difference $\chi_A - \chi_B$:

$$(\chi_A - \chi_B)^2 = \Delta_{AB}/23.05 \,, \tag{50}$$

where $\Delta_{AB} = D(A - B) - (1/2)\,[D(A - A) + D(B - B)]$ and D stands for the dissociation energy. This seminal idea has received since then wide and continuing attention which proves both its elusiveness and utility. Relation (50) is empirical in nature. A theoretically sound definition was provided by Mulliken [107] which relates electronegativity to the first ionization potential I_A and the electron affinity A_A:

$$\chi_A^M = (I_A + A_A)/2 \,. \tag{51}$$

Other measures of electronegativity involve the force concept as advocated by Allred and Rochow [108]

$$\chi_A = 0.36 Z_{eff}/r_c^2 + 0.74 \,, \tag{52}$$

where the effective nuclear charge Z_{eff} was obtained by Slater rules and r_c denotes a covalent radius. Numerical coefficients put the electronegativity χ_A (52) in line with Pauling's scale. Gordy on the other hand preferred the idea of the electrostatic potential [109]

$$\chi_A = 0.62(Z_{eff}/r_c) + 0.50 \,. \tag{53}$$

Again, numerical parameters were optimized against Pauling's classical definition of electronegativity (50). All these scales are more or less linearly related to each other. A conceptual step forward was made by Iczkowski and Margrave [110] who were the first to realize that atomic energy is a function of charge density

$$E_A(Q_A) = a_A + b_A Q_A + c_A Q_A^2 + \ldots \tag{54}$$

Further, the electronegativity is defined by

$$\chi_A(Q_A) = \partial E_A/\partial Q_A = b_A + 2c_A Q_A \,, \tag{55}$$

and we shall see that it coincides with Mulliken's formula (51). It is important to notice that the form (55) is easily amenable to generalizations being applicable to much more sophisticated expressions for total energy. The next conceptual jump was performed by Jaffé et al. [111] who introduced orbital electronegativity

$$\chi_{A\mu} = \partial E_A/\partial Q_{A\mu} \,, \tag{56}$$

where $Q_{A\mu}$ has the meaning of electron occupancy* of the AO $\phi_{A\mu}$. One can write in full analogy with Eq. (55)

$$\chi_{A\mu}(Q_{A\mu}) = b_{A\mu} + 2c_{A\mu} Q_{A\mu} \,, \tag{57}$$

---

* It should be pointed out however that orbital occupancies $Q_{A\mu}$ and gross atomic populations Q_A here involve exceptionally the sign of the electron charge in order to put Eqs (55–63) into the customary form.

which is related to orbital energy

$$E_{A\mu}(Q_{A\mu}) = a_{A\mu} + b_{A\mu}Q_{A\mu} + c_{A\mu}Q_{A\mu}^2, \tag{58}$$

substituting $Q_{A\mu} = 0, -1, -2$ in Eq. (58) and taking into account definition of the ionization potential and electron affinity

$$I_{A\mu} = E_{A\mu}(0) - E_{A\mu}(-1) \quad \text{and} \quad A_{A\mu} = E_{A\mu}(-1) - E_{A\mu}(-2), \tag{59}$$

one obtains straightforwardly

$$b_{A\mu} = (1/2)\,(3I_{A\mu} - A_{A\mu}), \qquad c_{A\mu} = (1/2)\,(I_{A\mu} - A_{A\mu}). \tag{60}$$

Mulliken's definition of electronegativity corresponds to singly occupied AO $\phi_{A\mu}$ which by using Eq. (57) yields:

$$\chi_{A\mu}(-1) = b_{A\mu} - 2c_{A\mu} = (1/2)\,(I_{A\mu} + A_{A\mu}). \tag{61}$$

Hence the Iczkowski-Margave energy formula (54) terminating after the quadratic term together with the electronegativity definition (55) is indeed equivalent to Mulliken formula (51). Applying Sanderson's argument that electronegativity of all atoms or groups in a molecule should be equal [112], Jaffe et al. obtained

$$b_{A\mu} + 2c_{A\mu}Q_{A\mu} = b_{B\nu} + 2c_{B\nu}Q_{B\nu}, \tag{62}$$

which yields gross atomic charges in a singly bonded A − B molecule:

$$Q_{A\mu} = \frac{(1/2)\,(b_{B\nu} - b_{A\mu}) - 2c_{B\nu}}{c_{A\mu} + c_{B\nu}}, \tag{63a}$$

$$Q_{B\nu} = \frac{(1/2)\,(b_{A\mu} - b_{B\nu}) - 2c_{A\mu}}{c_{A\mu} + c_{B\nu}}, \tag{63b}$$

It should be stressed that orbital electronegativities by Jaffé et al. are strongly dependent on both occupancy and orbital hybridization (through coefficients $b_{A\mu}$ and $c_{A\nu}$). Electronegativity of an empty orbital is greater than that of an orbital possessing a single electron because the former is more favourable for placing a negative charge. The electronegativity of the orbital containing two electrons of opposite spins should be zero because of the Pauli principle. The coefficient $c_{A\mu}$ ensures that electronegativity decreases as electron population increases. It is noteworthy that Jaffé et al. introduced a name bond electronegativity for $\chi_{A\mu}$ (57) and defined the ionic characters of bonds in terms of electronegativity differences. One should also mention that the orbital electronegativity concept was suggested first by Pritchard and Sumner [113]. Introduction of (hybrid) orbital electronegativity offers additional flexibility for a description of different bonds emanating from the same atom. Concomitantly, one can talk about variable electronegativity for different spatial directions thus yielding a more refined picture of modified atoms in molecules (MAM) [54, 114]. It is noteworthy that instead of the gross atomic charge Q_A one can employ net atomic charge q_A in formula (54) and (55) and define the energy E_A of a neutral atom as the origin of the energetic scale. Then the formulae (55–63b) are somewhat different but the main features remain the same. The equalization of the electronegativities of all atoms in a molecule gives a set of $n - 1$ simultaneous equations for atoms, which together with the conservation of charge condition enable estimates of atomic charges. Huheey presented a method for calculating group electronegativities of molecular fragments [115]. This approach is a combination of the Iczkowski-Margrave and Jaffe et al. concepts. He assumes also a complete equalization of electronegativity of all atoms to an average value, but this requirement was occasionally relaxed. Huheey's method proved useful in calculating total molecular energy and its partition into three contributions: (a) single-center charge transfer effect measured by electronegativity, (b) Madelung electrostatic term and (c) covalent interaction due to orbital overlapping. He found this analysis illuminating and practical [115]. Additionally, Huheey provided the largest library of group electronegativities [116]. He found a correlation between charges on the carbethoxy group estimated by electronegativity and polar substituent constants σ^* [117]. Interestingly, partial electronegativity equalization of 80% gave significantly better correlation. For our purpose particularly interesting are Huheey's results on some small linear molecules [118] which can be favourably compared with Politzer's spatial partitioning atomic charges (Table 6). Agreement of electronegativity charges with the latter is unexpectedly good. This is of utmost importance since electronegativity equalization calculations are easily made and they are not restricted to linear molecules as Politzer's method. Performance of such simple models underlines the relevance of interpretive quantum chemistry indicating that good thinking often yields qualitative but penetrating insight into molecular properties. We note in passing that MPA analysis gives results at variance with the two above-mentioned methods, HF being a classical case for its pathological sensitivity on basis sets (vide supra).

A renaissance of interest in electronegativity was triggered recently by development of the density functional theory (DFT) [120–122]. It has led to a number of useful concepts. Some of them will be briefly mentioned here. The absolute electronegativity χ and absolute hardness η are defined as:

$$\chi = -|\partial E/\partial N|_Z = -\mu, \qquad \eta = (1/2)\,|\partial^2 E/\partial N^2|_Z, \tag{64}$$

where E is the electronic energy of a molecule, atom or ion, N is a number of electrons and Z denotes a fixed set of nuclear charges. It appears that the absolute electronegativity is equal to the negative chemical potential μ which in turn is unique everywhere in a molecule. One easily finds out that the absolute electronegativity is equal to Mulliken's value

$$\chi = (1/2)\,(I + A) \quad \text{whereas} \quad \eta = (1/2)\,(I - A). \tag{65}$$

Further, for small changes in N one can write:

$$\mu = \mu^0 + 2\eta\,\Delta N, \tag{66}$$

Table 6. Comparison of net atomic charges obtained by Huheey's electronegativity equalization, Politzer's spatial partitioning and Mulliken population analysis

Molecule	Atom	Huheey[a]	Politzer[b]		MPA[c]	
HF	H	0.29	0.26[d];	0.27[e]	0.23[d];	0.48[e]
	F	−0.29	−0.26[d];	−0.27[e]	−0.23[d];	−0.48[e]
HCL	H	0.16	0.17[f]		−	
	Cl	−0.16	−0.17[f]		−	
LiF	Li	0.62	0.52[g]		−	
	F	−0.62	−0.52[g]		−	
LiH	Li	0.48	0.37[h]		−	
	H	−0.48	−0.37[h]		−	
NaF	Na	0.63	0.62[g]		−	
	F	−0.63	−0.62[g]		−	
NaCl	Na	0.72	0.60[g]		−	
	Cl	−0.72	−0.60[g]		−	
HCCH	H	0.14	0.14[i]		0.22[i]	
	C	−0.14	−0.14[i]		−0.22[i]	
HCCLi	H	0.03	0.10[i]		0.17[i]	
	C	−0.17	−0.23[i]		−0.41[i]	
	C	−0.32	−0.36[i]		−0.41[i]	
	Li	0.46	0.49[i]		0.71[i]	
HCCF	H	0.14	0.15[i]		0.25[i]	
	C	−0.17	−0.19[i]		−0.17[i]	
	C	0.11	0.09[i]		0.22[i]	
	F	−0.07	−0.05[i]		−0.31[i]	
HCCCl	H	0.12	0.15[i]		−	
	C	−0.20	−0.15[i]		−	
	C	0.05	−0.02[i]		−	
	Cl	0.04	0.03[i]		−	

[a]Ref. [118], [b]Spatial partitioning described in Ref. [91]; [c]Mulliken population analysis, [d]Nesbet basis set [92]; [e]Clementi basis set [92], [f]Politzer P and Meroney BF as cited in ref. [119], [g]Ref. [120], [h]Ref. [92], [i]Ref. [91].

in analogy to Eq. (55). At equilibrium two chemical potentials μ_A and μ_B in A–B molecule are equal giving rise to a charge shift

$$\Delta N = (\chi_A^0 - \chi_B^0)/2(\eta_A + \eta_B), \tag{67}$$

where $\chi_A^0 = -\mu_A^0$. This electron transfer leads to energy lowering

$$\Delta E = -(\chi_A^0 - \chi_B^0)^2/4(\eta_A + \eta_B), \tag{68}$$

indicating that electronegativity difference stimulates the electron transfer whereas the sum of hardness parameters has an inhibitory influence [121]. Electronegativities of atoms are totally equalized in molecules or fragments [120, 122]. This is related to exact electron densities and equilibrium nuclear positions, but in simpler theoretical approaches a requirement of full equalization is probably too stringent. The first to criticize full electronegativity equalization was Pritchard [123] and claimed that about 10% of the original difference remains persistently unbalanced. Other authors also noted a necessity of partial equalization within simple schemes [115–117, 124–126]. Gasteiger et al. [125, 126] devised a PEOE (partial equalization of orbital electronegativity) scheme based on an iterative damping procedure which prevents excessive charge transfer. Good correlations with ESCA shifts, pk_a values, and even with $J(C-H)$ coupling constants were obtained.

There is a number of applications of electronegativity in describing molecular properties. The interested reader is encouraged to consult Structure and Bonding 66 (1987) [127] which contains several scholarly written review articles. We would like to emphasize that simple electronegativity schemes are an invaluable tool in providing approximate atomic charges in large biological systems and in their dynamics simulations. Charge transfer in solvent-substrate interactions can also be efficiently taken into account. As just one example of the electronegativity equalization method we shall mention charge calculations on deoxyribose, the dipeptide and water dimers [128]. A desirable feature of the approach is the Madelung correction of atomic electronegativities. Results can be favourably compared with available ab initio data.

3.6 Deduction of Charges by Fitting Functionals and One-electron Properties

Point-charge models of electron density distributions received a considerable boost by the work of Hall et al. [129, 131–133]. Hall and Martin [129] employed the Dirichlet minimum energy theorem of classical electrostatics to obtain optimal point charges. If the true electric field is $\vec{E}(\vec{r})$ and its point charge approximation is $\vec{E}_{app}(\vec{r})$ then the functional

$$F = (1/8\pi) \int (\vec{E} - \vec{E}_{app})^2 \, d\vec{r}, \tag{69}$$

measures the accuracy of the approximation. Since it is positive, F should be kept at minimum. Functional (69) can be expressed in terms of electron density distribution

$$F = (1/2) \int \int (\varrho(1) - \varrho_{app}(1)) (\varrho(2) - \varrho_{app}(2)) (1/r_{12}) \, d\vec{r}_1 \, d\vec{r}_2 . \tag{70}$$

It appears that minimization of F is a fitting procedure of the electron density in a least squares sense with $1/r_{12}$ operator as a weighting function. Let us try to represent ϱ_{app} by point charges:

$$\varrho_{app}(\vec{r}) = \sum_A q_A \, \delta(\vec{r} - \vec{r}_A) . \tag{71}$$

Then optimal charges are obtained by a system of equations

$$(\partial F/\partial q_A) = -V_A(\vec{r}_A) + \sum_{\substack{B \\ (B \neq A)}} q_B/R_{AB} = 0 , \tag{72}$$

where V_A is the true electrostatic potential. In other words optimal charges reproduce exact potential at each charge which may or may not coincide with nuclear charges. Furthermore, derivatives of Eq. (72) with respect to $\partial/\partial r_A$ also vanish, implying that the electric field deduced from optimal charges perfectly matches the exact electric field produced by the true electron density. A possibility of accurate calculation of the electrostatic potentials exerted on the nuclei is intriguing because they are closely related to total molecular energies [130]. On the other hand the number of fractional point charges in Eq. (71) is by no means limited and their position can cover all critical regions in a molecule. Approximate density distribution can be expressed in orbital form

$$\varrho_{app}(\vec{r}) = \sum_A \sum_\mu Q_{A\mu}\, \phi^2_{A\mu}(\vec{r})\,, \tag{73}$$

which together with charge conservation $Q = \sum_A \sum_\mu Q_{A\mu}$ constraint yields a system of equations [131]:

$$(\partial F/\partial Q_{A\mu}) = \sum_B \sum_v \langle \phi_{Bv}\phi_{Bv} \mid \phi_{A\mu}\phi_{A\mu} \rangle Q_{Bv} - \sum_B \sum_C \sum_v \sum_\eta P_{v\eta}$$

$$\times \langle \phi_{Bv}\phi_{C\eta} \mid \phi_{A\mu}\phi_{A\mu} \rangle - \lambda = 0\,, \tag{74}$$

where $\langle \phi_{Bv}\phi_{C\eta} \mid \phi_{A\mu}\phi_{A\mu} \rangle$ are Coulomb integrals, λ is the Lagrange multiplier, and $P_{v\eta}$ are elements of the density matrix in the accurate distribution $\rho(\vec{r})$ $= \sum_B \sum_C \sum_v \sum_\eta P_{v\eta}\phi_{Bv}(\vec{r})\, \phi_{C\eta}(\vec{r})$. This type of orbital population analysis is intellectually pleasing but unfortunately the available results are based on rather modest basis sets STO–NG (N = 2, 3, 6) [131]. But its virtues have been demonstrated with Li–H where the Mulliken population gave an unrealistic positive charge on the H atom whereas optimal populations reversed the sign of charge transfer. An important outcome was the discussion of the hybridization influence on charge distribution in the H_2O molecule. Namely, the approximate density ϱ_{app} (73) involves only squares of AOs. Inclusion of an additional term describing single-center mixed densities $\phi_\mu\phi_v(\mu, v\varepsilon A)$ only slightly changes a set of eqs. (74) but greatly improves fitting by diminishing the value of the functional F (70).

Another interesting observation made by Hall [132, 133] was based on the use of floating spherical Gaussian orbitals (FSGO). The closed shell electron density distribution in SCF-FSGO formalism is

$$\varrho(\vec{r}) = \sum_{\mu v} P_{\mu v}G_\mu(\vec{r})\, G_v(\vec{r})\,. \tag{75}$$

It is a well known property of Gaussian functions that the product of their pairs forms another spherical Gaussian placed at $\vec{r}_{\mu v} = (\alpha_\mu\vec{r}_\mu + \alpha_v\vec{r}_v)/(\alpha_\mu + \alpha_v)$, where α_μ and α_v are their nonlinear parameters. Hence $\varrho(\vec{r})$ is essentially a sum of spherical

Gaussians. Concomitantly, Hall suggests fitting of $\varrho(\vec{r})$ by point charges $P_{\mu\nu}S_{\mu\nu}$ centered at $\vec{r}_{\mu\nu}$ positions:

$$\varrho_{app}(\vec{r}) = \sum_{\mu\nu} P_{\mu\nu}S_{\mu\nu}\delta(\vec{r} - \vec{r}_{\mu\nu}), \tag{76}$$

where $S_{\mu\nu}$ is a customary overlap integral. Thus spherical Gaussians are shrunk into delta functions, which preserves not only the monopole, but all spherical moments. The number of point charges $P_{\mu\nu}S_{\mu\nu}$ is $n(n + 1)/2$ where n is determined by the number of basis set functions. They do not coincide with nuclei in general. Hall has shown that ϱ_{app} (76) yields the extramolecular electrostatic potential outside the Van der Waals radii with a good accuracy [132, 133].

The number of point charges implied by the $\varrho_{app}(\vec{r})$ distribution (76) can be considerably reduced by their contraction to orbital populations of G_μ suggested by Shipman [134]:

$$Q_\mu = \sum_\nu 2P_{\mu\nu}S_{\mu\nu}\alpha_\mu/(\alpha_\mu + \alpha_\nu), \tag{77}$$

meaning that overlap charge is divided proportional to the magnitude of nonlinear parameters and properly normalized. On can easily check that partitioning (77) preserves total charge and dipole moments. However, the calculated extramolecular electrostatic potentials calculated by Shipman's charges are inferior to potentials estimated by Hall's multi-point charges indicating that a high price had to be paid for the achieved simplicity. A similar idea to that of Shipman was pursued by Huzinaga and Narita [135].

Spherical Gaussians and their populations could be fitted into accurate density by their substitution into ϱ_{app} (73) and subsequent minimization of the functional F (70). A detailed study of a number of Gaussians necessary for a good fitting of density and acceptable one-electron properties was performed by Hall and Smith [136]. A mixed Gaussian model suggested by Hall and Tsujinaga [137] is based on a simple physical picture of well localized electrons belonging to inner shells, lone and bond pairs which are well described by point charges. However, peripheral parts of a molecule should be described by diffuse spherical Gaussians which cannot be contracted to single point(s). A use of just one diffuse Gaussian greatly improved representation of molecular electrostatic potentials [137].

Fitting of the total molecular densities by using spheres of adjustable radii centered on the nuclear positions were discussed by several researchers [138–140]. They were applied for extracting atomic charges in molecular and ionic crystals and in analysis of ab initio density distributions. Shifting of the hydrogen sphere along the O–H bond was found advantageous in the H_2O molecule [141]. The projection of molecular charge density into spherical atoms was put into an elegant formalism by Yanez et al. [142, 143], employing spherical Gaussians. Atomic density basis functions were obtained by the least squares optimization. Scale factors were used in order to guarantee the adequate contraction or expansion of spherical densities in various chemical environments. This type of population analysis proved useful in rationalizing

[13]C chemical shifts [144] and inductive effects [145]. For the sake of completeness we mention the method of Larsson and Braga [146] which involves expansion of molecular orbitals in spherical hamonics around atomic centers. The expansion coefficients are functions of the distance from the nucleus and the quotient between this function and a corresponding atomic orbital is almost constant in the core region. The square of the quotient defines an atomic charge component. Unlike Mulliken's populations these charges are not basis set dependent.

In concluding discussion of atomic charges and orbital populations deduced by fitting procedures of the Dirichlet functional and one-electron density distributions a word on the generalization of Hall's method is in place. Instead of spherical Gaussians (75) one can use Cartesian GTOs. Then the electron density distribution can be expressed as

$$\varrho(\vec{r}) = \sum_{\mu\nu} P_{\mu\nu} D_{\mu\nu}(\nabla)\, G(\vec{r} - \vec{r}_{\mu\nu}), \tag{78}$$

where $D_{\mu\nu}(\nabla)$ is an appropriate function of differential operators $(\partial/\partial x)$, $(\partial/\partial y)$ and $(\partial/\partial z)$. The point charge model, which should closely match the accurate density, is obtained by replacing spherical Gaussians by the delta functions [154].

$$\varrho_{app}(\vec{r}) = \sum_{\mu\nu} P_{\mu\nu} D_{\mu\nu}(\nabla)\, \delta(\vec{r} - \vec{r}_{\mu\nu}). \tag{79}$$

Approximation (79) preserves all the spherical moments indicating that molecular electrostatic potentials are well reproduced at larger distances. This conjecture was confirmed by actual calculations which have shown that Mulliken and Shipman charges give the wrong sign of the potential in some important regions, whereas the extended Hall model (79) yields good results [147].

It was mentioned earlier that the electron density ϱ is relatively little changed by chemical bonding. These subtle changes relative to free atom densities can be magnified and made more visible by some one-electron properties operators which weigh different segments of molecular volume. It is well known that the molecular dipole is quite sensitive on the finer details of electron charge distributions. It can be expressed as a sum of two contributions — atomic monopole and atomic dipole terms. For an A−B diatomic molecule the dipole moment reads:

$$\mu = (Z_B - Q_B) R_{AB} + \mu_A + \mu_B = \mu_{mon} + \mu_A + \mu_B, \tag{80}$$

where it was tacitily assumed that the origin of the coordinate system coalesces with the nucleus A. The atomic monopole or charge transfer contribution is denoted by μ_{mon}. Local atomic dipole moments μ_A and μ_B are defined by Eq. (48). They correspond to atomic hybridization (polarization). If they could be neglected, then atomic charges could be straightforwardly calculated from the known (experimental or theoretical) dipole moments. The fact is that atomic dipoles are quite appreciable. This is illustrated by ab initio results on some diatomics (Table 7) obtained by charge distribution partitioning of Bader et al. [93] defined by the point along the internuclear axis exhibiting a density minimum (Section 3.4). A survey of the presented data shows

Table 7. Partitioning of molecular dipole moments in some diatomics (in Debyes)

Molecule AB	State	μ_{mon}	μ_A	μ_B	μ_{tot}
LiF	$X^1\Sigma^+$	6.34	0.14	−0.19	6.29
BeO	$X^1\Sigma^+$	7.76	0.20	−0.52	7.44
LiO	$X^2\Pi_i$	6.67	0.19	−0.05	6.81
BeN	$^2\Pi_i$	8.05	−0.84	−1.46	5.74
BF	$X^1\Sigma^+$	4.47	−4.46	−0.95	−0.94
CO	$X^1\Sigma^+$	5.00	−3.72	−1.01	0.27
BeF	$X^2\Sigma^+$	5.31	−2.04	−0.82	2.45
BeO	$^3\Sigma^-$	0.39	−5.26	−0.10	−4.97

that atomic dipoles can outweigh monopole term μ_{mon} like in BF and BeO ($^3\Sigma^-$) molecules strongly indicating that atomic charges cannot be derived from molecular dipole moments. Results for CO and BeF support this conclusion. A recent study of Bader et al. [148] has shown that μ_{mon} and atomic polarization dipoles μ_A are of equal importance in determining both static dipole moment an the dynamic molecular dipole induced by nuclear displacements.

The next widely used criterion for production of point charges is the molecular electrostatic potential [149–152]:

$$V(\vec{r}) = \sum_A Z_A/|\vec{r} - \vec{R}_A| - \sum_{\mu\nu} P_{\mu\nu} \int [\phi_\mu \phi_\nu/|\vec{r} - \vec{r}'|] \, d\vec{r}', \qquad (81)$$

which is fitted by the approximate potential of desired point charges.

$$V_{app}(\vec{r}) = \sum_A q_A/|\vec{r} - \vec{R}_A|, \qquad (82)$$

optimized by the least-squares fit:

$$\sum_{i=1}^{m} w_i[V(\vec{r}_i) - V_{app}(\vec{r}_i)]^2 - \lambda \left[\sum_A q_A - C\right] = \min, \qquad (83)$$

where λ is the Lagrange multiplier and C is the molecular total charge $C = \sum_A q_A$ being zero for neutral molecules. Positions $\vec{r}_i$ are related to m-sampling grid points which are chosen outside the molecular Van der Waals region. Weighting factors w_i are sometimes used to discriminate closer contacts with the positive probe charge which describes interaction with the extramolecular electrostatic potential, but usually they are put equal to one. Electrostatic potential $V(\vec{r})$ (81) is a well defined physical property which can be experimentally determined by using X-ray densities [43]. As a one-electron property it is an order of manitude more accurately determined than the corresponding wavefunction. Nevertheless, the electrostatic potential derived (ESPD) charges (82) are not free of criticism. In the inaugurational paper by Momany [149] an additional constraint is introduced in Eq. (83) which is related to a conservation of the components of the total dipole moment. This is unjustified in most molecules since atomic dipole moments arising from hybridization are not negligible. On the contrary, they yield a substantial contribution as discussed earlier. Hence, Momany's procedure is in fact too much constrained by the dipole moment. This requirement was abandoned by other researchers [150, 151] but the resulting dipole moments calculated by ESPD charges are in good agreement with the molecular dipoles computed as an expectation value of the corresponding wavefunctions particularly if larger basis sets are used. Surprisingly, this finding was used as a decisive support for ESPD charges. In reality, it actually speaks against them [114]. Let us consider a few examples which illustrate the point (Table 8). ESPD charges are compared with Politzer spatial partitioning results, modified Bader density integration technique which employs promolecule density as a references and charges obtained by the stockholder allocation of density. The last three methods yield consistent results for CO, HF and HCL. On the other hand ESPD charges are widely different, being too large by factors varying between 1.5–3.0 thus exaggerating the intramolecular charge transfer. It is easy to see that ESPD charges have little physical

Table 8. Comparison of ESPD charges with results of other partitioning methods for some small molecules

Molecule	Atom	Politzer[a]	MBP[b]	Stockholder[c]	ESPD[d]
CO	C	0.145	0.122	0.139	–
	O	−0.145	−0.122	−0.139	–
HF	H	0.26	0.174	0.228	0.462(0.442)
	F	−0.26	−0.174	−0.228	−0.462(0.442)
HCl	H	0.16	0.132	0.124	0.246
	Cl	−0.16	−0.132	−0.124	−0.246
HCCH	H	0.14	–	0.094	0.292(0.297)
	C	−0.14	–	−0.094	−0.292(−0.297)
HCN	H	0.18	–	0.133	–
	C	0	–	0.066	–
	N	−0.18	–	−0.201	−0.430(−0.514)[e]
OCO	C	0.46	–	0.413	0.908(0.868)
	O	−0.23	–	−0.208	−0.454(−0.434)
H$_2$CO	H	–	–	0.052[g]	−0.037(0.018)
	C	–	–	0.173	0.578(0.423)
	O	–	–	−0.276	−0.502(−0.459)
HCOOH[f]	H	–	–	0.079[g]	0.461(0.059)
	C	–	–	−0.272	0.781(0.674)
	O	–	–	−0.330	−0.598(−0.568)

[a] Politzer's integration density method [91];
[b] Modified Bader partitioning relative to promolecule density;
[c] Stockholder partitioning [86] obtained by experimental densities except in H$_2$CO and HCOOH molecules where theoretical extended DZ set was utilized;
[d] Electrostatic potential derived charges [151] calculated by the 6-31G** basis set. Values in parentheses refer to results of Cox and Williams [150];
[e] Nitrogen charge in CH$_3$CN;
[f] These atoms refer to the formyl group H − C = O;
[g] Eisenstein M, Hirshfeld FL (1983) J Comp Chem 4 : 15.

significance. As a matter of fact, the extramolecular electrostatic potentials cannot be calculated satisfactorily by atomic monopoles. Accurate reproduction of extra-molecular potentials requires inclusion of atomic monopoles, dipoles and quadrupoles [153, 154] or multipole expansion of overlap charges in addition [155]. A brief analysis presented above supports this conjecture. A statement of Cox and Williams [150] that ESPD charges are close to stockholder charges is obviously not correct. A fact that ESPD charges reproduce rather well molecular dipoles and quadrupoles is a crown argument that they are not adequate [152]. These too entities namely cannot be sufficiently accurately expressed by atomic charges placed at the nuclei. It can be easily shown that the diagonal elements of the molecular quadrupole tensor are of the form:

$$Q_{xx} = \frac{|e|}{2} \sum_A [(Z_A - Q_A)(3R_{Ax}^2 - R_A^2) + 2(3R_{Ax}\mu_{Ax} - \vec{R} \cdot \vec{\mu}_A) + Q_{xx}^A],$$

(84)

where R_{Ax} denotes nuclear coordinates, μ_{Ax} and Q_{xx}^A are components of local atomic dipole and quadrupole moments. The first term in (84) corresponds to the point-charge approximation. It appears that the second term is by no means negligible.

Another undesirable feature of ESPD charges is that they depend on the choice of grid points [149, 150] (see Table 8). It seems that sampling of Cox and Williams is more satisfactory [150]. However, the electrostatic potential alone can not give a unique resolution of charge density into atomic multipoles (vide supra). There is a number of different selections of atomic multipoles with the same ESP fit (see e.g. [168]).

4 Experimental Determination of Atomic Charges

A comprehensive presentation of the experimental methods used in determining atomic charges is unfortunately prohibited by limited space. Instead, some results will be briefly discussed since their complete omission would be inexcusable.

Experimental determination of atomic charges by ED, X-ray or a combined use of X-ray and neutron scattering techniques has been mentioned and compared with theoretical charges several times earlier. Therefore it should be pointed out here that a direct way of extracting atomic monopoles does not exist and that there is always a need of fitting the experimental density by some one-center or two-center [156] model representations, by some spatial or functional partitioning etc. Stockholder partitioning is not unreasonable in this respect, but a great care in defining the reference density of prepared-for-bonding atoms has to be exercised. In some cases the population of AOs can be obtained [43].

Considerable attention has been devoted lately to a production of atomic charges from infrared (IR) intensities [157–162]. It is based on the resolution of molecular dipole moments into bond contributions

$$\vec{\mu} = \sum_i \mu_i \vec{e}_i,$$

(84)

where μ_i is the absolute value of the bond moment "i" and e_i is a unit vector coinciding with the straight line passing through the bonded nuclei. Unlike the expression of the molecular dipole by the monopole (charge-transfer) and hybridization terms (80), partitioning to bond dipoles (84) is arbitrary. Nevertheless, it sometimes serves a useful purpose [163]. IR intensities are related to dipole moment derivatives with respect to the internal vibrational coordinates R_v

$$(\partial\vec{\mu}/\partial R_v) = \sum_i [\vec{e}_i^0(\partial\mu_i/\partial R_v) + \mu_i^0(\partial\vec{e}_i/\partial R_v)] , \qquad (85)$$

where μ_i^0 and $\partial\mu_i/\partial R_v$ are the so-called electrooptical parameters (EOP). Further, it is assumed that the equilibrium bond dipole can be expressed in the point charge approximation:

$$\mu_i^0 = q_i^0 r_i^0 , \qquad (86)$$

where q_i^0 is the partial charge residing on the bonded atoms (with $\pm$ signs) and r_i^0 is the equilibrium bond distance. Derivatives of bond dipoles

$$\partial\mu_i/\partial R_v = [(\partial q_i/\partial R_v) - q_i^0\delta_{iv}]/r_i^0 , \qquad (87)$$

are given by the atomic charge flux $\partial q_i/\partial R_v$ and atomic charge q_i^0. It should be pointed out that $R_v = r_i$ refers to the bond stretching which is most important as a rule. EOP charges within the framework stated above can be deduced from the static molecular dipole and IR intensities provided that accurate experimental data are available for various isotopes and within a family of related molecules. In addition a reliable force field must be used. Even if all these requirements are fulfilled the physical meaning of EOP charges is questionable. An immediate objection is the use of the atomic monopole approximation (86) whereas it is a common knowledge that the charge-transfer and atomic hybridization terms are equally important. A less obvious shortcomming is a constraint that bond dipoles are directed along the corresponding internuclear axis and that they rigidly follow bending motions of the latter. This is not justified as discussed by Bader et al. [164]. Hence the derived EOP charges have limited significance although the contrary was claimed by Gussoni et al. [158]. Adopting "experimental" EOP charges as a standard, a modified Mulliken population analysis was designed, which reproduces molecular dipole moments, by adding a specific overlap term [161, 162]. It was shown that corrected Mulliken populations, EOP and ESPD derived atomic charges give very similar values. This is in accordance with objections to EOP charges raised above. Since we have established that ESPD charges, which in turn reproduce molecular dipole (and quadrupole) moments in the monopole approximation quite closely, are about 1.5–3.0 times too large, it follows that EOP estimates are equally unreliable.

A more careful counterpart analysis within the atomic polar tensor (APT) formalism was performed yielding APT of an atom as a sum of three terms, only one of them being related to the equlibrium atom charge [165]. Cioslowski [166] introduced a generalized atomic polar tensor (GAPT) and defined atomic charge by:

$$q_{GAPT}^A = (1/3)\,[(\partial\mu_x/\partial x_A) + (\partial\mu_y/\partial y_A) + (\partial\mu_z/\partial z_A)]\,. \tag{88}$$

This is again open to criticism because neither dipole moment nor its (Cartesian) derivatives are dependent on the charge transfer term only. Furthermore, atomic density is not fixed leading to additional contributions. It is difficult to assess the validity of the relation (88). As a simple illustrative example we mention q_{GAPT} charges [167] in CO (± 0.406) which are, by a factor 3, larger than the relatively reliable charges offered by three different methods (Politzer, MBP and stockholder) presented in Table 8. An interesting discussion of derivatives of the components of the molecular dipole moments over the corresponding Cartesian coordinates was given by Dinur and Hagler [168]. It was shown that for linear and planar rigid molecules a derivative $\partial\mu_z/\partial z_A$ for the out-of-plane (axis) coordinate is equal to the effective or apparent charge q_A^{ef}, which in turn is a sum of the monopole q_A^0 and a derivative $\partial\sum\mu_{Az}/\partial z_A$, i.e. $q_A^{ef} = q_A^0 + \partial\sum_A\mu_{Az}/\partial z_A$. Here, μ_{Az} is the z-component of the atomic dipole. These two terms are inseparable indicating that the formula (88) for q_{GAPT}^A does not directly yield true atomic charges q_A^0. Effective charges of Dinur and Hagler in HF are ± 0.449 [168] being very close to ESPD charges thus showing once again that the latter have nothing to do with reliable atomic monopoles. Additional data for formaldehyde and formamide suport this conclusion. We note in passing that Dinur and Hagler perfomed fitting of the electrostatic potential of H_2O by widely different sets of atomic multipoles. It appears that several of them yield essentially the same ESP fit indicating that this type of atomic multipole has little physical meaning.

The most realistic atomic charge are probably given by ESCA technique [169]. It can be chown that binding energy shifts (ΔBE) of the localized inner core electrons are intimately related to electrostatic potential exerted at the ionized nucleus [169 – 171] which we give here in the atomic monopole approximation:

$$\Delta BE_A = k_1 Q_A + k_3 M_A + k_4\,, \tag{89}$$

where $M_A = \sum_B{}' (Z_B - n_B - Q_B)/R_{AB}$ is the Madelung potential and n_B is the number of the inner-shell electrons in the ground state of the neutral molecule. The valence shell gross atomic charges are denoted by Q_A etc. Empirical constants k_i are adjusted by the least square fit. They put electrostatic potentials in line with the energy shifts which are measured relative to a reference compound. Several remarks are in place here. Firstly, unlike the extramolecular molecular potential which is poorly represented by point-charges (unless specially derived for that purpose) the electrostatic potentials exerted at the nuclei are well reproduced by atomic net monopoles [172] in a fashion analogous to the formula (89). The weighting factors K_1, K_3 and K_4 were employed in order to remedy imperfections introduced by the Mulliken approximation. An important finding was made that the semiempirical SCC-MO atomic charges have the same performance as the ab initio DZ ones [172] which provided the basis of the SCC-MO atomic monopole electrostatic potential model (SCC-AMEP) for calculation of ESCA shifts. This simple approach proved very useful in rationalizing inner-shell binding energies in a large variety of molecules involving the first and second row atoms [173]. It should be pointed out that formula (89) is developed within the ground state potential approach. However, in spite of the fact that the photoionization is a very fast process it is not instantaneous. Consequently, the positive hole causes reorganization of the electron charge distribution which affects the final energy balance. In other words, the experimental ΔBE shifts involve a relaxation effect implying that the ground state atomic charges Q_A are contaminated by the charge distribution in the molecular ion. Hence, reorganization correction has to be explicitly taken into account which can be done in a rigorous ab initio fashion or semiempirically e.g. by equivalent core approximation. We defer a detailed discussion to the chapter 8 of this book [174]. Nevertheless, it should be stressed here that in principle it is possible to circumvent the relaxation trap by making use of the Manne-Åberg theorem [175] which states that the average energy of an inner-shell ionization spectrum involving the main peak and all shake-up and shake-off satellites equals the ionization energy of the frozen initial ground state. This energy is directly related to the ground state atomic charges. Experimentally this is a very difficult task but it is not impossible. Another route of attack of the relaxation problem is by measuring the Auger spectra [176, 177]. It appears that the Auger parameter α is closely related to reorganization energy. In our opinion this is the best way of obtaining experimental atomic charges. Of course, ESCA charges are extracted from XPS spectra via a chain of approximations:

$$\Delta BE \sim \text{electrostatic potential} \sim \text{monopole approximation.}$$

Nevertheless, they seem to be quite realistic.

5 Applications and Discussion

Intramolecular charge transfer is one of the essential ingredients of chemical bonding whenever heteroatoms are involved. Concomitantly, it affects more of less all chemical and physical facets of molecules including bond distances, bond energies and heats of formation, electric and magnetic properties etc. It is impossible to give here a comprehensive review. Instead we shall focus attention on a few outstanding features only.

5.1 Covalent vs Ionic Bonds

It is useful to make a distriction in chemistry between covalent and ionic bonds. In the simplest description, a covalent bond is characterized according to the Lewis concept by the strong bonding [22, 178] between pairs of atoms realized by pairs of electrons without charge transfer. The corresponding AOs and MAOs strongly overlap. On the other hand the ionic bond is formed by electrostatic attraction between a cation-anion pair of atoms. In the latter case, one of the atoms transfers an electron to a bonding partner. Perfect covalent bonding takes place only in homonuclear diatomics or in some systems where 100% of covalency is dictated by symmetry (e.g. diamond). In VB and MO theories these two extremes are limiting cases and majority of molecules possess bonds exhibiting characteristics of both types, the only difference being their different extent. If purely covalent bonds are scarce, the ideal ionic bonds are even harder to find. Since the intramolecular charge shift is dictated by electronegativity, one can say that bond ionicity increases with increasing electronegativity difference between the bonded atoms. In this sense alkali halides and in particular LiF would be the best candidates for ionic bonds. There are several definitions of ionic character involving electronegativity differences etc., but finally they are all reduced to differences in charges placed on atoms in question. Recently, a controversial topic of ionicity of carbon-lithium bonds was extensively discussed. Results of several researchers favor the ionic description of this bond. Collins and Streitwieser [75] obtained in $LiCH_3$ a net charge on Li of 0.79 with the 4-31G basis and 0.80 with the extended 6-31G* basis employing their integrated spatial population analysis. This should be compared with LiF net charges of 0.86 and 0.87, respectively. Similar values are predicted by Weinhold et al. [72] with a weighted natural population analysis based on orthogonalized atomic orbitals. In $LiCH_3$ the net charge on Li was 0.79 and 0.80 with 4-31G and 6-31G* basis sets, respectively, i.e. they obtained virtually the same values as Streitwieser. However, in LiF the ionicity is somewhat more pronounced as evidenced by Li charges which read 0.92 and 0.93, respectively. In hyperlithiated carbon compounds this method predicts negative charges on carbon which approach the valence shell capacity of 10. The total gross electronic populations were 9.13 for CLi_4, 9.08 for CLi_5 and 9.44 for CLi_6. An even larger ionicity of lithium bonds could be deduced from straightforward interpretation of the results of the zero flux atomic basin integrative method of Bader. Employing this approach Ritchie and Bachrach [179] calculated net charges on Li in $LiCH_3$ of 0.91 whereas the total electron population on carbon in CLi_4 and CLi_6 were 9.44 and 10.38, respectively. The latter value surely exceeds the tolerable limit of 10 electrons being inconsistent with the shell structure of atoms. This illustrates difficulties if topological densities are taken literally. As discussed in Section 3.4 Bader's method gives an insight into the charge transfer only if and when it is compared to the corresponding density distribution in promolecules or other suitable reference charge distribution. Needless to say, extent of the ionic character strongly depends on the computational method and population technique applied. For instance, Schiffer and Ahlrichs [180] analyzed lithium bonds in various compounds with the method of modified atomic orbitals discussed earlier. They emphasize that the covalent character of the LiC bond in $LiCH_3$ should be greater than in the LiCl diatomic molecule, which in turn is not a purely ionic bond according to their analysis. Jug and coworkers [76] have recently developed a charge concept which yield a net charge of 0.68 on Li in $LiCH_3$ with the 6-31G* basis. In LiF diatomic molecule a net charge of 0.80 is found whereas the total electron population on carbon atom in CLi_4 was 5.29, thus being much more moderate than in analyses mentioned above. These results are supported by the multipole moments preservation scheme [64, 65]. The dipole moment conservation overlap partitioning gave Li a net charge of 0.59 in $LiCH_3$ by the 4-31G basis set. The net charge of ± 0.87 was predicted in LiF. It is clear that there is not a common agreement about ionicity in lithium compounds.

Another much debated issue is related to the hypervalency problem in molecules involving second row atoms. Consider PF_3 and PF_5 molecules. Common sense suggests that PF_3 has three "normal" and predominantly covalent bonds, whilst the bonding in PF_5 requires either extensive d orbital participation or existence of three covalent and two ionic bonds. In the latter case the axial fluorine atoms would carry a net negative charge-1 whereas phosphorus would have $+2$ charge to make the balance zero. A similar situation arises in $(CH_3)_2SO$ [181] where the molecule takes either a zwitterionic S^+-O^- form or three single and one double more or less covalent bonds emanate from sulphur. The simpler basis sets like STO-3G and 3-21G prefer the former picture, but more involved basis sets such as e.g. 6-31G* underline the importance of extra-valence d-orbials in describing energetic and structural properties of hypervalent molecules. This conclusion seems to be generally accepted but Reed and Schleyer [182] point out that d orbitals in hypervalent compound play merely a role of additional central-atoms acceptor functions which polarize existing valence AOs. Their population is typically 0.3 electrons. Reed and Schleyer [182] considered a series of 32-valence-electron species like F_3NO, O_3ClF, F_3SN etc. and found by NPA population analysis that σ-bonds are significantly ionic. The natural bond orbital (NBO) method indicates a presence of the negative hyperconjugation of the n $\rightarrow \sigma^*$ type whereas d AOs of the central atoms are of secondary importance in the π bonding. Streitwieser et al. [183] claim that in phosphine oxide, PO_3^- and $P(CH_2)_3^-$ highly polar bonding takes place. Integrated projected population (IPP) and NPA analyses gave net charges at oxygen -1.56 and -1.23, respectively, in the second molecule, whereas the corresponding P net charges are 3.68 and 2.70 which is exorbitantly high. The IPP criterion gives the same charge of -1.56 e for oxygen in phosphine oxide. It is known that IPP is an approximation to IBP (integrated Bader population) atomic charges which in turn can not be directly used in discussion of ionicity (vide Sect 3.4). Hence the reported IPP data do not seem meaningful. NPA analysis gives more conservative estimates of charges but still they are too similar to IPP results which casts some doubts about their reliability. It is also possible that in Weinhold's method [72] the d orbital occupancy is underestimated by the very nature of the employed weighting procedure. In our method which makes use of orbital contraction for extended basis sets, but does not employ weighting factor for sulfur AOs, the d orbital populations are larger and the need for ionic bonding in sulfur-containing compounds is less pronounced [184]. Finally, quite recently Bachrach provided results of the topological (IBP) density analysis in some phosphines, phosphaalkenes and phosphaalkynes [185]. It was found that P atomic charges vary from 1.88 to 1.12, while carbon net charges range between -0.60 to -1.50. These values are too high, as expected, and little physical significance can be attributed to these numbers. Apparently, the problem of ionic character in molecules involving heavier atoms and particularly in hypervalent compounds is not settled as yet. An interesting energy criterion for estimating ionicity in molecules was put forward by Horn and Ahlrichs [186].

5.2 Energetic Properties of Molecules

Charge separation in (neutral) molecules, leading to a distribution of positive and negative net atomic charges, contributes to molecular stabilization via Coulomb interactions. Actually, it is related to total molecular SCF energies more than anticipated. It was shown by Politzer [187, 188] that there is a good approximate relationship

$$E_t = \sum_A k_A Z_A V_A, \tag{90}$$

which is capable of reproducing total molecular energy with an error of 0.5%. Here, k_A is the adjustable weighting factor, Z_A is the atomic number and V_A is the electrostatic potential at the site of the nucleus. A simplified expression of V_A involving net atomic charges of atoms $B \neq A$ and orbital populations of AOs placed at atoms A reads [189]

$$V_A = -\sum_{\mu}^{A} (\xi_{A\mu} Q_{A\mu}/n_{A\mu}) + \sum_{B}' (Z_B - Q_B)/R_{AB}, \tag{91}$$

where $\xi_{A\mu}$ and $n_{A\mu}$ are screening constants and principal quantum numbers, respectively. Formula (90) in conjunction with the point charge approximation (91)

Table 9. Comparison of the total molecular energy computed as the Hartree-Fock expectation value by using DZ basis set and the corresponding entities obtained in the point-charge approximation employing the same DZ set and by the semiempirical SCC-MO formal atomic charges (in a.u.)

Molecule	Point charge approximation		ab initio DZ
	SCC-MO	ab initio DZ	Average energy
H_2O	−76.086	−75.940	−76.004
NH_3	−56.129	−56.009	−56.171
N_2H_4	−111.238	−111.226	−111.126
H_2O_2	−150.848	−150.694	−150.737
CH_3OH	−115.201	−115.239	−115.006
$CHOOH$	−188.771	−188.882	−188.689
HCN	−93.005	−92.769	−92.829
C_2H_4	−78.047	−77.969	−78.005
N_2H_2	−110.073	−110.127	−109.942
H_2	−0.992	−0.858	−1.127
CH_4	−39.963	−39.989	−40.182
C_2H_2	−76.964	−76.842	−76.792
H_2CO	−114.001	−113.906	−113.821
C_2H_6	−78.921	−79.189	−79.198
CO_2	−187.221	−187.304	−187.538
CO	−112.705	−112.641	−112.676
NNO	−183.334	−183.521	−183.576
N_2	−108.799	−108.749	−108.870
Weighting factors	$k_H = 0.4150$	0.3679	
	$k_C = 0.413$	0.4309	
	$k_N = 0.4252$	0.4266	
	$k_O = 0.4191$	0.4233	

was tested against Snyder-Basch DZ average energy results [190] by using SCC-MO and ab initio DZ wavefunctions. In both cases Mulliken populations were employed. Results are compared in Table 9. Performance of the point-charge formula is surprisingly good. It is interesting that the quality of semiempirical and ab initio results is the same as evidenced by identical standard deviations (0.1 au). It is noteworthy that the adjustable weighting factors k_A are not far from 0.5 which is required by the virial theorem. It is reasonable to assume that a better population analysis would give even better agreement with the SCF results.

The role of atomic charges in determining the enthalpies of formation of organic compounds was studied by Benson [191]. The influence of charge shift in molecules on bond energies and some other energetic features are discussed in depth by Fliszar [192]. A formula relating stabilization energy to asymmetry in bond charge distribution between the participating atoms, which shows that the charge-transfer term yields a dominating contribution was developed by Matcha [193]. We note in passing that dissociation energies in alkali-halides are qualitatively well reproduced by the pure ionic bonding model [114]. Of course, this is a consequance of a fortuitous, but almost complete cancellation of the remaining terms in the Rittner potential [114]. Interestingly, a model of a single electron transfer from alkali to halide atoms works well for some other properties [114]. Finally, a Coulomb term is a necessary ingredient in all refined versions of the molecular mechanics methods [194 − 196].

5.3 Magnetic Properties

Diamagnetic shielding of the nuclei (Lamb shift) and diamagnetic susceptibility of molecules (Langevin term) have been thoroughly discussed by us in a number of papers [54, 197 − 200]. Both of these properties are not strongly dependent on atomic charge distributions and in most cases the independent atom model (IAM) alias spherical promolecule arrangement of neutral atoms suffices for their useful description. However, if the differences in electronegativity are highly pronounced, then intramolecular charge transfer has to be taken into account explicitly. The IAM model fails in alkali halides as expected and an ionic model has to be invoked for a satisfactory prediction of diamagnetic susceptibilities. The corresponding formulas are easily generalized and expressed within the modified atom model [MAM] to allow for a charge shift. It is worth mentioning that pure ionic bonding in alkali halides gives the best results, which makes the calculation feasible on the "back-of-an envelope". One should point out that both of these diamagnetic properties are very well understood in terms of the MAM model given in the monopole approximation [54].

An extremely simple point-charge description of electron density distribution was developed by Amos and Yoffe [201] by using the Frost model for a molecule. Each pair of electrons is placed in a spherical or sometimes elliptical Gaussian whose position and exponent are variationally optimized. Orbital centers are typically located near nuclei for inner shells and along bonds for valence electrons thus conforming to chemical intuition. This model was applied to calculating various molecular properties including susceptibility and qualitatively correct results were obtained in some small molecules [201]. We note in passing that exact additivity of magnetic susceptibility in molecules was discussed recently by Bader within his theory of zero-flux atoms [202].

NMR chemical shift is a sum of diamagnetic and paramagnetic contribution. We have mentioned that the Lamb shift is fairly accurately reproduced by atomic charges and that even the IAM model might suffice for its estimate. In some specific cases the paramagnetic term is also directly related to atomic charges. The famous example is provided by Spiesecke and Schneider [203] who measured ^{13}C NMR shifts in tropylium cation, benzene, cyclopentadienyl anion and cycloclatetraene dianion. It appears that ^{13}C shifts are linearly correlated with a local π charge obtained simply by counting π electrons per atom. This work was extended by Olah and Mateescu [204] who employed Hückel π charges. Full valence treatment of $\sigma + \pi$ electrons was dicussed by Fliszar [192] who pointed out that the linear relationships, which are occasionally observed, are a consequence of near cancellation of nonlocal diamagnetic and paramagnetic contributions. This is, however, only rarely the case.

Indirect spin-spin coupling of the directly linked magnetic nuclei is transmitted via the bonded spin paired electron couple. It depends on atomic charge density but the hybridization effect should be carefully delineated [125].

5.3 Electric Properties

Total molecular multipole moments have a substantial atomic monopole contribution but the effect of local anisotropy of electron density has to be explicitly included too (cf. Eqs. (48), (80) and (84)). Amos and Yoffe point charge model was applied for calculation of molecular polarizability in some small heteroatomic molecules and hydrocarbons [201]. A pretty good semiquantitative agreement with experiment was achieved for the latter compounds [201, 205].

The point-charge model was utilized in tackling electric field grandient tensor and applied in rationalizing the Mössbauer effect in ^{119}Sn and ^{57}Fe compounds with a reasonable success [206].

Electrostatic potentials play a crucial role in determining intra- and extramolecular interactions being important indicators of chemical reactivity, pharmacological properties and solvent-solute effects [187, 188, 207–213]. Atomic charges can give a rough picture and answer some questions at the qualitative level particularly if they are specially parametrized for that purpose thus assuming effective values, but subtle effects require a more detailed and refined description of the electron density distribution. This is a necessary prerequisite for better representation of both intra- and intermolecular actions. Some approaches are based on localized molecular orbitals and multipole expansion about their centroids [214] or by using transferable LMO fragments and by their representation by suitable sets of point charges [215] at and/or around the LMO center of charge. A number of other multipolar models are developed for tackling problems of macromolecular conformations, nucleic acid bases interactions, molecular recognition in crystals etc. [216–218] but they are of minor importance here since we are mostly interested in charge monopoles placed preferably at the nuclei. In this respect one should mention the use of empirical effective charges in discussing preferred orientation of organic molecules in dense media [219] and point-charge modelling of accurate HF intermolecular potentials suitable for treating a difficult but very important problem of solvation of large biological molecules [220]. This method proved particularly fruitful in its application to Monte-Carlo calculations. Needless to say, these charges serve only as a mathematical device and have not an immediate physical meaning. A simple model based on fractional charges located at atomic centers of solvent molecules was dicussed by Morokuma [221]. An interesting outcome of these calculations is that the electrostatic energy is by far the most important mode of interaction. The same conclusion was reached later by ab initio, large, basis-set studies of Van der Waals complexes [222]. Finally, it is noteworthy that a simple and transparent point-charge model was put forward recently [223] which yields reasonable estimates of angular geometries of Van der Waals systems possessing lone pairs and satisfactorily describes variation of the potential energy.

5.5 General Remarks

Atomic charge distribution in molecules is one of the most debated issues in theoretical chemistry. It was impossible to present results of several thousands of papers in a complete and well balanced manner. We are aware of the fact that this review is necessarily something of a torso, but omission of Bethe's crystal field theory [224] would be intolerable. Indeed, this simple model based on the point charge representation of ligands in complexes gave an essentialy correct description of the electronic properties of central ions. Consequently it has had a profound influence on the development of inorganic chemistry. It is a common knowledge today that a point-charge picture of various ligands is an oversimplification, but the crystal field model correctly reproduced symmetry of the Coulomb field exerted on the central atom which is an essential part of the physics of the problem.

Another wide application of the notion of atomic charge is its use in interpreting chemical reactivity. The latter is in very many cases qualitatively understood in terms of the frontier orbitals. However, if the HOMO-LUMO energy gap between reactants is large, then the chemical reaction is charge controlled [225].

One of numerous manifestations of the substituent effect is a change in electron distribution of the parent molecule, which is termed inductive effect. It depends on the electron acceptor or electron releasing properties of substituent atoms, and/or substituent group(s). Changes of atomic charges upon substitution summarize in a most concise way the inductive effect. It is gratifying that various methods yield charges which are linearly related to Taft's empirical polar parameters σ^* [192].

Although we are not concerned here with the solid state physics and chemistry it should be mentioned in passing that point charge description of coulomb interactions in ionic crystals was an indispensable tool in the early 1930 s [55, 226]. More recently the cluster model of ionic solids was developed where it was tacitly assumed that the observable properties of the crystal could be predicted from the behaviour of a small group of representative ions forming a fragment cluster [227]. The formal charge of ions in a cluster is an inevitable part of the model.

Finally, a word of caution is necessary in order to circumvent various pitfalls in interpreting atomic charges. They should not be unterstood in terms of classical charges for several reasons. The point charges have an infinite self-energy and their static arrangement is impossible in view of Earnshaw's theorem. Frozen distribution of atomic charges are not permitted in quantum theory either, because of Heisenberg's uncertainty principle. The concept of atomic charge (monopole) leans heavily on an idea that each atom has its own domain within a molecule. Electron charge density comprised in atomic region can be developed in a multipole moment series which in turn involves the atomic monopole as the first term. Its center is usually placed at or near the nucleus in question.

6 Epilogue

Summarizing the main results of the present critical survey ot the existing methods one can safely say that there is no unique way of allocating atomic charges in molecules. Furthermore, there is no general consensus about an optimal charge partitioning of the electron density which satisfactorily describes the charge transfer effect and possesses a universal applicability to both small and large molecules of arbitrary symmetry and geometry. Many of the considered methods have serious drawbacks. However, there are several prospective avenues of research which promise useful results, because the concept of atomic charge is of paramount importance in chemistry. Spatial integration techniques based in the first place on the Bader's theory of zero-flux atoms are intellectually pleasing, but they do not reflect a true charge transfer effect by themselves thus being unsatisfactory. Their values relative to judiciously chosen reference densities of neutral atoms seem to be more realistic but not in all cases. Orbitally based population analyses have still to find the best possible division of the shared density which will be verified on a large variety of different molecules. Another burden of the orbital approximation is the extreme sensitivity of the resulting atomic charges on the employed basis sets. Well balanced basis sets, implying that each atom should be described by its own AOs, or to put it in another way, that drawbacks in the local AOs on one atom should not be compensated by a larger selection of (weird) AOs on other atoms, still wait for criteria for their appraisal. We have suggested some minimal conditions which acceptable basis sets should satisfy [65, 200] but a lot of additional work is necessary in this area. Finally, electronegativity schemes for estimating atomic charges have to resolve the total/ partial electronegativity equalization dilemma and percentagewise assessment of the latter. This approach, provided that the Madelung term is included, is the best candidate for a means of producing useful charges in very large biological systems.

In spite of the fact that each definition of charges is somewhat arbitrary, several schemes yield fairly consistent and sensible values along the series of related molecules implying that the resulting atomic charges have some semblance of truth within their own scales. It should be stressed that atomic charges relative to the predetermined reference value corresponding to a specific moiety or molecule are more meaningful than the actual magnitude of each gross atomic population. This is remarkable because atomic charge is an important parameter of chemical bonding, possessing a high interpretive power (in identifying e.g. substituent effects) and being one of the crucial descriptors within the modified atom (MAM) model [22, 54, 114, 228, 229].

Several experimental methods for extracting atomic charges from observed data are considered. Taking into account obstacles in their interpretation it follows that the most straightforward insight into atomic charge distribution is offered by ESCA spectroscopy particularly if the reorganization effect upon ionization is properly taken into account. It appears that ESCA shifts are intimately related to the electrostatic potential exerted at the nucleus in question, which in turn is to a good approximation expressed in the atomic charge monopole form. In this sense atomic charges are measurable, but their accuracy depends on both experimental apparatus and AMEP (atomic monopole electrostatic potential) approximation.

We feel confident that future work will shed even more light not only on atomic charges but also on atomic orbital populations and their anisotropy by employing ESCA, X-ray, NMR and NQR techniques in conjuction with appropriate theoretical models.

Acknowledgement: A part of this work was done at the Organisch-chemisches Institut der Universität Heidelberg and one of us (ZBM) would like to thank the Alexander von Humboldt-Stiftung for financial support and Professor R. Gleiter for kind hospitality.

7 References

1. Hirshfeld FL (ed) (1977) Isr J Chem 16: Nos. 2 and 3; Coppens P, Hall MB (eds) (1982) Electron distributions and chemical bond, Plenum, New York
2. Bacon GE (1977) Neutron scattering in chemistry, Butterworths, Boston
3. Gillon B, Schweizer J (1989) In: Maruani J (ed) Molecules in physics, chemistry and biology, vol 3, Kluwer, Dordrecht
4. Bonham RA, Lee JS, Kennerly R, St John W (1978) Adv Quant Chem 11: 1
5. Hohenberg P, Kohn W (1964) Phys Rev B136: 349
6. Teller E (1962) Rev Mod Phys 34: 627
7. Balasz NL (1967) Phys Rev 156: 42
8. Dahl JP, Avery J (1984) Local density approximations in quantum chemistry and solid state physics, Plenum, New York
9. Hellmann H (1933) Z Phys 85: 180; Feynman RP (1939) Phys Rev 56: 340
10. see for example – Deb BM (1981) The force concept in chemistry, Van Nostrand Reinhold, London
11. For a recent paper see: Gilbert MM, Donn JJ, Peirce M, Sundberg KR, Ruedenberg K (1985) J Comp Chem 6: 209 and the references therein
12. Wahl AC (1966) Science 151: 961
13. Streitwieser A, Owens PH (1973) Orbital and electron density diagrams, MacMillan, New York
14. Van Wazer JR, Absar I (1975) Electron densities in molecules and molecular orbitals, Academic, New York
15. Bader RFW (1970) An introduction to the electronic structure of atoms and molecules, Clarke, Irwin
16. Daudel R, Leroy G, Peeters D, Sana M (1983) Quantum chemistry, John Wiley, Chichester
17. Bader RFW, Henneker WH, Cade PE (1967) J Chem Phys 46: 3341
18. Savariault JM, Lehmann MS (1980) J Am Chem Soc 102: 1298
19. Dunitz JD, Seiler P (1983) J Am Chem Soc 105: 7056

20. Breitenstein M, Donnohl H, Meyer H, Schweig A, Zittlau W (1982) in ref 2
21. Dunitz JD (1988) Bull Chem Soc (Japan) 61: 1
22. Maksić ZB (1990) In: Maksić ZB (ed) Theoretical models of chemical bonding, vol 2, Springer, Berlin Heidelberg New York, p 137
23. Kunze KL, Hall MB (1986) J Am Chem Soc 108: 5122
24. Kunze KL, Hall MB (1987) J Am Chem Soc 109: 7617
25. Low AA, Hall MB (1990) J Phys Chem 94: 628
26. Hirshfeld FL, Rzotkiewicz (1974) Mol Phys 27: 1319
27. Figgis BN, Reynolds PA (1985) Inorg Chem 24: 1864
28. Figgis BN, Forsyth JB, Reynolds PA (1987) Inorg Chem 26: 101
29. Schwarz WHE, Valtazanos P, Ruedenberg K (1985) Theoret Chem Acta 68: 471
30. Schwarz WHE, Mensching L, Valtazanos P, von Niessen W (1986) 30: 439
31. Schwarz WHE, Ruedenberg K, Mensching L, Miller LL, Jacobson R, Valtazanos P, von Niessen W (1989) Angew Chem 101: 605
32. Ruedenberg K, Schmidt MW, Gilbert MM, Elbert ST (1982) Chem Phys 71: 41; 71: 65; Ruedenberg K, Schmidt MW, Gilbert MM (1982) ibid, 71: 51
33. Breitenstein M, Dannöhl H, Meyer H, Schweig A, Seeger R, Seeger U, Zittlau W (1983) Int Rev Phys Chem 3: 335
34. Bader RFW, Chandra AK (1968) Can J Chem 46: 953
35. Hall MB (1982) in Ref [2]
36. Brillouin L (1934) Actualites Sci Ind 71: 159
37. Moller C, Plesset MS (1934) Phys Rev 46: 618
38. Coppens P, Stevens ED (1977) Isr J Chem 16: 157
39. Coppens P (1982) see Ref [2] p 61
40. Stevens ED (1982) see Ref [2] p 331
41. Mason R (1982) see Ref [2] p 351
42. Figgis BN, Reynolds PA, Wright S (1983) J Am Chem Soc 105: 434
43. Klein CL, Stevens ED (1988) In: Liebman JF, Greenberg A (eds) Structure and reactivity, VCH, New York, p 25
44. Low AA, Hall MB (1990) In: Maksić ZB (ed) Theoretical models of chemical bonding, vol 2, Springer, Berlin Heidelberg New York, p 543
45. Bader RFW, Nguyen-Dang TT, Tal Y (1981) Rep Prog Phys 44: 893
46. Bader RFW, Nguyen-Dang TT (1981) Adv Quant Chem 14: 63
47. Bader RFW (1985) Acc Chem Res 18: 9
48. Cremer D, Kraka E (1984) Angew Chem 96: 612; Cremer D, Kraka E (1984) In: Maksić ZB (ed) Conceptual quantum chemistry. Models and applications Part 2, Croat Chem Acta 57: 1231
49. Cremer D (1987) In: Maksić ZB (ed) Modelling of structure and properties of molecules, Ellis Horwood, Chicherter p 125
50. Kraka E, Cremer D (1990) In: Maksić ZB (ed) Theoretical models of chemical bonding, Part 2, Springer, Berlin Heidelberg New York, p 453
51. Boyd RJ, Edgecombe KE (1988) J Am Chem Soc 110: 4182
52. Morse P, Feshbach H (1953) Methods of theoretical physics Part 1, McGraw-Hill, New York
53. Gatti C, Fantucci P, Pacchioni G (1987) Theoret Chim Acta 72: 433
54. Maksić ZB (ed) (1990) Theoretical models of chemical bonding. Part 1 Atomic hypothesis and the concept of molecular structure, Springer, Berlin Heidelberg New York
55. Pauling L (1960) The nature of the chemical bond, 3rd edn, Cornell University Press, New York
56. Mulliken RS (1955) J Chem Phys 23: 1833
57. Löwdin PO (1953) J Chem Phys 21: 374
58. Doggett G (1969) J Chem Soc A 229
59. Hillier IH, Wyatt JF (1969) Int J Quant Chem 3: 67
60. Fenske RF, Caulton KG, Radtke DD, Sweeney (1966) Inorg Chem 5: 951
61. White WD, Drago RS (1970) J Chem Phys 52: 4717 and the references cited therein
62. Claverie P (1978) In: Pullman B (ed) Intermolecular interactions. From diatomics to biopolymers, Wiley, New York, p 283

63. Thole BT, van Duijnen P Th (1983) Theoret Chim Acta 63: 209
64. Jug K (1973) Theoret Chim Acta 31: 63
65. Jug K (1975) Theoret Chim Acta 39: 301
66. Christoffersen RE, Baker KA (1971) Chem Phys Lett. 8: 4
67. Politzer P, Stout Jr, EW (1971) Chem Phys Lett 8: 519
68. Gallup GA, Norbeck JM (1973) Chem Phys Lett 21: 495
69. Löwdin PO (1950) J Chem Phys 18: 63
70. Carlson BC, Keller JH (1957) Phys Rev 105: 102
71. Maksić ZB, Z Naturf (1981) 36a: 373
72. Reed AE, Weinstock RB, Weinhold F (1985) J Chem Phys 83: 735
73. Reed AE, Weinhold F (1983) J Chem Phys 78: 4066
74. Löwdin PO (1955) Phys Rev 97: 1474
75. Collins JB, Streitwiser A (1980) J Comp Chem 1: 81
76. Jug K, Fasold E, Gopinathan MS (1989) J Comp Chem 10: 965
77. Guerillot CR, Lissillour R, Le Beuze A (1979) Theoret Chim Acta 52: 1
78. Davidson ER (1976) Reduced density matrices in quantum chemistry, Academic, New York, p 3
79. Davidson ER (1967) J Chem Phys 46: 3320
80. Roby KR (1974) Mol Phys 27: 81
81. Roby KR (1974) Mol Phys 28: 1441
82. Cruickshand DWJ, Avramides EJ (1982) Phil Trans Roy Soc A 304: 533
83. Boys SF (1960) Rev Mol Phys 32: 296
84. Heinzmann R, Ahlrichs R (1976) Theoret Chim Acta 42: 33
85. Ehrhardt C, Ahlrichs R (1985) Theoret Chim Acta 68: 231
86. Hirshfeld FL (1977) Theoret Chim Acta 44: 129
87. Daudel R (1952) Comp Rend Acad Sci 235: 886; Roux M, Daudel R (1955) ibid 240: 90
88. Daudel R (1953) Comp Rend Acad Sci 237: 601; Daudel R (1968) The fundamentals of theoretical chemistry, Pergamon
89. Bader RFW, Stephens ME (1974) Chem Phys Lett 26: 445; J Am Chem Soc 97: 7391
90. Hartmann H, Jug K (1965) Theoret Chim Acta 3: 439
91. Politzer P, Harris RR (1970) J Am Chem Soc 92: 6451
92. Kern CW, Karplus M (1964) J Chem Phys 40: 1374
93. Bader RFW, Beddall PM, Cade PE (1971) J Chem Phys 93: 3095
94. Maclagan RGAR (1971) Chem Phys Lett 8: 114
95. Bader RFW, Beddall PM, Peslak J (1973) J Chem Phys 58: 557
96. Bader RFW, Messer RR (1974) Can J Chem 52: 2268
97. Stutchbury NCJ, Copper DL (1983) J Chem Phys 79: 4967
98. Kraka E (1984) Ph D Thesis, University of Köln, p 117
99. Spackman MA, Maslen EN (1986) J Phys Chem 90: 2020
100. Maslen EN, Spackman MA (1985) Austr J Phys 38: 273
101. Cade PE, Huo W (1973) At Data Nucl Data Tables 12: 415, (1975) ibid. 15: 1
102. McLean AD, Yoshimine M (1967) Tables of linear molecules wavefunctions, IBM J Res Develop suppl
103. Clementi E, Roetti C (1974) At Data Nucl Data Tables 14: 177
104. Streitwieser Jr A, Collins JB, McKelvey JM, Grier D, Sender J, Toczko AG (1979) Proc Natl Acad Sci USA 76: 2499
105. Glaser R, J Comp Chem (1989) 10: 118
106. Bachrach SM, Streitwieser A (1989) J Comp Chem 10: 514
107. Mulliken RS (1934) J Chem Phys 2: 782
108. Allred AL, Rochow EG (1958) J Inorg Nucl Chem 5: 264
109. Gordy W (1946) Phys Rev 69: 604
110. Iczkowski RP, Margrave JL (1961) J Am Chem Soc 83: 3547
111. Hinze J, Jaffé HH (1962) J Am Chem Soc 84: 540; Hinze J, Whitehead MA, Jaffé HH (1963) J Am Chem Soc 85: 148
112. Sanderson RT (1951) Science 114: 670
113. Pritchard HO, Sumner FH (1956) Proc Roy Soc (London) A235: 136

114. Maksić ZB, Eckert-Maksić M, Rupnik K (1984) In: Maksić ZB (ed) Conceptual quantum chemistry. Models and applications, Part 2, Croat Chem Acta 57: 1295
115. Huheey JE (1965) J Phys Chem 69: 3284; (1966) ibid 70: 2086
116. Huheey JE, Evans RS (1970) J. Inorg. Nucl. Chem. 32: 777
117. Huheey JE (1966) J Org Chem 31: 2365
118. Evans RS, Huheey JE (1973) Chem Phys Lett 19: 114
119. Politzer P (1971) Theoret Chim Acta 23: 203
120. Parr RG, Donnelly RA, Levy M, Palke WE (1978) J Chem Phys 68: 3801
121. Parr RG, Pearson RG (1983) J Am Chem Soc 105: 7512
122. Politzer P, Weinstein H (1979) J Chem Phys 71: 4218
123. Pritchard HO (1963) J Am Chem Soc 85: 1876
124. Mullay J (1985) J Am Chem Soc 107: 7271, (1986) ibid 108: 1770
125. Gasteiger J, Marsili M (1980) Tetrahedron 36: 3219; Guillen MD, Gasteiger J (1983) Tetrahedron 39: 1331
126. Mortier WJ, Van Genechten K, Gasteiger J (1985) J Am Chem Soc 107: 829
127. Sen KD, Jørgensen CK (eds) (1987) Electronegativity, Springer, Berlin Heidelberg New York
128. Mortier WJ, Ghosh SK, Shankar S (1986) J Am Chem Soc 108: 4315
129. Hall GG, Martin D (1980) Isr J Chem 19: 255
130. Politzer P, Parr RG (1974) J Chem Phys 61: 4258; Politzer P (1980) Isr J Chem 19: 224
131. Smith CM, Hall GG (1985) Int J Quant Chem 31: 685
132. Hall GG (1973) Chem Phys Lett 20: 501
133. Tait AD, Hall GG (1973) Theoret Chim. Acta 31: 311
134. Shipman LL (1974) Chem Phys Lett 31: 361
135. Huzinaga S, Narita S (1980) Isr J Chem 19: 242
136. Hall GG, Smith CM (1984) Int J Quant Chem 25: 881
137. Hall GG, Tsujinaga K (1986) Theoret Chim Acta 69: 425
138. Coppens P (1975) Phys Rev Lett 35: 98
139. Kurki-Suonio K, Salmo P (1971) Ann Acad Sci Fenn A6: 25
140. Francl MM, Hout RF, Hehre WJ (1984) J Am Chem Soc 106: 563
141. Rys J, King HF, Coppens P (1976) Chem Phys Lett 41: 383
142. Yanez M, Stewart RF, Pople JA (1978) Acta Cryst A34: 641
143. Escudero F, Yanez M (1982) Mol Phys 45: 617
144. Mo O, Yanez M (1979) Theoret Chim Acta 53: 337; Dorado M, Mo O, Yanez M (1980) J Am Chem Soc 102: 947
145. Catalan J, Escudero F, Laso J, Mo O, Yanez M (1980) J Mol Struct 69: 217
146. Larsson S, Braga M (1985) Theoret Chim Acta 68: 291
147. Martin D, Hall GG (1981) Theoret Chim. Acta 59: 281
148. Bader RFW, Larouche A, Gatti C, Carroll MT, MacDougall P J, Wiberg KB (1987) J Chem Phys 87: 1142
149. Momany FA (1978) J Phys Chem 82: 592
150. Cox SR, Williams DE (1981) J Comp Chem 2: 304
151. Chirlian LE, Francl MM (1987) J Comp Chem 8: 894
152. Williams DE, Ji-Min Y (1988) Adv At Mol Phys 23: 87
153. Williams DE (1988) J Comp Chem 9: 745
154. Bentley J (1981) In: Politzer P, Truhlar DG (eds) Chemical applications of atomic and molecular electrostatic potentials, Plenum, New York
155. Stone AJ (1981) Chem Phys Lett 83: 233; Stone AJ, Alderton (1985) Mol Phys 56: 1047
156. Matthews DA, Stucky GD, Coppens P (1972) J Am Chem Soc 94: 8001
157. Gussoni M, Castiglioni C, Zerbi G (1983) Chem Phys Lett 95: 483
158. Gussoni M, Castiglioni C, Zerbi G (1984) J Phys Chem 88: 600
159. Gussoni M, Castiglioni C, Zerbi G (1984) J Chem Phys 80: 1377
160. Ramos MN, Gussoni M, Castiglioni C, Zerbi G (1988) Chem Phys Lett 151: 397
161. Gussoni M, Ramos MN, Castiglioni C, Zerbi G (1987) Chem Phys Lett 142: 515
162. Ramos MN, Gussoni M, Castiglioni C, Zerbi G (1989) Croat Chem Acta 62: 595
163. Exner O (1975) Dipole Moments in Organic Chemistry, George Thieme, Stuttgart

164. Bader RFW, Larouche A, Gatti C, Caroll MT, MacDougall PJ, Wiberg KB (1987) J Chem Phys 87: 1142 and the reference cited therein
165. Rogers JD, Hilman JJ (1982) 77: 3615; King WT, Mast GB (1986) J Phys Chem 90: 2521
166. Cioslowski J (1989) Phys Rev Lett 62: 1469
167. Cioslowski J, Hay PJ, Ritchie JP (1990) J Phys Chem 94: 148
168. Dinur U, Hagler AT (1989) J Chem Phys 91: 2949; 2959
169. Siegbahn K, Nordling C, Johansson G, Hedman J, Heden PF, Hamrin K, Gelius U, Bergmark T, Werme LO, Manne R, Baer Y (1969) ESCA Applied to Free Molecules, North Holland, Amsterdam
170. Basch H (1970) Chem Phys Lett 5: 337
171. Schwartz ME (1970) Chem Phys Lett 6: 631
172. Maksić ZB, Rupnik K (1983) Z. Naturf 38a: 308
173. Maksić ZB, Rupnik K (1986) J Mol Struct 141: 309
174. Maksić ZB (1990) In: Maksić ZB (ed.) Theoretical models of chemical bonding, vol 3, Springer, Berlin Heidelberg, New York, the following chapter
175. Manne R, Aberg T (1970) Phys Lett 7: 282
176. Gaarenstroom SW, Winograd N (1977) J Chem Phys 67: 3500
177. Lang ND, Williams AR (1979) Phys Rev B20: 1369
178. For a recent discussion of a crucial role of overlapping in chemistry see: Salem L, Berthier G, Lefour JM, Koga T (1989) Chem Phys Lett 160: 167; Salem L, Berthier G, Lefour JM, Koga T, Durup J (1990) Chem Phys Lett 166: 632
179. Ritchie JP, Bachrach SM (1987) J Am Chem Soc 109: 5959
180. Schiffer H, Ahlrichs R (1986) Chem Phys Lett 124: 172
181. Hehre W, Radom L, Schleyer P von R, Pople JA (1986) Ab initio molecular orbital theory, John Wiley, New York, p 83
182. Reed AE, Schleyer P von R (1990) J Am Chem Soc 112: 1434 and the references cited therein
183. Rajca A, Rice JE, Streitwieser Jr A, Schaefer III HF (1987) J Am Chem Soc 109: 4189
184. Jug K, Fasold E, Int J Quantum Chem, submitted
185. Bachrach SM (1989) J Comp Chem 10: 392
186. Horn H, Ahlrichs R (1990) J Am Chem Soc. 112: 2121
187. Politzer P (1976) J Chem Phys 64: 4239; (1978) ibid 69: 491; (1979) ibid 70: 1067
188. Politzer P (1980) Isr. J Chem 19: 224
189. Maksić ZB, Rupnik K (1983) Theoret Chim Acta 62: 219
190. Snyder LC, Basch H (1972) Molecular wave functions and properties, John Wiley, New York
191. Benson SW, Luria M (1975) J Am Chem Soc. 97: 707; Benson SW (1978) Angew. Chemie 90: 868
192. Fliszar S (1983) Charge distributions and chemical effects, Springer, Berlin Heidelberg New York
193. Matcha RL (1983) J Am Chem Soc 105: 4859
194. Abraham RJ, Hudson B (1984) J Comp Chem 5: 562; (1985) ibid 6: 173
195. Hammarström LG, Liljefors T, Gasteiger J (1988) J Comp Chem 9: 24
196. Meyer AY (1990) In: Maksić ZB (ed) Theoretical models of chemical bonding, vol 1, Atomic hypothesis and the concept of molecular structure, Springer, Berlin Heidelberg New York, p 213 and the references cited therein
197. Maksić ZB, Rupnik K (1983) Theoret Chim Acta 62: 397
198. Maksić ZB, Mikac N (1978) J Mol Structure 44: 255; (1980) Mol. Phys 40: 455
199. Maksić ZB, Rupnik K (1983) Croat Chem Acta 56: 461
200. Maksić ZB, Supek S (1988) Theoret Chim Acta 74: 275
201. Amos AT, Yoffe JA (1975) Theoret Chim Acta 40: 221; J Chem Phys 63: 4723
202. Bader RFW (1989) J Chem Phys 91: 6989
203. Spiesecke H, Schneider WG (1961) Tetrahedron Lett 468
204. Olah GA, Mateescu GD (1970) J Am Chem Soc. 92: 1430
205. Yoffe JA (1978) Chem Phys Lett 54: 562; (1979) Theoret Chem Acta 52: 147
206. Sams JR (1972) In: McDowell CA (ed) MTP International review of science, Physical chemistry, Series One, vol 4, Butterworths, London, p 85
207. Scrocco E, Tomasi J (1973) Top Curr Chem 42: 95; Adv Quant Chem 11: 115

208. Tomasi J (1981) In: Politzer P, Truhlar DG (eds) Chemical applications of atomic and molecular electrostatic potentials, Plenum, New York, p 257
209. Tomasi J (1988) J Mol Structure (Theochem) 179: 273; Alagona G, Bonarccorsi R, Ghio C, Montagnani R, Tomasi J (1988) Pure Appl Chem 60: 231
210. Tomasi J, Alagona G, Bonaccorsi R, Ghio C, Cammi R (1990) In: Maksić ZB, Theoretical models of chemical bonding, vol 3, Molecular spectroscopy, electronic structure and intramolecular interactions, Springer, Berlin Heidelberg New York
211. Hadži D, Koller J, Hodošček M, Kocjan D (1987) In: Maksić ZB (ed) Modelling of structure and properties of molecules, Ellis Horwood, Chichester, p 286
212. Tomasi J, Alagona G, Bonaccorsi R, Ghio C (1987) In: Maksić ZB (ed) Modelling of structure and properties of molecules, Ellis Horwood, Chichester
213. Warshel A (1981) Acc Chem Res 14: 284
214. Goldblum A, Perahia D, Pullman A (1979) Int J Quant Chem 15: 121; Etchebest C, Lavery R, Pullman A (1982) Theoret Chim Acta 62: 17
215. Bonaccorsi R, Scrocco E, Tomasi J (1977) J Am Chem Soc 99: 4546
216. Langlet J, Claverie P, Caron F, Boeuve JC (1981) Int J Quant Chem 19: 229
217. Gresh N, Claverie P, Pullman A (1984) Theoret Chim Acta 66: 1
218. Price SL, Stone AJ (1987) J Chem Phys 86: 2859; Stone AJ, Price SL (1988) J Phys Chem 92: 3325
219. Gavezzotti A (1989) Chem Phys Lett 161: 67
220. Clementi E (1976) Lecture notes in chemistry No. 2, Springer, Berlin Heidelberg New York, and the references therein
221. Noell JO, Morokuma K (1975) Chem Phys Lett 36: 465; (1976) J Phys Chem 80: 2675
222. Rendell APL, Bacskay GB, Hush NS (1985) Chem Phys Lett 117: 400; Cummins PL, Rendell APL, Swanton DJ, Bacskay GB, Hush NS (1986) Int Rev Phys Chem 5: 139
223. Legon AC, Millen DJ (1989) Can J Chem 67: 1683
224. Bethe H (1929) Ann. Phys 5: 133
225. Klopman G (1968) J Am Chem Soc 90: 223; J Mol Struct (Theochem) 103: 121 and the references cited therein
226. Born M, Mayer JE (1932) Zeit. für Phys 75: 1
227. Recio JM, Luaña V, Pueyo L (1989) J Chem Ed 66: 307 and the references cited therein; Francisco E, Luaña V, Recio JM, Pueyo L (1988) J Chem Ed 65: 6
228. Maksić ZB (1988) J Mol Struct (Theochem) 170: 39
229. Maksić ZB (1988) In: Maruani J, ed., Molecules in physics, chemistry and biology, vol III, Kluwer, Dordrecht, p 49

Electron Spectroscopy for Chemical Analysis (ESCA) — Basic Features and their Model Description

Zvonimir B. Maksić

Theoretical Chemistry Group, Ruder Bošković Institute, 41001 Zagreb, Croatia, Yugoslavia and Faculty of Natural Sciences and Mathematics, University of Zagreb, Marulicev trg 19, 41000 Zagreb, Croatia, Yugoslavia

The conceptual framework of the inner-core binding energy shifts in molecules is considered. It is shown that the basic facets of electron spectroscopy for chemical analysis (ESCA) are well understood and can be interpreted by simple physical models based on classical electrostatics. This transparent and intuitively appealing approach provides a rationale for characteristic ESCA fingerprints of atoms in chemical environments and offers the underlying principle for empirical additivity of ESCA shifts exhibited by functional groups. The role of the final state after photoionization has been completed is discussed in extenso. Finally, the importance of ESCA in atomic charge analysis, determination of gas-phase acidities and basicities and its interplay with other spectroscopic techniques (PES, NMR, MW and Mössbauer) is briefly discussed.

1 Introduction

It is appropriate to commence this chapter with a quotation of the renowned physicist Victor Weisskopf: "There are three kinds of physicists, as we know, namely the machine builders, the experimental physicists and the theoretical physicists. If we compare these three classes, we find that machine builders are the most important ones. If we compare this (situation) with the discovery of America, then, I would say, the machine builders correspond to the captains and their ships – builders who really developed the techniques at that time. The experimentalists were those fellows on the ships that sailed to the other side of the world and then jumped upon the new islands and just wrote down what they saw. The theoretical physicists were those fellows who stayed in Madrid and told Columbus that he was going to land in India."

Development of the ESCA technique offers an excellent illustration of that statement. Refinements of the instruments used in nuclear spectroscopy in studying β-decay and internal conversions in radioactive decay by Siegbahn et al. [1–8] opened several new fields of research which subsequently gave an immense number of data on the electronic structure of atoms, molecules, solid matter, surfaces, liquids, etc. [9–17]. Additionally, ESCA (which is an acronym for electron spectroscopy for chemical analysis) has become one of the most modern tools in analytical chemistry.

The main facets of ESCA are very well understood and covered by numerous books, review articles and special issues of the scientific journals [9–20]. Our attention will be focused here on the binding energy shifts of inner-core electrons representing fingerprints of particular atoms in characteristic chemical environments. These ESCA chemical shifts are interesting per se since they monitor changes in charge distribution of the peripheral valence electrons. Perhaps more importantly they lend themselves to a simple model description which can be reduced to the concept of atomic charges in chemical moieties. It appears that functional groups have characteristic (electrostatic) contributions to ESCA shifts leading to a simple additivity rule of thumb. Interpretation of the basic ESCA lines in terms of the intramolecular electrostatic potential represents modelling par excellence in quantum chemistry because a very complex phenomenon of the inner-shell ionization is reduced to the electrostatic interaction of atomic point-charges in the initial (molecule) and the final (the resulting molecular ion) states. This conceptual approach together with ESCA data convincingly shows that the inner-core electrons are highly localized in the vicinity of their host nuclei or to put it in another way – inner shells retain their atomic nature. Providing this information and giving an important insight into the meaning of atomic charges in molecules, ESCA technique offers a significant piece of evidence that atoms do not lose their identity by chemical bonding. Hence it yields an important part of the observational basis for the concept of modified atoms in molecular environments. Further, being intimately related to the intramolecular charge drift, ESCA shifts are closely connected to many other physical and chemical properties of molecules. They are also related to some entities obtained by other spectroscopic methods as will be briefly discussed at the end of this chapter. To conclude this Introduction we cannot resist mentioning that the ESCA method illustrates rather nicely a symbiotic relationship between modern physics and chemistry. The former enables penetrating understanding of the electronic structure of molecules and as far as ESCA is concerned it gives important information about the electronic charge reorganization upon chemical bond formation, on atomic charges and electronegativities, etc. Amazingly enough, this rich harvest was started by a rather technical improvement of the resolution in the energy analysis of photoelectrons. We shall confine our discussion in what follows to first and second row atoms.

2 Theoretical Aspects of ESCA

Essentially, the basic phenomenon in ESCA is photoeffect discovered by Hertz [21] and explained by Einstein [22]. The corresponding law is simply expressed by:

$$E_k = h\nu - E_B - \varphi_W \tag{1}$$

where ν is a frequency of the incident photon, E_K is the kinetic energy of an expelled electron, E_B is its binding energy given with respect to the Fermi level and φ_W is the work function of the studied specimen. In what follows we shall be mostly interested in the inner-shell binding energies E_B of free molecules. They are given by

$$E_B = E_t^{f+} - E_t^i \tag{2}$$

where E_t^{f+} and E_t^i stand for the total energies of the final molecular ion state and initial ground state of the target molecule, respectively. The main ESCA peak is related to the hole state of the resulting ion whereas satellite peaks arise due to excitations of other electrons above this lowest hole state.

According to the Born-Oppenheimer approximation the binding energy can be divided into electronic and nuclear contributions

$$E_B = E_B^{ad} + \Delta E_{vib} + \Delta E_{rot} \tag{3}$$

where $\Delta E_{vib} = E_{vib}^{f+} - E_{vib}^i$ and $\Delta E_{rot} = E_{rot}^{f+} - E_{rot}^i$ are differences in vibrational and rotational excitation energies between final and initial states, respectively. It is assumed that the residual molecular ion is fully relaxed, i.e. it is left in the vibrational and rotational ground state. Needless to say, the electronic charge cloud is also considered as reorganized in the final state thus justifying a notation E_B^{ad}, where ad in the superscript denotes *adiabatic*. In our crude description of the inner-shell ionization phenomenon we shall neglect the change in vibrational and rotational motion by putting $\Delta E_{vib} = \Delta E_{rot} = 0$. Hence, E_B will refer to the adiabatic binding energy from now on.

A very useful concept is offered by the vertical ionization, which by the well-known Koopmans' theorem [23] establishes a one-to-one correspondence between the main ESCA peaks and inner-shell orbital enrgies:

$$E_B^v(m) = \langle H(n-1) \rangle - \langle H(n) \rangle = -\varepsilon_m. \tag{4}$$

Here, the ejected electron is denoted by m and the vertical energy of ionization $E_B^v(m)$ is obtained by using frozen orbitals ψ_j produced by the nonrelativitistic Hartree-Fock SCF calculations. The wavefunctions of the initial and final states are of the form:

$$\Psi^i = \hat{A} \prod_{j=1}^{n} \psi_j \quad \text{and} \quad \Psi^{f+}(m) = \hat{A} \prod_{j \neq m} \psi_j . \tag{5}$$

Since the orbitals of the final state are unrelaxed it is not surprising that the calculated ionization potentials given by the negative of the corresponding orbital energies $-\varepsilon_m$ are too high. For example, the vertical binding energies are exaggerated by more than 20 eV for Ne(1s) ionization. We shall dwell on reorganization energies much more in Sect. 3.2. However, neglect of the relaxation is not the only intrinsic error in the canonical Hartree-Fock orbitals. In a correct treatment, one should include relativistic and correlation effects. Concomitantly, the inner-core binding energy is given by three terms

$$E_B = E_t^{f+} - E_t^i + \Delta E^{corr} + \Delta E^{rel} , \tag{6}$$

where E_t^f and E_t^i are calculated at the Hartree-Fock level, ΔE^{corr} and ΔE^{rel} is the difference in the correlation and relativistic energy of the final and initial state, respectively. An idea about the magnitude of ΔE^{corr} and ΔE^{rel} contributions one can get from good calculations performed on small systems like Ne, CH_4 and H_2O (Table 1). Perusal of the presented data shows that the relativity and electron correlation contributions to 1s binding energies are percentagewise very small. More importantly, they are roughly constant for the same atom in a series of related molecules. Hence, they can be neglected when ESCA shifts are considered relative to a preselected reference molecule or easily absorbed in some adjustable parameters if necessary. A substantial difference between negative orbital energies $-\varepsilon_{1s}$ and experimental binding energies indicates that something very important is missing. This is the reorganization energy as we shall see later. A relaxation of the electron charge distribution has to be explicitly taken into account particularly if inner-core binding energies in molecules widely differing in size are compared (vide infra).

Table 1. Relativistic and correlation contributions to $E_B(1s)$ binding energies in some simple systems[a]

Quantity	Ne[b]	CH_4[c]	H_2O[d]
$-\varepsilon_{1s}$	891.7	305.0	559.5
ΔE^{corr}	0.6	−0.1	0.5
ΔE^{rel}	0.8	0.1	0.4
$E_B(exp)$	870.1	290.7	539.7

[a] In eV. [b] Ref. [24]. [c] Ref. [25]. [d] Ref. [26].

2.1 Sudden Approximation Calculations

As mentioned above, Hartree-Fock inner-shell orbital energies yield the binding energies within the frozen orbital or sudden approximation. Early ab initio calculations of the DZ quality gave ESCA shifts in agreement with scarce experimental data available at that time [27]. A number of subsequent ab initio studies have shown that a variation of $-\varepsilon_{1s}$ is linearly related to ESCA measured values provided that groups of similar molecules are considered [28–30]. Performance of Koopmans' sudden approximation increases with flexibility of the employed basis set as a rule. As an illustration we provide results of ab initio calculations of C(1s) binding energies in some halomethanes based on STO-3G, 4-31G, 6-31G and 6-31G** sets (Table 2). Survey of the results reveals that $-\varepsilon_{1s}$ values are too high as expected. However, they do follow a general trend. This is most easily seen by determining least square linear relations. The correlated single-electron energies $-\varepsilon_{1s}^k$ are in good accordance with experiment if basis sets of at least DZ quality are utilized. This is evident from the average absolute errors which for STO-3G and 4-31G sets read 1.4 eV and 0.2 eV, respectively. The minimal STO-3G is obviously unsatisfactory but the DZ 4-31G set describes the change in $E_B(C_{1s})$ very well within the family of substituted methanes. It seems that basis sets exhibit a saturation feature since further increase in sophistication (6-31G and 6-31G**) actually decreases a quality of results as evidenced by Δ_a values of 0.4 eV and 0.5 eV. It is gratifying that a relatively simple basis set like 4-31G is so successful, since 4-31G calculations are feasible in large molecular systems. It should be pointed out however that a good performance of the sudden

Table 2. Comparison of calculated C(1s) Binding energies in some halomethanes with experimental data. Koopman' theorem and several basis sets are used. Least square fit linear relations are given at the bottom[a, b]

Mole-cule	STO-3G		4-31G		6-31G		6-31G		Exp.
	$-\varepsilon_{1s}$	$-\varepsilon_{1s}^k$	$-\varepsilon_{1s}$	$-\varepsilon_{1s}^k$	$-\varepsilon_{1s}$	$-\varepsilon_{1s}^k$	$-\varepsilon_{1s}$	$-\varepsilon_{1s}^k$	$-\varepsilon_{1s}^k$
CH_4	300.2	290.4	300.4	290.6	305.0	290.5	305.1	290.4	290.7
CH_3F	305.4	295.2	307.4	293.2	307.9	293.2	307.8	293.3	293.5
CH_2F_2	304.9	294.7	310.8	296.0	310.7	295.6	310.2	295.8	296.3
CHF_3	311.5	300.7	314.4	298.9	314.6	299.3	312.6	298.4	299.0
CF_4	309.6	298.9	318.0	301.8	317.1	301.5	316.0	301.9	301.7
CH_3Cl	302.8	292.8	306.8	292.7	307.5	292.7	307.4	292.8	292.3
CH_2Cl_2	304.9	294.7	308.8	294.3	309.5	294.6	309.3	294.8	293.9

$-\varepsilon_{1s}^k(\text{STO-3G}) = 16.0 - 0.91\varepsilon_{1s} \quad \Delta_a = 1.4$
$-\varepsilon_{1s}^k(\text{4-31G}) = 40.0 - 0.82\varepsilon_{1s} \quad \Delta_a = 0.2$
$-\varepsilon_{1s}^k(\text{6-31G}) = 11.9 - 0.91\varepsilon_{1s} \quad \Delta_a = 0.4$
$-\varepsilon_{1s}^k(\text{6-31G**}) = 33.5 - 1.06\varepsilon_{1s} \quad \Delta_a = 0.5$

[a] In eV. A superscript k denotes correlated values. The average absolute error is denoted by Δ_a.
[b] Taken from Ref. [31]

approximation and 4-31G set is possible only because the relaxation energy is roughly constant along the studied series and can be included into the correlation coefficients. Generally this is not the case.

2.2 Calculations Involving the Initial and Final State

Explicit treatment involving calculations on both the initial ground state and the final hole state is not an easy task since McDonald's theorem [32] should be satisfied for the latter. This theorem says that the Hartree-Fock energy gives an upper limit to the true value for a higher state only if it is orthogonal to all lower states of the same symmetry. Since a 1s hole state is of a very high energy, there are many lower-lying states of the same symmetry and the orthogonalization process requires a formidable effort to say the least. It is gratifying, however, that the 1s orbital has little overlapping with valence orbitals implying that the effect of orthogonalization is small [33]. Hence the hole state calculations are performed as if they satisfy McDonald's theorem.

The first ΔSCF calculations on atoms were by Bagus who treated the F^-, Ne, Na^+ and Cl^-, Ar, K^- series [33]. A much better agreement with experiment is obtained with the ΔSCF procedure for the inner-shells ionizations than by employment of the Koopman's theorem. It should be pointed out that the atomic orbital basis sets were carefully optimized for both neutral and hole states. The pioneering ΔSCF computations on molecules were made by Schwartz [34] in 1970. At that time the experimental $E_B(1s)$ value was available only for CH_4. Schwartz's calculated $E_B(1s_C)$ of 291.0 eV was in excellent accordance with experiment (290.7 eV). He estimated, however, 1s binding energies for some other small systems such as BH_3, NH_3, H_2O and HF. Subsequent measurements have proved that the ΔSCF estimates were surprisingly good (Table 3). A single discrepancy found in BH_3 is actually a consequence of a fact that the experimental value refers to B_2H_6 system. Amusingly, agreement with measured data in molecules is much better than in atomic Ne, the case considered as a test for the adopted procedure. Frozen orbital binding energies calculated using Koopmans' theorem are too high as expected. Summarizing a brief discussion of the

Table 3. ΔSCF calculations of $E_B(1s)$ binding energies in some simple molecules (in eV)[a]

System	$-\varepsilon_{1s}$	ΔSCF	Exp.
BH_3	207.3	197.5	196.5[b]
CH_4	304.9	291.0	290.7
NH_3	422.8	405.7	405.6
H_2O	559.4	539.4	539.7
HF	715.2	693.3	
Ne	891.4	868.6	870.1

[a] Theoretical and experimental data are taken from Refs. [34] and [35], respectively. [b] This value is related to a dimer B_2H_6

results presented in Tables 1 and 3 one can say that explicit account of the reorganization energy at the appropriate level of sophistication suffices for a good description of the inner-core binding energies and their chemical shifts. Numerous computational studies performed later on showed that this conjecture is correct provided that adequate basic sets are used. We shall single out the work of Levy et al. [36] since they made CI correction to ΔSCF values in considering $E_B(1s_C)$ values in a series of fluorinated methanes. They found that the influence of the electron correlation is negligible in CH_4, which is in line with an earlier calculation of Meyer [25], but improves agreement with experiment along the series CH_3F, CH_2F_2, CHF_3 and CF_4. Levy et al. [36] conclude that the effect of CI increases with a number of F atoms and can not be ignored for the highest members of this family of molecules.

3 Development of a Simple Model for ESCA Shifts

3.1 Ground State Electrostatic Potential Model

It was realized quite early that the main contribution to ESCA shifts arose from differences in Coulomb interactions [9]. A theoretical analysis that demonstrated the crucial role of the electrostatic potentials exerted at the nucleus of the atom undergoing ionization was independently performed by Basch [28] and Schwartz [37]. Their treatments represent very nice examples of a deductive method in developing simple conceptual models. The orbital energy of a core electron m situated on the nucleus A is given in a molecule by

$$\varepsilon_m = \langle m | -(1/2)\,\Delta_1\,|m\rangle + \sum_B{}' \langle m | -Z_B/r_{B1}\,|m\rangle + \sum_j (2J_{mj} - K_{mj}), \quad (7)$$

where prime denotes the $B \neq A$ and J_{mj} and K_{mj} are Coulomb and exchange integrals, respectively. Rearrangement of these three terms yields a more convenient expression:

$$\varepsilon_m = \langle m | -(1/2)\,\Delta_m - Z_A/r_{A1}\,|m\rangle + \sum_{\substack{j = \text{core orb.} \\ \text{on A}}} (2J_{mj} - K_{mj}) - \sum_{\substack{j \neq \text{core orb.} \\ \text{on A}}} K_{mj}$$

$$+ \sum_{\substack{j = \text{val. orb.}}} 2J_{mj} + \sum_{\substack{j = \text{core orb.} \\ \text{on all B}}} 2J_{mj} + \sum_B{}' \langle m | -Z_B/r_{B1}\,|m\rangle. \quad (8)$$

It has been shown by explicit calculations that the first term involving the kinetic energy of the inner-core electron and its attraction to the host nucleus are constant to a high degree of accuracy being essentially an atomic entity [38]. This is corroborated by calculations of $\langle r \rangle$ expectation values in atoms by Bagus which change very little upon ionization in atoms [33] indicating that the inner-shell AOs remain practically invariant. Hence the first term vanishes when ESCA shifts are considered. The same holds for the second term. Further, exchange integrals K_{mj} between the core-electron in question and its counterparts placed on other atoms are very small. This is vindicated by actual calculations [37], but it is also intuitively clear at once in view of the minute overlapping of the respective AOs. In any case, their contribution to the ESCA shifts is negligible. It follows that the orbital energies and their change in different chemical moieties are determined by the last three terms in (8). The first of them arises due to the Coulomb repulsion between the core and valence electrons placed on the target atom in the first place as well as with all other valence shell electrons of the remaining atoms. This division is arbitrary but legitimate. The second term describes Coulomb repulsion between electron m and all other core electrons. The very last term in (8) yields attraction between a core electron m and nuclear charges B (B ≠ A). One can safely say that ESCA shifts are governed by a variation in classical Coulomb interactions between the inner-core electron in question and the rest of a molecule [28]. One can make a step further in simplifying the formula (8) and argue that inner-core integrals J_{mj} are given to a good accuracy by a point-charge approximation enabling absorption of the last-but-one term in (8) into $Z_B^{ef.} = Z_B - n_B$, where n_B denotes a number of inner-shell electrons. The same seems to hold for J_{mj} integrals between inner- and valence electrons

$$\sum_{\substack{j = val. \\ orb.}} 2J_{mj} \cong 2 \sum_{\substack{j = val. \\ orb.}} \langle j| \, 1/r_{A1} \, |j\rangle \, . \tag{9}$$

This is justified since e.g. $1s$ AOs are highly localized and for distant interactions act like delta-functions. Hence, the shifts in orbital energies are reduced to changes of the electrostatic potential exerted at the nucleus of the ionized atom:

$$\Delta(\varepsilon_m) \cong \Delta \left[2 \sum_{\substack{j = val. \\ orb.}} \langle j| \, 1/r_{A1} \, |j\rangle - \sum_{B}' Z_B^{ef}/R_{AB} \right] = \Delta V_A \, . \tag{10}$$

This is in fact a generalization of the earlier observation of Ha and O'Konski [37] that changes in $1s$ electron binding energies in various states of neutral atoms and their ions relative to the ground states parallel changes in $\langle r^{-1} \rangle$ expectation values. A similar analysis in molecules was made by Schwartz [34] who preferred to employ localized molecular orbitals (LMOs).We give just the final result:

$$\Delta(\varepsilon_m) \cong \Delta V_A + \Delta \sum_{j}^{A} [2 \langle L_j| \, 1/r_{A1} \, |L_j\rangle - 2J_{mj} + K_{mj}] \, . \tag{11}$$

The summation in formula (11) describes the difference between the actual interactions between electrons placed in localized MOs L_j involving the host atom A and the value this interaction would have if the inner-core electron m was collapsed to the nucleus A (in our imagination but not in reality) and exchange integrals were absent. Schwartz has shown that this term is pretty small and can be abandoned. It was tacitly assumed in these derivations that atomic units were employed. Therefore there is no formal distinction beteen potential and potential energy. Schwartz found by performing DZ ab initio calculations on a selection of 11 molecules that $\Delta(-\varepsilon_{1s})$ $= 1.11 \, \Delta V_A$ on average, or in other words that changes in the electrostatic potential at the nucleus were roughly proportional to changes in inner-shell orbital energies of $1s_C$, $1s_O$, $1s_F$ and $1s_N$. An important outcome of that analysis is that the electrostatic potentials can be calculated at the semiempirical level [40] thus considerably reducing costs in studying large molecular families and increasing the applicability to sizeable systems not amenable to ab initio procedures. A possibility of using semiempirical wavefunctions in producing electrostatic potentials gives a pragmatic value to the relation (10) in addition to its interpretative meaning. For if we need ab initio calculations for ΔV_A values, then relation (10) would not offer any provision since we can directly calculate $\Delta(-\varepsilon_i)$ shifts. It is gratifying however that the intramolecular electrostatic potential lends itself to further simplifications if approximations customary to semiempirical methods are utilized [40]. By expanding MOs in terms of atomic orbitals

$$\psi_j = \sum_A^A \sum_\mu c_{A\mu j}\, \chi_{A\mu}$$

the first summation in the formula (10) is given by

$$2 \sum_j \langle j|\, 1/r_{A1}\,|j\rangle \cong 2 \sum_j \sum_\mu^A c_{A\mu j}^2 \langle \chi_{A\mu}|\, 1/r_{A1}\, |\chi_{A\mu}\rangle$$

$$+ 2 \sum_j \sum_B{}' \sum_\nu^B c_{B\nu j}^2 \langle \chi_{B\nu}|\, 1/r_{A1}\, |\chi_{B\nu}\rangle \tag{12}$$

where j runs over all valence MOs and the neglect of the diatomic differential overlap approximation is employed which is characteristic of the CNDO etc. semiempirical schemes. If CNDO methodology is adopted, then it follows that all single center integrals $\langle \chi_{A\mu}|\, 1/r_{A1}\, |\chi_{A\mu}\rangle$ are equal to a common value K_A, since all radial functions of the STO orbital are assumed to be the same. One can also suppose that the two-center integrals are well approximated by $\langle \chi_{B\nu}|\, 1/r_{A1}\, |\chi_{B\nu}\rangle = 1/R_{AB}$, where R_{AB} is the interatomic distance. Surprisingly enough, this approximation is better than expected at first sight [41]. It follows that the approximate relation (12) takes a form

$$2 \sum_j \langle j|\, 1/r_{A1}\, |j\rangle \cong K_A Q_A + \sum_B{}' Q_B/R_{AB} \tag{13}$$

where Q_A and Q_B are total numbers of electrons situated on atoms A and B, respectively. Hence, the shifts in inner-core binding energies (10) are given by

$$\Delta(-\varepsilon_m) \cong \Delta[-K_A Q_A + \sum_B{}' (Z_B^{ef} - Q_B)/R_{AB}] . \tag{14}$$

Expression given in parentheses can be rephrased in terms of atomic charges $q_A = Z_A^{ef} - Q_A$ and by denoting $M_A = \sum_B{}' q_B/R_{AB}$ in analogy to the Madelung's term appearing in the solid state theory

$$\Delta(-\varepsilon_m) = \Delta[K_A q_A + M_A] . \tag{15}$$

Since shifts in binding energies are defined relative to a reference level in a predetermined molecule $\Delta(-\varepsilon_m)$ can be concisely written as

$$\Delta(-\varepsilon_m) = K_A q_A + M_A + L_A \tag{16}$$

where $L_A = K_A q_{Ar} + M_{Ar}$ and r denotes a reference molecule. This formula is in a full analogy with the expression developed by Siegbahn et al. [9, 10] on more intuitive grounds. It should be pointed out that the constant K_A can be interpreted equally well in terms of the repulsion between electrons of the inner-core and the valence shell, if the approximate relation (9) is not employed. This is however immaterial because K_A and L_A are considered as adjustable empirical parameters determined by the least square error procedure in fitting experimental binding energy shifts:

$$\Delta E_B(A) = K_A q_A + M_A + L_A \tag{17}$$

Although Eqs. (16) and (17) are formally identical, physically speaking they are not equivalent since by fitting experimental data one absorbs a good deal of the relaxation energy particularly in the first term. This is not necessarily the case in Eq. (16) where all entities are theoretically calculated.

It is noteworthy that Eq. (17) can be generalized in several different ways as we shall see later (vide infra). Nevertheless, a simple formula (17) provides a basis for the so called ground state potential model (GPM) which employs only the charge distribution of the neutral molecule [41]. In our terminology this approach will be referred to as the ground state atomic monopole electrostatic potential (GS-AMEP) model, which is very convenient for semiempirical exploitation. A posteriori, one can say that performance of the GS-AMEP model is surprisingly good provided that the estimated atomic charges are reliable. This approach, however, can not describe satisfactorily ESCA shifts in molecules widely differing in extent of the electron charge reorganization i.e. in systems varying grossly in size and shape. Explicit account of the relaxation should be made in this case, which is the subject of the next section.

3.2 Relaxation Potential Model

A lucid intuitive analysis of the relaxation energy influence on the electron binding energies in atoms was made by Snyder [41a]. His starting point was a Slater-Zener [41b] model of atoms based on the concept of the shell-structure, simple STO orbitals and the Aufbau principle. The total energy of an atom or ion is then given by

$$E_t = -\sum_n (N_n/n^2)(Z_A - s_n)^2 \tag{18}$$

where N_n is the number of electrons in orbitals of principal quantum number n with a shielding constant s_n in an atom of nuclear charge Z_A. The shielding constants are estimated by well known Slater rules and atomic units are employed. The binding energy of an electron in the m-th inner-shell reads:

$$E_B(m) = (-1/m^2)(Z_A - s_m)^2 + 2(1/m^2) Z_A(Z_A - s_m) - (2/m^2)$$

$$\times [0.85 N_{m-1} + \sum_{l=1}^{m-2} N_l] (Z_A - s_m) - (4/m^2)(N_m - 1)$$

$$\times (s_m - s'_m)(Z_A - s_m) + \sum_{n>m} (-2N_n/n^2)(s_n - s'_n)(Z_A - s_n) \tag{19}$$

where the prime denotes the screening constants of the cationic ion. These five terms have a simple physical meaning. If their negative values are considered, then they represent in the same order as in formula (19) the kinetic energy of the electron m, its attraction to the nucleus, its repulsion with all electrons in lower shells of n < m, its repulsion with the electron of the ionized shell m and finally its repulsion with electrons in higher shells n > m. To make a long story short, Snyder derived a formula for the relaxation energy in an ion with a 1s hole:

$$E_A^{relax}(1s) = (1.2 + 2.5N_2 + 1.5N_3)\,eV \tag{20}$$

where electron populations of n = 2 and n = 3 shells are denoted by N_2 and N_3 respectively, and atoms up to n = 3 shell are considered. This equation predicts that the relaxation energy is independent of the nuclear charge Z_A which is not quite correct. Nevertheless, the estimated magnitudes of E_A^{relax} for Ne^+ and Ar^+ 1s holes are within 5% of the Hartree-Fock values. The most important feature is however reflected in Eq. (20), namely that the reorganization energy is proportional to the charge density in upper (valence) shell(s). A meticulous analysis of the relaxation energy by Snyder gave an additional useful result. He concluded that relaxation should be a quadratic function of the charge of the inner-core hole. This is important in tackling the problem of the core hole localization in molecules possessing several equivalent sites like in homonuclear diatomics, acetylene, benzene, etc. For example, in O_2 molecule core hole(s) would be produced in $1\sigma_g$ or $1\sigma_u$ MOs, if the molecular orbital formalism is adopted. Such a hole should be delocalized in principle over both atoms and each oxygen would have an electron population of 7.5 electrons. However, Snyder's argument indicates that a distribution of a hole over N_c core centers is unfavorable. Indeed, it would produce a hole charge of $1/N_c$ for each center. The overall relaxation energy would be then given by $N_c(1/N_c)^2 = 1/N_c$ or in other words it would be N_c times smaller than in the localized state not to mention Coulomb repulsion. An early calculation of Bagus and Schaefer [43] on O_2 has conclusively shown that Snyder's conjecture was right and that in O_2^+ (1s; $^4\Sigma_g^-$) and O_2^+ (1s; $^2\Sigma_g^-$) states, core holes are localized.

Qualitatively, Snyder's approach can be generalized to molecules via the electrostatic potential model. For this purpose the seminal paper of Hedin and Johansson [44] should be considered first. They derived a simple expression for the correction to Koopmans' theorem which is intrinsically involved in the ΔSCF method. This correction is written in a form of a polarization or better termed as relaxation potential, which for the deep core levels gives roughly the same result as a second-order perturbation theory. The relaxation potential V_p is a sum

$$V_r = \sum_{k \neq m} (V_k^* - V_k) \tag{21}$$

It represents a change in the Hartree-Fock potential due to a removal of the inner-core electron m and describes the relaxation of the orbitals ψ_k given by $\psi_k^* = \psi_k + \delta\psi_k$. Briefly, the positive binding energy of the electron m is of the form

$$E_B(m) = -\varepsilon_m - (1/2)\langle m| V_r |m\rangle \tag{22}$$

where ε_m is the Hartree-Fock eigenvalue. An equivalent result was independently obtained by Liberman [45] $E_B(m) = (1/2) (\varepsilon_m + \varepsilon_m^*)$, where ε_m^* is the eigenvalue of the open-shell hole state. It appears that the relaxation energy is obtained by

$$E_R^{(m)} = (1/2) \langle m| V_r |m\rangle . \tag{23}$$

Hedin and Johansson [44] gave an interesting interpretation of the ionization process based on the Eq. (22). Ejection of electron m can be imagined as a two-step process. First, the interaction between electron m and outer shells is switched off adiabatically. The outer shells relax towards electron m and the increased Coulomb interaction is stored. The latter is one half of the change in the Coulomb potential. This means that the level of electron m is raised by extent of the matrix element $\langle m| (1/2) V_p |m\rangle$. In the second step electron m is expelled from the raised level to infinity meaning that the energy of ionization is lower then required by the Koopmans' theorem, as evident from Eq. (22).

3.2.1 Equivalent Core Approximation

Although the approach of Hedin and Johansson [44] is conceptually simple, it implies a hole-state calculation in order to obtain relaxed potential v_k^* in Eqs. (21–23). There is, however, a way to circumvent this obstacle by making use of the idea of "equivalent cores". A salient feature of this approximation is embodied in a fact that an electron in an inner orbital will almost completely shield an outer electron from one unit of nuclear charge. In other words, an outer orbital in an atom of nuclear charge Z_A with a hole in an inner-shell is closely mimicked by the corresponding outer orbital in the ground state of the atom $Z_A + 1$ of the next element in the periodic table. Objection that the inner orbital is contracted in atom $Z_A + 1$ is not warranted because this fact has very little effect on the Coulomb and exchange integrals between inner- and outer- electrons. As mentioned earlier a point-charge description of the inner orbital gives very good estimates of electrostatic interactions. Hence, the equivalent-core concept is well suited for semiempirical treatment of ESCA shifts, but it can be employed also within the ab initio SCF formalism as we shall see shortly. It is interesting to note that the equivalent-core concept represents a gedanken inverse process known in nuclear physics as the K-capture. In this type of nuclear transformation an $1s$ inner-core electron coalesces with the nucleus diminishing its charge by one unit.

We shall give now some illuminative examples which will illustrate adequacy of the equivalent-core (EC) approximation in simulating ΔSCF procedure. They serve a purpose of lending some credence to simple models based on the equivalent-core picture, which will be heavily used and discussed in the later stage. We commence with Shirley's equivalent core calculations [46] on some noble gas atoms by using Eq. (22). Orbital energies were taken from the work of Rosen and Lindgren [47], who published optimized relativistic Hartree-Fock-Slater estimates which in turn were essentially equal to relativistic HF results. Equivalent-core binding energies E_B^{ec} are compared with the core-hole calculations and measured values in Table 4. Perusal of data shows that equivalent core model works very well as evident from the good agreement with experiment. This accordance is in fact slightly better than that of the

Table 4. Comparison of noble-gas core-electron binding energies estimated by equivalent-core approximation and hole-state calculations with experimental data (in eV)

Orbital	E_B^{EC}	E_B^H	E_B (exp.)
Ne (1s)	868	870	870.2
Ar (1s)	3203	3209	3205.9
Ar (2s)	326	327	326.3
Ar ($2p_{1/2}$)	251	250	250.6
Ar ($2p_{3/2}$)	248	248	248.5
Kr (1s)	14351	14358	14326
Kr (2s)	1926	1933	1924.6
Kr ($2p_{1/2}$)	1730	1735	1730.9
Kr ($2p_{3/2}$)	1676	1681	1678.4
Kr (3s)	297	296	292.8
Kr ($3p_{1/2}$)	226	225	222.2
Kr ($3p_{3/2}$)	218	217	214.4
Kr ($3d_{3/2}$)	94	93	94.9
Kr ($3d_{5/2}$)	93	92	93.7

core-hole state calculations. A first application of the EC approximation in rationalizing ESCA shifts in molecules was made by Jolly and Hendrickson [48]. We shall illustrate the idea with two test molecules CH_4 and CO. The inner-core ionization energy of CH_4 is given by the energy ΔE_1 of the reaction

$$CH_4 \rightarrow C^*H_4^+ + e \qquad \Delta E_1 = E_B(C_{1s}, CH_4) \tag{24}$$

where the asterisk denotes a core-hole. We would like to eliminate the core-vacant state by using the corresponding EC model:

$$C^*H_4^+ + N^{5+} \rightarrow NH_4^+ + C^{*5+} \qquad \Delta E^{ec} = \delta_1. \tag{25}$$

This gedanken process is accompanied with a change in energy which is signified by δ_1. Addition of chemical "equations" (24) and (25) yields

$$CH_4 + N^{5+} \rightarrow NH_4^+ + C^{*5+} + e \qquad \Delta E_1 + \delta_1 \tag{26}$$

which gives the binding energy of the 1s electron if δ_1 is known or so small that it can be neglected. Suppose that we are interested in the ESCA chemical shift in CO relative to CH_4. Then one has to write down two relations analogous to (24) and (25) which give

$$CO + N^{5+} \rightarrow NO^+ + C^{*5+} + e \qquad \Delta E_2 + \delta_2 \tag{27}$$

where $\Delta E_2 = E_B(C_{1s}, CO)$. Combination of (26) and (27) provides a desired result:

$$CO + NH_4^+ \rightarrow CH_4 + NO^+ \qquad \Delta E \tag{28}$$

where the energy change of the characteristic reaction (28) gives $\Delta E = \Delta E_2 - \Delta E_1 + (\delta_2 - \delta_1)$. If the core-exchange energy δ does not change very much in different chemical moieties, then $\Delta E = E_B(C_{1s}, CO) - E_B(C_{1s}, CH_4)$ yields a difference in binding energies of the $1s$ electron in CO and CH_4. The energy of the characteristic reaction can be estimated by the experimental heats of formation or theoretically by ab initio calculations. Jolly [48, 49] has shown that a thermodynamical approach reproduces observed ESCA shifts with a good success. Subsequent theoretical studies have shown that core-exchange energies δ_1 are by no means small [50, 51]. Fortunately, they are not very sensitive on the chemical environment. A comment on the geometries in molecular (equivalent) core-hole states is in place. Molecular ground state structural parameters are commonly employed because it is supposed that heavy nuclei cannot follow very fast electron reorganization during a very short photoionization event. Average lifetimes of core hole states are typically within a range $10^{-13} - 10^{-17}$ s. For example, by using a formula $\tau_A = 3.8 \times 10^{-10}$ s$/Z_A^{3.93}$ one obtains average life times for the K-hole states of 0.33×10^{-14} s and 1.02×10^{-16} s for C and Ag atoms, respectively. Hence, it is reasonable to assume that nuclei remain frozen on the time scale of the ESCA experiment. An additional argument is provided by the fact that inner-shell electrons are placed in essentially non-bonding molecular orbitals. Nevertheless, some changes in bond distances occur giving rise to the fine structure of the spectra lines [3, 12, 52]. However, they can be safely neglected in approximate treatments which are discussed in particular in this chapter. For our purpose it is useful to keep in mind that the electron reorganization is completed during ejection of the core-electron whereas nuclei remain unrelaxed.

An important analysis of the EC approximation was made by Shirley [53]. He found that the quantum mechanical electrostatic potential at the nucleus [28, 37] is nearly equivalent to the thermodynamical EC scheme. Hence a development of the equivalent core formalism within the electrostatic potential model is fully justified. We shall commence a lapidary sketch of the plausible chain of approximation by recalling a relation of Liberman [45] discussed earlier:

$$E_B(m) = -(1/2)[\varepsilon_m + \varepsilon_m^*]. \tag{29}$$

Since we are interested in binding energy shifts relative to a reference level:

$$\Delta E_B(m) = (-1/2)[\Delta \varepsilon_m + \Delta \varepsilon_m^*] \cong (-1/2)[\Delta V_A(m) + \Delta V_A(m)^*] \tag{30}$$

where $V_A(m)$ denote the electrostatic potential at the nucleus A with an ionized inner-core electron m. By invoking EC approximation one obtains:

$$\Delta E_B(m) \cong (-1/2)[\Delta V_A(m) + \Delta V_{A+1}(m)^+] \tag{31}$$

where $A + 1$ denotes the atom with a nuclear charge $Z_A + 1$ of the equivalent core. A superscript $+$ signifies a fact that in the EC approach one electron is missing in the valence shell of the $Z_A + 1$ atom. This positive charge is subsequently delocalized over a molecule by the extra-atomic relaxation (charge drift) process (vide infra).

Keeping in mind that the relation (31) is approximate we shall employ equality sign from now on. An alternative form reads:

$$\Delta E_B(m) = -\Delta V_A(m) - \Delta E_A^{relax}(m) \tag{32}$$

where

$$E_A^{relax}(m) = (1/2)\,[V_{A+1}(m)^+ - V_A(m)] \tag{33}$$

is the relaxation energy given within the EC model. Formulas (31) and (32) explicitly involve reorganization energy and form a framework for the relaxation potential model(s). They can be easily reduced to the AMEP picture yielding:

$$\Delta E_B(m) = K_A(q_A + q_A^*) + (M_A + M_A^*) + (L_A + L_A^*) \tag{34}$$

in full analogy with Eq. (17), where asterisk denotes equivalent core entities. This approach will be termed relaxation atomic monopole electrostatic potential $(R - AMEP)$ model. Then the relaxation energy is obtained by

$$E_A^{relax}(m) = (1/2)\,[(K_A^* q_A^* - K_A q_A) + (M_A^* - M_A) + (L_A^* - L_A)] \tag{35}$$

Formulas (34) and (35) which represent a generalization of Snyder's Eq. (20) are convenient for semiempirical calculations. We shall, however, employ them in ab initio estimates of the C(1s) binding energies in fluorinated methanes and compare results with a performance of the $GS - AMEP$ model. This will illustrate importance of finding a suitable basis set in ESCA shift computations. Survey of the data obtained by utilizing Eqns. (17) and (34) presented in Table 5 shows that STO-3G basis set is too crude even for the simple AMEP model. On the other hand it is gratifying that the DZ 4-31G basis set has such a good performance particularly if the relaxation is taken into account. The average absolute error of 0.1 of the $R - AMEP$ model is close to experimental precision. It is interesting to note that increase in complexity of the basis set does not necessarily imply a better description of ESCA shifts as

Table 5. Performance of various basis sets in reproducing C(1s) ESCA shifts in fluoromethanes within the ground state and relaxation AMEP models (in eV)[a]

Mole-cule	STO-3G		4-31G		6-31G		6-31G**[b]		Exp.
	GS	R	GS	R	GS	R	GS	R	
CH_4	0.1	0.1	−0.2	−0.1	−0.2	−0.2	−0.2	−0.1	0
CH_3F	2.3	2.4	3.0	2.9	3.1	3.0	3.2	3.0	2.8
CH_2F_2	4.8	4.8	6.0	5.8	6.1	5.9	6.2	6.0	5.6
CHF_3	10.0	9.1	7.4	8.0	7.0	7.7	6.3	7.1	8.3
CF_4	9.4	9.5	11.3	11.1	11.3	11.2	11.4	11.4	11.0
CH_3Cl	2.1	2.1	1.7	1.6	1.7	1.6	1.9	1.7	1.6
CH_2Cl_2	3.8	3.7	3.3	3.2	3.4	3.2	3.7	3.4	3.2
Δ^{b}	0.8	0.7	0.3	0.1	0.4	0.2	0.6	0.3	

[a] Taken from ref. [31]. [b] Average absolute error

evidenced by the performance of 6-31G and 6-31G** basis sets (Table 5). We would like to stress in this connection that a choice of the adequate theoretical scheme (e.g. a suitable semiempirical method) is crucial for approximate but satisfactory rationalization of the inner-core binding energy variation in molecules and molecular crystals.

3.2.2. Transition Potential Model

EC approximation offers a good description of ESCA shifts in widely different intramolecular environments. However, it requires two calculations per ejected inner-core electron. A more economical approach is provided by the transition potential (TP) model. It is based on an idea of the transition operator. A usual ΔSCF scheme optimizes initial and final state spin-orbitals separately. On the contrary, transition opertator model employs a simple common set of spin orbitals which are optimized once. A salient feature is that the transition operator is constructed inter alia with spin-orbitals involved in the ionization which are occupied by half an electron. Employing transition operator formalism one obtains by Koopmans' theorem

$$\langle m| \hat{H}_T |m\rangle = \varepsilon_m^T \tag{36}$$

inner-core energies ε_m^T which intrinsically involve electron relaxation. It is shown that ε_m^T is close to the ΔSCF value [54–56] meaning that the hole-state problem is reduced to a simple Hartree-Fock calculation. This concept is easily translated into terms of the transition electrostatic potential model which is calculated with the average nuclear charge of the ionized atom between initial and final state. It is taken to be $(Z_A + Z_{A+1})/2 = Z_A + 1/2$. In other words, a pseudo-atom possessing a nuclear charge and a core electron population a half-way between the initial and final state $Z_{A+1/2}$ is employed in the calculation of wavefunctions. Reducing a transition electrostatic potential formalism to the AMEP model one obtains:

$$\Delta E_B(m) = K_A^{TP} q_A^{TP} + M_A^{TP} + L_A^{TP}. \tag{37}$$

It is also of interest to develop an expression for the relaxation energy E_R:

$$E_R = -\varepsilon_A - (-\varepsilon_A^{TP}) \cong (K_A q_A - K_A^{TP} q_A^{TP}) + (M_A - M_A^{TP}) + (L_A - K_A^{LTP}) \tag{38}$$

where entities without superscript refer to the initial ground state. Taking into account that $q_A = Z_A - n_A - Q_A$ and $q_A^{TP} = Z_A - n_A - Q_A^{TP} + 1/2$, where n_A is a number of inner-shell electrons in the ground state and Q_A denotes a number of valence shell electrons, one obtains

$$E_R = (K_A^{TP} - K_A) Q_A + K_A^{TP} \Delta Q_A - (M_A^{TP} - M_A) + L_A' \tag{39}$$

where $\Delta Q_A = Q_A^{TP} - Q_A$ is a charge transfer toward ionized atom and

$$L_A' = (Z_A - n_A) (K_A - K_A^{TP}) - (1/2) K_A^{TP} + (L_A - L_A^{TP}). \tag{40}$$

Formulas developed in Sects. 3.2.1 and 3.2.2 will be used in semiempirical calculations of the ESCA shifts discussed in the forthcomming text.

4 Qualitative Description of Inner-core Binding Energies

4.1 Semiempirical Calculations

Binding energy shifts in inner-core energies in a wide variety of sizeable molecules are most conveniently and economically studied at semiempirical level. However, a choice of a practical method is not trivial. The point is that methods based on zero differential overlap (ZDO) or on some sort of the neglect of diatomic overlap (NDO) approximation do not reflect most of the fine details of the electron charge distribution in molecules. For instance, massive CNDO/2 computations of ESCA shifts have shown that a single correlation for a particular atom is not possible and that separate calibrations are necessary for each family of characteristic molecules. As an example we note F(1s) shifts in substituted methanes and benzenes [59]. It seems that CNDO/2 wavefunctions should be necessarily deorthogonalized as evidenced by calculations of several one-electron properties [60–63]. Unfortunately, the effect of deorthogonalization was not examined in relation to ESCA shifts. On the other hand, the extended Hückel (EHT) method grossly overestimates intramolecular charge transfer thus being unsuitable for calculations of the ESCA shifts. We mention in passing that both CNDO/2 and EHT are notoriously unreliable in reproducing ESCA shifts of nitrogen atoms [64, 65]. To make a long story short, it appears that the best semiempirical approach in describing electron distribution in molecules is offered by the self-consistent charge (SCC-MO) method. It retains diatomic overlaps, which are of outmost importance in bonding phenomena, and mimicks the intramolecular charge drift by a suitable effective hamiltonian depending on the atomic electron charge populations [66]. Its performance has been tested by extensive calculations of various one-electron properties with considerable success [61, 62, 67]. One can say that the SCC-MO method satisfies two mutually almost exclusive requirements: practicability and reasonable accuracy of the results. Anticipating forthcomming discussion it can be safely stated that SCC-MO method reproduces ESCA shifts much better than not only CNDO/2 and EHT approaches but also more reliably than more modern semiempirical schemes like MINDO/3, MNDO and AM1 [68, 69].

Before dwelling on SCC-MO results in rationalizing and predicting ESCA shifts in molecules and molecular crystals, we would like to make two important points. The AMEP model is based on the idea of spherical atoms placed at the equilibrium positions and modified in a sense that they carry a formal atomic charge. This simple picture can be refined to include anisotropy of the electron density distributions at the sites of the nuclei by including higher electric multipoles. It was examined by Ellison and Larcom [70] and independently from a different standpoint by Maksić et al. [71]. Ellison and Larcom analysed Hartree-Fock orbital energies of inner-shell electrons much in a sense of the derivation of formulas (16) and (17), but in calculating two-center Coulomb and exchange integrals between $1s$ core and valence electrons they invoked multipole expansion approximation. The final formulas are formally the same with one notable difference: the Madelung term M_A involves atomic dipole and quadrupole contributions in addition to the usual monopole interaction [70]. Our starting point was the electrostatic potential exerted at the nucleus in question

[71]. Employing Taylor expansion of the $1/r$ operator placed at the host nucleus A one obtains:

$$1/|\vec{R}_B + \vec{r}'_B| = 1/R_B - (1/R_B^3)\,\vec{R}'_B \cdot \vec{r}'_B + (3/2R_B^5)$$

$$\times [x'_B y'_B X_B Y_B + x'_B z'_B X_B Z_B + y'_B z'_B Y_B Z_B] - (1/2R_B^5)$$

$$\times [(x'_B)^2\,(R_B^2 - 3X_B^2) + (y'_B)^2\,(R_B^2 - 3Y_B^2) + (z'_B)^2\,(R_B^2 - 3Z_B^2)] + \ldots$$

$$(41)$$

where capital letters signify distance and coordinates of an atom B relative to the ionized atom A, whilst prime denotes electron coordinates measured from the nucleus B. By using Slater type AOs and INDO approximation the electrostatic potential energy takes a form:

$$V_K = K_A Q_A - \sum_B{}' q_B/R_B + \sum_B{}' (10/\sqrt{3}\,\zeta_B R_B^3)\sum_\alpha \alpha_B P^B_{2s\,2p\alpha}$$

$$+ \sum_B{}' (6/\zeta_B^2 R_B^5)\sum_\alpha P^B_{2p\alpha\,2p\alpha}(R_B^2 - 3\alpha_B^2)$$

$$- \sum_B{}' (126/\zeta_B^2 R_B^5)\sum_{\alpha<\beta} \alpha_B \beta_B P^B_{2p\alpha\,2p\beta}\,. \qquad (42)$$

Here Greek letters denote coordinates x, y and z, ζ_B denotes effective nuclear charge, P stands for the elements of the first order density matrix and sum is extended over all atoms B provided $B \neq A$. It is easy to see that the formula (42) is rotationaly invariant. The first two terms correspond to analogous expressions (14–17). The third term represents atomic dipole contribution which is related to $2s$–$2p$ hybridization as evidenced by $P_{2s\,2p\alpha}$ matrix elements. The last two terms in (42) are related to atomic quadrupoles. Subsequent inclusion of dipole and quadrupole interactions has led to a slight decrease in standard deviations for C(1s) chemical shifts [71]. This is in line with a study of Ellison and Larcom [70] for carbon shifts. They found also a slight improvement of N(1s) shifts but considerable worsening of the results for O(1s) binding energies. The general conclusion is that inclusion of atomic anisotropy is not warranted [70, 71]. This conjecture should be taken cum grano salis since the executed calculations were of the very approximate CNDO/2 and INDO type. We shall see however that the atomic monopole (AMEP) approach can be justified a posteriori.

A more important observation, made first by Ellison and Larcom [72], is that $2s$ and $2p$ AOs have in fact different energies of Coulomb and exchange interactions with the inner-shell $1s$ electrons. Hence single-center populations of $2s_A$ and $2p_A$ subshells of a host atom should be separately parametrized with empirical weighting factors K_{As} and K_{Ap}. This refinement markedly improved performance of the CNDO/2 method in reproducing C(1s) and O(1s) ESCA shifts [72]. Since our approach is based on the electrostatic potential model exerted on the host nucleus, the difference in $2s$ and $2p$ AOs is reflected in different nonlinear screening factors leading to Eqs:

$$\Delta E_B(1s_A) = K_{A1} Q^A_{2s} + K_{A2} Q^A_{2p} + K_{A3} M_A + K_{A4}\,, \qquad (43)$$

$$\Delta E_B(1s_A) = K_{A1}(\zeta^A_{2s} Q^A_{2s} + \zeta^A_{2p} Q^A_{2p}) + K_{A2}(\zeta^{A*}_{2s} Q^{A*}_{2s} + \zeta^{A*}_{2p} Q^{A*}_{2p})$$

$$+ K_{A3}(M_A + M^*_A) + K_{A4}\,. \qquad (44)$$

Here Q_{2s}^A and Q_{2p}^A are populations of $2s$ and $2p$ sub-shells, respectively, and asterisk denotes the equivalent core entities. The last formula (40) explicitly involves relaxation. An analogous formula within the transition potential would read:

$$\Delta E_B(1s_A) = K_{A1}^{TP}Q_{2sA}^{TP} + K_{A2}^{TP}Q_{2pA}^{TP} + K_{A3}^{TP}M_A^{TP} + K_{A4}^{TP}. \tag{45}$$

Unlike many other authors we preferred to use the weighting factors of the Madelung terms diffrent from 1 in order to allow for a better description of the intramolecular charge transfer part of the relaxation energy. This is particularly important for the ground state (GS-AMEP) approach (43) and the corresponding earlier formula (17). The SCC-MO calculations of the intramolecular electrostatic potentials [73] based on the Clementi-Raimondi atomic orbitals [74] can be very favourably compared with the DZ ab initio values of Snyder and Basch [75]. They were given in a simple AMEP approximation and interstingly enough it was found that the coefficients of the Madelung term were practically equal to unity [73] for atoms H, C, N and O in widely different molecules. On the other hand, correlations with the experimental shifts show that the Madelung's empirical coefficient is significantly different from 1, indicating that it absorbs a portion of the charge drift relaxation [76]. Concomitantly, when the formula (17) is applied in rationalizing ESCA shifts it is taken for granted that its Madelung coefficient is empirically adjusted. Finally, it should be pointed out that in all SCC-MO computations of ESCA shifts Clementi-Raimondi [74] basis set is utilized, since it distinguishes $2s$ and $2p$ AOs. The best ground state experimental geometries are employed as a rule. In some cases MINDO/3 structural parameters were utilized. Although this method leaves much to be desired, inaccuracies in interatomic distance have a little effect on the calculated ESCA shifts. The reason lies in a fact that the first order correction of the $1/R_{AB}$ is given by $\Delta R_{AB}/R_{AB}^2$, where ΔR_{AB} is an error in estimating interatomic distance. Concomitantly, it should be small as a rule.

We shall illustrate performance of the AMEP model in rationalizing ESCA shifts by using SCC-MO wavefunctions. In particular, we shall discuss B(1s), C(1s), N(1s) and O(1s) inner-core binding energy variations in some widely different molecules encompassing characteristic bonding situations. Results for the boron ESCA shifts are displayed in Table 6. They are obtained by using ground state (GS), transition potential (TP) and equivalent core (EC) model calculations based on the SCC-MO wavefunctions [77, 78]. The corresponding formulae (17), (34), and (37) which involve three empirical parameters since we employ a separate weighting factor for Madelung term as mentioned earlier. One observes a good accordance with experiment with an average absolute error of 0.3 eV for all three approaches which is reasonable for the semiempirical level of calculation. Interestingly, inclusion of the relaxation energy does not improve the overall agreement with experiment. Generally speaking a trend of changs is well reproduced and the results are consistent with a simple intuitive picture that the $1s$ binding energy is proportional to the net positive charge of the host atom. Thus, the largest binding energy is found in BF_3 where the boron atom bears the largest positive formal charge. Since this level was chosen as a reference, the chemical shifts in core energy levels are negative. The first (one-center) term q gives the main contribution to the $1s$ binding energy. The small corrections arises from the charges residing on the neighbouring atoms via the Madelung term. Since

Table 6. Comparison of the theoretical gas phase B(1s) ESCA shifts, estimated by the ground state (GS), transition potential (TP) and equivalent core (EC) model calculations, with experimental data (in eV)*

Molecule	GS-AMEP	TP-AMEP	EC-AMEP	Exp.
BF_3	-0.2	-0.1	0.1	0
$B(OCH_3)_3$	-4.2	-5.3	-4.7	-4.4
B_2H_6	-7.0	-6.2	-6.3	-6.3
$B(CH_3)_3$	-5.8	-6.2	-6.1	-6.4
BH_3CO	-7.7	-6.7	-6.9	-7.6
$BH_3N(CH_3)_3$	-9.3	-8.8	-8.4	-9.1
$1,6-C_2B_4H_6$	-7.5	-7.5	-7.7	-7.4
$1,5-C_2B_3H_5$	-6.9	-7.1	-7.1	-6.8
$2,4-C_2B_5H_7$				
[1, 7]	-6.7	-7.3	-7.3	-6.7
[3]	-7.0	-7.4	-7.4	-7.2
[5, 6]	-7.6	-8.0	-8.0	-7.9
$B_2H_5N(CH_3)_2$	-7.3	-7.5	-7.0	-7.5
B_5H_9				
[1]	-8.3	-8.5	-8.6	-8.6
[2, 5]	-7.1	-7.2	-7.3	-6.7
Δ_{av}	0.3	0.3	0.3	

* Taken from Refs. [77] and [78]

boron is highly electropositive atom, its electronegative nearest neighbours are negatively charged. Hence they destabilize the 1s level B(1s) due to the Coulomb repulsion with the inner-shell electron. It follows that q and M terms exhibit opposite effects on the core electron. Furthermore, one can expect a rough linear dependence between the Madelung term and the net charge of the host atom. In fact, we found a nice straight line $M_B = -0.81q_B + 0.03$ eV, if the compounds $B(OCH_3)_3$ and $BH_3N(CH_3)_3$ are excluded. This linear relationship is the reason behind the success of the simple charge model q in correlating ESCA shifts. This conjecture holds quite generally although it should be strongly pointed out that Madelung term is important in a number of molecules.

Inspection of Table 6 reveals that intramolecular shifts in B_5H_9 and $2,4-C_2B_5H_7$ are also well reproduced just like the ESCA shifts between various molecules. In order to check the predictive ability of the AMEP model we performed calculations on the series BH_3, BH_2F and BHF_2 which were not used in the parametrization procedure. The B(1s) shifts relative to BF_3 obtained by the GS-AMEP approach are -2.2, -4.4, and -6.8 eV, respectively [77]. One notices that the B(1s) binding energies of the multiply hydrogenated molecules exhibit a typical additive effect. Our predicted B(1s) shift in BH_3 can be favourably compared with the DZ ab initio result of Snyder and Basch [75] of -6.944 eV. On the other hand, agreement with the ab initio data of Schwartz and Allen [79] is less satisfactory presumably due to the use of a less flexible basis set in their work.

Carbon 1s gas phase ESCA shifts for a rather large specimen of molecules are given in Table 7. They cover a large variety of bonding patterns arising due to several

Table 7. Comparison of the theoretical gas phase C(1s) ESCA shifts, estimated by the ground state (GS), transition potential (TP) and equivalent core (EC) model calculations, with experimental data (in eV) *

Molecule	GS-AMEP	TP-AMEP	EC-AMEP	Exp.
HCN	1.5	2.6	2.5	2.6
CO	3.0	4.7	4.9	5.3
CO_2	6.1	6.6	6.8	6.8
H_2CO	3.1	3.7	3.6	3.3
CHF_3	8.7	8.5	8.6	8.3
CF_4	11.1	10.8	11.1	11.0
CHOOH	4.7	4.8	5.0	5.0
CH_2F_2	6.0	6.0	5.9	5.6
CH_3OH	2.0	2.1	2.0	1.8
CH_3F	3.0	3.2	3.1	2.8
CH_4	−0.2	0.3	0.2	0.0
C_2H_2	−0.2	0.3	0.3	0.4
C*HCHCHCHO	1.7	1.1	1.3	0.8
CHC*HCHCHO	0.6	−0.2	0	−0.4
C_2H_4	0	−0.1	−0.1	−0.1
C_2H_6	0	0.1	0.1	−0.2
C_3H_6	−0.2	−0.3	−0.5	−0.3
$C*(CH_3)_4$	0.5	−0.5	−0.3	−0.4
CH_3C*HO	3.0	3.0	3.1	3.2
$C*H_3CHO$	0.6	0.5	0.3	0.6
OC*CO	3.5	3.3	3.3	4.2
$C*H_3COOH$	0.4	−0.2	0.1	0.7
$(C*H_3)_2CO$	0.5	0.4	0.3	0.5
$(CH_3)_2C*O$	2.9	2.6	2.9	3.1
C_2F_6	9.3	8.8	8.6	8.9
$C*H_3CH_2OH$	0.4	0.3	0.2	0.2
CH_3C*H_2OH	1.9	1.6	1.7	1.6
C_6H_6	0.2	−0.5	−0.4	−0.5
$C*H_2CF_2$	1.4	0.9	0.5	0.4
CH_2C*F_2	5.4	5.2	5.3	5.1
C_2H_4O	2.0	1.8	1.8	1.8
OCC*O	1.0	0.5	−	0.8
Δ_{av}	0.4	0.2	0.2	

* Taken from Refs. [77] and [78]

possible coordination numbers (n = 4, 3 and 2). The calculated values are obtained by the same formalism as applied to B(1s) shifts. The range of ESCA shifts is rather large (11 eV) and they are measured relative to CH_4 (290.8 eV). It is interesting to mention that C(1s) binding energies exhibit only a mild dependence on the coordination number or in other words on the hybridization state of the carbon atom. This is evidenced by the shifts in C_2H_6, C_2H_4 and C_2H_2 which are −0.2, −0.1 and 0.4 eV, respectively. There is clear tendency of increasing C(1s) binding energy in passing from sp^3 to sp hybridization but the effect is not highly pronounced. It is not unexpected because ESCA shifts strongly depend on intramolecular electrostatics

Table 8. Changes in C(1s) binding energies upon fluorination of methane (in eV)

Fluorination	ε_{1s}^a	ε_{1s}^b	ΔSCF^c	GS^d	TP^e	EC^e	Exp.
$CH_4 \rightarrow CH_3F$	4.8	2.9	3.4	3.2	2.9	2.9	2.8
$CH_3F \rightarrow CH_2F_2$	4.9	3.2	3.0	3.0	2.8	2.8	2.75
$CH_2F_2 \rightarrow CHF_3$	5.0	3.3	3.4	2.7	2.5	2.7	2.75
$CHF_3 \rightarrow CF_4$	–	3.3	–	2.4	2,3	2.5	2.70

[a] Koopmans theorem ab initio results [80]. [b] Koopmans theorem ab initio results [81]. [c] ΔSCF results of Brundle et al. [81]. [d] Ref. [77]. [e] Ref. [78]

and to a much lesser extent on covalent effects. Since the dominant feature is intramolecular charge transfer, it is expected that the largest shift should be found in CF_4. This is indeed the case since carbon bears the highest formal charge (0.73 |e|). Perusal of the results shows that a rule of thumb works very well: larger positive charge of the carbon atom undergoing ionization leads to higher binding energy. Let us focus our attention on the stepwise fluorination in a series: CH_4, CH_3F, CH_2F_2, CHF_3 and CF_4. One observes that there is an almost uniform increase in experimental C(1s) binding energy upon fluorination. However, a closer inspection reveals that increments are not exactly constant being 2.8, 2.75 and 2.70 eV, respectively. The SCC-MO calculations within GS, TP and EC potential model reproduce relatively well a decrease in increments (Table 8). On the other hand, ab initio calculations with the frozen MOs approximation [80] and ΔSCF level [81] were not able to give a right answer (Table 8) presumably due to modest basis sets employed. This example illustrates rather nicely that simple models can be very powerful in some instances. It is noteworthy that differences in the SCC-MO q_C charges between the subsequent members of the series $CH_4 \rightarrow CF_4$ are decreasing assuming values of 0.22, 0.20, 0.18 and 0.16 |e|. This is concomitant with ESCA shifts. The fluorination of ethane and ethylene fits also the simple picture that the C(1s) binding energy is proportional to the net positive charge of the considered carbon atom. The multiple fluorination is roughly additive but it should be pointed out that the effect of substitution at the adjacent carbon atom is smaller by an order of magnitude. Finally, it should be stressed that explicit treatment of relaxation cuts the errors by half on average and that TP and EC approaches have about the same performance.

Atoms possessing lone pair(s) are difficult to tackle by the semiempirical theory. We shall consider in some detail ESCA shifts of N(1s) and O(1s) binding energies. Results obtained by the same procedure for F(1s), Si(2p), S(2p) and Ge(3p) levels are available elsewhere [76, 83–85]. A simple comment will be made here only. It is related to F(1s) ESCA shifts. In spite of a fact that fluorine atom has three lone pairs, its core electron binding energies are surprisingly accurately reproduced even at the GS-AMEP level [76]. The reason lies perhaps in the fact that fluorine AOs are very compact thus complying better with approximations inherent in the AMEP model. Semiempirical estimates of N(1s) shifts are tested against the experimental data in Table 9. The overall qualitative agreement with measured values is good although the average absolute error is greater than in carbon atoms. In order to increase performance of the EC model we employed the four-parameter formula (44). In spite of that explicit treatment electron reorganization did not improve results over the

Table 9. Comparison of the theoretical gas phase N(1s) ESCA shifts, estimated by the ground state (GS), transition potential (TP) and equivalent core (EC) model calculations, with experimental data (in eV) *

Molecule	GS-AMEP	TP-AMEP	EC-AMEP**	Exp.
N_2	−1.3	0	−0.5	0
HCN	−3.6	−2.8	−3.1	−3.1
$(CN)_2C_2(CN)_2$	−3.8	−4.2	−4.0	−2.8
ONF_3	7.3	6.5	6.8	7.1
NF_3	4.6	4.4	4.5	4.3
N_2F_4	2.6	2.9	2.7	2.4
N*NO	−1.2	−1.6	−1.6	−1.3
NN*O	2.2	2.6	2.4	2.6
N_2H_4	−3.5	−3.6	−3.6	−3.8
NH_3	−4.1	−3.9	−4.2	−4.3
CH_3NH_2	−4.5	−4.4	−4.5	−4.8
$(CH_3)_3N$	−4.8	−5.3	−5.0	−5.2
CH_3NO_2	2.4	2.5	2.7	2.2
Δ_{av}	0.4	0.5	0.4	

* Taken from Refs. [78] and [82]. ** Results obtained by the formula (44)

GS-AMEP model. Although the average error is sizeable, it is noteworthy that the SCC-MO performs much better than MINDO/3, MNDO and AM1 methods [68, 69] not to mention somewhat obsolete EHT and CNDO/2 approaches [64, 65]. It is gratifying that several facets are nicely reproduced. For example, the destabilizing effect of the CH_3 substitution on the N(1s) level in NH_3 is well described. It appears that the inner-core binding energy is decreasing along the series NH_3, CH_3NH_2, $(CH_3)_2NH$ and $(CH_3)_3$ due to the increasd negative charge of the nitrogen atom. Further, the nonequivalent N atoms in NNO are clearly distinguished. A disturbing detail that the reference level in N_2 assumes a large negative value in the GS-AMEP model is remedied by taking into account relaxation energy. Indeed, it becomes virtually zero in the TP-AMEP framework as it should be by definition. The EC-AMEP falls somewhat short in this respect (Table 9).

The calculated O(1s) shifts are presented in Table 10. The quality of the results is comparable to the accuracy obtained for the nitrogen N(1s) levels with a notable exception — GS-AMEP model has a somewhat worse performance. In other words, relaxation of the electron distribution embodied in TP-AMEP and EC-AMEP models does improve agreement with experiment. For instance, EC-AMEP value for O_2 is close to zero (0.3 eV), whereas GS-AMEP model makes a three-times greater error. It is encouraging that different oxygen atoms are clearly distinguished within the same molecule implying that the semiempirical SCC-MO approach is potentially useful as an aid in assigning ESCA peaks. Needless to say, it is more successful in reproducing O(1s) levels than methods based on NDDO approximation [68].

Inner-core binding energy in solids contains implicitly a value of the work function and a contribution of the intermolecular relaxation. One has to make therefore separate parametrization of the AMEP model for molecular crystals [86–88]. We shall briefly discuss ESCA shifts in some biologically important compounds here.

Table 10. Comparison of the theoretical gas phase $O(1s)$ shifts, estimated by the ground state (GS), transition potential (TP) and equivalent core (EC) model calculations, with experimental data (in eV)*

Molecule	GS-AMEP	TP-AMEP	EC-AMEP**	Exp.
HCO*OH	−4.0	−4.1	−4.0	−4.3
C_3O_2	−2.9	−3.6	−3.5	−3.5
CO	−1.8	−1.7	−1.8	−1.0
CO_2	−2.3	−2.6	−2.4	−2.4
CH_3CHO	−4.4	−4.4	−4.4	−5.5
$(CH_3)_2CO$	−4.7	−4.8	−4.8	−4.1
HCOO*H	−2.9	−2.9	−2.8	−2.7
CH_3CO*OH	−4.0	−4.0	−3.9	−4.9
CH_3COO*H	−2.9	−2.8	−2.7	−3.1
H_2O	−4.2	−3.0	−3.3	−3.6
CH_3OH	−4.9	−4.0	−4.3	−4.4
C_2H_5OH	−5.0	−4.2	−4.5	−4.5
C_2H_4O	−5.2	−4.9	−5.0	−4.9
N_2O	−2.7	−2.9	−2.9	−2.1
C_4H_4O	−3.9	−4.1	−4.1	−3.6
SOF_2	−5.2	−4.6	−4.6	−3.7
SO_2	−3.7	−3.9	−3.7	−3.5
CH_3NO_2	−3.7	−3.7	−3.6	−4.1
O_2	0.9	0.4	0.3	0
Δ_{av}	0.6	0.5	0.4	

* Taken from Refs. [76] and [78]. ** Results obtained by the formula (44)

Uracil and some of its derivatives (Fig. 1) are considered first. The TP- and EC-AMEP results [86] are given in Table 11. Perusal of the data shows a good agreement with experiment. There are very few discrepancies. It should be mentioned in this connection

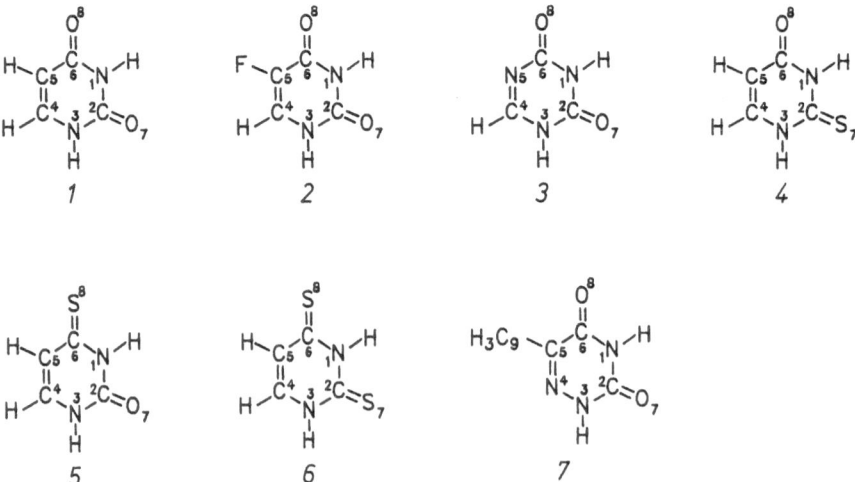

Fig. 1. Schematic representation of pyrimidines: uracil (*1*), 5-fluorouracil (*4*), 4-thiouracil (*5*), 2,4-dithiouracil (*6*) and 6-azathymine (*7*)

Table 11. Comparison of the theoretical solid state core electron ESCA shifts in uracil and its derivatives, obtained by the TP- and EC-AMEP model, with experimental data* (in eV)

Compound	Atom	TP-AMEP[a]	EC-AMEP[b]	Exp.
Uracil	C_6	− 3.8	− 3.8	− 3.9
1	C_5	− 5.1	− 5.1	− 5.1
	C_4	− 1.9	− 1.9	− 1.8
	C_2	− 0.9	− 0.9	− 1.0
	N_3	− 9.1	− 8.9	− 9.2
	N_1	− 9.5	− 9.2	− 9.2
5-Fluorouracil	C_4	− 1.2	− 1.2	− 1.4
2	C_2	− 0.8	− 0.8	− 0.8
	C_5	− 2.4	− 2.4	− 3.6
	C_6	− 3.2	− 3.3	− 3.0
	N_3	− 9.4	− 9.0	− 8.7
	N_1	− 9.7	− 9.3	− 8.7
5-Azauracil	C_4	− 0.9	− 0.9	− 1.3
3	C_2	− 0.7	− 0.6	− 0.8
	C_6	− 2.8	− 2.8	− 2.7
	N_1	− 9.0	− 8.8	− 8.7
	N_3	− 9.4	− 9.1	− 9.2
	N_5	−10.4	−10.2	−10.3
2-Thiouracil	C_4	− 2.2	− 2.2.	− 1.8
4	C_2	− 3.2	− 3.2	− 2.2
	C_6	− 4.3	− 4.3	− 3.8
	C_5	− 5.3	− 5.4	− 5.2
	N_3	− 9.5	− 9.2	− 9.0
	N_1	−10.0	− 9.6	− 9.0
4-Thiouracil	C_4	− 4.1	− 4.1	− 3.4
5	C_2	− 1.3	− 1.3	− 1.1
	C_6	− 4.4	− 4.4	− 4.2
	C_5	− 5.6	− 5.6	− 5.2
	N_3	− 9.5	− 9.2	− 9.2
	N_1	− 9.7	− 9.4	− 9.2
2,4-Dithiouracil	C_4	− 4.2	− 4.2	− 3.3
6	C_2	− 3.3	− 3.3	− 2.2
	C_6	− 4.6	− 4.6	− 4.1
	C_5	− 5.6	− 5.6	− 5.1
	N_3	− 9.6	− 9.4	− 9.2
	N_1	− 9.8	− 9.6	− 9.2
6-Azathymine	C_4	− 1.7	− 1.7	− 2.0
7	C_5	− 4.1	− 4.1	− 4.2
	C_9	− 4.4	− 4.3	− 4.3
	C_2	− 0.7	− 0.7	− 1.1
	N_3	− 9.1	− 8.8	− 9.0
	N_6	− 9.8	− 9.6	− 9.6
	N_1	− 8.6	− 8.4	− 8.7

* The C(1s) and N(1s) experimental shifts are determined relative to the gas-phase reference levels in CH_4 and N_2 (290.7 and 409.9 eV, respectively).
[a] Formula (37) is employed with the weighted Madelung term. [b] Formula (44)

that the standard deviation of the experimental levels is ~ 0.2 eV whereas the average absolute errors of EC and TP calculations are 0.3–0.4 eV. Although the inaccuracies in measurements and in the calculations are relatively high, some interesting conclusions can be drawn. Let's consider C(1s) shifts first. It is well established by now that the inner-shell binding energies are determined mainly by electron density residing on the host atom. In well-localized systems this density depends predominantly on the immediate chemical environment. It follows that the C(1s) BE shifts of carbon atoms participating in specific structural units should be similar. The situation is somewhat different in delocalized systems, where mobile π electrons are able to transfer the influence of distant fragments and substituents. Nevertheless, even in these cases BE's may be found which differ very little. For instance, $C_2(1s)$ shifts in *1, 2, 3, 4* and *8* vary within only ± 0.2 eV around the mean value which coincides with the BE shift in *2*. Further, $C_2(1s)$ shifts in *5* and *7* are close because the immediate neighbourhood is the same etc. however, the great utility of the ESCA technique lies in a fact that the same kind of atoms in differing environments will have significantly different inner-core levels. A number of electronegative atoms surrounding carbon atom studied has a decisive influence on the C(1s) shifts. It is therefore not surprising that the highest C(1s) binding energy is found at the carbonyl carbon C_2 directly bonded to the to ring nitrogens. The opposite effect is exerted by the electropositive sulfur.

Two basically different kinds of nitrogen atoms can be distinguished: pyrimidine-like $(-N=)$ and pyrrole-like $(-NH-)$. Their gross atomic populations differ greatly, because pyrimidine-like nitrogen possesses sp^2-sigma lone pair, which is energetically very convenient for accomodation of electrons. On the contrary, pyrrole-like nitrogen has a lone pair placed in the π-orbital. It is considerably delocalized over the ring as a consequence of the back-bonding effect. Consequently, the pyrimidine-like N atom should have higher electron density and lower N(1s) binding energy. This is nicely borne out by the AMEP model (Table 11). Finally, a word on the unresolved N_1 and N_3 peaks in *1, 2* and *6* is in place. Our calculations indicate that $N_3(1s)$ BE

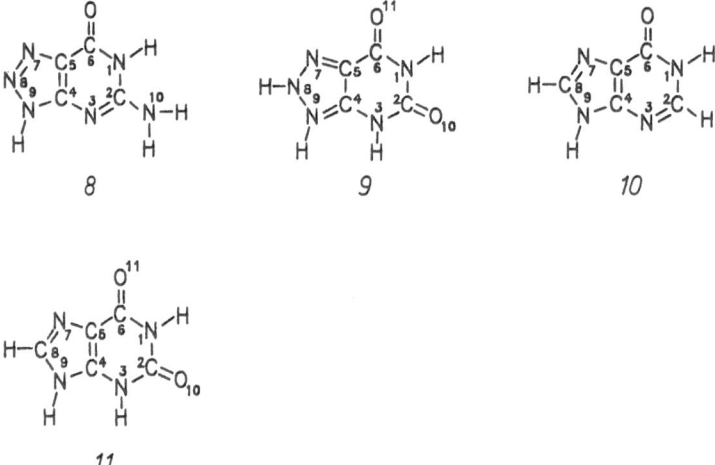

Fig. 2. Schematic representation of purines: 8-azaguanine (*8*), 8-azaxanthine (*9*), hypoxanthine (*10*) and xanthine (*11*)

is somewhat higher (by ~ 0.3 eV) and that the splitting is a consequence of the variation in the relaxation energy [86].

Results obtained in some biologically interesting purines (Fig. 2) are displayed in Table 12. TP- and EC-AMEP formulae (37) and (44) were applied. Semiempirical results are in good accordance with experiment as evidenced by low average errors

Table 12. Comparison of the theoretical solid state core electron ESCA shifts in some purines, obtained by the TP- and EC-AMEP models, with experimental data* (in eV)

Compound	Atom	TP-AMEP	EC-AMEP	Exp.
8-Azaguanine	C_6	− 1.9	− 1.8	− 1.8
8	C_5	− 4.3	− 4.3	− 4.9
	C_4	− 3.3	− 3.4	− 3.3
	C_2	− 2.5	− 2.6	− 2.0
	N_3	−10.6	−10.4	−10.3
	N_1	− 9.2	− 8.9	− 8.9
	N_7	− 9.3	− 9.4	− 9.7
	N_9	− 8.4	− 8.2	− 8.9
	N_8	− 8.9	− 8.9	− 8.9
	N_{10}	−10.0	−10.3	−10.3
8-Azaxanthine	C_6	− 1.6	− 1.6	− 1.9
9	C_2	− 0.9	− 0.9	− 1.0
	C_4	− 3.4	− 3.4	− 3.4
	C_5	− 4.2	− 4.3	− 4.7
	N_3	− 9.2	− 8.9	− 9.0
	N_1	− 9.4	− 9.1	− 9.0
	N_7	− 9.9	− 9.7	− 9.6
	N_8	− 7.9	− 7.7	− 8.1
	N_9	−10.2	−10.0	−10.6
Hypoxanthine	C_6	− 2.0	− 2.0	− 1.9
10	C_5	− 4.9	− 4.6	− 5.0
	C_4	− 3.7	− 3.8	− 3.6
	C_2	− 3.3	− 3.3	− 3.2
	C_8	− 3.8	− 3.8	− 4.7
	N_3	−10.5	−10.3	−10.1
	N_1	− 9.1	− 8.8	− 8.9
	N_7	−10.3	−10.3	−10.1
	N_9	− 9.3	− 9.0	− 8.9
Xanthine	C_8	− 4.0	− 4.0	− 4.0
11	C_6	− 1.9	− 1.9	− 1.9
	C_5	− 4.8	− 4.8	− 4.9
	C_4	− 3.5	− 3.5	− 3.5
	C_2	− 1.0	− 1.0	− 1.0
	N_3	− 9.1	− 8.9	− 9.1
	N_1	− 9.5	− 9.2	− 9.1
	N_7	−10.5	−10.5	−10.4
	N_9	− 9.6	− 9.3	− 9.1

* The $C(1s)$ and $N(1s)$ experimental shifts are determined relative to the gas-phase reference levels in CH_4 and N_2 (290.7 and 409.9 eV, respectively).
[a] Formula (37) is applied with the weighted Madelung term. [b] Formula (44)

of 0.2–0.3 eV. The ESCA C(1s) and N(1s) shifts exhibit a pattern similar to that found in pyrimidine bases discussed earlier (Table 11). The smallest absolute values of C(1s) shifts are found in carbon atoms surrounded by three highly electronegative heteroatoms in harmony with a simple rule of thumb. For example, $C_2(1s)$ binding energies in 8-azaguanine and hypoxanthine are 288.1 and 287.4 eV, respectively, according to EC-AMEP model. This is in line with the number of neighbouring N atoms (three and two, respectively) and the corresponding gross atomic electron populations of 3.81 and 3.87 e, thus being consistent with the aforementioned simple picture. Again, carbon atoms with similar immediate environment have very close 1s binding energies. For example, the $C_6(1s)$ energy variations are all within the ± 0.2 eV range measured from the C_6 value in 8-azaguanine. One can distinguish three types of nitrogens in a series of studied molecules: amino (NH_2), pyrimidine-like ($-N=$) and pyrrole-like ($-N-$) atoms. Interestingly, atoms N_{10} and N_3 in 8 have practically the same N(1s) binding energies of 399.5 eV which is accidental. The main reason behind it are very close gross atomic populations of 5.15 and 5.17 e, respectively. Nitrogen 1s binding energies in similar bonding situations are almost equal as expected. This is evident by comparison of $N_1(1s)$ BEs in 9 and 11, $N_7(1s)$ and $N_9(1s)$ in 10 are close to the corresponding values in 11 etc. Finally, we predict splitting of $N_8(1s)$ and $N_9(1s)$ values in 8 by 0.7 eV as well as a more stable $N_3(1s)$ level than both $N_1(1s)$ and $N_9(1s)$ levels in 11 (Table 12).

Nucleic acid bases (Fig. 3) play a crucial role in molecular biology. Therefore they were studied by the SCC-MO AMEP model in particular since previous calculations of the ESCA shifts were not satisfactory. The CNDO/2 estimates were disapointingly poor [89] whereas a comparison between experimental values and ab initio orbital energies has led to two widely different correlations for nitrogen (1s) shifts in cytosine and thymine [90]. In contrast, SCC-MO AMEP approach yields quite reliable estimates of the ESCA shifts in nucleic acid basis adenine being an exception which

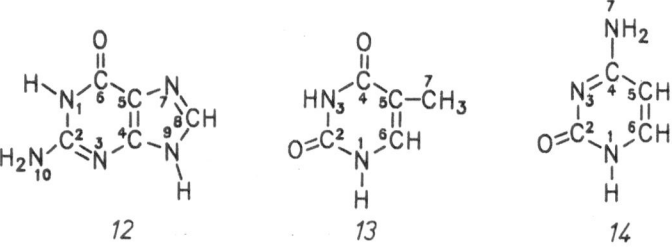

Fig. 3. Schematic representation of nucleic acid bases: guanine (12), thymine (13), cytosine (14) and adenine (15)

will be discussed later on. The results presented in Table 13 were obtained by formulas (17), (43) and (44). The average errors for guanine, thymine and cytosine are 0.3 eV for all three expressions. This is comparable to experimental accuracy which is believed to be ± 0.2 eV for intramolecular and ± 0.3 eV for intermolecular shifts. It appears that provision made for different $2s$ and $2p$ contributions or an explicit account of the relaxation offers no significant advantage over the standard GS-AMEP

Table 13. Comparison of the theoretical solid state by core electron ESCA shifts in nucleic acid bases, obtained by the GS-AMEP and EC-AMEP models, with experimental data* (in eV)

Compound	Atom	GS-AMEP[a]	GS-AMEP[b]	EC-AMEP[c]	Exp.
Guanine	C_2	− 2.5	− 2.4	− 2.5	− 2.2
12	C_4	− 3.4	− 3.4	− 3.7	− 3.5
	C_5	− 4.1	− 4.6	− 4.7	− 4.9
	C_6	− 2.2	− 2.4	− 2.4	− 2.0
	C_8	− 3.9	− 3.7	− 4.1	− 4.2
	N_1	− 8.9	− 9.0	− 9.0	− 9.1
	N_3	−11.0	−10.9	−10.9	−10.4
	N_7	− 9.9	−10.0	− 9.9	−10.4
	N_9	− 9.1	− 9.2	− 9.1	− 9.1
	N_{10}	−10.1	− 9.8	− 9.9	−10.4
Thymine	C_2	− 1.1	− 1.3	− 0.8	− 0.8
13	C_4	− 2.0	− 2.2	− 1.9	− 2.2
	C_5	− 4.8	− 4.9	− 5.1	− 4.9
	C_6	− 4.1	− 4.0	− 3.9	− 4.1
	C_7	− 5.4	− 5.4	− 5.4	− 5.6
	N_1	− 8.5	− 7.7	− 7.7	− 7.6
	N_3	− 8.8	− 8.9	− 8.9	− 8.8
Cytosine	C_2	− 1.6	− 1.7	− 1.6	− 1.3
14	C_4	− 3.1	− 2.9	− 3.1	− 2.8
	C_5	− 5.2	− 5.5	− 5.1	− 5.3
	C_6	− 4.0	− 3.7	−	− 4.2
	N_1	− 8.1	− 7.6	− 7.9	− 8.5
	N_3	− 9.9	−10.5	−10.0	−10.3
	N_7	−10.0	− 9.7	− 9.7	− 9.4
Adenine	C_2	− 3.9	− 3.7	− 4.2	− 5.0
15	C_4	− 3.3	− 3.4	− 3.6	− 4.1
	C_5	− 4.2	− 4.6	− 4.6	− 6.0
	C_6	− 3.1	− 3.1	− 3.4	− 2.9
	C_8	− 3.8	− 3.6	− 3.8	− 4.5
	N_1	−10.0	−10.3	−10.0	−11.3
	N_3	−10.0	−10.3	− 9.9	−10.8
	N_7	−10.3	−10.0	−10.4	−10.4
	N_9	− 9.0	− 9.2	− 9.1	− 9.0
	N_{10}	− 9.2	− 9.9	− 9.1	−10.3
Δ_{av}[d]		0.3	0.3	0.3	

* The C(1s) and N(1s) experimental shifts are determined relative to the gas-phase reference levels in CH_4 and N_2 (290.7 and 409.9 eV, respectively).
[a] Formula (17) applied with the weighted Madelung term. [b] Formula (43). [c] Formula (44).
[d] Average absolute errors for guanine, thymine and cytosine

model (17). The general features of the ESCA shifts here follow the same pattern found and discussed in great detail previously when purines and pyrimidines were considered. We would just like to point out that the $N_3(1s)$ level in guanine should be by 1 eV less stable than $1s$ local AOs at N_7 and N_{10} sites, what is not borne out by experiment. However, the overall agreement with measured data for guanine, thymine and cytosine is good. This is remarkable since earlier semiempirical work was unsuccessful as already mentioned. Calculations of Yonezawa et al. [91] gave much larger errors than our SCC-MO AMEP model. Bosa et al. [89] employed CNDO/2 method within both GS and RP formalisms. These two approaches gave widely different ordering of the inner-core levels both being far from the linear relations with experimental data. It was concluded that the ground state computations have low information content. This conclusion is not justified in view of a very good performance of the formulae (17) and (43) if SCC-MO wavefunctions are used. Hence unsatisfactory results obtained by the GS approach can be traced down to imperfections of the CNDO/2 method. Finally, large discrepancies between our results and observed values in adenine (Table 13) show that new experimental measurements would be well advised.

A general comment of the solid state ESCA shifts is appropriate: A difference between the gas and crystal values yields a sum of work function φ_W and specific effects appearing in crystals like intermolecular hydrogen bond influence, intermolecular Coulomb interactions etc. and extramolecular relaxation energy:

$$E_B^A(gas) - E_B^A(solid) = \varphi_W + E_A^{relax}(extra) + E_A^{ef}(cryst), \qquad (46)$$

where $E_A^{ef}(cryst)$ stands for effects characteristic for the crystal in question whereas $E_A^{relax}(extra)$ signifies extramolecular final state effect. It is difficult to delinate various terms in Eq. (46). Their sum however exhibits a very small variation at least for planar molecules. Our calculations show that $E_B^A(gas) - E_B^A(solid)$ values span a range between 4–6 eV. This holds for a set of molecules encompassing purines [87], nucleic acid bases as well as thiophene, pyrrole and furan [88]. The average values for the difference (46) are 4.9 and 5.6 eV for carbon and nitrogen atom, respectively. Now, the work function is found to be very insensitive to the size of the planar hydrocarbons [92] being practically constant (~ 4.0 eV). It is plausible that a similar value holds for nucleic acid bases, purines and pyrimidines discussed in this section. Thus one obtains a rough estimate that $E_A^{relax}(extra) + E_A^{ef}(cryst)$ varies between 1–2 eV and that it is larger for N atoms by 0.7 eV on average.

Since ESCA is a very fast technique and gives an "instantaneous snapshot" of a molecule, it is well suited for studying systems undergoing fast changes like e.g. topomerization. Keto-enol tautomerism of β-diketones seems to be such a case [93]. The problem of relative stabilities of keto- and enol-forms as well as the nature of the internal hydrogen bond in the latter has been a subject-matter of a long standing controversy. It is well established by now that the enol tautomers are thermodynamically more stable than the keto-forms [93]. It is also perfectly clear that appearance of the asymmetrical intramolecular hydrogen bond is more a rule than an exception. In this connection one should point out that some experimental techniques which operate on larger time scales like e.g. NMR and ED may lead to erroneous structures possessing symmetric internal hydrogen bonds in prototype molecules like malonaldehyde (MA), acetylacetone (ACAC) and in their derivatives. It is gratifying that correct answers have been provided by ESCA spectroscopy [94, 95] and by theoretical calculations of the inner-core binding energies [96, 96]. We illustrate SCC-MO GS-AMEP results for MA and ACAC with Table 14 where they are compared with the experimental data and ΔSCF calculations. It appears that both measured values and theoretical predictions are consistent with inner-core binding energies which should appear in the enol-forms of MA and ACAC. In particular, the difference between keto- and alcohol oxygen BEs is well reproduced by the GS-AMEP model. The higher binding energy of the alcohol oxygen $1s$ electron is in harmony with lower electron density at this atom. It is also noteworthy that GS-AMEP BEs for carbon atoms are in good accordance with 4-31G ΔSCF calculations in ACAC and with the corresponding experimental data [95]. This example illustrates rather nicely potential capability of ESCA in studying fast processes. Another useful application ESCA spectroscopy has found in investigations of systems undergoing large amplitude vibrations. Phosphorus pentafluoride provides an excellent example where NMR data indicated equivalence of all five fluorine atoms. On the other hand a short time scale ESCA measurements and subsequent theoretical SCC-MO AMEP calculations have convincingly shown that axial and equatorial fluorines are clearly distinguished the latter possessing significantly higher binding $F(1s)$ energies [97].

Table 14. Comparison of GS-AMEP inner-core binding energies in malonaldehyde and acetylacetone (ACAC) with available SCF and experimental values (in eV)*

Molecule	Atom	GS-AMEP[a]	GS-AMEP[b]	ΔSCF[c]	Exp.
	O_1	539.8	539.8	–	539.71[d]
	O_2	538.4	538.5	–	538.14
	C_1	291.8	291.3	–	–
	C_2	293.1	293.1	–	–
	C_3	293.0	293.3	–	–
	Δ_{ox}	1.4	1.3	–	1.6
	O	538.8	538.9	–	–
	C_1	292.4	291.8	–	–
	C_2	293.6	293.8	–	–
	O_1	539.6	539.5	538.2	538.7[c]
	O_2	537.9	538.0	537.3	537.4
	C_1	291.4	291.1	289.5	289.5
	C_2	293.0	293.2	292.6	293.5
	C_3	292.9	293.4	293.3	293.0
	C_4	291.3	291.1	291.0	290.7
	Δ_{ox}	1.7	1.5	0.9	1.3
	O	538.4	538.4	537.8	–
	C_1	292.0	291.5	291.1	–
	C_2	293.4	293.8	293.5	–
	C_3	291.4	291.1	290.9	–

* Difference in BEs of keto- and enol oxygens is denoted by Δ_{ox}
[a] GS-AMEP formula employing only charge of the host atom [96]
[b] GS-AMEP formula (43) distinguishing $2s$ and $2p$ AOs of the host atom [96]. [c] Ref. [95]. [d] Ref. [94]

4.2 The Role of the Relaxation Energy

Redistribution of the electron charge upon a creation of the inner-core positive hole is an important ingredient of the photoionization event. The accompanying relaxation energy affects ESCA shifts in molecules considerably differing in size. In some specific cases it completely determines ESCA shifts like in linear alkanes [98]. It was found that the measured core-level BEs decrease with an increasing number of carbon atoms in molecules. The whole shift ranges over 0.6 eV and it is governed by the reorganization energy. Further, relaxation energy is intimately related to the Auger parameter α and influences significantly proton affinities in molecules possessing large polarisable atoms (vide infra). Its knowledge would also allow determination of reliable ground state atomic charges via the intramolecular potential model which will be discussed at a later stage.

An important question arises therefore whether it is possible to measure relaxation energies. The answer is in principle positive and it is provided by the Manne-Åberg theorem [99]* which is related to distribution of energy of photoelectrons. Let us denote wavefunctions of the initial and final states (5) as follows

$$\Psi^i = \hat{A}\varphi_m\psi^G_m(n - 1) \tag{47a}$$

and

$$\Psi^{f+}(m) = \hat{A}\varphi_k\psi^{+j}_m(n - 1), \tag{47b}$$

where φ_m is the one-particle orbital of the inner-core to be ionized and φ_k denotes the out-going electron. The $n - 1$ electron wave-function in the ground state is given by $\psi^G_m(n - 1)$ whereas the $n - 1$ particle function in the final state of the molecular ion is signified by $\psi^{+j}_m(n - 1)$. Here a superscript j denotes possible excited states of the hole-state since a valence electron can be promoted to higher orbitals too. Now, in time-dependent quantum theory there are two limiting types of time-dependent perturbations which allow a reduction to time independent solutions. They encompass very rapid and very slow perturbations leading to sudden and adiabatic approximations, respectively. In the latter $\psi^{+j}_m(n - 1)$ are completely relaxed and consequently they are solutions of the $n - 1$ hamiltonian:

$$\hat{H}(n - 1)\,\psi^{+j}_m(n - 1) = E^j_m\psi^{+j}_m(n - 1). \tag{48}$$

We shall assume that $\psi^{+j}_m(n - 1)$ form a complete orthonormal set. On the other hand, in a sudden approximation core ionization is an instantaneous event and orbitals remain frozen implying that $\psi^G_m(n - 1)$ is a good wavefunction for the hole state as it is tacitly assumed in the Koopmans' theorem. In reality relaxation is completed during photoionization meaning that orbitals are not frozen. Therefore $\psi^G_m(n - 1)$ is a mixture of $\psi^{+j}_m(n - 1)$ states which can be formally expressed as:

$$\psi^G_m(n - 1) = \sum_{j=0}^{\infty} c_{mj}\psi^{+j}_m(n - 1). \tag{49}$$

---

* See also Ref. [100].

The coefficients in the series (49) are given by the overlap integrals $c_{mj} = \langle \psi_m^G(n - 1) | \psi_m^{+j}(n - 1) \rangle$. The probability of finding the $n - 1$ electron system in a specific state with an energy E_m^j is proportional to:

$$P_{mj} = |c_{mj}|^2 = |\langle \psi_m^G(n - 1) | \psi_m^{+j}(n - 1) \rangle|^2 . \qquad (50)$$

It is implicitly supposed here that the one-electron matrix element $\langle \varphi_m | \vec{r} | \varphi_k \rangle$ does not change appreciably with photoelectron energy over the range of possible final states. If one considers the energy associated with the frozen wavefunction $\Psi_m^G(n - 1)$ and the $n - 1$ hamiltonian then an interesting sum rule emerges:

$$\langle \psi_m^G(n - 1) | \hat{H} | \psi_m^G(n - 1) \rangle = \sum_{j=0}^{\infty} E_m^j |\langle \psi_m^G(n - 1) | \psi_m^{+j}(n - 1) \rangle|^2 . \qquad (51)$$

The ionization potential in the Koopman' frozen orbitals approximation is given by:

$$I_m^{Koop} = -\varepsilon_m^{Koop} = - [\langle \psi_m^G(n - 1) | \hat{H}(n - 1) | \psi_m^G(n - 1) \rangle - E_m^0], \qquad (52)$$

where E_m^0 is the ground state energy of n-electrons. By using Eq. (51) and taking into account closure relation:

$$\sum_{j=0}^{\infty} P_{mj} = \sum_{j=0}^{\infty} |c_{mj}|^2 = 1 \quad \text{one obtains:}$$

$$I_m^{Koop} = \sum_{j=0}^{\infty} P_{mj} I_m^j , \qquad (53)$$

where $I_m^j = E_m^j - E_m^0$. Alternatively

$$I_m^{Koop} = I_m^0 + \sum_{j=1}^{\infty} P_{mj}(I_m^j - I_m^0) , \qquad (54)$$

where I_m^0 is the basic ESCA ionization potential related to fully relaxed $n - 1$ single-electron orbitals. Hence the relaxation energy $E^{relax}(m)$ is by definition:

$$E^{relax}(m) = I_m^{Koop} - I_m^0 = \sum_{j=1}^{\infty} P_{mj}(I_m^j - I_m^0) . \qquad (55)$$

It follows that the Koopmans theorem ionization potential is in fact the average value of all ionization potentials involving shake-up and shake-off lines [16]. Hence, it is in principle possible to measure experimentally the Hartree-Fock eigenvalues (viz. (53)), although in practice this is rarely feasible with reasonable accuracy. The same holds for the relaxation energy (55) which is given by a difference of the average line determined by the Koopmans theorem (the so called "centroid" or "center of gravity" line) and the main peak in the ESCA spectrum. We conclude that it is possible to obtain experimental data on reorganization energies in some favourable cases.

A complementary way of extracting relaxation energies from the experimental data involves combined measurements of core-ionization potentials and Auger kinetic energies. Then the relaxation energy relative to a suitable compound is directly proportional to the corresponding change in the Auger parameter α [100–104]:

$$\Delta E^{relax} = (1/2)\, \Delta\alpha\,. \tag{56}$$

We note in passing that approximate estimates of the relaxation energy can be also obtained by the ESCA technique and gas-phase basicity and acidity measurements (see Sect. 5.3). It should be recalled that theoretical procedures for estimating electron redistribution energies were discussed in some detail in Sect. 3.2. We shall dwell here on their calculations at the semi-empirical level of sophistication. The EC expression $E_A^{relax}(m)$ (35) can be broken down into three terms:

$$E_A^{relax}(m) = E_A^r(contr) + E_A^r(flow) + E_A^r(mix)\,, \tag{57}$$

where

$$E_A^r(contr) = 13.6 Q_A (\zeta_A^* - \zeta_A)/n \qquad (eV)\,, \tag{58a}$$

$$E_A^r(flow) \;\; = -7.2(M_A^* - M_A) \qquad (eV)\,, \tag{58b}$$

$$E_A^r(mix) \;\; = 13.6 \zeta_A^*(Q_A^* - Q_A) \qquad (eV)\,. \tag{58c}$$

Here the asterisk denotes equivalent core and n is the principal quantum number of the valence electrons at the host atom A. Analogous expressions are readily obtained for the transition potential model. One simply replaces ζ_A^*, Q_A^* and M_A^* entities by the corresponding pseudoatom terms and divides each of them by 2. Partitioning of the total relaxation energy (57) is arbitrary but physically meaningful offering a transparent interpretation of the electron charge reorganization. The first term (58a) is local and arises due to contraction of the valence orbitals of the ionized atoms because of the created positive hole which increases the effective positive charge of the host nucleus. This term is proportional to the total electron density apportioned to atom A as first noticed by Snyder (20). The two-center term, Eq. (58b), is obviously a consequence of the migration of intramolecular electron density toward the positive hole, thus leading to a difference in Madelung potentials. It is called charge-flow contribution. The last term, Eq. (58c), is again monocentric, but includes both charge-transfer contribution $Q_A^* - Q_A$ and contraction of the host atom orbitals as evidenced by ζ_A^*. It is therefore called a mixed term and describes the fact that the portion of charge transferred to the host atom possessing a positive hole experiences a field produced by the effective nuclear charge $n\zeta_A^*$. Formulas (58a–c) will be employed in discussing various contributions to the total relaxation energy within the SCC-MO AMEP model. But first we shall test performance of the EC-AMEP model against the ab initio relaxation energies in some simple characteristic molecules.

The SCC-MO EC-AMEP relaxation energies obtained by the formula (35) employing suitable weighting factor for the Madelung terms are compared with some

Table 15. Comparison of the SCC-MO relaxation energies obtained by the EC-AMEP model with the corresponding ab initio values (in eV)

Molecule	EC-AMEP	AB INITIO
CH_4	13.7	12.1[a]; 14.40[b]; 14.31[c]; 14.10[h]; 13.56[j]
CH_3F	14.0	11.6[a]; 13.78[c]
CH_2F_2	14.1	11.8[a]; 13.69[c]
CHF_3	14.1	11.7[a]; 13.58[c]
CF_4	14.7	−; 13.53[c]
C^*O	12.4	12.31[b]; 10.79[i]
CO^*	19.7	21.38[b]; 19.91[i]
O_2	19.4	20.4[d]
H_2O	18.5	20.15[b]; 20.0[e]; 20.0[f]; 18.8[g]; 19.37[j]
CH_3O^*H	20.8	20.0[d]
N_2	16.4	16.74[d]
NH_3	17.0	17.0[d]; 16.7[j]

[a] Ref. [29]; [b] Ref. [105]; [c] Ref. [106]; [d] Empirical estimates taken from Ref. [107]; [e] Ref. [26]; [f] Ref. [34]; [g] Ref. [108]; [h] Ref. [25]; [i] Ref. [109]; [j] Ref. [110]

ab initio estimates in Table 15. One observes that actual values of the relaxation energy are basis set dependent. It appears that basis functions of at least DZ quality are necessary for reasonable description of the charge reorganization effect. Further, magnitude of the E^{relax} entity is large but its variation is relatively small in particular within the series of related molecules. This conclusion is substantiated by a family of fluorinated methanes where E^{relax} is slowly diminishing upon multiple fluorination according to calculations of Clementi and Routh [106]. Although SCC EC-AMEP model fails to reproduce these fine details, the general agreement with rigorous calculations is good thus lending credence to the semiempirical approach. In particular, accordance with empirical estimates of the relaxation energy in N_2 and NH_3 [107] is excellent. Discrepancies with ab initio results rarely exceed 1 eV. Hence, the changes of the E^{relax} obtained within the same framework of the SCC-MO EC-AMEP model should be fairly reliable.

It is of some interest to consider briefly partitioning of the relaxation energy in carbon and nitrogen atoms. The former, estimated by the EC-AMEP approach, are presented in Table 16. One finds out that the $E_C^r(tot.)$ values span a range between 12.4 and 18.5 eV, which is appreciable. The largest contributions are given by $E_C^r(contr)$ and $E_C^r(mix)$ terms. They are both positive the latter being greater. The monoatomic contraction term is amazingly constant with very few exceptions, which is a general case as we shall see later. The most pronounced variation (~ 5.4 eV) is found in the mixed term $E_C^r(mix)$. The largest change in $E_C^r(flow)$ is about a half of this value (2.7 eV). The lowest $E_C^r(mix)$ value is found in CO and concomitantly the total relaxation energy is small too. The point is that the positive hole-state can not aquire a lot of electron density from a single neighbouring atom which in turn is highly electronegative. Hence the difference $Q_C^* - Q_C$ in (58c) is relatively small. In contrast, the central carbon atom in tetramethylmethane has four chanels ((CH_3) groups) which donate electrons toward the positive hole. As a consequence, the positive hole is

Table 16. Partitioning of the relaxation energy of C(1s) hole-states in some simple organic molecules as obtained by the SCC-MO EC-AMEP model (in eV)*

Molecule	E_C^r(contr.)	E_C^r(mix.)	E_C^r(flow)	E_C^r(tot.)
CH_4	8.9	9.5	−4.7	13.7
HCN	8.7	8.7	−4.2	13.2
CO	8.5	7.5	−3.6	12.4
H_2CO	8.4	9.6	−4.5	13.6
CHOOH	8.2	10.7	−4.6	14.3
CH_3OH	8.6	10.4	−4.6	14.5
C_2H_2	8.9	9.7	−4.1	14.4
C_2H_4	8.9	10.7	−4.3	15.2
C_2H_6	8.9	10.4	−4.4	15.0
C_3H_6	8.9	11.4	−3.7	16.6
$C^*(CH_3)_4$	8.8	12.9	−3.9	17.8
C_6H_6	8.9	12.4	−3.9	17.4
OC^*CCO	8.4	11.7	−3.8	16.3
OCC^*CO	8.6	11.9	−2.0	18.5

* Ionized Carbon atoms are denoted by an asterisk. Total relaxation energy is denoted by E_C^r(tot.). Taken from Ref. [78]

almost completely screened as evidenced by the Q_C^* population of 4.96e. This leads to an appreciable $Q_C^* - Q_C$ difference and to a high E_C^r(mix) value. It is therefore not surprizing that the relaxation energy in $C^*(CH_3)_4$ is large. The same holds for the central carbon atom in carbon suboxide OCC^*CO. We note in passing that the difference in ESCA shifts for terminal and central carbons in OCCCO is to a large extent determined by a difference in relaxation. It is interesting to mention that a delocalized system like benzene has a high reorganization energy (17.4 eV) which again has its origin predominantly in a large E_C^r(mix) term. The positive hole is here also very well screened by $Q^* = 4.95$ e presumably due to the presence of mobile π electrons. It follows that the extent of screening of the positive hole and the accompanying relaxation energy depend on the number of the neighbouring atoms, their nature (relative electronegativity) and the conjugative ability of whole molecule or at least its pertinent fragment.

The variation of the E_N^r(contr) term in nitrogen is somewhat more pronounced than in carbon (Table 17). It apparently depends on a number of surrounding electronegative atoms. Consider a series of molecules N_2H_4, N_2F_4, NF_3 and ONF_3 where number of highly electronegative neighbours is none, two, three and four, respectively. The energies related to the host atom contraction decrese as 11.4, 10.5, 10.2 and 9.8 eV, respectively, for an obvious reason. However, the main reason for a variation in total relaxation energy comes from substantial changes in E_N^r(mix) term as anticipated. It ranges from 9.3 eV in N_2 to 14.4 in ONF_3. The high value in the latter molecule is caused by an effective shielding of the positive hole as evidenced by a difference $Q_N^* - Q_N = 0.93$ |e|. On the other hand the charge drift in N_2 is small as measured by $Q_N^* - Q_N$ which assumes a value 0.60 (in units e) thus giving rise to a low E_N^r(mix) value of 9.3 eV. A very effective screening of the inner-core hole is found in $(CH_3)_3N(0.91$ |e|) yielding a high mixed term. It is interesting to note that

Table 17. Partitioning of the relaxation energy of N(1s) hole-states in some simple molecules as obtained by the SCC-MO EC-AMEP model (in eV)*

Molecule	E_N^r(contr.)	E_N^r(mix.)	E_N^r(flow)	E_N^r(tot.)
N_2	11.1	9.3	-4.0	16.4
HCN	11.4	10.6	-3.9	18.1
ONF_3	9.8	14.4	-5.0	19.2
NF_3	10.2	13.1	-4.5	18.8
N_2F_4	10.5	12.8	-4.9	18.4
N*NO	11.0	12.5	-3.9	19.6
NN*O	10.6	11.6	-4.7	17.5
N_2H_4	11.4	12.0	-4.4	19.0
NH_3	11.5	10.3	-4.7	17.0
CH_3NH_2	11.5	11.7	-4.3	18.9
$(CH_3)_3N$	11.6	14.1	-3.8	21.9

* Ionized nitrogen atoms are denoted by an asterisk. Total relaxation energy is denoted by E_N^r(tot.). Taken from Ref. [78]

the ESCA shift between N_2 and $(CH_3)_3N$ is predominantly a consequence of the difference in their relaxation energies.

Let us briefly consider reorganization energies in some purines as representatives of the biological molecules discussed earlier (Table 18). One observes that the total relaxations in these planar conjugated molecules are relatively high as compared to small molecules for both C(1s) and N(1s) core-holes (Tables 16–18). This finding can be simply rationalized by the fact that E_A^r(mix) terms assume high values due to almost complete screening of the positive holes. For example, valence populations for the atom in question and its equivalent core counterpart $A(A_Q, Q_A^*)$ in hypoxanthine read: C_2 (3.87, 4.82), C_4 (3.83, 4.87), C_5 (3.93, 4.93), C_6 (3.87, 4.77), N_1 (5.02, 6.00), N_3 (5.18, 6.16), N_7 (5.22, 6.18), and N_9 (5.03, 5.99). It appears that differences $Q_A^* - Q_A$ are smaller but very close to unity as a rule. Two notable exceptions are carbons C_5 and C_4. The former achieves a perfect shielding whilst in the latter the charge flow overcompensates positive hole by transfering more than one electron. It is noteworthy that E_A^r(contr) term is again surprisingly constant for both A = C and N. This holds also for a wide selection of planar molecules including pyridines, nucleic acid bases etc. [86, 88]. Finally, we considered only relaxation energies obtained by the equivalent core approximation. Similar results are offerred by the transition potential formalism. It follows that the E_A^r(contr) and E_A^r(flow) terms are virtually the same within the EC and TP models. This is a consequence of the relations:

$$\zeta_A^{TP} \cong (\zeta_A^* + \zeta_A)/2 \quad \text{and} \quad M_A^{TP} \cong (M_A^* + M_A)/2 \tag{59}$$

which hold very closely. On the other hand, E_A^r(mix) term differs in the EC and TP methods by 1 eV. This discrepancy leads to a more or less constant difference between total relaxation energy provided by EC and TP procedures meaning that their contributions to ESCA shifts are equivalent. Indeed, this is a reason behind an equal

Table 18. Break-down of the reorganization energy upon inner-shell ionization in some purines as estimated by the SCC-MO EC-AMEP model (in eV)*

Compound	Atom	E_A^r(contr.)	E_A^r(mix.)	E_A^r(flow)	E_A^r(tot.)
8-Azaguanine	C_6	8.4	12.5	−3.7	17.3
	C_5	8.6	13.2	−3.8	18.1
	C_4	8.5	13.2	−3.8	17.9
	C_2	8.4	13.1	−3.8	17.7
8-Azaxanthine	C_6	8.5	13.0	−3.8	17.7
	C_2	8.3	12.3	−3.8	16.8
	C_4	8.4	12.4	−3.7	17.1
	C_5	8.6	13.1	−3.8	17.9
Hypoxanthine	C_6	8.6	13.3	−3.8	18.0
	C_5	8.7	13.3	−3.8	18.2
	C_4	8.5	12.5	−3.7	17.3
	C_2	8.6	12.6	−3.8	17.5
	C_8	8.6	12.7	−3.8	17.5
Xanthine	C_6	8.5	13.3	−3.8	18.0
	C_5	8.7	13.3	−3.7	18.2
	C_4	8.4	12.5	−3.6	17.3
	C_2	8.3	12.4	−3.8	16.9
	C_8	8.6	12.7	−3.7	17.6
8-Azaguanine	N_3	11.4	15.2	−3.4	23.3
	N_1	11.1	15.1	−3.8	22.3
	N_7	11.4	14.7	−3.6	22.5
	N_9	11.0	14.8	−3.9	22.0
	N_8	11.3	14.9	−3.5	22.7
	N_{10}	11.4	12.2	−3.9	19.7
8-Azaxanthine	N_3	11.1	14.9	−3.9	22.0
	N_1	11.1	14.9	−3.8	22.1
	N_7	11.3	15.2	−3.6	22.9
	N_8	10.9	14.8	−3.9	21.8
	N_9	11.4	15.3	−3.5	23.1
Hypoxanthine	N_3	11.5	15.1	−3.5	23.1
	N_1	11.1	15.1	−3.9	22.3
	N_7	11.5	14.9	−3.5	22.9
	N_9	11.1	14.9	−3.9	22.2
Xanthine	N_3	11.1	14.8	−3.8	22.0
	N_1	11.1	14.9	−3.8	22.2
	N_7	11.5	14.8	−3.5	22.8
	N_9	11.1	14.9	−3.9	22.1

* Total relaxation energy is denoted by E_A^r(tot.). Taken from Ref. [87]

performance of the EC and TP models. One can easily trace down a source of the difference for the E_A^r(mix) term. Taking into account its form (58c) and the first relation (59) one finds out that the difference in E_A^r(mix) term is proportional to

$$\zeta_A^* - \zeta_A^{TP} \cong (\zeta_A^* - \zeta_A)/2 .$$

4.3 Group Additivity of ESCA Shifts

It is possible to assign empirical values of ESCA shifts for some characteristic groups attached to a studied atom [111–113]. Entire ESCA inner-core binding energy shifts are then given by a sum of group contributions. We shall illustrate this empirical approach by the group shifts of phosphorus (Table 19). Let's consider P($2p$) shift in $O = PBr_3$. Summing up shifts for $-Br$ and $=O$ fragment (i.e. atoms in this case) one obtains $\Delta E_B(P(2p))$ of $1.58 + 3 \times 0.87 = 4.2$ (in eV) which can be favourably compared with the experimental value of 4.3 eV [112]. Further, the measured value of P($2p$) shift in P(phenyl)$_3$ is 1.3 eV whereas group additivity yields 0.9 eV. Hence, the concept of group shifts offers an approximate but quick estimate of ESCA shifts in bulky compounds which is of a great practical value.

It is easy to rationalize group additivity of ESCA shifts by the electrostatic potential model. One starts with a reference molecule which allows putting the additive constant L_A in Eq. (17) equal to zero and by invoking the electroneutrality of a molecule

$$q_A = - \sum_B{}' q_B \tag{59}$$

one obtains

$$\Delta E_B(A) = \sum_B{}' q_B(1 - K_A R_{AB})/R_{AB} \tag{60}$$

where a comma in the superscript implies $B \neq A$ as usual. A sum over B can be confined to a group G which is linked to the host atom A. Then

$$\Delta E_B(A) = \sum_G \left[\sum_{B \in G} q_B(1 - K_A R_{AB})/R_{AB} \right] = \sum_G \Delta E_B^G(A). \tag{61}$$

Formula (17) and consequently approximate expression (61) do not involve explicitly the relaxation energy. This is however not a drawback because one can

Table 19. P($2p$) Group shifts for groups attached to phosphorus* (in eV)

Group	Group shift ΔE_B^{gr}	Group	Group shift ΔE_B^{gr}
$-$phenyl	0.30	$-$Br	0.87
$-$C	0.33	$-$OC	0.89
$-$H	0.34	$-$OP	1.12
$-$S	0.57	$-$Cl	1.22
$-$N	0.60	$-S_2$	1.38
$-ONH_4$	0.70	$-$F	1.43
$-S_3$	0.70	$=$O	1.58
$-$OH	0.87	$=$S	1.78

* The group shifts are derived with the binding energy shift scale referred to the binding energy for red phosphorus (130.1 eV) [112]

start equally with the transition potential formula (37) and simply replace q_B in Eq. (61) by the charge q_A^{TP}. Hence, the group additivity of the ESCA shifts is theoretically well founded. This rule enables a "back of the envelope" calculation of the ESCA shifts which is very useful for a practicing chemist. From a theoretical point of view empirical additivity is of a great conceptual value. It gives a direct support of the idea of modified atoms in molecules [114–115].

5 Applications of ESCA

ESCA has a wide practical application in elemental analysis as an analytical tool, in surface studies, investigation of polymers and alloys, in industrial control, environmental monitoring and geology exploration etc. [12]. We shall dwell, however, on implications of ESCA in resolving problems of the electronic structure of molecules and its connections with other experimental techniques. It will turn out that there is a close link between core-level binding energy shifts and a number of chemical properties.

5.1 Atomic Charge Analysis

Interatomic charge drift in molecules is one of the most important manifestations of chemical bonding which has far reaching consequences. It affects practically all physical and chemical properties of molecules. It is therefore not surprizing that atomic charge is one of the crucial descriptors of modified atoms in chemical environments. However, the notion of the effective charge (monopole) of modified atom is not easy to define in a unique way. This question is addressed and in dept analysis of the existing methods and recipes for determining atomic charges is given in the preceding chapter [117]. A general conclusion of this discussion is that the optimal approach for extracting atomic charges from experimental and accurate theoretical electron density distributions is still to be found. Hence, it is of some interest to present a method of calculating atomic charges from ESCA data. In particular, the ACHARGE (atomic charge) analysis will be briefly presented.

We commence with the early work on fluorinated benzenes [118] and give an example provided by p-difluorobenzene. The starting point is the electrostatic formula in its GS-AMEP version (17). The next step is the choice of the reference level for ESCA shifts which allows putting $L_A = 0$ just as in the case of estimating group contributions of the inner-shell binding energy changes. It is convenient to write the resulting system of Eqs. in a matrix form:

$$\vec{q} = \tilde{C}^{-1}\vec{s} \tag{62}$$

where $\vec{q}$ and $\vec{s}$ are column vectors of atomic charges and the corresponding ESCA shifts, respectively. The form of the matrix $\tilde{C}$ depends on the nature of the studied molecule. Let us assume that a molecule has n different atoms which do not involve hydrogens. Then $\tilde{C}$ is the $n \times n$ matrix of the coefficients:

$$c_{AA} = K_A \quad \text{and} \quad c_{AB} = e^2/R_{AB} \quad (A \neq B). \tag{63}$$

The requirement of the conservation of charge is an additional condition which can be used in assessing the accuracy of the approach. Another possibility is to optimize $n + 1$ equations in the least-square fit sense. It is tacitly assumed here that ESCA shifts and coefficients K_A are known. The latter are usually identified as average $\langle 1/r \rangle_A$ values for valence atomic orbitals of the host atom. If the hydrogens are present like in the case of the above mentioned p-difluorobenzene, then the conservation of charge condition gives the missing n-th equation and the system (62) is fully determined yielding [118]:

$$\begin{pmatrix} q(C_1) \\ q(C_2) \\ q(F) \\ q(H) \end{pmatrix} = \begin{pmatrix} 27.18 & 32.68 & 14.61 & 21.93 \\ 16.34 & 43.52 & 10.19 & 28.03 \\ 14.61 & 20.38 & 35.18 & 17.57 \\ 2 & 4 & 2 & 4 \end{pmatrix}^{-1} \begin{pmatrix} 2.72 \\ 0.76 \\ -3.51 \\ 0 \end{pmatrix} \tag{64}$$

The last row shows that the total charge in the molecule is zero. The adopted values by Davis et al. [118] are $K_C = 22.0$ and $K_F = 32.5$ (in eV). The estimated atomic charges (in units of 10^{-2} |e|) $q(C_1) = 23$ (22), $q(C_2) = -3$ (−3), $q(F) = -18$ (−20) and $q(H) = 1$ (2) are in excellent agreement with the CNDO/2 values given within parentheses thus lending some credence to the ACHARGE scheme. The charge alternation of carbon atoms in the ring is nicely reproduced. There is a difficulty however when more hydrogen atoms are present which are chemically non-equivalent. Davis et al. assumed that all hydrogens have the same charge in fluorobenzenes. This is probably not a bad approximation in this family of compounds but it might pose a problem in general. ACHARGE analysis has been extended to fluorinated methanes, ethanes, ethylenes and perfluoro-cyclobutane and cyclobutene [119]. An assumption is made that carbon and hydrogen atoms in reference compounds CH_4, C_2H_6 and C_2H_4 were neutral. Nevertheless, good accordance with CNDO/2 charges is obtained and additive nature of inductive effect is established. Similar agreement between the empirical ESCA charges and CNDO/2 results was found for a variety of molecules by Stucky et al. [120]. They also discussed a low sensitivity of the empirical charges on the variation in K_A (single center $\langle r^{-1} \rangle$ integrals) parameters. It should be pointed out that the atomic charges in tetracyanoethylene and tetracyanoetylene oxide derived from the combined use of neutron and X-ray measurements are not in a very good accordance with the ESCA and CNDO/2 estimates. One should stress however that crystallographic densities are extracted under the assumption of spherical symmetry of core and valence electron distributions.

Inherent drawbeck of the ACHARGE method appart "invisibility" of hydrogens is embodied in a fact that ESCA shifts involve relaxation energy which is only partly reproduced by the GS-AMEP model. Hence, it would be useful to have both experimental ESCA shifts and experimental estimates of the electron reorganization effect. Fortunately, this is sometimes possible as we discussed earlier (cf. Eqs. (55) and (56)). Formally, nothing has to be changed in Eq. (62) the only exception being a replacement of $\Delta E_B(A)$ by $\Delta E_B(A) - \Delta E_A^{relax}$ in the vector $\vec{s}$. This type of approach has been applied to a number of simple diatomic halides by Thomas et al. [104, 121].

Table 20. Empirical ESCA charges in some halides involving an explicit account of the relaxation energy[a]

Molecule	Atom	K_A	R_{AB}	q_A	MPA[b]
HF	F	36.2	0.917	−0.26	−0.31
HCl	Cl	21.3	1.274	−0.15	−0.22
HBr	Br	18.0	1.414	−0.12	−0.13
HI	I	15.0	1.609	−0.12	−0.06
ClF	Cl	21.3	1.628	0.04	0.30

[a] Weighting parameters K_A, R_{AB} and charges q_A are in eV, Å and [e] units, respectively.
[b] Mulliken population analysis. Results of the DZ + P calculations [121]

Some representative results are displayed in Table 20. It is interesting that empirical ESCA charges are not excessive. In particular, the charge of the fluorine atom in HF is −0.26, which we believe is a very good estimate [122]. It is also remarkable that Mulliken population inspite of its imanent imperfections is not always bad as some people seem to think. Finally, the charge distribution in ClF is $Cl^{+\delta}F^{-\delta}$ ($\delta = 0.04$) contrary to the result obtained by the molecular Zeeman measurements [123]. We feel that the empirical ESCA charges are more reliable in this respect.

5.2 Interplay with Other Spectroscopic Methods

There is a close partnership between ESCA and valence photoelectron spectroscopy (PES) as thoroughly discussed by Jolly [124–126]. It can be shown that differences in core electron binding energy between molecules of the same element are proportional to the corresponding difference in localized lone pair ionization potentials. More specifically, the latter is eight-tenths of the core BE shift:

$$\Delta(\text{LOIP}) = 0.8\Delta E_B \tag{65}$$

where LOIP is a localized orbital ionization potential of the "isolated" lone pair. There is a good empirical and theoretical evidence that relation (65) holds to a good accuracy. This simple connection allows an insight into the bonding and nonbonding character of some MOs involving lone pair orbital. A simple example is given by $p\pi$ lone-pair orbital of H_2O which is strictly non-bonding. Its ionization potential of 12.61 eV may be taken as the gauge LOIP for the oxygen $2p\pi$ lone pair. If we would like to learn something about the nature of this lone pair in OF_2 molecule then $1s$ BEs of oxygen are necessary. It appears that O(1s) binding energy in OF_2 is 5.53 eV greater than that in H_2O. Hence, by using relation (65) one can easily predict the isolated LOIP related to $O(2p\pi)$ orbital. It should read $12.61 + 0.8 \times 5.53 = 17.0$ eV. The observed oxygen lone-pair ionization potential in OF_2 is 13.13 eV implying that it is lower by 3.9 eV than the expected value. Hence we conclude that the lone-pair MO of oxygen in OF_2 has strongly anti-bonding character. This type of reasoning sheds not only some light on details of the electronic structure of molecules but considerably facilitates the assignment of the bands in PES spectra as well [124, 125]. It is clear that ESCA and PES could and should be applied in a synergistic way.

There is a connection between ESCA shifts and changes in diamagnetic shielding of the nuclei of molecules inserted into strong homogeneous magnetic fields. The average screening of the NMR chemical shift has two contributions:

$$\sigma_A^{av} = \sigma_A^d + \sigma_A^p \tag{66}$$

the first being diamagnetic Lamb σ_A^d term and the second the paramagnetic σ_A^p component. The average diamagnetic screening is given by:

$$\sigma_A^d = (e^2/3mc^2) \langle O| \sum_i (1/r_{Ai}) |O\rangle \tag{67}$$

Apparently, the diamagnetic shielding represents electronic part of the electrostatic potential. Taking into account e.g. Eq. (10) one readily obtains a relation derived first by Basch [28]:

$$\Delta\sigma_A^d = -(\Delta E_B(A)/3mc^2) + (e^2/3mc^2) \sum_B{}' Z_B/R_{AB} . \tag{68}$$

It follows that knowledge of ESCA shifts and molecular geometry enables calculation of the diamagnetic contribution to the NMR shift. Hence, a joint use of ESCA and NMR techniques provides paramagnetic components $\Delta\sigma_A^p$ which is generally not possible by NMR observations alone.

It was realized by Ramsey as early as 1950 [127] that there is an intimate relationship between the paramagnetic term σ_A^p and the nuclear spin-rotation constant in molecular beam magnetic resonance. The nuclear spin-rotation constant describes the hyperfine interaction between the nuclear magnetic moment and the magnetic field created by the charge distribution of a freely rotating molecule. Ramsey's relation reads [128]:

$$\sigma_A^p = -(e^2/3mc^2) \sum_B{}' (Z_B/R_{AB}) + (m_p/3hmg_A) \sum_\alpha C_{A\alpha\alpha} I_{\alpha\alpha} \tag{69}$$

where m is the mass of the electron, m_p is the proton mass, g_A is the so called g-factor of a nucleus A, $C_{A\alpha\alpha}$ is the spin-rotation constant for nucleus A along the α-th molecular inertial axis and $I_{\alpha\alpha}$ is the corresponding principal moment of inertia. Other constants have their usual meaning. By combining Eqs. (66), (68) and (69) and taking into account that $C_{A\alpha\alpha}$ constants can be deduced from the microwave spectroscopy (MW), one obains an exact relation involving experimental data of NMR, ESCA and MW techniques:

$$\Delta\sigma_A^{av} = -(\Delta E_B(A)/3mc^2) + (\pi e/3mc\mu_N g_A) \Delta \sum_\alpha C_{A\alpha\alpha} I_{\alpha\alpha} . \tag{70}$$

It may be used as an aid in providing more accurate $C_{A\alpha\alpha}$ constants from more precisely known NMR and ESCA chemical shifts. Es a final comment we note that there is a linear relationship between Mössbauer isomer shift and inner-core binding energy shifts [16].

5.3 Core-binding Energy Shifts, Proton Affinities and Gas-phase Acidities and Basicities

Notions of basicity (proton affinity) and acidity (proton affinity of the anion) belong to the most fundamental concepts of chemistry. In general terms acidity and basicity are related to ability of molecules to accept negative or positive charge, respectively. In Brønsted definition this is attained by a loss or gain of a proton. On the other hand Lewis defines an acid as an electron acceptor whereas a base is an electron donor. Martin and Shirley were the first to notice a broad similarity between an ejection of the inner-core electron and protonation at the site of the host (ionized) atom [129]. They found it conceptually useful to disect the protonation process into two hypothetical steps:

$$(RR'R'')COH + H^+ \rightarrow (RR'R'')COH_2^+ \qquad \Delta H = -E(H^+) \tag{71}$$

where $(RR'R'')COH$ denotes a series of the studied alcohols and the product is written in a form which indicates that the excess of positive charge is located on the proton. In other words, there is no relaxation of electron density (or perhaps of the nuclei) and the corresponding change in enthalpy by a "rigid-molecule protonation" is given by $-E(H^+)$. The second "gedanken" step involves electron charge relaxation leading to a shielding of the added positive charge. Generally, the positive charge is distributed over the whole molecule which will be designated by $[(RR'R'')COH_2]^+$

$$(RR'R'')COH_2^+ \rightarrow [(RR'R'')COH_2]^+ \qquad \Delta H = -E_0^r(PA) \tag{72}$$

where $E_0^r(PA)$ is the relaxation contribution to the proton affinity of the oxygen atom. This is in full analogy with components consisting of frozen orbitals' (Koopmans') eigenvalue $-\varepsilon$ and reorganization energy E_A^r encountered in inner-core binding energy considerations:

$$E_B(A) = -\varepsilon_{1sA} - E_A^r \tag{73a}$$

As we allready know the corresponding shifts in the electrostatic potential approximation read:

$$\Delta E_B(A) = -\Delta V_A - \Delta E_b^r \tag{73b}$$

Since relative basicities are usually examined one obtains:

$$\Delta(PA)_b = -\Delta V_b(0) - \Delta E_b^r \tag{74}$$

where index b denotes a base in question and zero within parentheses in $\Delta V_b(0)$ is related to a vanishing total charge in a neutral molecule. Potential energy at the site of protonation $V_b(0)$ depends on the ground state charge distribution which is determined by the nature of substituents. Hence, it is termed inductive energy contribution. The second term in (74) is named a polarization or relaxation energy component. Although both terms in Eqs. (73) and (74) are interlocked it is advantageous to distinguish them because their magnitude varies in different families of molecules. They reflect static and dynamic aspect of the protonation process. In fact, it is sometimes possible to delineate induction and polarization contributions under some approximations (vide infra). It can appear also that one effect dominates proton affinity (PA) changes the other being negligible in a particular series of related compounds. This was exactly the case in simple alcohols with varying substituents differing in size (methyl, ethyl, isopropyl, tert. butyl) where the PA shifts can be attributed to changes in polarization of the final molecular ion state [129]. It was found indeed that $\Delta E_B(O_{1s}) \sim -\Delta(PA)_0 \sim \Delta E_b^r(0)$ as intuitively expected. The pioneering work of Martin and Shirley has been soon extended by other researchers showing that ESCA shifts are generally linearly related to proton affinities [130, 131]. It should be pointed out in this connection that the electrostatic formula for calculating PA trends (74) can be derived from the Hellmann-Feynman theorem [130]. Additional studies have convincingly shown that basicity (PA) is inversely proportional to the oxygen inner-core binding energies and that the coefficient is larger than one in magnitude (~ -1.5) [132, 133]. This indicates that proton affinity is less strongly influenced by changes in substituents than the ESCA shifts. Nice correlations between PA and ESCA shifts have been found for nitrogen in amines, for phosphine and its methyl derivatives etc. [132]. A work of Smith and Thomas [133] deserves some more scrutiny. They studied a large member of carboxylic acids and found a linear correlation between core-ionization energies and proton affinities for double-bonded oxygen with a slope of -1.6. Interestingly, both types of compounds where either initial-(induction) or final-state (relaxation) effects are predominating fit the correlation equally well. A different situation arises when acidities are in question. The acidity is defined as a negative of the proton affinity of the anion PA_a. The corresponding expression for relative PA_a values reads:

$$\Delta PA_a = -\Delta V_a(1) + \Delta E_a^r \tag{75}$$

where $\Delta V_a(1)$ is the relative potential energy at the added proton in the neutral molecule and ΔE_a^r is the relaxation effect accomanying abstraction of that proton. Now, it is clear that if the inductive effect favors ejection of the inner-core electron, then it hinders at the same time detachement of the proton. Concomitantly, if core-electron binding energies are governed by inductive effect then a positive correlation with acidity is expected. Conversely, if ΔE_B shifts are dominated by final-state polarization effect a negative correlation could be anticipated in view of the opposite sign of the relaxation in formulas (73b) and (75). Smith and Davis [133] have shown that these expectations are fully justified. The induction dominated core-ionization energies in substituted carboxylic acids RCOOH (R = CH_3, CFH_2, CF_2H and CF_3) had a negative (coefficient -1.1) correlation with the proton affinity of the anions. On the other hand, halogen derivatives in a series RCOOH (R = H, CFH_2, $CClH_2$ and $CBrH_2$) have positive correlation with hydroxyl oxygen $1s$

ionization energies since it is well known that large atoms like Cl and Br are highly susceptible to polarization [134].

It is noteworthy that comparison of gas-phase acidities and proton affinities can be employed in estimating relative importance of inductive and relaxation effects of various substituents in determining ease of protonation, proton abstraction or core-hole creation by photoionization at various sites in a molecule. For this purpose some assumptions are necessary. Smith and Thomas [133] deduced from their experimental correlations that the following relation approximately holds:

$$\frac{\Delta V_A}{\Delta V_a(1)} \cong \frac{\Delta E_A^r}{\Delta E_a^r} \cong -1.1 \,. \tag{76}$$

This is not incompatible with earlier consideration of Davis and Shirley [134]. It appears that knowledge of ESCA shifts and anion proton affinities enables estimates of inductive and relaxation contributions $\Delta V_a(1)$ and ΔE_a^r, respectively, via Eqs. (73b), (75) and (76). Experimental data confirm that previous CNDO/2 calculations of Davis and Shirley [134] were quite reliable in this respect (Table 21). It is remarkable that such crude wavefunctions accompanied by the electrostatic approximation yield so good relative $\Delta V_a(1)$ and ΔE_a^r values.

This type of analysis provides valuable information about origin of acidity. For example, it is well known that carboxylic acids and aromatic alcohols are more acidic than aliphatic alcohols. A common "explanation" for this difference is that resonance stabilization in anions of the latter compounds is much smaller than in the former families of molecules. However, ESCA and gas-phase acidity measurements revealed that resonance delocalization of negative charge in anions played only a minor role whereas the induction effects were of crucial importance [135].

Table 21. Comparison of the calculated and experimental relative values of inductive and relaxation contributions to the acidity in some carboxylic acids (in eV)

Molecule	Experimental[a]		Theoretical[b]	
	$\Delta V_a(1)$	ΔE_a^r	$\Delta V_a(1)$	ΔE_a^r
CH_3COOH	0	0	0	0
CFH_2COOH	10.3	−0.5	9.5	0.2
CF_2HCOOH	18.4	0.4	17.0	0.4
CF_3COOH	25.1	−0.1	26.6	1.4
CFH_2COOH	10.3	−0.5	9.5	0.2
$CClH_2COOH$	10.6	−2.2	9.8	−2.3
$HCOOH$	7.5	4.3	11.2	5.9
CH_3COOH	0	0	0	0
C_2H_5COOH	−0.1	−2.2	−0.2	−1.8

[a] Ref. [133]. [b] Ref. [134]

6 Concluding Remarks

The main aim of this chapter was to provide a conceptual framework for understanding the most outstanding features of ESCA spectra. It appears that atomic "fingerprints" and group additivity of ESCA shifts can be rationalized by the simple electrostatic model involving a notion of electric monopoles of atoms in chemical environment thus offering a simple rule of thumb. Further, it is useful to disect inner-core binding energy shifts into initial-state (neutral molecule) and final-state (molecular ion) contributions. The latter is conveniently described by equivalent-core and/or transition potential concepts. One can say without exaggerating that a reduction of rigorous theoretical expression for ESCA shifts to approximate but simple and transparent electrostatic model represents a masterpiece of modelling in quantum chemistry. It is noteworthy that an analogous approach is useful in rationalizing Auger spectra [136, 137].

ESCA spectroscopy provides plethora of information about electron charge distribution in molecules which in turn is related to all molecular properties. It can be applied together with some other spectroscopies (PES, NMR, MW, Mössbauer) in a complementary fashion. A combined use of ESCA with gas-phase acidity and basicity measurements yields an enlightening insight into interplay of inductive and polarization effects in ionized systems. We mention in passing that ESCA shifts are also correlated with Hammett substituent constants [138], and with enthalpies of some specific chemical reactions [139]. Additionally they are related to structural parameters of molecules [140] etc. Discussion of all these aspects is clearly outside the scope of this paper. It is gratifying, however, and intellectually pleasing that so many properties can be directly related via ESCA to simple electrostatics and to the concept of atomic charge which, last but certainly not least is essential ingredient of the model of modified atoms in molecules.

7 Acknowledgements

A part of this work was completed during the author's stay at the Organisch-chemisches Institut der Universität Heidelberg and he would like to express his thanks to the Alexander von Humboldt-Stiftung for financial support and to Professor R. Gleiter for his hospitality.

8 References

1. Hedman A, Siegbahn K, Svartholm N (1950) Proc Phys Soc London Sect A 63: 960
2. Siegbahn K (1982) Science 217: 111 and the references cited therein
3. Gelius U, Basilier E, Svensson S, Bergmark T, Siegbahn K (1974) J El Spectry Rel Phenom 2: 405

4. Siegbahn H, Siegbahn K (1973) J El Spectry Rel Phenom 2: 319
5. Siegbahn H, Asplund L, Kelfve P, Hamrin K, Karlson L, Siegbahn K (1974) J El Spectry Rel Phenom 5: 1059
6. Siegbahn H, Asplund L, Kelfve P, Siegbahn K (1975) J El Spectry Rel Phenom 7: 411
7. Siegbahn H, Svensson S, Lundholm M (1981) J El Spectry Rel Phenom 24: 205
8. Siegbahn K (1954) Ark Fys 7: 86; (1954) ibid, 8: 19
9. Siegbahn K, Nordling C, Fahlman A, Nordberg R, Hamrin K, Hedman J, Johansson G, Bergmark T, Karlsson SE, Lindgren I, Lindberg B (1967) ESCA — Atomic, molecular and solid state structure studied by means of electron spectroscopy, Almqvist & Wiksells, Uppsala
10. Siegbahn K, Nordling C, Johansson G, Hedman J, Heden PF, Hamrin K, Gelius U, Bergmark T, Werme LO, Manne R, Baer Y (1969) ESCA applied to free molecules, North Holland, Amsterdam
11. Shirley DA (ed.) (1972) Electron spectroscopy, North-Holland, Amsterdam
12. Carlson TA (1975) Photoelectron and auger spectroscopy, Plenum, New York
13. Brundle CR, Baker AD (eds) (1977–1981) Electron spectroscopy, Vols 1–4, Academic, New York
14. Cardona M, Ley L (eds:) (1978–79) Photoemission in solids, Topics in applied physics, Vols 26 and 27, Springer, Berlin Heidelberg New York
15. Siegbahn H, Karlsson L (1982) In: Flüge S (ed) Encyclopedia of physics, vol 31, Springer, Berlin Heidelberg New York, p 215
16. Gelius U (1974) Phys Scr 9: 133
17. Gelius U (1974) J El Spectry Rel Phenom 5: 985
18. Faraday Disc Chem Soc (1975) Vol 60
19. J El Spectry Rel Phenom (1988) Vol 47 dedicated to Siegbahn K on the occasion of his 70th birthday
20. J El Spectry Rel Phenom (1990) vol 51, Parts A and B
21. Hertz H (1887) Ann Phys 33: 301
22. Einstein A (1905) Ann Phys 17: 132
23. Koopmans T (1934) Physica 1: 104
24. Verhaegen G, Berger JJ, Desclaux JP, Moser CM (1971) Chem Phys Lett 9: 479; Moser CM, Nesbet RK, Verhaegen G (1971) Chem Phys Lett 12: 230
25. Meyer W (1937) J Chem Phys 58: 1017
26. Meyer W (1971) Int J Quant Chem 5: 341
27. Basch H, Snyder LC (1969) Chem Phys Lett 3: 333
28. Basch H (1970) Chem Phys Lett 5: 337
29. Brundle CR, Robin M, Basch H (1970) J Chem phys 53: 2196
30. Adams DB, Clark DT (1973) Theoret Chim Acta 31: 171
31. Kovaćek K, Maksić ZB (unpublished results)
32. McDonald JKL (1933) Phys Rev 43: 830
33. Bagus PS (1965) Phys Rev A 139: 619
34. Schwartz ME (1970) Chem Phys Lett 5: 50
35. Carlson TA (1975) Photoelectron and auger spectroscopy, Plenum, New York
36. Levy B, Millie Ph, Ridard J, Vinh J (1974) J El Spectry Rel Phenom 4: 13
37. Schwartz ME (1970) Chem Phys Lett 6: 631
38. Watson RE (1960) Phys Rev 118: 1036
39. Ha TK, O'Konski CT (1969) Chem Phys Lett 3: 603
40. Schwartz ME, Switalski JD, Stronski RE (1972) in Ref. 11, p 605
41. Davis DW, Shirley DA (1974) J El Spectry Rel Phenom 3: 137
42. (a) Snyder LC (1971) J Chem Phys 55: 95; (b) Slater JC (1930) Phys Rev 36: 57
43. Bagus PS, Schaefer III HF (1971) J Chem Phys 55: 1474
44. Hedin L, Johansson A (1969) J Phys B 2: 1336
45. Liberman D (1964) Bull Am Phys Soc 9: 731
46. Shirley DA (1972) Chem Phys Lett 16: 220
47. Rosen A, Lindgren I (1968) Phys Rev 176: 114
48. Jolly WL, Hendrickson DN (1970) J Am Chem Soc 92: 1863

49. Jolly WL (1972) In: Shirley DA (ed) Electron spectroscopy, North-Holland, Amsterdam, p 629
50. Clark DT, Adams DB (1972) J Chem Soc Farad II 68: 1819
51. Adams DB (1976) J. El. Spectry Rel Phenom 9: 251
52. Clark DT, Muller J (1976) Theoret Chim Acta 41: 193; Goscinski O, Palma A (1977) Chem Phys Lett 47: 322; Müller J, Agren H, Goscinsky O (1979) Chem Phys 38: 349
53. Shirley DA (1972) Chem Phys Lett 15: 325
54. Goscinski O, Pickup BT, Purvis G (1975) Chem Phys Lett 30: 87
55. Goscinski O, Howat G, Åberg T (1975) J Phys B 8: 11
56. Goscinski O, Hehenberger M, Roos B, Siegbahn P (1975) Chem. Phys Lett 33: 427
57. Howat G, Goscinski O (1975) Chem Phys Lett 30: 87
58. Siegbahn H, Medeiros R, Goscinski O (1976) J El Spectry Rel Phenom 8: 149
59. Davis DW, Shirley DA, Thomas TD (1972) J Chem Phys 56: 671
60. Shillady DD, Billingsley FP, Bloor JE (1971) Theoret Chim Acta 21: 1
61. Bloor JE, Maksić ZB (1971) Mol Phys 22: 35; (1972) J Chem Phys 57: 3572; (1973) Mol Phys 26: 397
62. Bloor JE, Maksić ZB (1975) Proc Sec Int Symp NQR Spectroscopy, Pisa, p 1
63. Chung-Phillips A (1989) J Comp Chem 10: 17
64. Davis DW, Shirley DA (1974) J El Spectry Rel Phenom 3: 137
65. Schwartz ME, Switalski JD (1972) J Am Chem Soc 94: 6298
66. White WD, Drago RS (1970) J Chem Phys 52: 4717
67. Maksić ZB, Bloor JE (1972) Croat Chem Acta 44: 435; Graovac A., Maksić ZB, Rupnik K, Veseli A (1977) Croat Chem Acta 49: 695
68. Maksić ZB, Supek S (1989) J Mol Structure (Theochem) 198: 427
69. Maksić ZB, D. Kovaček, Šuste T (1991) J Mol structure (Theochem) (in print)
70. Ellison FO, Larcom LL (1972) Chem Phys Lett 13: 399
71. Maksić ZB, Kovačević K, Metiu H (1974) Croat Chem Acta 46: 1
72. Ellison FO, Larcom LL (1971) Chem Phys Lett 10: 580
73. Maksić ZB, Rupnik K (1983) Z Naturforsch 38a: 308
74. Clementi E, Raimondi DL (1963) J Chem Phys 38: 2686
75. Snyder LC, Basch H (1972) Molecular wave functions and properties, John Wiley, New York
76. Maksić ZB, Rupnik K (1977) Croat Chem Acta 50: 307
77. Maksić ZB, Rupnik K (1980) Theoret Chim Acta 54: 145
78. Maksić ZB, Rupnik K (unpublished results)
79. Schwartz ME, Allen LC (1970) J Am Chem Soc 92: 1466
80. Schwartz ME, Coulson CA, Allen LC (1970) J Am Chem Soc 92: 447
81. Brundle CR, Robin MB, Basch H (1970) J Chem Phys 53: 2196
82. Maksić ZB, Rupnik K (1980) Z Naturforsch 35a: 988
83. Maksić ZB, Rupnik K, Mileusnić N (1981) J Organomet Chem 219: 21
84. Maksić ZB, Rupnik K (1983) Croat Chem Acta 56: 461
85. Maksić ZB, Rupnik K (1980) Croat Chem Acta 53: 413
86. Maksić ZB, Rupnik K, Veseli A (1983) J El Spectry Rel Phenom 32: 163
87. Maksić ZB, Rupnik K, Veseli A (1983) Z Naturforsch 38a: 866
88. Maksić ZB, Rupnik K (1981) Nouv J Chim 5: 515
89. Bossa M, Gianturco FA, Maraschini F (1975) J El Spectry Rel Phenom 6: 27
90. Barber M, Clark DT (1970) Chem Comm 22; (1970) ibid. 24
91. Ishida K, Kato H, Nakatsuji H, Yonezawa T (1972) Bull Chem Soc (Japan) 45: 1574
92. Riga J, Pireaux JJ, Caudano R, Verbist JJ (1977) Phys Scr 16: 346
93. Maksić ZB, Eckert-Maksić M, Kovaček D (1989) Croat. Chem. Acta 62: 623; Eckert-Maksić M, Maksić ZB, Margetić D (1989) Croat Chem Acta 62: 645 and the references cited therein
94. Brown RS (1977) J Am Chem Soc 99: 5497
95. Clark DT, Harrison A (1981) J El Spectry Rel Phenom 23: 39
96. Maksić ZB, Rupnik K, Eckert-Maksić M (1979) J El Spectry Rel Phenom 16: 371
97. Maksić ZB, Rupnik K (1979) J El Spectry Rel Phenom 16: 481 and the references cited therein

 98. Pireaux JJ, Svensson S, Basilier E, Malmqvist PA, Gelius U, Caudano R, Siegbahn K (1976) Phys Rev A 14: 2133
 99. Manne R, Åberg T (1970) Chem Phys Lett 7: 282
100. Lundqvist BI (1969) Phys Cond Matter 9: 236
101. Wagner CD, Biloen R (1973) Surf Sci 35: 82
102. Siegbahn H, Goscinski O (1976) Phys Scr 13: 225
103. Thomas TD (1980) J El Spectry Rel Phenom 20: 117
104. Bomben KD, Gimzewski JK, Thomas TD (1983) J Chem Phys 78: 5437
105. Hillier IH, Saunders VR, Wood MH (1970) Chem Phys Lett 7: 323
106. Clementi E, Routh A (1972) Int J Quant Chem 6: 525
107. Jen JS, Thomas TD (1974) J El Spectry el Phenom 4: 43
108. Goscinski O, Hehenberger M, Roos B, Siegbahn P (1975) Chem Phys Lett 33: 427
109. Clark DT, Cromarty BJ, Sgamellotti A (1977) Chem Phys Lett 51: 356
110. Clark DT, Cromarty BJ, Sgamellotti A (1978) J El Spectry Rel Phenom 13: 85
111. Gelius U, Heden PF, Hedman J, Lindberg BJ, Manne R, Nordberg R, Nordling C, Siegbahn K (1970) Phys Scr 2: 70
112. Hedman J, Klasson M, Nordling C, Lindberg BJ (1972) In: Shirley DA (ed) Electron spectroscopy, North-Holland, Amsterdam, p 681
113. Lindberg BJ, Hedman J (1975) Chem Scr 7: 155
114. Maksić ZB (1990) In: Maksić ZB (ed) Theoretical models of chemical bonding, part 1, Atomic hypothesis and the concept of molecular structure, Springer, Berlin Heidelberg New York, p 283
115. Maksić ZB, Eckert-Maksić M, Rupnik K (1984) Croat Chem Acta 57: 1295; Maksić ZB (1988) J Mol Structure (Theochem) 170: 39
116. Maksić ZB (1989) In: Maruani (ed) Molecules in physics, chemistry and biology, vol 3, Kluwer, Dordrecht
117. Jug K, Maksić ZB (1991) preceding chapter of this book
118. Davis DW, Shirley DA, Thomas TD (1972) J Am Chem Soc 94: 6565
119. Davis DW, Banna MS, Shirley DA (1974) J Chem Phys 60: 237
120. Stucky GD, Matthews DA, Hedman J, Klasson M, Nordling C (1972) J Am Chem Soc 94: 8009
121. Saethre LJ, Thomas TD, Gropen O (1985) J Am Chem Soc 107: 2581
122. See Section 3.4 of Ref. [117]
123. Ewing JJ, Tigelaar HL, Flygare WH (1972) J Chem Phys 56: 1957; McGurk J, Norris CL, Tigelaar HL, Flygare WH (1973) J Chem Phys 58: 3118
124. Jolly WL (1981) J Phys Chem 85: 3792
125. Jolly WL, Eyermann CJ (1982) J Phys Chem 86: 4834
126. Jolly WL (1983) Acc Chem Res 16: 370
127. Ramsey NF (1950) Phys Rev 78: 699
128. Flygare WH (1964) J Chem Phys 41: 793
129. Martin RL, Shirley DA (1974) J Am Chem Soc 96: 5299
130. Davis DW, Rabalais JW (1974) J Am Chem Soc 96: 5305
131. Benoit FM, Harrison AG (1977) J Am Chem Soc 99: 3980
132. Mills BE, Martin RL, Shirley DA (1976) J Am Chem Soc 98: 2380
133. Smith SR, Thomas TD (1978) J Am Chem Soc 100: 5459
134. Davis DW, Shirley DA (1976) J Am Chem Soc 98: 7898
135. Siggel MR, Thomas TD (1986) J Am Chem Soc 108: 4360
136. Siegbahn H, Goscinski O (1976) Phys Scr 13: 225
137. Kelfve P, Blomster B, Siegbahn H, Siegbahn K, Sanhueza E, Goscinski O (1980) Phys Scr 21: 75
138. Lindberg B, Svensson S, Malmquist PA, Basilier E, Gelius U, Siegbahn K (1976) Chem Phys Lett 40: 175
139. Stams DA, Thomas TD, MacLaren DC, Ji D, Morton TH (1990) J Am Chem Soc 112: 1427
140. Gutsev GL, Boldyrev AI (1989) J El Spectry Rel Phenom 49: 1

Experimental Momentum-Space Chemistry by (e, 2e) Spectroscopy

K. T. Leung

Department of Chemistry, University of Waterloo, Waterloo, Ontario N2L 3G1, Canada

The notion that chemists might benefit by looking at molecular orbitals and chemical bonding phenomena from the complementary momentum-space perspective was first suggested by Coulson and Duncanson some half a century ago. With the rapid development of (e, 2e) spectroscopy in the past two decades, electron momentum distributions of individual ionic states in the valence shell can be measured direclty to high precision. In addition to accessing the nature, symmetry and parentage of a characteristic orbital in a particular ionic state, the experimental momentum distribution provides a direct and stringent evaluation of the quality of ab-initio self-consistent-field wavefunctions. This has provided a powerful technique to investigate chemical bonding in the laboratory. A phenomenological look at (e, 2e) spectroscopy in its first two decades of development will be presented with a special emphasis on its applications in quantum chemistry. Some general observations regarding the use of (e, 2e) data in electronic structure determination and wavefunction evaluation will be made. A summary of momentum-space chemical concepts will be given in order to provide qualitative understanding of the observed features in the experimental spherically averaged momentum density using the three-dimensional density topography of calculated wavefunctions. Finally, a bibliographical update of recent (e, 2e) literature will be provided.

1 Introduction

One of the major challenges in quantum chemistry is the search for the exact solution to the equation of motion for electrons in atoms and molecules [1]. The complex motion of electrons in an atomic or molecular system is conveniently described by the Schroedinger equation. Since this equation cannot be solved exactly for systems with more than one electron, iterative procedures are used to obtain the best approximate solution. One such method of obtaining an approximate wavefunction is the self-consistent-field method which is based on the energy variational principle [2, 3]. Once the wavefunction is determined, the expectation values of any physical observables can be calculated using quantum mechanics [4]. To determine experimentally the precision of such a computation, and in particular the quality of the approximate wavefunction itself, in modelling the system of interest, two general approaches may be used. The first approach involves the comparison of expectation values of physical observables (such as the dipole moment, polarizability, ionization energy, oscillator strengths, etc.) generated using the approximate wavefunction with the experimental values. This approach usually involves further approximation to the form of a particular operator of interest [5, 6]. The second approach is to compare the wavefunction probability density, itself an observable function, directly [7, 8]. It is the experimental determination of this wavefunction probability density, expressed in the momentum-space representation [9], that we will discuss in this article. In particular, the technique of (e, 2e) spectroscopy [10], provides direct and precise measurements of the ionization energy spectrum and momentum-space density functions of individual ionic states, giving a comprehensive "anatomy" of the total momentum density into its individual orbital components in regions where chemical bonding occurs.

The possibility of applying a simple (e, 2e) scattering technique to study the electronic structure of matter was first proposed by Smirnov and Neudatchin and later by Glassgold and coworkers [11]. In 1969, the first (e, 2e) experiment, performed on a thin carbon film, was accomplished by Amaldi et al. [12]. Two decades of intense development in (e, 2e) studies have produced over 240 publications. Over 66 systems including free atoms and molecules, and thin films have been investigated by this method. It is also of interest to note that the number of experimental groups has increased from the original four (Giardini-Guidoni [13], Weigold [14], Brion [15], and Coplan [16]) in the 1970s to over ten groups who participated in the recent (e, 2e) conference in 1989 [17]. The steady growth of (e, 2e) spectroscopy is clearly evident from its yearly publication history, shown in Fig. 1. Detailed references and brief data symmary of (e, 2e) work recorded up to the end of 1983 can be found in an (e, 2e) bibliography compiled by Leung and Brion [18]. Between 1984 and 1988, there were about 105 additional papers, representing 75% of all the (e, 2e) work published before 1984 (see Appendix). The thrust of most of the reported (e, 2e) studies can be generally classified into two groups. The first involves the investigation of collision physics and electron scattering phenomena while the second focuses on the extraction of electronic structural information of (homological series of) atoms and molecules. Insights and results generated from the former group have established the theoretical foundation of which the structural investigations is based in the latter. It is the latter group of

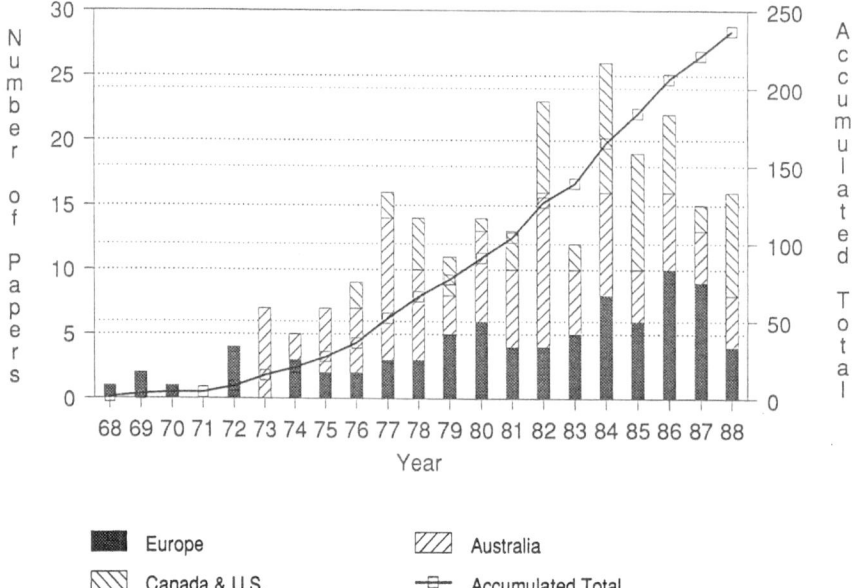

Fig. 1. Publication history of (e, 2e) literature from 1969 to 1988. The numbers of publications per year per continent are shown by stack bars (left axis) while the acummulated total up to a given year is indicated by a heavy line (right axis)

studies that are of more interest to chemists and will be the theme of the present article.

Of the atomic systems investigated by (e, 2e) spectroscopy, the noble gases are the most extensively studied systems. Experimentally, the valence-shell ionization energy spectrum of the He 1s orbital has been used as a standard to determine the energy resolution, and the momentum distribution of the Ar 3p orbital has routinely been used to calibrate the angular (or momentum) resolution. The early work on the noble gases, with representative work by Weigold et al. [14] and by Botticelli et al. [19], established the theoretical foundation of symmetric (e, 2e) scattering [20-22]. The more recent work by Lahmam-Bennani and coworkers [23, 24], on the other hand, provided new insights to the scattering mechanisms and to the dynamics in asymmetric (e, 2e) reactions, particulary in the Bethe ridge region [25]. Of the remaining atomic systems, the study on the hydrogen atom deserves special mention because not only did this work definitively reaffirm the validity of the scattering approximations commonly employed in (e, 2e) studies, it also gave for the first time a precise proof of the exact solution to the Schroedinger equation for a one-electron system [26]. The studies on heavier atoms, such as Xe [27] and Pd [28] for instance, investigated relativistc effects in momentum distributions arising from solving the Dirac-Fock wave equation. As well, the study on Cd reported the momentum distribution of a d-orbital (Cd 4d) for the first time [29] while the work on Na gave the first measurement of the momentum distribution of an excited state (Na ^{3}P) [30].

The evolution of molecular (e, 2e) spectroscopy paralleled the early development of photoelectron spectroscopy. It is evident that the energy resolution obtained in (e, 2e) spectroscopy limits the scope of molecular studies to those of relatively simple systems in which the majority of the accessible ionic states are well resolved.

Nevertheless, (e, 2e) studies on molecules as large as SF_6 [31], para-$Cl_2C_6H_4$ [32] and $Cr(CO)_6$ [33] have been reported. We may conveniently group all of the reported molecular (e, 2e) studies into three general classes (see Appendix):

1) Simple molecules (H_2, halogen, simple diatomics and triatomics, and first row hydrides), their valence-isoelectronic homological members, their halogen-substituted and methylsubstituted derivatives.
2) Hydrocarbons including acetylene, benzene and other simple linear and cyclic members.
3) Other molecules including fluorides, SF_6 and $Cr(CO)_6$.

Like photoelectron spectroscopy, studies of simple homological groups with isoelectronic or valence-isoelectronic structure are common in (e, 2e) spectroscopy. These studies facilitate better understanding of any general features observed in the momentum distributions of a particular member. Furthermore, halogen-substituted [34] and methyl-substituted [35] derivatives can be used to infer the general effect of charge delocalization and other effects caused by selective change of a ligand in the momentum distributions of individual orbitals.

Although the very first (e, 2e) experiment was performed on a thin carbon film [12], it was not until the early 1980s that more intense efforts on condensed matter studies by (e, 2e) spectroscopy were made. In particular, high resolution studies in the valence band [36] and detailed modelling of multiple scattering effects in solids [37] were achieved by Ritter and coworkers. Other (e, 2e) studies were also made on Al films by Williams and coworkers [38] and on Ag films by Nakel and coworkers [39]. Recent advances in parallel detection technology [40] are expected to accelerate the development of these studies on thin films and further enhance our understanding of the physics of (e, 2e) processes and of the electronic structure of matter in the condensed phase. The possibility of measuring directional momentum density of chemisorbed species oriented on surfaces may also be realized in the near future [41].

Like most other important developments in science, it is evident that numerous articles reviewing various aspects of (e, 2e) spectroscopy have been written [10, 42 – 47] (see, for example, Table A4 in Appendix). Numerous summaries of studies carried over a period of time have also been given by different laboratories at various stages in the past twenty years. It would therefore appear to be a somewhat tenuous attempt to redo what has already been done so well and exhaustively. This article will only attempt to give a "handbook" overview of the relevant concepts commonly employed in the field, with special focus on its applications in electronic structural investigation and in the elucidation of momentum-space chemical concepts verified and/or discovered by (e, 2e) spectroscopy. No attempt is made to provide a complete data summary of all the (e, 2e) work. Instead, recent data recorded in our laboratory will be used as examples to illustrate the concepts discussed [48, 49]. Any oversights in citations and references are unintentional, and I hope this can be remedied by an (e, 2e) bibliographical update given in the Appendix, which together with an earlier (e, 2e) bibliography [18] I shall resist from claiming complete.

Before proceeding any further, it may be amusing to comment on the names/acronyms of the field itself. The first series of experiments had its origin in nuclear physics and indeed much of the technology was borrowed from nuclear physics. As such, the name (e, 2e) spectroscopy was created in an analogous fashion as $(\gamma, 2\gamma)$, (p, 2p) etc., following the nuclear physicist's notation of denoting a reaction [50]. Since then, depending upon in which laboratory the work was performed and when the paper was published, variations of the names were evident. Some of the more common ones include electron coincidence spectroscopy, electron impact spectroscopy, and binary (e, 2e) spectroscopy, but the name (e, 2e) spectroscopy generally remains the accepted norm for the field. In the (e, 2e) workshop held in Vancouver, Canada in 1986, yet another name, electron momentum spectroscopy (EMS), was suggested in order to create perhaps a more convenient acronym. Ironically, the initial adoption of EMS as the neon sign of the field happened when workers in other areas began to reference (e, 2e) (not EMS) work. It is, however, the original name, i. e. (e, 2e) spectroscopy, which gives me a more concrete representation of the physical process, i. e. the (e, 2e) reaction, itself. I shall therefore, at the risk of being called a traditionalist, indulge in using the traditional name throughout this article to avoid further confusion.

What follows then, is an attempt to look at (e, 2e) spectroscopy with specific focus on its contributions to quantum chemistry in its first twenty years (1969-1988) of development, from an experimentalist's perspective.

2 Principle and Theory

The technique involves a simple reaction: $M(e, 2e)M^+$. This equation is written in nuclear physicist's notation (devised by Bothe) where the reactants are to the left of the comma while the products are on the right hand side [50]. It corresponds to the more familiar chemist's notation: $M + e^- \rightarrow e^- + e^- + M^+$. In general, any ionization process caused by electron impact under a variety of scattering kinematic conditions involving an incoming electron and a neutral target as the initial reactants and two outgoing electrons and an ion as the products can be collectively called an (e, 2e) reaction [10]. In the present case, we shall ignore all the processes in the low energy regime and limit the scope of the present discussion to high energy (e, 2e) reactions, in which the kinetic energies of the electrons are typically of the order of hundreds of eV. The high energy electron scattering processes can be further divided into two groups according to whether the two outgoing electrons emerge with the same (symmetric scattering) or different (asymmetric scattering) momentum magnitudes and polar angles. Before the advances of synchrotron radiation sources in the early 1970s, small-angle asymmetric (e, 2e) reactions, also known as dipole (e, 2e) spectroscopy, provided an important technique of measuring photoionization cross sections as a function of (pseudo-) photon energy [51, 52]. The recent extension of asymmetric (e, 2e) scattering into the large momentum transfer (the Bethe ridge) region [23, 24, 53, 54], has demonstrated its effectiveness in providing information similar to those obtained using the symmetric geometry. In the case of symmetric scattering, the coplanar geometry has been used for investigating aspects of the scattering theory [55] and is the geometry of choice to study thin film phenomena in a transmission configuration [56]. The noncoplanar geometry, on the other hand, is commonly used in the determination of the electronic structure of atoms and molecules. The remainder of this article will concentrate mainly on symmetric noncoplanar (e, 2e) reactions with specific focus on their applications in electronic structure and wavefunction density determination for free atoms and molecules.

The scattering theory of (e, 2e) reaction was developed mainly by McCarthy and coworkers and has been discussed in great detail in many review articles [10, 22, 57]. Only key equations relevant to the extraction of wavefunction information are discussed here.

2.1 Scattering Kinematics

In an (e, 2e) experiment [10, 20, 57], the scattering kinematics are completely determined. A target atom or molecule is ionized by high energy electron impact using a fast incident electron with kinetic energy E_0 and momentum p_0. The scattered electron (with kinetic energy E_1 and momentum p_1) emerges along with an ejected electron (with kinetic energy E_2 and momentum p_2) and a positive ion (with recoil energy E_{recoil} and recoil momentum q).

$$M + e^- \rightarrow M^+ + e^- + e^-. \tag{1}$$

$$(p_0, E_0) \ (q, E_{recoil}) \ (p_1, E_1) \ (p_2, E_2)$$

Neglecting the very small ion recoil energy, the conservation of energy dictates that:

$$E = E_0 - (E_1 + E_2), \tag{2}$$

where E is the ionization or binding energy of the bound electron. Clearly, if the kinetic energies of the outgoing electrons (i.e. E_1 and E_2) are kept constant, the ionization energy can be sampled by varying E_0. Also, if the very small thermal motion of the target before the impact is ignored, then the conservation of momentum requires that:

$$q = K - p_2, \tag{3}$$

where the momentum transfer (from the incident electron to the bound electron) K is defined as $p_0 - p_1$. If one considers that the scattering takes place in a close, binary encounter [10, 57] condition (which can be realized by maximizing the momentum transfer), then the ion can be virtually regarded as a spectator. Under these conditions, the momentum p of the bound electron before the impact-ionization is equal in magnitude, but opposite in sign, to the ion recoil momentum q. i.e.

$$p = p_2 - (p_0 - p_1). \tag{4}$$

Under the symmetric noncoplanar scattering conditions, the magnitude of the momentum p can be effectively sampled by varying the relative azimuthal angle φ.

$$p = \{[2p_1 \cos \theta_1 - p_0]^2 + [2p_1 \sin \theta_1 \sin (\varphi/2)]^2\}^{1/2}, \tag{5}$$

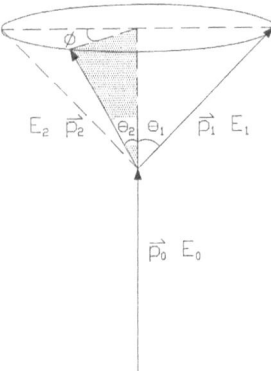

Fig. 2. Schematic diagram of the symmetric noncoplanar (e, 2e) scattering geometry

where $\varphi = \pi - (\varphi_1 - \varphi_2)$. A schematic diagram of the symmetric noncoplanar scattering geometry ($p_1 = p_2$, $\theta_1 = \theta_2 = 45°$ and φ variable) [10, 57] depicting the definitions of the solid angles used in Eq. 5 is given in Fig. 2.

2.2 Differential Cross Section

Consider an (e, 2e) reaction which changes an N-electron neutral system in its ground state denoted by Ψ_0^N, to a N-1 electron ionic system in an eigenstate denoted by Ψ_F^{N-1}. In atomic units ($h/2\pi = m_e = e = 1$), the (e, 2e) differential cross section (also known as triple or five-fold differential cross section commonly denoted by $d^3 \sigma/d\Omega_1 \, d\Omega_2 \, dE_1$ or $d^5 \sigma/d\Omega_1 \, d\Omega_2 \, dE_1$ respectively in the literature)[58] can be written as [10, 57]:

$$\sigma_F(\mathbf{p}_0, \mathbf{p}_1, \mathbf{p}_2) = (2\pi)^4 \sum_{\text{av}} (p_1 p_2/p_0) \, |M_F(\mathbf{p}_0, \mathbf{p}_1, \mathbf{p}_2)|^2 , \tag{6}$$

where $\sum_{\text{av}}$ represents an average over the initial degenerate states and a sum over the unresolved final states. The cross section gives the transition probability of the initial two-body system (the incident electron and the neutral target) to the final three-body system (the scattered electron, the ejected electron and the residual ion). The scattering amplitude, $M_F(\mathbf{p}_0, \mathbf{p}_1, \mathbf{p}_2)$, is given by:

$$M_F(\mathbf{p}_0, \mathbf{p}_1, \mathbf{p}_2) = \langle \chi_1 \chi_2 \Psi_F^{N-1} | T \, (E) | \Psi_0^N \chi_0 \rangle , \tag{7}$$

where the χ's represent the wavefunctions of incident and outgoing electrons, and T (E) represents the transition operator. Antisymmetrization is assumed implicitly throughout unless stated otherwise.

2.3 Plane Wave Impulse Approximation

Under the high energy and maximized momentum transfer kinematic conditions, the incident electron knocks out a bound electron in an essentially clean manner. The incident, scattered and ejected electrons interact only very weakly with the residual ion. The (e, 2e) collision may then be considered as a close, binary encounter, direct knockout reaction. At sufficiently high energies, the wavefunctions of the incident and outgoing electrons can be represented by plane waves. The plane wave wavefunction $|\mathbf{p}\rangle$ of an electron of momentum $\mathbf{p}$ is given by $(2\pi)^{-3/2} \exp(i\mathbf{p} \cdot \mathbf{r})$. The transition operator T(E) becomes a three-body operator, depending only on the coordinates of the two colliding electrons and the centre of mass of the residual ion. It can be further approximated by the electron-electron scattering operator t(E) [10, 57, 58]. Under these conditions, the scattering amplitude becomes a product of an antisymmetrized electron-electron collision amplitude and an overlap function of the target and ion states in the momentum-space representation. i. e.

$$M_F(\mathbf{p}_0, \mathbf{p}_1, \mathbf{p}_2) = \langle \mathbf{k}' \, |t(E)| \, \mathbf{k} \rangle \, \langle \mathbf{p}\Psi_F^{N-1} \, |\Psi_0^N \rangle, \tag{8}$$

where $\mathbf{k}' = (\mathbf{p}_1 - \mathbf{p}_2)/2$, $\mathbf{k} = (\mathbf{p}_0 - \mathbf{p})/2$, and $\mathbf{p} = \mathbf{p}_1 + \mathbf{p}_2 - \mathbf{p}_0$.

In the plane wave impulse approximation, the differential cross section is therefore given by:

$$\sigma_F(\mathbf{p}_0, \mathbf{p}_1, \mathbf{p}_2) = (2\pi)^4 \, (p_1 p_2/p_0) \, \sigma_{MOTT} \, \Sigma_{av} \, |\langle \mathbf{p}\Psi_F^{N-1} \rangle \, \Psi_0^N \rangle|^2. \tag{9}$$

The half-off-shell Mott scattering cross section, σ_{MOTT}, is given by [10, 58]:

$$\sigma_{MOTT} = |\langle \mathbf{k}' | \, t(E) \, |\mathbf{k}\rangle|^2,$$
$$= (2\pi^2)^{-2} \, \{2\pi\alpha/[\exp(2\pi\alpha) - 1]\} \, \{K^{-4} + K'^{-4}$$
$$- (KK')^{-2} \cos[\alpha \ln(K^2/K'^2)]\}, \tag{10}$$

where $K = |\mathbf{k} + \mathbf{k}'|$, $K' = |\mathbf{k} - \mathbf{k}'|$ and $\alpha = |\mathbf{p}_1 - \mathbf{p}_2|^{-1}$. The Mott cross section varies rapidly with θ and this must be taken into account in the interpretation of the symmetric coplanar measurements [10]. In the symmetric noncoplanar geometry, however, σ_{MOTT} reduces to:

$$\sigma_{MOTT} = (2\pi^2)^{-2} \, \{2\pi\beta/[\exp(2\pi\beta) - 1]\} \, K^{-4}, \tag{11}$$

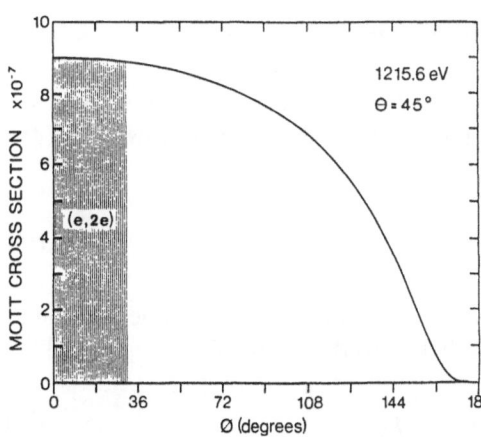

Fig. 3. Mott scattering cross section as a function of the relative azimuthal angle φ for the symmetric noncoplanar geometry. The impact energy E_0 is 1215.6 eV while the energies of the outgoing electrons are each 600 eV (i.e. $E_1 = E_2$). The polar angles of the outgoing electrons $(\theta = \theta_1 = \theta_2)$ are 45°. The normal angular range for a symmetric noncoplanar (e, 2e) experiment is indicated by the shaded region. Clearly, the Mott cross section is essentially constant over this experimental range

where $\beta = \{2p_1 \sin \theta_1 \sin [(\pi - \varphi)/2]\}^{-1}$ and $K^2 = p_0^2 + p_1^2 - 2p_0p_1 \cos \theta_2$. A plot of σ_{MOTT} as a function of φ corresponding to $E_0 = 1215.6$ eV, $E_1 = E_2 = 600$ eV and $\theta = 45°$ in the symmetric noncoplanar geometry is given in Fig. 3. Clearly for the range of φ normally accessible in the experiment, the Mott cross section is virtually constant (variation over the effective experimental range $\varphi = 0°$ to $35°$ is less than 2%). Since absolute cross sections are not measured in most symmetric noncoplanar experiments, the Mott term as well as the constant kinematic factors can usually be ignored.

2.4 Target Hartree-Fock Approximation

The important term in the equation for the cross section (Eq 9) for the symmetric noncoplanar (e, 2e) reaction is the modulus square of the overlap amplitude, $\langle \mathbf{p}\Psi_F^{N-1} | \Psi_0^N \rangle$. The ground state of the target wavefunction $|\Psi_0^N\rangle$ and the F-th eigenstate of the ion wavefunction $|\Psi_F^{N-1}\rangle$ can each be expanded into a linear combination of configuration states, $|\Phi\rangle$'s, with expansion coefficients a_A and t_{jB}^F respectively. i.e.

$$|\Psi_0^N\rangle = \Sigma_A a_A |\Phi_A\rangle . \tag{12}$$

$$|\Psi_F^{N-1}\rangle = \Sigma_B t_{jB}^F C_{jsB} \psi_j \dagger |\Phi_B\rangle . \tag{13}$$

The ion wavefunction is regarded as an expansion of a hole orbital $\psi_j\dagger$ coupled by the Clebsch-Gordon coefficients C_{jsB} with a target configuration $|\Phi_B\rangle$, to give an ion state with symmetry s. The overlap amplitude then becomes:

$$\langle \mathbf{p}\Psi_F^{N-1} | \Psi_0^N \rangle = \Sigma_{ABj} a_A t_{jB}^F C_{jsB} \langle \mathbf{p} | \psi_j \rangle \langle \Phi_B | \Phi_A \rangle . \tag{14}$$

If the same one-electron potential is used for the ion as for the target, then orthonormality of the configurations gives:

$$\langle \mathbf{p}\Psi_F^{N-1} | \Psi_0^N \rangle = n_s \Sigma_{Aj} a_A t_{jA}^F C_{jsA} \psi_j(\mathbf{p}) , \tag{15}$$

where n_s is the number of equivalent electrons (or the dimension of the symmetry group). The momentum-space orbital $\psi_j(\mathbf{p})$ is given by the Fourier transform of the position-space orbital $\psi_j(\mathbf{r})$. i.e.

$$\psi_j(\mathbf{p}) = (2\pi)^{-3/2} \int d\mathbf{r} \exp(-i\mathbf{p}.\mathbf{r}) \psi_j(\mathbf{r}) . \tag{16}$$

In the special case when the Hartree-Fock ground state Φ_0 is a good description of the target state $|\Psi_0^N\rangle$, i.e. $a_A \simeq 0$ for $A \neq 0$, the overlap amplitude becomes [10]:

$$\langle \mathbf{p}\Psi_F^{N-1} | \Psi_0^N \rangle = n_s \Sigma_j a_0 t_{j0}^F C_{js0} \psi_j(\mathbf{p}); \tag{17}$$

and for a close-shell target,

$$\langle \mathbf{p}\Psi_F^{N-1} \mid \Psi_0^N \rangle = n_s^{1/2} a_0 \Sigma_j \, t_{j0}^F \psi_j(\mathbf{p}) \, . \tag{18}$$

Normally, any set of independent particle orbitals, $\{\psi_j\}$, can be redefined so that for a given ion state $|\Psi_F^{N-1}\rangle$ only one term is sufficient in the sum (Eq. 18). The associated orbital is called the characteristic orbital ψ_c so that:

$$\langle \mathbf{p}\Psi_F^{N-1} \mid \Psi_0^N \rangle = n_s^{1/2} a_0 t_{c0}^F \psi_c(\mathbf{p}) \, , \tag{19}$$

and

$$\sigma_F(\mathbf{p}_0, \mathbf{p}_1, \mathbf{p}_2) = (2\pi)^4 \, (p_1 p_2 / p_0) \, \sigma_{MOTT} \, [(n_s a_0^2 S_{c0}^F) \, \Sigma_{av} \, |\psi_c(\mathbf{p})|^2] \, , \tag{20}$$

where the spectroscopic factor $S_{c0}^F = |t_{c0}^F|^2$ can be regarded as the probability that the F-th ion state contains the configuration with a hole orbital $\psi_c\dagger$ in the target. The differential cross section is therefore proportional to $\Sigma_{av} \, |\psi_c(\mathbf{p})|^2$.

In the Born-Oppenheimer approximation, a molecular wavefunction is a product of separate electronic, vibrational and rotational functions. In cases where the final rotational and vibrational states cannot be resolved, the wavefunction must be integrated over the rotational and vibrational states. The differential cross section is given by:

$$\sigma_F(\mathbf{p}_0, \mathbf{p}_1, \mathbf{p}_2) \propto \int d\Omega \int dv \, |\psi_c(\mathbf{p})|^2 \, . \tag{21}$$

The vibrational integral $\int dv$ has been shown to be accurately approximated by assuming that the nuclei are fixed at their equilibrium positions [10, 59]. The differential cross section, in the plane wave impulse and target Hartree-Fock approximations, is therefore proportional to the spherically averaged modulus square of the characteristic orbital evaluated at the equilibrium geometry in the momentum-space representation. i.e.

$$\sigma_F(\mathbf{p}_0, \mathbf{p}_1, \mathbf{p}_2) \propto \int d\Omega \, |\psi_c(\mathbf{p})|^2 \, . \tag{22}$$

2.5 Validities of the Approximations

The final expression for the (e, 2e) differential cross section (Eq. 20) depends on the validities of the plane wave impulse approximation (PWIA) and the target Hartree-Fock approximation (THFA). The PWIA requires scattering kinematics with high momentum transfer and high energy. The symmetric noncoplanar scattering kinematics ensure maximized momentum transfer. Earlier works [60] on the PWIA have demonstrated its validity for incident energies $E_0 > 400 \, eV$. Although the use of more sophisticated scattering approximations such as the distorted wave impulse approximation [61-63] and the averaged eikonal approximation [10, 64] has provided significant improvement in the prediction of the shape of the angular correlation in coplanar (e, 2e) studies, no noticeable differences can be found in the noncoplanar experiments except in the high ($\geq 1.5 a_0^{-1}$) momentum region [65]. It has, however, been suggested [66] that in order for the PWIA to predict the correct s-to-p intensity ratio (see later) when all the many-body states are considered, an incident energy $E_0 \geq 1000$ eV or about 20 times that of the ionization energy of interest is necessary.

Experimentally, the use of a higher incident energy imposes more demanding requirements on the performance of the spectrometer. At higher incident energies, the (e, 2e) cross-section decreases and the range of φ becomes smaller for a given range of p. While the smaller range of φ improves the statistics due to a lower relative accidental coincidence rate, it also necessitates a smaller acceptance angle as well as better electron optics for angular selection in order to maintain sufficient angular (or momentum) resolution. A smaller acceptance angle may in turn reduce the count rate.

The target Hartree-Fock approximation (THFA) defines a unique set of orbitals (the canonical ones) apart from a unitary transformation. The Hartree-Fock orbitals for the ground ionic state are in general not the same as for the excited ionic states or for the target (i.e. neutral) ground state. (The overlap amplitude has been calculated in some cases [67] and is found to be over 90% for valence orbitals). This orthogonality problem is "eliminated" in the THFA by defining all single-particle orbitals to be those of the target ground state (see Eq. 15). Such a treatment has been found to be essentially adequate for the outer valence orbitals in most of the (e, 2e) studies reported to date. In the case where electron correlation effects are important [68 − 73], the more general expression (Eq. 9) must be used. Moreover, the ion wavefunction (Eq. 13) must be expanded to include the 2-particle-1-hole, 3-particle-2-hole, etc. processes. In this case, the interpretation of the cross section becomes complicated [10]. The resulting momentum distribution contains an admixture of orbital momentum densities and is not characteristic of any single orbital. The weighting coefficients will depend upon whether initial or final state configurational interactions or both are important.

3. Experimental Method and Instrumentation

The experiment requires a complete control and determination of the scattering kinematics of the incident and outgoing electrons. The energies and momenta of the electrons are analyzed using standard electrostatic electron optical techniques [74, 75]. A block diagram of a conventional double-analyzer (e, 2e) spectrometer is shown in Fig. 4. Briefly, the spectrometer consists of one primary electron optical system (PEOS) and two secondary electron optical systems (SEOS's). The PEOS provides a finely focussed incident electron beam to effect high energy electron-impact ionization of gaseous targets. The SEOS's momentum-analyze the outgoing electrons after the (e, 2e) reactions in the symmetric noncoplanar scattering conditions. Angular correlation of the outgoing electrons is achieved by physically rotating one of the SEOS's to the appropriate relative azimuthal angle. Channel electron multipliers (or channeltrons) are conventionally used as the detectors for the electron energy analyzers but the use of multichannel position-sensitive detectors [75, 76] has become increasingly popular to provide simultaneous parallel detection of an array of electrons with different momenta. Due to the normally long data acquisition period, the

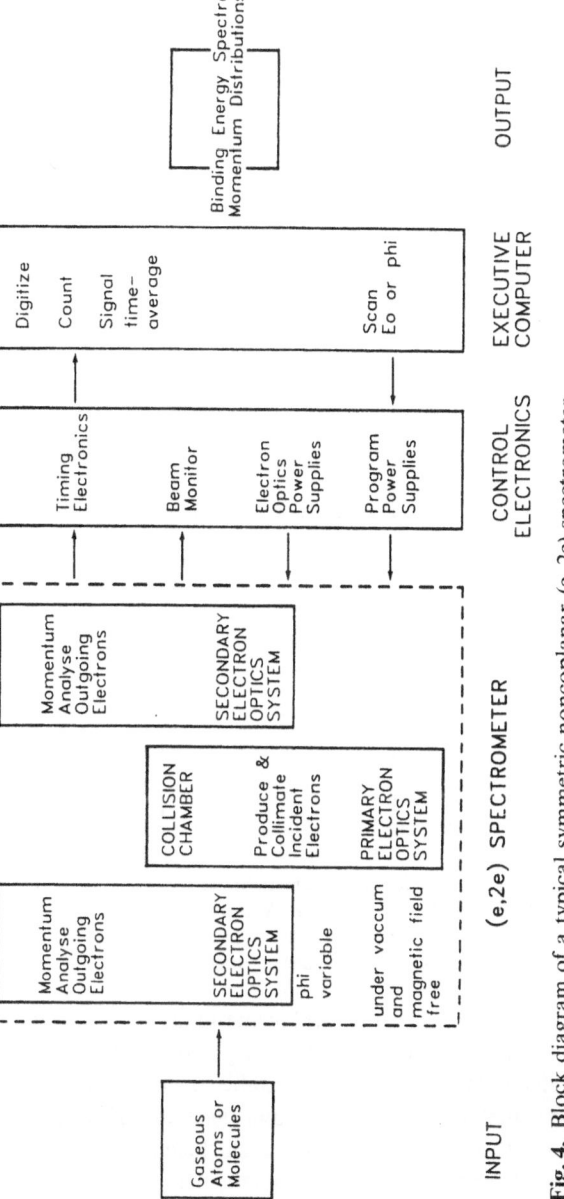

Fig. 4. Block diagram of a typical symmetric noncoplanar (e, 2e) spectrometer

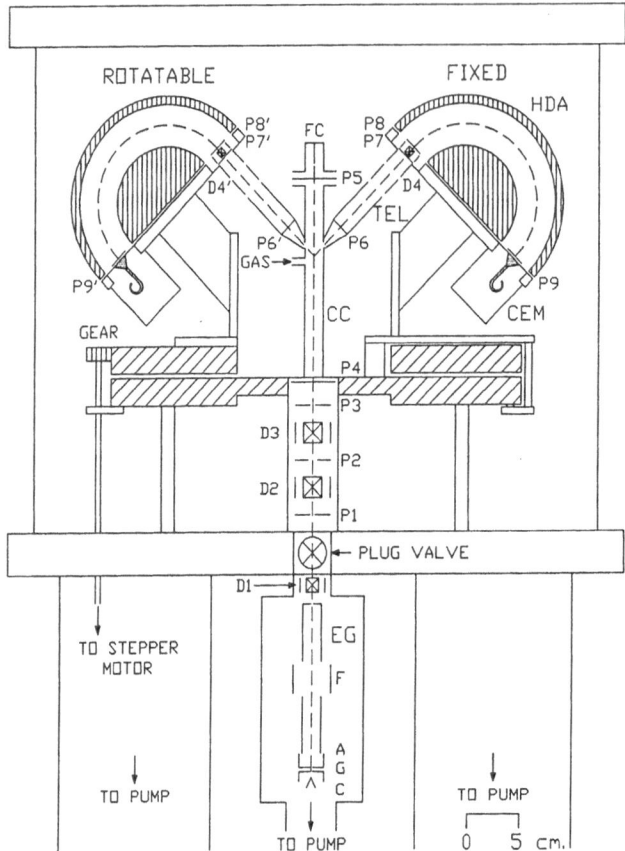

Fig. 5. Schematic diagram of an (e, 2e) spectrometer at the University of Waterloo. Legend: *EG*: Electron Gun; *C*: Cathode; *G*: Grid; *A*: Anode; *F*: Focus; *D*'s: Deflectors; *P*'s: spray Plates; *CC*: Collision Chamber; *FC*: Faraday Cup; *TEL*: Triple Elements Lens; *HDA*: Hemispherical Deflector Analyzer; *CEM*: Channel Electron Multiplier

spectrometer is under the constant control of an online computer, which also performs data collection and parameters scanning of E_0 and φ. The power supplies for the various optical elements are made to be stable over the lengthy data acquisition period. The spectrometer is housed in a high vacuum chamber (with typical base pressure in the 10^{-7} to 10^{-8} Torr range). It is shielded from spurious electronic noise as well as the Earth's and/or any local magnetic field.

Most conventional (e, 2e) spectrometers employ the double-analyzer design described above, with differences in the type of energy analyzers (hemispherical vs cylindrical mirror analyzers), retardation input lenses (3-element or 5-element, tube or apertures lenses), and detection systems (single channel electron multipliers or multichannel position sensitive detectors) [75]. Other miscellaneous differences are found in the beam steering optics and angle control mechanisms. A schematic diagram of the (e, 2e) spectrometer at the University of Waterloo is shown in Fig. 5.

3.1 Coincidence Rate and Signal-to-Noise

Rather standard signal processing electronics (such as preamplification, amplification, discrimination, etc.), common in nuclear physics experiments, are used in (e, 2e) spectroscopy to convert an electron as detected by a channeltron via capacitive decoupling to a workable signal for coincidence determination. Almost all of the existing spectrometers use the so-called single-delay timing method to determine whether the two detected electrons originate from the same electron impact ionization event [10]. In this method, one of the two detected signals is used as the start pulse for a time-to-amplitude converter (TAC) while the other signal is used as a stop pulse after a time delay. A schematic diagram of the timing electronics is shown in Fig. 6. A typical timing spectrum obtained at a fixed E_0-φ pair (corresponding to a fixed ionization energy and orbital momentum respectively) is shown in Fig. 7. Clearly two kind of coincidences are possible: A "true" coincidence is generated by an (e, 2e) reaction while an "accidental" coincidence is generated by randomly scattered, uncorrelated electrons which happen to arrive at the detectors at the same time. In the timing spectrum, it is evident that the "true" coincidence peak is riding on top of a flat "accidental" coincidence background. Two single channel analyzers (SCA's) are used to window the peak (true plus accidental) and background (accidental) regions. The total number of counts per unit time detected in the peak and background windows are denoted by N_{T+A} and N_A respectively. If the widths of the peak and background windows are denoted by t_T and t_A respectively and the ratio between the widths of these two timing windows (t_A/t_T) is given by r, then the true coincidence is given by:

$$N_T = N_{T+A} - N_A/r, \tag{23}$$

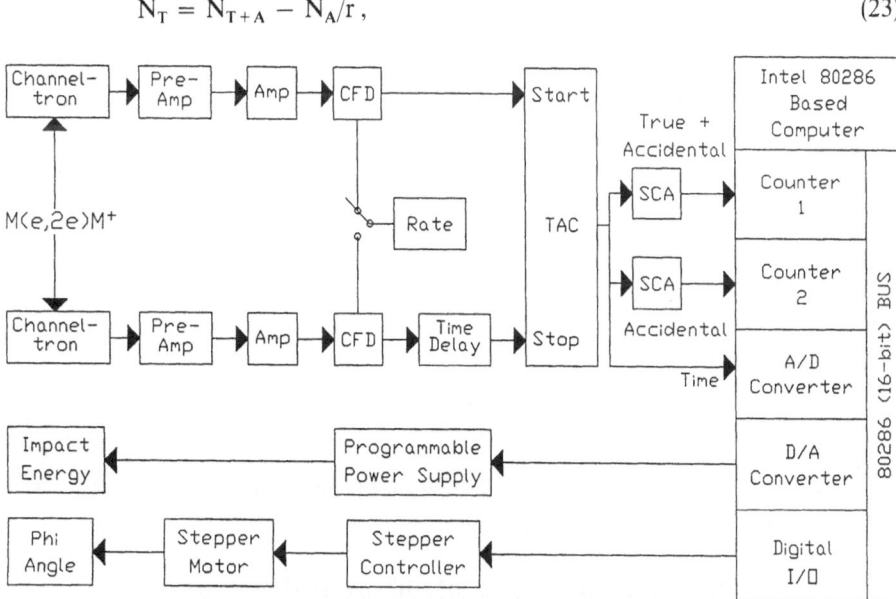

Fig. 6. Schematic diagram of the timing electronics. Legend: *PreAmp*: Preamplifier; *Amp*: Amplifier; *CFD*: Constant Fraction Discriminator; *TAC*: Time-to-Amplitude Converter; *SCA*: Single Channel Analyser; *A/D*: Analog-to-Digital; *D/A*: Digital-to-Analog; *I/O*: Input-Output

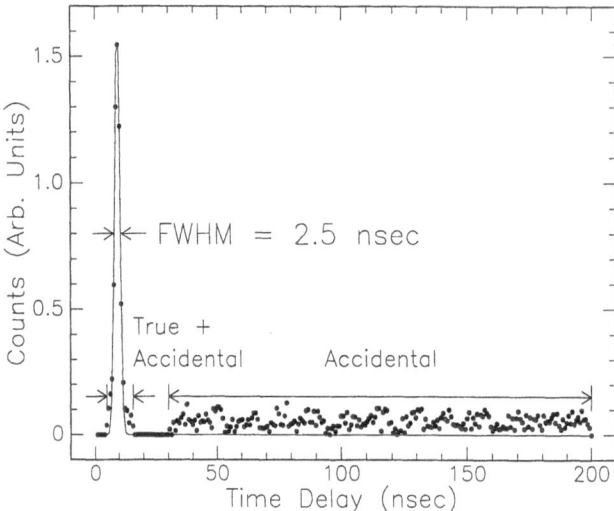

Fig. 7. A typical timing spectrum. A Gaussian lineshape with a Full-Width-at-the-Half-Maximum (FWHM) of 2.5 nsec is fitted to the "true" coincidence peak. This FWHM represents the typical timing resolution of our spectrometer

and its uncertainty is given by:

$$\Delta N_T = (N_{T+A} + N_A/r^2)^{1/2}. \tag{24}$$

The experimental true coincidence count rate, N_T, is related to the triple or five-fold differential cross section discussed earlier [10]. i.e.

$$N_T = nI[d^3\sigma/d\Omega_1 \, d\Omega_2 \, dE_1] \, \Delta\Omega_1 \, \Delta\Omega_2 \, \Delta E_1, \tag{25}$$

where n is the number of target species in the collision region and I is the number of incident electron per second. $\Delta\Omega_1$ and $\Delta\Omega_2$ are the acceptance solid angles of the two secondary electron optics systems. The experimental accidental coincidence count rate, N_A, is given by: [10]

$$N_A = N_1 N_2 t_T. \tag{26}$$

The singles (elastics) count rates, such as N_1 (which is assumed to be the same as N_2), are defined by:

$$N_1 = nI[d^2\sigma/d\Omega_1 \, dE_1] \, \Delta\Omega_1 \, \Delta E_1. \tag{27}$$

where $d^2\sigma/d\Omega_1 \, dE_1$ is the double or three-fold differential cross section. The signal-to-noise is therefore given by [10]:

$$N_T/N_A = (nI \, t_T \, \Delta E)^{-1} \, [d^3\sigma/d\Omega_1 \, d\Omega_2 \, dE_1]/[d^2\sigma/d\Omega_1 \, dE_1]^2, \tag{28}$$

where the energy resolution of the secondary electron optics system is $\Delta E(= \Delta E_1 = \Delta E_2)$. If one were to ignore the target dependent triple and double differential cross sections, the signal-to-noise ratio N_T/N_A is proportional to $(nI \, \Delta E_1 \, t_T)^{-1}$ while the true coincidence signal N_T is proportional to $(nI \, \Delta E)\Delta\Omega_1 \, \Delta\Omega_2$. It is clear that the best operating condition for the energy resolution, incident beam current and target number density is a compromise between the signal-to-noise and the true coincidence rate. Increasing the acceptance solid angles will increase the true coincidence count rate at the expense of the angular resolution. A smaller, true timing window will, however, improve the signal-to-noise without affecting the true coincidence count rate.

3.2 Modes of Operation

In a symmetric noncoplanar (e, 2e) spectrometer, there are two common modes of operation. In the ionization energy mode, the two SEOS's are positioned at a fixed φ and each system is set to detect outgoing electrons with fixed kinetic energies equal to half of the incident energy less the ionization energy of interest. It is clear that by scanning the incident energy employed in the PEOS and keeping the detection energies of the two outgoing electrons fixed, one is effectively scanning the ionization energy (Eq. 2). In the angular correlation mode, the PEOS is set at a fixed incident energy (corresponding to a fixed ionization energy) while the relative azimuthal angle φ is varied. Using the φ-to-p relation (Eq. 5), the momentum distribution at a particular ionization energy is obtained.

3.3 Performance Considerations

A typical (e, 2e) spectrometer is commonly operated with an impact energy of 1.0–2.0 keV. An energy resolution of 1.0–2.0 eV FWHM, momentum resolution better than 0.1 a u. FWHM and timing resolution of 2–10 nsec FWHM can be routinely obtained. Over the past two decades, the momentum and timing resolutions have been advanced significantly while the energy resolution and coincidence rate continue to be the experimental challenges. A standard method of improving the energy resolution is to monochromatize the incident electron beam in order to minimize the thermal spread resulting from thermionic emission. This method however cannot be employed until the achievable coincidence data rate is enhanced.

The rather low coincidence rate remains to be the singularly most important experimental challenge to be overcome. Conventional multi-energy detection technique, involving position-sensitive detection in the energy dispersive plane of an energy analyzer, has been incorporated in (e, 2e) experiments by Weigold and coworkers [77], and recently by Williams and coworkers [78]. Perhaps more exciting developments involving multi-angle detection are found in at least two laboratories. The first multi-angle (and the first true multichannel) (e, 2e) machine was developed by Coplan and coworkers in 1978, using a spherical analyzer coupled with two banks of five single-channel multipliers arranged at appropriate locations to sample a selected number of azimuthal angles [79, 80]. The same group has advanced this concept further with their second generation machine equipped with improved input optics and with two banks of seven multipliers [81]. The multi-angle detection technique using a complete cylindrical mirror analyzer and a position-sensitive detector is being developed by Brion and coworkers [82]. Perhaps the ultimate instrumentation development is the toroidal analyzer currently under development by Weigold and coworkers [83]. This analyzer is capable of sampling electrons emerging with a preselected range of energies and angles simultaneously. All of these efforts seek to greatly enhance the achievable data rate and should help to fully integrate (e, 2e) spectroscopy to condensed matter as well as exotic gas-phase studies [83].

4 Applications

Two general types of electronic structural information can be obtained by (e, 2e) spectroscopy, namely, the ionization or binding energy spectrum measured at a fixed momentum, and the electron momentum distribution measured at a fixed ionization energy of a particular ionic state. In principle, the technique is capable of studying the complete ionization energy spectrum including both the core and valence shells. However, due to the very high energies required in order to satisfy the kinematic approximations involved, almost all of the (e, 2e) studies published to date are concerned only with the valence region. It should also be noted that because of the random orientations of the gaseous targets, the measured momentum distribution corresponds to momentum density averaged over all angular orientations (i. e. spherical averaging occurs). Despite these limitations, important information and useful insights concerning the electronic structure can be inferred from these measurements. Instead of providing a case-by-case summary of all the reported (e, 2e) data, some general observations and an overview of the more common applications to quantum chemistry by (e, 2e) spectroscopy are discussed.

4.1 Valence-Shell Ionization Energies

(e, 2e) spectroscopy measure the complete valence-shell ionization energy spectrum (typically 10–50 eV) of atoms and molecules. Before the advent of synchrotron radiation sources, the inner valence region was not easily accessible by photoelectron spectroscopy [84]. According to Koopmans' theorem [85], ionization may be considered as the ejection of an orbital in an noninteracting, sudden fashion, which corresponds to one observed peak in the ionization energy spectrum. This independent particle ionization (or one peak per orbital) picture generally applies quite well to the outer valence orbitals (< 20 eV). In the inner valence region (> 20 eV), however, significant population splittings due to many-body (or satellite) states have been observed in almost all of the systems (both atoms and molecules) studied to date by (e, 2e) spectroscopy. This breakdown of the independent particle picture is generally attributed to electron correlation effects in the initial states and/or final states [86]. Indeed, one of the pioneering (e, 2e) studies [14] was the first not only to demonstrate

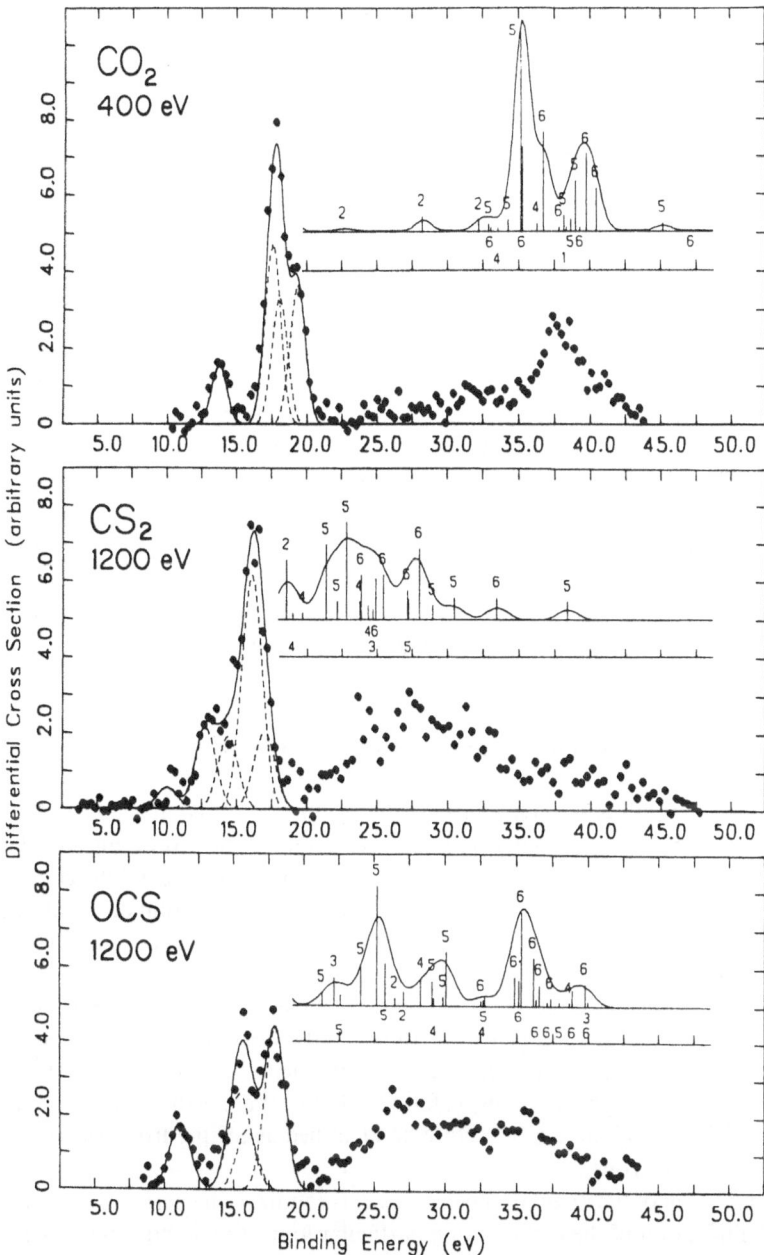

Fig. 8. Binding or Ionization Energy spectra for the valence-isoelectronic series: CO_2, CS_2 and OCS [92]. The many-body structures in the inner valence regions (>20 eV) are compared with the corresponding polestrength diagrams generated by Green's function calculations. The poles are labelled numerically in an increasing fashion from the outermost (labelled by 1) to the innermost (usually labelled by 6) orbitals of the corresponding molecules. Individual poles are convoluted with the experimental energy resolution and the sum gives the overall calculated band-shape for these many-body states. Clearly, reasonably good agreements are obtained between the experiments and calculations

the presence of many-body states in the inner valence region of argon, but also to identify the orbital characters of these many-body states by direct measurements of their momentum distributions. Numerous other examples can be found, for instance, in the isoelectronic series: CO [87, 88], N_2 [89, 90] and C_2H_2 [91], the valence isoelectronic series: CO_2, OCS and CS_2 [92], and other methyl-substituted compounds [35, 93, 94].

The spectroscopic factor or pole-strength, which is defined to be the probability that an eigenstate contains a principal configuration of a hole in the characteristic orbital, can be derived from the peak intensities in the ionization energy spectra integrated over all angles [95]. In particular, the Green's function calculations developed by Cederbaum and coworkers [96, 97] have been used to compare with the experimental (e, 2e) spectrum, giving new insights concerning the breakdown of the pole-strength of a particular valence orbital over a range of ionization energy. The spectroscopic factor therefore provides a quantitative means of characterizing the extent of an orbital contribution at a particular ionization energy. The interplay between results from an (e, 2e) experiment and a Green's function calculation (and other calculations such as the symmetry-adapted cluster expansion method) [98] will eventually lead to a better understanding of configurational interaction and electron correlation effects [86]. An example depicting the many-body structure in the inner valence region of the valence-isoelectronic series: CO_2, OCS and CS_2^{92} is shown in Fig. 8.

4.2 Identification of Many-body States and Orbital Ordering

Although the energy resolution in (e, 2e) spectroscopy may not be comparable to that of photoelectron spectroscopy, many-body structure and orbital ordering can be identified by the orbital-specific momentum distributions measured at carefully chosen ionization energies. As each orbital has its own characteristic momentum distribution, a measurement of the angular dependence of the ionization spectra can identify the dominant orbital compositions of many-body structure in the ionization spectra. In essence, the momentum distribution itself represents the signature of the orbital. Excellent examples of this advantage can be found in the heavier members of noble gases: Ar, Kr and Xe, whereby the characteristic momentum distributions of the satellite states in the inner valence region are found to follow those of the corresponding ns orbitals [99]. These satellites can therefore be identified as arising from ionization of the respective ns electrons.

Another application of the angular dependence of the ionization energy spectrum is the correct identification of the location of a particular ionization band. An excellent example can be found in a recent study of tetramethylsilane (Fig. 9) [49]. Previous photoelectron studies using both resonance [100] and synchrotron light sources [101] revealed great inconsistency (of the order of $1-2$ eV) in the energy position of the band arising from ionization of the innermost valence $5a_1$ orbital. This ambiguity was caused in part by the poor energy resolution obtained in the X-ray photoemission study [100]. The presence of a strong close-lying $(4t_2)^{-1}$ band observed by a recent high resolution synchrotron radiation work [101] further obscured the $(5a_1)^{-1}$ band. The $5a_1$ and the $4t_2$ orbitals correspond to a s-type (with its maximum at zero momentum) and a p-type (with its maximum at a nonzero momentum value)

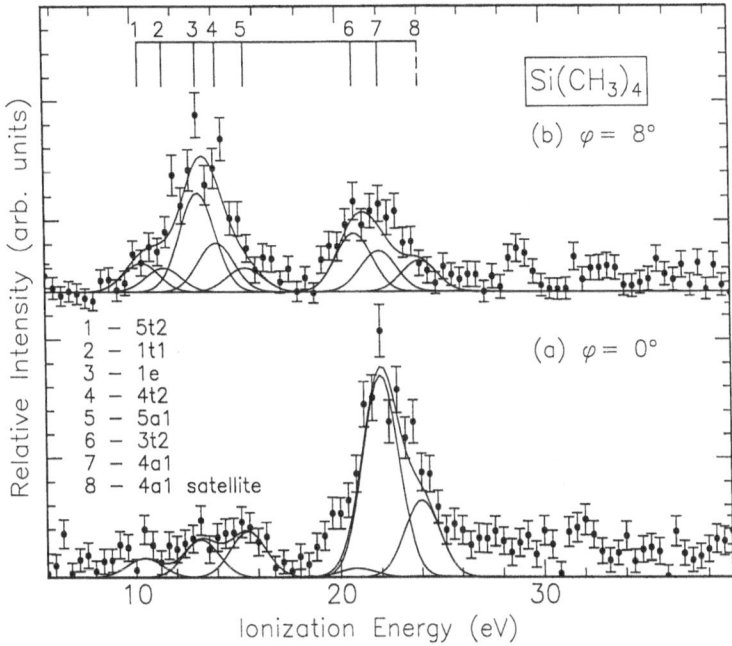

Fig. 9. Ionization energy spectra of tetramethysilane measured at $\varphi = 0°$ and at $\varphi = 8°$ [49]. The relative intensities of the inner-valence $(4t_2)^{-1}$ (corresponding to a p-type) band and $(5a_1)^{-1}$ (corresponding to a s-type) band are markedly different at the two angles due to the symmetries of the ionized orbitals. This contrast in intensity is used to identify the energy locations of these states in the respective spectra

distribution respectively. By making use of the different angular dependence in the intensities of the $(5a_1)^{-1}$ and $(4t_2)^{-1}$ bands at appropriate azimuthal angles, clear identification of the energy positions of the $(5a_1)^{-1}$ and the $(4t_2)^{-1}$ bands can be made.

By comparing the shapes of the experimental momentum distributions with those of the theoretical distributions of individual orbitals and/or by observing the angular dependence of individual ionic states in the ionization energy spectrum, one can unambiguously identify the proper orbital ordering. Indeed, there are two notable examples: H_2CO [72, 102], and OCS [92], in which (e, 2e) data have been used to correct misassigned orbital ordering in the outer valence region. In formaldehyde, despite the somewhat poor energy resolution, the proper ordering of the $5a_1$ and $1b_2$ orbitals was assigned using the different angular dependence of the (e, 2e) ionization energy spectrum [102]. An unambiguous identification of the ordering of the 2π and 9σ orbitals was made in OCS using their characteristic orbital momentum distributions [92]. It should be noted that, while both the photoionization asymmetry parameter and partial photoionization cross section can give excellent insights into photoionization physics [84], the characters of the initial states are often hidden in the dipole matrix elements and are not easily identifiable. (e, 2e) spectroscopy therefore appears to be a more direct tool for the analysis of orbital ordering.

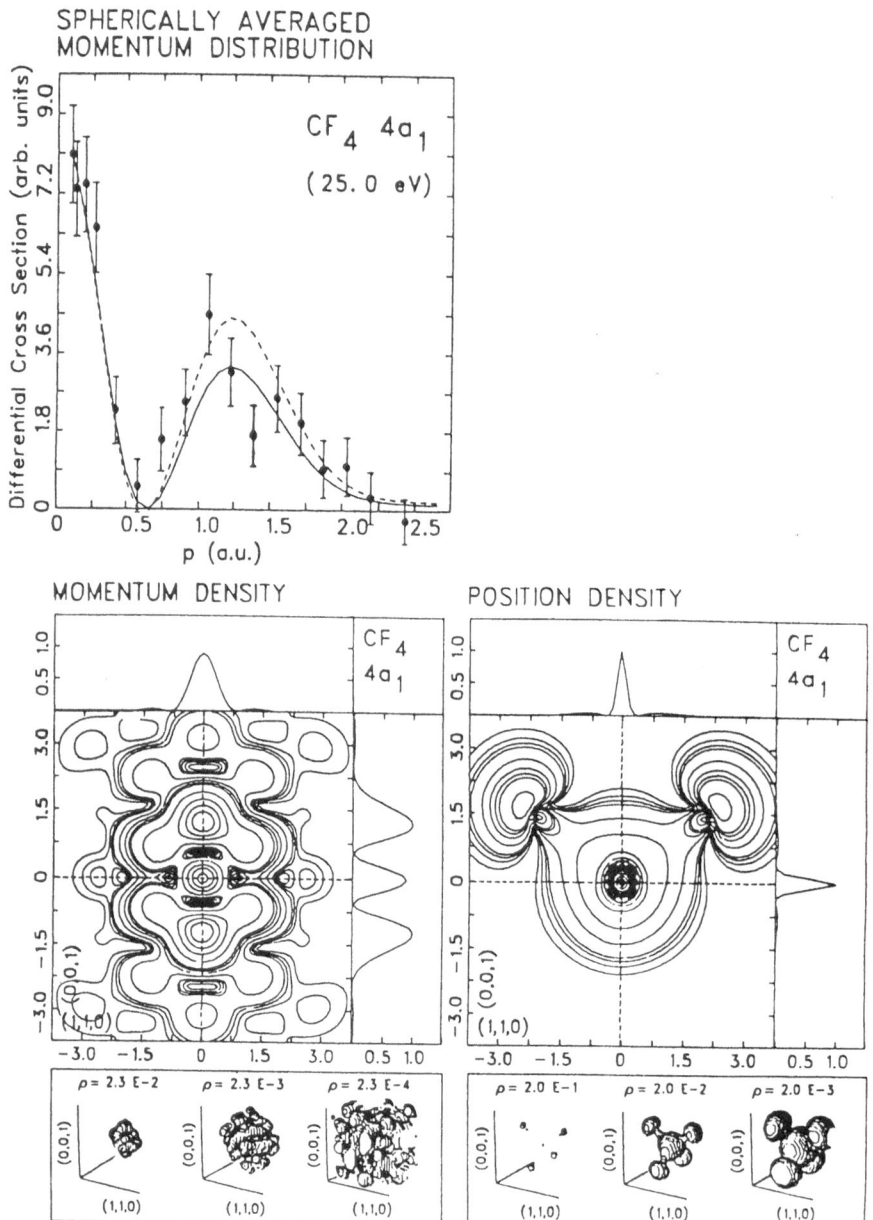

4.3 Qualification of Orbital Symmetry

In general, an orbital momentum distribution is phenomenologically classified according to where its maximum occurs. In atoms, the symmetry of an s orbital can be distinguished directly from a p (and other higher l) atomic orbital using their characteristic momentum distributions. The so-called "s-type" distribution has its maximum at zero momentum while a "p-type" distribution has its

◀ **Fig. 10.** Spherically averaged momentum distributions of the $4a_1$ orbital of CF_4 [104]. Theoretical momentum distributions are evaluated using Snyder and Basch wavefunction [106] (*solid line*) and the Gaussian 4-31 G* wavefunction [109] (*dashed line*), and are height-normalized at the maximum to the experimental distribution. The indicated energy corresponds to the sitting ionization energy and not to the vertical ionization potential. The instrumental momentum resolution (0.1 a. u. FWHM) has been folded into the calculations. Two-dimensional density contour maps and three-dimensional constant density surface plots of the $4a_1$ orbital in momentum-space (lower left) and position-space (lower right) are generated using the Snyder and Basch wavefunction [106]. The contour plane (of the position-space map) corresponds to one of the F-C-F planes. The contour values are 0.2, 0.4, 0.6, 0.8, 2, 4, 6, 8, 20, 40, 60 and 80% of the maximum density. The density values indicated in the constant density surface plots correspond (from left to right) to 20, 2 and 0.2% of the maximum density.

maximum at a nonzero momentum value. While the orbital symmetry of a molecular orbital is more difficult to quantify, some approximate guidelines may be used. Any molecular orbital which consists of totally symmetric basis functions, i.e. with $l = 0$, will have finite intensity at zero momentum. For a molecular orbital which consists of basis functions with nonzero angular momenta, the situation is less clear and detailed consideration of its three-dimensional density is necessary. It can be shown (see later) that since all momentum-space orbital wavefunctions have inversion symmetry, any symmetry orbital with a node (i.e. zero intensity) at the position-space origin will have a node in the momentum-space wavefunction as well, which will therefore give rise to a p-type momentum distribution after spherical averaging.

From the above consideration, because atomic s, diatomic σ_g and polyatomic a_1 orbitals all have totally symmetric components in their wavefunctions, their corresponding momentum distributions therefore follow a s-type profile. Atomic p, d and any other higher l orbitals, diatomic σ_u and π orbitals, and polyatomic e and t orbitals, on the other hand, have p-type momentum distributions. Some classic examples are the momentum distributions of the outer valence ns and np orbitals in the noble gases [99] and those of π and σ orbitals of linear diatomics and triatomics (see Appendix).

Besides the simple s-type and p-type momentum distributions, there are the "sp-type" momentum distributions, with maxima both at zero momentum and at some nonzero momentum value. While the secondary maxima may be caused by the so-called bond oscillation effect in three-dimensional momentum density (see later), these secondary maxima are generally not sufficiently intense (after spherical averaging) to be observed except in the case of core molecular orbitals. The momentum distributions of core molecular orbitals have not been observed in (e, 2e) spectroscopy except in a recent study [103]. As such, the maximum at zero momentum and the maximum at a finite momentum value for a valence-shell orbital momentum distribution may be attributed respectively to the totally symmetric and nonsymmetric components of the wavefunction itself. Even in an apparently well-defined case such as the momentum distribution of the $4a_1$ orbital of CF_4 [104], shown in Fig. 10, it is difficult to quantitatively infer the extent of s-p hybridization from the relative intensities of the maxima corresponding to s and p components in the experimental momentum distribution. This is due to the complexities in most momentum-space molecular wavefunctions and clearly more work is needed to extract the wavefunction properties directly (see later). Another example of a sp-type orbital is the outermost σ bonding orbital of the valence-isoelectronic triatomic series: CO_2, OCS and CS_2 [92].

4.4 Stringent Test of Ab-initio Wavefunctions

Perhaps the most powerful and more direct use of the experimental momentum distributions is the evaluation of ab-initio wavefunctions. An ab-initio wavefunction is commonly obtained by a self-consistent-field (SCF) procedure, which optimizes the wavefunction by minimizing its total energy (the variational method) [2, 3]. Such a procedure of energy minimization often underestimates the chemically significant outer (large r) parts of the valence orbitals since these regions do not contribute significantly to the total energy. (e, 2e) spectroscopy provides momentum distributions of valence orbitals in the low momentum (0-2 a.u.) region. Momentum distributions of the valence orbitals can therefore be used as a sensitive and complementary test for the quality of theoretical wavefunction in regions where chemical bonding occurs. Even though absolute cross sections are not determined in most (e, 2e) studies, the (relative intensity) measurements of experimental momentum distributions of a series of valence orbitals, relatively normalized to each other, can be a very powerful tool in identifying any orbital-specific inadequacy in a particular class of ab-initio wavefunctions.

The general theory of (e, 2e) spectroscopy used for electronic structural determination (discussed earlier) is based upon two important approximations: the plane wave impulse approximation (PWIA) and the target Hartree-Fock approximation (THFA) [10]. The validity of the PWIA obtained under appropriate scattering conditions is best illustrated by the excellent agreement of the experimental momentum distribution of the atomic H $1s$ orbital measured at several impact energies with the exact solution [26]. The THFA reduces the ion-neutral overlap function, which contains the electronic structural information, to the momentum-space density of a canonical orbital. While

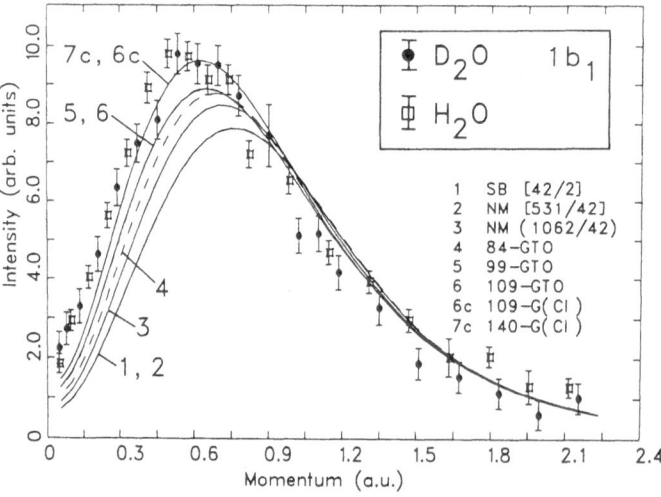

Fig. 11. Spherically averaged momentum distributions of the $1b_1$ orbital of D_2O (*solid circles*) and H_2O (*open squares*) [70]. Detailed calculations of the momentum distributions at the single configuration (Hartree-Fock) and multi-configuration (CI) levels of various degrees of sophistication are shown [70]. The sitting ionization energy is 12.2 eV and the instrumental momentum resolution (0.15 a.u. FWHM) has been folded into the calculations. Clearly, the CI wavefunctions give the best agreement to the experiment

this independent particle approximation is generally valid for most outer valence electrons and is commonly employed in the interpretation, better agreement can be found when configurational interaction is considered, i. e. beyond the single configuration Hartree-Fock approach. The failure of the Hartree-Fock quality wavefunction in modelling the momentum distributions is most evident in the outermost lone-pair orbitals in a number of small molecules, such as the first-row hydrides: HF [105], H_2O [70] (Fig. 11) and NH_3 [71]. The recent calculations reported by Bawagan et al. have clearly demonstrated the importance of the use of correlated wavefunctions for correctly predicting the experimentally observed momentum distributions of these molecules [70, 71, 105]. These comparative studies show that (e, 2e) spectroscopy may be used to provide unique information for modelling the "residual" electron density in the non-bonding (and other "diffuse") region.

Within the limits of the target Hartree-Fock approximation, however, it has been shown that wavefunctions based upon the widely used double-zeta basis, such as, for example, the Snyder and Basch basis [106], are generally inadequate to provide a satisfactory agreement with the (e, 2e) data. Other standard basis sets [107, 108] commonly available in public-domain Gaussian-type programs [109], such as 4-31G, and 6-31G basis sets, generally give agreement not significantly improved or worse (relative to the double-zeta basis) [106] when compared with the experiment. It should be noted that the inclusion of more primitive functions to describe the core shell (i.e. going from 4-31G to 6-31G), will lower the total energy, but should have little effects on the valence-shell orbital momentum distributions. An example is shown in Fig. 12, which compares the experimental momentum distributions of the two Jahn-Teller components of the 3e' orbital in cyclopropane with the double-zeta quality wavefunctions [48]. Within the statistics of the experiment, the data also show that there is little discernable difference between the two experimental curves, indicating that either the present data is not sufficiently sensitive to the Jahn-Teller effect or the two Jahn-Teller components in fact have the same momentum distribution.

The inclusion of additional s and p "diffuse" functions (commonly indicated by the " + " sign in basis set notation) to double-zeta quality basis will increase the momentum distribution in the low momentum region while the inclusion of additional d "polarization" functions (commonly indicated by the "*" sign in basis set notation) will enhance the momentum distribution in the higher momentum region. Depending upon the particular molecular orbital of interest, the inclusion of "diffuse" and "polarization" functions will in general provide a somewhat better agreement to the experimental momentum distribution, with rather small improvement in the total energy. These features are illustrated by the generally small deviations of theoretical momentum distributions from the experiment for the 3e' orbital of cyclopropane in Fig. 13a. Fig. 13b shows even smaller deviations of a wide range of calculations with respect to the 6-31G wavefunction for the same orbital.

It is clear, however, that for a particular theoretical wavefunction, there is little correlation (other than in the general sense) between the degree of overall agreement for the experimental momentum distributions of valence shell, and any one particular wavefunction property such as the total energy, dipole moment or Virial ratio. For instance, while the momentum distributions of a theoretical wavefunction with a lower total energy usually give a better overall agreement with the experimental momentum distributions, it is difficult to predict a priori the overall degree of

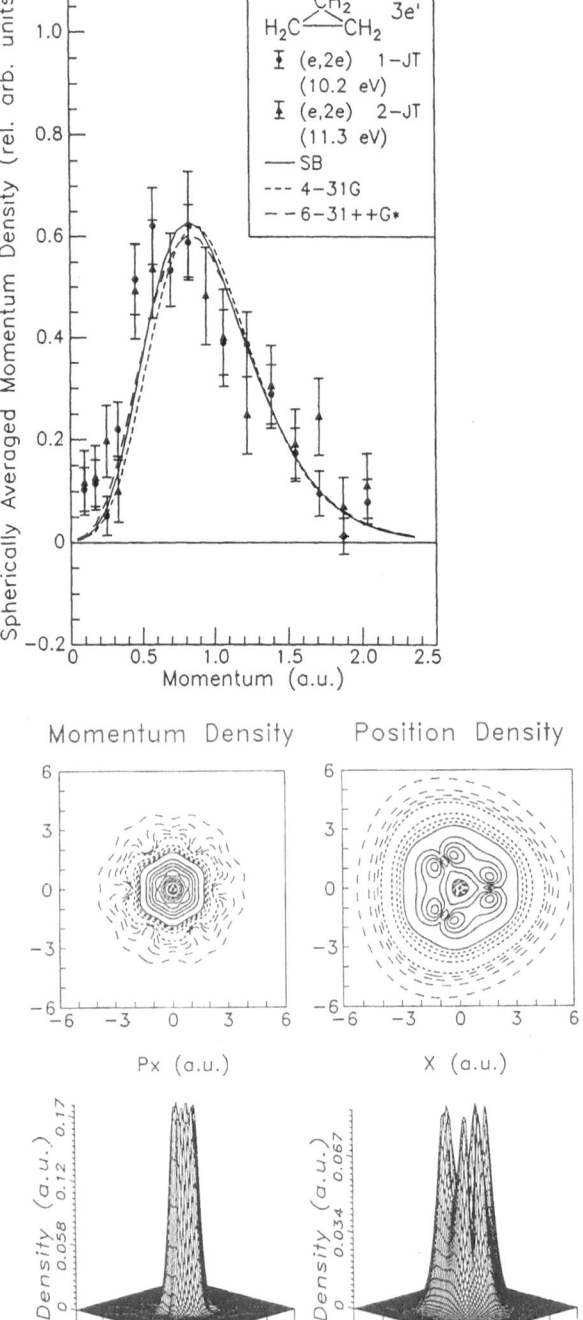

Fig. 12. Spherically averaged momentum distributions of the two Jahn-Teller (JT) components of the 3e' orbital of cyclopropane [48]. Theoretical momentum distributions are evaluated using Snyder and Basch (SB) wavefunction [106] (*solid line*) and the 4-31G (shorter dashed line) and 6-31 + +G* (*longer dashed line*) Gaussian wavefunctions [109], and are relatively normalized using a R-factor approach [48]. The indicated energies correspond to the sitting ionization energies. The instrumental momentum resolution (0.1 a. u. FWHM) has been folded into the calculations. Two-dimensional density contour maps and three-dimensional perspective plots of the orbital in momentum-space (lower left) and position-space (lower right) are generated using the Snyder and Basch wavefunction [106]. The contour plane (of the position-space map) corresponds to the C-C-C plane. The contour values are 0.1, 0.3, 0.5, 0.7 (*dashed lines*), 1, 3, 5, 7 (dotted lines), 10, 30, 50, 70 and 90% (*solid lines*) of the maximum density

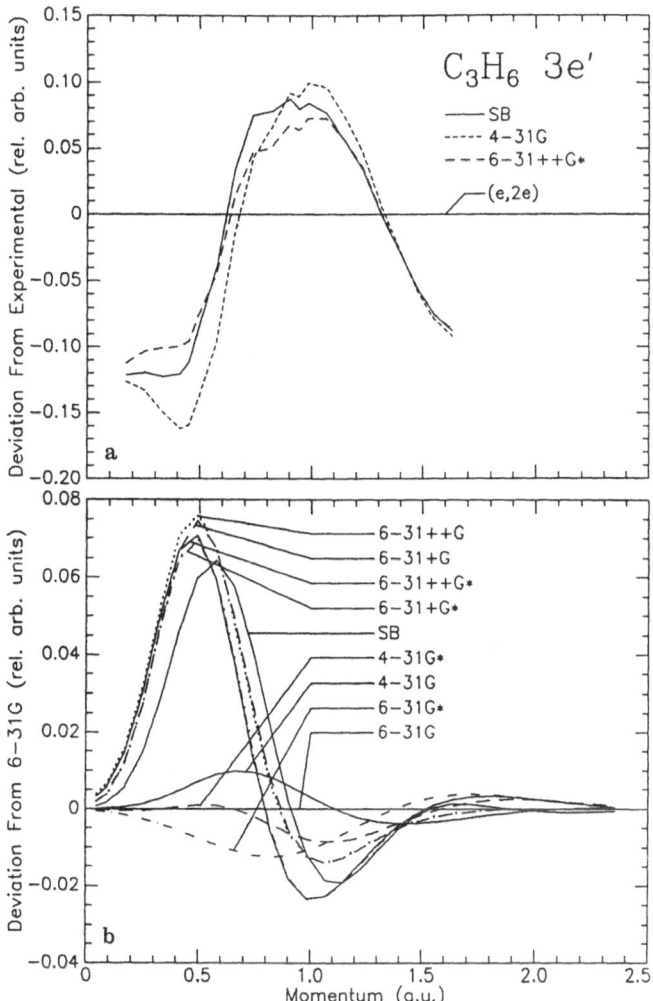

Fig. 13. Deviations of theoretical momentum distributions from the experiment (a) and from the 6-31G calculation (b) for the *3e'* orbital of cyclopropane [48]. Theoretical momentum distributions are evaluated using Snyder and Basch (SB) wavefunction [106] (*solid line*) and various Gaussian wavefunctions of different quality [109]

improvement nor the degree of agreement for a particular orbital in the valence shell. The need to develop the so-called universal wavefunction, i.e. a wavefunction which gives excellent agreement to all observable properties, is clearly evident [73]. This development can clearly be assisted by the experimental momentum distributions obtained by (e, 2e) spectroscopy.

From an experimentalist's perspective, it is somewhat unsatisfactory to quantify the degree of agreement between the orbital momentum distributions obtained by experiment and by various theoretical wavefunctions of different quality using the common practice of visual inspection. Using a modified procedure based on a R-factor analysis commonly employed in the study of I-V curves in low energy electron diffraction [110], we have recently developed a somewhat more quantitative approach to characterize the degree of agreement [48]. The R-factor is defined to be the sum

Table 1. Summary of LCAO MO wavefunction and density equations in position and momentum space

Wavefunction	Position space[a]	Momentum space[a]	Comment
(I) molecular orbital	$\psi_m(\mathbf{r}) = \sum_a C_{ma}\varphi_a(\mathbf{r})$	$\psi_m(\mathbf{p}) = \sum_a C_{ma}\exp(-i\mathbf{p}\cdot\mathbf{R}_a)\varphi_a(\mathbf{p})$	LCAO MO formalism ψ_m MO φ_a AO
(II) atomic orbital	$\varphi_a(\mathbf{r}) = \sum_{i=1}^{n_a} b_{i\lambda}\chi_{ai}(\mathbf{r}-\mathbf{R}_a)$	$\varphi_a(\mathbf{p}) = \sum_{i=1}^{n_a} b_{i\lambda}\chi_{ai}(\mathbf{p})$	AO is again a linear combination of basis function with correct symmetry.
(III) basis function	$\chi_{ai}(\mathbf{r}_a) = N_{n_l l_i m_i} R_{n_l l_i}(\mathbf{r}_a) Y_{l_i m_i}(\Omega)$ where $\mathbf{r}_a = \mathbf{r}-\mathbf{R}_a$; $r_a = |\mathbf{r}-\mathbf{R}_a|$	$\chi_{ai}(\mathbf{p}) = N_{n_l l_i m_i} P_{n_l l_i}(p) Y_{l_i m_i}(\Omega)$	Angular part of the basis function is the same in both r-space and p-space.
(IV) radial part	$R_{n_l l_i}(\mathbf{r}_a)$: STF, GTF, etc.	$P_{n_l l_i}(\mathbf{p}) = (2/\pi)^{1/2}\,(-i)^{l_i}\int r_a^2\,dr_a\, j_{l_i}(pr_a)\, R_{n_l l_i}(r_a)$	For $P_{n_l l_i}(\mathbf{p})$ see Ref. 120. Note that j_{l_i} is the spherical Bessel function.
(V) density	$\varrho_m(\mathbf{r}) = \psi_m^*(\mathbf{r})\,\psi_m(\mathbf{r})$ $= \varrho_m^{QC}(\mathbf{r}) + \varrho_m^I(\mathbf{r})$	$\varrho_m(\mathbf{p}) = \psi_m^*(\mathbf{p})\,\psi_m(\mathbf{p})$ $= \varrho_m^{QC}(\mathbf{p}) + \varrho_m^I(\mathbf{p})$	orbital density decomposition
(VI) quasi-classical part	$\varrho_m^{QC}(\mathbf{r}) = \sum_a |C_{ma}|^2\,|\varphi_a(\mathbf{r}-\mathbf{R}_a)|^2$	$\varrho_m^{QC}(\mathbf{p}) = \sum_a |C_{ma}|^2\,|\varphi_a(\mathbf{p})|^2$	one-centre part
(VII) interaction part	$\varrho_m^I(\mathbf{r}) = \sum_{a\neq b} C_{ma}^*\,C_{mb}\,\varphi_a^*(\mathbf{r}-\mathbf{R}_a)\,\varphi_b(\mathbf{r}-\mathbf{R}_b)$	$\varrho_m^I(\mathbf{p}) = \sum_{a\neq b} C_{ma}^*\,C_{mb}\exp[-i\mathbf{p}\cdot(\mathbf{R}_b-\mathbf{R}_a)]\,\varphi_a^*(\mathbf{p})\,\varphi_b(\mathbf{p})$	two-centre part

[a] $\mathbf{r}$ and $\mathbf{p}$ correspond to the position and momentum of the electron respectively. $\mathbf{R}_i$ refers to the equilibrium position of the atomic centre i

of the squares of the residuals (which is defined as the difference between the experiment and theory) divided by the sum of the square of the experimental data over a specified momentum range, in this case 0.06 to 1.5 a.u. (It should be noted that the present limit of 1.5 a.u. for the integration range is due to an anticipated breakdown in the scattering approximation at higher momentum). By minimizing the R-factor, all the theoretical curves can be appropriately normalized. The minimal R-factors so obtained for different theoretical curves are used as indices for the quality of agreement with the experiment. This kind of analysis will eventually take the effect of statistical accuracy of the data into account and is hoped to provide a more quantitative comparison between experiment and theory [48].

In summary, comparison of the experimental (e, 2e) momentum distributions with theoretical distributions can reveal inadequacies of individual orbitals for a theoretical wavefunction, especially in the variationally less sensitive outer valence region. These experimental results can be used as important feedback for further improvement of the theoretical wavefunction itself.

5 Momentum-Space Chemistry

The development of (e, 2e) spectroscopy and in particular the measurements of electron momentum distributions of individual orbitals prompted a renewed interest in the understanding of chemical bonding directly in momentum-space. Theoretical quantum chemistry [1, 111], has almost totally relied upon the position representation, to formulate problems in chemical structure and reactivity. The momentum representation, although being an equivalent representation in atomic and molecular problems, is generally believed to lead to much greater mathematical difficulties. Most of the attempts [112–115] to use momentum-space wavefunctions directly have been limited primarily to simple atomic systems, because of the complexities involved in solving the integral equation of motion in momentum-space. A more practical approach is therefore to first solve the quantum mechanical problem using the (differential) Schroedinger equation in position-space and then apply the Fourier transform to the solution to obtain the momentum-space wavefunction [112].

In Table 1, a summary of the basic equations of a molecular wavefunction in the LCAO-MO formalism written in position and momentum representations is given. Detailed derivations of these equations can be found elsewhere [116–120]. It is obvious from Table 1 that the Fourier transform preserves the general LCAO-MO formalism; i.e. a position-space LCAO molecular wavefunction can be expressed in momentum-space as a linear combination of momentum-space atomic wavefunctions. The major difference is the introduction of a phase factor (which depends on the nuclear positions R_a) into the complex molecular orbital (MO) coefficients (Eq. (I), Table 1). The density function can be considered as the sum of an one-centre "quasi-classical" term and a two-centre "interaction" term in both position-space and momentum-space representations.

The momentum-space wavefunction is obtained by the Fourier transform of the position-space wavefunction (Eq. 13). By writing $\exp(-i\mathbf{p} \cdot \mathbf{r})$ in a power series expansion and replacing the momentum with its corresponding (position-space) operator ∇_r, the following relation is obtained,

$$\Psi_j(\mathbf{p}) = (2\pi)^{-3/2} \int d\mathbf{r} \; [(1 - i(\nabla_r \cdot \mathbf{r}) - 1/2 \, (\nabla_r \cdot \mathbf{r})^2 + ...)] \, \Psi_j(\mathbf{r}). \qquad (29)$$

The momentum-space (p-space) wavefunction involves an integral of a sum of powers of the gradient of the position-space (r-space) wavefunction over all space. In the case of atomic orbitals where large changes in the r-space wavefunction occur only near the nucleus, the small r part of the r-space wavefunction contributes most significantly to the large p part of the p-space wavefunction (an inverse spatial reversal relation). In the case of molecular orbitals, however, the spatial reversal relation, although still qualitatively correct, must be applied with caution since any region where there is a rapid change in the r-space wavefunction may contribute significantly to the large momentum part of the p-space wavefunction. Such rapid change can occur between atomic centres, for instance across a nodal plane. A notable example of this is the $CO_2(1\pi_g)$ orbital [104]. It should be noted that this physical association of the gradient of the r-space wavefunction with the p-space molecular density also gives a directional reversal relation. The longitudinal change across any r-space density lobe is obviously less rapid (contributing to the small momentum part) than changes in the other perpendicular direction (contributing to the large momentum part). The corresponding directional properties of the p-space density lobe will therefore be reversed.

There are many other interesting relations between the position-space density and the corresponding momentum-space density. Many of the momentum-space notions, first discussed by Coulson and Duncanson [121] in the early 1940s and later extended by Epstein et al. [8, 122, 123], are based upon the Fourier transform properties. In addition, another important property called the Virial property was first discussed by Epstein and Tanner [123] and can be used to interpret the changes of the density difference (or bond density) [124, 125] during bond formation. While the current gas-phase (e, 2e) experiments do not allow three-dimensional mapping of the momentum density function, these relations are useful in giving a qualitative interpretation of the origins of many structures observed in the experimental spherically averaged momentum density as well as in the theoretical momentum density topography depicted by density [126, 127] and density difference [124] maps. With the advent of (e, 2e) spectroscopy and its extension to condensed matter studies, true density mapping may become possible in the near future. Using these concepts, the nature of chemical bonding in free molecules and on surfaces can be inferred qualitatively from the measurements. Extensive discussions of these are made in the literature [121, 123, 125, 127–129], only a brief summary will therefore be given.

5.1 Symmetry, P-space Origin, and P-space Virtual Boundary

All symmetry properties of the position-space wavefunctions are preserved under the Fourier transform. This is evident from Eq. (III) in Table 1 which indicates that the spherical harmonics (which give the angular dependence of atomic orbitals) are

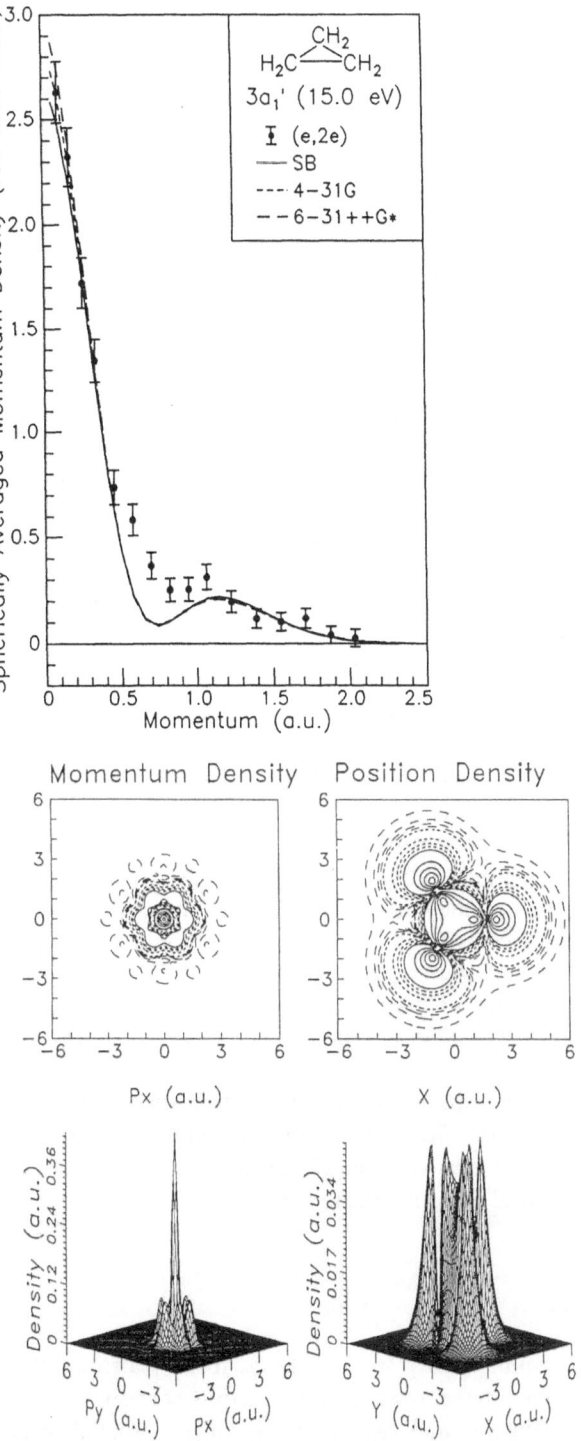

Fig. 14. Spherically averaged momentum distributions of the $3a_1'$ orbital of cyclopropane [48]. Theoretical momentum distributions are evaluated using Snyder and Basch (SB) wavefunction [106] (*solid line*) and the 4-31G (*shorter dashed line*) and 6-31++G* (*longer dashed line*) Gaussian wavefunctions [109], and are relatively normalized using a R-factor approach [48]. The indicated energy corresponds to the sitting ionization energy. The instrumental momentum resolution (0.1 a.u. FWHM) has been folded into the calculations. Two-dimensional density contour maps and three-dimensional perspective plots of the orbital in momentum-space (lower left) and position-space (lower right) are generated using the Snyder and Basch wavefunction [106]. The contour plane (of the position-space map) corresponds to the C-C-C plane. The contour values are 0.1, 0.3, 0.5, 0.7 (*dashed lines*), 1, 3, 5, 7 (*dotted lines*), 10, 30, 50, 70 and 90% (*solid lines*) of the maximum density. The momentum density map clearly reflects the additional inversion symmetry

invariant under the transform. In addition, the momentum-space wavefunction has an additional inversion symmetry, which arises from the fact that the electron has no net translational motion in the centre-of-mass frame. This inversion symmetry is apparently represented by the presence of phase shifts in the MO coefficients (Table 1). The presence of the inversion symmetry in momentum-space can be illustrated by the density maps in the plane of the three-member carbon ring (Fig. 14), which shows that the D_{3h} molecular symmetry of cyclopropane in position-space is changed to D_{6h} in momentum-space [48].

One important difference between the r-space and p-space LCAO-MO formalisms is in the way of expressing the localities of the density function. In r-space, these localities are expressed by referencing electrons to their own atomic centres, represented by r_a. In p-space, these direct references are lost (i.e. there are no nuclear positions in p-space) and the nuclear position vectors (R_a) introduce an extra phase into the MO coefficients (Eq. (I), Table 1). The resulting momentum density function has inversion symmetry because it is composed of a linear combination of symmetric (with respect to inversion) p-space atomic orbital density functions. However, the nuclear geometry information is not lost but is retained in the interaction term (Eq. (VII), Table 1) in the p-space formalism. The inversion symmetry in p-space may also be regarded as a direct manifestation of the invariance of a SCF wavefunction under the time reversal operation.

Since electrons experience sharp attractive potentials near the nuclear centres, the nuclear centres in r-space act, by cosmological analogy, like "black holes", i.e. increasing the momenta as the electrons fall towards them and causing the corresponding p-space density to inflate outward towards the p-space virtual boundary at $p = \infty$ (which corresponds to the electron momenta at the nuclei). The p-space origin, on the other hand, can be thought of as being like a cosmological "white hole" since the electrons with zero momentum (i.e. free electrons) are being pushed out into the more attractive high momentum region near the nuclear centres. The low momentum regime therefore corresponds to the chemically important valence region in r-space. One can therefore picture the effect of the Fourier transform as "turning the wavefunction inside out", without changing the symmetry properties. This in effect inflates the chemically less interesting core (inner r-space) region localized around each nuclear centre towards the p-space virtual boundary and collapses the chemically significant valence (outer r-space) region towards the p-space origin, where the (e, 2e) technique is most sensitive. The (e, 2e) method therefore provides a new and unique vantage point from which to view the chemically significant part of the molecular wavefunction in chemical bonding phenomena.

5.2 Inverse Spatial Reversal

In addition to the inversion symmetry in p-space orbitals, there is also amplitude inversion, namely, a contraction in the r-space wavefunction corresponds to an expansion in the p-space wavefunction and vice versa. This inverse spatial reversal property arises from the pr_a dependence of the spherical Bessel function in Eq. (IV), Table 1. This property has been demonstrated in many earlier works [123] and is well illustrated in the study of noble gases using (e, 2e) spectroscopy [99]. The property

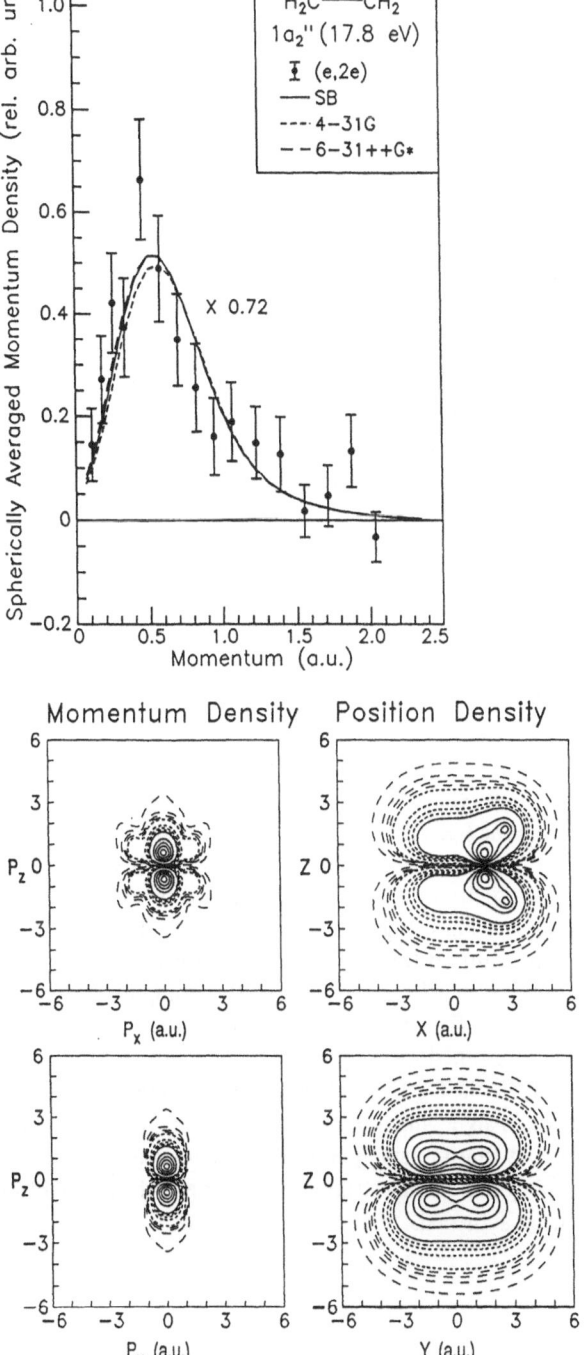

Fig. 15. Spherically averaged momentum distributions of the $1a_2''$ orbital of cyclopropane [48]. Theoretical momentum distributions are evaluated using Snyder and Basch (SB) wavefunction [106] (*solid line*) and the 4-31G (*shorter dashed line*) and 6-31++G* (*longer dashed line*) Gaussian wavefunctions [109], and are relatively normalized using a R-factor approach [48]. The indicated energy corresponds to the sitting ionization energy. The instrumental momentum resolution (0.1 a.u. FWHM) has been folded into the calculations. Two-dimensional density contour maps in momentum-space (lower left) and position-space (lower right) are generated using the Snyder and Basch wavefunction [106]. The contour planes (of the position-space map) correspond to two mutually orthogonal planes with respect to the C-C-C plane. The contour values are 0.1, 0.3, 0.5, 0.7 (*dashed lines*), 1, 3, 5, 7 (*dotted lines*), 10, 30, 50, 70 and 90% (*solid lines*) of the maximum density. The momentum density maps clearly reflect the molecular density directional reversal property

can be readily explained using the Heisenberg uncertainty principle; namely, a more localized r-space wavefunction with a small spatial uncertainty Δr corresponds to a large p-space uncertainty Δp, giving a more diffuse p-space wavefunction.

5.3 Molecular Density Directional Reversal

The relative longitudinal and transverse spatial extents (or lobes) of an orbital in r-space are interchanged in p-space and vice verse [123]. For instance, the position density of a diatomic σ bonding orbital is directed along the internuclear axis in r-space (the longitudinal direction), whereas the corresponding momentum density of such a bonding orbital is oriented in the transverse direction, i. e. perpendicular to the internuclear direction. Similar concepts can be used to draft out the directions of lobes in an antibonding orbital. In essence, the density directional reversal property is a combined effect of the symmetry property and the inverse spatial reversal property applied to molecular orbitals. This property is illustrated in Fig. 15 by the change in the out-of-plane pseudo-π like density distribution of the $1a_2''$ orbital of cyclopropane in the respective representations.

5.4 Molecular Density Oscillations

The molecular momentum density of a bonding (antibonding) orbital exhibits cosinusoidal (sinusoidal) modulations with p-space spatial periodicity of $2\pi/|\mathbf{R}_b - \mathbf{R}_a|$ along the bonding direction [123]. This effect is clearly evident in a core-shell orbital such as the $1a_1'$ orbital of cyclopropane [48] shown in Fig. 16. The p-space periodicity calculated directly from the carbon-carbon separation of 2.83 a.u. is 2.22 a.u., which is in good agreement with the periodicity of 2.32 a.u. as measured directly from the P_Y-P_X plane of the momentum density map (Fig. 16). This oscillation effect is a direct consequence of the phase factor introduced by the Fourier transform in the p-space LCAO-MO formalism. The modulation effects come from the complex coefficient product term $C_{ma}^* C_{mb} \exp(-i\mathbf{p} \cdot (\mathbf{R}_b - \mathbf{R}_a))$ in the interaction density $\varrho_m^I(\mathbf{p})$ in Eq. (VII), Table 1. The periodicity is determined by the $\mathbf{p} \cdot (\mathbf{R}_b - \mathbf{R}_a)$ term and the relative sign of the product of MO coefficients $C_{ma}^* C_{mb}$. (Note that C_{ma} and C_{mb} for an antibonding σ orbital are opposite in sign, which introduces an extra 90° phase shift). The significance of the oscillations is that in principle the relative nuclear separations can be inferred from the p-space periodicities, if a complete mapping of the momentum density in three dimensions can be achieved.

5.5 Virial Property

The formation of a chemical bond is accompanied by an increase in the average momentum $\langle p \rangle$ [123] by transferring momentum density from the p-space origin to the high momentum region in the transverse direction. This is a direct consequence

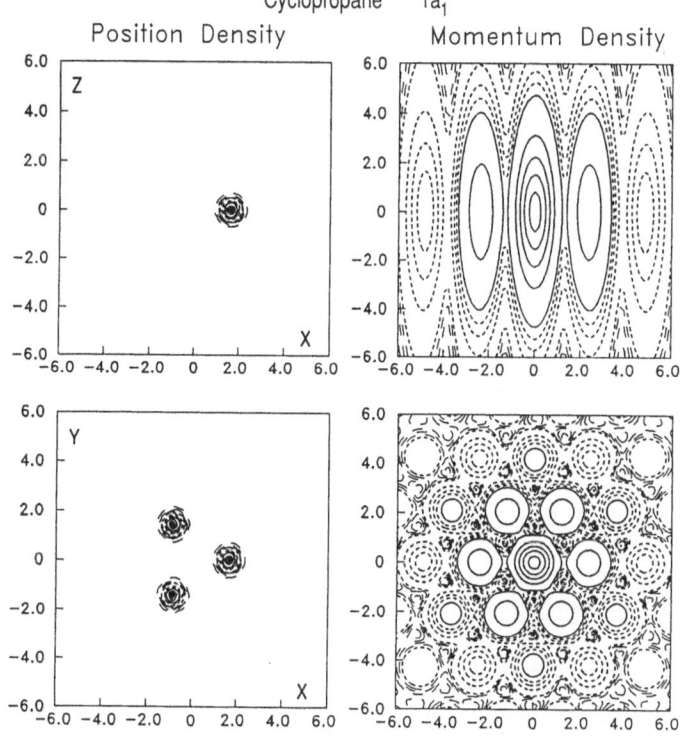

Fig. 16. Calculated two-dimensional density contour maps of the $1a_1$ orbital of cyclopropane in position-space (left) and momentum-space (right) [48]. The contour maps are generated using the Snyder and Basch wavefunction [106]. The (Z-X) contour plane (of the position-space map) is orthogonal to the C-C-C(Y-X) plane. The contour values are 0.1, 0.3, 0.5, 0.7 (*dashed lines*), 1, 3, 5, 7 (*dotted lines*), 10, 30, 50, 70 and 90% (*solid lines*) of the maximum density. The intense oscillatory structures in the momentum density maps reflect the molecular density oscillation property in momentum-space

of the Virial theorem. Consider, for example, diatomic molecules in the Born-Oppenheimer approximation:

$$2\langle T \rangle + \langle V \rangle + RdE/dR = 0 \tag{20}$$

where E, T and V are respectively the total, kinetic and potential energies and R is the internuclear separation. For a system in its equilibrium nuclear geometry, one has the special case where the average potential energy is the negative of twice the kinetic energy, i.e. $\langle V \rangle = -2\langle T \rangle$. Hence any process which lowers the total energy (noting that $\langle E \rangle = -\langle T \rangle$) will raise the average momentum $\langle p \rangle$ since $\langle T \rangle = 1/2\langle p^2 \rangle$. This therefore produces a general transfer of the momentum density into the high momentum region. The direction of such density redistribution is governed by the Fourier transform properties mentioned above. The Virial property leads to a different partitioning of the bonding and antibonding regions in p-space. Figure 17 illustrates the changes of the bond density in momentum-space during the $(1\sigma_g)$ bond formation of H_2 [130]. An estimate of the spherically averaged bond density of H_2 by (e, 2e) spectroscopy is shown in Fig. 18 [130]. Clearly, the familiar picture

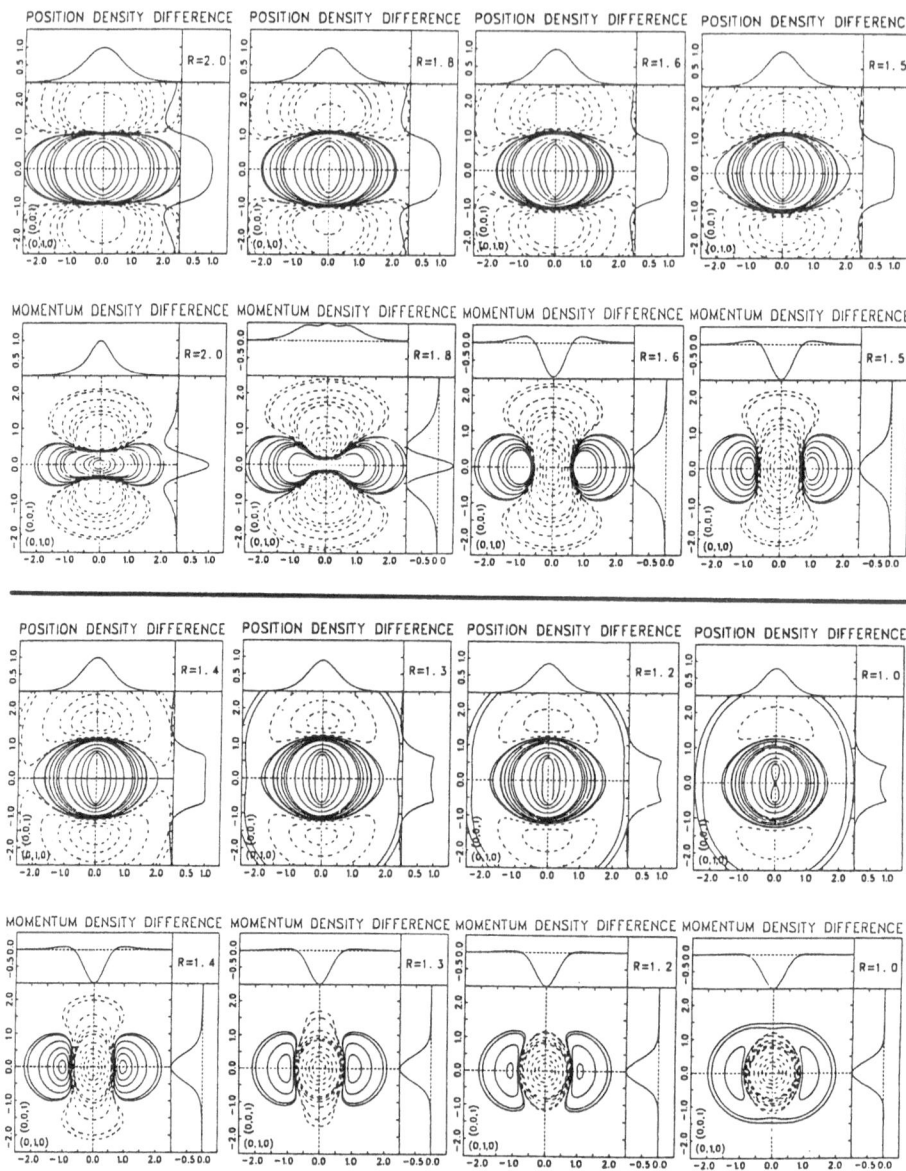

Fig. 17. Directional density difference (bond density) maps of H_2 in momentum-space and position-space as a function of internuclear separation R. The equilibrium separation corresponds to R = 1.4 a.u. The extended Hartree-Fock wavefunctions at different R's are used to generate the maps [130]. Contours of negative density difference are shown as dashed lines. The internuclear (bond-parallel) direction is along the (0, 0, 1) direction. While the position-space maps appear to show relatively little change as the two H atoms approach the equilibrium geometry, the corresponding momentum-space maps indicate rather substantial changes

of a chemical bond in position-space is replaced by a rather different model in momentum-space, which corresponds to a probability redistribution from the central cylinder-like low momentum region to the ring-like structure in the higher momentum region, in agreement with the Virial property.

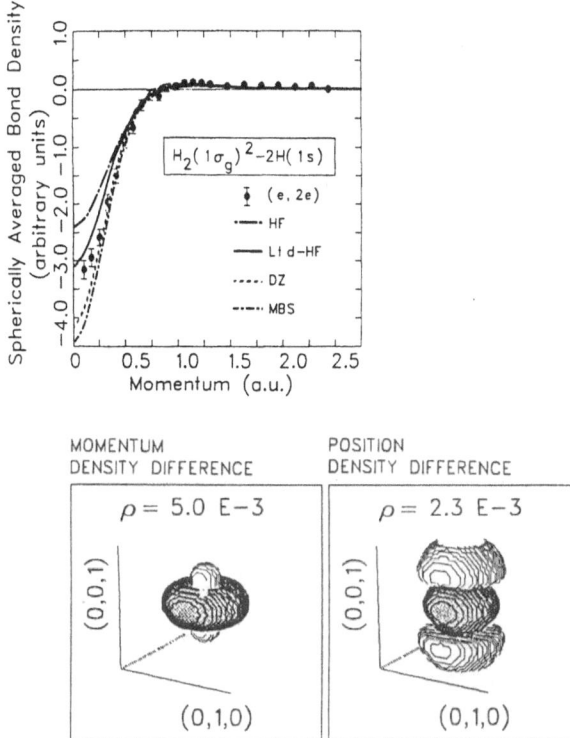

Fig. 18. Sperically averaged bond density (upper) and three-dimensional surface plot of the density difference (bond density) of H_2 in momentum-space (lower left) and position-space (lower right). The experimental spherically averaged) bond density is obtained from the difference between the (experimental spherically averaged) momentum density of the $1\sigma_g$ orbital of H_2 and the exact (spherically averaged) momentum density of the H 1s orbital. The extended Hartree-Fock wavefunction (HF) at the equilibrium internuclear geometry is used to generate the bond density surfaces [130]. The bond density surfaces correspond to $\pm 2\%$ of the maximum bond density values in the respective spaces. Positive bond density surfaces are shaded

6 Concluding Remarks

Like most other electron-based spectroscopies, (e, 2e) spectroscopy has evolved to a critical stage in the past two decades. The validities of many of the scattering approximations have been established, which in turn secure the significance of the measured angular dependence in the (e, 2e) cross section for applications in quantum chemistry. Perhaps unlike many of the electron-based experiments which seek to function as a pseudo-photon analog, (e, 2e) experiments provide truly unique measurements/information that no other photon-based experiments can offer. While momentum distributions corresponding to the total wavefunction can be obtained by techniques such as Compton scattering, the more refined momentum distributions of individual ionic states can only be provided by (e, 2e) spectroscopy. Not only can the (e, 2e) technique provide orbital-specific information regarding the quality of a particular wavefunction, it can also identify any inadequacy of an ab-initio wavefunction in the chemically interesting bonding region. This unique capability of (e, 2e) spectroscopy, along with many of its important applications in quantum chemistry, has clearly been exploited extensively in the past decade. Computations of increasingly precise wavefunction were carried out using experimental momentum distributions as an optimizing gauge. Better understanding of many-body effects in impact-ionization phenomenon of the valence-shell can also be obtained from (e, 2e) data and from calculations such as the Green's function method.

The 1989 (e, 2e) workshop [17] saw many new and exciting developments in (e, 2e) spectroscopy. In particular, the inherent limitation of low coincidence counting rate is slowly being resolved by new experimental advances, particularly in the areas of parallel position sensitive detection and multiplexing electron optics. Intense on-going research is clearly evident in the experimental improvement of the achievable energy resolution so that many more complex molecules (with close-lying states) can be investigated. The extensions of (e, 2e) spectroscopy to condensed matter studies and to studies of more exotic species (such as species in excited states or inner shells, weakly bound molecular complexes, etc.) are progressing steadily. Attempts at studying (e, 2e) reactions on single-crystal surfaces using a reflective glancing incidence geometry have already been made [41]. This kind of experiment enhances the prospect of determining directional momentum density of species oriented on surfaces, and seeks to develop a new way of studying chemisorption using the (e, 2e) technique.

Acknowledgements: This work was supported by the Natural Sciences and Engineering Research Council of Canada. I am pleased to acknowledge many illuminating discussions with my collaborators: Professors M. Coplan and P. W. Langhoff. I wish to especially thank Professor C. E. Brion for introducing me to (e, 2e) research, which has never ceased to excite and challenge me.

7 References

1. Lowdin PO, Pullman (ed) (1983) New horizons of quantum chemistry, Reidel D, Boston
2. Pople JA, Beveridge DL (1970) Approximate molecular orbital theory, McGraw-Hill, New York
3. Richards WG, Cooper DL (1983) Ab-initio molecular orbital calculations for chemists, Clarendon, Oxford and references therein
4. McWeeny R, Sutcliffe BT (1969) Methods of molecular quantum mechanics, Academic London
5. Schaefer HF III (1972) The electronic structure of atoms and molecules, Addison-Wesley, Reading, Massachusetts
6. Lipscomb WN (1972) In: Buckingham AD (ed) Theoretical chemistry, physical chemistry series one, vol 1. Butterworth, London, p 167
7. Bader RFW (1975) In: Buckingham AD, Coulson CA (eds) Theoretical chemistry. Physical chemistry series two, vol 2. Butterworths, London p 79
8. Epstein IR (1975) In: Buckingham AD, Coulson CA (eds) Theoretical chemistry. Physical chemistry series two, vol 2. Butterworths, London p 107
9. Williams BG (1977) Physica scripta 15: 92
10. McCarthy IE, Weigold E (1976) Phys rep 27C: 275
11. Smirnov YF, Neudatchin VG (1966) JETP lett 3: 192. See also: Glassgold AE, Ialongo (1968) Phys rev 175: 151; and references therein
12. Amaldi U, Egidi A, Marconero R, Pizzella P (1969) Rev sci instrum 40: 1001
13. Camilloni R, Giardini-Guidoni A, Tiribelli R, Stefani G (1972) Phys rev lett 29: 618
14. Weigold E, Hood ST, Teubner PJO (1973) Phys rev lett 30: 475

15. Hood ST, Hamnett A, Brion CE (1976) Chem phys lett 39: 252
16. Coplan MA, Moore JH, Tossell JA (1978) J chem phys 68: 329
17. Coplan M (ed) (1989) Conf Proc (e, 2e) Spectroscopy Workshop, Univ. of Maryland, College Park, Maryland
18. Leung KT, Brion CE (1985) J electron spectrosc relat phenom 35: 327
19. Botticelli A, Camilloni R, Giardini-Guidoni A, Missoni G, Stefani G, Tiribelli R, Vinciguerra D (1974) Annali di chimica 64: 189
20. McCarthy IE (1973) J phys B 6: 2358; J phys B 8 (1975) 2133
21. McCarthy IE, Weigold E (1976) Adv in phys 25: 489; and references therein
22. McCarthy IE (1982) in momentum wave-function − 1982, AIP Conf. Proc. vol 86, AIP, New York (1982), p 5; and references therein
23. Lahmam-Bennani A, Wellenstein HF, Duguet A, Rouault M (1983) J phys B 16: 121
24. Dal Cappello MC, Dal Cappello C, Tavard C, Cherid M, Lahmam-Bennani A, Duguet A (1989) J phys (Paris) 50: 207; and references therein
25. Inokuti M (1971) Rev mod phys 43: 297
26. Lohmann B, Weigold E (1981) Phys lett 86A: 139
27. Cook JPD, Mitroy J, Weigold E (1984) Phys rev lett 52: 1116
28. Frost L, Mitroy J, Weigold E (1986) J phys B 19: 4063
29. Frost L, Weigold E, Mitroy J (1983) J phys B 16: 223
30. Weigold E, Zhang D, Zheng Y (1989) ICPEAC 16th, Abstracts of Contributed Papers, In: Dalgarno A, Freund RS, Lubell MS, Lucatorto TB (eds). New York, New York p 152
31. Giardini-Guidoni A, Fantoni R, Tiribelli R, Vinciguerra D, Camilloni R, Stefani G (1979) J chem phys 71: 3182. See also: Niessen W von, Weigold E, Zheng Y, ICPEAC 16th, Abstracts of Contributed Papers, Dalgarno A, Freund RS, Lubell MS, Lucatorto TB (1989) (eds). New York, New York p 340
32. Bawagan AO, Brion CE, Coplan MA, Tossell JA, Moore JH (1986) Chem phys 110: 153
33. Chornay DH, Coplan MA, Tossell JA, Moore JH, Baerends EJ, Rozendaal A (1985) Inorg chem 24: 877
34. Gorunganthu RR, Coplan MA, Leung KT, Tossell JA, Moore JH (1989) J chem phys 91: 1994
35. See for example Bawagan AO, Brion CE (1988) Chem phys 123: 51
36. Gao C, Ritter AL, Dennison JR, Holzwarth NAW (1988) Phys rev B 37: 3914; and references therein
37. Jones R, Ritter AL (1986) J electron spectrosc relat phenom 40: 285
38. Dey S, Hayes P, Williams JF (1987) J phys D 20: 504
39. Schule E, Nakel W (1982) J phys B 15: L639
40. Lower J, Weigold E (1989) ICPEAC 16th Abstracts of Contributed Papers, Dalgarno A, Freund RS, Lubell MS, Lucatorto TB (eds). New York, New York p 814
41. Zhu H, Best PE (1989) Conf. Proc. (e, 2e) Spectroscopy Workshop, Coplan M (ed) Univ. of Maryland, College Park, Maryland
42. Giardini-Guidoni A, Fantoni R, Camilloni R, Stefani G (1981) Comments at mol phys 10: 107
43. Coplan MA, Tossell JA, Moore JH (1982) in: Momentum wave-functions − 1982, AIP Conf. Proc. V86, AIP New York p 82
44. Weigold E (1984) Comments at mol phys 15: 223
45. McCarthy IE (1986) Aust j phys 39: 587
46. Brion CE (1986) Int j quantum chem 29: 1397
47. McCarthy IE, Weigold E (1988) Rep prog phys 51: 299
48. Banjavcic MP, Daniels TA, Leung KT (1990) Chem phys, to be submitted
49. Daniels TA, Banjavcic MP, Leung KT (1990) Chem phys, to be submitted
50. Segre E (1977) Nuclei and particles, 2nd ed, Benjamin, Massachusetts p 501
51. Wiel MJ van der, Brion CE (1974) J electron spectrosc relat phenom : 439
52. Brion CE (1982) in: Physics of electronic and atomic collisions, Datz S (ed), North-Holland, Amsterdam p 579; and references therein
53. Avaldi L, Camilloni R, Fainelli E, Stefani G (1987) J phys B 20: 4163
54. Lahmam-Bennani A, Avaldi L, Fainelli E, Stefani G (1988) J phys B 21: 2145

55. Ugbabe A, Weigold E, McCarthy IE (1975) Phys rev A 11: 576
56. D'Andrea A, Del Sole R (1978) Surf sci 71: 306
57. McCarthy IE (1980) Coherence and correlation in atomic collisions, Kleinpoppen H, Williams JF (eds). Plenum, New York p 1
58. Mott NF, Massey HSW (1965) The theory of atomic collisions, Clarendon, Oxford
59. Leung KT, Sheehy JA, Langhoff PW (1989) Chem phys lett 157: 135; and references therein
60. Weigold E, Hood ST, McCarthy IE (1975) Phys rev A 11: 566; and references therein
61. Camilloni R, Giardini-Guidoni A, McCarthy IE, Stefani G (1978) Phys rev A 17: 1634
62. Fuss I, McCarthy IE, Noble CJ, Weigold E (1978) Phys rev A 17: 604
63. Weigold E, Noble CJ, Hood ST, Fuss I (1979) J phys B 12: 291
64. Camilloni R, Giardini-Guidoni A, McCarthy IE, Stefani G (1980) J phys B 13: 397
65. Dixon AJ, McCarthy IE, Noble CJ, Weigold E (1978) Phys rev A 17: 597
66. McCarthy IE, Weigold E (1983) Comtemp phys 24: 163
67. Bagus PS, Schrenk M, Davis DW, Shirley DA (1974) Phys rev A 9: 1090
68. McCarthy IE, Ugbabe A, Weigold E, Teubner P (1974) Phys rev lett 33: 459
69. Dixon AJ, Hood ST, Weigold E (1978) Phys rev lett 40: 1262
70. Bawagan AO, Brion CE, Davidson ER, Feller D (1987) Chem phys 113: 19
71. Bawagan AO, Mueller-Fiedler R, Brion CE, Davidson ER, Boyle C (1988) Chem phys 120: 335
72. Bawagan AO, Brion CE, Davidson ER, Boyle C, Frey RF (1988) Chem phys 128: 439
73. Clark SAC, Weigold E, Brion CE, Davidson ER, Frey RF, Boyle CM, Niessen W von, Schirmer J (1989) Chem phys 134: 229
74. Harting E, Read FH (1976) Electrostatic lenses, Elsevier, Amsterdam
75. Granneman EHA, Wiel M van der (1980) handbook on synchrotron radiation, vol I, Koch EE (ed). Chapter VI; and references therein
76. Wiza JL (1979) Nucl inst meth 162: 587; and references therein
77. Cook JPD, McCarthy IE, Stelbovics AT, Weigold E (1984) J phys B 17: 2339
78. Hayes P, Bennett MA, Flexman J, Williams JF (1988) Rev sci instrum 59: 2445
79. Moore JH, Coplan MA, Skillman TL, Brooks ED (1978) Rev sci instrum 49: 463
80. Skillman TL, Brooks ED, Coplan MA, Moore JH (1978) Nucl instrum and method 115: 267
81. Goruganthu RR, Coplan MA, Moore JH, Tossell JA (1988) J chem phys 89: 25
82. Brion CE, private communication
83. Weigold E (1989) Conf. Proc. (e, 2e) Spectroscopy Workshop, Coplan M (ed). Univ. of Maryland, College Park, Maryland
84. Berkowitz J (1979) Photoabsorption, photoionization, and photoelectron spectroscopy, Academic, New York
85. Koopmans T (1933) Physica 1: 104
86. Schaefer HF (ed) (1977) Methods of electronic structure theory, Plenum, New York
87. Dey S, Dixon AJ, Lassey KR, McCarthy IE, Teubner PJO, Weigold E, Bagus PS, Viinikka EK (1977) Phys rev A 15: 102
88. French CL, Brion CE, Bawagan AO, Bagus PS, Davidson ER (1988) Chem phys 121: 315; and references therein
89. Weigold E, Dey S, Dixon AJ, McCarthy IE, Lassey KR, Teubner PJO (1977) J electron spectrosc relat phenom 10: 177
90. Cook JPD, Pascual R, Niessen W von, Weigold E (1989) ICPEAC 16th, Abstracts of contributed papers, Dalgarno A, Freund RS, Lubell MS, Lucatorto TB (eds), New York, New York p 322
91. Dixon AJ, McCarthy IE, Weigold E, Williams GRJ (1977) J electron spectrosc relat phenom 12: 239
92. Leung KT, Brion CE (1985) Chem phys 95: 241; and references therein
93. Minchinton A, Cook JPD, Weigold E, Niessen W von (1985) Chem phys 93: 21
94. Clark SAC, Bawagan AO, Brion CE (1989) Chem phys 137: 407; and references therein
95. McCarthy IE (1985) J electron spectrosc relat phenom 36: 37
96. Cederbaum LS, Domcke W (1977) Adv chem phys 36: 205; and references therein
97. Niessen W von, Schirmer J, Cederbaum LS (1984) Comp phys rep 1: 57; and references therein

 98. Nakatsuji H, Hirao K (1978) J chem phys 68: 2053
 99. Leung KT, Brion CE (1983) Chem phys 82: 87; and references therein
100. Perry WB, Jolly WL (1974) J electron spectrosc relat phenom 4: 219
101. Bice JE, Tan KH, Bancroft GM, Tse JS (1987) Inorg chem 26: 4106
102. Hood ST, Hamnett A, Brion CE (1976) Chem phys lett 41: 428
103. Lahmam-Bennani A, Grisogono AM, Pascual R, Weigold E (1989) ICPEAC 16th, Abstracts of contributed papers, Dalgarno A, Freund RS, Lubell MS, Lucatorto TB (eds). New York, New York p 341
104. Leung KT, Brion CE (1984) Chem phys 91: 43; and references therein
105. Brion CE, Hood ST, Suzuki IH, Weigold E, Williams GRJ (1980) J electron spectrosc relat phenom 21: 71
106. Snyder LC, Basch H (1972) Molecular wavefunctions and properties, Wiley, New York
107. Poirier R, Kari R, Csizmadia IG (1985) Handbook of gaussian basis sets, Elsevier Science, New York
108. Davidson ER, Feller D (1986) Chem rev 86: 681
109. Hehre WJ, Lathan WA, Ditchfield R, Newton MD, Pople JA (1980) Nat resource comput software cat vol 1, GAUSS 76
110. Van Hove MA, Weinberg WH, Chan CM (1986) Low energy electron diffraction, Springer-Verlag, New York
111. Carbi R (ed) (1982) Current aspects of quantum chemistry − 1981, Elsevier Scientific, New York
112. Podolsky G, Pauling L (1929) Phys rev 34: 109
113. Fock V (1935) Z phys 98: 98
114. Levy M (1950) Proc roy soc A 204: 145
115. Lombardi JR (1982) J phys chem 86: 3513
116. Epstein IR (1971) Chem phys lett 9: 9
117. Levin VG (1972) Phys lett 39 A: 125
118. Levin VG, Neudatchin VG, Pavlitchenkov AV, Smirnov YF (1975) J chem phys 63: 1541
119. Komarov FF, Temkin MM (1976) J phys B 9: L255
120. Kaijer P, Smith VH jr (1977) Adv quant chem 10: 37
121. Coulson CA, Duncanson WE (1941) Proc Cambridge Phil Soc 37: 55, 67, 74, 397, 406; (1942) 38: 100; (1943) 39: 180
122. Epstein IR (1973) Acc chem res 6: 145
123. Epstein IR, Tanner AC (1977) Compton scattering. In: Williams BG (ed), McGraw-Hill, New York p 209
124. Henneker WH, Cade PE (1968) Chem phys lett 2: 575
125. Leung KT, Brion CE (1983) Chem phys 82: 113
126. Camilloni R, Stefani G, Fantoni R, Giardini-Guidoni A (1979) J electron spectrosc relat phenom 17: 209
127. Cook JPD, Brion CE (1982) Momentum wave-function − 1982, AIP Conf Proc V86, AIP Press, New York p 278
128. Minchinton A, Brion CE, Cook JPD, Weigold E (1983) Chem phys 76: 89
129. Rawlings DC, Davidson ER (1985) J phys chem 89: 969
130. Leung KT, Brion CE (1984) J am chem soc 106: 5859; and references therein

Appendix: An (e, 2e) Bibliographical Update [1984−88]

The following update follows the general format of an earlier (e, 2e) bibliography published in J. Electron. Spectrosc. Relat. Phenom. (1985) 35: 327−52, hereafter called I. The present update covers all the (e, 2e) and related work published between 1984 and 1988 (inclusive). Unlike the earlier bibliography, no attempt is made here to give a complete data summary. Instead, the works published are categorized in a rather simple (and somewhat arbitrary) fashion to indicate some patterns in the scope of species studied. In particular, Table A1 and Table A2 list respectively all the atomic and molecular systems studied to-date (1969–88) by (e, 2e) spectroscopy. Table A3 lists the (e, 2e) work done on condensed matter, e.g. thin films. The reference codes used in Tables A1 to A3 can be found in Table A5 below and in Table 3 of I. Finally, Table A4 gives a list of recent theory and review articles published in the 1984–88 period. Other theory and review articles published before 1984 can be found in Table 2 of I.

The references are arranged alphabetically in Table A5 by the assigned reference codes, which follow the same general format of I. In particular, the reference code is normally indicated by no more than 6 characters with the characters appearing before the year (as indicated by the two integers) to represent the first characters of the last names in the author list. The occasional "&" sign indicates that the number of authors exceeds three and a complete list will not be given. The items in each reference are given in the following order: Authors, Journal, Volume (Year) Page. This is then followed by the title of the article in a separate line.

Together with I, it is hoped that the present update provides a more complete bibliography of the current (e, 2e) literature.

Table A1. Atoms [1969–88]

Hydrogen atom	H	WH&77, WN&79a, WN&79b, LW81, MW81, MW83a, LM&84, W86
Noble gases	He	HM&73, MU&74, WUT75, DMW76, GT&77b, MCS77, CG&78, FM&78, SCG78, WK&79, L81, WK&81, LB83a, LB83b, LR83, L84, LW&83b, LW&84b, BJP86, LW&85, IKG86, MM86, SC&86, BS87, CDL87, DC&87, FO87
	Ne	BC&74, UWM75, WHM75, GT&77b, DM&78, FM&78, SCG79, CG&80, LB83a, DT&84, DL&85, LW&85, BB86, LDD86
	Ar	FW73a, FW73b, HM&73, WHT73, BC&74, HM&74, UWM75, WHM75, FM&78, MUP78, W78, ML81, LB83a, LW&83, LW&83a, LW&84a, MAM84, AK85, LW&85, MW85, BB86, BT86, SA&86, AC&88, BBT88, HB&88a
	Kr	BC&74, WHM75, GT&77b, FG&81, LB83a, AK85, LDD86
	Xe	UWM75, GT&77b, HHB77a, DM&78, GF&80c, LB83a, CMW84, AK85, BT86, CM&86, BBT88
Others	Na	LW82
	K	LW82
	Mg	PM&88
	Cd	LWM83, MR84
	Hg	MF82
	Pb	FMW86

Table A2. Molecules [1969–88]

Hydrogen	H_2			M73, WH&73, DM&75, WM&77, MC&81, WK&81, LB83b, LB84a, L85, LS85
	D_2			DM&75
		HF		BM&79, BH&80
		HCl		BH&80, SB&80
		HBr		BM&82
		HI		BM&82
Halogen	CL_2			FG&87
	Br_2			FG&88
	I_2			GP&88
Other diatomics	CO			M75, DD&77b, TM&82, FB&88
	N_2			CS&76, WD&77
	NO			BC&82, FGT82, TM&82
	O_2			SWB80, TM&82
Other triatomics	CO_2			GT&77a, CB82a, LB85a
	N_2O			FG&80, MFW82
	OCS			CW&81, LB85b
	CS_2			LB&85
Water	H_2O			DD&77a, HHB77a, CC&84, BL&85, BB&87, LSL89
	D_2O			BB&87
	H_2S			CB79, CBH80, FBD88
			CH_3OH	MBW81
			$(CH_3)_2O$	CBC89
	H_2CO			HHB76b, BB&88
Ammonia	NH_3			CS&76, HHB76a, WMW77, TL&84, BB87, BM&88
	PH_3			HHB77b
			NH_2CH_3	TL&84, BB88
			$NH(CH_3)_2$	BB88
			$N(CH_3)_3$	BB87, BB88
	NF_3			BB87, BB88
	$N(C_2H_5)_3$			RC&87
Methane	CH_4			HW&73, M73, WD&76, HHB77a, CC&81

Table A2. (continued)

Methane	SiH$_4$			CW&89
		CH$_3$F		CC&81
		CH$_3$Cl		MG&82, MC&84
		CH$_3$Br		MC&84, MC&85
		CH$_3$I		MC&84, MC&85
	CH$_3$CN			MB&83
Flourides	CF$_4$			CC&82, LB&84b
	SiF$_4$			FG&86
			CF$_3$H	CC&82, CC&83a
Acetylene	C$_2$H$_4$			CM&78,
				DH&78,
				CM&79,
				GC&89
		C$_2$H$_3$F		FG&82,
				CC&83a,
				GC&89
		C$_2$H$_3$Cl		CC&83b, GC&89
		C$_2$H$_3$Br		CC&83b, GC&89
		C$_2$H$_3$I		GC&89
Other hydrocarbons	C$_2$H$_2$			DM&77,
				CMT78,
				CM&79,
				GC&88
			C$_2$HCH$_3$	GC&88
			C$_2$(CH$_3$)$_2$	CM&79, GC&88
			C$_2$(CF$_3$)$_2$	GC&88
	C$_2$H$_6$			DD&76
	C$_3$H$_6$			CM&79, TMC79,
				BW&89
	C$_4$H$_4$			
	C$_6$H$_6$			FM&81
	C$_6$H$_4$Cl$_2$			BB&86
	C$_3$N$_3$H$_3$			HC&85
Polyatomics	SF$_6$			GF&79
Metal Carbonyls	Cr(CO)$_6$			CC&85

Table A3. Condensed Matter [1969–88]

C film	A69, AE&69, CG&72a, CG&72b, KLP75, RDD84, RDJ84, JR86, GR&88, GW&89
Si (theory)	DD78
Al film	PKN79, DHW87
Al oxide film	LNS72, KLP75, HB&88b
Cu (theory)	LNS72
Ag film	SN82
Au (theory)	FMS82
Th (theory)	FMS82

Table A4. Reviews and General Theory [1984–88]

NZ74	Manifestations of collective properties of the degenerate electron gas in the (e, 2e) quasi-elastic knock-out process.
DD78	Theory of (e, 2e) reaction near solid surfaces: Aplication to Si.
SSK84	The nodal structure of the momentum distribution of molecules.
SWS84	The shell structure of atoms and ions in momentum space.
TMC84	Studies of molecular orbital momentum distributions by (e, 2e) spectroscopy.
W84	Electron momentum spectroscopy – some recent developments.
AC&85	The e-e correlation in (e, 2e) reaction: a semiclassical approach.
K85	Differences in the ionization spectra of atoms in the (gamma, e) and (e, 2e) reactions.
M85	Interpretation of intensities in electron-momentum and photoelectron spectroscopies.
RD85	Molecular electron density (distributions in position and momentum space.
TCC85	Electron spectroscopy and (e, 2e) collision processes.
W85	Electron momentum spectroscopy – a new way of looking at the dynamic structure of matter.
AC&86	Coulomb interaction in the final state of electron impact ionization: Effects on the triple differential cross section.
B86	Looking at orbitals in the laboratory: The experimental investigation of molecular wavefunctions and binding energies by electron momentum spectroscopy.
BS86	Higher-order effects in the large-angle coplanar symmetric (e, 2e) processes at high energies.
JP86	Recent progress in the theory of (e, 2e) reactions.
M86	Electron momentum spectroscopy.
MHS86	The (e, 2e) collisions near ionization threshold-electron correlations.
PS86	Notes on dipolar (e, 2e) reactions.
AC&87a	The interpretation of binary (e, 2e) spectroscopy using spin-coupled theory.
AC&87b	Electron momentum spectroscopy by asymmetric (e, 2e) collisions belonging to the Bethe ridge.
J87	Theory of (e, 2e) reactions.
LCD87	A critical evaluation of the various methods for absolute scale determination in (e, 2e) experiments.
MR87	A comparison of various forms of the half-on-shell Coulomb T matrix applied to (e, 2e) collisions.
W87	Electron momentum spectroscopy of molecules: a review of recent development.
JF&88	(e, 2e) collisions in the presence of a laser field.
LA&88	The asymmetric (e, 2e) collisions at intermediate and high imnpact energy: success and limits of first-order models.
DD&89	Description of (e, 2e) triple differential cross sections: Situation of molecular targets.

Table A5. References [1984–88]

A69	Amaldi U (1969) Ann ist super sanita 5: 680–92.
AC&85	Avaldi L, Camilloni R, Popov YV, Stefani G (1985) NATO ASI Ser., Ser. B 134: 633.
AC&86	Avaldi L, Camilloni R, Popov YV, Stefani G (1986) Phys rev A 33: 851–60.
AC&87a	Allan NL, Cooper DL, Gerratt J, Raimondi M (1987) J electron spectrosc relat phenom 42: 127–48.
AC&87b	Avaldi L, Camilloni R, Fainelli E, Stefani G (1987) J phys B 20: 4163–72.
AC&88	Avaldi L, Camilloni R, Fainelli E, Stefani G (1988) J phys B 21: L359–64.
AK85	Amusia MY, Kheifets AS (1985) J phys B 18: L679–84.
B 86	Brion CE (1986) Int j quantum chem 29: 1397–428.

Table A5. (continued)

BB86	Brothers MJ, Bonham RA (1986) J phys B 19: 3801–14.
BB87	Bawagan AO, Brion CE (1987) Chem phys lett 137: 573–7.
BB88	Bawagan AO, Brion CE (1988) Chem phys 123: 51–63.
BBT88	Brion CE, Bawagan AO, Tan KH (1988) Can j chem 66: 1877–89.
BB&86	Bawagan AO, Brion CE, Coplan MA, Tossell JA, Moore JH (1986) Chem phys 110: 153–60.
BB&87	Bawagan AO, Brion CE, Davidson ER, Feller D (1987) Chem phys 113: 19–42.
BB&88	Bawagan AO, Brion CE, Davidson ER, Boyle C, Frey RF (1988) Chem phys 128: 439–55.
BJP86	Byron Jr, FW, Joachain CJ, Piraux B (1986) J phys B 19: 1201–10.
BL&85	Bawagan AO, Lee LY, Leung KT, Brion CE (1985) Chem phys 99: 367–82.
BM&88	Bawagan AO, Mueller-Fiedler R, Brion CE, Davidson ER, Boyle C (1988) Chem phys 120: 335–57.
BS86	Baliyan KS, Srivastava MK (1986) Pramana 27: 409–12.
BS87	Baliyan KS, Srivastava MK (1987) Phys rev A 35: 908–10.
BT86	Brion CE, Tan KH (1986) Aust j phys 39: 565–85.
BW&89	Banjavcic MP, Watt BH, Pope TD, Daniels TA, Hammond RP, Leung KT (1989) Chem phys lett 160: 371–6.
CBC89	Clark SAC, Bawagan AO, Brion CE (1989) Chem phys 137: 407–26.
CC&84	Cambi R, Ciullo G, Sgamellotti A, Brion CE, Cook JPD, McCarthy IE, Weigold E (1984) Chem phys 91: 373–81. See also: Chem phys 98 (1985) 166.
CC&85	Chornay DH, Coplan MA, Tossell JA, Moore JH, Baerends EJ, Rozendaal A (1985) Inorg chem 24: 877–82.
CDL87	Cherid M, Duguet A, Lahmam-Bennani A (1987) J phys B 20: L187–91.
CM&86	Cook JPD, McCarthy IE, Mitroy J, Weigold E (1986) Phys rev A 33: 211–21.
CW&89	Clark SAC, Weigold E, Brion CE, Davidson ER, Frey RF, Boyle CM, Niessen W von, Schirmer J (1989) Chem phys 134: 229–39.
DC&87	Duguet A, Cherid M, Lahmam-Bennani A, Franz A, Klar H (1987) J phys B 20: 6145–56.
DD78	D'Andrea A, Del Sole R (1978) Surf sci 71: 306–26.
DD&89	Dal Cappello MC, Dal Capello C, Tavard C, Cherid M, Lahmam-Bennani A, Duguet A (1989) J phys (Paris) 50: 207–17.
DFG84	Di Martino V, Fantoni R, Giardini-Guidoni A (1984) Symp at surf phys 34–8 (Edited by Howorka F, Lindinger W, Maerk TD; Inst Atomphys. Univ. Innsbruck; Innsbruck, Austria).
DHW87	Dey S, Hayes P, Williams JF (1987) J phys D 20: 504–10.
DL&85	Daoud A, Lahmam-Bennani A, Duguet A, Dal Cappello C, Tavard C (1985) J phys B 18: 141–53.
DT&84	Dal Cappello C, Tavard C, Lahmam-Bennani A, Dal Cappello MC (1984) J phys B 17: 4557–64.
FBD88	French CL, Brion CE, Davidson ER (1988) Chem phys 122: 247–69.
FB&88	French CL, Brion CE, Bawagan AO, Bagus PS, Davidson ER (1988) Chem phys 121: 315–33.
FG&86	Fantoni R, Giardini-Guidoni A, Tiribelli R, Cambi R, Rosi M (1986) Chem phys lett 128: 67–75.
FG&87	Frost L, Grisogono AM, McCarthy IE, Weigold E, Brion CE, Bawagan AO, Mukherjee PK, Niessen W von, Rosi M, Sgamellotti A (1987) Chem phys 113: 1–18.
FG&88	Frost L, Grisogono AM, Weigold E, Brion CE, Bawagan AO, Tomasello P, Niessen W von (1988) Chem Phys (1988) 119: 253–64.
FMW86	Frost L, Mitroy J, Weigold E (1986) J phys B 19: 4063–74.
FO87	Furtada FM, O'Mahony PF (1987) J phys B 20: L405–9.
GC&88	Goruganthu RR, Coplan MA, Moore JH, Tossell JA (1988) J chem phys 89: 25–33.

Table A5. (continued)

GC&89	Goruganthu RR, Coplan MA, Leung KT, Tossell JA, Moore JH (1989) J chem phys 91: 1994–2001.
GF&86	Giardini-Guidoni A, Fantoni R, Tiribelli R, Cambi R, Rosi M (1986); Contrib.-symp at surf phys Howorka F, Lindinger W, Maerk TD (eds), Inst. Atomphys. Univ. Innsbruck: Innsbruck, Austria.
GP&88	Grisogono AM, Pascual R, Weigold E, Bawagan A, Brion CE, Tomasello P, Niessen W von (1988) Chem phys 124: 121–30.
GR&88	Gao C, Ritter AL, Dennison JR, Holzwarth NAW (1988) Phys rev B 37: 3914–23.
GW&89	Gao C, Wang YY, Ritter AL, Dennison JR (1989) Phys rev lett 62: 945–8.
HB&88a	Hayes P, Bennett MA, Flexman J, Williams JF, (1988) Rev sci instrum 59: 2445–52.
HB&88b	Hayes P, Bennett MA, Flexman J, Williams JF (1988) Phys rev B 38: 13371–6.
HC&85	Hiser S, Chornay DJ, Coplan MA, Moore JH, Tossell JA (1985) J chem phys 82: 5571–6.
IKG86	Ippolitov II, Katyurin SV, Glinkin OB (1986) Izv vyssh uchebn zaved, fiz 29: 45–50.
J87	Joachain CJ (1987) Few-body syst suppl 2: 294–308.
JF&88	Joachain CJ, Francken P, Maquet A, Martin P, Veniard V (1988) Phys rev lett 61: 165–8.
JP86	Joachain CJ, Piraux B (1986) Comments at mol phys 17: 261–83.
JR86	Jones R, Ritter AL (1986) J electron spectrosc relat phenom 40: 285–97.
K85	Kheifets AS (1985) Zh eksp teor fiz 89: 459–69.
KLP75	Krasil'nikova NA, Levin VG, Perisantseva NM (1975) Zh eksp teor fiz 69: 1562–8. [Sov phys JETP 42 (1976) 796–9].
L84	Lahmam-Bennani A (1984) Phys rev A 29: 962–5.
L85	Liu JW (1985) Phys rev A 32: 3784–6. (See also: Phys rev A 38 (1988) 1659).
LA&88	Lahmam-Bennani A, Avaldi L, Fainelli E, Stefani G (1988) J phys B 21: 2145–57.
LB84a	Leung KT, Brion CE (1984) J am soc 106: 5859–64.
LB84b	Leung KT, Brion CE (1984) Chem phys 91: 43–58.
LB85a	Leung KT, Brion CE (1985) Chem phys 93: 319–31.
LB85b	Leung KT, Brion CE (1985) Chem phys 95: 241–58.
LB&85	Leung KT, Brion CE, Fatyga BW, Langhoff PW (1985) Chem phys 96: 227–40.
LCD87	Lahmam-Bennani A, Cherid M, Duguet A (1987) J phys B 20: 2531–44.
LDD86	Lahmam-Bennani A, Duguet A, Dal Cappello C (1986) J electron spectrosc relat phenom 40: 141–61.
LM&84	Lohmann B, McCarthy IE, Stelbovics AT, Weigold E (1984) Phys rev A 30: 758–67.
LNS72	Levin VG, Nendachin VG, Smirnov YF (1972) Phys stat sol. 49: 489.
LS85	Liu JW, Smith VH, Jr (1985) Phys rev A 31: 3003–11.
LSL89	Leung KT, Sheehy JA, Langhoff PW (1989) Chem phys lett 157: 135–41.
LWM83	Frost L, Weigold E, Mitroy J (1983) J phys B 16: 223–331.
LW&83a	Lahmam-Bennani A, Wellenstein HF, Duguet A, Rouault M (1983) J phys B 16: 121–30.
LW&83b	Lahmam-Bennani A, Wellenstein HF, Dal Cappello C, Rouault M, Duguet A (1983) J phys B 16: 2219–30.
LW&84a	Lahmam-Bennani A, Wellenstein HF, Duguet A, Daoud A (1984) Phys rev A 30: 1511–3.
LW&84b	Lahmam-Bennani A, Wellenstein HF, Dal Cappello C, Duguet A (1984) J phys B 17: 3159–72.
LW&85	Lahmam-Bennani A, Wellenstein HF, Duguet A, Lecas M (1985) Rev sci instrum 56: 43–51.
M84	McCarthy IE (1984) Los Alamos Natl. Lab, LA-10227-C. Proc workshop high energy excitations condens matter, vol 2: 479–92.
M85	McCarthy IE (1985) J electron spectrosc relat phenom 36: 37–58.
M86	McCarthy IE (1986) Aust j phys 39: 587–600.
MC&85	Minchinton A, Cook JPD, Weigold E, Niessen W von (1985) Chem phys 93: 21–38.

Table A5. (continued)

MHS86	Mazeau J, Huetz A, Selles P (1986) ICPEAC 14th invited paper, Lorents DC, Meyerhof WE, Peterson JR (eds), North-Holland, Amsterdam p 141–51.
MM86	McCarthy IE, Mitroy J (1986) Phys rev A 34: 4426–7.
MMW84	McCarthy IE, Mitroy JD, Weigold E (1984) Flinders Univ. South Aust., Inst. At. Stud., FIAS-R-137.
MR84	Martin NLS, Ross KJ (1984) J phys B 17: 4033–40.
MR87	McCarthy IE, Roberts MJ (1987) J phys B 20: L231–4.
MW85	McCarthy IE, Weigold E (1985) Phys rev A 31: 160–6.
NZ74	Nendatchin VG, Zhivopistsev FA (1974) Phys rev lett 32: 995–7.
PKN79	Perisantseva NM, Krasil'nikova NA, Nendachin VG (1979) Zh eksp teor fiz 76: 1047–57. [Sov phys JETP 49 (1979) 530–5].
PM&88	Pascual R, Mitroy J, Frost L, Weigold E (1988) J phys B 21: 4239–47.
PS86	Popov YV, Shabalina EK (1986) J phys B 19: L855–8.
RC&87	Rosi M, Cambi R, Fantoni R, Tiribelli R, Bottomei M, Giardini-Guidoni A (1987) Chem phys 116: 399–410.
RD85	Rawlings DC, Davidson ER (1985) J phys chem 89: 969–74.
RDD84	Ritter AL, Dennison JR, Dunn J (1984) Rev sci instrum 55: 1280–6.
RDJ84	Ritter AL, Dennison JR, Jones R (1984) Phys rev lett 53: 2054–7.
SA&86	Stefani G, Avaldi L, Lahham-Bennani A, Duguet A (1986) J phys B 19: 3787–800.
SC&86	Smith AD, Coplan MA, Chornay DJ, Moore JH, Tossell JA, Mrozek J, Smith Jr VH, Chant NS (1986) J phys B 19: 969–80.
SSK84	Simas AM, Smith Jr VH, Kaijser P (1984) Int j quantum chem 25: 1035–44.
SWS84	Simas AM, Westgate WM, Smith JrVH (1984) J chem phys 80: 2636–42.
TCC85	Tavard C, Dal Cappello C, Dal Cappello MC (1985) Stud phys theor chem 35: 79–84.
TMC84	Tossell JA, Moore JH, Coplan MA (1984) Int j quantum chem, quantum chem symp 18: 483–95.
W84	Weigold E (1984) Comments at mol phys 15: 223–49.
W85	Weigold E (1985) Aust phys 22: 256–9.
W87	Weigold E (1987) Flinders Univ. South Aust. Inst. At. Stud., FIAS-R-188.
WUT75	Weigold E, Ugbabe A, Tenbner PJO (1975) Phys rev lett 35: 209–12.
W86	Weigold E (1986) ICPEAC 14th Invited paper, Lorents DC, Meyerhoff WE, Peterson JR (eds), North-Holland, Amsterdam 125–40.

Theoretical Parameters of NMR Spectroscopy

Jozef Kowalewski and Aatto Laaksonen

Division of Physical Chemistry, Arrhenius Laboratory, University of Stockholm, S-10691 Stockholm, Sweden

The theory of magnetic shielding and nuclear spin-spin coupling in nuclear magnetic resonance (NMR) spectroscopy is reviewed. For each of the parameters, we review first the physical basis starting with the Hamiltonians for the relevant interactions. Then, we present a noncomprehensive review of the methods used for numerical calculations. The emphasis is on general characterization of the various approaches and on the nonempirical work, but the semiempirical and empirical work judged as the most important by the reviewers is also quoted. Some representative numerical results are presented.

1 Introduction

The nuclear magnetic resonance (NMR) phenomenon was discovered more than forty years ago [1–5]. The chemical potential of NMR spectroscopy was realized soon, in particular after the discoveries of the phenomena of magnetic shielding [6, 7] and the indirect nuclear spin-spin coupling [8–10]. The state of the understanding of NMR in the early 1960s is summarized in the beautiful and still amazingly fresh book by Abragam [11]. During the 1970s and 1980s, new NMR parameters caught the chemists' attention: the relaxation times, nuclear Overhauser enhancements, and various types of connectivities in two-dimensional experiments [12]. Nevertheless, the chemical shifts and spin-spin couplings retained their role as parameters of great interest for both experimentalists and theoreticians. In this chapter, the theory for these two parameters will be reviewed. The emphasis will be placed on explaining the origin of the two effects. The "state of the art" calculations will be presented briefly and without claiming a complete survey. The most important correlations between the two NMR parameters and other molecular properties of theoretical interest will be discussed.

Nuclear magnetic resonance spectra of liquids are usually analyzed in terms of the following phenomenological Hamiltonian (in frequency units):

$$\hat{H} = -\sum_N \frac{\gamma_N}{2\pi} (1 - \sigma_N) \, B_0 \hat{I}_{zN} + \sum_{N<N'} J_{NN'} \hat{\vec{I}}_N \cdot \hat{\vec{I}}_{N'} \tag{1.1}$$

The terms $(\gamma_N/2\pi) \, B_0 \, \hat{I}_{zN}$ correspond to the Zeeman splitting of the nuclear spin energy levels in the magnetic field B_0. B_0 is, however, not exactly the magnetic field experienced by the nuclei, since the nuclei are, to a small and on the chemical situation dependent extent, shielded by the electrons. This shielding is the origin of the term σ_N in Eq. (1.1), which is called the shielding constant. The term $\sigma_N B_0$ can be treated as an induced magnetic field due to the electronic motion and opposing the external field B_0. The terms $J_{NN'} \hat{\vec{I}}_N \hat{\vec{I}}_{N'}$ describe the phenomenon of nuclear spin-spin coupling, $J_{NN'}$ is called the coupling constant. It can sometimes be advantageous to write the spin-spin coupling term as $K_{NN'} \mu_N \mu_{N'}$ where μ_N and $\mu_{N'}$ are the nuclear magnetic moments, $K_{NN'}$ is called the reduced nuclear spin-spin coupling constant. The relation between $J_{NN'}$ and $K_{NN'}$ is:

$$K_{NN'} = 4\pi^2 J_{NN'}/h\gamma_N\gamma_{N'} \tag{1.2}$$

The Hamiltonian of Eq. (1.1) is called phenomenological because it can be used to explain the energy levels, transition frequencies, and spectral intensities in terms of parameters, shielding constants (the differences of which are called chemical shifts), and spin-spin coupling constants, without relating those to the more fundamental properties of the system. The subject of this chapter is to give the physical background explaining the origin of σ_N and $J_{NN'}$ and to relate these properties to the electronic structure as given by the wavefunctions for molecules. The energies corresponding to effects included in the Hamiltonian of Eq. (1.1) are very small compared to the electrostatic energies of the interactions between nuclei and electrons. This suggests that perturbation theory should be the quantum mechanical strategy of choice for obtaining first principle expressions analogous to Eq. (1.1). Once such expressions are obtained, we should be able to identify in the terms corresponding to the shielding constants and the coupling constants. Let us recapitulate the perturbation theory [13] briefly. Assume that the Hamiltonian may be written as:

$$\hat{H} = \hat{H}^{(0)} + \lambda \hat{H}^{(1)} + \lambda^2 \hat{H}^{(2)} + \dots \tag{1.3}$$

For a free molecule in the absence of an external field, the unperturbed Hamiltonian $\hat{H}^{(0)}$ would be the Born-Oppenheimer Hamiltonian containing the electrostatic interactions between the nuclei and electrons and the kinetic energy of the electrons:

$$\hat{H}^{(0)} = \hat{T} + \hat{V} = \sum_i \frac{\hat{p}_i^2}{2m} + \hat{V} = -\frac{\hbar^2}{2m} \sum_i \nabla_i^2 + \hat{V} \tag{1.4}$$

The index i refers to electrons. λ in Eq. (1.3) is a small constant. In the perturbation theory one assumes further that all the eigenfunctions $\Psi_n^{(0)}$ of $\hat{H}^{(0)}$ (the unperturbed wavefunctions) are known. This assumption is in fact not very realistic for many-electron systems. The task of the quantum chemistry in general is to develop methods for obtaining approximate $\Psi_n^{(0)}$; in the discussion of calculations of the properties of interest we are indeed going to assess the effects of the level of approximation used to compute the unperturbed wavefunctions on the calculated shieldings and coupling constants. In the derivation of the formal expressions for the properties of interest we are going to assume all the unperturbed wavefunctions and the corresponding eigenvalues (unperturbed energies) to be known. The total energy and the wavefunction for the ground state may then be written as:

$$E_0 = E_0^{(0)} + \lambda E_0^{(1)} + \lambda^2 E_0^{(2)} + \dots \tag{1.5}$$

$$\Psi_0 = \Psi_0^{(0)} + \lambda \Psi_0^{(0)} + \lambda^2 \Psi_0^{(2)} + \dots \tag{1.6}$$

In addition, the first.order correction to the ground state wavefunction, $\Psi_0^{(1)}$ may be expressed in terms of the eigenstates of $\hat{H}^{(0)}$:

$$\Psi_0^{(1)} = \sum_n \frac{\langle n | \hat{H}^{(1)} | 0 \rangle}{E_0 - E_n} \Psi_n^{(0)} \tag{1.7}$$

The energy correct to the second order may be expressed in terms of $\Psi^{(0)}$ and $\Psi^{(1)}$:

$$E_0 = \langle \Psi^{(0)} | \hat{H}^{(0)} | \Psi^{(0)} \rangle \tag{1.8}$$

$$\lambda E^{(1)} = \langle \Psi^{(0)} | \lambda \hat{H}^{(1)} | \Psi^{(0)} \rangle \tag{1.9}$$

$$\lambda^2 E^{(2)} = \langle \Psi^{(0)} | \lambda \hat{H}^{(1)} | \lambda \Psi^{(1)} \rangle + \langle \Psi^{(0)} | \lambda^2 \hat{H}^{(2)} | \Psi^{(0)} \rangle \tag{1.10}$$

This scheme may be generalized to include several independent perturbations. With:

$$\hat{H} = \hat{H}^{(0)} + \lambda \hat{H}^{(1,0)} + \mu \hat{H}^{(0,1)} \tag{1.11}$$

where μ is another small constant, we may write the wavefunction as:

$$\Psi = \Psi^{(0)} + \lambda\Psi^{(1,0)} + \mu\Psi^{(0,1)} + \ldots \tag{1.12}$$

In the expansion of energy we shall now get terms analogous to Eqs. (1.9) and (1.10) as well as an additional cross-term:

$$\lambda\mu E^{(1,1)} = \langle\Psi^{(0)}|\,\lambda\hat{H}^{(1,0)}\,|\mu\Psi^{(0,1)}\rangle + \langle\Psi^{(0)}|\,\mu\Psi^{(0,1)}\,|\lambda\hat{H}^{(1,0)}\rangle \tag{1.13}$$

The outline of this chapter is as follows. In Sect. 2, we are going to derive a formal theory for the magnetic shielding. In Sect. 3, the computational methods for obtaining the shielding constants will be reviewed. In Sect. 4, the formal theory of the spin-spin coupling will be formulated. In Sect. 5, the review of calculations of the nuclear spin-spin coupling constants will be provided. This chapter was completed in the spring of 1988.

2 The Formal Theory of Magnetic Shielding

The first goal of this section is to derive a Hamiltonian suitable for describing a molecular system that is under the influence of an external magnetic field and that contains nuclear spins. The second goal is to use this Hamiltonian and the perturbation theory-type approaches to obtain a qualitative physical picture of the magnetic shielding phenomenon. The approach towards these goals will involve some detours introducing important concepts (the vector potential, the current density) which are judged as less familiar to the general chemical readership. The approach used in this section is based on the books of Atkins [13], Slichter [14], and Ando and Webb [15]. In the following, the operator sign "$\hat{}$" will be omitted.

2.1 The Vector Potential

In order to introduce the concept of the vector potential, we are going to recapitulate some elements of vector analysis [16]. A vector can undergo three types of multiplication:

$$\bar{a} \text{ scalar} = \text{vector} \tag{2.1}$$

$$\bar{a} \cdot \bar{b} = \text{scalar} \tag{2.2}$$

$$\bar{a} \times \bar{b} = \text{vector} \tag{2.3}$$

Assume now that $\bar{a}$ is a vector operator (in the following, the operator sign "$\hat{}$" will be omitted), which in the following will be denoted $\bar{\nabla}$.

$$\bar{\nabla} = \bar{i}\frac{\partial}{\partial x} + \bar{j}\frac{\partial}{\partial y} + \bar{k}\frac{\partial}{\partial z} \tag{2.4}$$

An analogue to Eq. (2.1) can be obtained by letting $\bar{\nabla}$ operate on a scalar function:

$$\text{grad } f = \bar{\nabla}f = \bar{i}\,\frac{\partial f}{\partial x} + \bar{j}\,\frac{\partial f}{\partial y} + \bar{k}\,\frac{\partial f}{\partial z}\ \text{(vector)} \tag{2.5}$$

If f is identified with the ordinary electrostatic potential, then -grad f is the electric field vector. $\bar{\nabla}$ can also operate on a vector function $\bar{A}$ (a vector function has, for any set of arguments, a magnitude and a direction; alternatively, we can say that a vector function $\bar{A}(x, y, z)$ in the three-dimensional space is for every x, y, z a vector with components A_x, A_y, A_z). In analogy with Eq. (2.2), we have:

$$\text{div } \bar{A} = \bar{\nabla} \cdot \bar{A} = \frac{\partial A_x}{\partial x} + \frac{\partial A_y}{\partial y} + \frac{\partial A_z}{\partial z}\ \text{(scalar)} \tag{2.6}$$

while the analogue of Eq. (2.3) is:

$$\text{curl } \bar{A} = \bar{\nabla} \times \bar{A} = \bar{B}\ \text{(vector)} \tag{2.7}$$

with:

$$B_x = \nabla_y A_z - \nabla_z A_y = \frac{\partial A_y}{\partial y} - \frac{\partial A_y}{\partial z} \tag{2.8}$$

$$B_y = \nabla_z A_x - \nabla_x A_z = \frac{\partial A_x}{\partial z} - \frac{\partial A_y}{\partial x} \tag{2.9}$$

$$B_z = \nabla_x A_y - \nabla_y A_x = \frac{\partial A_y}{\partial x} - \frac{\partial A_x}{\partial y} \tag{2.10}$$

Let us now look at two cases when $\bar{\nabla}$ is operating twice:

$$\bar{\nabla} \times (\bar{\nabla}f) = \text{curl(grad } f) = 0 \tag{2.11}$$

We can treat Eq. (2.11) as a special case of of the more general $\bar{R} \times \bar{R}f = 0$ (where $\bar{R}$ is any vector), because $\bar{R} \times \bar{R} = 0$.

$$\bar{\nabla} \cdot (\bar{\nabla} \times \bar{A}) = \text{div (curl } \bar{A}) = 0 \tag{2.12}$$

which can be considered as a special case of $\bar{R} \cdot (\bar{R} \times \bar{T}) = 0$, because $\bar{R} \times \bar{T}$ is perpendicular to $\bar{R}$. We need these two properties in order to come to two importants theorems, which will be given without proof. Theorem 1 is related to Eq. (2.11) and can be stated as follows: If $\bar{\nabla} \times \bar{A} = 0$ then there exists a scaler function f such that $\bar{A} = \text{grad } f$. Theorem 2 is related to Eq. (2.12): If $\bar{\nabla} \times \bar{B} = 0$, then there exists a vector function $\bar{A}$ such that $\bar{B} = \text{curl } \bar{A}$. Now, we need a quick look at the basic equations of electromagnetism, the Maxwell equations [13, 17]. We define the electric displacement vector $\bar{D}$ as:

$$\bar{D} = \varepsilon_0\bar{F} + \bar{P} \tag{2.13}$$

where ε_0 is the electric permittivity of the medium, $\bar{F}$ is the electric field, $\bar{P}$ the polarization. The magnetic flux density $\bar{B}$ is defined:

$$\bar{B} = \mu_0\bar{H} + \mu_0\bar{M} \tag{2.14}$$

where μ_0 is the magnetic permittivity of the medium, $\bar{H}$ the magnetic field, and $\bar{M}$ the magnetization. The Maxwell equations can be written as:

$$\text{div }\bar{D} = \varrho \tag{2.15}$$

$$\text{div }\bar{B} = 0 \tag{2.16}$$

$$\text{curl }\bar{F} = -\frac{\partial\bar{B}}{\partial t} \tag{2.17}$$

$$\text{curl }\bar{H} = \bar{J} + \frac{\partial\bar{D}}{\partial t} \tag{2.18}$$

$\bar{J}$ is the current density. If $\bar{B}$ is time-independent, then Eq. (2.17) together with theorem 1 says that the electric field must be expressible as a gradient of a scalar function — which is the ordinary electrostatic potential, or the scalar potential. Looking at Eq. (2.18), we see that the curl of the magnetic field does not need to be zero, even if $\bar{D}$ is time-independent. On the other hand, Eq. (2.16) and theorem 2 say that the magnetic flux density must be expressible as a curl of a vector function:

$$\bar{B} = \text{curl }\bar{A} = \bar{V} \times \bar{A} \tag{2.19}$$

The vector function $\bar{A}$ is called the vector potential. Since $\bar{B}$ is given by derivatives of $\bar{A}$, many different vector potentials can produce the same magnetic flux density; for example, adding a constant vector to $\bar{A}$ does not change $\bar{B}$. Also, adding to $\bar{A}$ a vector function which has the property of being expressible as a gradient of a scalar function will not change $\bar{B}$. If $\bar{A}' = \bar{A} + \text{grad } f$, then:

$$\bar{V} \times \bar{A}' = \bar{V} \times (\bar{A} + \bar{V}f) = \bar{V} \times \bar{A} + \bar{V} \times \bar{V}f = \bar{V} \times \bar{A} \tag{2.20}$$

Obviously, the choice of the $\bar{A}$ which gives the required physically relevant $\bar{B}$ is somewhat arbitrary. For our purpose, we shall find that an additional condition:

$$\bar{V} \cdot \bar{A} = 0 \tag{2.21}$$

will be convenient. It is easy to convince oneself that the vector potential:

$$\bar{A} = \frac{1}{2} B_0(-\bar{i}y + \bar{j}x) \tag{2.22}$$

gives rise to the magnetic flux density B_0 in z-direction. More generally, a uniform magnetic flux density can be derived from the vector potential:

$$\bar{A} = \frac{1}{2}\bar{B} \times \bar{r} \tag{2.23}$$

where $\bar{r}$ is the radius vector. Both Eq. (2.22) and Eq. (2.23) fulfill Eq. (2.21). It should be made clear that replacing $\bar{r}$ in Eq. (2.23) by $\bar{r} - \bar{R}_0$ (where $\bar{R}_0$ is a constant vector) also gives a correct vector potential. This replacement corresponds to the change of origin of the coordinate system in which the vector potential is defined, or to the gauge transformation.

2.2 The Hamiltonian in the Presence of the Magnetic Field and Its Relation to Molecular Properties

The long and tedious way we had to go to make the introduction of the concept of vector potential plausible is motivated by the important role this concept plays in constructing the Hamiltonian for a molecule in the magnetic field. Without attempting a proof, we state that in the presence of a magnetic field described by a vector potential $\bar{A}$, the Hamiltonian is obtained by replacing the usual momentum operator:

$$\bar{p} = -i\hbar\bar{\nabla} \tag{2.24}$$

by the canonical momentum in the presence of a magnetic field, $\bar{\pi}$:

$$\bar{\pi} = \bar{p} + e\bar{A} \tag{2.25}$$

The usual electrostatic Hamiltonian of Eq. (1.4) is therefore replaced by:

$$H = \sum_i \frac{\pi_i^2}{2m} + V \tag{2.26}$$

Since $\pi^2 = \bar{\pi} \cdot \bar{\pi}$, we can write:

$$\bar{\pi}^2\varphi = (\bar{p} + e\bar{A}) \cdot (\bar{p} + e\bar{A})\,\varphi = p^2\varphi + e\bar{A} \cdot \bar{p}\varphi + e\bar{p}(\bar{A}\varphi) + e^2A^2\varphi \tag{2.27}$$

where the third term has to be treated carefully:

$$\bar{p}(\bar{A}\varphi) = -i\hbar\bar{\nabla}(\bar{A}\varphi) = i\hbar(\bar{\nabla} \cdot \bar{A})\,\varphi - i\hbar(\bar{\nabla}\varphi)\,\bar{A} = -i\hbar\bar{A} \cdot \bar{\nabla}\varphi = \bar{A} \cdot \bar{p}\varphi \tag{2.28}$$

This result allows us to write:

$$\bar{\pi}^2 = \bar{p}^2 + 2e\bar{A} \cdot \bar{p} + e^2A^2 \tag{2.29}$$

and:

$$H = \sum_i \frac{p_i^2}{2m} + \sum_i \frac{e}{m}\bar{A} \cdot \bar{p}_i + \frac{e^2}{2m}A^2 + V = H^{(0)} + \lambda H^{(1)} + \lambda^2 H^{(2)} \tag{2.30}$$

with $H^{(0)}$ given by Eq. (1.4) and:

$$\lambda H^{(1)} = \sum_i \frac{e}{m} \bar{A} \cdot \bar{p}_i \tag{2.31}$$

$$\lambda^2 H^{(2)} = \frac{e^2}{2m} A^2 \tag{2.32}$$

Introducing the vector potential corresponding to the constant magnetic field, given by Eq. (2.23), we obtain:

$$\lambda H^{(1)} = \sum_i \frac{e}{m} \frac{1}{2} (\bar{B} \times \bar{r}_i) \cdot \bar{p}_i = \sum_i \frac{e}{2m} \bar{B} \cdot (\bar{r}_i \times \bar{p}_i) = \sum_i \frac{e}{2m} \bar{B} \cdot \bar{L}_i \tag{2.33}$$

L_i is the angular momentum operator for the ith electron. We can treat the $H^{(2)}$ term in a similar fashion. Assuming $\bar{B}$ in the z-direction and remaining careful when taking the scalar product of two vector products [13], we can get:

$$\lambda^2 H^{(2)} = \sum_i \frac{e^2}{8m} B^2 (x_i^2 + y_i^2) \tag{2.34}$$

The coordinates x_i and y_i for the electron i in Eq. (2.34) are taken relative to the origin of the vector potential; we shall return to this point below. λ in the left-hand sides of Eqs. (2.33) and (2.34) may now be identified with the magnetic flux density. Then, we can use the perturbation expansion of energy and obtain a quantum mechanical expression for the field-dependence of the total energy. This expression can be compared with the classical Taylor-expansion of the energy of a system in a magnetic field; such a comparison leads to quantum-mechanical expressions for the magnetic susceptibility [13]. In order to obtain something useful for the discussion of our NMR parameters, we have to include the nuclear spins. Fortunately, this is not too difficult. We need to recognize that a nuclear spin acts as a magnetic dipole with its own vector potential $\bar{A}_{nuc}$. The vector potential related to the dipolar flux density is:

$$\bar{A}_{dipole} = \frac{\mu_0}{4\pi} \frac{\bar{\mu} \times \bar{r}}{r^3} \tag{2.35}$$

Remembering that the magnetic dipole moment of a nucleus is related to its spin by:

$$\bar{\mu}_N = \gamma_N \bar{I}_N \tag{2.36}$$

we can write the vector potential of the nucleus:

$$\bar{A}_{nuc} = \frac{\mu_0 \gamma_N}{4\pi} r_N^{-3} \bar{I}_N \times \bar{r}_N \tag{2.37}$$

Now, the total vector potential that we need to deal with becomes a sum of the external field vector potential $\bar{A}_{ex}$ and $\bar{A}_{nuc}$:

$$\bar{A} = \bar{A}_{ex} + \bar{A}_{nuc} \tag{2.38}$$

and our Hamiltonian in the presence of the external field and a magnetic nucleus becomes:

$$H = \sum_i \frac{1}{2m} (\bar{p}_i + e\bar{A}_{ex} + e\bar{A}_{nuc})^2 + V \tag{2.39}$$

2.3 The Current Density

Before making use of the Hamiltonian given by Eq. (2.39) to work out theoretical expression for the shielding factor, we shall introduce a useful concept of current density. In analogy with the classical electromagnetism, it appears plausible that the induced magnetic field (and σB_0 is just such an induced field) is related to currents. We choose in this and next subsection to deal with a single electron case and define the current density vector:

$$\bar{j}(\bar{r}) = -\frac{e}{2m} (\psi^*\bar{\pi}\psi + \psi\bar{\pi}^*\psi^*) = -\frac{e}{2m} (\psi^*\bar{p}\psi - \psi\bar{p}\psi^*) - \frac{e^2}{m} \bar{A}\psi^*\psi$$

$$= \frac{e\hbar i}{2m} (\psi^*\bar{\nabla}\psi - \psi\bar{\nabla}\psi^*) - \frac{e^2}{m} \bar{A}\psi^*\psi \tag{2.40}$$

since p/m or π/m implies the velocity and $\psi^*\psi$ the charge density, we may interpret Eq. (2.40) as giving the rate of change of the charge density, or the current density. If the wavefunction for a molecule is real (as it can always be written for an orbitally non-degenerate state) the first term in the last expression disappears. If in addition the magnetic field is absent (the vector potential is zero everywhere), then $\bar{j}(\bar{r})$ is zero everywhere and no current flows. For an orbitally non-degenerate state, but in the presence of a magnetic field, $\bar{j}(\bar{r})$ becomes non-zero for two reasons. First, we have the term proportional to $\bar{A}$, which is now non-zero. Furthermore, the field induces an imaginary component in the wavefunction, making ψ and ψ^* different and rendering the first term in the last expression non-zero. The origin of this effect is to be sought in the fact that the first-order correction to the wavefunction due to the external magnetic field can be written:

$$\lambda\psi^{(1)} = -\sum_n \frac{\langle n| \lambda H^{(1)} |0\rangle}{E_n - E_0} \tag{2.41}$$

Let $H^{(1)}$ be given by Eq. (2.31) or Eq. (2.33). Since both $\bar{p}$ and $\bar{L}$ are imaginary operators and the unperturbed wavefunctions are real, the $\psi^{(1)}$ must be purely imaginary. For simplicity, we may write

$$\psi^{(1)} = -\sum_n c_n\psi_n^{(0)} \tag{2.42}$$

remembering that the coefficients c_n are imaginary. The wavefunction correct to the first order in the field may be written as $\psi = \psi^{(0)} + \lambda\psi^{(1)}$ and we may use it to calculate the current density correct to the first order. We note that $\bar{A}$ is already first order in the field and replace $\bar{A}\psi^*\psi$ by $\bar{A}(\psi^{(0)})^2$ and write

$$\bar{j}(\bar{r}) = \frac{e\hbar i}{2m}\left\{(\psi^{(0)} - \lambda\psi^{(1)})\,\bar{\nabla}(\psi^{(0)} + \lambda\psi^{(1)})\right.$$

$$\left. - (\psi^{(0)} + \lambda\psi^{(1)})\,\bar{\nabla}(\psi^{(0)} - \lambda\psi^{(1)}) - \frac{e^2}{m}\,\bar{A}(\psi^{(0)})^2\right\}$$

$$= \left[\frac{e\hbar i}{2m}\left\{\psi^{(0)}\,\bar{\nabla}\psi^{(0)} - \lambda\psi^{(1)}\,\bar{\nabla}\psi^{(0)} + \lambda\psi^{(0)}\,\bar{\nabla}\psi^{(1)} - \lambda^2\psi^{(1)}\,\bar{\nabla}\psi^{(1)}\right.\right.$$

$$\left. - \psi^{(0)}\,\bar{\nabla}\psi^{(0)} - \lambda\psi^{(1)}\,\bar{\nabla}\psi^{(0)} + \lambda\psi^{(0)}\,\bar{\nabla}\psi^{(1)} + \lambda^2\psi^{(1)}\,\bar{\nabla}\psi^{(1)}\right\}$$

$$\left. - \frac{e^2}{m}\,\bar{A}(\psi^{(0)})^2\right]$$

We can now divide the total current density into a part which only contains $\psi^{(0)}$ and a part which also contains $\psi^{(1)}$:

$$\bar{j}^d(\bar{r}) = - \frac{e^2}{m}\,\bar{A}(\psi^{(0)})^2 \tag{2.44}$$

and

$$\bar{j}^p(\bar{r}) = \frac{e\hbar i\lambda}{m}\left\{\psi^{(0)}\,\bar{\nabla}\psi^{(1)} - \psi^{(1)}\,\bar{\nabla}\psi^{(0)}\right\}$$

$$= \frac{e\hbar i\lambda}{m}\sum_n (c_n - c_n^*)\,\{\psi_n^{(0)}\,\bar{\nabla}\psi_0^{(0)} - \psi^{(0)}\,\bar{\nabla}_n^{(0)}\} \tag{2.45}$$

Both contributions to the current density are proportional to the magnetic flux density. It can also be shown that the two contributions correspond to currents in opposite directions. $\bar{j}^p$ is directed so that it counteracts the external flux density, while $\bar{j}^d$ enhances it. We shall find this physical distinction to be important for the discussion of the magnetic shielding.

2.4 Relation of the Magnetic Shelding to Currents

Let us now return to our complete Hamiltonian, Eq. (2.39) for a single electron system. Let us write the canonical momentum $\bar{\pi}$ including only the external field vector potential $\bar{A}_{ex}$:

$$\bar{\pi} = \bar{p} + e\bar{A}_{ex} \tag{2.46}$$

and let us introduce a modified unperturbed Hamiltonian:

$$H^{(0)'} = \frac{\pi^2}{2m} + V \tag{2.47}$$

Further, let us neglect the term in Eq. (2.39) that is quadratic in $\bar{A}_{nuc}$ (we shall return to this term later). We can then rewrite Eq. (2.39) as:

$$H = \frac{1}{2m} (\bar{\pi} + e\bar{A}_{nuc})^2 + V =$$

$$= H^{(0)'} + \frac{e}{2m} (\bar{\pi} \cdot \bar{A}_{nuc} + \bar{A}_{nuc} \cdot \bar{\pi}) = H^{(0)'} + H^{(1)'} \qquad (2.48)$$

and thereby divide the total Hamiltonian into an "unperturbed part" (which already contains the effect of the external field) and a "perturbation" reflecting the effect of the nuclear magnetic moment. Since the wavefunctions in the presence of the magnetic field are not easily available, we cannot use this formalism for performing calculations. However, it is useful for conceptual understanding of the phenomenon. We can formally write an expression for the first order energy correction due to $H^{(1)'}$:

$$E^{(1)} = \int \psi^* H^{(1)'} \psi \, d\tau = \frac{e}{2m} \int \psi^* (\bar{\pi} \cdot \bar{A}_{nuc} + \bar{A}_{nuc} \cdot \bar{\pi}) \psi \, d\tau \qquad (2.49)$$

where ψ is the wavefunction in the presence of the external field. Letting the divergence of $\bar{A}_{nuc}$ disappear in agreement with Eq. (2.21) allows us to rewrite $E^{(1)}$ as:

$$E^{(1)} = \frac{e}{2m} \int \bar{A}_{nuc} \cdot \{\psi^* \bar{\pi} \psi + \psi^* \bar{\pi} \psi\} \, d\tau \qquad (2.50)$$

Observing that $\bar{p}$ is Hermitian and performing some fairly simple manipulations, we may find that:

$$E^{(1)} = \frac{e}{2m} \int \bar{A}_{nuc} \cdot \{(\psi^* \bar{p} \psi - \psi \bar{p} \psi^*) + 2e\bar{A}_{ex} \psi^* \psi\} \, d\tau = - \int \bar{A}_{nuc} \cdot \bar{j}_{ex}(\bar{r}) \, d\tau \qquad (2.51)$$

where $\bar{j}_{ex}$ is the current density due to the external field. By setting into Eq. (2.51) the form of $\bar{A}_{nuc}$ given by Eq. (2.37), we obtain:

$$E^{(1)} = -\gamma_N \frac{\mu_0}{4\pi} \bar{I}_N \cdot \int \frac{\bar{r} \times \bar{j}_{ex}}{r^3} \, d\tau \qquad (2.52)$$

The energy of a magnetic dipole in a magnetic field is classically given by:

$$E = -\bar{\mu} \cdot \bar{B}' = -\gamma_N \bar{I}_N \cdot \bar{B}' \qquad (2.53)$$

We can identify the energy correction of Eq. (2.52) with the energy of Eq. (2.53) if we treat $\bar{B}'$ as an induced field:

$$\bar{B}' = \frac{\mu_0}{4\pi} \int \bar{r} \times \bar{j}_{ex}/r^3 \, d\tau \qquad (2.54)$$

or

$$\sigma \bar{B}_0 = \frac{\mu_0}{4\pi} \int \bar{r} \times \bar{j}_{ex}/r^3 \, d\tau \qquad (2.55)$$

Thus, in order to find σ, we need the part of the right-hand side of Eq. (2.55) which is linear in the magnetic field. This is obtained if $\bar{j}_{ex}$ is replaced by the first-order current density of Eq. (2.43).

We are not going to pursue these derivations any further. We note however that σ is related to induced currents and that we are going to have two distinct contributions to σ, connected with the diamagnetic and paramagnetic contributions to $\bar{j}$. One should at this stage mention that the induced magnetic field of Eq. (2.54) does not necessarily need to be parallel to the external field $\bar{B}_0$. Thus, the left-hand side of Eq. (2.55) should really be written $\underline{\underline{\sigma}} \, \bar{B}_0$ where $\underline{\underline{\sigma}}$ is now a shielding tensor. The fact that the NMR spectra of liquids may be interpreted in terms of Eq. (1.1) is due to the rapid isotropic tumbling of molecules. Mathematically, this corresponds to the statement that we observe an average shielding or one-third of the trace of the shielding tensor:

$$\sigma_{obs} = \frac{1}{3} (\sigma_{xx} + \sigma_{yy} + \sigma_{zz}) \tag{2.56}$$

Before leaving this section, we would like to point out that the concept of current density is also useful in discussing the theory of magnetic susceptibility [13].

2.5 The Ramsey Theory of Magnetic Shielding

The theory of magnetic shielding originally formulated by Ramsey [18] starts also with Eq. (2.39). In the original paper, the magnetic shielding constant for a certain nucleus in a molecule is obtained assuming that all other nuclei have zero magnetic moment and that the nucleus concerned has a magnetic moment of magnitude μ and of the same direction as the externally applied magnetic flux density B. The energy of the molecular system, for an orientation specified by subscript λ, is then calculated by perturbation theory to second order and by collecting terms whose dependence on μ and B is linear in the product μB. Let us denote this energy W_λ and the desired magnetic shielding constant σ_λ (if a magnetic field of unit strength is applied, the magnetic field at the nucleus has a component $- \sigma_\lambda$ parallel to the magnetic field). σ_λ can the be obtained from:

$$W_\lambda = \sigma_\lambda B \mu \tag{2.57}$$

At the final stage, Ramsey [18] performs the averaging over all molecular orientations. For the purpose of this review, we judge it more instructive to give expressions for the individual tensor components, $\sigma_{\alpha\beta}$. In the notation of Ando and Webb [15], the expression for the tensor component of the magnetic shielding for nucleus N becomes:

$$\sigma_{N\alpha\beta} = \langle \Psi_0 | H_{N\alpha\beta}^{(1,1)} | \Psi_0 \rangle - \sum_{k \neq 0}^{\infty} (E_k - E_0)^{-1}$$
$$\times [\langle \Psi_0 | H_\alpha^{(1,0)} | \Psi_k \rangle \langle \Psi_k | H_{N\beta}^{(0,1)} | \Psi_0 \rangle$$
$$+ \langle \Psi_0 | H_{N\alpha}^{(0,1)} | \Psi_k \rangle \langle \Psi_k | H_\beta^{(1,0)} | \Psi_0 \rangle] \tag{2.58}$$

The operators are given by the following expressions:

$$H_{N\alpha\beta}^{(1,1)} = \frac{\mu_0 e^2}{8\pi m} \sum_i (\bar{r}_i \cdot \bar{r}_{iN}\delta_{\alpha\beta} - \bar{r}_{i\alpha}\bar{r}_{iN\beta})\, r_{iN}^{-3} \tag{2.59}$$

$$H_{\alpha}^{(1,0)} = \frac{\mu_0 e\hbar}{8\pi m} \sum_i L_{i\alpha} \tag{2.60}$$

$$H_{N\alpha}^{(0,1)} = \frac{\mu_0 e}{4\pi m} \sum_i L_{iN\alpha} r_{iN}^{-3} \tag{2.61}$$

where μ_0 is the permittivity of free space, r_{iN} refers to the separation of electron i from nucleus N and r_i to the distance between the electron i and the origin (unspecified so far) of the vector potential for the external magnetic flux density, $\delta_{\alpha\beta}$ is the Kronecker delta. $L_{i\alpha}$ and $L_{iN\alpha}$ are the α components of the electron angular momentum operators defined with respect to r_i and r_{iN}, respectively. The tensor component then becomes:

$$\sigma_{N\alpha\beta} = \sigma_{N\alpha\beta}^d + \sigma_{N\alpha\beta}^p \tag{2.62}$$

where the diamagnetic and paramagnetic contributions are given by:

$$\sigma_{N\alpha\beta}^d = \frac{\mu_0 e^2}{8\pi m} \langle \Psi_0 | \sum_i (\bar{r}_i \cdot \bar{r}_{iN}\delta_{\alpha\beta} - r_{i\alpha}r_{iN\beta})\, r_{iN}^{-3} | \Psi_0 \rangle \tag{2.63}$$

and:

$$\sigma_{N\alpha\beta}^p = -\frac{\mu_0 e^2}{8\pi m^2} \sum_{k\neq 0} (E_k - E_0)^{-1} \left[\langle \Psi_0 | \sum_i L_{i\alpha} | \Psi_k \rangle \langle \Psi_k | \sum_i L_{iN\beta} r_{iN}^{-3} | \Psi_0 \rangle \right.$$
$$\left. + \langle \Psi_0 | \sum_i L_{iN\alpha} r_{iN}^{-3} | \Psi_k \rangle \langle \Psi_k | \sum_i L_{i\beta} | \Psi_0 \rangle \right] \tag{2.64}$$

If the origin of the external vector potential is located at the nucleus of interest, the diamagnetic term becomes analogous to the expression for the diamagnetic shielding in atoms, given by Lamb [19], while the paramagnetic term arises from the lack of spherical symmetry of the electrical potential. It should be made clear that the two terms are closely related and that the separation into the diamagnetic and paramagnetic contribution is largely artificial.

3 Calculations of Magnetic Shielding

The aim of this section is to present an overview of the methods currently used in theoretical calculations of magnetic shielding parameters. Because of the large amount of approaches, it is not possible to go deeply into the details in the various methods. We are limiting ourselves to a description of the main lines of the most frequently used schemes. For readers interested in the details of computational aspects, we refer to recent excellent reviews of the subject [15, 20, 21]. Also, the field is reviewed annually [22].

We start with a short historical review, followed by an introduction to some of the most successful methods developed and used during the last ten years. There, we are mainly concentrating on ab initio methods.

3.1 Simplifications of the Ramsey Theory

As the entire field of computational quantum chemistry, the calculations on real molecules started when powerful computers became available, although the theoretical basis was developed decades earlier.

The first theoretical treatment of chemical shielding is that of Ramsey [18], based on the Rayleigh-Schrödinger perturbation theory. In the theory of Ramsey, the electronic wavefunctions for the ground and the excited states are calculated in the absence of magnetic fields. According to Ramsey (see previous section Eqs. (2.62)–(2.64)), the component $\sigma_{\alpha\beta}$ of the shielding tensor $\sigma_{N\alpha\beta}$ of a given nucleus N is given as a two-term expression, $\sigma_{N\alpha\beta} = \sigma_{N\alpha\beta}^d + \sigma_{N\alpha\beta}^p$, where the superscripts d and p denote diamagnetic and paramagnetic contribution, respectively.

A principally important difficulty is related to the fact that the diamagnetic term depends only on the ground-state wavefunction for the molecule, while the para-magnetic contribution depends also on the excited states. Very little is known about the high energy electronic states for most of the molecules and almost nothing about the continuum states. On the basis of calculations on simple systems [23, 24], the contribution from the continuum states is likely to be rather significant. As a consequence, it is difficult to get a balanced description of the two terms. This difficulty was already recognized by Ramsey, who suggested that if one were able to estimate an average value of the excitation energy, then the summation over excited states would not be required, because of the closure relation:

$$\sum_n |n\rangle\langle n| = 1 \tag{3.1}$$

Using the closure relation, one can express the paramagnetic part of the shielding in terms of the ground-state wavefunctions only. This approach has been used in the early work by Saika and Slichter [25] who reported simple valence-bond estimates of the fluorine shieldings. Generally, there is no rigorous way to determine the average excitation energy term; this has been chosen as an empirical parameter in calculations made in past.

Karplus and Das [26] proposed a simplification of Eqs. (2.63) and (2.64) by re-presenting the ground state and the excited states in terms of molecular orbital wavefunctions. This approximation replaces the infinite sum with the sum over the finite number of orbitals. Within this framework the Ramsey equations become:

$$\sigma_{N\alpha\beta}^d = \frac{\mu_0 e^2}{8\pi m} \sum_j^{occ} \langle \psi_j | \bar{r} \cdot \bar{r}_N \delta_{\alpha\beta} - r_\alpha r_{N\beta}) \, r_N^{-3} | \psi_j \rangle \tag{3.2}$$

and:

$$\sigma^{P}_{N\alpha\beta} = -\frac{\mu_0 e^2}{8\pi m^2} \sum_{j}^{occ} \sum_{k}^{unocc} (E_k - E_j)^{-1}$$

$$\times [\langle \psi_j | L_\alpha |\psi_k\rangle \langle \psi_k| L_{N\beta} r_N^{-3} |\psi_j\rangle$$

$$+ \langle \psi_j| L_{N\alpha} r_N^{-3} |\psi_k\rangle \langle \psi_k| L_\beta |\psi_j\rangle] \tag{3.3}$$

The virtual orbitals of the MO calculations provide only a poor description of the excited states. Thus, Karplus and Das [26] also introduce the average excitation energy approximation and apply the closure relation.

A balanced description of the two terms in Eq. (2.62) is especially important in view of the underlying physics. The paramagnetic and the diamagnetic currents, giving rise to the paramagnetic and diamagnetic parts of the shielding, are in opposing directions and the σ^p and the σ^d are therefore expected to have opposite signs. An additional complication arises because of the fact that Eqs. (2.63) and (2.64), as well as Eqs. (3.2) and (3.3) contain a still arbitrary origin of the external vector potential, i.e. are gauge dependent. Physically, this gauge dependence is irrelevant and the changes in the two terms upon the gauge transformation should cancel exactly. Unless complete basis sets are used in the calculations, this cancellation is, however, not exact and acts as an additional error source, which also tends to become more serious as the size of the system increases. A usual choice is to make the origin of the external vector potential coincident with the studied nucleus. With this choice of the gauge, Eq. (3.2) and Eq. (3.3) have proved to give qualitatively correct results for diatomic molecules, when used with very large basis sets [20].

An elegant way to circumvent the problem has been proposed by Pople [27], who followed the early work of London [28]. Pople suggested that the molecular orbitals should be expanded in terms of the so-called gauge-independent atomic orbitals (GIAOs):

$$\theta_\mu(B) = \chi_\mu \exp\left[-\left(\frac{ie}{\hbar}\right) \bar{A}_\mu \cdot \bar{r}\right] \tag{3.4}$$

with:

$$\bar{A}_\mu = \frac{1}{2} \bar{B} \times \bar{r}_\mu \tag{3.5}$$

Pople combined the GIAO approach with a series of approximations and derived simplified expressions for the shielding constants similar to those obtained by Karplus and Das [26] without the GIAO formalism. Karplus and Pople [29] subsequently combined their efforts and reported together an important application of their simple approaches to the case of carbon-13. They find that the carbon-13 shieldings in conjugated molecules are likely to be dominated by the local paramagnetic contribution, for which they derive the following expression [29]:

$$\sigma^{AA}_p \propto (\Delta E)^{-1} \langle r^{-3}\rangle_{2p} \sum_B Q_{AB} \tag{3.6}$$

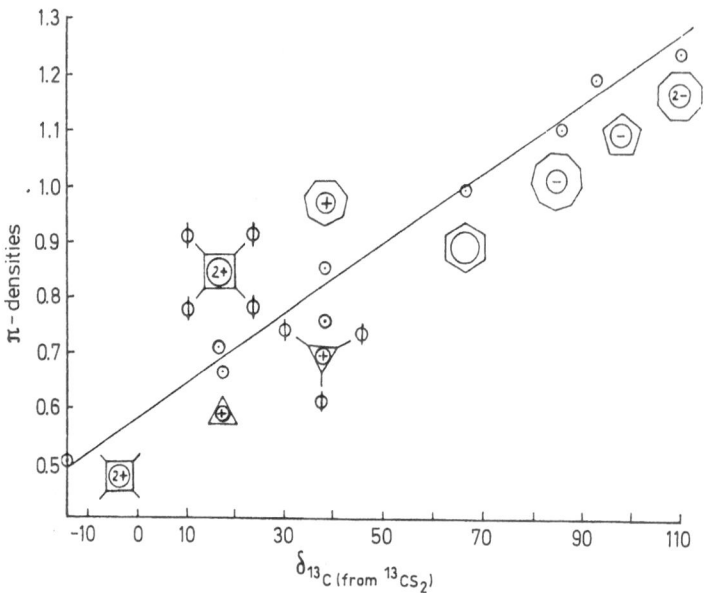

Fig. 1. Correlation of carbon-13 chemical shift and π-electron density in aromatic systems. (Reprinted with permission from Olah and Matescu (1970). J. Am. Chem. Soc. 92: 1430. Copyright by the American Chemical Society.)

Here, ΔE is the average excitation energy and Q_{AB} is a combination of products of various bond orders. $\langle r^{-3} \rangle_{2p}$ is the mean inverse cube of the distance from the nucleus for the carbon $2p$ atomic orbital. This factor leads directly to a correlation with electron density, as the atomic orbitals are expected to expand as electrons are added to the atom. This correlation is widely used [30, 31]; an illustrative example taken from a paper by Olah and Matescu [32] is shown in Fig. 1.

3.2 The Coupled Hartree-Fock and Related Approaches

Various SCF perturbation theories or coupled Hartree-Fock (CHF) methods have today a well-established position in the calculations of second-order molecular properties. The SCF perturbation method, developed by Stevens et al. [33, 34] and further by Pople et al. [35] and Ditchfield et al. [36], avoids the SOS formalism within the LCAO–MO SCF framework. The resulting equations are given as:

$$\sigma_{N\alpha\beta} = \frac{\mu_0 e^2}{8\pi m} \sum_{\mu}^{N} \sum_{v}^{N} P_{\mu v}^{(0)} \langle \chi_\mu | (\bar{r} \cdot \bar{r}_N \delta_{\alpha\beta} - r_\alpha r_{N\beta}) r_N^{-3} | \chi_v \rangle$$

$$- \frac{\mu_0 e}{8\pi m^2} \sum_{\mu}^{N} \sum_{v}^{N} (P_{\mu v}^{(1,0)})_\alpha \langle \chi_\mu | L_{N\beta} r_N^{-3} | \chi_v \rangle \tag{3.7}$$

$$P_{\mu v}^{(0)} = 2 \sum_{j}^{occ} C_{\mu j}^{(0)} C_{v j}^{(0)} \tag{3.8}$$

$$(P_{\mu v}^{(1,0)})_\alpha = 2 \sum_{j}^{occ} [C_{\mu j}^{(0)} (C_{v j}^{(1,0)})_\alpha - (C_{\mu j}^{(1,0)})_\alpha C_{\mu j}^{(0)}] \tag{3.9}$$

In the approach of Pople et al. [35] and Ditchfield et al. [36] an external small (finite) perturbation is explicitly included in the Hamiltonian and the Hartree-Fock equations are solved. The magnetic shielding constant $\sigma_{N\alpha\beta}$ is then calculated by derivating the energy with respect to the perturbation:

$$\sigma_{N\alpha\beta} = \left[\frac{\partial^2 E(\mu_{N'}, B)}{\partial\mu_{N\beta}\partial B_\alpha} \right]_{\mu_{N\beta} = B_\alpha = 0} \tag{3.10}$$

Due to the finite nature of the perturbations, the method is frequently called the finite perturbation theory (FPT) or the finite perturbation method (FPM).

Another important theoretical contribution to calculations of second-order molecular properties made during the 1970s is that of Nakatsuji [37]. Nakatsuji showed that the best second-order sum-over-state energy at the Hartree-Fock level of approximation is that obtained by the coupled Hartree-Fock (CHF) method. The Nakatsuji SOS-CI method was later used by Fukui et al. [38] to calculate magnetic shielding constants in first-row hydrides (See Tables 1—4). Fukui et al. obtain identical results with both methods (CHF and SOS-CI) when the same basis sets are used. The very attractive feature in SOS-CI is that it is a non-iterative procedure. The rather time-consuming diagonalisation of the complex matrix is done only once. The use of SCFPT is free from the Rayleigh-Schrödinger sum-over-states formalism but the problem caused by the gauge dependence still remains unless a complete basis set is used.

As discussed above an elegant method to circumvent the gauge dependence problem

Table 1. Some selected theoretical values of magnetic shielding parameters in hydrogen fluoride. (All values in ppm). Experimental values quoted are found in the theoretical papers

	GIAO-FPT	GIAO-CHF	IGLO	SOS-CI/ SCF-FPT	CHF	MBPT	*EXP.*
	a)	b)	c)	d)	e)	e)	
^{19}F							
σ	412.4	411.6	413.5	342.5	413.5	428.9	410 ± 6
$\sigma_\perp$	—	376.1	379.4	—	379.3	402.8	
$\sigma_\parallel$	—	482.7	481.7	—	481.8	481.3	
$\Delta\sigma$	—	106.7	102.3	—	102.5	78.6	108
^1H							
σ	30.18	30.6	28.05	—	27.5	27.9	28.72
							28.51 ± 0.2
							28.37
							28.8 ± 0.5
							29.2 ± 0.5
$\sigma_\perp$	—	24.2	20.0	—	19.6	19.6	
$\sigma_\parallel$	—	43.5	44.1	—	44.1	44.5	
$\Delta\sigma$	—	19.3	24.1	—	24.5	24.9	24

[a] Chesnut DB, Foley CK (1986) J. Chem. Phys. 84: 853 (same program as in Ditchfield R (1974) Mol. Phys. 27: 789. Basis set: $\langle 4,3,1/2\rangle$;

[b] Fukui H, Miura K, Shinbori H (1985) J. Chem. Phys. 83: 907 Basis set: 4-31G;

[c] Schindler M, Kutzelnigg W (1982) J. Chem. Phys. 76: 1919 Basis set: $\langle 7,7,3,1/4,2\rangle$;

[d] Fukui H, Yoshida H, Miura K (1981) J. Chem. Phys. 74: 6988 errata: (1982) J. Chem. Phys., 77: 5229. The SOS-CI and SCF-FPT give identical results. Basis set: 4-31G/gauge: fluorine;

[e] Iwai M, Saika A (1982) J. Chem. Phys. 77: 1951 Basis set: $\langle 8,6,3/6,4,1\rangle$/gauge: fluorine

Table 2. Some selected theoretical values of magnetic shielding parameters in water. (All values in ppm). Experimental values quoted are found in the theoretical papers

	GIAO/FPT a)	RPA b)	EOM-RPA c)	IGLO d)	SOS-CI/ SCF-FPT e)	CHF e)	SCF-CI f)	*EXP.*
^{17}O								
σ	332.0	327.1	·324.8	327.4	267.3	266.2	295.8	334
$\Delta\sigma$			56.1					
^{1}H								
σ	32.0	29.9	30.2	30.6	342.5	28.1	29.3	30.1

[a] Chesnut DB, Foley CK (1986) J. Chem. Phys. 84: 853 (same program as in Ditchfield R (1974) Mol. Phys. 27: 789) Basis set: $\langle 4,3,1/2 \rangle$;

[b] Lazeretti P, Zanasi R (1983) Phys., Rev. a27: 1301 101 uncontracted gaussians (near Hartree-Fock quality) gauge: centre of mass;

[c] Lamanna UT, Guidotti C, Arrighini GP (1977) J. Chem. Phys. 67: 604 47 STO's (near-Hartree-Fock quality), gauge: the studied nucleus;

[d] Schindler M, Kutzelnigg W (1982) J. Chem. Phys. 76: 1919 Basis set: $\langle 8,6,3/5,2 \rangle$;

[e] Fukui H, Yoshida H, Miura K (1981) J. Chem. Phys. 74: 6988 errata: (1982) J. Chem. Phys. 77: 5229 The SOS-CI and SCF-FPT give identical results. Basis set: 4-31G/gauge: oxygen;

[f] Fukui H, Miura K, Tada F (1983) J. Chem. Phys. 79: 6112 The same method is used to calculate shift in HF: Fukui et al. (1983) Int. J. Quant. Chem. 23: 633 CHF/$\langle 5,3/3 \rangle$ SCF-CI/4-31G gauge: centre of mass

is to expand each MO in terms of gauge invariant atomic orbitals (GIAO) [39]. In Eq. (3.4) the real atomic orbitals are multiplied by a complex factor which depends on the gauge of the vector potential. The name GIAO has been critized to be mis-

Table 3. Some selected theoretical values of magnetic shielding parameters in ammonia. (All values in ppm). Experimental values quoted are found in the theoretical papers

	GIAO-FPT a)	IGLO b)	SOS-CI/ SCF-FPT c)	EXP.
^{15}N				
σ	265.2	265.4	244.1	264.5
$\sigma_\perp$	—	278.0	—	
$\sigma_\parallel$	—	240.3	—	
$\Delta\sigma$	—	−37.7	—	−40
^{1}H				
σ	32.7	31.5	—	30.7 (liq)

[a] Chesnut DB, Foley CK (1986) J. Chem. Phys. 84: 853 (same program as in Ditchfield R (1974) Mol. Phys. 27: 789) Basis set: $\langle 4,3,1/2 \rangle$;

[b] Schindler M, Kutzelnigg W (1982) J. Chem. Phys. 76: 1919 Basis set: $\langle 7,5,3/4,2 \rangle$;

[c] Fukui H, Yoshida H, Miura K (1981) J. Chem. Phys. 74: 6988 errata: (1982) J. Chem. Phys. 77: 5229 The SOS-CI and SCF-FPT give identical results Basis set: 4-31G/gauge: nitrogen.

Table 4. Some selected theoretical values of magnetic shielding parameters in methane. (All values in ppm). Experimental values quoted are found in the theoretical papers

	GIAO-FPT a)	IGLO b)	SOS-CI/ SCF-FPT c)	LORG-RPA d)	EXP.
^{13}C					
σ	193.0	193.9	221.0	196	197.4
$\sigma_\perp$	—	278.0	—		
$\sigma_\parallel$	—	240.3	—		
$\Delta\sigma$	—	−37.7	—		
^{1}H					
σ	30.6	31.3	—		30.6
$\sigma_\perp$	—	33.8	—		
$\sigma_\parallel$	—	27.0	—		
$\Delta\sigma$	—	−6.8	—		

[a] Chesnut DB, Foley CK (1986) J. Chem. Phys. 84: 853 (same program as in Ditchfield R (1974) Mol. Phys. 27: 789 Basis set: $\langle 4,3,1/2 \rangle$;
[b] Schindler M, Kutzelnigg W (1982) J. Chem. Phys. 76: 1919 Basis set: $\langle 8,6,2/5,2 \rangle$;
[c] Fukui H, Yoshida H, Miura K (1981) J. Chem. Phys., 74: 6988 errata: (1982) J. Chem. Phys. 77: 5229 The SOS-CI and SCF-FPT give identical results. Basis set: 4-31G/gauge: carbon;
[d] Hansen AE, Bouman TD (1985) J. Chem. Phys. 82: 5035 Basis set: $\langle 3,3,1/2,1 \rangle$

leading and it was proposed by Pople [40] that the notation "gauge dependent atomic orbitals" would be more appropriate, since the method is based on determining different gauge origins for different atomic orbitals. Within the FPT method the equations for the magnetic shielding constant given by Ditchfield are

$$\sigma_{N\alpha\beta} = \frac{\mu_0 e^2}{8\pi m} \sum_\mu^N \sum_\nu^N P_{\mu\nu}^{(0)} \langle \chi_\mu | (\bar{r}_N \cdot \bar{r}_\nu \delta_{\alpha\beta} - r_{N\alpha} r_{\nu\beta}) r_N^{-3} | \chi_\nu \rangle$$

$$+ \frac{\mu_0 e h^2}{8\pi m} \sum_\mu^N \sum_\nu^N \left\{ (P_{\mu\nu}^{(0)} \left[\left\langle \frac{\partial \theta_\mu}{\partial B_\alpha} \Big| \frac{L_{N\beta}}{r_N^3} \Big| \chi_\nu \right\rangle + \left\langle \chi_\mu \Big| \frac{L_{N\beta}}{r_N^3} \Big| \frac{\partial \theta_\nu}{\partial B_\alpha} \right\rangle \right] \right.$$

$$\left. + \frac{\partial P_{\mu\nu}(B_\alpha)}{\partial B_\alpha} \left\langle \chi_\mu \Big| \frac{L_{N\beta}}{r_N^3} \Big| \chi_\nu \right\rangle \right\} \tag{3.11}$$

Calculations based on Eq. (3.11) gave, for the first time, good agreement with experiment [39]. Ditchfield further divided the expression for diamagnetic and paramagnetic terms to consist of contributions from local, non-local, and interatomic parts to analyze the various mechanisms behind the magnetic shielding:

$$\sigma_N = \sigma_N^d(\text{loc.}) + \sigma_N^p(\text{loc.}) + \sigma_N^d(\text{non-loc.}) + \sigma_N^p(\text{non-loc.})$$

$$+ \sigma_N^d(\text{inter.}) + \sigma_N^p(\text{inter.}) \tag{3.12}$$

It should be mentioned that the diamagnetic and paramagnetic terms in Eq. (3.12) are not strictly related to the corresponding terms in Ramsey's theory.

Schindler and Kutzelnigg introduced a method called individual gauge for localized orbitals (IGLO) [41]. In this method the CHF equations are solved in terms of localized MOs with an optimum gauge for each localized molecular orbital. The tensor components of the magnetic shielding parameter are obtained as a sum over all orbital contributions. Schindler and Kutzelnigg introduce a new set of molecular orbitals ψ_k related to φ_k:

$$\varphi_k = \varphi_{ko} + \varphi_{kl} + \dots \tag{3.13}$$

(MOs expanded in powers of the flux density) via:

$$\varphi_k = e^{iA_k}\psi_k \tag{3.14}$$

where:

$$A_k = \frac{e}{2\hbar}\,(\overline{R}_k \times \overline{B}) \cdot \overline{r} \tag{3.15}$$

is a local operator proportional to the flux density and $\overline{R}_k$ is the local gauge origin. Compared to a standard CHF scheme the advantage in the IGLO-CHF lies in the fact that the localized contributions to diamagnetic and paramagnetic terms are small in magnitude; thus, the error in the cancelling terms is also smaller than in the total terms with a common gauge origin in the CHF approach. Because localized MOs are used in the calculations, it is possible to investigate the contributions coming from various parts of the molecule and especially from delocalized electrons or lone pairs. Because the theory of Kutzelnigg and Schindler is not based on common gauge origin, new "resonance" and "exchange" contributions arise, which are treated in an approximate way. Only the local contributions can be divided into diamagnetic and paramagnetic terms as in the Ramsey theory. Levy and Ridard [42] have proposed method which uses local origins for pairs of orbitals without introducing complex phase factors as it is done in the GIAO method by Ditchfield [39] and the IGLO method by Kutzelnigg and Schindler [41]. Another interesting development in calculations of magnetic shielding parameters makes use of the so-called equations of motion (EOM) method [43]. The EOM method has been previously used to calculate second-order properties by Lamanna et al. [44]. The random phase approximation (RPA) is one approximate solution to the EOM method. The EOM-RPA method, which is equivalent to other CHF methods (i.e. SCFPT, FPT, CHF, Nakatsuji's SOS-CI) was first developed with a common gauge origin by Lazzeretti et al. [45] and later by Hansen and Bouman [46], who introduced their localized orbitals in a local origin (LORG) method within the EOM-RPA scheme. As in IGLO method, a separate analysis of contributions from various parts of a molecule is possible in LORG approach. It is clear that the introduction of local gauge origins in the methods to calculate magnetic shielding parameters (such as the GIAO) of Ditchfield [39] with atomic orbitals multiplied with a complex phase factor as local origins, the IGLO method by Kutzelnigg and Schindler [41] with localized molecular orbitals multiplied with a complex phase factor as gauge origins, method of Levy and Ridard [42] with pairs of atomic orbitals as local origins and the LORG method by Hansen and Bouman [46] with localized molecular orbitals in the local origins) have been successful. Compared to standard CHF methods with common gauge origins, a good agreement with experiment is obtained already when moderate size basis sets are used, while very large basis sets are required in standard methods to achieve the same accuracy. As discussed in the literature, the reason for the improvement is more cosmetic than physical. The methods with local origins are capable of removing basis set errors to a large extend. In the standard methods with a common gauge origin, the total magnetic shielding constant is calculated as a difference between two large terms (the size of the terms grows with the size of the molecule), the paramagnetic term being much more sensitive to the basis set quality. In the local origin methods this error is prevented from accumulating. The other advantage of the local origin method is that it facilitates an analysis of the local and non-local contributions. Since chemical shifts (as well as NMR coupling constants) are valuable tools in probing various aspects connected to chemical bonding, a closer analysis of the origin of these mechanisms is of great value especially if the transferability of these mechanisms from molecule to molecule upon substitution is of interest. In Tables 1–4, some selected values from theoretical calculations of magnetic shielding parameters and their anisotropies in first-row hydrides are gathered. Anisotropies can be measured for example by solving the molecule in an ordered medium such as liquid crystals. The anisotropy, $\Delta\sigma$, is defined as $\sigma_{\parallel} - \sigma_{\perp}$ for axially symmetric sites and as $1/2(3\sigma_{zz} - \sigma_{yy} - \sigma_{xx})$, $\sigma_{zz} > \sigma_{yy} > \sigma_{xx'}$, otherwise.

The calculations referenced in the tables are probably rather representative of today's methods in the field. It is clear from the table that the local gauge origin methods are able to give results very close to the experiment. The advantage of the local origin methods over the conventional common origin methods becomes, however, better demonstrated in calculations of larger molecules than the ones showed in Tables 1–4. The reason for quoting first row hydrides is to compare the state of accuracy in theoretical calculations of magnetic shielding with corresponding nuclear spin-spin results below. In the case of small molecules, the theoretical results obtained with today's method are very close to the experimental ones within the (still quite large) uncertainty in measured values: the relative values of chemical shifts are measured very accurately, while it is difficult to obtain accurate values for the absolute shielding parameters. On the other hand, there is the same dilemma in calculations of chemical shielding parameters as there is in theoretical predictions of the NMR coupling constants. Namely, in most of the cases the measurements are carried out in a medium, while the quantum-chemical calculations correspond to gas-phase conditions for one single molecule at 0 K. Medium effects are completely lost in theoretical calculations. The remaining discrepancy between the theoretical and experimental values is most probably reduced when vibrational corrections are included in the calculations. Lazzeretti et al. [47] have calculated the ^{13}C shielding surfaces for the methane molecule. They find the results very sensitive to bond stretching, σ being decreased with increasing bond length.

3.3 Electron Correlation Effects in Magnetic Shielding

As will be discussed in Sect. 5, the electron correlation contributions to the nuclear spin-spin coupling constants are of crucial importance. Agreement with experiment is obtained first when correlation contributions are added to the Fermi contact term in theoretical calculations of NMR coupling constants. In the case of magnetic shielding, there have been so far very few attempts to extend the calculations beyond the Hartree-Fock level. On the other hand, the good agreement of the calculated results with the experiment indicates that the electron correlation effects for magnetic shielding are not likely to be very large in small molecules. Generally, the possible correlation contributions to magnetic shielding are assumed to arise from the paramagnetic mechanisms.

Daborn et al. [48] have developed a method to calculate magnetic second-order properties at the CI level. The CI perturbation theory is developed in the framework of SCFPT theory. Because of the imaginary nature of the magnetic perturbation, it is computationally more efficient to use infinitesimal rather than finite field methods. Daborn et al. [48] use LiH as a test case, which is not so interesting because the electron correlation effects are small in this molecule. Iwai and Saika [49] have calculated magnetic shielding parameters for simple diatomic molecules (H_2, HF, and F_2) using many-body perturbation theory (MBPT) and compare the results with CHF calculations. Iwai and Saika also find small correlation corrections to magnetic shielding parameters.

3.4 Approximate Methods

Using the local gauge origin methods it is possible to make reliable calculations of magnetic shielding parameters for medium-size molecules, at least when trends in large series of molecules are investigated. As an example of extensive studies of ^{13}C shifts in organic molecules could be mentioned [41, 46, 50]. For large molecular systems one is still more or less forced to use semi-empirical schemes, among which complete neglect of differential overlap (CNDO), intermediate neglect of differential overlap (INDO) and modified INDO (MINDO) methods are well established [51].

The INDO parametrized FPT scheme has been very successful both in calculations of spin-spin coupling constant (see below) and chemical shielding and is more or less routinely used by many experimental groups to interpret NMR spectra. An extensive review of semi-empirical calculations of magnetic shielding and details of various computational schemes including program listings is found elsewhere [15]. Also, Xα schemes have been developed to calculate magnetic shielding [52, 53]. These results do not so far compare very well with the experiments.

3.5 Relativistic Theories

For heavy atoms large relativistic corrections are needed in the Ramsey theory. Relativistic theories for magnetic shielding are developed by Kolb et al. [54], Pyykkö [55] and Pyper [56]. The theory of Kolb et al. is based on relativistic random-phase approximation (RRPA) in terms of Dirac-Fock wavefunctions. In Pyykkö's [55] Dirac-Fock theory the main relativistic effects are included in the standard second-order theory. Pyper's [56] coupled Dirac-Fock theory consists of five terms, of which two are purely relativistic. No purely relativistic calculations of magnetic shielding parameters on molecules have been reported so far. Pyykkö et al. [57] have studied relativistic contribution in hydrogen halides to the NMR "heavy atom chemical shift".

The correct non-relativistic limit is obtained with relativistic extended Hückel (REX) level and the relativistic contributions are then obtained by comparing the REX and extended Hückel (EHT) results using second-order perturbation theory.

4 The Origin of Nuclear Spin-Spin Coupling

The nuclear spin-spin coupling term in the phenomenological Hamiltonian is bilinear in the nuclear spins. Classically, we can envision an interaction of this form, the direct dipole-dipole coupling between the nuclear magnetic moments. This interaction averages, however, to zero in isotropic fluids [58] and thus cannot explain field-independent splittings in the liquid phase spectra.

If we want to obtain a quantum-mechanical expression for $J_{NN'}$, we can follow an approach similar to that used for the shielding. We start by formulating a suitable Hamiltonian and proceed with the perturbation expansion of energy. We follow here the approach taken in the earlier reviews by Kowalewski [59, 60].

4.1 The Basic Interactions

In order to arrive at terms in the molecular Hamiltonian appropriate for describing the nuclear spin-spin coupling, we need to introduce the concept of magnetic hyperfine interaction. In its dominant interaction (Coulombic attraction) with the electrons in an atomic or molecular system, the nucleus can be represented as a point charge. The interactions between the nuclei and electrons resulting from the non-point charge character of the nucleus are denoted as hyperfine interactions. The hyperfine interactions can be electric or magnetic. The most important electric hyperfine interaction is that between the electric quadrupole moment of the nucleus and the field gradient at the site of the nucleus. The magnetic hyperfine interaction is that between the nuclear and electronic magnetic moments. In the following, we are going to concentrate on the magnetic hyperfine interaction.

The most correct way to derive the form of the magnetic hyperfine interaction is actually to start with the relativistic counterpart of the Schrödinger equation, the Dirac equation (spin is really a relativistic phenomenon!). A derivation of this type can be found in the book by Weissbluth [61] or in the review by Blinder [62]. We shall not repeat this derivation here, but only state the form of the final Hamiltonian in the non-relativistic limit and comment it. The magnetic hyperfine Hamiltonian contains three terms of first order in the nuclear magnetic moment or spin:

$$H = H_{1b} + H_2 + H_3 \tag{4.1}$$

with:

$$H_{1b} = \frac{\mu_0 \mu_B h}{2\pi i} \sum_N \gamma_N \bar{I}_N \cdot \sum_k r_{kN}^{-3} (\bar{r}_{kN} \times \bar{\nabla}_k) \tag{4.2}$$

$$H_2 = \frac{\mu_0 \mu_B h}{2\pi} \sum_N \gamma_N \sum_k [3(\bar{S}_k \cdot \bar{r}_{kN})(\bar{I}_N \cdot \bar{r}_{kN}) r_{kN}^{-5} - (\bar{S}_k \cdot \bar{I}_N) r_{kN}^{-3}] \qquad (4.3)$$

$$H_3 = \frac{4\mu_0 \mu_B h}{3} \sum_N \gamma_N \sum_k \delta(\bar{r}_{kN}) \bar{S}_k \cdot \bar{I}_N \qquad (4.4)$$

μ_B is the Bohr magneton, the index k refers to electrons, $\bar{S}_k$ is the electron spin operator, and $\delta(\bar{r}_{kN})$ is the Dirac delta function. The first term, H_{1b}, corresponds to the interaction between the nuclear magnetic moment of the nucleus and the magnetic field related to the orbital motion of the electrons. The origin of this term can also be found in the Hamiltonian of Eq. (2.39); upon squaring the expression in the parantheses, we obtain a term that is independent of the external field and linear in the nuclear vector potential or magnetic moment. If we include more than one magnetic nucleus in Eq. (2.39), we can actually obtain one more term of interest:

$$H_{1a} = \frac{\mu_0^2 e \hbar \mu_B}{(4\pi)^2} \sum_{N,N'} \gamma_N \gamma_{N'} \sum_k r_{kN}^{-3} r_{kN'}^{-3} [(\bar{I}_N \cdot \bar{I}_{N'})(\bar{r}_{kN} \cdot \bar{r}_{kN'}) - (\bar{I}_N \cdot \bar{r}_{kN'})(\bar{I}_{N'} \cdot \bar{r}_{kN})]$$

$$(4.5)$$

The H_{1a} term is of second order in the nuclear magnetic moments and differs in this way from the terms included in Eq. (4.1). The remaining two terms in Eq. (4.1) have their origin in the interaction between the nuclear magnetic moment and the magnetic moment of the electron spin. A simplified derivation of these terms is given in the book by Atkins [13]. Essentially, both terms can be obtained assuming the nuclear vector potential as given in Eq. (2.35) and evaluating carefully the interaction between the magnetic field $\bar{B} = \bar{\nabla} \times \bar{A}$ and the magnetic moment of the spinning electron. The first of these terms, denoted H_2, describes the dipole-dipole interaction between the two magnetic moments, while the last one, denoted H_3, is the so-called contact term, derived originally by Fermi [63]. This term describes the interaction between the nuclear spin and the electrons at the site of the nucleus. Neglecting the electrons originating from other atoms, only the s-electrons have non-zero density at the nuclear site and can contribute to the contact interaction.

One should mention in this connection that the form of the magnetic hyperfine Hamiltonian of Eq. (4.1) is approximate. It is valid for the point nucleus in the non-relativitic limit. The relativistic corrections to the magnetic hyperfine Hamiltonian were first discussed by Breit in 1930 [64] and more recently by Pyykkö et al. [65].

4.2 Ramsey's Theory for Nuclear Spin-Spin Coupling

Ramsey followed his fundamental contributions to the theory of magnetic shielding [18] by equally important contributions to the theory of spin-spin coupling. He showed in his classical paper [66] how perturbation theory using the Hamiltonians of Eqs. (4.1)–(4.5) as perturbations can lead to the quantum-mechanical expressions corresponding to the spin-spin coupling term in the phenomenological Hamiltonian. Here, we follow again the approach taken by Kowalewski [59].

The three terms in Eq. (4.1) can contribute to the total energy correction of second order in nuclear magnetic moments by terms that in general can be written:

$$E_{NN'}^{AB} = k_{NN'}^{AB} \sum_n \frac{\langle 0| \bar{A}_N \cdot \bar{I}_N |n\rangle \langle n| \bar{B}_{N'} \cdot \bar{I}_{N'} |0\rangle}{E_0 - E_n} \tag{4.6}$$

where the superscript AB refers to the combination of mechanisms giving rise to the energy term. $k_{NN'}^{AB}$ is a constant. Vector operators $\bar{A}_N$ and $\bar{B}_{N'}$ depend on the electronic coordinates, momenta and spins and their exact form may be derived from Eqs. (4.2)–(4.4) for any combination of perturbations. The summation is carried out over all the excited states of the molecule.

Equation (4.6) is not exactly of the form that the coupling term has in Eq. (1.1) but rather:

$$E_{NN'} = h\bar{I}_{N'} \cdot \underline{J}_{NN'} \cdot \bar{I}_{N'} \tag{4.7}$$

where it is assumed that the integrations implied by Eq. (4.6) involve electron spin and coordinates only, allowing the nuclear spins to be factored out. Moreover, in going from Eq. (4.6) to Eq. (4.7) the matrix elements of vector operators $\bar{A}_N$ and $\bar{B}_{N'}$ are treated as vectors and the general relation:

$$(\bar{A}_N \cdot \bar{I}_N) (\bar{B}_{N'} \cdot \bar{I}_{N'}) = \bar{I}_N \cdot (\bar{A}_N \bar{B}_{N'}) \cdot \bar{I}_{N'} \tag{4.8}$$

is utilized. $(\bar{A}_N \bar{B}_{N'})$ is again a second-rank tensor, expressed as a direct product of two vectors [16]. Ramsey [66] has shown that in the absence of preferred direction, the free tumbling of molecules will average out the eventual anisotropy leading to the spin-spin interaction of form given in Eq. (1.1) with:

$$J = \frac{1}{3} (J_{xx} + J_{yy} + J_{zz}) \tag{4.9}$$

The analogy between the tensorial nature of the spin-spin coupling constant and the shielding constant should be noticed. Neglecting the anisotropy of the interaction, we can now rewrite Eq. (4.6) in terms of matrix elements of $\bar{A}_N$ and $\bar{B}_{N'}$ and the scalar product of $\bar{I}_N$ and $\bar{I}_{N'}$:

$$E_{NN'}^{AB} = \frac{1}{3} k_{NN'}^{AB} \left(\sum_n \frac{\langle 0| \bar{A}_N |n\rangle \langle n| \bar{B}_{N'} |0\rangle}{E_0 - E_n} \right) \bar{I}_N \cdot \bar{I}_{N'} \tag{4.10}$$

Each of the different contributions to $J_{NN'}$ may be written:

$$J_{NN'}^{AB} = \frac{2k_{NN'}^{AB}}{3h} \sum_n \frac{\langle 0| \bar{A}_N |n\rangle \langle n| \bar{B}_{N'} |0\rangle}{E_0 - E_n} \tag{4.11}$$

where the factor of two arises because of the presence of two equivalent energy terms with N and N′ permuted. The sum of the effects from different mechanisms will give the total value of the coupling constant. Ramsey [66] has shown that all the cross-term contributions to the motionally averaged coupling constant vanish and that the three parts of the perturbation can be treated separately. The reason for vanishing cross-terms between H_{1b} and both H_2 and H_3 is that, in the absence of spin-orbit coupling, the excited states $|n\rangle$ giving non-zero matrix elements of H_{1b} are of the same multiplicity as the ground state (normally singlet), while only the excited triplet states are coupled to the singlet ground state by the spin-dependent operators H_2 and H_3. The cross-term between H_2 and H_3 does not vanish because of multiplicity; the motional average is zero because the two operators involved have different transformation properties under rotations. Upon substitution of the forms of vector operators of Eqs. (4.2)–(4.4) into Eq. (4.11), the following expressions are obtained:

$$J_{NN'}^{(1b)} = -\frac{2}{3h}\left(\frac{\mu_0\mu_B\hbar}{2\pi}\right)^2 \gamma_N\gamma_{N'} \sum_n \langle 0| \sum_k r_{kN}^{-3}(\bar{r}_{kN} \times \bar{V}_k)\,|n\rangle$$
$$\cdot \langle n| \sum_j r_{jN'}^{-3}(\bar{r}_{jN'} \times \bar{V}_j)\,|0\rangle/(^1E_n - E_0) \tag{4.12}$$

$$J_{NN'}^{(2)} = -\frac{2}{3h}\left(\frac{\mu_0\mu_B\hbar}{2\pi}\right)^2 \gamma_N\gamma_{N'} \sum_n \langle 0| \sum_k 3r_{kN}^{-5}(\bar{S}_k \cdot \bar{r}_{kN})\,\bar{r}_{kN} - r_{kN}^{-3}\bar{S}_k\,|n\rangle$$
$$\cdot \langle n| \sum_j 3r_{jN'}^{-5}(\bar{S}_j \cdot \bar{r}_{jN'})\,\bar{r}_{jN'} - r_{jN'}^{-3}\bar{S}_j\,|0\rangle/(^3E_n - E_0) \tag{4.13}$$

$$J_{NN'}^{(3)} = -\frac{2}{3h}\left(\frac{4\mu_0\mu_B\hbar}{3}\right) \gamma_N\gamma_{N'} \sum_n \langle 0| \sum_k \delta(\bar{r}_{kN})\,\bar{S}_k\,|n\rangle$$
$$\cdot \langle n| \sum_j \delta(\bar{r}_{jN'})\,\bar{S}_j\,|0\rangle/(^3E_n - E_0) \tag{4.14}$$

where the summation is carried out over the excited singlet states in Eq. (4.12) and over excited triplet states in Eq. (4.13) and Eq. (4.14). Finally, the first-order contribution corresponding to the Hamiltonian of Eq. (4.5) should be included:

$$J_{NN'}^{(1a)} = -\frac{\mu_0^2 e\mu_B}{48\pi^3} \gamma_N\gamma_{N'} \langle 0| \sum_k (\bar{r}_{kN} \cdot \bar{r}_{kN'})\,r_{kN}^{-3}r_{kN'}^{-3}\,|0\rangle \tag{4.15}$$

The terms given by Eqs. (4.12) and (4.15) are often denoted as paramagnetic and diamagnetic orbital contributions, J^{OP} and J^{OD}, respectively. There exists an analogy between those terms and the paramagnetic and diamagnetic contributions to the magnetic shielding; the coupling contributions can be thought of as arising through paramagnetic and diamagnetic currents, induced in the molecular electronic distribution by the nuclear magnetic moment of one of the nuclei, coupling to the magnetic moment of the other nucleus. $J^{(2)}$ will below occasionally be denoted J^{SD}. The largest interest among theoreticians working with calculations of coupling constants has been concentrated on the Fermi contact term, $J^{(3)}$ or J^{FC}. It can be shown [59] that this term is isotropic. In complete analogy with the case of magnetic shielding, the expressions derived by Ramsey [66] cannot be used without further approximations, which will be the subject of the next section. Before turning to that, one should mention that the relativistic analogue of the Ramsey theory [66] for spin-spin coupling was presented by Pyykkö [67].

5 Calculations of Nuclear Spin-Spin Coupling Constants

The calculations of nuclear spin-spin coupling constants reported in the 1970s and early 1980s have been subject to reviews by Kowalewski [59, 60] and the book by Ando and Webb [15]. More recently, the field was covered on an annual basis by Laaksonen [68] and by Overill [69]. Among recent reviews, one should also mention the paper by Jameson [70].

In this section, we review briefly and without claims on completeness, the computational methods used in numerical calculations of the coupling constants and some of their applications. The calculations of J^{OP}, J^{SD}, and J^{FC} can be classified into different categories depending on the type of perturbational treatment or on the level of approximation used in the calculations of wavefunctions. The methods based on the straightforward approximations to the Ramsey's equations given in the last section (sum-over-states methods, uncoupled methods) will be reviewed in Sect. 5.1. The coupled Hartree-Fock (CHF) type methods, in which calculations seek to attain self-consistency in the molecular orbital description, will be reviewed in Sect. 5.2. The methods aiming at removing some of the deficiencies still present in the CHF methods will be discussed in Sect. 5.3. The calculations of the diamagnetic orbital terms are straightforward in principle, as they only involve the unperturbed ground-state wavefunctions and expectation values over such functions; this mechanism will be discussed in Sect. 5.4. Summarizing comments about the state of the art of the nonempirical calculations will be given in Sect. 5.5.

The calculations of the wavefunctions can be of widely differing quality. The most sophisticated nonempirical computational machinery applied to calculations of coupling constants involves multiconfigurational self-consistent field (MC SCF) or configuration interaction (CI) techniques. Somewhat more approximate methods involve nonempirical SCF MO approach. Even more approximate are various semiempirical calculations, among which the INDO scheme was most widely and most successfully used.

5.1 Sum-over-States and Related Methods

The methods covered in this subsection use various approximations for the unperturbed ground state $|0\rangle$ and expand the first-order correction to the wavefunction in a truncated set of approximate excited state wavefunctions. Most of the work in this category deals with the Fermi contact contributions only and the calculations deal with approximations to Eq. (4.16). An important early contribution to the understanding of coupling constants was presented by Pople and Santry [71]. They assumed that the set of excited states could be created by promoting one electron from an occupied MO to a virtual one, that the MOs were obtained as linear combinations of a minimal valence shell basis set of atomic orbitals, that only one-center integrals over Dirac delta function needed to be retained, and that the excitation energies could be expressed as differences of orbital energies. This last approximation is only consistent with an independent electron-type model, which the authors also assumed. The series of approximations leads to:

$$J_{NN'} = \frac{\mu_0^2 \mu_B^2}{9\pi^2} h\gamma_N\gamma_{N'}s_N^2(0)\,s_{N'}^2(0)\,\pi_{NN'} \tag{5.1}$$

where the quantity π, called the atom-atom polarizability, is given by:

$$\pi_{NN'} = -4 \sum_i^{occ} \sum_a^{unocc} \frac{c_{iN} c_{aN} c_{iN'} c_{aN'}}{\varepsilon_a - \varepsilon_i} \tag{5.2}$$

$s_N(0)$ is the amplitude at the origin of the valence shell s orbital associated with nucleus N, and c_{iN} is the LCAO coefficient of this atomic orbital in the ith MO. ε_i and ε_a are the orbital energies for the occupied MO i and the virtual MO a. Probably the greatest success of the Pople-Santry method was the successful rationalization of the observed trends in the geminal couplings (couplings through two chemical bonds) in the CH_2 fragment, presented by Pople and Bothner-By [72]. According to the authors, the experimental findings may be summarized by the following four points:

1) in hydrocarbons, $^2J_{HH}$ increases as the hybridization of the carbon atom becomes more s-like;
2) the substitution of an electronegative atom in the position α to the CH_2 group leads to a positive shift in the coupling constant;
3) the substitution of an electronegative atom in the position β to the CH_2 group leads to a negative shift in the coupling constant;
4) the presence of a π-electron system next to the CH_2 group generally leads to a negative shift in the coupling constant.

Pople and Bothner-By explained these trends by considering a four-electron model of the CH_2 fragment with a set of two bonding and two antibonding orbitals. One of the bonding and one of the antibonding MOs were assumed symmetric with respect to the plane of symmetry perpendicular to the CH_2 fragment; the other two MOs were assumed antisymmetric. The first point in the list above was explained by increased energy separation within the bonding and the antibonding sets with the increasing s-character, giving increasing dominance of the positive contribution from the lowest energy excitation. The substituent effects were rationalized by the facts that the withdrawal of electrons from the symmetric bonding orbital (generally inductive effect) should lead to a positive change in the coupling constant, while the withdrawal of electrons from the antisymmetric bonding orbital (generally hyperconjugative effect) should lead to a negative change.

The Pople-Santry approach can also be applied to the non-contact contributions to the coupling constants. The same set of approximations that leads to Eq. (5.1) will then provide an equation containing the expectation values of the r^{-3} operator over the valence shell p-orbitals for the coupled nuclei. Pople and Santry [71] used this fact and proposed that the non-contact contributions to the couplings involving protons should vanish, because no p-type orbitals on hydrogen atoms are included in the minimal basis set. This conclusion was proved wrong at a much later date (see below). At an even earlier date, an even simpler MO method was proposed by McConnell [73]. He introduced the average excitation energy approximation and the closure relation already in the Eq. (4.6). Assuming, in analogy with Pople and Santry [71], the LCAO MO model and one-center approximation for the integrals, one obtains:

$$J_{NN'}^{(3)} = \frac{\mu_0^2 \mu^2}{9\pi^2} h\gamma_N\gamma_{N'} (^3\Delta E)^{-1} P_{s_N s_{N'}}^2 s_N^2(0) s_{N'}^2(0) \tag{5.3}$$

where $P^2_{s_N s_{N'}}$ is the element of the bond order matrix between the valence shell atomic orbitals s_N and $s_{N'}$:

$$P_{s_N s_{N'}} = 2 \sum c_{iN} c_{iN'} \tag{5.4}$$

Actually, the McConnell formula may be considered an approximation to the Pople-Santry model, as Eq. (5.1) will reduce to Eq. (5.3) if the average excitation energy is introduced. Contrary to Eq. (5.1), Eq. (5.3) has an obvious drawback in predicting a positive sign for all coupling constants between the nuclei with magnetogyric ratios of the same sign, which is not in agreement with experimental evidence. Nevertheless, the McConnell model and its modifications have been shown to be useful for rationalization of trends in certain coupling constants between directly bonded nuclei, e.g. $^1J_{CH}$ [74–77] and $^1J_{CC}$ [76–80]. Basically the model was applied to relate the coupling constant to the extent of $2s$ orbital contribution in the bonding hybrids on the carbon atom (or atoms) participating in the bond, obviously an attractive idea for chemists.

Among the early, simple and important theoretical work, one must also mention the valence bond (VB) theory formulated by Karplus and Anderson [81]. In analogy with the work of McConnell [73], they used the average excitation energy approximation and concluded that the occurrence of non-vanishing coupling between the non-bonded atoms implies deviations from perfect electron pairing in the ground state of the molecule. Karplus [82] developed a set of further approximations to this treatment which allowed considering the molecules in terms of a limited set of bonds, orbitals and electrons. Applying this approach to a six-electron fragment, Karplus [82, 83] derived form of the angular dependence of the vicinal (through three chemical bonds) proton-proton coupling constant in ethane:

$$^3J_{HH'} = A + B\cos\varphi + C\cos 2\varphi \tag{5.5}$$

A, B, and C are constants and φ is the dihedral HCCH angle. Equation (5.5), known as the Karplus relation, and its much later modifications (e.g., the work of Haasnoot et al. [84] are still today one of the fundaments of the conformational studies in solution. During the 1970s and 1980s, the VB methods for calculating coupling constants played a much less important role than the MO treatments, on which we are going to concentrate in the following.

The simple Pople-Santry sum-over-states approach has subsequently been refined and we are going to mention here some of the highlights in this development. The nonempirical and semi-empirical SCF MO calculations appeared in the late 1960s [85–88]. The method was pushed to the limits by Kowalewski et al. [89, 90] in the early 1970s. The authors used the unperturbed ground state wavefunctions of CI quality and expanded the first-order corrections into all singly and doubly excited (with respect to the SCF ground state) triplet configurations. The working equation of the method is:

$$\sum_{\mu=1}^{K} \langle \lambda | H^{(0)} - E_0^{(0)} | \mu \rangle c_{\mu,N}^{(1)} = -\langle \lambda | \sum \delta(\bar{r}_{kN}) S_{zk} | 0 \rangle, \tag{5.6}$$

$\lambda = 1, \ldots, K$

where $c_{\mu,N}^{(1)}$ is the coefficient of the excited triplet configuration μ in the first-order correction to the wavefunction and K is the number of triplet configurations, which can be of the order of several thousands. Equation (5.6) was solved iteratively, following the methods developed by Roos and Siegbahn [91, 92], without ever constructing the large Hamiltonian matrix. Kowalewski and coworkers reported large basis set calculations for the hydrogen molecule [89], where the results agreed very well with experiment, and for the first-row hydrides, ethene, and ethyne [90]. In spite of the great computational effort, the calculations for these polyatomic systems were, by and large, unsuccessful.

Before leaving the sum-over-states type methods, we would like to mention the work of Pyykkö and Wiesenfeld [93] covering a large number of coupling constants in a spectacular, "around the periodic table" fashion. The method is similar to the Pople-Santry approach [71], but the orbital energies and molecular orbital coefficients are obtained from relativistic extended Hückel calculations and the hyperfine integrals are taken from relativistic atomic calculations. The relativistic effects were found to be very important for couplings involving heavy atoms.

5.2 The Coupled Hartree-Fock Method

Generally speaking, the coupled Hartree-Fock method means that the Hartree-Fock equations for the molecular orbitals are solved in the presence of a perturbation. There are several ways to implement the CHF method in practice. We are going to present one of these approaches, the finite perturbation (FP) method, in some detail.

The FP method was originally proposed by Slater and Kirkwood [94]. In connection with the spin-spin coupling constants, the method was first proposed by Pople and coworkers [35, 95, 96]. It is convenient to carry out the presentation in terms of the reduced coupling constant, $K_{NN'}$, defined in Eq. (1.2). Consider a molecule with two nuclear magnetic moments, μ_N and $\mu_{N'}$, directed along the z-axis. In the presence of the Fermi contact interaction alone, the Hamiltonian for the molecule may be written as:

$$H = H_0 + \mu_N H_N + \mu_{N'} H_{N'} \tag{5.7}$$

where:

$$H_N = (4/3)\, \mu_0 \mu_B \sum_k \delta(\bar{r}_{kN})\, S_{zk} \tag{5.8}$$

From the power expansion of the energy in the presence of the two magnetic moments, the expression for the reduced coupling constant may be written as:

$$K_{NN'}^{(3)} = \left[\frac{\partial^2 E(\mu_{N'} \mu_{N'})}{\partial \mu_N\, \partial \mu_{N'}} \right]_{\mu_N = \mu_{N'} = 0} \tag{5.9}$$

Using the Feynman-Hellmann theorem. proved to be valid for the SCF functions used here [35], the second derivative of the energy may be set equal to the first derivative of the expectation value of H_N with respect to $\mu_{N'}$, evaluated in the presence of $\mu_{N'}$:

$$K_{NN'}^{(3)} = \left[\frac{\partial}{\partial \mu_{N'}} \langle \Psi(\mu_{N'})| \hat{H}_N |\Psi(\mu_{N'})\rangle \right]_{\mu_{N'}=0} \tag{5.10}$$

Thus, the calculation of $K_{NN'}$ requires calculation of the wavefunction in the presence of one of the nuclear magnetic moments, $\mu_{N'}$. Subsequent evaluation of the expectation value of H_N and its derivative is much simpler and can easily be done for any number of magnetic nuclei in a molecule.

In the FP approach of Pople and coworkers [35, 95, 96] $\Psi(\mu_{N'})$ is calculated in the presence of a unnaturally large (finite) nuclear magnetic moment and the derivative is evaluated numerically. $\Psi(\mu_{N'})$ is computed as a spin-unrestricted SCF LCAO MO wavefunction. This is necessary because the presence of $\mu_{N'}H_{N'}$ (the matrix elements of which enter the Fock matrix with different signs for α and β electrons) induces a non-vanishing spin-density in the molecule. The spin-density matrix ϱ may be defined as a difference between the first-order density matrices associated with the α and β electron spins:

$$\underline{\varrho} = \underline{\underline{P}}^{\alpha} - \underline{\underline{P}}^{\beta} \tag{5.11}$$

and the expectation value of H_N appearing in Eq. (5.10) may be written in terms of $\underline{\varrho}$:

$$\langle \Psi(\mu_{N'})| \, H_N \, |\Psi(\mu_{N'})\rangle = \frac{2\mu_0\mu_B}{3} \sum_{\lambda,\nu} \varrho_{\lambda\nu}(\mu_{N'}) \, \langle\lambda| \, \delta(\bar{r}_N) \, |\nu\rangle \tag{5.12}$$

where λ and ν denote the atomic orbitals. Now, the $\mu_{N'}$ dependence of the right-hand side of Eq. (5.12) is limited to the spin-density matrix. Derivation with respect to $\mu_{N'}$ gives the reduced coupling constant:

$$K_{NN'}^{(3)} = \frac{2\mu_0\mu_B}{3} \sum_{\lambda,\nu} \langle\lambda| \, \delta(\bar{r}_N) \, |\nu\rangle \left[\frac{\partial}{\partial\mu_{N'}} \varrho_{\lambda\nu}(\mu_{N'}) \right]_{\mu_{N'}=0} \tag{5.13}$$

which is generally valid for any type of SCF LCAO MO procedure. Pople et al. [35, 95, 96] then introduce a number of approximations consistent with the semi-empirical methods, such as CNDO and INDO, and calculate the derivative using the method of finite differences. They arrive at:

$$K_{NN'}^{(3)} = \frac{4\mu_0^2\mu_B^2}{3} \, s_N^2(0) \, s_{N'}^2(0) \, \frac{\varrho_{s_N s_N}(h_{N'})}{h_{N'}} \tag{5.14}$$

where $h_{N'}$ is the quantity actually added to, or subtracted from, the diagonal matrix elements of the Hartree-Fock operator. A similar FP approach for calculating non-contact terms was developed by Nakatsuji et al. [97].

The FP CNDO and, to an even larger extent, FP INDO calculations had a great impact on the development of the understanding of the nuclear spin-spin coupling constants. An important series of papers by Maciel and coworkers covered calculations of a large number of coupling constants between directly bonded carbons and protons, $^1J_{CH}$ [98], between two directly bonded carbons, $^1J_{CC'}$ [99], and between protons separated by two or more bonds [100–102]. Besides providing a lot of new information, the calculations corroborated some of the earlier results. For example, the linear relation between $^1J_{CH}$ and the square of the bond-order $P_{s_C s_H}$ was demon-

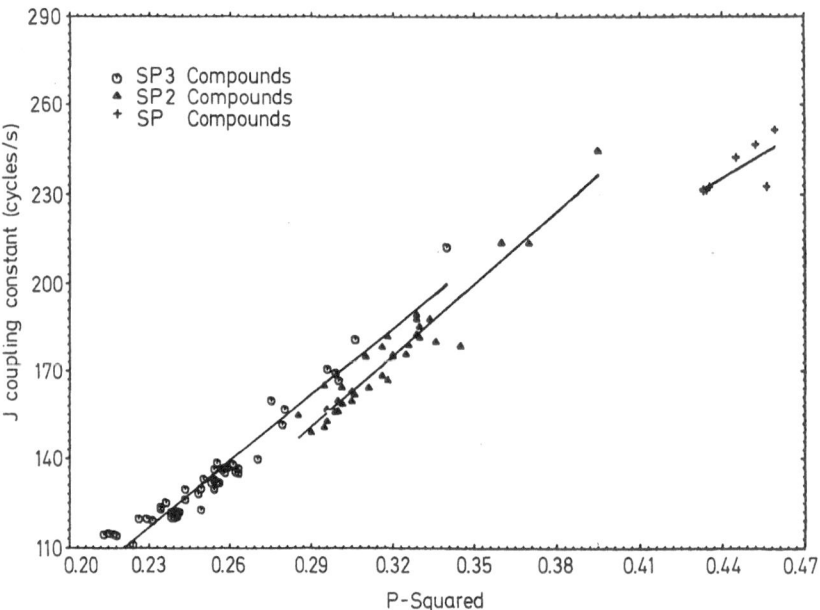

Fig. 2. Plot of calculated values of J_{CH} vs valculated value of $P^2 S_C S_H$. (Reprinted with permission from Maciel, McIver, Ostlund, and Pople (1970). J. Am. Chem. Soc. 92: 1. Copyright by the American Chemical Society.)

strated [98], cf. Fig. 2, and the Karplus relation for $^3J_{HH}$ was confirmed [101], cf. Fig. 3. Among other classical papers in the field one should mention the work of Wasylishen and Schaefer on nitrogen compounds [103, 104] and on long-range proton couplings in toluene [105]. The long-range proton-proton couplings were studied

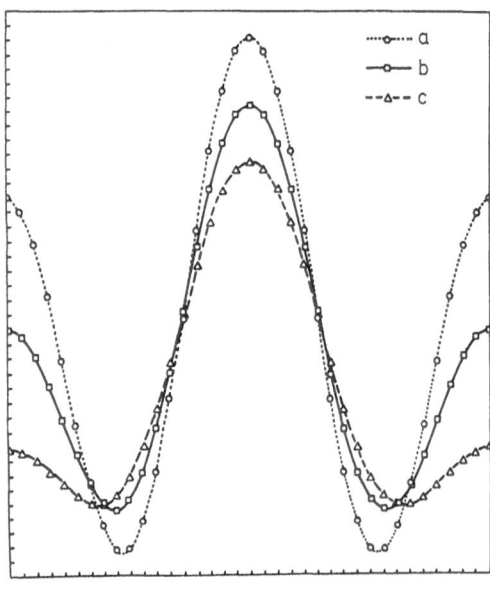

Fig. 3. (a) Plot of $^3J_{HH}$ vs the HCCH dihedral angle for one proton pair in ethane. (b) Plot of $^3J_{H*h*}$ versus the $H*CCH*$ dihedral angle for one proton pair in propene, $H_2CH*—H*C=CH_2$. (c) Plot of $^3J_{H*H*}$ versus the $H*CCH*$ dihedral angle in acetaldehyde, $H_2CH*—H*Co$. (Reprinted with permission from Maciel, McIver, Ostlund, and Pople (1970) J. Am. Chem. Soc. 92: 4497. Copyright by the American Chemical Society.)

dilligently also by other authors, among which one should mention Barfield and coworkers [106–108], who also worked on three-bond carbon-proton and carbon-carbon couplings [109].

As a method of performing coupled Hartree-Fock calculations, the FP scheme is not particularly efficient. Another, equivalent but more efficient, computational scheme denoted self-consistent perturbation theory (SCPT) has been proposed by Blizzard and Santry [110, 111] for the Fermi contact and other mechanisms. As in the FP method, Eq. (5.13) is also the basic expression of the SCPT procedure. The derivative of the spin-density matrix is, however, evaluated from the first-order correction to the molecular orbitals. Among important applications of the SCPT method in connection with the semi-empirical INDO scheme, we would like to mention the calculations of the vibrational corrections to the coupling constants in ammonia by Solomon and Schulman [112].

Still another computational scheme, already mentioned in Sect. 3, was proposed by Nakatsuji [37]. The method has an advantage of being non-iterative and was implemented in the context of ab initio calculations of spin-spin couplings by Galasso [113]. Numerous nonempirical coupled Hartree-Fock calculations employing the above-mentioned approaches have been reported. Among the early work, one should mention the paper by Ostlund et al. [114] using the FP method and the work of Snyder and Ditchfield [115]. The early results were not very encouraging, mainly because of deficiencies of the small basis sets used and further progress occurred first in late 1970s, when Guest and coworkers [116] reported large basis set SCPT calculations of the one-bond proton-first-row nucleus and geminal proton-proton coupling constants in the first-row hydrides. They showed that the calculations were basis-set sensitive and proved that the assumption of negligible non-contact contributions to the proton-proton couplings, as proposed by Pople and Santry [71], was wrong (see below). Soon afterwards, Kowalewski and coworkers [117] reported large basis set FP calculations of the Fermi contact contribution to the coupling constants in the hydrogen molecule and the first-row hydrides. Kowalewski, Laaksonen and coworkers reported subsequently similar calculations for the second-row hydrides [118], for ethane [119], and other substituted methanes [120]. More recent work included usually all three second-order contributions (and often the diamagnetic orbital term, see below). Lazzeretti and coworkers reported large basis-set CHF calculations of the coupling tensors in the first-row hydrides [121], phosphine and the corresponding anion and cation [122], and some other charged systems [123]. Raynes, Lazzeretti, and coworkers [124, 125] studied coupling constants in methane as a function of molecular geometry and estimated the vibrational corrections. Schulman and Lee [126] studied the one-bond nitrogen-nitrogen coupling constant in hydrazine. Galasso reported calculations for the first- and second-row hydrides [113] and for systems with two first- or second-row atoms [127–131].

5.3 Beyond the Coupled Hartree-Fock Approach

There are two main reasons to extend the calculations beyond the coupled Hartree-Fock method. The first and main reason is to include the dynamic correlation between the electrons, i.e. to account for instantaneous rather than averaged repulsion of the electrons. The effect of correlation is to keep electrons apart. The spatial correlation of the electrons may be thought of as consisting of several parts viz. the left-right correlation (keeping the electrons in a bond close to different nuclei), the in-out or radial correlation (when one electron is close to the bond axis or an atom, the other one will tend to be far away), and the angular correlation (the electrons tend to stay on opposite sides of the bond or the atom). Consider now a simple so-called Dirac vector model for coupling through two bonds, such as the proton coupling in water, as given in the book by Harris [132]. According to this model, the coupling will be efficient if the two electrons are close to different nuclei and if the bonding electrons tend to have parallel spins at the intervening atom. Both these phenomena can clearly be affected by correlation.

It is straightforward in principle to extend the FP version of the CHF method to include the correlation effects, by performing configuration interaction (CI) calculations based on the UHF molecular orbitals. This approach has been taken by Kowalewski, Laaksonen, and coworkers [117–120]. They found that the correlation effects were very important for the quantitative interpretation of the couplings. In particular, the couplings through two bonds are severely overestimated by the CHF calculations (see below). It can also be of interest to note that for the hydrogen molecule the FP CI [117] and the earlier SOS CI [89] calculations gave identical results — for a two-electron system both types of calculations involve full configuration interaction and become equivalent.

The other reason for making calculations extending beyond the limits of the CHF method is that, as pointed out by Guest and coworkers [116], the coupled Hartree-Fock method is unsuitable for systems that show instabilities with respect to spin-unrestricted (UHF) solutions, e.g. some multiply bonded systems. A related methodology that could handle this type of systems was proposed by Laaksonen et al. [133, 134]. They demonstrated that the problem of the UHF instabilities could be circumvented by employing multiconfigurational (MC) SCF procedure with a small number of configurations and reported numerical calculations for ethene, ethyne, hydrogen cyanide, and some other multiply bonded molecules.

Also other methods for including the correlation effects in calculations of nuclear spin-spin coupling constants have been proposed. One of them, described by Iwai and Saika [135, 136] is the finite perturbation — many body perturbation theory (MBPT) where the effects of correlation are evaluated by diagrammatic perturbation techniques. Another group of methods contains the techniques using the second quantization formalism proposed, in the context of coupling constants, by Oddershede and coworkers [137, 138] and the equation of motion (EOM) methods, suggested by Galasso [139]. Both these types of approach were originally developed for interpretation of electron spectra and are thus designed to give a balanced description of the ground state and the excited states. It seems that at least some of the problems with the uncoupled sum-over-states methods is related to the lack of balance in the description of $\Psi^{(0)}$ and $\Psi^{(1)}$. Thus, the second quantization and EOM methods appear promising. A wide range of names and acronyms is in use, some of which seem to be equivalent to the coupled Hartree-Fock (time-dependent Hartree-Fock, TDHF, random-phase approximation, RPA, first-order polarization propagator approximation, FOPPA), while others (higher random-phase approximation, HRPA, second-order polarization propagator approximation, SOPPA) include at least some correlation effects and are able to handle UHF unstable systems. The SOPPA method was compared with HRPA by Geertsen and Oddershede [140] and with CHF by Scuseria [141], while the HRPA was compared with CHF by Galasso and Fronzoni [142, 143]. Besides this method-oriented work, Fronzoni and Galasso [144] applied also the HRPA method to calculate one-bond coupling constants involving silicon. The general conclusions from the SOPPA and HRPA work confirms the importance of the correlation effects for the Fermi contact term. For the other mechanisms, the correlation effects seem to be of lesser importance.

5.4 The Diamagnetic Orbital Term

In the classical Pople-Santry paper [71] it was stated that the orbital diamagnetic term should be negligible for the coupling involving protons and this remained a generally accepted idea until Lee and Schulman [145] and Matsuoka and Aoyama [146] disproved it by their calculations in 1980. During the last decade, calculations of J^{OD} have become increasingly popular. Galasso and coworkers [147] investigated the basis-set dependence of J^{OD} in water and found that fairly small basis set could adequately describe the property. A similar conclusion for the same molecule was also reached by Overill and Guest [148] who in addition found that the role of correlation was minor and reported a large number of calculations for the first- and second-row hydrides. Overill and Saunders [149] investigated in a similar way a series of molecules with multiple bonds. In many of the cases, J^{OD} was found to have the opposite sign to J^{OP}, and to be of similar magnitude. As mentioned above, the J^{OD} term was also included in most of the other nonempirical calculations of coupling constants reported after 1982 and the conclusions on non-negligibility and basis set-intensivity of J^{OD} are now generally agreed on.

5.5 The State of the Art Numerical Results

Some of the best available numerical results for the one-bond and two-bond coupling constants in the first-row hydrides are gathered in Tables 5 and 6; similar results for various coupling constants in ethyne are collected in Table 7.

The general conclusion to be drawn from Table 5 is that the theory is indeed very successful in reproducing the experimental nuclear spin-spin coupling constants between directly bonded nuclei in simple molecules. It must be kept in mind, however, that the excellent agreement with experiments is first obtained after considerable computational effort and taking into consideration the complicated combinations of several partly cancelling terms.

The data summarized in Table 6 are taken from essentially the same set of papers as the results of Table 5. We can note that the trend in the geminal couplings can be successfully reproduced but that the calculated two-bond coupling constants deviate more from their experimental counterparts than the couplings between directly bonded atoms. The two-bond couplings are clearly a more difficult case, where the basis set effects combined with the cancellations of fairly large contributions make life difficult for theorists.

Table 5. State of the art calculations of the reduced one-bond coupling constants $^1K_{HX}$ in the first-row hydrides. All values are in 10^{19} m^{-2} kg s^{-2} A^{-2}

Term	Method	HF	H2O	NH$_3$	CH$_4$
FC	CHF[a]	42.3	48.8	57.5	48.2
	CHF[b]	—	52.0	57.7	49.3
	SOPPA, EOM[c]	—	40.0	—	37.7
	FP CI[d]	30.7	40.2	43.9	37.7
SD	CHF[a]	−0.9	−0.4	−0.2	−0.1
	CHF[b]	—	−0.2	−0.1	−0.1
	SOPPA, EOM[c]	—	0.0	—	0.0
OP	CHF[a]	17.4	7.5	2.4	0.5
	CHF[b]	—	7.5	2.4	0.4
	SOPPA, EOM[c]	—	6.8	—	0.4
OD	SCF[e]	0.0	0.1	0.1	0.1
	SCF[f]	—	0.0	—	—
	CI[e]	—	0.1	—	—
Vibr. corr.[g]		—	−0.1	4.4	2.4
Total calc.[h]		47.2	47.3	50.6	40.6
Experimental		46.9	48	50	41.3

[a] Guest et al. [116], basis set $\langle 5,3,1/3,1 \rangle$ for H$_2$O, NH$_3$, CH$_4$; $\langle 8,4,2/4,2 \rangle$ for HF;
[b] Lazzeretti and Zanasi [121], basis set $\langle 6,5,1/3,1 \rangle$;
[c] SOPPA for H$_2$O by Geertsen and Oddershede [138], basis set as in b); EOM for CH$_4$ by Galasso [139], basis set $\langle 4,3,1/3,1 \rangle$;
[d] Kowalewski et al. [117], basis set as in a);
[e] Overill and Guest [148], basis set $\langle 5,3,1/3,1 \rangle$;
[f] Galasso et al. [147], basis set $\langle 8,6,2/6,2 \rangle$;
[g] Geertsen and Oddershede [138] for H$_2$O; Solomon and Schulman [112] for NH$_3$; Lazzeretti et al. [125] for CH$_4$;
[h] FP CI for FC [117] + CHF for SD & OP [116] + SCF for OD + vibr. corr

Table 6. State of the art calculations of the two-bond coupling constants $^2J_{HH}$ in the first row hydrides. All values are in Hertz

Term	Method	H$_2$O	NH$_3$	CH$_4$
FC	CHF[a]	−23.8	−24.3	−25.4
	CHF[b]	−25.0	−25.1	−26.3
	SOPPA, EOM[c]	−13.5	−12.2	−15.5
	FP CI[d]	−12.1	−12.2	−13.6
SD	CHF[a]	1.4	0.8	0.4
	CHF[b]	1.5	0.9	0.6
	SOPPA, EOM[c]	1.0	—	0.3
OP	CHF[a]	7.1	4.7	2.7
	CHF[b]	8.1	5.6	3.4
	SOPPA, EOM[c]	7.9	—	2.9
OD	SCF[e]	−7.2	−5.3	−3.5
	SCF[f]	−7.2	—	—
	CI[e]	−7.0	—	—

Table 6   continued

Vibr. corr.[g]	2.4	−0.7	−0.6
Total calc.[h]	−8.4	−12.7	−14.6
Experimental	−7.2	−10.4	−12.4

[a] Guest et al. [116], basis set $\langle 5,3,1/3,1 \rangle$ for H_2O, NH_3, CH_4; $\langle 8,4,2/4,2 \rangle$ for HF;

[b] Lazzeretti and Zanasi [121], basis set $\langle 6,5,1/3,1 \rangle$;

[c] SOPPA for H_2O by Geertsen and Oddershede [138], basis set as in b); EOM for CH_4 by Galasso [139], basis set $\langle 4,3,1/3,1 \rangle$;

[d] Kowalewski et al. [117], basis set as in a);

[e] Overill and Guest [148], basis set $\langle 5,3,1/3,1 \rangle$;

[f] Galasso et al. [147], basis set $\langle 8,6,2/6,2 \rangle$;

[g] Geertsen and Oddershede [138] for H_2O; Solomon and Schulman [112] for NH_3; Raynes et al. [124] for CH_4;

[h] FP CI for FC (117) + CHF for SD & OP [116] + SCF for OD + vibr. corr

Table 7. State of the art calculations of coupling constants in ethyne (acetylene). All values are in Hertz

Term	Method	$^1J_{OC}$	$^1J_{CH}$	$^2J_{CH}$	$^3J_{HH}$
FC	FP MC[a]	214.3	291.7	28.4	17.7
	FP CI[a]	173.8	235.0	41.1	6.0
	CHF[b]	355.8	449.3	−51.6	75./
	SOPPA[b]	175.7	246.5	42.4	9.8
	EOM[c]	164.5	229.5	35.3	8.9
SD	CHF[b]	29.3	3.7	−2.6	3.1
	SOPPA[b]	7.4	0.5	0.6	0.5
	EOM[c]	5.5	0.3	0.5	0.3
OP	CHF[a]	15.3	−3.6	6.6	3.1
	CHF[b]	14.8	−3.6	7.7	4.3
	SOPPA[b]	3.5	−0.8	4.7	3.4
	EOM[c]	3.2	−0.4	3.9	4.1
OD	SCF[d]	0.0	0.4	−1.3	−3.5
	SCF[b]	0.0	0.3	−1.4	−3.6
	SCF[c]	0.0	0.3	−1.3	−3.6
Total calc.					
	FP CI[e]	189.1	231.8	46.4	5.6
	SOPPA[b]	186.6	246.5	46.3	10.2
	EOM[c]	173.2	229.7	38.4	9.7
Experimental		171.5	248.7	49.2	9.6

[a] Laaksonen et al. [134], basis set $\langle 4,2/2 \rangle$;

[b] Geertsen and Oddershede [140], basis set $\langle 5,3,2/3,1 \rangle$;

[c] Galasso [139], basis set $\langle 4,3,1/3,1 \rangle$;

[d] Overill and Saunders [149], basis set as in a);

[e] FC from a) + OP from a) + OD from d); SD neglected

The subject of Table 7, ethyne (acetylene) is a more difficult case because of its multiple bond and triplet instabilities. Clearly, the CHF method is not able to handle the system properly. On the other hand, the SOPPA, EOM, and FP CI methods are quite successful. The three tables together certainly should give a fairly optimistic picture of the state of the art in calculations of nuclear spin-spin coupling constants, and, at the same time, should be a warning against oversimplifications, which certainly can lead to erroneous conclusions.

6 References

1. Bloch F, Hansen WW, Packard M (1946) Phys. Rev. 69: 127
2. Bloch F (1946) Phys. Rev. 70: 460
3. Bloch F, Hansen WW, Packard M (1946) Phys. Rev. 70: 474
4. Purcell EM, Torrey HC, Pound RV (1946) Phys. Rev. 69: 37
5. Proctor WG, Yu FC (1950) Phys. Rev. 77: 717
6. Dickinson WC (1950) Phys. Rev. 77: 736
7. Gutowsky HS, McCall DW (1951) Phys. Rev. 82: 748
8. Gutowsky HS, McCall DW, Slichter CP (1951) Phys. Rev. 84: 589
9. Hahn EL, Maxwell DE (1951) Phys. Rev. 84: 1246
10. Hahn EL, Maxwell DE (1952) Phys. Rev. 88: 1070
11. Abragam A (1961) Principles of nuclear magnetism, Oxford University Press, Oxford
12. Ernst RR, Bodenhausen G, Wokaun A (1987) Principles of nuclear magnetic resonance in one and two dimensions, Clarendon, Oxford
13. Atkins PW (1983) Molecular quantum mechanics, 2nd ed. Oxford University Press, Oxford
14. Slichter CP (1980) Principles of magnetic resonance, 2nd ed, Springer, Berlin Heidelberg New York
15. Ando I, Webb GA (1983) Theory of NMR parameters, Academic, London
16. Arfken G (1985) Mathematical methods for physicists, 3rd ed., Academic, Orlando
17. Feynman RP, Leighton RB, Sands M (1964) The Feynman lectures on physics, vol 2, Addison-Wesley, Reading
18. Ramsey NF (1950) Phys. Rev. 77: 567; (1950) 78: 699; (1952) 86: 243
19. Lamb WE (1941) Phys. Rev. 60: 817
20. Ditchfield R (1974) In: Lewy GC (ed) Topics in carbon-13 NMR spectroscopy, Ch. 1, J. Wiley, New York
21. Jameson CJ (1987) In: Mason J (ed) Multinuclear NMR, Plenum New York
22. Jameson CJ (1987) Nuclecar magnetic resonance, The Royal Society of Chemistry, 16: 1
23. Snyder LC, Parr RG (1961) J. Chem. Phys., 34: 827
24. Hameka HF (1964) J. Chem. Phys. 40: 3127
25. Saika A, Slichter CP (1954) J. Chem. Phys. 22: 26
26. Karplus M, Das TP (1961) J. Chem. Phys., 34: 1683
27. Pople JA (1957) Proc. Royal Soc. 239: 541
28. London F, (1937) J. Phys. Radium 8: 397; (1937) J. Chem. Phys., 5: 837
29. Karplus M, Pople JA (1963) J. Chem. Phys. 38: 2803
30. Farnum DG (1975) Adv. Phys. Org. Chem. 11: 123
31. Young RN (1979) Progr. NMR Spectrosc. 12: 261
32. Olah GA, Matescu GD (1970) J. Am. Chem. Soc. 92: 1430
33. Stevens RM, Pitzer RM, Lipscomb WN (1963) J. Chem. Phys. 38: 550
34. Stevens RM, Lipscomb WN (1964) J. Chem. Phys. 40: 2238; (1964); 41: 3710; (1965) 42: 3660; (1965) 42: 4302

35. Pople JA, McIver JW, Ostlund NS (1968) J. Chem. Phys. 49: 2960
36. Ditchfield RM, Miller DP, Pople JA (1970) J. Chem. Phys. 53: 613; (1971) 54: 4186
37. Nakatsuji H (1974) J. Chem. Phys. 61: 3728
38. Fukui H, Yoshida H, Miura K (1981) J. Chem. Phys. 74: 6988
39. Ditchfield RM (1972) J. Chem. Phys. 56: 5688
40. Pople JA (1962) Discuss., Faraday Soc. 34: 7; (1963) J. Chem. Phys. 37: 53; (1963) J. Chem. Phys. 37: 60; (1964) Mol. Phys. 7: 301
41. Kutzelnigg W (1980) Isr. J. Chem. 19: 193; Schindler M, Kutzelnigg W (1982) J. Chem. Phys. 76: 1919; (1983) Mol. Phys. 48: 781
42. Levy B, Ridard J (1981) Mol. Phys. 44: 1099
43. Rowe DJ (1968) Rev. Mod. Phys. 40: 153
44. Lamanna UT, Guidotti C, Arrighini GP (1977) J. Chem. Phys. 67: 604
45. Lazzeretti, Zanasi (1983) Phys. Rev. A27: 1301; (1985) A32: 2607
46. Hansen AE, Bouman TD (1985) J. Chem. Phys. 82: 5035
47. Lazzeretti P, Zanasi R, Sadlej AJ, Raynes WT (1987) Mol. Phys. 62: 605
48. Daborn GT, Ferguson WI, Handy NC (1980) Chem. Phys., 50: 255; Daborn GT, Handy NC (1981) Chem. Phys. Lett., 81: 201
49. Iwai M, Saika A (1982) J. Chem. Phys. 77: 1951
50. McMichael Rohling C, Allen LC, Ditchfield R (1984) Chem. Phys. 87: 9
51. Pople JA, Beveridge DL (1970) Approximate molecular orbital theory, McGraw-Hill, New York
52. Bieger W, Seifert G, Esching H, Grossmann G (1985) Chem. Phys. Lett. 115: 275
53. Frier DG, Fenske RF, You XZ (1985) J. Chem. Phys. 83: 3526
54. Kolb D, Johnson WR, Shorer P (1982) Phys. Rev. A26: 19
55. Pyykkö P (1983) Chem. Phys. 74: 1
56. Pyper NC, (1983) Chem. Phys. Lett. 96: 204; (1983) Chem. Phys. Lett. 96: 211
57. Pyykkö P, Görling A, Rösch N (1987) Mol. Phys. 61: 195
58. Harris RK (1986) Nuclear magnetic resonance spectroscopy, ch 4, Longmans, London
59. Kowalewski J (1977) Progr. NMR Spectr. 11: 1
60. Kowalewski J (1982) Ann. Rep. NMR Spectr. 12: 81
61. Weissbluth M (1978) Atoms and molecules, Academic, New York
62. Blinder SM (1965) Adv. Quant. Chem., vol 2, Academic In: Löwdin P-O (ed), New York
63. Fermi E (1930) Z. Phys., 60: 320
64. Breit G (1930) Phys. Rev. 35: 1447
65. Pyykkö P, Pajanne E, Inokuti Œ (1973) Int. J. Quant. Chem. 7: 785
66. Ramsey NF (1953) Phys. Rev. 91: 303
67. Pyykkö P (1977) Chem. Phys. 22: 289
68. Laaksonen A (1986) Nuclear magnetic resonance, The Royal Society of Chemistry, 15: 81
69. Overill RE (1987) Nuclear magnetic resonance, The Royal Society of Chemistry, 16: 84
70. Jameson CJ (1987) In: Mason J (ed) Multinuclear NMR, ch 4 Plenum New York
71. Pople JA, Santry DP (1964) Mol. Phys. 8: 1
72. Pople JA, Bothner-By AA (1965) J. Chem. Phys. 42: 1339
73. McConnell HM (1956) J. Chem. Phys. 24: 460
74. Muller N, Pritchard DE (1959) J. Chem. Phys., 31: 768
75. Muller N, Pritchard DE (1959) J. Chem. Phys. 31: 1471
76. Maksic ZB (1971) Int. J. Quant. Chem. 5: 301
77. Maksic ZB, Eckert-Maksic M, Randic M (1971) Theor. Chim. Acta 22: 70
78. Frei K, Bernstein HJ (1963) J. Chem. Phys. 38: 1216
79. Newton MD, Schulman JM (1972) J. Am. Chem. Soc. 94: 767 (1972) 94: 773
80. Schulman JM, Newton MD (1974) J. Am. Chem. Soc., 96: 6295 (1974)
81. Karplus M, Anderson DH (1959) J. Chem. Phys. 30: 6
82. Karplus M (1959) J. Chem. Phys. 30: 11
83. Karplus M (1963) J. Amer. Chem. Soc. 85: 2870
84. Haasnoot CAG, de Leeuw FAAM, Altona C (1980) Tetrahedron, 36: 2783
85. Loewe P, Salem L (1965) J. Chem. Phys. 43: 3402
86. Armour EAG, Stone AJ (1967) Proc. Roy. Soc., A302: 25
87. Ditchfield R, Murrell JN (1968) Mol. Phys. 14: 481
88. Ditchfield R (1969) Mol. Phys. 17: 33

89. Kowalewski J, Roos B, Siegbahn P, Vestin R (1974) Chem. Phys. 3: 70
90. Kowalewski J, Roos B, Siegbahn P, Vestin R (1975) Chem. Phys. 9: 29
91. Roos B (1972) Chem. Phys. Lett. 15: 153
•92. Roos B, Siegbahn P (1977) In: Schaefer HF (ed) Methods of electronic structure theory, Ch. 7, Plenum, New York
93. Pyykkö P, Wiesenfeld L (1981) Mol. Phys. 43: 557
94. Slater JC, Kirkwood JG (1931) Phys. Rev. 37: 682
95. Pople JA, McIver JW, Ostlund NS (1967) Chem. Phys. Lett. 1: 465
96. Pople JA, McIver JW, Ostlund NS (1968) J. Chem. Phys., 49: 2965
97. Nakatsuji H, Kato H, Morishima I, Yonezawa T (1970) Chem. Phys. Lett. 4: 607
98. Maciel GE, McIver JW, Ostlund NS, Pople JA (1970) J. Amer. Chem. Soc. 92: 1
99. Maciel GE, McIver JW, Ostlund NS, Pople JA (1970) J. Amer. Chem. Soc. 92: 11
100. Maciel GE, McIver JW, Ostlund NS, Pople JA (1970) J. Amer. Chem. Soc. 92: 4151
101. Maciel GE, McIver JW, Ostlund NS, Pople JA (1970) J. Amer. Chem. Soc. 92: 4497
102. Maciel GE, McIver JW, Ostlund NS, Pople JA (1970) J. Amer. Chem. Soc. 92: 4506
103. Wasylishen RW, Schaefer T (1972) Can. J. Chem. 50: 2989
104. Wasylishen RE, Schaefer T (1973) Can. J. Chem. 51: 3087
105. Wasylishen RE, Schaefer T (1972) Can. J. Chem., 50: 1852
106. Barfield M (1971) J. Amer. Chem. Soc. 93: 1066
107. Barfield M, Sternhell S (1972) J. Amer. Chem. Soc. 94: 1905
108. Barfield M, Dan AM, Fallick CJ, Spear RJ, Sternhell S, Westman PW (1975) J. Amer. Chem. Soc., 97: 1482
109. Barfield M (1980) J. Am. Chem. Soc. 102: 1
110. Blizzard AC, Santry DP (1970) J. Chem. Soc. Chem. Commun. 87
111. Blizzard AC Santry DP (1971) J. Chem. Phys. 55: 950 (1971). Erratum: (1973) J. Chem. Phys. 58: 4714
112. Solomon P, Schulman JM (1977) J. Am. Chem. Soc. 99: 7776
113. Galasso V (1983) Theor. Chim. Acta 63: 35
114. Ostlund NS, Newton MD, McIver JW, Pople JA (1969) J. Magn. Reson. 1: 298
115. Ditchfield R, Snyder LC (1972) J. Chem. Phys. 56: 5823
116. Guest MF, Saunders VR, Overill RE (1978) Mol. Phys. 35: 427
117. Kowalewski J, Laaksonen A, Roos B, Siegbahn P (1979) J. Chem. Phys. 71: 2896
118. Kowalewski J, Laaksonen A, Saunders VR (1981) J. Chem. Phys. 74: 2412
119. Laaksonen A, Kowalewski J, Siegbahn P (1980) Chem. Phys. Lett. 69: 109
120. Laaksonen A, Kowalewski J (1981) J. Amer. Chem. Soc. 103: 5277
121. Lazzeretti P, Zanasi R (1982) J. Chem. Phys. 77: 2448 (1982)
122. Lazzerretti P, Rossi E, Taddei F, Zanasi R (1982) J. Chem. Phys. 77: 408
123. Lazzeretti P, Rossi E, Taddei F, Zanasi R (1982) J. Chem. Phys. 77: 2023
124. Raynes WT, Lazzeretti P, Zanasi R, Fowler PW (1985) J. Chem. Soc. Chem. Commun. 1538
125. Lazzeretti P, Zanasi R, Raynes WT (1986) J. Chem. Soc. Chem. Commun. 57
126. Schulman JM, Lee WS (1982) J. Magn. Res. 50: 142
127. Galasso V (1983) J. Mol. Struct. (Theochem) 93: 201
128. Galasso V (1984) J. Chem. Phys. 80: 365
129. Galasso V (1984) Chem. Phys. 83: 407
130. Galasso V (1984) Chem. Phys. 83: 407
131. Galasso V (1984) Chem. Phys. Lett., 108, 435
132. Harris RK ref 58, p 212
133. Laaksonen A, Saunders VR (1983) Chem. Phys. Lett., 95: 375
134. Laaksonen A, Kowalewski J, Saunders VR (1983) Chem. Phys., 80, 22
135. Iwai M, Saika A (1983) Phys. Rev. A 28: 1924
136. Saika A (1985) Bull. Magn. Reson. 7: 100
137. Oddershede J, Jorgensen P, Beebe NHF (1977) Chem. Phys., 25: 451
138. Geertsen J, Oddershede J (1984) Chem. Phys. 90: 301
139. Galasso V (1985) J. Chem. Phys. 82: 899
140. Geertsen J, Oddershede J (1986) Chem. Phys., 104: 67
141. Scuseria GE (1986) Chem. Phys. Lett. 127: 236
142. Fronzoni G, Galasso V (1989) J. Mol. Struct. (Theochem) 122: 327
143. Galasso V, Fronzoni G (1986) J. Chem. Phys. 84: 3215

144. Fronzoni G, Galasso V (1986) Chem. Phys. 103: 29
145. Lee WS, Schulman JM (1979) J. Chem. Phys. 70: 1530
146. Matsuoka O, Aoyama T (1980) J. Chem. Phys. 71: 5718
147. Galasso V, Lazzeretti P, Rossi E, Zanasi R (1983) J. Chem. Phys. 79: 1554
148. Overill RE, Guest MF (1983) Chem. Phys. Lett. 98: 229
149. Overill RE, Sannders VR (1984) Chem. Phys. Lett. 106: 197.

Theoretical Approaches to ESR Spectroscopy

David Feller [+] **and Ernest R. Davidson** [*]

[*] Department of Chemistry, Indiana University, Bloomington, Indiana 47405
[+] The Molecular Science Research Center, Richland, Washington 99352

1 Introduction

Atoms and molecules which contain unpaired electrons possess magnetic moments due to the electron's intrinsic spin and, in cases where the total spin S $\neq$ 0, to the electron's orbital angular momentum. Experimental techniques which measure the interaction of the intrinsic spin with an external magnetic field fall into the category of "electron spin resonance" (ESR) spectroscopy. Techniques which deal with orbital angular momemtum are often differentiated through the use of the label "electron paramagnetic resonance" (EPR) spectroscopy. In this article we shall use the term ESR to refer to both.

Many molecules exist as closed shell singlets in their ground states and are, therefore, inaccessible to ESR techniques. Nonetheless, a significant number of molecules have nonsinglet ground states, such as the class of highly reactive species known as free radicals. Recent advances in experimental techniques now permit radical cations and anions to be studied in condensed phases. In addition to these, ESR is able to observe most ions (in gas, liquid and solid phases), paramagnetic point point defects in crystals and many transition metal and rare-earth ions. Concentrations as low 0.01 micromolar are detectable. The sphere of applicability is sufficiently large to make ESR a major investigative tool in the arsenal of modern chemists.

The principal information provided by the ESR experiment concerns the distribution of the unpaired spin density within a molecule. The unpaired spin density, ϱ_s, is conventionally defined as the difference between the α and β densities, normalized to unity. If N_α and N_β are the number of α and β spin electrons, then

$$\varrho_s = (\varrho_\alpha - \varrho_\beta)/(N_\alpha - N_\beta) . \tag{1}$$

Because the magnetic environment of the unpaired electron is sensitive to the perturbing effects of neighboring magnetic nuclei, an ESR spectrum can be used as a fingerprint to signal the presence of a particular molecular species. Much as the related technique of nuclear magnetic resonance (NMR) has been widely used as a probe of molecular structure, ESR can also be used to determine the spin multiplicity and number of equivalent atoms in a molecule.

For a single electron in the presence of an external magnetic field there are two energy levels whose splitting is proportional to the strength of the field. This phenomenon, which is responsible for the so-called anomalous Zeeman splitting of spectral lines, was first observed in the Stern-Gerlach experiment. A transition between the two energy levels can be induced if an oscillating magnetic field of the correct frequency is applied at right angles to the original field. The resonance condition for a particular magnetic field strength H_0 is given by

$$\Delta E = h\nu_0 = g_e\beta_e H_0 \tag{2}$$

where h is Planck's constant (6.63×10^{-27} erg sec), g_e is the free electron g-value (2.0023) and β_e is the Bohr magneton (9.27×10^{-21} erg/G). Most ESR experiments utilize a fixed frequency, $\nu_0 \cong 9.3$ GHz, for the oscillating magnetic fiel in the so-called X-band portion of the microwave spectrum. ΔE is then $\varLambda 4$ J mol^{-1}. Resonance

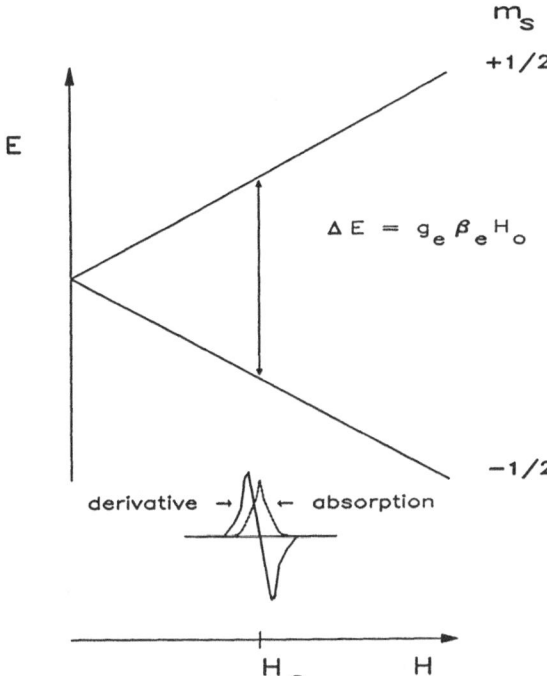

$$m_s$$

$$+1/2$$

E

$$\Delta E = g_e \beta_e H_o$$

$$-1/2$$

derivative → ← absorption

$$H_o \qquad H$$

Fig. 1. The energy splitting for a free electron in a magnetic field, H, and the absorption plot resulting from a transition induced by microwave radiation of the proper frequency

will then occur at a magnetic field strength of approximately 3300 G. Since the $m_s = -^1/_2$ state is lower in energy (and thus more heavily populated), inducing transitions to the $m_s = +^1/_2$ state will require that energy be absorbed from the applied oscillating magnetic field. A derivative plot of the absorption curve is shown in Fig. 1 beneath a diagram of the $2S + 1$ equispaced energy levels. Use of derivative plots is a common technique designed to improve the poor signal-to-noise ratio which results from the very small net absorption. The excess number of electrons in the lower state at 300 K is only 1 in 700.

Except for the lineshape there is little information contained in the sort of simple spectrum seen in Fig. 1. Fortunately, most real spectra exhibit more complexity (and more useful information). The majority of elements possess at least one naturally occuring isotope with a nonzero magnetic moment, I. These occur if the atomic number is odd and the atomic mass is even (I is integral, eg. ^2H, ^6Li), or when the atomic number is even and the atomic mass is odd (I is half integral, e.g. ^1H, ^{13}C, ^{17}O). The magnetic field from such a nucleus will couple with the electron's magnetic moment so as to split the original spectral lines into $2I + 1$ nearly equally spaced lines, as shown in Fig. 2 for an $I = ^1/_2$ nucleus. Because of the opposite signs of the nuclear and electronic charges the lowest energy conformation, in a strong magnetic field, corresponds to a situation where the spins are opposed to each other. While the magnetic environment of the unpaired electron is dominated by the external field (~ 3000 G), the field at the nucleus ($\sim 10^6$ G) is still dominated by the electronic contribution. Transitions occur with $\Delta m_i = 0$, $\Delta m_s = \pm 1$.

A detailed derivation of the many terms present in the relativistic, Born-Oppenheimer Hamiltonian needed to treat every aspect of the ESR phenomenon is beyond

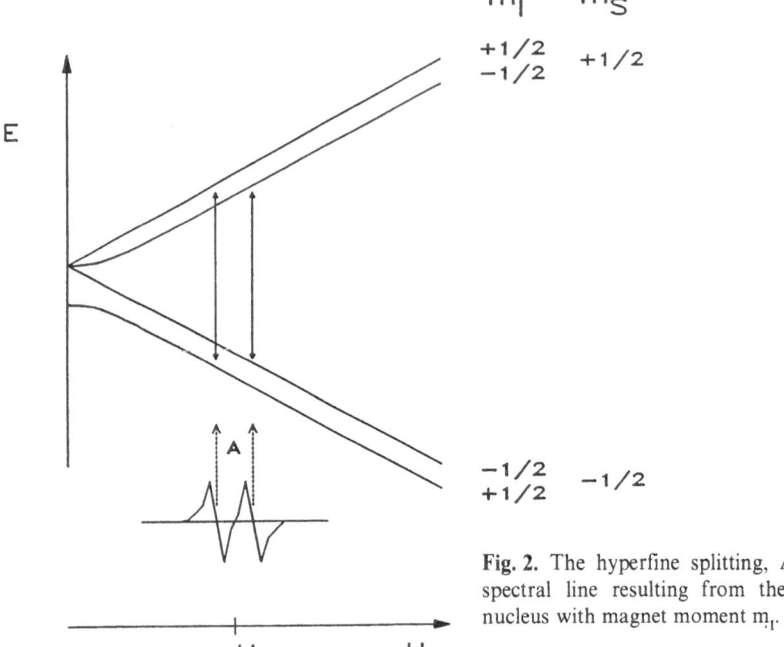

Fig. 2. The hyperfine splitting, A, of the single spectral line resulting from the presence of a nucleus with magnet moment m_{I_i}.

the scope of the present work. For such a discussion the reader is referred to the monographs by Bethe and Salpeter [1] and Harriman [2] who develop the theoretical framework beginning with the relativistic Dirac equation's description of the interaction of a free electron with a uniform magnetic field. Other authors [3] have taken the point of view that the hyperfine interaction is not a relativistic effect, but can be understood in classical terms by assuming an intrinsic magnetic moment for the electron. A recent examination of these issues by Kutzelnigg [4] concludes that the hyperfine interaction is not intrinsically a relativistic effect. He points out that the δ function term in the Hamiltonian, to be described in the following paragraph, arises as an artifact of trying to use first order perturbation theory in terms of two component spinors instead of the four component Dirac spinors to describe the interaction.

Of the several terms in the Hamiltonian arising from the nuclear magnetic moment, the term of primary interest is the so-called macroscopic Hamiltonian:

$$H_{macro} = \sum_{v} I_{v} \cdot A \cdot S, \tag{3}$$

where I_{v} is the nuclear magnetic moment for nucleus v and S is the total electronic spin and A is the hyperfine splitting tensor. For a manifold of m_{S}, m_{I} states the expectation value of the macroscopic Hamiltonian can be shown identical to the expectation value of the following microscopic Hamiltonian:

$$H_{micro} = (8\pi/3)(g_{e}/g_{0}) \sum_{k} \sum_{v} g_{v}\beta_{v}S_{k} \cdot I_{v}\delta(r_{kv})$$

$$- (g_{e}/g_{0}) \sum_{k} \sum_{v} g_{v}\beta_{v}[S_{kv} \cdot I_{v}/r_{kv}^{3} - 3(S_{k} \cdot r_{kv})(I_{v} \cdot r_{kv})/r_{kv}^{5}] \tag{4}$$

where the summation indices run over electrons (k) and nuclei (v) and variable r in the above equation is the distance separating the j'th electron and v'th nuclei. The parameters g_e and g_0 are the so-called g values of the electron in the radical and the free electron (2.0023). Finally, g_v is the nuclear g factor and β_v is the nuclear magneton. The connection between the macroscopic and microscopic Hamiltonians comes from the definition of the various components in the macroscopic A:

$$A_{ij} = \delta_{ij} A_{iso}(v) + A_{aniso}(i, j, v) \qquad i, j = x, y, z \tag{5}$$

where

$$A_{iso}(v) = (8\pi/3) (g_e g_0) g_v \beta_v \rangle \delta(r_v) \rangle_{spin} \tag{6}$$

and two examples of diagonal and off-diagonal elements are as follows:

$$A_{aniso}(x, y, v) = -(g_e/g_0) g_v \beta_v \langle 3xy/r^5 \rangle_{spin} \tag{7}$$

$$A_{aniso}(z, z, v) = (g_e g_0) g_v \beta_v \langle (r^2 - 3z^2)/r^5 \rangle_{spin} \tag{8}$$

The subscript "spin" implies averaging over the unpaired spin density (normalized to unity). The anisotropic hyperfine tensor is a traceless quantity, i.e. $\sum A_{aniso}(i, i) = 0$.

As indicated by its name, the first term (also known as the Fermi [5] contact term) is isotropic in space and possesses no classical counterpart. Contributions to the hyperfine properties arising from this term are proportional to the amount of unpaired spin density at a particular nucleus, $\langle \delta(0) \rangle_{spin}$. Thus, A_{iso} provides an indication of the unpaired spin density distribution throughout a molecule. The second term is anisotropic in space and is denoted here as A_{aniso}. For a diatomic molecule oriented along the z-axis A_{aniso} is proportional to $\langle (3z^2 - r^2)/r^5 \rangle_{spin}$. The anisotropic part reflects the asymmetry of the spin density about each nuclei and can be viewed as arising from the interaction of the two magnetic dipoles. It is often referred to as the "dipolar" contribution because of this relationship to classical physics.

Besides the hyperfine splitting constant, ESR measurements yield information about the "g tensor", a 3-dimensional generalization of g_e for molecular systems. Even in the atomic case the g factor will not always equal 2.0023 because of the effects of electronic orbital angular momentum. The Landé formula gives the g factor appropriate for a given atomic energy level. In a molecule with zero orbital angular momentum the g factor may still deviate from g_e due to the effects of excited states (with nonzero angular momentum) interacting with the ground state wave function via the spin orbit operator. The smaller the excitation energy, the larger the effect.

Lastly, approximate information about the %s and %p character of the molecular wave function can be derived from ESR data. The expectation value of an operator θ is given by Tr(θP) where, in the case of spin properties, P is the spin density. P is rigorously defined as the difference between the α and β spin densities. However, if we are operating within the restricted Hartree-Fock (RHF) approximation, the closed shell α and β contributions to P cancel each other and we are left with the entire unpaired spin density arising from one singly occupied orbital (for a doublet). With the additional assumptions of a minimal AO basis and no contributions to the hyperfine properties on one atom arising from atomic orbitals centered on neighboring

atoms, the isotropic and anisotropic hyperfine properties of a diatomic molecule
are given by:

$$A_{iso} = \varrho_{s,s} \langle \phi_s | \delta(o) | \phi_s \rangle \tag{9}$$

$$A_{aniso} = \varrho_{pz,pz} \langle \phi_{pz} | (3z^2 - r^2)/r^5 | \phi_{pz} \rangle \tag{10}$$

where z is the internuclear axis and the unpaired electron resides in a σ orbital.

It is then argued that the two quantities $(\phi_s | \delta(o) | \phi_s \rangle$ and $\langle \phi_{pz} | (3z^2 - r^2)/r^5 | \phi_{pz} \rangle$
are characteristic of the atom on which the hyperfine properties were measured.
Thus, they are independent of the molecule under study and may be obtained from
any convenient tabulation of the interaction integrals involving one $s, p, d \dots$ etc.
electron with a free atom. Provided they are available and the experimental values
of A_{iso} and A_{aniso} are available, the corresponding ϱ values can be calculated. This
method of predicting spin densities from the hyperfine properties is often referred
to as the free atom comparision method (FACM). In practice the values of the integrals are often taken from the Hartree-Fock Slater work of Herman and Skillman
[6]. Some of the potential difficulties inherent in this approach will be discussed later.

2 Theoretical Descriptions

2.1 Unrestricted Hartree-Fock

As indicated above, the isotropic hyperfine constant for a nucleus N, $A_{iso}(N)$, is
proportional to the unpaired spin density at that nucleus, $\langle \delta(0) \rangle_{spin}$. For a doublet
state (one unpaired electron) the RHF approximation requires each pair of electrons
to occupy the same spatial orbital and ϱ_s is simply given by the square of the singly
occupied orbital, i.e. $|\phi_\alpha|^2$ for the high spin case. If some of the atoms in a molecule
happen to lie in a nodal plane of ϕ_α, RHF spin densities will predict isotropic hyperfine
splitting contants which are identically zero. An example of this occurs in the methyl
radical, CH_3 ($^2A_2''$), which possesses D_{3h} symmetry. The three hydrogens lie in the
same plane as the carbon atom, with HCH angles of 120°. Since the unpaired spin
resides in a p orbital on carbon perdendicular to the plane of the hydrogens, there is
no spin density on the hydrogens. While RHF wave functions yield a proton isotropic
hyperfine value of zero, experiment shows A_{iso} to be on the order of -69 MHz.

The presence of net unpaired spin density on the three methyl protons can be
achieved theoretically by relaxing the "restricted" Hartree-Fock constraint. In the
exact wave function the unpaired spin density shows a tendency to concentrate in
some portions of a molecule and not in others. The concentrated spin density will
affect or "polarize" spin throughout the rest of the molecule. The Pauli exclusion
principle functions to keep electrons with the same spin separated. So that, if we
consider one CH fragment of the methyl radical, the greater effective repulsion be-
tween $\pi\alpha$ and $\sigma\beta$ will cause the $\sigma\beta$ electron to shift slightly towards the proton and the
$\sigma\alpha$ electron to shift slightly towards carbon, balancing the charge. The overall result
will be that the proton will experience excess β spin. This "spin polarization" pheno-

memon can be viewed as a mechanism for the propagation of spin relaxation forces throughout the system. Obviously, any zeroth order wave function intended for use in isotropic hyperfine calculations must possess this minimal flexibility if it is to have any chance of success.

One of the easiest ways to introduce sufficient flexibility into the wave function to allow for these effects is to relax the restriction of identical spatial orbitals for α and β spins. This leads to the unrestricted Hartree-Fock or UHF wave function, given by:

$$\Phi_{UHF} = \mathscr{A}[\phi_1\phi_2\phi_3 \ldots \phi_A\phi'_1\phi'_2\phi'_3 \ldots \phi'\alpha_1\alpha_2\alpha_3 \ldots \alpha_A\beta_1\beta_2\beta_3 \ldots \beta] \qquad (11)$$

where A is the antisymmetrizer, the ϕ_i's are associated with the α spins and the ϕ'_i's are associated with the β spins. Φ is an eigenfunction of S_z, $m_s = {}^1/_2(A-B)$, but not of S^2. In general Φ will contain varying amounts of higher multiplet contaminants, with the dominant one corresponding to a state having 2 additional unpaired electrons. For example, the leading contaminant for a doublet state is a quartet. This defect can be remedied by a variety of methods for either annihilating the contaminants [7, 8] or projecting out a pure spin state. The choice of when to apply the projection, i.e. before or after optimization of the orbitals, leads to somewhat different results [9, 10].

Before discussing other ab initio methods and how they perform, it is useful to briefly consider several more approximate approaches which have been applied to the hyperfine problem. One of the simplest methods capable of producing a spin density for π radicals is due to Pariser and Parr and modified by Pople. The PPP method considers only π electrons. The $\sigma-\pi$ interaction is folded into an effective one-electron operator. Since no π orbitals are present, the PPP method cannot be used to directly explain the hyperfine splitting of the protons in CH_3, since those splitting are due to spin polarization of the σ electrons. However, a relationship due to McConnell [11, 12] of the form:

$$a_H = Q\varrho_\pi \qquad (12)$$

where a_H is the isotropic hyperfine coupling constant, ϱ_π is the Mulliken spin population in the $2p_\pi$ orbital adjacent to the proton and Q is a proportionality constant in the range -62 to -76 MHz can be used in conjunction with PPP to predict such splittings. The relationship is based on the premise that the negative spin density at the proton is solely determined by the adjacent carbon's out-of-plane spin population.

Conversely, if the hyperfine coupling constant is known experimentally, the McConnell relationship can be used to predict the carbon spin population. For example, Feller et al. [13] used the somewhat more elaborate, two term expression due to Gold [14]:

$$a_H = -69\varrho_\pi - 5.2\varrho_{nn} \qquad (13)$$

where ϱ_{nn} is the spin population on the next nearest nighbor carbons, and the experimental value for $a_H (-26\,\text{MHz})$ to predict the unpaired spin density on the central carbon in planar trimethylenemethane, $C(CH_2)_3$, $\varrho_\pi = 0.39$. This result is in reasonable agreement with the configuration interaction (CI) value of 0.35. A normalization relationship of $\varrho_{nn} = 1 - 3\varrho_{\pi}$ was used. As this example shows, the McConnell relationship is not necessarily limited to use with PPP. Any method that produces a spin population is acceptable. In practice although the relationship provides only qualitative agreement with experiment when used with spin densities derived from PPP, the accuracy is often sufficient for interpreting the ESR spectra of a large number of planar radicals.

If UHF is one of the simplest ways to introduce spin polarization effects, it has been argued that even greater savings in computer time may be achieved by avoiding the expense of computing all the electron-electron repulsion integrals needed for an ab initio wave function. The semiempirical UHF INDO method [15] has been suggested as being more reliable than its ab initio counterpart [16]. Compared with PPP, the INDO method is more elaborate. However, while INDO explicitly includes the valence σ electrons, it still does not include the inner shells (such as $1s$ orbitals on carbon) and so, it cannot be used to directly compute A_{iso}. Proportionality constants, derived by fitting to experimental data, can be used to relate the carbon and hydrogen spin populations to isotropic hyperfine coupling constants. For carbon the proportionality constant is simply $|2s_c|^2$, the density of the $2s$ orbital at the nucleus. The value employed in INDO calculations, 2.042 a.u.$^{-3}$, differs substantially from numerical Hartree-Fock values (~ 2.77). Some of the difficulties inherent in lumping a large variety of correlation effects, vibrational effects and spin contamination effects into this one constant have been pointed out by Chipman [21].

Experience has shown that isotropic hyperfine constants are exceedingly difficult quantities to compute. For first row atoms Li–Ne there are large and nearly cancelling contributions arising from the $1s$ and $2s$ orbitals. Even in the case of protons the sometimes opposing effects from incomplete basis sets and inflexible wave functions must be carefully balanced if an unbiased approximation of the full CI answer is

Table 1. Ab Initio Calculations of the Methyl Radical, CH_3, Isotropic Hyperfine Constants (in MHz)[a]

Basis	UHF	PUHF	S-CI	SD-CI	MR SD-CI	Full CI
Minimal STO[b]			−109/370	−122/394		
DZ STO			−102/64	−107/72		
STO-2G	−106/250					−78/208
STO-3G	−152/279	−50/96				−109/229
STO-6G	−212/434	−69/150	−109/369			−151/355
[3s, 2p/2s]	−142/110	−46/90				−99/211
[4s, 2p/2s]	−140/184	−46/64	−88/87			
(9s, 5p/4s)	−121/177	−39/62				
Extended[c]	−121/161		−75/78	−60/48	−72/81	

[a] The isotropic hyperfine results are listed as proton value/carbon value. The estimated values for a planar, nonvibrating molecule, taken from ref. [21] are $a_H \sim -70$ MHz, $a_C \sim 76$ MHz. Calculations were carried out at the experimental R_{CH} bond length of 2.039 a_0 (107.9 pm);

[b] Using energy-optimized exponents. If Slater rule exponents are used instead, A_{iso} on hydrogen is reduced to −82 MHz, ref. [26]

[c] The extended basis set calculation employed a (19s, 10p, 3d/10s, 2p) even-tempered Gaussian basis contracted to [10s, 5p, 3d/5s, 2p]. S-components of the cartesian d's were transformed out. The reference space contained 8 space orbital products (25 spin-adapted configurations). E(SCF) = −39.5751, E(CI) = −39.7704

sought. For instance, while UHF implemented with most commonly used basis sets provides poor agreement with experiment, it is possible to use this method in conjunction with very small, minimal basis sets and obtain reasonable results.

As seen in Table 1, an STO-2G UHF wave function yields proton A_{iso} of -106 MHz for CH_3 compared to the experimental value near -70 MHz. In this admittedly extreme example, the typical UHF overestimation of the spin density at the nucleus is being offset by the lack of tight s primitives in the basis set. As the number of Gaussian primitives used to expand each contracted basis function increases from 2 to 6 (STO-2G to STO-6G) the UHF isotropic hyperfine values increases rapidly, overshooting the experimental value by more than a factor of 6. Finally, as the basis set approaches the Hartree-Fock limit, the UHF values converge to approximately -121 MHz (hydrogen) and 161 MHz (carbon), slightly more than double the experimental values. In general UHF wave functions will display much larger isotropic hyperfine parameters than RHF. If the RHF wave function happens to underestimate A_{iso} then UHF may fortuitously show good agreement with experiment. If RHF is already larger than experiment, UHF will provide even worse agreement.

The projected UHF (PUHF) method, in which a pure spin state is projected from a UHF wavefunction following orbital optimization, is expected to produce smaller isotropic hyperfine values because its spin polarization contribution to the unpaired spin density is a factor of 3 less than the corresponding quantity in a UHF wave function. The so called "direct contribution", which is present in even the RHF model, is zero for the methyl radical. While early findings [17] with small basis sets suggested that PUHF might be capable of at least semiquantitative reliability, later results [18] showed that UHF followed the trends across a group of hydrocarbon radicals more accurately than PUHF. Neither wave function comes particularly close to the exact solution represented by the full CI, which includes all possible excitations consistent with the desired and spatial symmetries. Table 1 shows that UHF overestimates the spin density by about as much as PUHF underestimates it.

Because of the relative speed with which UHF and PUHF wave functions can be obtained, these methods have been applied to much larger systems than are possible with more elaborate methods. For example, molecules on the order of the cyclopentane [19] and cyclohexane cations [20] are amenable to study.

2.2 Singles CI

Within the conventional CI framework the simplest model which accounts for unpaired spin polarization adds single excitations from the Hartree-Fock configuration to the wave function. Examples of how well this model performs in the case of the methyl radical can be seen in Table 1 under the column denoted S–CI. Single excitations from a RHF wave function fall into four categories: (a) excitations from doubly occupied orbitals into the singly occupied orbital, $\phi_D^2 \phi_s^1 \rightarrow \phi_D^1 \phi_s^2$; (b) excitations from doubly occupied orbitals into virtuals, $\phi_D^2 \phi_s^1 \rightarrow \phi_D^1 \phi_v^1 \phi_s^1$ with the first two spins coupled as a triplet; (c) the same excitations with the first two spins coupled as a singlet and (d) $\phi_s^1 \rightarrow \phi_v^1$. CI coupling between the HF configuration and excitations in categories (a), (c) and (d) are forbidden by Brillouin's theorem. However, while their first order contribution to the wave function is zero, they can interact with configurations of type (b) in second order as well as with double excitations, and thus affect the spin density. In the case of the methyl radical it has been shown that inclusion of the Brilloin-forbidden configurations, as opposed to just the Brilloin-allowed singles, produces only a slight change in A_{iso} [21].

In general, single excitation CI produces isotropic hyperfine parameters which fall somewhere between the UHF and PUHF results. Thus, S-CI often comes closer to matching the full CI value, as seen in Table 1. Because of the ease with which single excitation CI wave functions can be computed there has been a renewed interest in this method in recent years. Nevertheless, agreement with experiment is still highly dependent on the particular basis set with which the calculation was performed.

The effects of basis set choice on single excitation CI's are particularly noticeable in the extreme case of a minimal basis. For example, a minimal STO basis (or the STO-6G contracted Gaussian analogue) produced a carbon A_{iso} in CH_3 which was nearly four times larger than the [$4s$, $2p/2s$] double zeta value. Without the flexibility to correlate the $1s$ core, the minimal basis S-CI result severely underestimates the negative core contribution to the property.

Further examples of the level of agreement with experiment to be found with singles CI comes from the work of McCanus et al. [22] on $H_2CN\cdot$, $H(HO)CN\cdot$ and $\cdot CONH_2$. Using a double zeta plus polarization (DZP) basis they found: H_2CN, $a(^1H) = 162$ (244), $a(^{13}C) = -70 (-81)$, $a(^{14}N) = 19$ (29); $H(OH)CN$, $a(^1H) = 103$ (152), $a(^{13}C) = -71 (-57)$, $a(^{14}N) = 17$ (29); $CONH_2$, $a(^1H_a) = 55$ (86), $a(^1H_b) = -5 (-4)$, $a(^{13}C) = 518$ (442), $a(^{14}N) = 67$ (61). The experimental values, given in parentheses, were obtained in aqueous solution using in situ radiolysis.

These same authors have used an extension of the S-CI model to study the $^2A_2$ allyl radical, $CH_2-CH-CH_2$ [23]. Allyl is the smallest odd alternate hydrocarbon. It is expected to show negative spin density on the central carbon and positive values on the terminal carbons. At the RHF level allyl displays two artifactual broken symmetry solutions to the HF equations at the optimal C_{2v} geometry. The energies corresponding to these broken symmetry solutions ($CH_2=CH-CH_2\cdot$ and $\cdot CH_2-CH$ $=CH_2$) are lower than the energy of the C_{2v} constrained solution. A π-space multiconfiguration SCF (MCSCF) wave function for 3 electrons in 3 orbitals, which simultaneously optimizes the shapes of the orbitals as well as their CI expansion coefficients, can overcome this defect in the RHF method. By adding single excitations from each of the four π configurations they found: $a(^{13}C_{central}) = -66 (-48)$, $a(^{13}C_{terminal})$ $= 51$ (61), $a(^1H_{central}) = 8$ (12), where the experimental values are given in parentheses. A DZ Gaussian basis was employed.

Acknowledging the importance of the basis set's ability to reproduce the experimental spin density at the nucleus, Konishi and Morokuma [24, 25] have proposed the use of special Slater-type orbital (STO) basis sets which are augmented with a single very tight s function chosen so that the S-CI atomic spin densities match the experimental values.

2.3 Single and Double Excitation CI

The earliest ab initio polyatomic calculations to include the more numerous double excitations, such as $\phi^2 \rightarrow \phi'^2$ or $\phi^2 \rightarrow \phi'^1\phi''^1$, were carried out on CH_3 with minimal and DZ STO basis sets [26]. As seen in Table 1, the effects of including double excitations is, as with single excitations, dependent on the nature of the basis set. The column denoted singles, doubles CI (SD-CI) shows that with both the minimal and DZ STO basis sets a slight increase in both isotropic hyperfine parameters is experienced. With the extended Gaussian basis SD-CI shows a slight decrease.

The sensitivity of A_{iso} to the basis set was emphasized in parallel sets of calculations [27] on CH ($^2\pi$) which alternatively imposed and relaxed the electron-nuclear cusp constraint, $\partial\psi/\partial r_{ia}|_{r=0} = -Z_a\psi|_{r=0}$, where r is the distance between the i'th electron and nucleus a with charge Z. The exact wave function will satisfy this condition, but approximate solutions, in general, will not. An extended (s, p, d, f) STO basis set which produced identical S-CI energies with and without the cusp constraint, changed the carbon A_{iso} from 69 MHz (cusp relaxed) to 16 MHz (cusp constraint imposed).

Most ab initio calculations on polyatomic molecules are now carried out with Gaussian functions which, due to their exp ($-r^2$) behavior, possess no cusp. Thus, there is no analogous constraint which can be imposed. Nevertheless, an indication of how well the Gaussian set performs in the vicinity of the nucleus can be gained from an examination of the value of the Hartree-Fock total electron density at the nucleus. Accurate values of this property are available from either numerical work or large STO calculations [28]. By using a sufficiently large number of Gaussians the exact value of the property can be reproduced to any desirable accuracy. With 9 energy-optimized s-type Gaussians the error is on the order of 5% in carbon.

So far we have dealt solely with the isotropic component of the hyperfine interaction. The anisotropic components, arising from the interaction of the electron and nuclear magnetic dipole moments, represent a far easier computational task since the direct contribution from the Hartree-Fock wave function is often quite close to experiment. In the general case, the hyperfine interaction tensor, A, is a traceless 3×3 tensor whose elements are given by

$$A_{ij} = g_e g_N \beta_N \langle (3r_i r_j - r^2\delta_{ij})/r^5 \rangle_{spin} \tag{14}$$

where i, j = x, y, z. The results of various A_{aniso} calculations for the methyl radical are presented in Table 2. The agreement with experiment [29] in this case is fairly typical, although many studies have reported even better results for other molecules.

Table 2. Ab Initio Calculations of the Methyl Radical Anisotropic Hyperfine Constants (in MHz)[a]

	A_{xx} (C)	A_{xx} (H)	A_{zz} (H)
Minimal STO[b] SD-CI	103	− 6	49
STO-6G Full CI	171	−17	66
[4s, 2p/2s] Singles CI	146	− 5	43
Extended Basis Set CI[c]	154	1	40

[a] The molecule lies in the yz plane, with one hydrogen extended along the z-axis. The three-fold axis is aligned along the x-axis. Experimental anisotropic hyperfine parameters are 126 ± 2 MHz (A_{xx} on ^{13}C), 1.4 ± 0.8 MHz (A_{xx} on ^{1}H) and 35 MHz (A_{px} on ^{1}H), ref. [29];
[b] Using energy-optimized exponents, ref. [30];
[c] A (19s, 10p, 3d/10s, 2p) → [10s, 5p, 3d/5s, 2p] basis. See footnote 3 of Table 1

As with the isotropic hyperfine parameter, it is possible to obtain results in fortuitous agreement with experiment when very small basis sets and highly truncated CI's are employed. For example, as seen in Table 2, a SD-CI[30] with a minimal STO basis gives better agreement than a full CI in what should be a nearly identical Gaussian basis (STO-6G). Single excitation CI with a DZ basis set performs just about as well as more elaborate wave functions.

2.4 Perturbation Theory, Cluster Expansions and Valence Bond

Several other theoretical methods have been applied to the methyl radical spin properties. Millie et al. [31] used a [6s, 4p, 2d/3s, 1p] contracted basis and first order perturbation theory to obtain $a_H = -58$ MHz and $a_C = 29$ MHz with the canonical HF HF orbital set. However, the results proved strongly dependent on the choice of orbitals. A unitary transformation to a set of localized MO's produced $a_H = 36$ MHz, $a_C = 87$ MHz. Ohta et al. [32] have used an extension of the conventional cluster expansion to open shell which they call the symmetry-adapted cluster (SAC) expansion. When used with a small [4s, 2s/2p] basis the SAC method and it's derivatives [33] produces numbers in the range $a_H = -59$ to -63, $a_C = 63$ to 80. Yet another approach was employed by Raimondi et al. [34] who used a multistructure valence-bond approach to CH_3. With a minimal STO basis possessing exponents chosen according to Slater rules the computed isotropic hyperfine values were $a_H = -100$ MHz and $a_C = 417$ MHz. Variation in the number of valence-bond structures over a range of 200 to 800 had a little affect on the properties, but it is likely that, as with CI, the impact of correlation effects is artificially supressed by the use of a minimal basis set.

CH_3 has a very low barrier to out-of-plane bending motion of the three hydrogens. This could lead to a sizable difference between the experimentally observed hyperfine properties and those derived from ab initio methods. Chang et al. [35] used a minimal STO basis and a SD-CI wave function to calculate a vibrational correction for the proton of -6 MHz. Ohta et al. [36] have computed the effects of the out-of-plane vibrational motion using "pseudo-orbital" theory. This theory, which includes only the spin-polarization single excitation operator in the cluster expansion formalism, predicted corrections of -27 MHz (^{13}C) and -5 MHz (^{1}H). When added to the low temperature, experimental values ($a_C = 107$ MHz, $a_H = -65$ MHz)[37] these corrections lead to estimates of $a_C = 80$ MHz and $a^H = -69$ MHz for the hypothetical, planar, nonvibrating radical. A more detailed analysis due to Chipman suggests values of $a_C \sim 76$ MHz and $a_H \sim -70$ MHz for the planar species, with vibrational corrections of -31 and -5 MHz respectively. Chipman also showed that the anisotropic components of the hyperfine interaction were relatively insensitive to out-of-plane bending, with most corrections being less than 2 MHz.

2.5 Extended Basis Set, Highly Correlated Wave Functions

As often happens in science, the development of new experimental techniques in ESR spectroscopy and the accompanying expansion of knowledge concerning small molecules has spurred theoreticians to the use of ever more sophisticated wave func-

Table 3. Examples of HF SD-CI Spin Properties[a] for Diatomic Molecules Obtained With SVP or DZP Basis Sets

		Isotropic		Anisotropic	
		CI	Expt.	CI	Expt.
BeF:[b]	Be	243	294	6	4[c]
	F	126	230	80	102[c]
MgF:[b]	Mg	−303	−337	−5	−6
	F	505	206	115	126
BeH:[d]	Be	201	199	8	9
	H	164	194		7
BO:[e]	B	968	1025	52	27
	O	15	14		
AlH⁺:[d]		1422	1583	99	97
AlO:[e]	Al	776	767	84	106
BeCl:[f]	Be	220	273	7	10
	Cl	18	34	14	20

[a] Results given in MHz;
[b] BeF and MgF results taken from Ref. [38 b];
[c] Experimental values estimated from the line shape assuming no rotational averaging;
[d] BeH and AlH⁺ results taken from Ref. [38 a];
[e] BO and AlO results taken from Ref. [38 d];
[f] BeCl results taken from Ref. [38 c]

tions. Low temperature, rare gas matrix isolation techniques have made it possible to observe the ESR spectra of many radicals and ions which would be extremely difficult to observe in the gas phase due to their highly reactive nature. Where comparision data is available, the effects of the matrix have ususally been shown to be small, i.e. < 10%. For example, gas phase measurements on the formal radical, HCO ($^2A'$), found a_H = 389 (microwave), 390 (laser magnetic resonance). While a matrix isolation study in CO at 4.2 K found a_H = 380 MHz.

Following the appearance of the first SD-CI molecular hyperfine calculations in 1971 very few subsequent calculations were attempted for a period of about 8 years. By the end of the decade the first of a large body of collaborative work between theoreticians and experimentalists using matrix isolation techniques began to appear. The calculations employed split valence (SV) or DZ quality basis sets augmented with a single set of cartesian d-type functions on first and second row atoms. Agreement with experiment was generally good, as can be seen in the results for the diatomics BeF, MgF, BeH, BO, AlH⁺ and AlO shown in Table 3 [38]. While there were instances where the theoretical and experimental isotropic parameters differed by a factor of two, e.g. fluorine in BeF and MgF, the impression one is given from Table 3 is that this level of theory (modest basis set, modest CI) is capable of at least semi-quantitative accuracy in the majority of cases.

A study of spin properties in ten small radicals was published in 1984 [39] which reopened the question of what level of theory was needed to reliably reproduce hyperfine parameters. To a large extent the quality of a theoretical calculation is determined by just two factors, the flexibility of the 1-particle basis set and the amount of correlation recovery. This study was the first to explore the impact of extended Gaussian basis sets and CI wave functions more elaborate than the simple HF SD-CI variety. Multireference SD-CI's (MR SD-CI's), which involve single and double excitations from a list of important configurations (the "reference" space) were employed along with transformations of the molecular orbitals to approximate natural orbitals.

For any basis set the best approximate solution of the Schrödinger equation, within the time-independent, nonrelativistic, Born-Oppenheimer approximation, is obtained with a full CI. However, the number of configurations representing higher order excitations increases so rapidly that full CI is intractable for all but the smallest systems. The MR SD-CI approach attempts to partially circumvent this problem by including only those higher order excitations which are singles and doubles from a few of the most important configurations. For example, if a double excitation were the second most important configuration in the wave function, judged on the basis of either its energetic contribution or the size of its CI coefficient, including single and double excitations from it would introduce some triple and quadruple excitations (relative to the Hartree-Fock) into the wave function.

Even this approach, when implemented with large bass sets, can very quickly generate millions of configurations. Since the vast majority of those configurations make a negligible contribution to the energy and properties, a selection procedure may be invoked to reduce the computational task even further. Analysis of a simple second

Table 4. Spin Properties Obtained With Extended Gaussian Basis Sets[a]

H_2^+

Basis Set	Energy	$\langle \delta_H \rangle$	$\langle (3z_{H^2} - r_{H^2})/r_{H^5} \rangle$
$(10s) \rightarrow [6s]$	−0.5909	0.2037	0.1277
$(10s, 1p) \rightarrow [6s, 1p]$	−0.6019	0.2033	0.1520
$(10s, 2p) \rightarrow [6s, 2p]$	−0.6024	0.2034	0.1533
$(10s, 2p, 1d) \rightarrow [6s, 2p, 1d]$	−0.6026	0.2033	0.1625
$(18s, 5p, 4d)$	−0.6026	0.2090	0.1656
Exact	−0.6026	0.2096	0.1620

HCO ($^2A'$)

Basis	Wave Function	Energy	a_C	a_O	a_H
STO-3G	RHF	−111.7293	420	− 6	230
	UHF	−111.7326	496	− 6	247
4-31G	UHF	−113.0706	437	−34	272
$(8s, 4p, 1d/8s, 1p)$	HF SD-CI	−113.6130	390	−31	294
$(12s, 6p, 2d/10s, 2p)$	MR SD-CI	−113.6175	376	−34	342
Experiment			377		381

[a] The H_2^+ calculations were done at R_{HH} = 2.00 a_o (105.8 ppm). Energies and H_2^+ properties are given in atomic units. The HCO calculations were done at the experimental geometry, R_{CO} = 2.220 a_o (117.5 pm), R_{CH} = 2.126 a_o (112.5 pm) and HCO angle = 124.95°. Isotropic hyperfine values for HCO are given in MHz. Results taken from Ref. [39]

order Raleigh-Schödinger selection procedure suggest that for many molecules somewhere on the order of 10% to 15% of the original, unselected configurations in a MR SD-CI are sufficient to reproduce the unselected energy and properties to better than 99%.

Overall the agreement with the experimental A_{iso} was within 10% for 6 of the 10 radicals appearing in the 1984 study. Representative findings are presented in Table 4 for H_2^+, where the exact values are available from theory, and HCO, where good experimental values are available for ^{13}C and ^1H. However, for H_2CN, H_2CO^+, H_2BO and H_3C-CHO^+ the deviation from experiment was much larger. Even relatively large CI was unable to come within 35% of the experimental proton isotropic hyperfine value obtained in a neon matrix.

In a follow-up study [40], which concentrated on the four troublesome radicals, the effects of better geometry optimization, vibrational averaging, larger reference spaces and the use of multiconfiguration SCF (MCSFC) molecular orbitals were all examined, but none of these factors seem to provide a dramatic improvement. While the source of the difficulty was to remain unknown for several years the MR SD-CI method continued to be applied to small molecular systems with generally good agreement with experiment. Table 5 shows some representative findings for 6 molecules [41, 42, 43].

In some cases, such as the $^2\Sigma^+$ SiO$^+$ cation [44] and the $^4\Sigma^-$ BC radical [45] matters were complicated by the failure of the SCF wave function to provide even a qualitatively correct description of the system. For SiO$^+$ the RHF solution places the unpaired electron on oxygen, while experiment indicated considerable unpaired spin density on silicon. It proved necessary to resort to an MCSCF wave function for the zeroth order description of the molecule. A comparison of the RHF and MCSCF singly occupied orbitals in Fig. 3 shows that the latter method has shifted the unpaired

Table 5. Examples of MR SD-CI Spin Properties Obtained with Extended Gaussian Basis Sets

Molecule		$A_{iso}{}^a$	$A_{aniso}{}^b$		
			xx	yy	zz
$C_2O_2^+$:^{c}	C	672 (589)^d	−20 (−12)	49 (38)	−29 (−29)
	O	−19 (−25)	28 (25)	−47	20
CO$^+$:^{c}	C	1512 (1573)			103 (98)
	O	13 (19)			−59 (−66)
SiO$^+$:^{e}	Si	−757 (−798)			−137 (−128)
	O	12			−137
BF$^+$:^{f}	B	1662 (1746)			51 (53)
	F	212 (226)			199 (199)
BC:^{g}	B	60 (86)			16 (14)
	C	10 (14)			−8
N_4^+:^{h}	N_1	6 (−10)			8 (10)
	N_2	362 (280)			17 (16)

^{a} Given in MHz;	^{e} Taken from Ref. [42];
^{b} In the principal magnetic axis system;	^{f} Taken from Ref. [43];
^{c} Taken from Ref. [41];	^{g} Taken from Ref. [45];
^{d} The experimental value is given in parentheses;	^{h} Taken from Ref. [47]

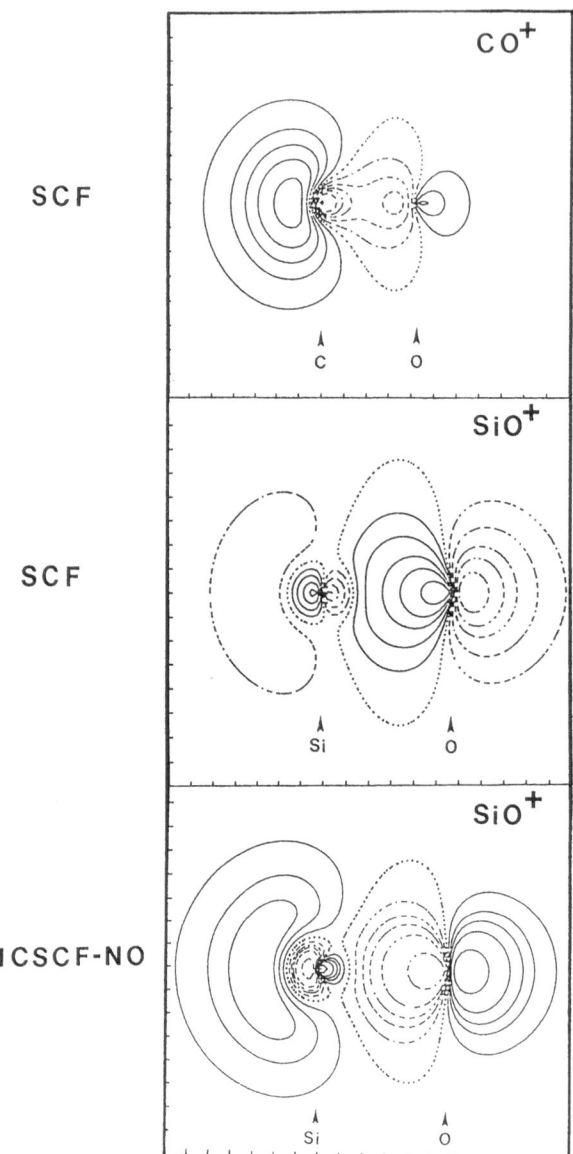

Fig. 3. Probability contours for the singly occupied orbitals in CO^+ and SiO^+ $(^2\Sigma^+)$ at the RHF and MCSCF-NO levels, indicating the qualitative error present at the simpler level of theory. The outer contour encloses 90% of the probability density. Each successive contour contains 20% less probability density

spin density in the right direction and does provide a qualitatively correct zeroth order description of the molecule. For BC two distinct SCF minima were identified, corresponding to localization of the unpaired electron on either boron or carbon. As seen in Fig. 4, an MCSCF wave function places the unpaired spin about equally on both atoms. Interestingly, although the related $^4\Sigma^-$ BC_3 radical also displayed two SCF minima, the energy separation was more than 50 times larger than in BC, so the existence of a second minimum presented no problem in the larger molecule [46]. While very extensive MCSCF/CI calculations on BC struggled to approach 75% agreement with experiment, for BC_3 the SCF wave function was already within 20%.

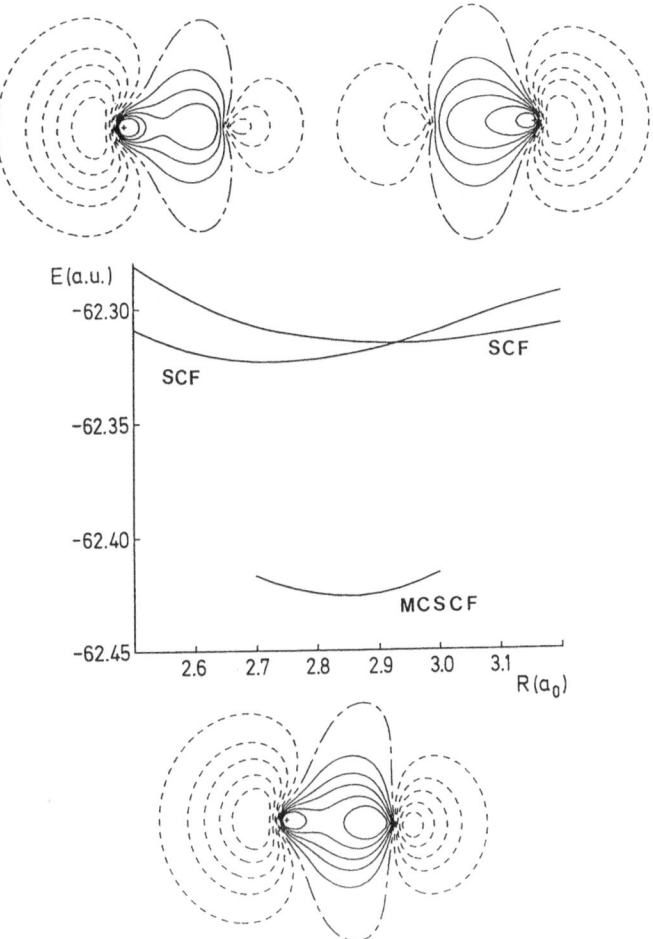

Fig. 4. Potential energy curves for the $^4\Sigma$-ground state of BC computed at the SCF and MCSCF levels. The SCF has two solutions corresponding to localization of the unpaired electron on either boron or carbon, as indicated by the contour plots for the 5σ molecular orbital. The MCSCF singly occupied orbital is a compromise of the two SCF solutions

Besides illustrating the need for more flexible zeroth order wave functions, the SiO^+ study highlighted limitations in the free atom comparision method (FACM), described previously. This method is frequently used in an effort to "experimentally" determined the $(\%s)/(\%p)$ composition of the orbital containing the unpaired electron. Nonetheless, the approach fails to consider important core polarization effects or the impact of off-diagonal contributions to $Tr(\theta P)$. Specifically, for SiO^+ FACM suggested that the unpaired electron on silicon resided in an orbital which was more hybridized (17%-s, 56%-p) than its first row counterpart in CO^+ (42%-$\hat{s}$, 43%-p). This runs counter to the observation that second row elements usually exhibit less hybridization than elements in the first row.

FACM begins with an implicit minimal basis set assumption, which is difficult to mimic with ab initio methods based on extended basis set, CI wave functions. While the CI spin density can easily be projected onto a minimal basis, if free atom AO's are used the computed Mulliken gross population is only 0.94 and hyperfine properties computed from this projected spin density differ qualitatively from the actual CI properties. Since the bonding in SiO^+ is mostly ionic, $\sim Si^{+2}-O^-$, it might be argued that a better set of AO's could be obtained from Si^{+2} and O^-. However, these AO's still do not provide a minimal set capable of adequately minimicking the CI. In order to obtain a meaningful minimal basis population analysis it proved necessary to develop a hybrid HF core AO/natural orbital valence basis set. The core orbitals were taken from Hartree-Fock calculations on the free atoms. The valence AO's on Si ($3s$ and $3p_z$) were taken from the 7σ singly occupied NO in such a way that they were orthogonal to the Si core and could still represent 7σ exactly.

As opposed to the FACM analysis, which assumes only one term is dominant in computing A_{iso} or A_{aniso}, the detailed analysis showed that additional terms involving the core orbitals, which have less than 1% of the spin population, make large contributions to both components. When only the leading term is considered A_{iso}/A_{aniso} is 1.8. This ratio falls to 0.6 when the entire sum is considered! The large core effect was not due to spin polarization effects, since even the leading NO configuration gave a value of $A_{iso} = -789$ MHz (^{29}Si), compared to the experimental value of -797 MHz. It has been shown to be due to coreother valence overlap, $\langle 0_{2p}|Si_{1s}\rangle$ and $\langle 0_{2p}|Si_{2s}\rangle$, where it is understood that all references are to the p_z components. Because of this overlap the singly occupied orbital in SiO^+ has the form:

$$7\sigma = c_1 Si_{3s} + c_2 Si_{3p} + c_3[0_{2p} + c_4 Si_{2s} + c_5 Si_{2p} + ...] \qquad (15)$$

where

$$c_4 = -\langle 0_{2p}|Si_{2s}\rangle \quad \text{and} \quad c_5 = -\langle 0_{2p}|Si_{2s}\rangle . \qquad (16)$$

If all p_z orbitals have their positive lobes on the right, then from Fig. 3 we can see that all coefficients are positive. When individual contributions to A_{iso} were examined it was found that the second largest contribution came from the off-diagonal term $2c_1 c_3 c_4 \langle Si_{3s}|\delta|Si_{2s}\rangle$ which was -0.90 in magnitude, compared to $+1.27$ for the $\langle Si_{3s}|\delta|Si_{3s}\rangle$ diagonal term. On the other hand, the experimentally derived spin populations for many first row diatomics, such as CO^+ and BO are often within 10–20% of the CI values. Thus, spin populations deduced from experiment using the FACM procedure may differ substantially from CI population analyses when core-other valence effects are large or when the bonding is mostly ionic and the "free atom" values correspond to neutral species.

As mentioned previously, the two dominant factors affecting the quality of all theoretical calculations are basis quality and correlation recovery. In spite of a growing body of experience with extended basis sets and MR SD-CI wave functions, the inconsistent performance of this relatively high level of theory was perplexing. In the case of the linear N_4^+ ($^2\Sigma_u$) radical cation [47] the Hartree-Fock values of A_{iso} may be nonzero because the unpaired electron occupies a sigma orbital. Sigma orbitals are permitted by symmetry to have nonzero density at the nuclei. The MR SD-CI values are given in Table 5, where we have used the labelling convention $N_1-N_2-N_2-N_1$. These values are hardly changed from the HF values, $A_{iso} = 14$ MHz (N_1) and $A_{iso} = 329$ MHz (N_2). Examination of the isotropic hyperfine properties was even extended over a large portion of the potential energy surface around the Møller-Plesset (4'th order) minimum, but no closer agreement with experiment was obtained. In fact, replacement of the original ($11s$, $6p$, $1d$) basis with a much larger ($15s$, $8p$, $3d$) basis actually resulted in slightly *worse* agreement. Since the anisotropic components in N_4^+ were reasonably close to experiment, efforts focused on the more difficult isotropic parameter.

Progress in systematically tracking down the sources of error in ab initio calculations has come about through a series of very recent studies of A_{iso} in first row atoms B–F. Prior to this it was not known whether the failure to obtain reliable isotropic hyperfine properties resulted primarily from basis set deficiencies or the lack of sufficiently complete correlation recovery (or some combination of both). These systems are small enough that it is possible to recover a very high percentage of the empirical correlation energy. They also permit us to avoid the question of "direct" (SCF) vs. polarization (CI) contributions to the property. For partially filled p shell atoms, the RHF solution permits no unpaired spin density at the nucleus. The same was true in CH_3. Thus, the entire answer is due to the effects of correlation.

Previous calculations of atomic A_{iso} offer a wide range of agreement with experiment, as seen in Table 6. The singles CI wave function of Schaefer et al. [48], performed with an (s, p, d, f) STO basis set, produced results in semiquantitive agreement with experiment [49, 50, 51, 52]. However, as will be discussed in a later section, experience has shown single excitation CI to be more unpredictable than more extensive CI. Numerical unrestricted Hartree-Fock (UHF) [53] which in principle is capable of an exact solution of the UHF equations, yields values anywhere from 2 to 4 times larger than experiment. Better agreement is seen in the MCSCF findings of Chipman using a numerical procedure to avoid the finite basis set problem [54]. This work emphasizes the importance of a small number of Brillouin-allowed single excitations in the correct spin density at the nucleus.

Table 6. Theoretical Determinations of Atomic Isotropic Hyperfine Constants in MHz

Method	Boron	Carbon	Nitrogen	Oxygen	Fluorine
(s, p, d, f) STO Basis S-CI[a]	10.5	15.6	7.9	−18.5	197.6
Numerical UHF[b]	24.3	86.6	60.4	−118.2	559.6
Numerical MCSCF[c]	22.3	26.6	10.9	−27.6	261.7
CGTO MR SD–CI[d]			9.9		
GTO CASSCF/MR SD–CI[e]			10.4		
GTO MR SD–CI[f]	6.4	17.8	10.1	−30.2[h]	285.3
A_{iso} (exp.)[g]	11.6		10.45	−34.5	301.7

[a] H. Schaefer et al., Ref. [48]. An ($8s$, $5p$, $6d$) was used on B, C and N. An ($8s$, $5p$, $6d$, $4f$) was used on O and F;

[b] P. Bagus et al., Ref. [53]

[c] D. Chipman, Ref. [54];

[d] ($13s$, $8p$, $4d$) → [$8s$, $4p$, $4d$], contracted Gaussian basis, Engels et al., Ref. [55];

[e] ($10s$, $5p$, $4d$, $4f$) uncontracted Gaussian basis, Bauschlicher et al. Ref. [56];

[f] ($23s$, $12p$, $10d$, $4f$, $2g$) uncontracted Gaussian basis, Feller and Davidson, Ref. [57];

[g] See text. The theoretical values were converted to MHz using the following nuclear g values: 1.7923 (^{11}B), 1.4048 (^{13}C), 0.4038 (^{14}N), −0.7575 (^{17}O), 5.2576 (^{19}F). $A(MHz) = g_N c \langle \delta(o) \rangle_{spin}$ where $c = 800.2370$

[h] ($23s$, $12p$, $10d$, $5f$, $3g$) Feller and Davidson, Ref. [65]

A recent examination of the effects of basis sets and CI methodology on the spin properties of the nitrogen atom was reported by Engels et al. [55] using a four configuration reference space and a *(13s, 8p, 4d) → [8s, 4p, 4d]* contracted basis. They found A_{iso} = 9.1 MHz for ^{14}N. This work provides a detailed analysis of the charge density at the nucleus and the spin polarization in the *1s* and *2s* shells as a function of the CI parameters. Another study of the nitrogen atom by Bauschlicher et al. [56] produced the best results to date. They found A_{iso} = 10.4 MHz, in nearly exact agreement with experiment, using a combination of MCSCF and MR SD-CI techniques along with a *(10s, 5p, 4d, 4f)* Gaussian basis. They attempted to calibrate their methods, to the extent possible, by comparing against full CI results obtained with somewhat smaller basis sets.

A larger systematic study by Feller and Davidson [57], conducted simultaneously with the work of Bauschlicher et al., reached similar conclusions. Using a sequence of progressively larger *(s, p, d, f, g)* even-tempered [58, 59] basis sets, it was determined that relatively large basis sets (by common molecular standards) and extensive correlation recovery was needed to reproduce experimental isotropic hyperfine constants to within 10%. For nitrogen the contributions to A_{iso} from progressively higher l functions were computed to be 2.5 (*d*'s), 0.4 (*f*'s) and 0.0 (*g*'s) out of a final value of 10 MHz. The importance of these functions, however, increases with atomic number. For instance, the d contribution on oxygen is 3.7 MHz. General conclusions, such as these, are difficult to formulate because of the complications arising from the near cancellation of large core ($\sim$ −50 MHz) and valence ($\sim$ +60 MHz) contributions to A_{iso}. Because of this cancellation, it is often possible to choose the exponents for a small basis in such a way that it will exactly reproduce the experimental value for any one system.

The even-tempered sequence of basis sets is *not* the most rapidly convergent set for this property since, for a given basis set size, it lacks the tightest *s* primitives found in an independently optimized set. Nonetheless, the even-tempered sets provide the only means available in the literature for conveniently generating arbitrarly large sets which approach a complete basis in the limit of $\beta \rightarrow 1$ as N → ∞. The results appear in Table 6 under the heading Gaussian Type Orbital (GTO) MR SD-CI.

With the exception of boron, where the theoretical value is only half of the experimental value, the agreement between the MR SD-CI and experimental numbers appearing in Table 6 is good. No explanation has been offered for the lack of agreement on boron. All indications are that the theoretical treatment should as good or better for boron than for the other atoms.

Besides the conclusion regarding large basis sets, it was also apparent that if a conventional CI approach was to be used it would have to be extensive enough to recover a large fraction of the total correlation energy. Failure to do so invites a skewing of the isotropic hyperfine property in favor of the negative core contribution or the positive valence contribution. The wave functions computed by Feller and Davidson recovered from 94 to 98% of the empirical correlation energy.

While configuration selection schemes based on various perturbation theory criteria can help reduce the number of configurations which must be treated varitaionally, the computational task is still quite large. For example, in the case of the nitrogen atom a 42 configuration reference space generated a total of 2 million single and double excitations. All of the singles were kept, due to their well-known importance for first order properties. Approximately 160,000 double excitations were also selected based on the second order Raleigh-Schödinger estimate of their energy contribution. Thus, several million configurations had to be treated via perturbation theory and several hundred thousand had to treated variationally, as opposed to 10^{12} in the full CI.

Table 7. Convergence of A_{iso} With Respect to Reference Space Size in MR SD–CI's on B_2 and H_2CO^+

B_2

Refer.	Size[a]	No. Config.[b]	$(9s, 5p) \to [4s, 2p]$			
			E_{CI}	$|C_{max}|^c$	$\sqrt{\Sigma c^2}$	A_{iso}^c
1	(1)	1,692/1,692	−49.2237	0.264	0.831	22.25
4	(8)	10,338/10,338	−49.2422	0.091	0.916	26.37
15	(33)	28,151/31,107	−49.2458	0.049	0.961	30.64
34	(108)	47,457/71,731	−49.2467	0.022	0.980	30.99
50	(162)	59,215/100,989	−49.2470	0.017	0.987	30.99

Refer.	Size	No. Config.	$(9s, 5p, 2d) \to [4s, 2p, 2d]$			
			E_{CI}	$|C_{max}|$	$\sqrt{c^2}$	A_{iso}
1	(1)	7,944/7,944	−49.2732	0.231	0.845	6.33
4	(8)	52,304/52,304	−49.2952	0.081	0.906	12.16
15	(39)	108,736/187,122	−49.3012	0.044	0.943	16.94
38	(104)	149,049/449,251	−49.3027	0.023	0.965	16.76

Refer.	Size	No. Config.	$(15s, 8p, 3d, 1f) \to [10s, 8p, 3d, 1f]$			
			E_{CI}	$|C_{max}|$	$\sqrt{c^2}$	A_{iso}
1	(1)	48,471/77,366	−49.3509	0.191	0.871	−5.81
4	(4)	132,612/532,388	−49.3760	0.077	0.906	−1.88
11	(23)	90,162/1,353,760	−49.3786	0.054	0.938	0.23
23	(63)	178,459/3,246,881	−49.3837	0.030	0.954	4.72

H_2CO^+

Refer.	Size	No. Config.	$(9s, 5p, 1d/5s, 1p) \to [4s, 2p, 1d/3s, 1p]$				
			E_{CI}	$|C_{max}|$	$\sqrt{c^2}$	$A_{iso}(C)$	$A_{iso}(H)$
1	(1)	19,404/22,541	−113.8365	0.098	0.908	−68.7	212.1
6	(12)	83,897/205,108	−113.8514	0.049	0.926	−87.3	272.8
23	(60)	137,268/904,558	−113.8575	0.027	0.941	−99.4	294.5
47	(133)	186,515/1,750,102	−113.8591	0.019	0.952	−104.7	296.7

[a] The number of space orbital products (spin-adapted configurations) in the reference space;

[b] The number of spin-adapted configurations selected/the number of configurations generated;

[c] The absolute value of the largest coefficient contribution to the CI wave function which is outside the reference space;

[d] The square root of the sum of the squares of the CI coefficients corresponding to the reference configurations;

[e] In MHz. All calculations were performed at the experimental geometry for B_2 ($^3\Sigma_g^-$), $R_{BB} = 3.005\,a_0$ (159.0 pm). The experimental ^{11}B A_{iso} was 15 MHz, ref. [47]. H_2CO^+ calculations were performed at $R_{CO} = 119.8$ pm, $R_{CH} = 111.7$ pm and HCH angle $= 123°$. The experimental proton A_{iso} was 371.9 MHz

For a given basis set, there are two parameters which largely control the quantity of correlation energy recovered by configuration selected MR SD-CI's. The two parameters are the size of the reference space and the value of the selection threshold, denoted T_E if selection is based on energy considerations. Two approximate guidelines of reference space quality have been suggested. One is the magnitude of the largest CI coefficient outside the reference space, denoted $|C_{max}|$. As the number of configurations in the reference space increases $|C_{max}|$ goes to zero. Another measure which is commonly used is the square root of the sum of the squares of the reference space coefficients in the CI, denoted $\sqrt{\Sigma c^2}$. As the reference space increase $\sqrt{\Sigma c^2}$ approaches unity. Experience has show that values of $|C_{max}|$ on the order of 0.03 to 0.05 and values of $\sqrt{\Sigma c^2}$ on the order of 0.94 or larger are necessary to achieve results within a few percent of the full CI results.

An indication of the sensitivity of A_{iso} to the size of the reference space can be seen in the B_2 $(^3\Sigma)$ and H_2CO^+ $(^2B_2)$ results shown in Table 7. For the [4s, 2p] and [4s, 2p, 2d] B_2 basis sets a selection threshold on the order of 5×10^{-9} was used in order to insure convergence of the results with respect to $T_E = 0$ calculations. However, with the [10s, 8p, 3d, 1f] extended basis it proved necessary to increase the threshold to something on the order of 10^{-7} with the largest reference space (23 space orbital products). Therefore, the last B_2 entry under the extended basis set must be considered unconverged. All indications from *smaller* basis sets and tighter thresholds with smaller reference spaces on *this* basis suggest that the 4.72 MHz value underestimates the zero threshold value by 2 to 4 MHz.

The need for very small selection thresholds $(T_E \sim 10^{-8} - 10^{-9})$ to reach convergence in A_{iso} is most pronounced with large basis sets and reference spaces. Since the size of the CI wave function increases very rapidly with the size of the basis set and reference space, it has only recently become possible to perform sufficiently large CI's to probe this behavior. Similar findings with regards to T_E were reported for a variety of one-electron properties, such as dipole and quadrupole moments [60]. While the error in the total energy is second order in the error in ψ, one-electron properties possess first order errors and are, thus, expected to show an increased sensitivity to the quality of the wave function.

Figure 5 shows the interplay of all three parameters: basis set size, selection threshold and reference space size. Contour values of constant correlation energy and A_{iso} for the nitrogen atom are plotted against the $|C_{max}|$ and T_E axes. Compared to the experimental value of 10.5 MHz, it is seen that theoretical calculations can obtain exact agreement even when the less extensive double zeta plus polarization (DZP) and triple zeta plus double polarization (TZP) basis sets are used, but *only in selected regions* of the contour map. The full CI results are approached in the lower lefthand limit of the plots. Unfortunately, the DZP basis shown good agreement only in the upper regions, which indicates that if good agreement is to be obtained with this basis it must be gotten with a fortuitous combination of small reference space and large selection threshold.

Further evidence of the need, in some cases, for large basis sets and extensive correlation recovery comes from the H_2CO^+ cation. Failure to achieve reasonable agreement on the isotropic hyperfine constants of this molecule (i.e. to within 30%) was what initially led to a reexamination of the requirements for theoretical calculations of these properties. Table 8 provides some indication of the extraordinarily high level

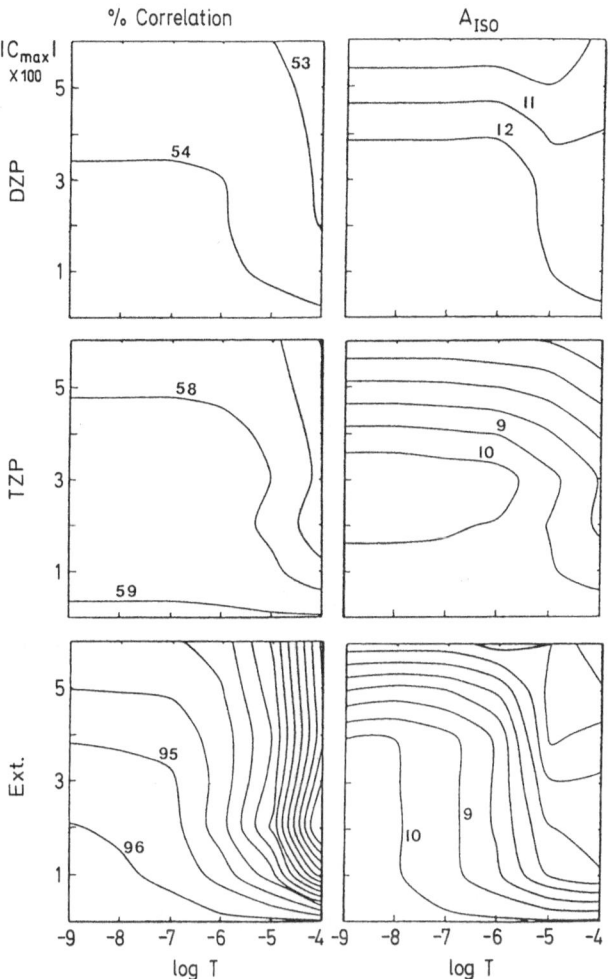

Fig. 5. Contour maps of constant correlation energy and A_{iso} (MHz) for three basis sets, ranging in size from DZP to very extended, for the nitrogen atom. The vertical axis indicates the magnitude of the largest CI coefficient outside the reference space. The horizontal axis indicates the value of the selection threshold used in performing perturbation theory selection

of CI which appears to be necessary in this case. Once again, as with the (s, p, d, f) extended basis calculation on B_2, the last calculation listed in Table 8 is unconverged with respect to zero selection threshold. The properties reported for the largest reference space were obtained with a threshold of $T_E = 1.3 \times 10^{-6}$. Even though the CI included over 140,000 spin-adapted configurations, these represent a tiny fraction of the 24 million possible single and double excitations. Second order perturbation theory indicates that more than 95% of the correlation energy available in the 24 million configurations has been variationally extracted with the 140,000 configurations. Evidence from other molecules suggests that further decreases in T_E below 10^{-6} can be expected to lead to substantial increases in A_{iso}, but the current level of available hardware and software were unable to handle the additional hundreds of thousands of configurations.

Table 8. MR SD–CI Results for H_2CO^+ ($^2B_2$) Selecting on Energy Contributions to the Lowest Root of Reference Space[a]

Primitive Basis Set	Ref.	Size[e]	E_{CI}	^{13}C	^{17}O	^1H
(8s, 4p, /4s, 1p)	1	(1)[f]	−113.5391	−53.2	−30.8	159.7
(10s, 5p, 1d /8s, 1p)[b]	1	(1)[f]	−113.5391	−53.2	−30.9	168.1
(12s, 6p, 2d /10s, 2p)[c]	13	(23)	−113.8911	−64.4	−35.0	204.6
(19s, 10p, 3d, 1f/10s, 2p, 1d)[d]	1	(1)	−113.9498	−57.3	−27.0	303.5
	50	(142)	−113.9588	−79.2	−39.1	252.5
Experiment				−108.7		371.9

[a] Calculations for all basis sets except the last one were performed at the following geometry: R_{CO} = 121.0 pm, R_{CH} = 110.0 pm and HCH angle = 124°. The the (s, p, d, f) basis set calculation was performed at the optimal 10-orbital/11-electron full CI geometry: R_{CO} = 119.8 pm, R_{CH} = 111.7 pm and HCH angle = 123°;
[b] The contracted basis was [*8s, 4p, 1d/6s, 1p*], ref. [39] and [40];
[c] The contracted basis was [*8s, 4p, 2d/6s, 2p*]. If selection is based on the two lowest roots of the reference space A_{iso} on hydrogen = 230 MHz, ref. [39] and [40];
[d] The contracted basis was [*10s, 5p, 3d, 1d/5s, 2p, 1d*]. The CI included 143,970 spin-adapted configurations selected from 24.5 million singles and doubles. $|C_{max}|$ = 0.017. With a selection threshold T_E = 1.3 × 10^{-6} approximately 95% of the estimated singles and doubles correlation energy was variationally recovered. D. Feller and E. R. Davidson, unpublished work;
[e] The number of space orbital products (spin-adapted configurations) in the reference space;
[f] All singles CI

Table 9. A Comparison of Basis Set and Numerical Isotropic Hyperfine Results (in MHz)[a]

	RHF				Singles CI			
		A (iso)		A (aniso)		A (iso)		A (aniso)
	Energy	B		B	Energy	B		B
B_2[b]	−49.0903	0.0		−19.48	−49.0903	−11.25		−22.66
	−49.0910	0.0		−19.60	−49.1083	−16.32		−22.64
		C	P	C P		C	P	C P
CP[c]	−378.4841	614	104	68 74	−378.4983	580	−105	54 128
	−378.4854	618	105	68 74	−378.4997	578	−103	57 127
		Si	O			Si	O	
SiO$^+$[d]	−363.4506	−26	−167		−363.4711	132	−98	
	−363.4625	−19	−167		−363.4831	124	−106	

[a] The basis set results are listed over the numerical results;
[b] Taken from Ref. [61];
[c] Taken from Ref. [62];
[d] Taken from Ref. [42]

The question of the impact of finite Gaussian basis sets on hyperfine property evaluations can be tackled in an entirely different manner than the one just discussed, i.e. repeated calculations with ever increasing basis sets. Numerical Hartree-Fock for atoms and diatomic molecules can, in principle, provide exact solutions of the HF equations. Thus, at least at the HF level it should be possible to provide a quantitative measure of the errors introduced by the Gaussian basis sets reported in print. Table 9 presents a comparision of extended basis sets [61, 62] and numerical results on B_2, CP and SiO^+. The numerical findings are based on the partial-wave procedure of McCullough and coworkers [63].

3 Conclusion

In the 20 years since the appearance of the first ab initio calculation of molecular hyperfine properties, rapid improvements in ESR techniques have produced a growing body of information on small radicals. This, in turn, has spurred renewed interest among theoreticians in spite of the fact that "small molecules" work is less in vogue today than it once was. Ab initio methods are increasingly applied to larger and larger systems, in an attempt to become more relevant to the whole of chemistry. Much of the research effort in large systems has focused on geometries and energy-related quantities. However, these represent a small fraction of the information which can be extracted from the wave function. Other properties, such as the hyperfine parameters, pose a much more severe challenge.

While there are other areas of Quantum Chemistry known to pose special computational difficulties, few present the spectrum of problems, ranging from a need for balanced core/valence correlation to a need for accurate geometries in species poorly desribed at the RHF level, that characterize hyperfine properties. In some instances, such as SiO^+ and H_2CO^+, experimental findings exposed inadequacies in molecular wave functions which otherwise appeared to be of high quality. In other cases, such as AlF^+ and BC, theoretical calculations were able to rule out spectra as originating from the presumed radical. The symbiotic nature of this relationship serves to the betterment of both branches of science.

As indicated by a soon to be released review article on condensed phase ESR of hydrocarbon radical cations [64] the number of applications of ab initio techniques to experimental problems arising from hyperfine property measurements has steadily increased. However, even the most fundamental requirements in the wave function, necessary for reliably and systematically obtaining A_{iso} values with true predictive accuracy, are not universally understood, as recent papers in print would indicate. Before ab initio calculations of A_{iso} become as widespread (or as successful) as geometry optimizations it will be necessary to develop new methods which avoid the necessity of recovering over 90% of the total correlation energy. Hopeful signs of progress in this regard may be seen in the very recent work of Chipman [66] and Carmichael [67]. Careful calibration of any new methods will be essential since nearly any answer can be obtained in a specific case by unwittingly shifting the balance in favour of core or valence contributions.

4 References

1. Bethe HA, Salpeter EE (1957) in: Quantum mechanics of one- and two-electron atoms, Academic, New York
2. Harriman JE (1978) in: Loebl EM (ed) Theoretical foundations of electron spin resonance, Physical Chemistry Series, vol 37, Academic, New York
3. See for example, Ramsey NF (1956) in: Molecular beams, Oxford University Press, Clarendon, England
4. Kutzelnigg W (1988) Theor. Chim. Acta 73: 173
5. Fermi E (1930) Z. Physik, 60: 320
6. Herman F, Skillman S (1963) in: Atomic structure calculations, Prentice-Hall, Englewood Cliffs, New Jersey
7. Amos AT, Hall GG (1961) Proc. Roy. Soc. A 263: 483
8. Amos AT, Snyder LC (1964) J. Chem. Phys. 41: 1773
9. Harriman JE (1964) J. Chem. Phys. 40: 2827
10. Phillips DH, Schug JC (1974) J. Chem. Phys. 61: 1031
11. McConnell HM (1956) J. Chem. Phys. 24: 764
12. McConnell HM, Chestnut DB (1957) J. Chem. Phys. 27: 984
13. Feller D, Borden WT, Davidson ER (1981) J. Chem. Phys. 74: 2256
14. Gold A (1969) J. Am. Chem. Soc. 91: 4961
15. Pople JA, Beveridge DL (1970) in: Approximate molecular orbital theory, McGraw-Hill, New York
16. Sieiro C, de la Vega JMG (1985) J. Mol. Struct. 120: 383
17. Chipman DM (1979) J. Chem. Phys. 71: 761
18. Chipman DM (1983) J. Chem. Phys. 78: 4785
19. Huang MB, Lunell S, Lund A (1983) Chem. Phys. Lett. 99: 201
20. Lunell S, Huang MB, Claesson O, Lund A (1985) J. Chem. Phys. 82: 5121
21. Chipman DM (1983) J. Chem. Phys. 78: 3112
22. McCanus H, Fessenden RW, Chipman DM (to be published).
23. McCanus H, Fessenden RW, Chipman DM (to be published).
24. Konishi H, Morokuma K (1971) Chem. Phys. Lett. 12: 408
25. Konishi H, Morokuma K (1972) J. Am. Chem. Soc. 94: 5603
26. Chang SY, Davidson ER, Vincow G (1970) J. Chem. Phys. 52: 1740
27. Poling SM, Davidson ER, Vincow G (1971) J. Chem. Phys. 54: 3005
28. Fraga S, Saxena KMS, Lo BWN (1971) Atomic Data 3: 323
29. Rogers MT, Kispert LD (1967) J. Chem. Phys. 46: 221; and Shiga T, Yamaoka H, Lund A (1974) Z. Naturforsch. Teil A 29: 653
30. Vincow G, Chang SY, Davidson ER (1971) J. Chem. Phys. 54: 4121
31. Millie P, Levy B, Berthier G (1972) Int. J. Quantum Chem. 6: 155
32. Ohta K, Hirao K, Yonezawa T (1980) J. Chem. Phys. 73: 1770
33. Nakatsuji H, Ohta K, Yonezawa T (1983) J. Chem. Phys. 87: 3068
34. Raimondi M, Tantardini GF, Simonetta M (1974) Mol. Phys. 30: 797
35. Chang SY, Davidson ER, Vincow G (1970) J. Chem. Phys. 52: 5596
36. Ohta K, Nakatsuji H, Meada I, Yonezawa T (1982) Chem. Phys. 67: 49
37. Fessenden RW (1967) J. Phys. Chem., 71: 74
38. a. Knight LB, Martin RL, Davidson ER (1979) J. Chem. Phys. 71: 3991; b. Knight LB, Wise MB, Childers AG, Davidson ER, Daasch WR (1980) J. Chem. Phys. 73: 4189; c. Knight LB, Wise MB, Childers AG, Daasch WR, Davidson ER (1981) J. Chem. Phys. 74: 4256; d. Knight LB, Wise MB, Davidson ER, McMurchie LE, (1982) J. Chem. Phys. 76: 126
39. Feller D, Davidson ER (1984) J. Chem. Phys. 80: 1006 and unpublished results
40. Feller D, Davidson ER (1985) Theoret. Chim. Acta 68: 57
41. Knight LB, Steadman J, Miller PK, Bowman DE, Davidson ER, Feller D (1984) J. Chem. Phys. 80: 4593
42. Knight LB, Ligon A, Woodward RW, Feller D, Davidson ER (1985) J. Amer. Chem. Soc., 107: 2857

43. Knight LB, Ligon A, Cobranchi ST, Cobranchi DP, Earl E, Feller D, Davidson ER (1986) J. Chem. Phys. 85: 5437
44. Knight LB, Ligon A, Woodward RW, Feller D, Davidson ER (1985) J. Am. Chem. Soc. 107: 2857
45. Knight LB, Feller D, Davidson ER (1989) J. Chem. Phys. 90: 690
46. Knight LB, Feller D, Davidson ER (to be published)
47. Knight LB, Johannessen KD, Cobranchi DC, Earl EA, Feller D, Davidson ER (1987) J. Chem. Phys. 87: 885
48. Schaefer HF III, Klemm RA, Harris FE (1968) Phys. Rev. 176: 49
49. Harvey JSM, Evans L, Lew H (1972) Can. J. Phys. 50: 1719
50. Holloway WW, Novick R (1958) Phys. Rev. Lett. 1: 367; Holloway WW, Luscher E, Novick R (1962) Phys. Rev. 126: 2109
51. Harvey JSM (1961) Proc. Roy. Soc. (London) A285: 581
52. Hirsch JM, Zimmerman GH, Larson DJ, Ramsey NF (1977) Phys. Rev. A 16: 484
53. Bagus PS, Liu B, Schaefer HF III (1970) Phys. Rev. A, 2: 555
54. Chipman D, (1989) Phys. Rev. 39: 415
55. Engels B, Peyerimhoff S, Davidson ER (1987) Mol. Phys. 62: 109
56. Bauschlicher CW, Langhoff SR, Partridge H, Chong DP J. Chem. Phys. (1988) 89: 2985
57. Feller D, Davidson ER (1988) J. Chem. Phys. (to be published)
58. Feller D, Ruedenberg K (1979) Theoret. Chim. Acta 52: 231
59. Schmidt MW, Rudenberg K (1979) J. Chem. Phys. 71: 3951
60. Feller D, Boyle CM, Davidson ER (1987) J. Chem. Phys. 86: 3424
61. Knight LB, Gregory BW, Cobranchi ST, Feller D, Davidson ER (1987) J. Am. Chem. Soc. 109: 3521
62. Knight LB, Petty JT, Cobranchi ST, Feller D, Davidson ER (1988) J. Chem. Phys. 88: 3441
63. Richman KW, Shi Z, McCullough EA (1987) Chem. Phys. Lett. 141: 186 and private communication
64. Lund A, Lindgren M, Lunell S (1988) in: Maruani J (ed) Hydrocarbon radical cations in condensed phases; Molecules in physics, chemistry and biology, Reidel, Dordrecht
65. Feller D, Davidson ER (1989) J. Chem. Phys. 90: 1024
66. Chipman D (1989) J. Chem. Phys. 91: 5455
67. Carmichael I (1989) J. Phys. Chem. 93: 190

Rovibrational Averaging of Molecular Electronic Properties

Cynthia J. Jameson

Department of Chemistry, University of Illinois Chicago, Illinois 60680, USA

The measured value of a molecular electronic property is a rovibrational average for a given v, J state or a thermal average. These electronic properties may exhibit measurable isotope effects and are found to be v, J-state-dependent and temperature dependent. Rovibrational effects are intimately related to both the molecular potential surface and the electronic property surface. When both surfaces are available it becomes possible to correct the observed values of the property for rovibrational effects and thereby to elicit the value of the property at the equilibrium geometry of the molecule, for comparison with the ab initio theoretical value. Alternatively, temperature coefficients and isotope shifts may be calculated from both theoretical surfaces, for direct comparison with experiment. In some favourable cases it is possible to obtain an empirical estimate of the sensitivity of the property to a change in geometry from a measurement of a temperature coefficient or an isotope effect. A review of the theory and its applications to the nuclear magnetic shielding, spin-rotation constant, nuclear hyperfine constant, electric field gradient, spin-spin coupling, dipole moment, magnetizability, electric dipole polarizability and hyperpolarizabilities reveals some interesting general trends.

1 Introduction

Rovibrational averaging of a molecular electronic property manifests itself in various ways. Different values of the electronic property for each J or v level may be measured where individual J- or v-labeled states are observed, with different values for different isotopomers. Where thermal averages over all populated v, J states are observed, a temperature dependence of the electronic property in the limit of the isolated molecule or different thermal average values of the electronic property for different isotopomers may be observed. The property may be zero except upon isotopic substitution, as in the dipole moment of CH_3D. In NMR spectroscopy where ultrahigh resolution (mHz) is becoming more accessible, temperature dependent and mass-dependent effects have become ubiquitous [1]. Even in modest high resolution the effects of rovibrational averaging are observed. See for example Fig. 1 in which the individual peaks for the ^{19}F in the various isotopomers $^{m'}SeF_6$ are observed with intensities directly proportional to the natural abundance of the Se isotopes [2].

Rovibrational averaging is intimately connected with ab initio calculations of electronic properties. As calculations become more sophisticated and/or as measurements become more precise, the gap between the ab initio calculated value of a pro-

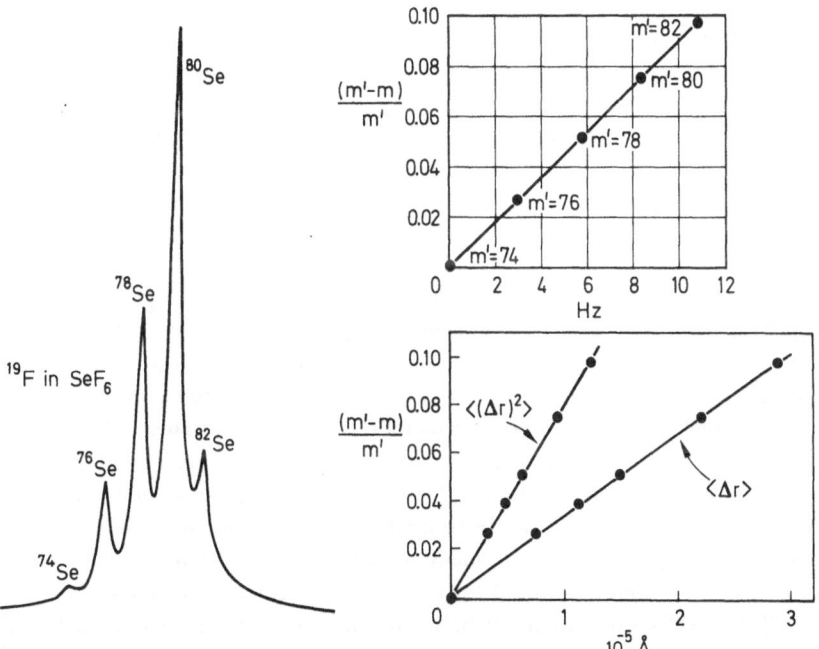

Fig. 1. ^{19}F NMR spectrum in liquid SeF_6 at 300 K at 188.3 MHz. The intensities of the peaks for the isotopomers are consistent with the natural abundance of Se isotopes. $^{77}SeF_6$ satellite peaks (not shown) are split by 1421 Hz. These isotope shifts between $^{m'}SeF_6$ and $^{m}SeF_6$ are plotted for m = 74, also the mean bond displacements and mean square amplitudes (at 300 K) of the Se—F bonds in SeF_6 molecule, are proportional to (m'—m)/m' (see text). Reproduced from Jameson et al. (1986) J. Chem. Phys. 85: 5480, 5484, with permission [2, 80]

perty and its experimental value becomes sufficiently small so that theoretical values calculated at the equilibrium geometry of a molecule cannot be compared directly with the measured value of the property even at 0 K. Rovibrational averaging is also intimately connected with determination of molecular geometry. The observed quantities in various methods of determining molecular structure are themselves particular rovibrational averages of various powers of atom-atom distances or averages of differences between various powers of atom position coordinates. Without the appropriate computations based on the averages which are directly observed, different techniques such as microwave spectroscopy, electron diffraction, and NMR spectroscopy lead to structures that are not consistent. Results from each technique have to be corrected by appropriate rovibrational averaging, back to the geometry corresponding to the minimum in the potential energy. Only the equilibrium geometry resulting from the various measurements can be directly compared.

The effective hamiltonian for a free molecule in a non-degenerate electronic state has been derived by Michelot [3, 4], which allows a study of the various interactions within the molecule and the direct and indirect effects induced by molecular vibration. In particular we could consider with this hamiltonian the terms characteristic of the interaction between a nuclear magnetic moment and an external magnetic field, the field induced at the nucleus by molecular motion, and the interaction of the magnetic moment induced by this molecular motion with an external magnetic field. The effective hamiltonian is obtained by a second order computation in degenerate perturbation theory and explicitly includes the parameters characteristic of spin-vibration and spin-rotation interactions together with those associated with vibrational and rotational Zeeman effects, and it can be written in a form of an effective hamiltonian for a given vibronic state.

Several levels of approximation are possible: (1) the Born-Oppenheimer approximation, (2) the adiabatic corrections, and (3) non-adiabatic corrections. At the level of the Born-Oppenheimer approximation, as the nuclei move about an equilibrium position corresponding to a minimum of the potential function, it is assumed that in this vicinity the electronic wavefunctions does not vary greatly with r. At level 1 the molecular potential of a given electronic state is invariant through isotopic substitution. Some electronic properties vanish at level 1 and at least level 2 approximation must be used. An example of this is the dipole moment of HD. To go beyond level 1, one may add a term $V_{ad}(r)$ to the potential energy, (see for example [4]) which depends upon nuclear mass and which accounts for the noninvariance upon isotopic substitution of the molecular potential of a given electronic state. Some properties are only partly accounted for at level 2 and level 3 must be used for a complete picture. Examples are the differences in the electron spin densities at the protons in aromatic radical ions such as $C_6H_6^-$ upon monodeuteration. Non-adiabatic corrections are introduced when the computation of the effective hamiltonian is extended to second order [3, 4]. The well-known relations between some electronic properties such as the spin-rotation tensor and the paramagnetic part of the shielding tensor [5], the molecular g tensor and the paramagnetic part of the magnetizability [6] only hold strictly for clamped nuclei. Nonadiabatic corrections make the relationships between these tensors only approximate for the vibrating molecule [3].

Many of the observed molecular electronic properties are coefficients of terms in the effective molecular hamiltonian. These coefficients may consist of the sum of a purely nuclear term and a term which depends upon electronic variables. This review comments on the latter only. Some rovibrational effects on properties are largely accounted for at level 1, i.e., within the context of the Born-Oppenheimer approximation, for example the isotope effects on NMR chemical shifts and coupling constants, the vibrational or rotational state dependence of eqQ and C. We shall concentrate our attention on these because they are the ones which are best understood from a theoretical point of view and also most clearly characterized from a quantitative experimental point of view. As a consequence of the Born-Oppenheimer approximation the concept of a mass-independent molecular electronic property surface can be used. This is analogous to the potential energy surface; the value of any property in a given electronic state can be expressed in terms of a Taylor series in the instantaneous displacement from equilibrium geometry.

$$P = P_e + \sum_i P_i q_i + \sum_{ij} \frac{1}{2!} P_{ij} q_i q_j + \sum_{ijk} \frac{1}{3!} P_{ijk} q_i q_j q_k + \cdots \tag{1}$$

The coordinates describing the nuclear displacements may be in dimensionless normal coordinates q_i, normal coordinates Q_i, curvilinear symmetry coordinates S_i or curvilinear internal coordinates $\mathfrak{R}_i$. The value of the property for nuclei fixed at the equilibrium geometry is denoted by P_e and the derivatives of the property surface P_i, P_{ij}, etc. are electronic properties which describe the sensitivity of P to small displacements of the nuclei away from equilibrium.

Molecular electronic property surfaces are emerging into their own rightful place in chemistry next to potential energy surfaces. They reflect basic features of the electronic distribution of the molecule, being sensitive to different portions of the distribution, depending on the nature of the property operator. Even those properties that are determined largely by the value of the wavefunction in the immediate vicinity of a nucleus are found to be sensitive to small changes in internuclear distances. One of the first formulations of coupled Hartree-Fock theory as applied to nuclear displacements, the influential paper by Geratt and Mills [7] considered dipole moment derivatives. Methods of computation of derivatives of a property surface have been reviewed for μ and α by Amos and also Helgaker [8]. The $\mu(\mathfrak{R})$ and $\alpha(\mathfrak{R})$ surfaces are of extreme importance in the calculations of transition probabilities, i.e., intensities in IR and Raman spectra. They are also relevant to the calculations of the purely vibrational components of the electric dipole polarizability α and the hyperpolarizabilities β and γ, the vibrational components of which have their origin in the perturbation of the nuclear motion by the electric field [9, 10], a deformation of the nuclear frame simultaneously with electronic deformation. In some cases these vibrational components of the hyperpolarizabilities can be of greater importance than the electronic hyperpolarizability itself [11].

The obvious method of determining property surfaces in ab initio calculations is by calculation of the property itself at the equilibrium geometry and at suitable values of curvilinear symmetry internal coordinates. The values of the property may be computed as expectation values of the corresponding perturbation operator or as the derivatives of the perturbed energy or as the derivatives of some other property. For example, the molecular electric dipole polarizability components can be computed

from the molecular wavefunctions in finite static electric fields as the second derivatives of the energy with respect to the components of the electric field, or as the first
derivatives of the electric dipole moment components with respect to the static electric
field components. Alternatively, it is sometimes convenient to evaluate the first few
derivatives at the equilibrium geometry using analytic gradient techniques [12]. Unfortunately, some properties require highly correlated wavefunctions, which makes the
evaluation of derivatives difficult. The expansion in (1) is not appropriate when dealing
with properties of molecules with very large amplitude internal motions such as to
render meaningless the notion of an equilibrium geometry [13].

Examples of property surfaces are shown in Fig. 2, 3, 4 and 5. Fig. 2 shows the ab
initio surfaces for the nuclear spin-spin coupling J(CO) in the NMR spectroscopy of
CO molecule, calculated by [14]. Figure 3 shows the experimental surfaces (solid
curves) for $\mu(HF)$ from [15], $\mu(HCl)$, $\mu(HBr)$ and $\mu(HI)$ from [16], in comparison with
ab initio values calculated at various internuclear separations by [17] at the MCSCF
and other levels of computation. In Fig. 4 reproduced from [18] is the ^1H nuclear
magnetic shielding surface for H_2^+ molecule calculated by Hegstrom [19] and the
experimental potential energy surface. In Fig. 5 the empirical nuclear quadrupole
coupling function for $^{127}I_2$ in the ground and B electronic states have been obtained
from the rotational and vibrational dependence in up to about 80 vibrational states
[20].

Theoretical connections have been made between many molecular electronic properties at the equilibrium geometry [5, 6, 21, 22]. This would mean that by appropriate
rovibrational corrections the same ab initio surface can be used for evaluating more
than one property. See, for example, [23].

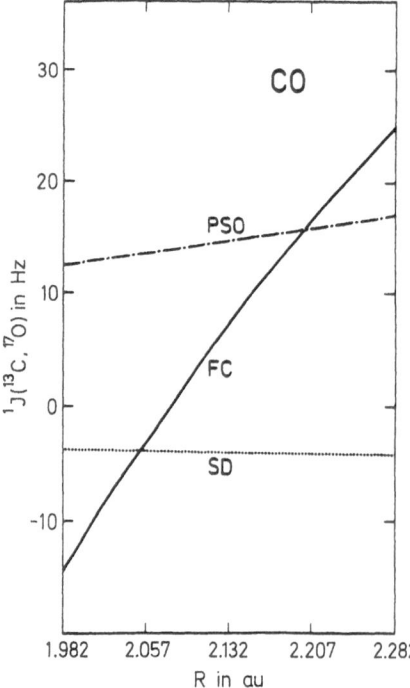

Fig. 2. Geometry dependence of the contributions to
the nuclear spin-spin coupling constant of carbon
monoxide, that is, the paramagnetic spin-orbit
(PSO), the Fermi contact (FC), and the spin-dipolar
terms (SD). The results are calculated in the second
order polarization propagator approximation. Reproduced from Geertsen J, Oddershede J, Scuseria
GE (1987) J. Chem. Phys. 87: 2138, with permission
[14]

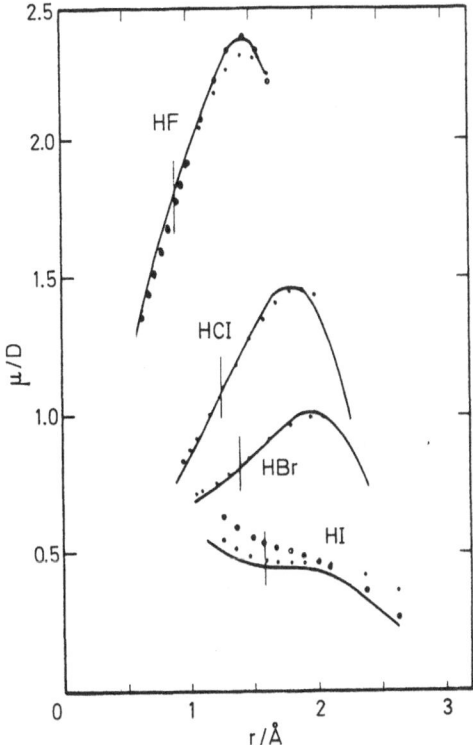

Fig. 3. Calculated and experimental dipole moment functions of the hydrogen halides. *Vertical bars* mark the equilibrium distances. *Solid curves* are from Sileo and Cool (HF) [15] and Ogilvie et al. (HCl, HBr, HI) [16], and the *calculated points* shown are for various ab initio calculations by Werner et al. [17]. Reproduced from Werner HJ, Reinsch EA, Rosmus P (1981) Chem. Phys. Lett. 78: 311, with permission [17]

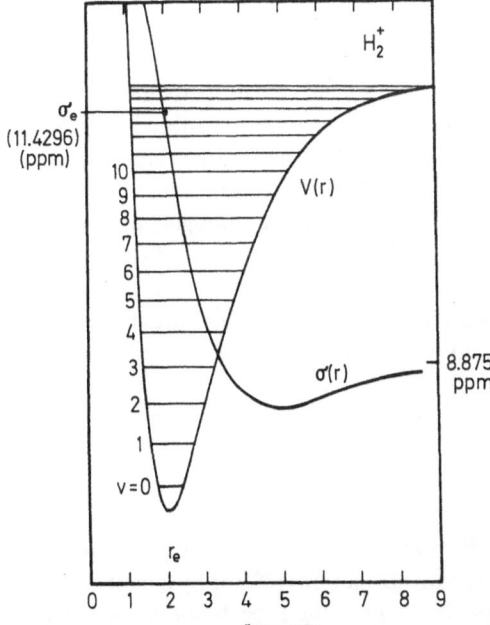

Fig. 4. The proton nuclear magnetic shielding surface and the potential energy surface for H_2^+, plotted from values given in [19]. Reproduced from Jameson CJ, Osten HJ (1986) Ann. Reports NMR Spectrosc. 17:1, with permission [18]

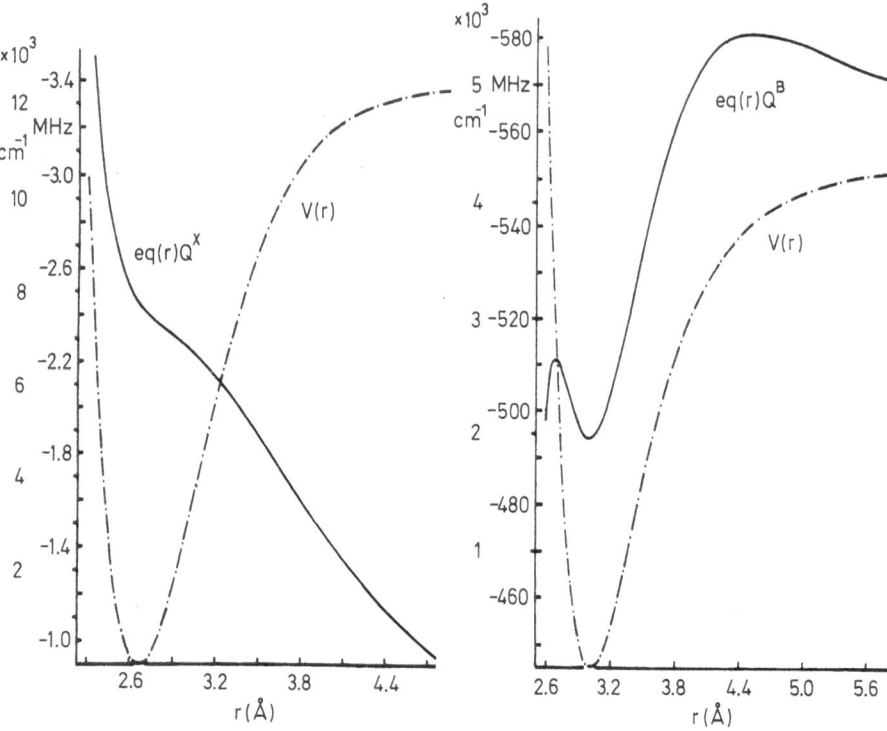

Fig. 5. Empirical nuclear quadrupole coupling function and potential energy curves for the X and B electronic states of $^{127}I_2$. Reproduced from Spirko V, Blabla J (1988) J. Mol. Spectrosc. 129: 59, with permission [20]

The properties discussed in this review can be expressed entirely in terms of nuclear position coordinates and thus the property surface can be expanded around the equilibrium geometry in terms of property derivatives with respect to nuclear coordinates of some type (Cartesian, normal, etc.). In those cases where the property has significant non-Born-Oppenheimer contributions, the concept of a property surface is still useful, provided that by means of perturbation theory, the first order vibronic function incorporates the change in electronic state that is due to small nuclear displacements q_i so that the electron cloud may be thought of as being modulated by the vibrational motion of the nuclei [24]. The concept of property surfaces has been extended by Buckingham et al. to include property surfaces expanded in terms of derivatives with respect to nuclear velocities or nuclear momentum. The latter have been used to describe VCD and other phenomena [24]. (See also [25–27].)

2 Theoretical Considerations

The expectation value of a property P in an arbitrary vibration-rotation state of a particular isotopic form of a molecule, $\langle P \rangle$, may be derived by Rayleigh-Schrodinger perturbation theory [28], by use of contact transformations [29], or by a method based on the Hellmann-Feynman theorem [30]. We require the expectation values of P in the state which is an eigenfunction of the molecular vibration-rotation hamiltonian operator

$$H = H_{(0)} + \lambda H_{(1)} + \lambda^2 H_{(2)} + \ldots$$

in which $H_{(0)}$ is the hamiltonian for non-degenerate harmonic oscillators and a rigid rotor, having known solution $X^{(0)}$. λ is a smallness parameter, $H_{(1)}$ and $H_{(2)}$ are perturbations each of which includes terms due to anharmonicity and vibration-rotation interaction. The property P has a corresponding operator which can be expressed as in (1), in which the sums over the reduced dimensionless coordinates are unrestricted. To first order in λ, the expectation value can be evaluated as

$$
\begin{aligned}
\langle P \rangle_n &= \langle X_n^{(0)} + \lambda X_n^{(1)} | P | X_n^{(0)} + \lambda X_n^{(1)} \rangle \\
&= \langle X_n^{(0)} | P | X_n^{(0)} \rangle + 2\lambda \langle X_n^{(0)} | P | X_n^{(1)} \rangle
\end{aligned}
\tag{2}
$$

or

$$\langle P \rangle = P_e + \sum_i \left[\frac{1}{2} P_{ii} - \sum_j \frac{1}{2} P_j \frac{\varphi_{iij}}{\omega_j} \right] \left\langle v_i + \frac{1}{2} \right\rangle + \text{rotational part} + \cdots \tag{3}$$

where ω_j and φ_{iij} are the quantities in

$$V = \frac{1}{2!} \sum_i \omega_i q_i^2 + \frac{1}{3!} \sum_{ijk} \varphi_{ijk} q_i q_j q_k + \cdots \tag{4}$$

Equation (3) was obtained by Toyama, Oka, and Morino [28]. The same result can be obtained by contact transformation, which see below.

The hamiltonian H itself is transformed by $T_1 H T_1^{-1}$ into [29]

$$H'_{(0)} + \lambda H'_{(1)} + \lambda^2 H'_{(2)} + \ldots \tag{5}$$

where

$$T_1 = e^{i\lambda S_1} \tag{6}$$

from which

$$H'_{(0)} = H_{(0)} \tag{7}$$

$$H'_{(1)} = H_{(1)} + i[S_1, H_{(0)}] \tag{8}$$

$$H'_{(2)} = H_{(2)} + \frac{i}{2} [S_1, (H_{(1)} + H'_{(1)})] \tag{9}$$

The function S_1 is chosen in such a way that all off-diagonal matrix elements of $H'_{(1)}$ vanish, i.e., such that the commutator of S_1 with $H_{(0)}$ is the negative of any term in $H_{(1)}$ which is non-diagonal. To obtain wavefunctions correct through second order, it is necessary to carry out a second contact transformation on H' to remove the off-diagonal elements of $H'_{(2)}$, i.e.,

$$H^\dagger = T_2 H' T_2^\dagger = H_{(0)}^\dagger + \lambda H_{(1)}^\dagger + \lambda^2 H_{(2)}^\dagger + \dots \tag{10}$$

where

$$T_2 = e^{i\lambda^2 S_2}. \tag{11}$$

It then follows that

$$H_{(0)}^\dagger = H'_{(0)} \tag{12}$$

$$H_1^\dagger = H'_{(1)} \tag{13}$$

$$H_{(2)}^\dagger = H'_{(2)} + [S_2, H'_{(0)}] \tag{14}$$

Since $H_{(0)}$ is diagonal in a harmonic oscillator basis set, and since $H'_{(0)} = H_{(0)}$, and $H'_{(1)}$ has been made diagonal in the same harmonic oscillator basis set, $H'_{(0)} + H'_{(1)}$ is diagonal in the harmonic oscillator basis set and the latter are therefore first order wavefunctions of H', i.e.,

$$\langle X^{(0)} | H'_{(0)} + H'_{(1)} | X^{(0)} \rangle = \langle X^{(0)} | T(H_{(0)} + H_{(1)}) T^{-1} | X^{(0)} \rangle \tag{15}$$

Therefore $T^{-1} X^{(0)}$ are first order wavefunctions of H. Similarly,

$$\langle X^{(0)} | H_{(0)}^\dagger + H_{(1)}^\dagger + H_{(2)}^\dagger | X^{(0)} \rangle = \langle X^{(0)} | T_2 T_1 (H_{(0)} + H_{(1)}$$
$$+ H_{(2)}) T_1^{-1} T_2^{-1} | X^{(0)} \rangle \tag{16}$$

and therefore $T_1^{-1} T_2^{-1} X^{(0)}$ are second order wavefunctions of H. To use wavefunctions of second order in evaluating the expectation value of P, we shall need to evaluate the matrix elements

$$\langle X^{(0)} | T_2 T_1 P T_1^{-1} T_2^{-1} | X^{(0)} \rangle,$$

which are, in fact, the harmonic oscillator matrix elements of the doubly contact-transformed P operator.

$$P = P_{(0)} + \lambda P_{(1)} + \lambda^2 P_{(2)} + \dots \tag{17}$$

where

$$P_{(0)} = P_e + \sum_i P_i q_i \tag{18}$$

$$P_{(1)} = \sum_{ij} \frac{1}{2!} P_{ij} q_i q_j \tag{19}$$

$$P_{(2)} = \sum_{ijk} \frac{1}{3!} P_{ijk} q_i q_j q_k \tag{20}$$

The doubly contact transformed operator is then found [31]

$$T_2 T_1 P T_1^{-1} T_2^{-1} = P^\dagger = P^\dagger_{(0)} + P^\dagger_{(1)} + P^\dagger_{(2)} + \dots \tag{21}$$

where

$$P^\dagger_{(0)} = P_{(0)} \tag{22}$$

$$P^\dagger_{(1)} = P_{(1)} + i[S_1, P_{(0)}] \tag{23}$$

$$P^\dagger_{(2)} = P_{(2)} + i[S_1, P_{(1)}] + \frac{i^2}{2}[S_1, [S_1, P_{(0)}]] + i[S_2, P_{(0)}] \tag{24}$$

$$P^\dagger_{(3)} = P_{(3)} + i[S_1, P_{(2)}] + i[S_2, P_{(1)}] - \frac{1}{2}[S_1, [S_1, P_{(1)}]] +$$
$$+ i^2[S_2, [S_1, P_{(0)}]] - \frac{i}{6}[S_1, [S_1, [S_1, P_{(0)}]]] \tag{25}$$

From the commutators of S_1 and S_2 with the normal coordinate operators q_j, the commutators of S_1 and S_2 with $P_{(0)}$, $P_{(1)}$, and $P_{(2)}$ can be evaluated. The results are: [31]

$$P^\dagger_{(0)} = P_{(0)} = P_e + \sum_i P_i q_i \tag{26}$$

$$P^\dagger_{(1)} = \sum_{ij} \frac{1}{2!} P_{ij} q_i q_j + \sum_{i \le j} \sum_k \{P_k S^k_{ij} q_i q_j + P_k S^{ijk}(1 + \delta_{ik} + \delta_{jk}) P_i P_j\} \tag{27}$$

The expression for $P^\dagger_{(2)}$ has been derived by Marcott, Golden and Overend [32]. The quantities S^k_{ij} and S^{ijk} involve the cubic force constants and harmonic frequencies. These and other quantities appearing in $P^\dagger_{(2)}$ are given by Amat et al. [29, 33]. The leading terms in the formula for the expectation value of a property of a diatomic molecule was derived by Buckingham [34], the general formula with terms up to 6th order by Herman and Short [35], Toyama, Oka, and Morino provided the leading terms in the formula for a polyatomic molecule [28]. The vibrational terms for a property of an asymmetry rotor was given to general n^{th} order by Krohn, Ermler, and Kern, [36] to which rotational terms were added by Fowler et al. [30] (with an earlier version in [37]).

The expectation value of P for non-degenerate modes is then given by [30]

$$\langle X^{(0)}| P^\dagger |X^{(0)} \rangle =$$

$$\langle P \rangle = P_0 + \sum_s A_s \left(v_s + \frac{1}{2}\right) + \sum_{s \le s'} B_{ss'} \left(v_s + \frac{1}{2}\right)\left(v_{s'} + \frac{1}{2}\right) \tag{28}$$

where

$$P_0 = P_e + \sum_s \frac{1}{64} P_{ssss} - \sum_s 7P_{sss}\varphi_{sss}/288\omega_s + \sum_{s \ne s'} 3P_{sss}\varphi_{sss'}\omega_{s'}/32(4\omega_s^2 - \omega_{s'}^2) \tag{29}$$

$$A_s = P_{ss}/2 - \sum_{s'} P_{s'}\varphi_{sss'}/2\omega_{s'} \tag{30}$$

$$B_{ss} = P_{ssss}/16 - \sum_{s'} P_{sss'}\varphi_{sss'}(8\omega_s^2 - 3\omega_{s'}^2)/8\omega_{s'}(4\omega_s^2 - \omega_{s'}^2) \tag{31}$$

$$B_{ss'} = P_{sss's'}/4 - \sum_{s''} (P_{sss''}\varphi_{s's's''} + P_{s's's''}\varphi_{sss''})/4\omega_{s''}$$
$$- \sum_{s''} P_{ss's''}\varphi_{ss's''}\omega_{s''}(\omega_{s''}^2 - \omega_s^2 - \omega_{s'}^2)/D_{ss's''} \tag{32}$$

$$D_{ss's''} \equiv (\omega_s + \omega_{s'} + \omega_{s''})(\omega_s + \omega_{s'} - \omega_{s''})$$
$$\times (\omega_s - \omega_{s'} + \omega_{s''})(-\omega_s + \omega_{s'} + \omega_{s''}) \tag{33}$$

In the first applications of these formulae the values of A_s, B_{ss}, and $B_{ss'}$ were calculated by Krohn et al. for the ^{17}O and D quadrupole coupling tensor and the dipole moment in H_2O, D_2O, and HDO [36]. Off-diagonal matrix elements, which correspond to transition integrals for P being the dipole moment operator, have also been derived by Overend [38–40] using contact transformations. The advantage of this type of approach for any electronic property is that the vibrational wavefunction need not be evaluated explicitly and only a knowledge of the derivatives of the property at the equilibrium geometry is necessary for calculations of vibrational state dependence, rotational state dependence, temperature and mass dependence of any electronic property.

Keeping only terms quadratic in the normal coordinates in the expansion (1) [28]

$$\langle P \rangle = P_e + \sum_s \left[\frac{P_{ss}}{2} - \sum_{s'}^{\text{tot. sym.}} P_{s'} \varphi_{sss'}/2\omega_{s'} \right] \left(v_s + \frac{1}{2} \right) + \text{rot. contrib.} \quad (34)$$

is directly obvious from

$$\langle q_{s'} \rangle = - \sum_s (\varphi_{sss'}/2\omega_{s'}) \left(v_s + \frac{d_s}{2} \right) \quad (35)$$

d_s is the degeneracy of the sth mode. The rotational contribution to a single reduced non-vanishing only when s' is totally symmetric, and

$$\langle q_s q_{s'} \rangle = \left(v_s + \frac{d_s}{2} \right) \delta_{ss'} . \quad (36)$$

normal coordinate $q_{s'}$, to be added to (35) is [28]

$$+ \frac{1}{4\pi c\omega_s} \left(\frac{1}{hc\omega_s} \right)^{1/2} \sum_\alpha \frac{a_{s'}^{\alpha\alpha}}{(I_{\alpha\alpha}^{(e)})^2} \langle J_\alpha^2 \rangle \quad (37)$$

where J_α^2 is the rotational angular momentum about the α inertial axis, and the coefficients $a_{s'}^{\alpha\alpha}$ are the derivatives of the α moment of inertia with respect to the s_i^{th} normal coordinate. Simple forms of the rotational contribution have been obtained for $\sum_\alpha a_{s'}^{\alpha\alpha}/I_{\alpha\alpha}^{(e)}$ in highly symmetric molecules with only one totally symmetric mode [41]

$$\sum_\alpha \frac{a_{s'}^{\alpha\alpha}}{I_{\alpha\alpha}^{(e)}} = \frac{6}{r_e(nm_X)^{1/2}} \quad \begin{array}{l} \text{for } n = 6, 4, \text{ and } 3 \\ \text{in } AX_6(O_h), AX_4(T_d), \text{ and } AX_3(D_3) \end{array} \quad (38)$$

$$= \frac{4}{r_e(2m_X)^{1/2}} \quad \text{for } AX_2(D_{\infty h})$$

$$= \frac{4}{r_e \mu^{1/2}} \quad \text{for diatomic molecules.}$$

The complete expressions for a symmetric top, involving degenerate vibrational modes, and for a spherical top are given by Fowler [30, 42]. The rotational contributions for asymmetric tops complete to second order in the contact transformation have also been derived by Fowler [43]:

$$\langle P \rangle_J - \langle P \rangle_0 = \frac{1}{\sqrt{3}} (B_e/\omega_1)^{3/2} \left[4P_1 + \sum_i P_{1ii} \right] J(J + 1) \tag{39}$$

where P_1, P_{1ii} are the coefficients in the expansion (1).

The above expressions then allow the calculation of the rovibrational average of an electronic property P observed for any $|vJK\rangle$ state. For the thermal average property measured in NMR or electron diffraction, where the observed property is an average over the different populated rovibrational states, one needs only the thermal averages of $\langle J(J + 1) \rangle$ and $\langle v + 1/2 \rangle$. This is accurately obtained for small number of vibrational modes by summing up over the rovibrational states weighted by populations and degeneracies:

$$\langle P \rangle^T = \frac{\sum\limits_{v, J, K} (2J + 1) g_{Ns} \langle P \rangle_{vJK} \exp(-E_{vJK}/kT)}{\sum\limits_{v, J, K} (2J + 1) g_{Ns} \exp(-E_{vJK}/kT)} \tag{40}$$

including nuclear spin statistics. Nuclear spin statistics are important at low temperatures, as shown in Fig. 6 where the thermal average of nuclear shielding components of ortho and para H_2 and D_2 are compared with HD [44].

For diatomic molecules the potential energy and the molecular electronic property surfaces are conveniently expressed in terms of the dimensionless reduced coordinate

$$\xi = (R - R_e)/R_e \tag{41}$$

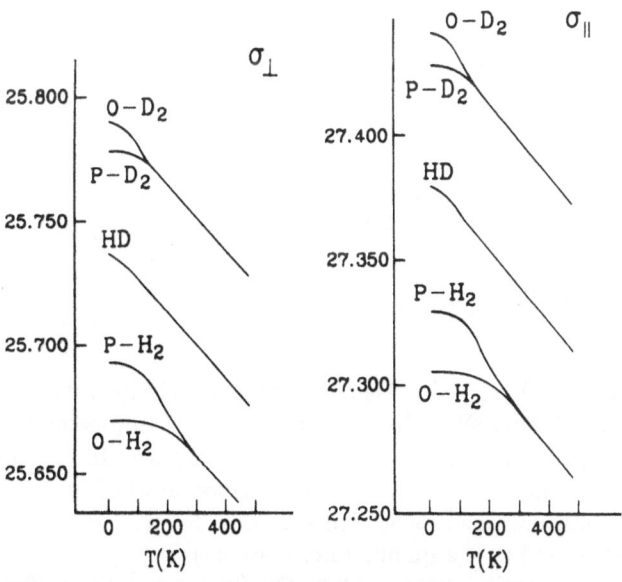

Fig. 6. Calculated nuclear magnetic shielding components for H or D in H_2, HD, and D_2 [44]. Reproduced from Ditchfield R (1981) Chem. Phys. 63: 185, with permission

as follows [34]:

$$V(R) = a_0 \xi^2 (1 + a_1 \xi + a_2 \xi^2 + a_3 \xi^3 + ...) \tag{42}$$

where

$$a_0 = \omega_e^2 / 4B_e \tag{43}$$

$$P = P_e + P'\xi + \frac{P''}{2!}\xi^2 + \frac{P'''}{3!}\xi^3 + \frac{P^{iv}}{4!}\zeta^4 + ... \tag{44}$$

The rovibrational averages $\langle \xi \rangle_{vJ}$, $\langle \xi^2 \rangle_{vJ}$, ... can be obtained by second order perturbation theory:

$$\langle \xi \rangle_{vJ} = -3a_1 \frac{B_e}{\omega_e} \left(v + \frac{1}{2} \right) \tag{45}$$

$$\langle \xi^2 \rangle_{vJ} = 2 \frac{B_e}{\omega_e} \left(v + \frac{1}{2} \right) \tag{46}$$

etc. Collecting terms, it is then possible to write,

$$\langle P \rangle = P_e + P_{rot}J(J + 1) + P_I \left(v + \frac{1}{2} \right) + P_{II} \left(v + \frac{1}{2} \right)^2 \tag{47}$$

where, keeping only terms up to $(B_e/\omega_e)^2$ [34],

$$P_{rot} = 4(B_e/\omega_e)^2 P' \tag{48}$$

$$P_I = (B_e/\omega_e) [P'' - 3a_1 P'] \tag{49}$$

and [35],

$$P_{II} = (B_e/\omega_e)^2 \left\{ \frac{P^{iv}}{4} - \frac{5}{2} a_1 P''' - \left(\frac{15}{2} a_1^2 - 3a_2 \right) P'' \right.$$
$$\left. + \left(-\frac{45}{2} a_1^3 + 39a_1 a_2 - 15a_3 \right) P' \right\} \tag{50}$$

Note that P_{rot} and $P_{II} \propto \mu^{-1}$ and $P_I \propto \mu^{-1/2}$. Therefore differences in the rotational and vibrational dependence of molecular electronic properties of isotopomers of diatomic molecules will largely reflect this μ^{-1} and $\mu^{-1/2}$ dependence. In (50) are the terms P_{II} which come in at the second contact transformation in which it should be noted that P' and P'' appear only with higher order terms in the potential energy (a_2 is a quartic force constant and a_3 is a quintic force constant).

The advantage of using an expansion such as (42) by Dunham is that the resulting $\langle P \rangle$ are expressible in the same a_i coefficients so that the relative order of magnitude of contributions to $\langle P \rangle$ are perceptible, as in (49)–(50). However, the numerical V(R) which is consistent with all the measured frequencies for the molecule may not be accurately described by (42). For more accurate average values $\langle \xi \rangle$ and $\langle \xi^2 \rangle$, one may need to perform a numerical integration using the numerical rotation-vibration wavefunctions which are the solutions of the Schrödinger equation with the numerical V(R). When a numerical integration over the numerical vibrational rotational wavefunctions (which reproduce the experimental spectroscopic energies of the ground state) of HD is used, the vibrational correction to the NMR spin spin coupling constant J(HD) at 0 K is found to differ by 0.09 Hz (out of a total rovibration-al correction equal to 1.81 Hz) from that obtained using the Dunham expansion [45]. Such accuracy in the rovibrational averaging procedure is warranted when the property derivatives P′ and P″ have been calculated to sufficiently high accuracy by ab initio methods.

3 Observed Vibrational and Rotational State Dependence of Electronic Properties

Outside of the dipole moment, the largest number of examples of vibrational depend-ence of an electronic property are probably those of the nuclear quadrupole coupling constants obtained from the Stark hyperfine structure in molecular beam electric resonance (MBER) spectroscopy, or from high resolution microwave. This property can also be obtained from pure nuclear quadrupole resonance spectroscopy or from NMR but these are of course thermal averages rather than state-labeled values. The nuclear quadrupole coupling constant eqQ is a measure of the electrical interaction between a nonspherical nucleus and a surrounding electronic environment which is likewise non-spherical. Since the trace of the electric field gradient tensor is identically zero, then for a nucleus in a linear molecule only one value needs to be specified, the component usually referred to as q_{zz} or simply q. The eqQ values of ^{35}Cl in HCl and DCl have been measured for several vibrational and rotational states and an analysis in the form of (47)–(50) leads to

$$\frac{1}{q_e}\left(\frac{dq}{d\xi}\right)_e = 1.46 \pm 0.06 \text{ and } \frac{1}{q_e}\left(\frac{d^2q}{d\xi^2}\right)_e = -3.3 \pm 0.4 [46],$$

which are in good agreement with Huo's theoretical values 1.49 and −3.07 respectively. Measurement of the rotational dependence of the ^{79}Br eqQ by MBER allowed P_{rot} in (47) to be obtained, from which by (48), $\dfrac{1}{q_e}\left(\dfrac{dq}{d\xi}\right)_e = 1.556(3)$ [47]. An earlier attempt to analyze the earlier much less precise data from millimeter wave spectra of the hydrogen halides and their isotopic species by using these same equations led to P_e, P_{rot}, and P_{II} values for each isotopomer [48], from which first and second derivatives could have been obtained. Unfortunately the experimental errors in the derived values P_{rot} and P_I themselves were too large. In I_2 the nuclear quadrupole coupling constant in the ground state has been determined for a large number of rovibrational states especially in the range $v = 26$–86 [20]. These are shown in Fig. 7. Although the data

Table 1. The linear and quadratic terms in the vibrational contributions
to the electric field gradient. The numbers shown are the coefficients of
$\left(v + \dfrac{1}{2}\right)$ in (44)–(46)[a]

Nucleus	$\dfrac{1}{q_e}\left(\dfrac{dq}{d\xi}\right)_e \langle\xi\rangle$	$\dfrac{1}{2q_e}\left(\dfrac{d^2q}{d\xi^2}\right)_e \langle\xi^2\rangle$	Sum
Cl in HCl	0.03701	−0.01090	0.02611
Br in HBr	0.03572	−0.00802	0.0277
O in HO⁻	0.08422	0.00329	0.08749
S in HS⁻	0.06294	−0.00923	0.05371
Cl in FCl	0.00624	−0.00624	0.0

[a] Calculated by Lucken [56] using theoretical derivatives given by Cade
[47]

were not fitted specifically to the form of (47), there is clearly a quadratic dependence
on $\left(v + \dfrac{1}{2}\right)$.

The sign and magnitude of the vibrational dependence of the electric field gradient
in diatomic molecules is determined mainly by the first derivative and thus by the
anharmonic contribution. In Table 1 the vibrational contributions are compared in
selected diatomic molecules, for which the first and second derivatives had been
calculated by Cade [49]. The anharmonic term clearly dominates in all but FCl
molecule.

The ^{19}F spin-rotation constant in F_2 has been measured by Ozier and Ramsey
[50] as a function of rotational quantum number:

$$C/kHz = -156.85 \pm 0.10 + (0.0024 \pm 0.0010)\, J(J + 1) . \tag{71}$$

From (48) we identify

$$0.0024 \pm 0.0010 = 4(B_e/\omega_e)\, C'$$

from which

$$C' = (dC/d\xi)_e = -650 \text{ kHz} .$$

The interesting part of the spin rotation constant is that due to the electrons. According to the theory of Reid and Chu [51] the spin rotation constant for a state v,J
is the sum of three contributions

$$C_{vJ} = C_{vJ}^{nucl} + C_{vJ}^{T} + C_{vJ}^{el} . \tag{52}$$

The nuclear contribution results from the magnetic field generated at the nucleus of
interest (whose g factor is g_N) by the motion of all the other nuclei N'. For a diatomic
molecule only the perpendicular component is non-vanishing and it simplifies to

$$\frac{C_{vJ}^{nucl}}{Hz} = \frac{e\mu_n g_N}{2\pi c\mu}\, Z_{N'} \langle R^{-3}\rangle_{vJ} \tag{53}$$

μ_n being the nuclear magneton and μ the reduced mass. This contribution is directly proportional to the direct nuclear spin-nuclear spin interaction constant which also can be measured independently from the experiment. The second term arises from the Thomas precession caused by the acceleration of the nucleus in question in the internal electric field, which for H_2 molecule is [51]

$$\frac{C_{vJ}^T}{Hz} = \frac{\hbar}{4\pi m_p^2 c^2} \left\langle \frac{1}{R} \frac{dV(R)}{dR} \right\rangle_{vJ} \tag{54}$$

where $V(R)$ is in (42). This term is zero at the equilibrium geometry and is a very small contribution for state v,J. The third term arises from the mixing of excited electronic states into the ground state by the molecular rotation and is therefore directly related to the paramagnetic part of the nuclear magnetic shielding, of which only the perpendicular component is non-vanishing for a linear molecule. For a diatomic molecule at its equilibrium geometry the relationship can be written as

$$C_e^{el} = \frac{3}{2\pi} \frac{\mu_n}{\mu_B} \frac{g_N \hbar}{2\mu R_e^2} \sigma_e^p. \tag{55}$$

The relationship holds strictly only at the equilibrium configuration. As an approximation, we can use the following relation between the rovibrationally averaged properties

$$C_J^{el} = C_e^{el} + C^{el'}\langle \xi \rangle_J + \frac{1}{2} C^{el''}\langle \xi^2 \rangle_J + \dots$$

$$\cong \frac{3}{2\pi} \frac{\mu_n}{\mu_B} \frac{g_N \hbar}{2\mu} \left[\sigma_e^p + \sigma^{p'}\langle \xi/R^2 \rangle_J + \frac{1}{2} \sigma^{p''}\langle \xi^2/R^2 \rangle_J + \dots \right] \tag{56}$$

From the MBMR hyperfine spectrum of H_2 the values $C_{J=1} = 113.904\,(30)$ kHz, $C_{J=3} = 111.10\,(25)$ kHz, and $C_{J=5} = 105.37\,(32)$ kHz were measured by Verberne and Ozier [52]. From these, after making the corrections using (53) and (54), the electronic parts alone are $-92.271\,(30)$, $-90.41\,(25)$, and $-87.74\,(32)$ kHz respectively for $J = 1, 3, 5$ states. These are then written in the expansion (56) and the average $\langle \xi^n/R^2 \rangle_{J,v=0}$ are obtained. It was found that either the set $\sigma^{p'} = +2.810 \times 10^{-6}$ and $\frac{1}{2} \sigma^{p''} = +0.29 \times 10^{-6}$ or the set $\sigma^{p'} = +2.963 \times 10^{-6}$ and $\frac{1}{2} \sigma^{p''} = 0$ fit the experimental values of C_J^{el} equally well [52], that is, the leading rovibrational correction term goes as $\langle \xi/R^2 \rangle$ rather than $\langle \xi^2/R^2 \rangle$. This is consistent with (48) in which the rotational state dependence of an electronic property of a diatomic molecule depends only on the first derivative; the first correction to (48) involves the 3rd derivative, as already indicated in (39). A later analysis by Raynes and Panteli of the rotational dependence of the spin rotation data combined with isotope effects on shielding data again provides only the first derivative [53]. Their further conclusion is that even for the H_2 molecule the theoretical proton magnetic shielding surfaces are not sufficiently accurate to reproduce the observed NMR isotope effects within the experimental errors.

The ^{199}Hg nuclear spin-electron spin interaction constants b in ^{199}HgH, ^{199}HgD, and ^{199}HgT have been determined as a function of vibrational and rotational state in the ground electronic state, from an analysis of the hyperfine splittings observed in the electronic spectra [54]. Fermi contact and dipolar terms contribute to this property:

$$b = g_e g_{Hg} \mu_B \mu_n \left\langle \frac{8\pi}{3} \langle \varrho_{spin} \rangle - \left\langle \frac{(3 \cos^2 \theta - 1)}{2r^3} \right\rangle \right\rangle_{vJ} \tag{57}$$

$\langle \varrho_{spin} \rangle$, the electron spin density at the Hg nucleus and $\langle (3 \cos^2 - 1)/2r^3 \rangle$ are averages over the molecular electronic function. The electronic spectra show a considerable decrease in b with increasing vibrational quantum number and is most pronounced in HgH, followed by HgD, and HgT. There was also a decrease in b with increasing rotation, which too was greater in HgH than in HgD or HgT [54]. The same parameter b can be obtained from the hyperfine structure in the ESR spectra of HgH and HgD in an argon matrix at 4 K [55]. Examination of the H–D isotope effect leads to the conclusion that the average spin density on Hg is slightly smaller in HgH than in HgD. The vibrational and rotational dependence of b and the isotope effects all are consistent with the $(\partial| \varrho_{spin} |/\partial R)_e$ being negative.

When the dependence on v of an electronic property of a diatomic molecule is measured, P_I and P_{II} in (47) can be determined and from P_I the quantity

$$(P'' - 3a_1 P')$$

can be obtained. In principle this can be combined with P' obtained from the rotational state dependence so both P' and P'' can be determined. The analysis is much more involved than this for polyatomic molecules.

The vibrational dependences of the nuclear quadrupole coupling constants of symmetric tops in which the quadrupolar nucleus lies on the symmetry axis, have been observed in the gas phase. A summary of data prior to 1983 is provided by Lucken [56]. The magnitude of eqQ is larger or smaller for $v_s = 1$ than for the ground vibrational state, depending on the vibrational mode. This is not surprising. As we can see in (28), for the sth mode the change is proportional to

$$\frac{1}{2} \left(\frac{P_{ss}}{2} - \sum_{s'} P_{s'} \frac{\varphi_{sss'}}{2\omega_{s'}} \right).$$

Since the first and second derivatives of the electric field gradient with respect to the reduced normal coordinates are of various signs, as are $\varphi_{sss'}$, the above expression can be of either sign. In most cases the property is known only for the $v = 0$ and $v = 1$ states, so that the determination of P_I depends on the assumption that the plot of the property vs. $\left(v + \frac{1}{2} \right)$ is a straight line, which has been shown in some cases to be a poor approximation (e.g., see Fig. 7). At best only combinations of first and second order derivatives of the property with respect to normal coordinates, e.g., $\frac{1}{2}$ $[P_{33} - P_1 \varphi_{331}/\omega_1 - P_2 \omega_{332}/\omega_2]$, can be obtained from the experimental

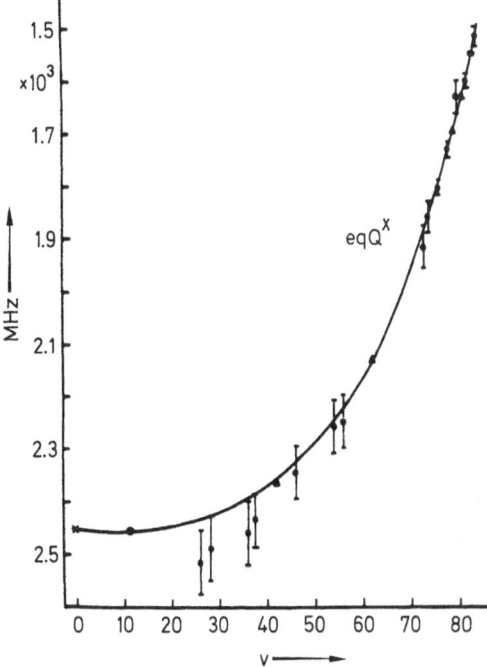

Fig. 7. Vibrational dependence of ^{127}I nuclear quadrupole coupling constant in I_2. Reproduced from Spirko V, Blabla J (1988) J. Mol. Spectrosc. 129: 59, with permission [20]

measurements of the vibrational dependence, even when linear plots of the property against $\left(v + \dfrac{1}{2}\right)$ are obtained. These are shown in Table 2 for AsF_3. Further drastic approximations (neglect of terms containing cubic force constants φ_{331}, φ_{332}, φ_{441}, and φ_{442} and neglect of off-diagonal L matrix elements in the normal coordinate analysis) lead to rough estimates of second derivatives with respect to bond extension and angle deformation [56]:

$$\frac{1}{q_e}\left(\frac{\partial^2 q}{\partial r^2}\right)_e \cong -3.5\,\text{Å}^{-2} \qquad \frac{1}{q_e}\frac{1}{r_e^2}\left(\frac{\partial^2 q}{\partial \alpha^2}\right)_e \cong -0.12\,\text{Å}^{-2}$$

for the electric field gradient at the ^{75}As nucleus in AsF_3.

Table 2. Empirical values of derivatives of $P = eqQ$ of ^{75}As in AsF_3

	s	ω_s/cm^{-1} [a]	$\left(\dfrac{P_{ss}}{2} - \overset{A_1}{\underset{s'}{\sum}} P_{s'}\dfrac{\varphi_{sss'}}{2\omega_s'}\right)\Big/\text{MHz}$ [b]
A_1	1	740.55	1.804(80)
A_1	2	336.50	0.988(6)
E	3	702.20	1.34(17)
E	4	262.30	0.23(15)

[a] Ref. [175]. [b] Obtained from $eqQ\,(v_s = 1) - eqQ\,(v = 0)$

4 Mass-independent Electronic Property Surfaces and Derivatives of Surfaces

While (28) gives the simplest form of the connection between the properties of the potential surface $\omega_{s'}$, $\varphi_{sss'}$, etc. and the rovibrationally averaged molecular electronic property, $\langle P \rangle$, they are not the relations of choice when discussing isotope effects, since the reduced dimensionless normal coordinates are mass-dependent. Furthermore, it is sometimes easier to understand the physical interpretation of a property derivative when it is expressed in terms of curvilinear internal displacement coordinates such as bond stretches and angle deformations. For these reasons, it becomes necessary to consider the transformation of the expansion (1) into these curvilinear internal coordinates $\mathfrak{R}_i$. For example, in the CH_4 molecule, $\mathfrak{R}_i$ ($i = 1$–4) denotes one of the four bond extensions Δr_1, Δr_2, Δr_3, and Δr_4 (which we will denote simply as r_1, r_2, r_3, r_4) and $\mathfrak{R}_i$ ($i = 5$–10) denotes one of the six interbond angle changes α_{12}, α_{13}, etc. $\mathfrak{R}_i$ have a nonlinear relationship to the dimensionless normal coordinates [57]

$$\mathfrak{R} = \bar{L}q \quad \text{or} \quad \mathfrak{R} = LQ \tag{58}$$

which stand for the relationships

$$\mathfrak{R}_i = \sum_r \bar{L}_i^r q_r + \frac{1}{2} \sum_{r,s} \bar{L}_i^{rs} q_r q_s + \frac{1}{3!} \sum_{r,s,t} \bar{L}_i^{rst} q_r q_s q_t + \cdots \tag{59}$$

where the sums are unrestricted. Expressions for L elements are given by Hoy, Mills, and Strey [57] and $\bar{L}$ elements are related to these since the reduced (dimensionless) normal coordinates q_s are related to the normal coordinates Q_s by

$$q_s = (\omega_s/\hbar)^{1/2} Q_s. \tag{60}$$

Additional $\bar{L}$ elements are given by Fowler and Raynes [58] and Louinila et al. [59]. For molecules possessing high symmetry, relationships between $\bar{L}$ elements lead to simplifications in (58). For $\mathfrak{R}_i = \Delta r_i$ in AX_n-type molecules of symmetry O_h, T_d, D_{3h}, $D_{\infty h}$ with only one totally symmetric mode of vibration of frequency ω_1, [41]

$$\bar{L}_i^1 = \left[\frac{\hbar}{2\pi c \omega_1 n m_X} \right]^{1/2}, \quad i = 1 \text{ to } n \tag{61}$$

and $\bar{L}_i^s = 0$ for $s \neq 1$. For example, in CH_4-type molecules including up to quadratic terms in the normal coordinates,

$$\langle \Delta r_1 \rangle = \bar{L}_1^1 \langle q_1 \rangle + \frac{1}{2} \bar{L}_1^{2a2a} \langle q_2^2 \rangle$$

$$+ \frac{1}{2} \bar{L}_1^{3x3x} \langle q_3^2 \rangle + \frac{1}{2} \bar{L}_1^{4x4x} \langle q_4^2 \rangle + \cdots \tag{62}$$

$$\langle (\Delta r_1)^2 \rangle = (\bar{L}_1^1)^2 \langle q_1^2 \rangle + (\bar{L}_1^{3x})^2 \langle q_3^2 \rangle + (\bar{L}_1^{4x})^2 \langle q_4^2 \rangle + \cdots \tag{63}$$

In terms of these curvilinear internal coordinates, (1) can be rewritten

$$P = P_e + \sum_i \left(\frac{\partial P}{\partial \mathcal{R}_i}\right) \mathcal{R}_i + \frac{1}{2!} \sum_{ij} \left(\frac{\partial^2 P}{\partial \mathcal{R}_i \partial \mathcal{R}_j}\right)_e \mathcal{R}_i \mathcal{R}_j$$

$$+ \frac{1}{3!} \sum_{ijk} \left(\frac{\partial^3 P}{\partial \mathcal{R}_i \partial \mathcal{R}_j \partial \mathcal{R}_k}\right) \mathcal{R}_i \mathcal{R}_j \mathcal{R}_k + \dots \tag{64}$$

where the sums are unrestricted. Symmetry relations between the derivatives $(\partial P/\partial \mathcal{R}_i)_e$, etc., also reduce the number of unique derivatives, depending on the property. The magnetizability, the electric dipole polarizability, and higher polarizability tensors of the molecule depend on the symmetry of the molecule as a whole in its equilibrium geometry. The number of unique non-vanishing components for various symmetry groups have been determined by Buckingham et al. [60] for molecular moments and polarizabilities in static fields. The numers of independent derivatives of the dipole moment, quadrupole moment, and dipole polarizability for various molecular types of given geometry and total charge have been tabulated by Fowler and Buckingham [21]. The symmetry properties of dipole moment, electric dipole polarizability and the first hyperpolarizability components have been given in detail for the various symmetry point groups by Cyvin, Rauch, and Decius [61]. Other properties are site-specific, such as the electric field gradient, the nuclear magnetic shielding, the spin rotation constant. These properties have the symmetry of the molecule if the site is at the center of the molecule; a different nuclear site leads to a lower symmetry. For example, in CH_4 the nuclear magnetic shielding and electric field gradient at the proton and the $^{13}C-^1H$ spin-spin coupling constant all have C_{3v} symmetry. Furthermore, because of this C_{3v} symmetry, the electric field gradient tensor at the proton has only one independent component. In CH_4 the $^1H-^1H$ spin-spin coupling is of C_{2v} symmetry. Buckingham et al. [62, 63] have tabulated the number of independent components of the spin-spin coupling tensor and also the nuclear shielding tensor [64–66] for some important point group symmetries. Furthermore, the number of non-vanishing unique first derivatives (and higher) of the property depends on this site symmetry. For example, in the pyramidal AB_3 molecule, the first derivatives of the electric field gradient at nucleus A with respect to the normal coordinates of the E vibrational modes are zero, but this does not apply to the electric field gradient derivatives at the B nuclei. There are similar site symmetry consequences for the property derivatives with respect to internal coordinates. For example, there is only one $(\partial \sigma^C/\partial r)_e$ for ^{13}C nuclear magnetic shielding in CH_4, but for ^1H shielding the derivatives $(\partial \sigma^{H_1}/\partial r_1)_e$ and $(\partial \sigma^{H_1}/\partial r_2)_e$ are different [41]. Thus, one may write as in [67]

$$\langle \sigma(^{13}C) \rangle = \sigma_e + \sigma_r(\langle r_1 \rangle + \langle r_2 \rangle + \langle r_3 \rangle + \langle r_4 \rangle)$$

$$+ \frac{1}{2} \sigma_{rr}(\langle r_1^2 \rangle + \langle r_2^2 \rangle + \langle r_3^2 \rangle + \langle r_4^2 \rangle) + \dots \tag{65}$$

where

$$\sigma_r \equiv (\partial \sigma^C/\partial r_1)_e = (\partial \sigma^C/\partial r_2)_e = (\partial \sigma^C/\partial r_3)_e = (\partial \sigma^C/\partial r_4)_e \tag{66}$$

whereas

$$\langle \sigma(^1H) \rangle = \sigma_e + \sigma_r \langle r_1 \rangle + \sigma_s(\langle r_2 \rangle + \langle r_3 \rangle + \langle r_4 \rangle)$$

$$+ \frac{1}{2}\sigma_{rr}\langle r_1^2 \rangle + \frac{1}{2}\sigma_{ss}(\langle r_2^2 \rangle + \langle r_3^2 \rangle + \langle r_4^2 \rangle) + \dots \tag{67}$$

where, for the proton participating in C–H bond 1 of four C–H bonds in CH_4,

$$\sigma_r \equiv (\partial \sigma^{H_1}/\partial r_1)_e . \tag{68}$$

$$\sigma_s \equiv (\partial \sigma^{H_1}/\partial r_2)_e = (\partial \sigma^{H_1}/\partial r_3)_e = (\partial \sigma^{H_1}/\partial r_4)_e . \tag{69}$$

The relations between the mass-dependent property derivatives P_i in (1) and the mass-independent derivatives P_r in (64) can be written in terms of the **L** tensor elements (which contain the mass factors). See for example [58].

$$P_i = \sum_r P_r L_r^i$$

$$P_{ij} = \sum_r P_r L_r^{ij} + \sum_{rs} P_{rs} L_r^i L_s^j$$

$$P_{ijk} = \sum_r P_r L_r^{ijk} + \sum_{rs} P_{rs}(L_r^i L_s^{jk} + L_r^j L_s^{ki} + L_r^k L_s^{ij}) + \sum_{rst} P_{rst} L_r^i L_s^j L_t^k \tag{70}$$

In early treatments only the first order terms in the **L** tensor were included in the above equations [68]. Symmetry relations between the **L** tensor elements lead to simple relationships between the derivatives of the properties. For example, for the properties P which have the symmetry of T_d in the isotopomers $CH_{4-n}D_n$,

$$P_1 = -4P_r L_1^1 \quad \text{(only one } A_1 \text{ mode) in } CH_4 \text{ and } CD_4 \tag{71}$$

or

$$P_n = -P_r(L_1^n + 3L_2^n), \quad n = 1, 2, 3 \subset A_1 \text{ in } CH_3D \text{ and } CD_3H \tag{72}$$

or

$$P_n = -2P_r(L_1^n + L_3^n), \quad n = 1\text{–}4 \subset A_1 \text{ in } CH_2D_2 . \tag{73}$$

In CH_4 and CD_4, the third derivatives are (leaving out the 3rd order terms in the **L** tensor) [69]:

$$P_{12a2a} = 8(P_{rr} + P_{rs}) \bar{L}_1^1 \bar{L}_1^{2a2a} + 9(P_{\alpha\alpha} + P_{\alpha\omega}) \bar{L}_5^{2a} \bar{L}_5^{12a} + \dots \tag{74}$$

and for u = 3x, 4x

$$P_{1uu} = 8(P_{rr} + P_{rs}) \bar{L}_1^1 \bar{L}_1^{uu} + 4(P_{\alpha\alpha} - P_{\alpha\omega}) \bar{L}_6^u \bar{L}_6^{1u} + 16P_{r\alpha} \bar{L}_1^u \bar{L}_6^{1u} + \dots \tag{75}$$

The analogs of (62) and (74) and (75) for the lower symmetry mixed-isotope species can be derived in a straightforward way for each molecular type and property site symmetry. For some explicit examples see Raynes [67]. The *first order* terms take on

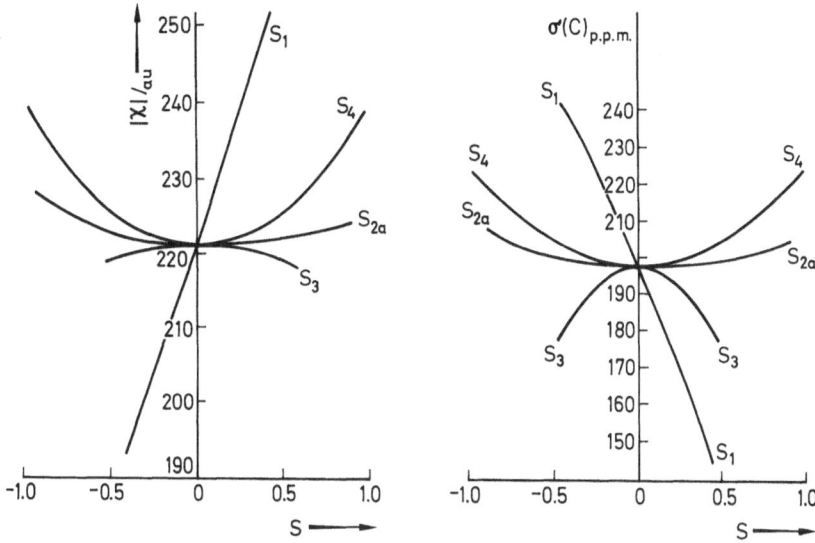

Fig. 8. Property surfaces of CH_4 molecule. The absolute value of the magnetizability and the ^{13}C nuclear magnetic shielding are plotted as functions of the symmetry coordinates defined in the text. S_1 and S_3 are in Angstroms and S_{2a} and S_{4z} are in radians. Reproduced from Lazzeretti P, Zanasi R, Sadlej AJ, Raynes WT (1987) Mol. Phys. 62: 605, with permission [70]

particularly simple forms in highly symmetric situations. For example in AX_n molecules (O_h, T_d and D_{3h} for $n = 6, 4, 3$) [41],

$$\langle P \rangle = P_e + (h/4\pi^2 \omega_1 cnm_x)^{1/2} \langle q_1 \rangle \cdot P' + \ldots \tag{76}$$

where $\quad P' \equiv [(\partial P/\partial r_1)_e + (\partial P/\partial r_2)_e + \ldots + (\partial P/\partial r_n)_e]$ (77)

where m_x is the mass of the X atom and $\langle q_1 \rangle$ is given by (35) and (37) where ω_1 is the totally symmetric vibrational frequency.

In ab initio calculations of surfaces the sensible approach is of course to use not bond displacements but simultaneous displacements which reflect the symmetry of the molecule, i.e. symmetry internal displacement coordinates. The choice of symmetry coordinates is the same as for the potential energy surface. Figure 8 shows sections through the χ (magnetizability) surface and the σ^C (^{13}C nuclear magnetic shielding) surface calculated by Lazzeretti et al. [70] in the symmetry coordinates given below:

$$S_1 = \frac{1}{2}(r_1 + r_2 + r_3 + r_4) \tag{78}$$

where r_i describes the increase in the length of bond i relative to its equilibrium value.

$$S_{2a} = (12)^{-1/2}(2\alpha_{12} - \alpha_{13} - \alpha_{14} - \alpha_{23} - \alpha_{24} + 2\alpha_{34}) \tag{79}$$

and

$$S_{2b} = \frac{1}{2}(\alpha_{13} - \alpha_{14} - \alpha_{23} - \alpha_{34}) \tag{80}$$

where α_{ij} denotes the increase in the interbond angle between bonds i and j from its equilibrium value.

$$S_{3x} = \frac{1}{2}(r_1 - r_2 + r_3 - r_4), \tag{81}$$

$$S_{3y} = \frac{1}{2}(r_1 - r_2 - r_3 + r_4), \tag{82}$$

and

$$S_{3z} = \frac{1}{2}(r_1 + r_2 - r_3 - r_4), \tag{83}$$

$$S_{4x} = 2^{-1/2}(\alpha_{24} - \alpha_{13}), \tag{84}$$

$$S_{4y} = 2^{-1/2}(\alpha_{23} - \alpha_{14}) \tag{85}$$

and

$$S_{4z} = 2^{-1/2}(\alpha_{34} - \alpha_{12}). \tag{86}$$

Similar surfaces in terms of the symmetry coordinates have been plotted for O and H nuclear shielding and magnetizability in H_2O [71], and for electric dipole polarizability and molecular quadrupole moment in CO_2 [26].

The largest contributions to the value of the property at the equilibrium geometry P_e, and to its derivative $(\partial P/\partial \boldsymbol{\mathcal{R}})_e$ may come from distinctly different regions within the molecule. For example the ^{19}F nuclear magnetic shielding in HF is dominated by the diamagnetic term which depends on r^{-1} of the electrons whereas $(\partial \sigma/\partial \boldsymbol{\mathcal{R}})_e$ is dominated by the paramagnetic term which depends on r^{-3} [18]. A mapping of the nuclear shielding density function shows the contributions to the shielding from various regions in the molecule, and a mapping of the distribution of $(d\sigma/d\boldsymbol{\mathcal{R}})_e$ shows which regions of the electronic distribution contribute to the change in shielding as the bond is extended [72]. A similar conclusion has been reached in the case of μ and $(\partial \mu/\partial \boldsymbol{\mathcal{R}})_e$ [73].

5 Temperature Dependence

5.1 Calculations

Calculation of the thermal average of a molecular property using (40) including nuclear spin statistics, is feasible for a molecule with only a few vibrational modes, e.g., H_2O [58]. In this example Fowler, Riley and Raynes used ab initio SCF property surfaces for the electric dipole moment, magnetizability, proton magnetic shielding and

[17]O magnetic shielding to obtain the derivatives P_r, P_{rs}, etc. of the various properties [71]. An accurate empirical anharmonic force field was used to determine the quantities ω_s, $\varphi_{sss'}$, $a_s^{\alpha\alpha}$, and E_{vJK}. The thermal averages $\langle P \rangle_T$ were calculated using (28)–(33) and (40) [58]. Rovibrational states up to $J = 25$ and levels less than 5000 cm^{-1} above the zero-point levels were included for each of the 18 isotopomers (^{1}H, ^{2}H, ^{3}H, ^{16}O, [17]O, ^{18}O). The literal summing up of terms can be time-consuming when the density of rovibrational states is high. If nuclear spin statistics are ignored, if rotation is averaged classically, and if the harmonic oscillator density of states is used, then

$$\left\langle v_s + \frac{1}{2} \right\rangle^T = \frac{1}{2} \coth(hc\omega_s/2kT) \tag{87}$$

$$\langle J_\alpha^2 \rangle^T = I_{\alpha\alpha} kT \tag{88}$$

Equations (35)–(37) become

$$\langle q_s \rangle = -\frac{1}{2\omega_s} \sum_{s'} \varphi_{s's's} \frac{1}{2} \coth(hc\omega_{s'}/2kT) \tag{89}$$

$$\langle q_s q_{s'} \rangle = \frac{d_s}{2} \coth(hc\omega_s/2kT) \, \delta_{ss'} \tag{90}$$

and the rotational contribution to a single reduced normal coordinate q_s to be added to (89) is

$$\langle q_s \rangle_{\text{cent}} = +\frac{kT}{4\pi c\omega_s} \left(\frac{1}{hc\omega_s}\right)^{1/2} \sum_\alpha \frac{a_s^{\alpha\alpha}}{I_{\alpha\alpha}^{(e)}} \tag{91}$$

(See Toyama et al. [28], and Jameson [41, 68]). In this case, keeping only terms quadratic in the normal coordinates in the expansion (1) [28, 68]

$$\langle P \rangle^T = P_e + \sum_s \left[\frac{P_{ss}}{2} - \sum_{s'}^{\text{tot. sym.}} P_{s'} \varphi_{sss'}/2\omega_{s'} \right] \cdot \frac{1}{2} \coth(hc\omega_s/2kT) + \cdots$$

$$+ \sum_{s'}^{\text{tot. sym.}} P_{s'} \frac{h}{2} \left(\frac{1}{hc\omega_s}\right)^{3/2} kT \sum_\alpha \frac{a_{s'}^{\alpha\alpha}}{I_{\alpha\alpha}^{(e)}} + \cdots \tag{92}$$

With these approximations the form of the temperature dependence of an electronic property becomes more obvious than in (40). For $x \geq 2.5$ the function coth x changes by less than 1 part in 100 when x changes by 1 part in 7. Thus, around room temperature ($T = 300 \text{ K} = 208.5 \text{ cm}^{-1}$) only those vibrational modes having frequencies such that $\omega_s/2kT \leq 2.5$, i.e., $\omega_s < 1000 \text{ cm}^{-1}$ make substantial vibrational contributions to the temperature coefficient of any property. Inclusing only terms up to P'' and taking the coth approximation for the thermal average of $\langle v + \frac{1}{2} \rangle$, (as in (87)) for diatomic molecules:

$$\langle P \rangle^T = P_e + 4(B_e/\omega_e)^2 P' \frac{kT}{B_e} + \left(\frac{B_e}{\omega_e}\right)(P'' - 3a_1 P') \frac{1}{2} \coth(hc\omega_e/2kT) + \cdots \tag{93}$$

Table 3. Relative contributions to the P′ term of ($\langle P \rangle^{400\text{K}}$ − $\langle P \rangle^{300\text{K}}$) for a diatomic molecule[a]

ω_e/cm^{-1}	% Rotation	% Anharmonic vibration
600	52	48
800	61	39
1000	70	30
1200	79	21
1400	86	14
1600	92	8

[a] Using $a_1 \cong -2$, which is a typical value

The above equation was derived by Buckingham [34].

The relative contributions of rotation, harmonic vibration, and anharmonic vibration to the temperature dependence of an electronic property of a diatomic molecule depend on ω_e and a_1, as well as the property derivatives. For 300 K to 400 K the relative contributions to the P′ term of the temperature dependence of $\langle P \rangle^T$ can be estimated, as shown in Table 3. For diatomic molecules (such as N_2 or CO) with fairly high vibrational frequencies the temperature dependence of a molecular electronic property can be dominated by rotation [68, 74].

The relative contributions of various terms to thermal average shielding in a diatomic molecule can be seen in Tables 4 and 5. The sign and magnitude of the vibrational dependence is determined by (P″ − $3a_1$P′). The values of a_1 being typically about −2, the vibrational dependence of many properties is contributed mainly by the first derivative and thus by the anharmonic contribution [68]. We note in Table 4 that the third and higher order terms in (44) have very little effect on the thermal average shielding. Terms up to the second derivative comprise 92–99% of the total calculated value. Although the rotational contribution to $\langle \Delta r \rangle^T$ is usually one order of magnitude smaller than the vibrational contribution, it plays an important role in the temperature dependence of an electronic property. As an example, the temperature dependence of nuclear magnetic shielding in some diatomic molecules calculated using (93) are shown in Table 5. The temperature dependence of the shielding depends nearly entire-

Table 4. The cumulative contributions, linear, quadratic and higher order rovibrational contributions to the thermal average shielding at 300 K, $\langle \sigma \rangle^{300} - \sigma_e$, in ppm [44]

Nucl.	Mol.	$(d\sigma/d\xi)_e \langle \xi \rangle$	$\left\{ \begin{array}{c} (d\sigma/d\xi)\langle \xi \rangle + \\ \frac{1}{2}(d^2\sigma/d\xi^2)_e \langle \xi^2 \rangle \end{array} \right\}$	Up to 6th order
[1]H	H_2	−0.5554	−0.3964 (97%)	−0.3814
[1]H	HF	−0.7021	−0.3721 (95%)	−0.3911
[19]F	HF	−7.42	−11.113 (99%)	−11.23
[1]H	LiH	−0.1228	−0.1238 (99%)	−0.1228
[7]Li	LiH	+0.1386	+0.0726 (92%)	+0.0786

Table 5. Temperature dependence of nuclear shielding, $\langle\sigma\rangle^{400} - \langle\sigma\rangle^{300}$, in ppm[a] [18]

Nucl.	Mol.	Rotational		Anharm. vib.	Harm. vib.	Total
^{1}H	H_2	−0.0282	(100%)	~0	~0	−0.0282
^{1}H	^{7}LiH	−0.0091	(79%)	−0.0008	−0.0016	−0.0115
^{7}Li	^{6}LiH	+0.0253	(96%)	+0.0023	−0.0013	+0.0263
^{1}H	HF	−0.0253	(100%)	~0	~0	−0.0253
^{19}F	HF	−0.2560	(100%)	~0	~0	−0.2560
^{13}C	$^{13}C^{16}O$	−0.1062	(98%)	−0.0016 ·	−0.0002	−0.1081
^{17}O	$^{12}C^{17}O$	−0.1231	(98%)	−0.0020	−0.0003	−0.1254
^{15}N	$^{15}N_2$	−0.1405	(99%)	−0.0013	−0.0004	−0.1423

[a] Calculated using (93)

ly on rotation, with vibrational effects accounting for 0–2%. This is easily understood by consulting (93). The linear temperature dependence of the rotational contribution dominates the change of the average shielding with temperature, especially for these diatomic molecules which have strong bonds characterized by high vibrational frequencies. The $(hc\omega/2kT)$ term is large, for which the coth function is very nearly a constant equal to 1.0. For F_2 and ClF molecules the rotational contribution still dominates although the anharmonic vibration contribution to the temperature dependence becomes significant [68].

5.2 Experimental Examples

The most extensive data on the temperature dependence of molecular electronic properties come from magnetic resonance. Measurements made in the gas phase in the zero-pressure limit in molecules which are incapable of undergoing rearrangement or conformational changes, can only be explained in terms of rovibrational averaging as described theoretically in the foregoing. Although the first attribution of a temperature-dependent vibrational averaging of a molecular electronic property was probably that by Petrakis and Sederholm [75], of the proton chemical shift variation with temperature in gas samples at 10 atm, it is now obvious that a large part of what was observed was due to intermolecular effects [76]. Buckingham called attention to the latter and also provided a general theoretical explanation for temperature dependence in diatomics which is that shown in (93) [34]. Extrapolation to the zero-pressure limit eliminates the intermolecular effects, leaving only the temperature dependence due to rovibrational averaging. The first such report was in ^{19}F nuclear magnetic shielding in F_2 and ClF [68], followed shortly by BF_3, CF_4, SiF_4, and SF_6 [77]. An example is shown in Fig. 9. Further examples have been reviewed [18, 76]. These results have been interpreted in terms of (92) which has a simple form for AX_n type molecules which have O_h, T_d or D_{3h} symmetry, as follows [41]:

$$\langle P\rangle^T = P_e + (\partial P/\partial r_1 + \partial P/\partial r_2 + \ldots \partial P/\partial r_n)_e$$

$$\times \left\{ (h/4\pi^2\omega_1 cnm_x)^{1/2} \sum_{s=1}^{3N-6} \frac{\varphi_{1ss}}{2\omega_1} \frac{1}{2} \coth\left(\frac{hc\omega_s}{2kT}\right) \right.$$

$$\left. + 3kT/4\pi^2c^2 nm_x r_e\omega_1^2 + O(L_r^{ij}) \right\} + \ldots \qquad (94)$$

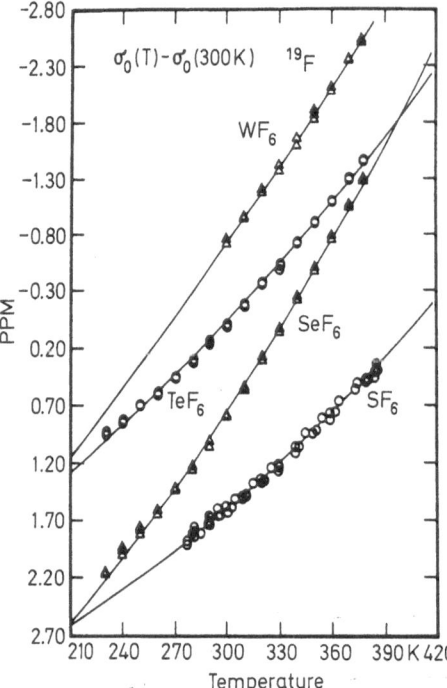

Fig. 9. Comparison of the observed temperature dependence $[\sigma_0(T) - \sigma_0(300 \text{ K})]$ of ^{19}F in octahedral fluorides with the calculated curves $(\partial\sigma/\partial r)_e [\langle\Delta r\rangle^T - \langle\Delta r\rangle^{300}]$, using $(\partial\sigma/\partial r)_e \approx -1930, -2690, -1770,$ and -2500 ppm/Å for SF_6, SeF_6, TeF_6, and WF_6, respectively. (Except for TeF_6, the individual plots are arbitrarily offset for display.) $(\partial\sigma/\partial r)_e$ in these MF_6 cases are of the form $[(\partial\sigma^F/\partial r)_1)_e + (\partial\sigma^F/\partial r_t)_e + 4(\partial\sigma^F/\partial r_c)_e]$. Reproduced from Jameson CJ, Jameson AK (1986) J. Chem. Phys. 85:5484, with permission [80]

The terms of order $O(L_r^{ij})$ not explicitly shown above include terms in L_r^{ij}, L_j^{ijk}, ... which appear in (70) upon nonlinear transformation of the derivatives of the property with respect to normal coordinates into derivatives with respect to curvilinear internal coordinates. Alternatively the entire quantity in curly brackets in (94) may be calculated using another method, due to Bartell [78, 79], which gives equivalent results [80] when the same anharmonic force field is used. The terms denoted by + ... involve second and higher derivatives of the property surface. For the central nucleus (A) in these symmetrical AX_n molecules, $[(\partial P/\partial r_1)_e + (\partial P/\partial r_2)_e + ... + (\partial P/\partial r_n)_e]$ is $n(\partial\sigma^A/\partial r_1)$. On the other hand, for the peripheral nucleus F in CF_4 for example, it is $(\sigma_r + 3\sigma_s) \equiv [(\partial^{F1}/\partial r_1)_e + 3(\partial\sigma^{F1}/\partial r_2)_e]$, with $(\partial\sigma^{F1}/\partial r_1)_e$ being the dominant contribution. For the F nucleus in the MF_6 molecules in Fig. 9 it is $(\sigma_r + \sigma_t + 4\sigma_c) \equiv [(\partial\sigma^{F1}/\partial r_1)_e + (\partial\sigma^F/\partial r_t)_e + 4(\partial\sigma^F/\partial r_c)_e]$, where the contributions from the bonds *cis* and *trans* to the F nucleus of interest are expected to be smaller than $(\partial\sigma^{F1}/\partial r_1)_e$. This linear combination of derivatives has been considered as an empirical parameter and the experimental data have been fitted by (94), neglecting second and higher derivatives. There is no guarantee that the neglected terms are much smaller than the terms involving the first derivative. Even though the absolute precision and accuracy of the experimental data are high (a few parts per billion) they do not warrant fitting to more than one empirical parameter. For this particular observable, the very large number of examples are typically of the form shown in Fig. 9, that is, the shapes of curves which best describe the experimental data points mimic the shapes of $\langle\Delta r\rangle^T$ vs. temperature for the molecule. Of course the functions $\langle(\Delta\alpha)^2\rangle^T$ also have similar shapes.

The temperature dependence of spin-spin coupling constants have been reported in two molecules, the B-F coupling in BF_3 [81] and the C-H coupling in CH_4 [82]. The latter has been interpreted using (94), keeping only the terms in the first derivatives. The empirical value of $(J_r + 3J_s) \equiv [(\partial J(CH_1)/\partial r_1)_e + 3(\partial J(CH_1)/\partial r_2)_e]$ was found to be 368(10) Hz $Å^{-1}$, that is, an increase of the coupling with increasing bond length [82]. This had been predicted to be a somewhat general trend (except where lone pairs are involved) from a study of isotope effects on coupling constants and from consideration of the change in electron density at the nuclei upon bond extension [83].

The other temperature-dependent electronic properties which have been widely reported are the nuclear spin hyperfine constants in ESR spectra of organic free radicals. Large temperature coefficients, characteristic of nuclear hyperfine constants at β position from the atom with the unpaired spin involve the averaging over torsional states of a highly torsional-angle-dependent electronic property. Oriented molecule studies in single crystals provide information on the second derivative of the hyperfine constant with respect to the torsional angle ϕ [84]:

$$a_\beta^H = B_0 + B_2 \cos^2 \phi \tag{95}$$

which could lead to temperature-dependent hyperfine constants a_β^H [85] under isotropic conditions in solution or in gas phase. A large number of examples have been reviewed [86]. This is similar to the 3-bond spin-spin coupling constants observed in NMR which are highly dependent on the dihedral angle. Averaging over the torsional angle in each conformation leads to a temperature dependent spin-spin coupling, but the largest temperature dependence comes from the temperature-dependent populations of the various conformers [87].

Table 6. Calculated and experimental results of rovibrational averaging of hyperfine constants in $\dot{C}H_3$ [90]

	calculated[a]	experimental[b]
$a^H(^{12}CH_3)$	-23.85	23.04 ± 0.01 [178, 148]
$\frac{\gamma^H}{\gamma^D} a^D(^{12}CD_3)$	-24.14	23.30 ± 0.07 [178]
ratio $\varkappa$	$1 - 0.012$	$1 - 0.011 \pm 0.003$
$a^C(^{13}CH_3)$	52.23	38.34 ± 0.01 [178]
$a^C(^{13}CD_3)$	49.92	35.98 ± 0.01 [178]
ratio	1.0462	1.0656 ± 0.0004
$(da^H/dT) (^{12}CH_3)$	1.48	1.3 ± 0.2 [150], 1.5 ± 0.2 [149]
$\frac{\gamma^H}{\gamma^D} (da^D/dT) (^{12}CD_3)$	2.04	2.3 [149]
$(da^C/dT) (^{13}CH_3)$	11.6	13.5 [149]
$(da^C/dT) (^{13}CD_3)$	15.9	

[a] Rovibrational averaging calculations by [90] using ab initio theoretical property surface calculated by [89];
[b] Signs were not determined for a values

Of greater interest is the temperature dependence of simple systems such as the a^H in $\dot{C}H_3$, for example. Analyses of the temperature dependence of the ^1H and ^{13}C hyperfine constants indicate that the most important vibrational mode is the symmetric out-of-plane bend of the planar $\dot{C}H_3$ radical. The dependence of these properties on the out-of-plane angle θ has been calculated by various methods. The ab initio SCF-CI calculations with a small optimized basis set [88] yield

$$a^C(\theta) = 148.8 + 1395\theta^2 - 2560\theta^4 \tag{96}$$
$$a^H(\theta) = -54.27 + 2500\theta^2 - 9570\theta^4 \tag{97}$$

Vibrational averaging using harmonic oscillator functions in a normal coordinate approximated by the out-of-plane angle θ, and using the experimental $\omega = 580 \text{ cm}^{-1}$ lead to calculated temperature coefficients of the right order of magnitude. An ab initio UHF calculation by Meyer for $\dot{C}H_3$, using a large basis set [89] yields spin densities which have been used by Schrader and Morokuma to calculate the temperature and mass dependence of a^C and a^H [90]. Results are shown in Table 6. The agreement between experiment and calculation is sufficiently good so that the phenomenon can be considered well understood.

6 Mass Dependence : Isotope Effects

Isotope effects on molecular electronic properties are a direct consequence of rovibrational averaging.

6.1 Calculations

6.1.1 Centrifugal Distortion

Complete to second order in the contact transformation, the rotational contribution to a rovibrationally averaged electronic property includes terms quadratic in the angular momentum and terms quartic in the angular momentum. These have been derived by Fowler for asymmetric tops [43]. There are contributions involving property derivatives P_s, $P_{ss'}$ and $P_{ss's'}$. For spherical tops (39) shows the term in P_1 and P_{1ss} where q_1 is the totally symmetric normal coordinate. The general leading term comes from the centrifugal distortion of a single reduced normal coordinate, (37) or its thermal average, (91). Note that this leading term does not depend on the anharmonic part of the potential. Thus, the centrifugal distortion contribution can be written simply in terms of the usual matrices which arise in a harmonic vibrational analysis such as [28]:

$$\langle \Delta r \rangle_{rot}^T = k \tilde{T} \tilde{U} F_S^{-1} G_S^{-1} U B \Omega X \tag{98}$$

where Ω denotes a diagonal matrix with the elements

$$\Omega_{ii}^{(\alpha\alpha)} = \frac{1}{I_{\beta\beta}^{(e)}} + \frac{1}{I_{\gamma\gamma}^{(e)}} \tag{99}$$

involving the equilibrium moments of inertia and X is a vector which has the Cartesian coordinates of the atoms at the equilibrium configuration as its elements. F_S^{-1} and G_S^{-1} are the inverse of the usual Wilson F and G matrices in symmetry coordinates. U is the transformation matrix from internal into symmetry coordinates and B is the transformation from Cartesian displacements into internal coordinates.

For highly symmetrical molecules the rotational effect on the mean bond length can be expressed in even simpler form, for example [28, 91]:

For a CO_2 type molecule

$$\langle \Delta r \rangle_{rot}^T = \frac{kT}{r_e(F_{11} + F_{12})} \tag{100}$$

For a AX_n type (planar, T_d or O_h):

$$\langle \Delta r \rangle_{rot}^T = \frac{3kT}{nr_e F_{11}} \tag{101}$$

where F_{11} and F_{12} are mass-independent quadratic force constants in internal coordinates. Since these equations involve only mass-independent terms, there will be *no rotational contribution* to the isotope effects on any molecular electronic property of these molecular types provided that isotopic substitution preserves the symmetry of the molecule, as in going from CH_4 to CD_4 (but not in CH_4 to CH_3D) [91].

6.1.2 Vibration in Diatomics

Of the quantities in (93), a_1 is mass-independent and the others depend on the reduced mass of the diatomic molecule as follows:

$$\omega_e^* = (\mu/\mu^*)^{1/2} \omega_e$$
$$B_e^* = (\mu/\mu^*) B_e \tag{102}$$

Furthermore, the thermal average $\left\langle v + \frac{1}{2} \right\rangle^T$ is also mass-dependent since the populations of the vibrational energy levels over which this average is taken depend on the vibrational frequency ω_e. We note that B_e/ω_e^2 is mass-independent, so the rotational contribution to the thermal average shielding in the diatomic molecule is independent of mass.

Therefore the isotope shift, the isotope effect on the property which is the nuclear magnetic shielding σ, is given by [91]

$$\langle \sigma \rangle - \langle \sigma \rangle^* = [(d^2\sigma/d\xi^2)_e - 3a_1(d\sigma/d\xi)_e] (B_e/\omega_e)$$
$$\times \{\coth(hc\omega_e/2kT) - (\mu/\mu^*)^{1/2} \coth[hc(\mu/\mu^*)^{1/2} \omega_e/2kT]\} \tag{103}$$

We have calculated the rovibrational corrections to shielding and the isotope shift using (93) and (103) for diatomic molecules, using shielding derivatives from the

Table 7. Anharmonic and harmonic vibrational contributions to the secondary isotope shift, in ppm[a]

Nucl.	Mol.	Anharm.	Harm.	Total	Expt.	Ref.
^{1}H	HD$-$H$_2$	-0.0723	$+0.0090$	-0.0632[b]	-0.036 ± 0.002	[179]
^{1}H	^{7}LiH$-^6$LiH	-0.0007	-0.0013	-0.0020		
^{7}Li	LiD$-$LiH	$+0.0438$	-0.0250	$+0.0188$		
^{1}H	DF$-$HF	-0.1719	$+0.0796$	-0.0923		
^{19}F	DF$-$HF	-1.7364	-0.7883	-2.5248	-2.5 ± 0.5	[180]
^{13}C	^{13}C^{18}O$-^{13}$C^{16}O	-0.0414	-0.0058	-0.0472[c]	-0.0476 ± 0.0016	[111]
^{17}O	^{13}C^{17}O$-^{12}$C^{17}O	-0.0438	-0.0063	-0.0501	-0.110 ± 0.001	[111]
^{15}N	^{15}N$_2$$-^{15}$N^{14}N	-0.0432	-0.0148	-0.0581	-0.0601 ± 0.002	[112]

[a] Ref. [18] using (103);
[b] For isotopomers of H$_2$ the classical treatment of rotation leads to significant error. A proper average taking into consideration the nuclear spin statistics gives better results: -0.045 [44];
[c] This may be compared with -0.06 ppm obtained from calculations including cubic terms

literature for H$_2$ [92], LiH [93], HF [94], CO [95] and N$_2$ [96]. Differences in shielding between pairs of isotopically related molecular species give the isotope shifts in Table 1. There is no rotational contribution to the isotope shift in diatomic molecules in the classical limit. The relative contributions of the anharmonic and harmonic vibrational effects vary so that neglect of the harmonic contribution in calculating the isotope shift can lead to error. Among these examples, the largest error would occur in the case of ^{19}F in the DF-HF system, in which the harmonic vibrational contribution is 31% of the observed isotope shift.

In diatomic molecules the mean bond displacement and the mean square amplitudes can be written from (93) as

$$\langle \Delta r \rangle^T = \langle \Delta r \rangle^T_{rot} + \langle \Delta r \rangle^T_{vib} \tag{104}$$

$$\langle \Delta r \rangle^T_{rot} = \frac{4 B_e r_e}{hc\omega_e^2} kT \tag{105}$$

$$\langle \Delta r \rangle^T_{vib} = -(3/2) \, a_1 \, (B_e/\omega_e) \, r_e \, \coth \, (hc\omega_e/2kT) \tag{106}$$

$$\langle (\Delta r)^2 \rangle^T_{vib} = (B_e/\omega_e) \, r_e^2 \, \coth(hc\omega_e/2kT) \tag{107}$$

From (106) we see that by making use of the implicit mass dependence of B_e and ω_e in (102) we can write [91]

$$\langle \Delta r \rangle - \langle \Delta r \rangle^* = -(3/2) \, a_1 r_e \{ (B_e/\omega_e) \coth \, (hc\omega_e/2kT)$$

$$-(\mu/\mu^*)^{1/2} \, (B_e/\omega_e) \coth[hc(\mu/\mu^*)^{1/2} \, \omega_e/2kT] \} \tag{108}$$

There are several approximations which can be invoked to simplify this expression further:

(i) if $\coth(hc\omega_e/2kT)$ is very close to 1.0 and $\coth[hc(\mu/\mu^*)^{1/2}\omega_e/2kT]$ is also very close to 1.0, i.e. for high vibrational frequencies, then $\langle \Delta r \rangle - \langle \Delta r \rangle^*$ reduces to

$$\langle \Delta r \rangle - \langle \Delta r \rangle^* \approx [1 - (\mu/\mu^*)^{1/2}] \, \langle \Delta r \rangle \tag{109}$$

Similarly

$$\langle(\Delta r)^2\rangle - \langle(\Delta r)^2\rangle^* \approx [1 - (\mu/\mu^*)^{1/2}] \langle(\Delta r^2)\rangle \tag{110}$$

(ii) $[1 - \mu/\mu^{*1/2}]$ can be further approximated by $(\mu^* - \mu) 2\mu^*$ except when isotopes of hydrogen are involved so that

$$\langle\Delta r\rangle - \langle\Delta r\rangle^* \approx \langle\Delta r\rangle \left(\frac{\mu^* - \mu}{2\mu^*}\right) = \langle\Delta r\rangle \frac{1}{2} \left(\frac{m' - m}{m'}\right) \left(\frac{m_A}{m_A + m}\right) \tag{111}$$

$$\langle(\Delta r)^2\rangle - \langle(\Delta r)^2\rangle^* \approx \langle(\Delta r)^2\rangle \left(\frac{\mu^* - \mu}{2\mu^*}\right) = \langle(\Delta r)^2\rangle \frac{1}{2} \left(\frac{m' - m}{m'}\right) \left(\frac{m_A}{m_A + m}\right) \tag{112}$$

where m_A is the mass of the observed nucleus A and the isotopes m' and m of the other nucleus X in the diatomic molecule. Thus, for a reference diatomic molecule $A^m X$ and its isotopomer $A^{m'} X$, neglecting higher order terms and using (111)–(112)

$$\begin{array}{c} \langle P\rangle \quad - \quad \langle P\rangle^* \\ \text{in } A^m X \qquad \text{in } A^{m'} X \end{array} \cong \left\{\left(\frac{dP}{dr}\right)_e \langle\Delta r\rangle_{vib} + \frac{1}{2}\left(\frac{d^2 P}{dr^2}\right)_e \langle\Delta r^2\rangle\right\}$$

$$\times \frac{1}{2}\left(\frac{m' - m}{m'}\right)\left(\frac{m_A}{m_A + m}\right) \tag{113}$$

The vibrational correction for the reference molecule, $\langle P\rangle - P_e$, is given to second order by the quantity in curly brackets in (113).

6.1.3 Extension to Polyatomic Molecules

The mass dependence implicit in (28)–(33) are in the L tensor elements. The derivatives with respect to dimensionless normal coordinates P_i, P_{ij}, P_{ijk}, are related to the mass-independent derivatives by these L tensor elements as shown in (70) and the force constants in terms of dimensionless normal coordinates: ω_i^{-1}, φ_{ijk}, ... are likewise expressible in terms of mass-independent potential energy derivatives F_{ii}, F_{ijk}, ... and these L tensor elements. Thus, the calculations of isotope effects on any electronic property is straightforward when both the property surface and the potential energy surface are known. Equivalently, following (64), the isotope shift can be written in terms of mass-independent property derivatives and mass-dependent averages of curvilinear internal coordinates [18]:

$$\sigma - \sigma^* = \sum_i \left(\frac{\partial\sigma}{\partial\mathbf{\mathcal{R}}_i}\right)_e [\langle\mathbf{\mathcal{R}}_i\rangle^T - \langle\mathbf{\mathcal{R}}_i\rangle^{T*}] + \frac{1}{2}\sum_{ij}\left(\frac{\partial^2\sigma}{\partial\mathbf{\mathcal{R}}_i \partial\mathbf{\mathcal{R}}_j}\right)_e$$

$$\times [\langle\mathbf{\mathcal{R}}_i\mathbf{\mathcal{R}}_j\rangle^T - \langle\mathbf{\mathcal{R}}_i\mathbf{\mathcal{R}}_j\rangle^{T*}] + \ldots \tag{114}$$

The isotope shift (at 0 K) of several properties of H_2O molecule have been calculated. The isotope shifts at 0 K of the electric dipole moment, electric quadrupole moment, the octapole moment and the diamagnetic part of the magnetizability were calculated by Ermler and Kern [97]. The effects of isotopic substitution on the electric dipole moment, the magnetizability and the nuclear magnetic shieldings [58] and the quadrupole moment, rotational g factor and spin rotation parameters [23] were also calculated by Fowler and Raynes. While it is not possible to do calculations at these levels for most systems of interest, it is essential to have at least a qualitative understanding of the extensive isotope shift data and to be able to combine these data with observed temperature coefficients for a partial knowledge of the nature of an electronic property surface. Since isotope shifts are usually easier to measure than temperature coefficients in the zero-pressure limit, it is worthwhile to find a simple model for the estimation of the isotope effects on an electronic property and a further simple model for the change in $\langle \Delta r \rangle$ and in $\langle (\Delta r)^2 \rangle$ upon a change in mass. The model has been described elsewhere [98] and we consider only a brief description here.

Consider a symmetrical molecule $A^m X_n$. For a property which is most sensitive to bond stretches rather than out-of-plane bends or bond angle deformations, the leading terms in (64) are

$$\langle P \rangle = P_e + \sum_i \left(\frac{\partial P}{\partial r_i} \right)_e \langle \Delta r_i \rangle^T + \frac{1}{2} \sum_{ij} \left(\frac{\partial^2 P}{\partial r_i^2} \right)_e \langle \Delta r_i^2 \rangle^T + \dots \qquad (115)$$

when $\left(\dfrac{\partial P}{\partial r_i} \right)_e$ are all equal, such as when P is the nuclear shielding of nucleus A, then the n terms can be combined,

$$\langle P \rangle = P_e + n \left[\left(\frac{\partial P}{\partial r_1} \right)_e \langle \Delta r_1 \rangle^T + \frac{1}{2} \left(\frac{\partial^2 P}{\partial r_1^2} \right)_e \langle \Delta r_1^2 \rangle^T \right] + \dots \qquad (116)$$

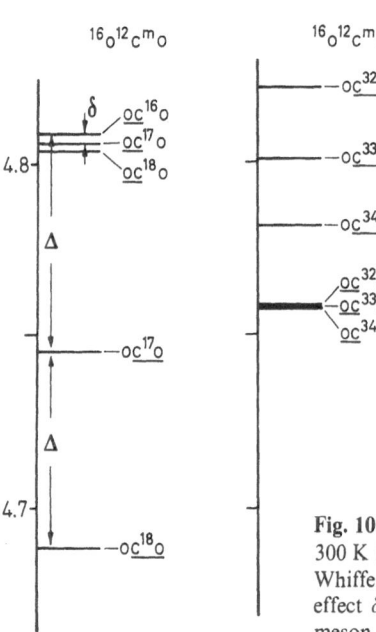

Fig. 10. Mean bond displacements in CO_2, and OCS calculated at 300 K [129] from the anharmonic force fields of Lacy [176] and Whiffen [177], illustrating the primary effect Δ, the secondary effect δ, and the additivity of the effects. Reproduced from Jameson CJ, Osten HJ (1984) J. Chem. Phys. 81: 4293, with permission [99]

Upon isotopic substitution of $^{m'}X$ for one ^mX in the A^mX_n molecule, there will be a change in symmetry, the normal coordinate analysis will be altered so that all the mean bond displacements $\langle \Delta r_1 \rangle$, $\langle \Delta r_2 \rangle$, ... $\langle \Delta r_n \rangle$ will be changed. However, the primary change will be in $\langle \Delta r_1 \rangle$ where $^{m'}X$ has replaced ^mX in the bond. There will be only very small secondary changes in the neighboring bonds [69, 99]. See for example Fig. 10. Thus, the difference between $\langle P \rangle$ in A^mX_n and $\langle P \rangle$ in $A^mX_{n-1}{}^{m'}X$ will be determined primarily by

$$\left(\frac{\partial P}{\partial r_1}\right)_e [\langle \Delta r_1 \rangle - \langle \Delta r_1 \rangle^*] + \frac{1}{2}\left(\frac{\partial^2 P}{\partial r_1^2}\right)_e [\langle (\Delta r_1)^2 \rangle - \langle (\Delta r_1)^2 \rangle^*]. \quad (117)$$

Other important terms are

$$(n-1)\left(\frac{\partial^2 P}{\partial r_1\, \partial r_2}\right)_e [\langle \Delta r_1\, \Delta r_2 \rangle - \langle \Delta r_1\, \Delta r_2 \rangle^*]$$

and various others such as

$$\left(\frac{\partial^2 P}{\partial r_1\, \partial \alpha}\right)_e [\langle \Delta r_1\, \Delta \alpha \rangle - \langle \Delta r_1\, \Delta \alpha \rangle^*].$$

We shall ignore these for the time being. With successive isotopic substitution at r_2, r_3, etc. the primary contributions to the isotope shift in $\langle P \rangle$ will be sequentially incremented by the terms in (117) except that Δr_2, Δr_3, etc. replace Δr_1. Thus, the isotope shift will be largely given by

$$\begin{array}{cc} \langle P \rangle & - \quad \langle P \rangle^* \\ \text{in } A^mX_n & \text{in } A^mX_{n-s}{}^{m'}X_s \end{array} \cong s \left\{ \begin{array}{l} \left(\dfrac{\partial P}{\partial r_1}\right)_e [\langle \Delta r_1 \rangle - \langle \Delta r_1 \rangle^*] \\[2mm] + \dfrac{1}{2}\left(\dfrac{\partial^2 P}{\partial r_1^2}\right)_e [\langle (\Delta r_1)^2 \rangle - \langle (\Delta r_1)^2 \rangle^*] \end{array} \right\}$$

$$(118)$$

Furthermore, from (113) we can write the mass factors approximately in the form,

$$\begin{array}{cc} \langle P \rangle & - \quad \langle P \rangle^* \\ \text{in } A^mX_n & \text{in } A^mX_{n-s}{}^{m'}X_s \end{array} \cong s \left(\frac{\langle P \rangle - P_e}{n}\right) \frac{1}{2}\left(\frac{m'-m}{m'}\right)\left(\frac{m_A}{m_A - m}\right) \quad (119)$$

Some properties are most sensitive to out-of-plane bends, for example a^C in planar $\dot{C}H_3$, in which case the deuteration isotope shift will be largely given by

$$\begin{array}{cc} \langle a^C \rangle & - \quad \langle a^C \rangle^* \\ \text{in } \dot{C}H_3 & \text{in } \dot{C}D_3 \end{array} \cong 3 \left\{ \frac{1}{2}\left(\frac{\partial^2 a^C}{\partial \theta^2}\right)_e [\langle \theta^2 \rangle - \langle \theta^2 \rangle^*] \right\} \quad (120)$$

The above expressions indicate an additivity of the isotope effect upon systematic isotopic replacement of equivalent atoms. Indeed, the bulk of the observations on various properties show additivity, deviations from additivity where they are observed at all in NMR are small [100–103]. If we push the approximation further and consider *only* the $\left(\frac{\partial P}{\partial r}\right)_e$ term, and write very roughly for A^mX_n

$$\langle P \rangle - P_e \cong n \left(\frac{\partial P}{\partial r}\right)_e \langle \Delta r \rangle \tag{121}$$

then very roughly

$$\langle P \rangle \quad - \quad \langle P \rangle^* \cong s \left(\frac{\partial P}{\partial r}\right)_e \langle \Delta r \rangle \frac{1}{2} \left(\frac{m' - m}{m'}\right) \left(\frac{m_A}{m_A - m}\right) \tag{122}$$
$$\text{in } A^mX_n \quad \text{in } A^mX_{n-s}{}^{m'}X_s$$

It turns out that even for molecular types other than AX_n it is still possible to write for any single substitution of ^mX by $^{m'}X$, even when X is not an end atom, that

$$\langle \Delta r \rangle - \langle \Delta r \rangle^* \cong k \langle \Delta r \rangle \left(\frac{m' - m}{m'}\right) \tag{123}$$

$$\langle (\Delta r)^2 \rangle - \langle (\Delta r)^2 \rangle^* \cong k' \langle (\Delta r)^2 \rangle \left(\frac{m' - m}{m'}\right) \tag{124}$$

This is illustrated in Fig. 1 in SeF_6 with masses m = 74 and m' = 74–82 for Se. The $\langle \Delta r \rangle$ and $\langle (\Delta r)^2 \rangle$ calculated using a reasonably good anharmonic force field for the SeF_6 molecule show the direct dependence on $\left(\frac{m' - m}{m'}\right)$. The observed isotope shifts show the same dependence on this mass factor, indicating some support for (122). Equation (122) provides a simple way of estimating $\left(\frac{\partial P}{\partial r}\right)_e$ from the observed isotope effect on property $\langle P \rangle$ if we have some measure of $\langle \Delta r \rangle$ in A^mX_n molecule.

Estimation of $\langle \Delta r \rangle$

A method of estimating $\langle \Delta r \rangle$ has been proposed [98]. For diatomic molecules this is not a problem since there are usually enough known spectroscopic constants to calculate $\langle \Delta r \rangle$ without having to use estimates. We note that including only terms up to quadratic allows us to write

$$\langle \Delta r \rangle^T_{vib} = -(3/2) (a_1/r_e) \langle (\Delta r)^2 \rangle^T \tag{125}$$

An especially useful form of an approximate potential for the diatomic molecule is the Morse function [104]:

$$V = D_e \{1 - \exp[-a(r - r_e)]\}^2 \tag{126}$$

For the Morse potential the ratio $\frac{1}{3} (d^3V/d\xi^3)_e/(d^2V/d\xi^2)_e = -ar_e$ can be identified with a_1, the same a_1 that we have previously defined so that, for diatomic molecules,

$$\langle \Delta r \rangle^T_{vib} = (3/2)\, a \langle (\Delta r)^2 \rangle^T \tag{127}$$

If the harmonic approximation is used for $\langle (\Delta r)^2 \rangle$ it is possible to express $\langle \Delta r \rangle_{vib}$ as [98]

$$\langle \Delta r \rangle_{vib} \approx \left(\frac{3h}{8\pi}\right)(-F_3 F_2^{-3/2})\, \mu^{-1/2} \tag{128}$$

where

$$F_3 \equiv (1/3)\,(\partial^3 V/\partial r^3)_e \qquad F_2 \equiv (\partial^2 V/\partial r^2)_e$$

and the temperature dependence has been suppressed. Herschbach and Laurie found that F_3 and F_2 are approximately exponential functions of internuclear distance, each described by a family of curves which are determined by the location of the bonded atoms in rows of the periodic table [105].

$$(-1)^n F_n = 10^{-(r_e - a_n)/b_n} \tag{129}$$

Thus we write

$$\langle \Delta r \rangle_{vib} \approx (3h/8\pi)\, \mu^{-1/2}\, 10^{-D} \tag{130}$$

where

$$D \equiv (r_e - a_3)/b_3 - 3(r_e - a_2)/2b_2 \tag{131}$$

The constant in (130) is 19.35×10^{-3} if μ is in amu. The a_2, b_2, a_3 and b_3 are reproduced in Table 7 of [18]. A comparison of (130) with the $\langle \Delta r \rangle_{vib}$ calculated at 300 K using (107) and the known spectroscopic constants of a large set of diatomic molecules [106] shows good overall agreement for $\langle \Delta r \rangle_{vib}$ values from 3×10^{-3} Å to 25×10^{-3} Å [98]. What this means is that it is possible to estimate $\langle \Delta r \rangle$ for a diatomic molecule by knowing only the bond length. Similarly

$$\langle (\Delta r)^2 \rangle \approx (h/4\pi)\, \mu^{-1/2}\, 10^{+d} \tag{132}$$

where $d \equiv (r_e - a_2)/2b_2$.

To apply (122) to polyatomic molecules we need to be able to estimate $\langle \Delta r \rangle$ for a bond in a polyatomic molecule. For the majority of molecules, insufficient spectroscopic information is available to permit full dynamic calculations, so it is necessary to develop an approximation method in order to estimate the mean bond displacements for these molecules. It appears that stretching cubic constants for polyatomic molecules can be deduced from the bond length in the same way as for diatomic molecules. The Herschbach and Laurie parameters used for such estimation for diatomic molecules are found to describe F_3 for polyatomic molecules as well [98, 105].

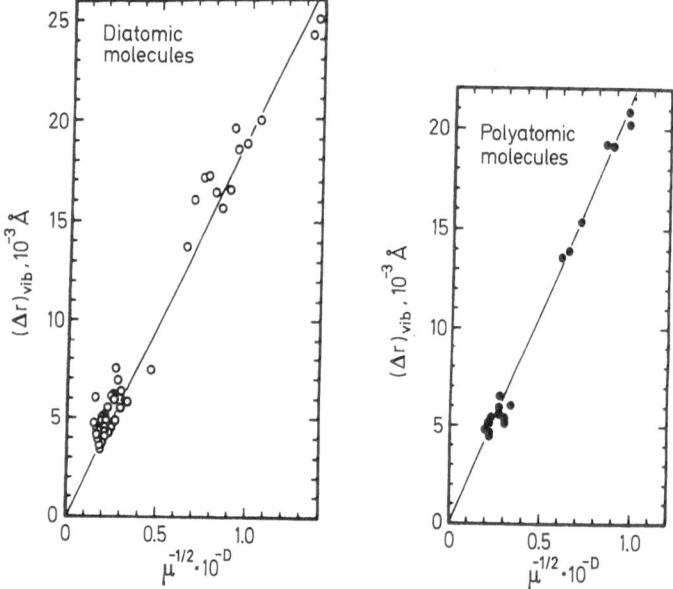

Fig. 11. Test of the approximation in (130) (straight line, slope $= 19.35 \times 10^{-3}$ is from (130)), against the calculated mean bond displacements for diatomic molecules. The same method is applied to polyatomic molecules; the straight line is a least-squares fit to the data (slope $= 22 \times 10^{-3}$). D is defined in (131). Reproduced from Jameson CJ, Osten HJ (1948) J. Chem. Phys. 81: 4300, with permission [98]

Therefore we can use (130) to predict mean bond displacements in polyatomic molecules if the Morse anharmonic stretching accounts for most of the mean bond displacement [98].

In Fig. 11 the plots of $\mu^{-1/2} \, 10^{-D}$ vs. $\langle \varDelta r \rangle_{\text{vib}}$ calculated for selected polyatomic molecules are compared with the results for some diatomic molecules [98]. A least-squares fit gives a slope of 22×10^{-3} rather than the 19.35×10^{-3} factor containing the fundamental constants in (130). The slope for the polyatomic molecules is somewhat larger because the other contributions due to bending and non-bonded interactions can be significant. Bartell has shown that, for molecules of the AX_n type, the Morse stretching contribution to $\langle \varDelta r \rangle$ is only 64% in CH_4 [107], and only about 40% in SF_6 [108]. However, (130) with a modified factor of 22×10^{-3} for polyatomics gives a satisfactory estimate of $\langle \varDelta r \rangle$ from r_e. Therefore we shall use

$$\langle \varDelta r \rangle_{\text{vib}}/\text{Å} \approx 22 \times 10^{-3} \mu^{-1/2} \, 10^{-D} \tag{133}$$

where D is given by (131).

The rotational contribution to $\langle \varDelta r \rangle$ is likewise easily estimated from r_e alone for an AX_n molecule [107]:

$$\langle \varDelta r \rangle_{\text{rot}} = 2\bar{E}_{\text{rot}}/(nF_2 r_e) \tag{134}$$

where $\bar{E}_{\text{rot}}$ is the classical average, kT for linear molecules, and $(3/2)$ kT for nonlinear ones, and F_2 is expressed according to (129). However, we do not even need $\langle \varDelta r \rangle_{\text{rot}}$

for estimating most isotope shifts. For bent triatomic molecules it has been shown that rotation does not play a significant role in the isotope shifts of the central atom upon end atom substitution [109]. We have shown that for diatomic molecules there is no rotational contribution to the isotope shift; for symmetrical substitution in the highly symmetric molecular types (e.g., CH_4, BF_3, SF_6, CO_2) there is also no rotational contribution to the isotope shift. Even for unsymmetrical substitution in CH_4 isotopomers, the rotational contribution plays no role in the D-induced ^{13}C isotope shift. For the isotope shift of the end atom the rotational contribution is also not too important. For example, the dynamical factor relevant to the proton isotope shift in CH_4–CH_3D is -2.687×10^{-4} Å from rotation and 5.53×10^{-3} Å from vibration, only 5% contribution comes from rotation. For the proton isotope shift in $H_2^{16}O$–$H_2^{17}O$ the dynamical factor is -4.176×10^{-7} Å from rotation and 3.102×10^{-5} Å from vibration, only 1.4% rotational contribution [98].

On the basis of (122) and $\langle \Delta r \rangle$ estimated by (130) or (133), empirical estimates of $(\partial\sigma/\partial r)_e$ have been obtained from isotope shifts in various molecules [53, 98, 101, 109–112]. Some of these are compared with theory in Table 8. In these apparently favorable cases, there is reasonably good agreement between the empirical value of the derivative and the value obtained from theoretical (ab initio) calculations. In the inversion of the data to obtain an empirical first derivative, only the leading term involving the first derivative with respect to bond stretch was used and in many instances, only the estimation methods for Δr and $(\langle \Delta r \rangle - \langle \Delta r \rangle^*)$ as in (111) and (130) or (133) proposed by Jameson and Osten [98] were used instead of a full vibrational analysis. Some cancellation of second order terms may occur for this property as shown by Fowler and Raynes in the case of CH_4 [67, 69]. Fleischer et al. have noted that for a wide variety of small polyatomic molecules the nuclear shielding surface is remarkably close to a linear function of bond length at the equilibrium structure [113]. Ditchfield has found this to be the case also for the diatomic molecules H_2,

Table 8. Comparison of empirical estimates of $(\partial\sigma/\partial r)_e$ with theoretical values, ppm Å^{-1}

Nucleus	Molecule	$(\partial\sigma/\partial r)_e$ empirical	Ref.	$(\partial\sigma/\partial r)_e$ theor. (ab initio)	Ref.
^{1}H	H_2	$-12.1, -11.5$	[53, 98]	-20.7	[44]
	CH_4	-38	[110]	-25.42	[181]
^{11}B	BH_4^-	-26.7	[98]	-27.0	
^{13}C	CO	-456	[111]	$-573.9, -535$	[182, 183]
				$-640, -413.7$	[184, 95]
	CH_4	-35	[110]	$-51.1, -52.62$	[181, 70]
^{15}N	N_2	-910	[112]	$-1090, -1132.5$	[184, 182]
				-640	[96]
	NH_3	-124	[101]	$-130.3, -129.2$	[181, 185]
				-144	[113]
	NH_4^+	$-65, -60$	[98, 101]	-67.9	[181]
^{17}O	CO	-1150	[111]	$-1166, -1077$	[182, 183]
				$-1240, -479$	[184, 95]
	H_2O	-294	[109]	$-270.9, -285$	[58, 186]
		-296	[101]	$-267.1, -275$	[181, 185]
^{31}P	PH_3	-180	[117]	$-150.8, -155,$	[181, 113, 185]
				-148.4	

LiH, and HF [44]. If the full shielding surface shown in Fig. 4 for H_2^+ is typical, the great sensitivity of nuclear magnetic shielding in molecules to bond stretching, and the nearly uniform negative sign of shielding derivatives with respect to bond stretch are easily understood.

6.2 General Trends in Secondary Isotope Effects

More observations of this mass dependence of electronic properties have been reported in NMR and ESR parameters than in other properties, thus many more of the general empirical observations have been made there than in other fields. The available high precision and resolution and the universality of the magnetic properties in all molecules make this possible. Isotope shifts of the order of a few parts per billion are now routinely observed in solution. NMR spectra showing isotope effects of the order of a few parts per million are shown in Fig. 12. These properties are site specific. The observations of a very large number of secondary isotope effects (i.e., the change in the observed property by isotopic substitution of an off-site (neighboring) atom) on nuclear magnetic shielding [114–116] have led to some interesting general trends.

The following trends were noted in the review of NMR isotope shifts by Batiz-Hernandez and Bernheim [114]:

1) The magnitude of the isotope shift is dependent on how remote the isotopic substitution is from the nucleus under observation.

2) The magnitude of the shift is largest where the fractional change in mass upon isotopic substitution is largest.

3) The magnitude of the shift is approximately proportional to the number of equivalent atoms in the molecule that have been replaced by isotopes.

4) Heavy isotopic substitution shifts the NMR signal of a nearby nucleus toward a higher magnetic field.

5) The magnitude of the shift is a function of the resonant nucleus, comparisons between nuclei reflecting the differences in their range of chemical shifts. The explanation of these observed trends have been given in [18, 91].

Secondary isotope effects on NMR spin-spin coupling constants have also been observed. These are fairly small so nearly all reports involve the large fractional mass change accompanying replacement of H with D. A review of the experimental data and interpretation in terms of the theory presented here was given by Jameson and Osten [83]. More recent experiments by Wasylishen et al. have revealed additional data such as isotope effects on J(NH) in NH_4^+ ion [100], J(PH) in PH_4^+ and PH_2^- [117] and J(SnH) in SnH_4 [102] and SnH_3^- [117]. The latter two isotope effects are large, -1.7 and 3.0 Hz respectively, and the signs are as predicted by the Jameson-Osten model.

Fig. 12. NMR spectra of ^{31}P in $PH_{2-n}D_n^-$ at 20 °C and of ^{119}Sn in $SnH_{3-n}D_n^-$ at -50 °C, both in liquid ammonia, showing the isotope effects on the average nuclear magnetic shielding. The shifts of the centers of the multiplets from the undeuterated species are additive. Reproduced from Wasylishen RE, Burford N (1987) Can. J. Chem. 65: 2707, with permission [117]

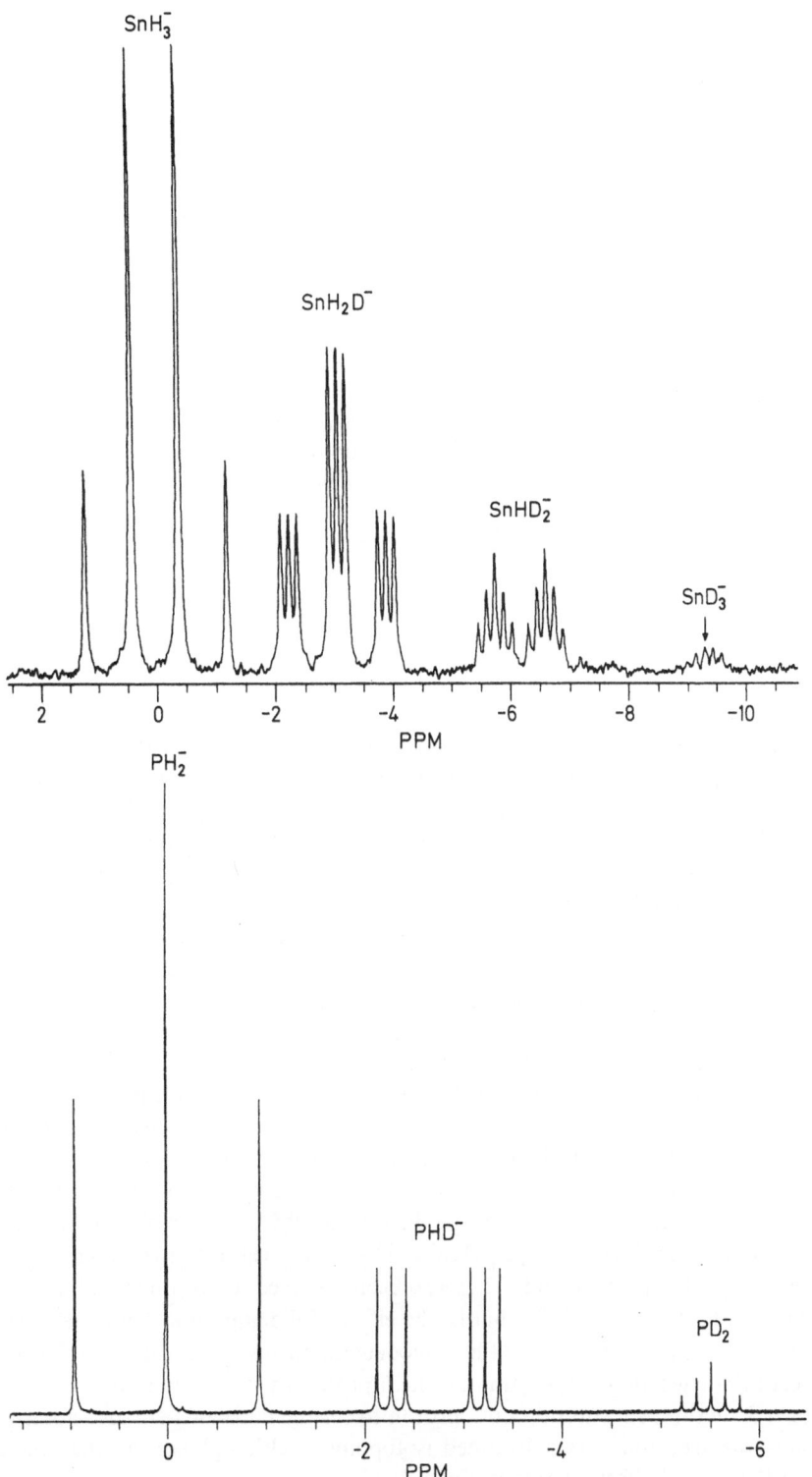

Table 9. Secondary isotope effects in the gas phase quadrupole coupling constants of symmetric tops

Nucleus	Heavy-Light	Ref.	$|eqQ|_{Heavy} - |eqQ|_{Light}$	% Change
^{35}Cl	$^{13}CD_3Cl-CD_3Cl$	[187, 188]	$-1.373\,MHz$	-1.8
^{75}As	AsD_3-AsH_3	[189]	$+2.12$	$+1.3$
^{121}Sb	SbD_3-SbH_3	[189]	$+5.01$	$+1.1$
^{35}Cl	$DCl-HCl$	[46]	-0.226	-0.33
^{79}Br	$^{13}CD_3Br-CD_3Br$	[190, 191]	-1.76	-0.31
^{79}Br	$DBr-HBr$	[47]	-1.673	-0.31
^{127}I	$DI-HI$	[192]	-4.912	-0.27
^{127}I	CD_3I-CH_3I	[193, 194]	-5.36	-0.27
^{79}Br	CD_3Br-CH_3Br	[191]	-1.42	-0.25
^{79}Br	$SiD_3Br-SiH_3Br$	[195]	-0.823	0.25
^{35}Cl	CD_3Cl-CH_3Cl	[188]	-0.1802	-0.24
^{14}N	$DCN-HCN$	[196]	-0.0061	-0.13
^{14}N	CD_3CN-CH_3CN	[197]	$+0.0049$	$+0.12$
^{35}Cl	$DC\equiv CCl-HC\equiv CCl$	[198]	-0.01	-0.01
^{79}Br	$DC\equiv CBr-HC\equiv CBr$	[199]	-0.04	-0.006
^{127}I	$^{13}CH_3I-CH_3I$	[194]	-0.016	-0.0008

Other properties also show some of the same general trends. For example, in Table 9 the isotope shifts in nuclear quadrupole coupling constants upon deuteration at a neighboring site appear to be uniformly in the same direction, that is, the magnitude becomes smaller upon heavy atom substitution. The exceptions are the apex atoms As and Sb in the pyramidal hydrides. The general trend 1) is also noted in Table 9. The magnitude of the effect is dependent on the remoteness of the isotopic substitution from the nuclear site, comparing the DCl, HCl pair with $DC\equiv CCl$, $HC\equiv CCl$. Also, the $D_3{}^{13}CCl-D_3{}^{12}CCl$, difference is larger than the difference D_3CCl-H_3CCl in the chlorine nuclear quadrupole coupling constant. Trend 2) should be observed for all molecular electronic properties sensitive to bond stretching since the factor $(m'-m)/m'$ appears in both $\langle\Delta r\rangle$ and $\langle(\Delta r)^2\rangle$.

Trend 3), universally observed in nuclear magnetic shielding, is clearly seen in the examples in Fig. 12 where the shift is in increments of -2.76, -5.52 ppm for one and two replacements of H by D in $PH_2{}^-$; similarly the increments are -3.1, -6.2, and -9.3 ppm for one, two, and three replacements of H by D in $SnH_3{}^-$. Additivity is also observed in the secondary isotope effects on NMR spin-spin coupling [83]. Some examples are J(PH) in $PH_{3-n}D_n$ [118] and $PH_{4-n}D_n^+$ [117], J(SnH) in $SnH_{4-n}D_n$ [102] and $SnH_{3-n}D_n^-$ [117]. Trend 3) has also been observed in proton hyperfine constants in naphthalene anion [119], that is, the effect of dideuteration can be deduced by adding up the effects observed in the monodeuterated anions, as shown in Fig. 13. The effect of α-deuterium substitution on the β proton hyperfine constants, and the effect of β deuterium substitution on α proton hyperfine constants are also additive for mono-, di-, tri-, and tetra-substituted isotopomers, although smaller than the α on α and the β on β effects shown in Fig. 13.

The *primary* isotope effects can not be determined easily in properties such as.

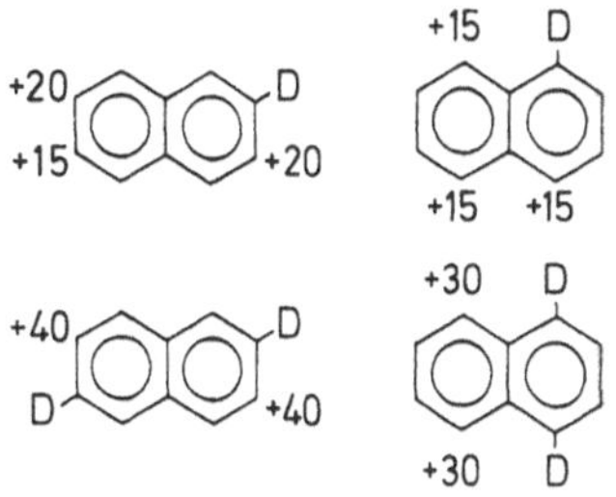

Fig. 13. Assignments of the experimental proton hyperfine splittings in the naphthalene-d anion radical and the corresponding splittings for the naphthalene-d_2 anion. Numbers are deviations of the splittings from the undeuterated naphthalene anion in milligauss, which are additive and are in agreement with the experimental data. Reproduced from Lawler RG, Bolton JR, Karplus M, Fraenkel GK (1967) J. Chem. Phys. 47: 2149, with permission [119]

nuclear magnetic shielding or nuclear quadrupole coupling constants because isotopic replacement of the nucleus in question involves not only a mass change but also a change in a nuclear parameter which is directly a part of the observed quantity, i.e., the nuclear magnetic dipole moment and electric quadrupole moment. Ratios of these quantities for nuclear isotopes are not very accurately known, with the exception of γ_H/γ_D which is 6.514 398 04 (120) [120]. Where γ_H/γ_D is involved, it is possible to make a determination of the primary isotope effect. Examples are discussed in Sect. 8.

The explanation of these general trends comes easily if we consider only the one term

$$\left(\frac{\partial P}{\partial r}\right)_e [\langle \Delta r \rangle - \langle \Delta r \rangle^*]$$

as in (122). Trends 1) to 3) can be attributed to dynamic factors. General trend 1) comes about because a secondary change in the average bond length at the site of the property upon isotopic substitution at a remote site is known to be negligibly small. At the same time, a site-specific property will clearly be most sensitive (large $(\partial P/\partial r)_e$) to changes in internal coordinates in its immediate vicinity; the change in a property upon infinitesimal bond extension at a remote site will be small in comparison, except when delocalized pi systems are involved. General trend 2) is clearly seen in the mass factors $[(m'-m)/m']\,[m_A/(m_A + m)]$. General trend 3) is also clearly seen in (122) where the number s of isotopic substitutions explicitly appears as a factor when second and higher order terms are neglected. Deviations from additivity of isotope shifts are generally very small and have been found to be regular, in ratios $0:3:4:3:0$ in $CH_{4-n}D_n$ types [99] or $0:1:1:0$ in $PH_{3-n}D_n$ types [101], similar to deviations from additivity of zero-point energies [121]. The exceptions occur when intermolecular affects are involved [103].

The other general trends, such as 4) and 5) in nuclear shielding, are purely dependent on the nature of the electronic property surface at the equilibrium geometry of the molecule. Where the surface has a steep and nearly linear dependence on bond stretching at the equilibrium geometry, the single first derivative $(\partial P/\partial r)_e$ may dominate the observed mass dependence (and temperature dependence) of the property. On the other hand, some properties have an overwhelmingly large dependence on an out-of-plane bend in a planar molecule, in which instances (120) is used, i.e.,

$$\left(\frac{\partial^2 P}{\partial \theta^2}\right)_e [\langle \theta^2 \rangle - \langle \theta^2 \rangle^*]$$

will dominate the observed rovibrational effects. In either case the isotope shifts nearly directly provide information on the nature of $(\partial P/\partial r)_e$ or $(\partial^2 P/\partial \theta^2)_e$ even when a full rovibrational analysis is not feasible. Table 8 shows that in these favorable cases the empirical $(\partial \sigma/\partial r)_e$ obtained compares well with ab initio theoretical values where available. Other examples are given in [18].

Where both the isotope effect and the temperature dependence have been observed, it is sometimes possible to find a single empirical derivative which reproduces both effects within experimental error, using only the first term in (116) and (118). This has been found for $(\partial \sigma^F/\partial r)_e$ in SeF_6 and TeF_6 [80], $(\partial \sigma^V/\partial r)_e$ in $[V(CO)_6]^-$ and $(\partial \sigma^{Co}/\partial r)_e$ in $[Co(CN)_6]^{3-}$ [122]. In $\dot{C}H_3$ and $\dot{C}D_3$, using the observed temperature dependence of a^H, Moss obtains [123]

$$a_e^H = -24.46\ (29)\ \text{gauss}, \qquad (\partial^2 a^H/\partial \theta^2)_e = +152\ (24)\ \text{gauss rad}^{-2}$$

whereas, using the isotope effect he gets with (120), a very similar result

$$a_e^H = -24.19\ (13), \qquad (\partial^2 a^H/\partial \theta^2)_e = +111\ (13)\,.$$

Ab initio derivatives [89] with appropriate rovibrational averaging give the results shown in Table 6.

7 Higher Order Contributions

How important are the terms involving the second and higher derivatives? We have already seen in Tables 1, 4, and 5 the contributions of the terms in the second derivative of the electronic property to the rovibrational averages in diatomic molecules. In Table 4 we also see the combined contributions from 3rd to 6th derivatives. Calculations indicate that these contribute altogether only 1–5% of the rovibrational corrections to the proton shielding in H_2, HF, and LiH at 300 K, and 8% for ^7Li in LiH.

Let us now consider second derivative terms in polytomic molecules. A typical equation for a rovibrationally averaged property such as σ^C in CH_4 is (from (65) and (62)–(63)) [67]:

$$\langle P \rangle = P_e$$

$$(a) \qquad + 4P_r \bar{L}_1^1 \langle q_1 \rangle$$

$$(b) \qquad + 2P_r [\bar{L}_1^{2a2a} \langle q_2^2 \rangle + \bar{L}_1^{3x3x} \langle q_3^2 \rangle + \bar{L}_1^{4x4x} \langle q_4^2 \rangle]$$

$$(c) \qquad + 2P_{rr} [(\bar{L}_1^1)^2 \langle q_1^2 \rangle + (\bar{L}_1^{3x})^2 \langle q_3^2 \rangle + (\bar{L}_1^{4x})^2 \langle q_4^2 \rangle]$$

$$(d) \qquad + 2P_{rs} [3(\bar{L}_1^1)^2 \langle q_1^2 \rangle - (\bar{L}_1^{3x})^2 \langle q_3^2 \rangle - (\bar{L}_1^{4x})^2 \langle q_4^2 \rangle]$$

$$(e) \qquad + \frac{9}{4}(P_{\alpha\alpha} + P_{\alpha\omega})(\bar{L}_5^{2a})^2 \langle q_2^2 \rangle + (P_{\alpha\alpha} - P_{\alpha\omega})[(\bar{L}_6^{3x})^2 \langle q_3^2 \rangle$$

$$+ (\bar{L}_6^{4x})^2 \langle q_4^2 \rangle]$$

$$(f) \qquad + 8P_{r\alpha} [\bar{L}_1^{3x} \bar{L}_6^{3x} \langle q_3^2 \rangle + \bar{L}_1^{4x} \bar{L}_6^{4x} \langle q_4^2 \rangle] \qquad (135)$$

The individual contributions of terms (a) through (f) have been calculated for the CH_4 molecule, for the ^{13}C nuclear magnetic shielding in $^{13}CH_4$ and for the electric dipole polarizability and magnetizability in $^{12}CH_4$. Results are shown in Table 10. We see that the second order terms (b) through (f) cannot generally be neglected. In these examples the first derivative terms alone, (a) + (b), comprise 137% or 92% of the total calculated temperature dependence of σ^C or α and 120%, 69%, or 72% of the total mass dependence of σ^C, α, or χ and 122%, 69%, or 70% of the total zero-point correction to σ^C, α, or χ.

Early work in this field recognized that rotation plays an important role in the temperature dependence of an electronic property, even as the observed temperature and mass dependence of various properties are fitted to only the terms in the first derivatives [68–91]. A further approximation ignored the curvilinear corrections (terms of the type (b)) in (135) [41]. By the use of Bartell's method, curvilinear corrections (b) were also explicitly included [80, 124–131] but still only a fit to the terms in the first derivatives was used because there are far too many second derivatives to be determined empirically from experimental data. We now find (in Table 10) that

Table 10. Contributions of various terms in Eq. (135)[a]

P	Ref.	(a) P_r 1st order	(b) P_r 2nd order	(c) P_{rr}	(d) P_{rs}	(e) $P_{\alpha\alpha}$	(e) $P_{\alpha\omega}$	(f) $P_{r\alpha}$	Total
Zero-point corrections $\langle P\rangle^{0K} - P_e$									
$\dfrac{\sigma^C}{ppm}$	[69]	−2.781 (rot = 0)	−1.587	−1.696	−0.067	2.573	−0.057	0.024	−3.591
$\dfrac{\alpha}{au}$	[67]	0.366 (rot = 0)	0.209	0.131	0.003	0.125	0.001	0.004	0.839
$\dfrac{\chi}{10^{-29}JT^{-2}}$	[69]	−0.283	−0.162	0.021	0.001	−0.217	0.005	−0.003	−0.638
Temperature dependence $\langle P\rangle^{369K} - \langle P\rangle^{180K}$									
$\dfrac{\sigma^C}{ppb}$	[67]	−60.86 (rot = −66.2)	−11.14	−0.05	−0.03	15.61	3.87	0.11	−52.49
$\dfrac{\alpha}{10^4au}$	[67]	79.90 (rot = 86.8)	14.34	0.04	0.01	7.42	0.29	0.15	102.10
Mass dependence $\langle P\rangle^{0K}_{CH_4} - \langle P\rangle^{0K}_{CH_4}$									
$\dfrac{\sigma^C}{ppm}$	[69]	0.763	0.406	0.453	−0.018	−0.686	0.059	0.001	0.978[b]
$\dfrac{\alpha}{au}$	[200]	−0.10	−0.053	−0.035	0.001	−0.033	−0.001	0.0	−0.221
$\dfrac{\chi}{10^{-29}JT^{-2}}$	[69]	0.077	0.041	−0.006	0.001	0.058	−0.006	0.0	0.165

[a] Rotational contribution ≈ 0 except where shown;
[b] Experimental value = 0.774

when only the terms in P_r are included, these terms account for 70–137% of the total rovibrational effects for a variety of electronic properties of CH_4. Calculations in H_2O also show that while the terms in P_r provide a large part of the contributions to the zero-point values, the temperature dependence, and the mass dependence of various molecular properties, the higher order terms can not be neglected. For example the isotope effect ($\langle P \rangle^{0K}_{OD2} - \langle P \rangle^{0K}_{OH_2}$) for σ^O is made up of -2.57, -0.01, -1.31, $+0.45$, -0.24 ppm from the terms in $(\partial\sigma/\partial r)_e$, $(\partial^2\sigma/\partial r^2)_e$, $(\partial\sigma/\partial\alpha)_e$, $(\partial^2\sigma/\partial\alpha^2)_e$, and 3rd and higher derivatives, respectively, giving a calculated total of -3.68 ppm [132].

It is fortunate that the P_r term constitutes such a large part of the observed rovibrational effects in some molecular electronic properties. Calculations to higher orders cannot be carried out in most molecules for which experimental data for electronic properties are available, whereas it is still necessary to make some sense out of the large body of information and look for some (albeit qualitative) predictive conclusions. However, the terms in the second derivatives are not insignificant. When the terms in the second derivatives are significant, a fitting of experimental data to one parameter leads to an empirical overestimation of P_r the first derivative of the property surface. On the other hand, fitting to more than one parameter is hazardous and could lead to unrealistic magnitudes specially when the contributions are not all the same sign. Examples of this pitfall are the fits to $(\partial\sigma^N/\partial r)_e$ and $(\partial\sigma^N/\partial\alpha)_e$ in NH_3 giving unphysical empirical estimates for these quantities [109], as is now obvious from ab initio calculations of these derivatives [133]. The empirical $(\partial\sigma^N/\partial r)_e$ values from the temperature dependence of the shielding of the two nitrogens in NNO are likewise flawed [129], being much too large compared to what one can obtain from the fit to the mass dependence [134].

These case studies of H_2O and CH_4 have included molecular electronic properties where the anharmonicity of the totally symmetric vibration gives an important contribution, since the properties happen to be very sensitive to bond extension and less so to angle deformations. Thus, the terms in the first derivatives of the property surfaces form the bulk of the rovibrational effects and the harmonic approximation is not at all applicable. In other molecular properties the terms in the first derivatives of the surface are not the most important ones. When the greatest sensitivity of the property is to non-totally symmetric displacements such as out-of-plane deformations, then the bulk of the observed rovibrational effects can be semi-quantitatively interpreted in terms of $\langle\theta^2\rangle$ rather than $\langle\Delta r\rangle$ terms. In these cases the harmonic approximation is probably adequate unless some double- or multiple-minimum anharmonic potential describes the angle deformation.

8 Additional Examples

The best interpretation of experiment requires 1) experiments in the limit of zero pressure so as to exclude intermolecular effects, 2) measurements of rovibrational averages in several vibrational states and/or rotational states, 3) a good anharmonic potential energy surface with up to third derivatives with respect to normal coordinates and, 4) a sufficient number of theoretically calculated values of the electronic property at selected displacements from the equilibrium molecular structure such that derivatives P_r, P_α, $P_{r\alpha}$, etc. up to fourth derivatives of the electronic property with respect to curvilinear internal coordinates can be obtained by fitting. For some properties the measurements in specific rovibrational states are not feasible. Instead, averages given in (40) have been measured as a function of temperature and/or for different isotopomers. If simultaneously available, the temperature dependence and isotope shifts of the property can provide redundant or (less commonly) complementary information. When it is assumed that only one derivative overwhelmingly dominates both, then it becomes possible to interpret both observations with a single electronic factor such as $(\partial P/\partial r)_e$ [122] or $(\partial^2 P/\partial\theta^2)_e$ [90]. This forms the basis of the simultaneous interpretation of nuclear magnetic shielding temperature dependence in the zero-pressure limit and isotope shifts in the early work [41, 68, 91] which have been reviewed recently [18].

8.1 Primary Isotope Effects

The electron spin-nuclear spin hyperfine constant of an unpaired electron in a sigma bond provides a good example. In HgH molecule the electron spin density surface shifts the unpaired electron density from Hg to H as the bond lengthens, which is not unexpected since the unpaired spin is entirely on H in the limit of dissociation

$$(HgH)^{\cdot} \to Hg + \dot{H}$$

This shift corresponds to derivatives of the following signs:

$$\left(\frac{\partial \varrho_{Hg}}{\partial r}\right)_e < 0 \quad \text{and} \quad \left(\frac{\partial \varrho_H}{\partial r}\right)_e > 0 \,.$$

If only the first derivative is included in the expansion in (1) then the effects of rovibrational averaging are as follows: ϱ_{Hg} and thus the nuclear hyperfine interaction constant a^{Hg} will decrease with increasing vibrational quantum number and decrease with increasing rotation. This is indeed observed in HgH, HgD, and HgT spectra, being most pronounced in HgH [54]. Since $\langle\Delta r\rangle_{HgH} > \langle\Delta r\rangle_{HgD}$ as discussed earlier then $[\langle a^{Hg}\rangle_{HgD}/\langle a^{Hg}\rangle_{HgH}] < 1$. This secondary isotope effect is indeed observed. ESR experiments reveal this ratio to be 0.992 and it was deduced that $\langle\varrho_{Hg}\rangle_{HgD} = 8.95$ au whereas $\langle\varrho_{Hg}\rangle_{HgH} = 8.88$ au [55]. Furthermore, we also expect $\langle\varrho_H\rangle_{HgH} > \langle\varrho_H\rangle_{HgD}$, a primary isotope effect. Since the observable a^H or a^D contains the factor γ_H or γ_D, the ratio $\varkappa$ is used:

$$\varkappa = \frac{\gamma_D}{\gamma_H} \cdot \frac{a^H}{a^D} = \frac{a^H}{6.514 \, a^D} = \frac{\langle\varrho_H\rangle_{HgH}}{\langle\varrho_D\rangle_{HgD}}$$

$\varkappa$ in this case is predicted to be greater than 1. The ESR experiments showed $\varkappa = 1.019$ [55].

A related property is the nuclear spin-spin coupling J observed in NMR. The terms in the effective hamiltonian for a and J are shown below:

$$H_2 = 2\mu_B \hbar (\mu_0/4\pi) \sum_N \gamma_N \sum_k [3(\mathbf{S}_k \cdot \mathbf{r}_{kN})(\mathbf{I}_N \cdot \mathbf{r}_{kN}) r_{kN}^{-5} - (\mathbf{S}_k \cdot \mathbf{I}_N) r_{kN}^{-3}]$$

$$\text{(136)}$$

$$H_3 = (16\pi\mu_B\hbar/3)(\mu_0/4\pi) \sum_N \gamma_N \sum_k \delta(\mathbf{r}_{kN}) \mathbf{S}_k \cdot \mathbf{I}_N \tag{137}$$

With these perturbation terms in the molecular hamiltonian $\mathbf{a}^N$ is the tensor which appears in the $\mathbf{S} \cdot \mathbf{a}^N \cdot \mathbf{I}_N$ terms in the first order correction to the energy for a non-singlet electronic state. $\mathbf{J}$ is the tensor in the $\mathbf{I}_N \cdot \mathbf{J}(NN') \cdot \mathbf{I}_{N'}$ terms in the second order correction to the energy. Only the trace of the tensor is observed in the gas or in solution. The reduced spin-spin coupling constant K is defined as

$$K(NN') = 4\pi^2 J(NN')/h\gamma_N\gamma_{N'} \tag{138}$$

We note that each half of the interaction in the second order perturbation theory expression for J is similar to that of the interaction for a. This has been used to advantage in creating models for interpretation of both properties [135–140, 83]. In the simplest case, the transfer of spin density from X to H upon bond extension in the above discussion of a^H and a^X translates to an increase in the reduced one-bond spin spin coupling constant K upon bond extension [83]. When only this first derivative is considered, this would mean that

$$\varkappa = \frac{\gamma_D}{\gamma_H} \frac{J(XH)}{J(XD)} > 1 .$$

The model depends on the dominance of the derivative of the Fermi contact mechanism, as has been shown theoretically in J(CO) [14] in Fig. 2. A review of the available data does show $\varkappa > 1$ where the above model is applicable. The model has been extended to one-bond couplings in which terms of opposite sign contribute in the Fermi-contact mechanism [83]. The presence of a lone pair on one of the coupled atoms has long been known to be associated with $K < 0$ for one-bond couplings. The sign of the derivative $(\partial K/\partial r)_e$ then depends on which term changes greatly upon bond extension. It was proposed by Jameson and Osten that in the usual situation (as for J(CH) in CH_4) $\varkappa > 1$, but in those other cases where a lone pair was involved, $\varkappa < 1$. Theoretical calculations on J(HF) lead to $\varkappa < 1$ as well, in agreement with these arguments [141]. Indeed, nearly all cases of $\varkappa < 1$ involve a lone pair on one of the coupled atoms, as in H_2Se, PH_3, $PhPH_2$, Ph_2PH [83]; and the more recently reported couplings in H_2O, PH_2^-, SnH_3^-, NH_3 also exhibit $\varkappa < 1$ [117, 142, 101]. In contrast, the lack of the lone pair in related molecules is accompanied by $\varkappa > 1$, in support of the Jameson-Osten model. These are shown in Table 11. NH_4^+ ion is an exception which has also exhibited exceptional ^1H secondary isotope shifts of the nuclear magnetic shielding constant, attributed to differential intermolecular (hydrogen bonding) effects [103].

A substantially different situation of rovibrational averaging attends the hyperfine constant of an unpaired electron largely in a pi orbital on an adjacent nucleus. Typical

Table 11. Comparison of primary isotope effect on spin spin couplings in related molecules

With lone pair	$\varkappa$	Ref.	No lone pair	$\varkappa$	Ref.
PH_2^-	0.98	[117]	PH_4^+	1.003	[117]
PH_3	0.93	[118]			
SnH_3^-	0.90	[142, 117]	SnH_4	1.001	[102]
			SnH_3^+	1.004	[102]
NH_3	0.992	[101]	NH_4^+	0.996	[101]
H_2O	0.996	[101]			

examples of this are $\dot{C}H_3$ and $[\dot{N}H_3]^+$ which are planar molecules. More complex examples are benzene anion or cation or other planar aromatic radicals. McConnell [143] has described the spin density on the proton in terms of a simple model,

$$\varrho_H \propto a^H = Q_{CH}\varrho^\pi \tag{139}$$

where $Q_{CH} < 0$; the spin density on H is due to a coupling between the sigma and pi electron system, with a spin polarization parameter Q_{CH} relating the spin density on H to the spin density on the adjacent carbon pi orbital. The coupling parameter is expected to decrease with increasing CH bond length and increase with out-of-plane bending. The first derivatives $(\partial\varrho_H/\partial q)_e$ where q is a normal coordinate are of either sign depending on whether the vibration assists spin polarization or not. If $(\partial\varrho_H/\partial r_{CH})_e < 0$ is the important term, then one expects to find

$$\varkappa = \frac{\langle\varrho_H\rangle}{\langle\varrho_D\rangle} < 1$$

and

$$\frac{d\langle\varrho_H\rangle}{dT} < 0 \quad \text{and} \quad \frac{d\langle a^H\rangle}{dT} < 0. \tag{140}$$

It has also been suggested that the observed temperature dependence is primarily due to out-of-plane vibrations, i.e., that the $(\partial^2\varrho_H/\partial\theta^2)_e \langle\theta^2\rangle^T$ term in (64) is responsible [144, 145], where θ is the out-of-plane angle and $(\partial^2\varrho_H/\partial\theta^2)_e < 0$. In the McConnell model these derivatives would correspond to

$$(\partial|Q_{CH}|/\partial r_{CH})_e < 0 \quad \text{and} \quad (\partial^2|Q_{CH}|/\partial\theta^2)_e < 0 \tag{141}$$

neglecting $(\partial\varrho^\pi/\partial r)_e$ and $(\partial^2\varrho^\pi/\partial\theta^2)_e$. Both the bond stretching and the out-of-plane bending lead to a primary isotope effect $\varkappa < 1$ and negative temperature coefficients $d\langle\varrho_H\rangle/dT < 0$. The $\varkappa < 1$ was first observed in $[\dot{N}H_3]^+$ and $[\dot{N}D_3]^+$ by Cole [146]; other examples are $\dot{C}H_3/\dot{C}D_3$ [147] and $CH_3\dot{C}H_2/CH_3\dot{C}D_2$ [147], in which $\varkappa$ was found to be respectively 0.986, 0.989, and 0.990. It is also found that $d\langle\varrho_H\rangle/dT < 0$ for $\dot{C}H_3$ [148–150], benzene anion [147], benzene cation [151], and $C_7H_7^-$ [152]. The relatively lower frequencies of out-of-plane bends compared to stretching frequencies and the relatively large temperature coefficients of a compared to other observables seem to indicate that (at least in the planar systems) the out-of-plane vibrational averaging makes significant contributions. The dominance of the out-of-plane bend over the other modes in the zero-point vibrational contributions to a^H has been demonstrated in calculations for $\dot{C}H_3$ [153] and $\dot{N}H_3^+$ [154]. Table 6 shows that the $\varkappa < 1$ and the temperature dependence in the typical case of $\dot{C}H_3$ and its isotopomers are reproduced reasonably well by calculations. It is encouraging that the calculations support the above semi-quantitative descriptions, lending credibility to their use in other cases where calculations as accurate as those in Table 6 are not available.

The sign of the temperature coefficient of a^H for protons attached to proximate $\dot{C}$ is by no means uniformly negative. An interesting correlation is that $d|a^H|/dT < 0$ when the spin density is positive (usual) on both the proximate and adjacent carbons, whereas $d|a^H|/dT > 0$ are found for protons attached to sites of *negative* spin density although the adjacent carbons have large positive spin densities. This correlation is consistent with (141) and (139), and may be assisted by $(\partial \varrho^\pi/\partial r)_e$, $(\partial^2 \varrho^\pi/\partial \theta^2)_e > 0$.

β and γ C–H protons are known to exhibit large positive temperature coefficients. Early examples are $CH_3\dot{C}H_2$, in which it is also observed that $\varkappa = 1.011$ in the comparison $CH_3\dot{C}H_2/CD_3\dot{C}H_2$ and $\varkappa = 1.178$ in $CH_3\dot{C}H_2/CHD_2\dot{C}H_2$ [147]. In these cases Stone and Maki suggested that hindered internal rotations could lead to the large temperature coefficients of β proton hyperfine constants [85]. Single crystal studies reveal the torsional angle (ϕ) dependence [84]:

$$a^H_{CH_\beta} = a_0 + a_2 \cos^2 \phi . \tag{142}$$

Averaging over the torsional states alone, using a reasonable estimate of a torsional barrier accounts for most of the temperature dependence. Of course, complete agreement with experiment is not expected unless the calculations include all the vibrational modes. A review of temperature-dependent hyperfine constants in organic free radicals has been given by Sullivan and Menger [86].

8.2 Beyond the Born-Oppenheimer Approximation

The necessity of some nonadiabatic corrections have been suggested for isotope effects on various electronic properties, especially electric dipole moments. Some ab initio calculations of property surfaces which include nonadiabatic corrections are in Table A1 (see Appendix). In the interpretation of vibrational circular dichroism using theories based on the usual expansion of a property in terms of nuclear displacements, the calculations have to go beyond the Born-Oppenheimer approximation to include the distortion of the electronic wavefunction by nuclear displacements. Whereas the adiabatic approximation allows the vibronic wavefunctions to be written as $\phi_n^{(0)}(\mathbf{r}) \cdot \chi_v(\mathbf{R})$, the vibronic wavefunction which includes nonadiabatic corrections to first order is

$$\Psi_{nv}(\mathbf{r}, \mathbf{R}) \approx \phi_n^{(0)}(\mathbf{r}) \cdot \chi_v(\mathbf{R})$$

$$- \sum_{m \neq n} \sum_{v''} \sum_s \frac{\phi_m^{(0)} \chi_{v''} \langle \phi_m^{(0)}| (\partial H_0/\partial q_s)_{R_0} |\phi_n^{(0)}\rangle \langle \chi_{v''}| q_s |\chi_v\rangle}{E_m^{(0)} - E_n^{(0)} + \hbar\omega_{v''v}}$$

$$\tag{143}$$

$\chi_v(R)$ are the second order wavefunctions of the rovibrational hamiltonian, $|T_1^{-1} T_2^{-1} X^{(0)}\rangle$ in (16). The appropriate average of a property $\langle P \rangle$ is then $\langle \Psi_{nv}| P |\Psi_{nv}\rangle$, which includes not only the averages over the zeroth order vibronic function (1st term in (143)) as explicitly given in (28)–(33) but also the contributions from the distortion of the electronic wavefunction by nuclear displacement. However, the alternate formalism by Buckingham, Fowler, and Galwas [24], in which a molecular electronic property surface (the magnetic dipole moment **m**) is expanded in terms of nuclear velocities, leads to a theory which provides the magnetic dipole vibrational transition integrals in terms of derivatives $(\partial \mathbf{m}/\partial \dot{\mathbf{R}}_N)$ which can be calculated within the clamped nucleus approximation [24].

A clear indication of non-Born-Oppenheimer behavior was noted in some free radicals. Hyperfine constants in orbitally degenerate or nearly degenerate electronic states show unusually large temperature coefficients and deuteration effects. Examples of these are $C_6H_6^-$ anion [155–157], $C_7H_7^-$ [158, 152], naphthalene anion [119]. The comparison of a^H with a^D as in $C_6H_6^-/C_6D_6^-$, show $\varkappa < 1$ as expected [157]. Furthermore, the proton hyperfine constants at the undeuterated positions in a partly deuterated species are altered, and the $\varkappa$ and the temperature coefficients for H positions made nonequivalent by deuteration are different from each other and are larger or smaller than in the parent anion [157]. It has been suggested [159] that these cases are subject to the effects of the Jahn-Teller theorem [160]. Vibronic interactions have to be taken into account because the observed effects on deuterium substitution are different and greater than would be expected in the context of the Born-Oppenheimer approximation. Symmetric substitution (as in benzene 1,3,5-d$_3$ and benzene-d$_6$ anions) do not lift the degeneracy.

8.3 Rovibrational Averaging Effects on the Determination of Molecular Geometry of Semi-rigid Molecules

Various definitions and interrelations of the geometric structures of polyatomic molecules have been reviewed by Kuchitsu and Oyanagi [161] and by Lide [162], and earlier by Herschbach and Laurie [163]. See also [163a] the desired geometry is of course the equilibrium geometry, that is, that which corresponds to the global minimum in the potential energy surface of the molecule as defined in the Born-Oppenheimer sense. This is never directly observed. Depending on the experiment, various rovibrationally averaged quantities related to geometry can be obtained, some of which are discussed here.

8.3.1 Diatomic Molecules

The averages which can be obtained from experiment are typically taken from spectroscopy in the form of the molecular rotational constant, or the direct nuclear spin-spin coupling, or a bond distance from electron diffraction. The rotational constant is inversely proportional to the moment of inertia so that the observed average is $\langle r^{-2} \rangle^{-1/2}$ for a given vibrational level. On the other hand, the direct nuclear spin spin coupling constant is proportional to components of the dipolar interaction tensor so that the observed average is $\langle r^{-3} \rangle^{-1/3}$ for a given vibrational level if observed in a molecular beam resonance spectrum, or a thermal average if observed in an NMR spectrum of oriented molecules. Thus, from spectroscopy one gets average nuclear *positions* in the principal inertial axes. On the other hand, from electron diffraction one gets thermal average internuclear *distances*, $\langle r \rangle$. Written in terms of the averages $\langle \xi \rangle$ and $\langle \xi^2 \rangle$ defined in (45)–(46) these are [163]:

$$\langle r^{-2} \rangle^{-1/2} = r_e \left\{ 1 + \langle \xi \rangle - \frac{3}{2} \langle \xi^2 \rangle + ... \right\} \tag{144}$$

$$\langle r^{-3} \rangle^{-1/3} = r_e \{ 1 + \langle \xi \rangle - 2 \langle \xi^2 \rangle + ... \} \tag{145}$$

$$\langle r \rangle = r_e \{ 1 + \langle \xi \rangle + ... \} \tag{146}$$

The structure obtained from rotational constants is called the "r_0" structure corresponding to the zero-point average, $\langle r^{-2} \rangle_{v=0}^{-1/2}$. On the other hand, $\langle r \rangle$ is essentially identical to one of the widely used types of distances derived from electron diffraction, the "r_g" structure, the center of gravity of the radial distribution function [164]

$$r_g = \frac{\int_0^\infty r P(r) \, dr}{\int_0^\infty P(r) \, dr} \tag{147}$$

which represents the average of instantaneous internuclear distance over all the vibrations of the molecule, is a function of the distribution of molecules over vibrational states.

8.3.2 Polyatomic Molecules

From rotational constants or effective moments of inertia I_0, one gets the r_0 structure which corresponds to the zero point average nuclear positions, as in diatomic molecules. On the other hand, these effective moments of inertia I_0 can be corrected to give a set of moments I^* from which a structure called the r_z structure can be calculated. It was shown by Morino et al. [165] and by Herschbach and Laurie [163] that the removal of the harmonic part of the vibrational dependence of the effective moments I_0 yields the moments (I^*) which correspond to a rigid molecule in which each atom is frozen at its average position. The differences in the r_z structure of various isotopomers of several triatomics have been calculated and these reveal additivity in multiple substitutions [161].

The simplest way of comparing the various types of averages in polyatomic molecules is by expressing the averages in terms of local cartesian displacements: Δz is an instantaneous displacement of $r(A-X)$ projected on the equilibrium $A-X$ axis taken as a local z axis, Δx and Δy denote the displacements perpendicular to the equilibrium internuclear axis z [165]. The thermal average internuclear *distance* from electron diffraction

$$r_g = \langle r \rangle = r_e + \langle \Delta r \rangle^T \cong r_e + \langle \Delta z \rangle^T + \frac{(\langle \Delta x^2 \rangle^T + \langle \Delta y^2 \rangle^T)}{2 r_e} + \delta r_{cent} \tag{148}$$

where δr_{cent} is the displacement due to centrifugal stretching, the rotational contribution already discussed in Sect. 6.1.1. The $\langle \Delta r \rangle^T$ in (148) for C–H in CH_4 is the same as in (62). Since anharmonic force fields are not accurately known for most molecules while harmonic force constants are more readily available, partial rovibrational corrections are usually attempted, leading to the "r_α" structure which is the distance between the average nuclear positions at thermal equilibrium

$$r_\alpha = r_g - \frac{(\langle \Delta x^2 \rangle^T + \langle \Delta y^2 \rangle^T)}{2r_e} - \delta r_{cent} \qquad (149)$$

$$\cong r_e + \langle \Delta z \rangle^T \qquad (150)$$

For C–H in CH_4, one obtains

$$r_\alpha = r_e + \bar{L}_1^1 \langle q_1 \rangle_{vib} \qquad (151)$$

when the harmonic corrections

$$\frac{\langle \Delta x^2 \rangle^T + \langle \Delta y^2 \rangle^T}{2r_e} \quad \text{or} \quad \frac{1}{2}\{\bar{L}_1^{2a2a}\langle q_2^2 \rangle + \bar{L}_1^{3x3x}\langle q_3^2 \rangle + \bar{L}_1^{4x4x}\langle q_4^2 \rangle\}$$

are made on the electron diffraction distances. When extrapolated to 0 K r_α becomes practically identical with the r_z structure [161].

$$r_\alpha(0 \text{ K}) = r_e + \langle \Delta z \rangle^{0K} \cong r_z \qquad (152)$$

Obviously, the harmonic corrections which need to be made on r_g values from electron diffraction or $\langle r^{-2} \rangle^{-1/2}$ values from rotational constants [165, 163] are different from the harmonic corrections which need to be made on $\langle r^{-3} \rangle^{-1/3}$ values from NMR dipolar coupling constants [166, 59], but all should lead to the same r_α structure. A knowledge of the anharmonic force field further converts [via (151) for CH_4, for example] the r_α structure to the desired equilibrium r_e structure. Actually, the harmonic corrections usually made to obtain the r_α structure are usually only approximate since the correct harmonic force field can only come out of a complete anharmonic vibrational analysis which reproduces all the observed frequencies for all isotopomers of the molecule.

8.4 Rovibrational Averaging in van der Waals Molecules

Rotational constants of van der Waals complexes can be obtained from their microwave spectra in molecular beam resonance spectroscopy. However, large amplitude anharmonic vibrations against the weak intermolecular bonds of these molecules usually mean that the average structural parameters of the various isotopes will differ significantly. In many cases, average nuclear quadrupole coupling constants and dipole moment components provide additional information. For the purposes of determining an average structure it is commonly assumed that the ^{35}Cl quadrupole coupling constant observed, in HCl–Ar for example, is unchanged from that in the isolated HCl molecule so that the observed coupling constant in HCl–Ar is merely a vibrationally averaged projection of $eq_{zz}Q$ of the diatomic along the inertial axis of HCl–Ar:

$$\langle eqQ \rangle \approx \langle eq_{zz}Q \rangle_{diat} \langle (3\cos^2\theta - 1)/2 \rangle \tag{153}$$

Since the diatomic molecule has been studied separately, the observed $\langle eqQ \rangle$ in the complex then provides the angle that HCl makes with the inertial axis of the complex, in the form of an average $\cos^{-1} \langle \cos^2\theta \rangle^{1/2}$. The equilibrium angle that corresponds to the nonvibrating complex at the minimum of its potential surface can be obtained only if the potential surface derivatives are known. In the examples of semi-rigid molecules discussed in the preceding sections, the values $\langle \Delta r \rangle$, $\langle (\Delta r)^2 \rangle^{1/2}$, $r_e \langle (\Delta \alpha)^2 \rangle^{1/2}$, etc. are very small, of the order of $10^{-3} r_e$ to $10^{-5} r_e$, so that the rovibrational corrections to obtain the equilibrium geometry are small. The same is no longer true in van der Waals complexes. The difference between the average and equilibrium geometry of these complexes is not usually small. In Ar–ClH the equilibrium polar angle θ is $0°$ while the average angles obtained via (153) from the ^{35}Cl quadrupole coupling constant are: $< ArClD$ is $33.8°$ and $< ArClH$ is $41.5°$ [167, 168].

It has been found that this assumption of no change in q_{zz} in the monomer upon complex formation is not appropriate. This has been demonstrated, for example, when the rotational constants and $\langle eqQ \rangle$ values for ^{14}N and D in 11 isotopomers of $(HCN)_2$ were determined from pulsed FT microwave spectroscopy [169]. If the dimer is treated as rigid linear monomers tilted away from the line of centers of mass, the average angles of tilt and the $\langle eq_{zz}Q(^{14}N) \rangle$ in (153) can be independently determined without making the assumption that the latter are the same as in the monomer. The difference between the derived $\langle eq_{zz}Q \rangle$ value in (153) and the $\langle eq_{zz}Q \rangle$ value for the isolated HCN monomer is found to be 4.9% and 2.3%, accounting for about 40% of the observed $[\langle eqQ \rangle_{complex} - \langle eq_{zz}Q \rangle_{monomer}]$ [(169].

The effects of isotopic substitution on the components of the dipole moment of a van der Waals complex along the inertial axes provide additional information. Here too it is not necessary to assume that $\langle \mu_z \rangle$ in the complex is the same as that for the isolated monomer. In cases where the quadrupole coupling constants as well as the dipole moment components of the complex are determined for a large number of isotopomers, the isotopic substitution trends in the quadrupole coupling and in the dipole moments are consistent [170].

A complete vibrational analysis of a van der Waals complex is made tractable by a separation of the large amplitude motions associated with the intermolecular or hydrogen bonds and the small amplitude motions associated with the semirigid monomers. The Born-Oppenheimer separation of slow coordinate ϱ from fast coordinates q_1 and q_2 can also be used for the separation of low and high frequency modes. The energy levels of the fast coordinates q_1 and q_2 can be solved at a frozen value for the slow coordinate ϱ. This is done for different values of ϱ to determine $E_v(\varrho)$. The latter is then used as the potential energy to solve for motion in ϱ. This has been applied to the $(HF)_2$ dimer. $(HF)_2$ is a hydrogen-bonded dimer with a planar asymmetric top structure which tunnels between two equivalent maxima on the potential surface. The slow coordinate ϱ is $(\theta_2 - \theta_1)$, a tunneling coordinate [171].

The empirical potential energy surface was determined by Barton and Howard [172] by inversion of the spectroscopic data from radiofrequency and microwave spectra of Dyke et al. [173]. Interpretation of high resolution infrared spectra of the dimer in the H-stretching region require corrections to the zero[th] order Born-Oppenheimer approximation, however [171]. The extra terms which appear in the Schrödinger equation for the slow coordinate have been derived in general [174]. In part these result from the change in the form of the high frequency normal coordinates with a change in the tunneling coordinate ϱ. The dynamic interaction of the tunneling momentum with the high frequency vibrations adds to the effective barrier when the high frequency stretching vibrations are excited [171]. Of course, for molecules with large amplitude internal motions, the notion of an equilibrium geometry is no longer appropriate. An alternative approach to the description of nuclear motion which has been found particularly useful in the study of van der Waals molecules is given by Sutcliffe and Tennyson [13].

9 Conclusions

We have shown that rovibrational corrections almost always need to be included in order to obtain from experiment the value of the electronic property at the structure corresponding to the potential energy minimum, and vice versa, rovibrational corrections have to be included in ab initio calculations of electronic properties before comparing with experiment. Where vibrational state dependence and rotational state dependence can be measured precisely (rarely) it is possible to obtain individually from experiment the derivatives of a property with respect to normal coordinates. The observed temperature dependence and isotope shifts of an electronic property contain information about the property derivatives but inversion of the data to find empirical property surface derivatives is not always possible. We have found some cases in which isotope shift data provide empirical derivatives which are in reasonably good agreement with the best available ab initio calculations.

When the property has a nearly linear dependence on bond stretching, or an overwhelmingly large $(\partial^2 P/\partial \theta^2)_e$ for example, then general trends appear in its temperature and mass dependence which make it possible to make qualitatively insightful predictions and correlations with other related properties of the same molecule. For quantitative interpretation of experiments however, it is virtually necessary to carry out a fully correlated ab initio calculation of the property surface in the immediate vicinity of the potential minimum as well as perform a complete rovibrational averaging on the potential surface (usually a semi-empirical one) in order that the terms in second or higher order can be included, since these are not always negligibly smaller than the leading terms.

10 References

1. Allerhand A, Dohrenwend M (1985) J. Am. Chem. Soc. 107: 6684
2. Jameson CJ, Jameson AK, Oppusunggu D (1986) J. Chem. Phys. 85: 5480
3. Michelot F (1982) Mol. Phys. 45: 971
4. Michelot F (1982) Mol. Phys. 45: 949

5. Flygare WH (1964) J. Chem. Phys. 41: 793
6. Eshbach JR, Strandberg WP (1952) Phys. Rev. 85: 24
7. Geratt J, Mills IM (1968) J. Chem. Phys. 49: 1719
8. Danishoh Jørgensen P, Simons J (eds) (1985) Geometrical derivatives of energy surfaces and molecular properties, Reidel, Dordrecht
9. Bishop DM (1987) J. Chem. Phys. 86: 5613
10. Bishop DM, Cheung LM, Buckingham AD (1980) Mol. Phys. 41: 1225
11. Adamowicz L, Bartlett RJ (1986) J. Chem. Phys. 84: 4988
12. Amos RD (1987) Adv. Chem. Phys. 67 (Part 1): 99
13. Sutcliffe BT, Tennyson J (1981) In: Avery J, Dahl JP, Hansen AE (eds) Understanding Molecular Properties, Reidel, Dordrecht
14. Geertsen J, Oddershede J, Scuseria GE (1987) J. Chem. Phys. 87: 2138
15. Sileo RN, Cool TA (1976) J. Chem. Phys. 65: 117
16. Ogilvic JF, Rodwell WR, Tipping RH (1980) J. Chem. Phys. 73: 5221
17. Werner HJ, Reinsch EA, Rosmus P (1981) Chem. Phys. Lett. 78: 311
18. Jameson CJ, Osten HJ (1986) Ann. Reports NMR Spectrosc. 17: 1
19. Hegstrom RA (1979) Phys. Rev. A 19: 17
20. Spirko V, Blabla J (1988) J. Mol. Spectrosc. 129: 59
21. Fowler PW, Buckingham AD (1985) Chem. Phys. 98: 167
22. Lazzeretti P, Zanasi R (1986) Phys. Rev. A 33: 3727
23. Fowler PW, Raynes WT (1982) Mol. Phys. 45: 667
24. Buckingham AD, Fowler PW, Galwas PA (1987) Chem. Phys. 112: 1
25. Lazzeretti P, Zanasi R, Stephens PJ (1986) J. Phys. Chem. 90: 6761
26. Morrison MA, Hay PJ (1979) J. Chem. Phys. 70: 4034
27. Nafie LA, Freedman TB (1983) J. Chem. Phys. 78: 7108; Nafie LA (1983) J. Chem. Phys. 79: 4950
28. Toyama M, Oka T, Morino Y (1964) J. Molec. Spectrosc. 13: 193
29. Amat G, Nielsen HH, Tarrago G (1971) Rotation-vibration of polyatomic molecules. Dekker, New York
30. Fowler PW (1981) Mol. Phys. 43: 591
31. Secroun C, Barbe A, Jouve P (1973) J. Mol. Spectrosc. 15: 1
32. Marcott C, Golden WG, Overend J (1978) Spectrochim. Acta A 34: 661
33. Amat G, Goldsmith M, Nielsen HH (1957) J. Chem. Phys. 27: 838
34. Buckingham AD (1962) J. Chem. Phys. 36: 3096
35. Herman RM, Short S (1968) J. Chem. Phys. 48: 1266
36. Krohn BJ, Ermler WC, Kern WC (1974) J. Chem. Phys. 60: 22
37. Riley G, Raynes WT, Fowler PW (1979) Mol. Phys. 38: 877
38. Overend J (1978) J. Chem. Phys. 64: 2878
39. Overend J (1982) In: Person WB, Zerbi G (eds) Vibrational intensities in infrared and Raman spectroscopy. Elsevier, Amsterdam, p 190, p 203
40. Person WB, Overend J (1977) J. Chem. Phys. 66: 1442
41. Jameson CJ (1977) J. Chem. Phys. 67: 2814
42. Fowler PW (1984) Mol. Phys. 51: 1423
43. Fowler PW (1982) Mol. Phys. 46: 913
44. Ditchfield R (1981) Chem. Phys. 63: 185
45. Oddershede J, Geertsen J, Scuseria GE (1988) J. Phys. Chem. 92: 3056
46. Kaiser EW (1970) J. Chem. Phys. 55: 1686
47. Johnson DW, Ramsey NF (1977) J. Chem. Phys. 67: 941
48. Tokuhiro T (1967) J. Chem. Phys. 47: 109
49. Cade PE (1967) J. Chem. Phys. 47: 2390
50. Ozier I, Ramsey NF (1966) Bull. Am. Phys. Soc. 11: 23
51. Reid RV, Chu AH (1974) Phys. Rev. A9: 609
52. Verberne J, Ozier I, Zandee L, Reuss J (1978) Mol. Phys. 35: 1649
53. Raynes WT, Panteli N (1983) Mol. Phys. 48: 439
54. Eakin DM, Davis SP (1970) J. Mol. Spectrosc. 35: 27
55. Knight Lв, Weltner W (1971) J. Chem. Phys. 35: 2061
56. Lucken EAC (1983) Adv. Nucl. Quad. Reson. 5: 83

57. Hoy AR, Mills IM, Strey G (1972) Mol. Phys. 24: 1265
58. Fowler PW, Raynes WT (1981) Mol. Phys. 43: 65
59. Lounila J, Wasser R, Diehl P (1987) Mol. Phys. 62: 19
60. Buckingham AD (1967) Adv. Chem. Phys. 12: 107
61. Cyvin SJ, Rauch JE, Decius JC (1965) J. Chem. Phys. 43: 4083
62. Buckingham AD, Love I (1970) J. Magn. Reson. 2: 338
63. Buckingham AD, Pyykkö P, Robert JB, Wiesenfeld L (1982) Mol. Phys. 46: 177
64. Buckingham AD, Malm SM (1971) Mol. Phys. 22: 1127
65. Buckingham AD, Pople JA (1963) Trans. Faraday. Soc. 59: 2421
66. Raynes WT (1978) Nucl. Magn. Reson. 7: 1
67. Raynes WT (1988) Mol. Phys. 63: 719
68. Jameson CJ (1977) J. Chem. Phys. 66: 4977
69. Raynes WT, Fowler PW, Lazzeretti P, Zanasi R, Grayson M (1988) Mol. Phys. 64: 143
70. Lazzeretti P, Zanasi R, Sadlej AJ, Raynes WT (1987) Mol. Phys. 62: 605
71. Fowler PW, Riley G, Raynes WT (1981) Mol. Phys. 42: 1463
72. Jameson CJ, Buckingham AD (1980) J. Chem. Phys. 73: 5684
73. Hunt KLC (1989) J. Chem. Phys. 90: 4909
74. Jameson CJ, Jameson AK, Wille S, Burrell PM (1981) J. Chem. Phys. 74: 853
75. Petrakis L, Sederholm CH (1961) J. Chem. Phys. 35: 1174
76. Jameson CJ (1980) Bull. Magn. Reson. 3: 3
77. Jameson CJ, Jameson AK, Cohen SM (1977) J. Chem. Phys. 67: 2771
78. Bartell LS (1963) J. Chem. Phys. 38: 1827
79. Bartell LS (1979) J. Chem. Phys. 70: 4581
80. Jameson CJ, Jameson AK (1986) J. Chem. Phys. 85: 5484
81. Jameson AK, Moyer J, Jameson CJ (1978) J. Chem. Phys. 68: 2873
82. Bennett B, Raynes WT (1987) Mol. Phys. 61: 1423
83. Jameson CJ, Osten HJ (1986) J. Am. Chem. Soc. 108: 2497
84. Heller C, McConnell HM (1960) J. Chem. Phys. 32: 1535
85. Stone EW, Maki AH (1962) J. Chem. Phys. 37: 1326
86. Sullivan PD, Menger EM (1977) Adv. Magn. Reson. 9: 1
87. Gutowsky HS, Belford GG, McMahon PE (1962) J. Chem. Phys. 36: 3353
88. Chang SY, Davidson ER, Vincow G (1970) J. Chem. Phys. 52: 5596
89. Meyer W (1969) J. Chem. Phys. 51: 5149
90. Schrader DM, Morokuma K (1971) Mol. Phys. 21: 1033
91. Jameson CJ (1977) J. Chem. Phys. 66: 4983
92. Raynes WT, Davies AM, Cook DB (1971) Mol. Phys. 21: 123
93. Stevens RM, Lipscomb WN (1964) J. Chem. Phys. 40: 2238
94. Stevens RM, Lipscomb WN (1964) J. Chem. Phys. 41: 184
95. Stevens RM, Karplus M (1968) J. Chem. Phys. 49: 1094
96. Laws EA, Stevens RM, Lipscomb WN (1971) J. Chem. Phys. 54: 4269
97. Ermler WC, Kern CW (1971) J. Chem. Phys. 55: 4851
98. Jameson CJ, Osten HJ (1984) J. Chem. Phys. 81: 4300
99. Jameson CJ, Osten HJ (1984) J. Chem. Phys. 81: 4293
100. Wasylishen RE, Friedrich JO (1984) J. Chem. Phys. 80: 585
101. Wasylishen RE, Friedrich JO (1987) Can. J. Chem. 65: 2238
102. Leighton KL, Wasylishen RE (1987) Can. J. Chem. 65: 1469
103. Sanders JKM, Hunter BK, Jameson CJ, Romeo G (1988) Chem. Phys. Lett. 143: 471
104. Morse P (1929) Phys. Rev. 34: 57
105. Herschbach DR, Laurie VW (1961) J. Chem. Phys. 35: 458
106. Mills IM (1974) In: Dixon RN (ed) Theoretical chemistry. Royal Society of Chemistry, London, p 110
107. Bartell LS (1963) J. Chem. Phys. 38: 1827
108. Bartell LS, Doun SK, Goates SR (1979) J. Chem. Phys. 70: 4585
109. Osten HJ, Jameson CJ (1985) J. Chem. Phys. 82: 4595
110. Osten HJ, Jameson CJ (1984) J. Chem. Phys. 81: 4288
111. Wasylishen RE, Friedrich JO, Mooibroek S, Macdonald JB (1985) J. Chem. Phys. 83: 548
112. Friedrich JO, Wasylishen RE (1985) J. Chem. Phys. 83: 3707

113. Fleischer U, Schindler M, Kutzelnigg W (1987) J. Chem. Phys. 86: 6337
114. Batiz-Hernandez H, Bernheim RA (1967) Progr. NMR Spectrosc. 3: 63
115. Hansen PE (1983) Ann. Reports NMR Spectrosc. 15: 105
116. Hansen PE (1988) Prog. NMR Spectrosc. 20: 207
117. Wasylishen RE, Burford N (1987) Can. J. Chem. 65: 2707
118. Jameson AK, Jameson CJ (1978) J. Magn. Reson. 32: 455
119. Lawler RG, Bolton JR, Karplus M, Fraenkel GK (1967) J. Chem. Phys. 47: 2149
120. Smaller B, Yasaitis E, Anderson HL (1951) Phys. Rev. 81: 896
121. Bigeleisen J, Goldstein P (1963) Z. Naturforsch. A 21: 205
122. Jameson CJ, Rehder D, Hoch M (1987) J. Am. Chem. Soc. 109: 2589
123. Moss RE (1965) Mol. Phys. 10: 339
124. Jameson CJ (1980) Mol. Phys. 40: 999
125. Jameson CJ, Osten HJ (1985) Mol. Phys. 56: 1083
126. Jameson CJ, Osten HJ (1985) J. Chem. Phys. 83: 5425
127. Jameson CJ, Osten HJ (1985) Mol. Phys. 55: 383
128. Jameson CJ, Osten HJ (1984) J. Chem. Phys. 81: 4915
129. Jameson CJ, Osten HJ (1984) J. Chem. Phys. 81: 2556
130. Jameson CJ (1987) J. Am. Chem. Soc. 109: 2586
131. Jameson CJ, Rehder D, Hoch M (1988) Inorg. Chem. 27: 3490
132. Fowler PW, cited in Jameson CJ, Osten HJ (1986) Ann. Reports NMR Spectrosc. 17: 1
133. Schindler M (1987) J. Am. Chem. Soc. 109: 5950
134. Wasylishen RE, private communications
135. Karplus M (1960) J. Chem. Phys. 33: 1842
136. Karplus M, Fraenkel GK (1961) J. Chem. Phys. 35: 1312
137. Karplus M (1969) J. Chem. Phys. 50: 3133
138. Jameson CJ, Gutowsky HSG (1969) J. Chem. Phys. 51: 2790
139. Jameson CJ, Damasco MC (1970) Mol. Phys. 18: 491
140. Jameson CJ (1969) J. Am. Chem. Soc. 91: 6232
141. Murrell JN, Turpin MA, Ditchfield R (1970) Mol. Phys. 18: 271
142. Wasylishen RE, Burford N (1987) J. Chem. Soc. Chem. Commun. 1987: 1414
143. McConnell HM (1956) J. Chem. Phys. 24: 632, 764
144. Reddoch AH, Dodson CL, Paskovich DH (1970) J. Chem. Phys. 52: 2318
145. Schrader DM, Karplus M (1964) J. Chem. Phys. 40: 1593
146. Cole T (1961) J. Chem. Phys. 35: 1169
147. Fessenden RW, Schuler RH (1963) J. Chem. Phys. 39: 2147
148. Zlochower IA, Miller WR, Fraenkel GK (1965) J. Chem. Phys. 42: 3339
149. Garbutt GB, Gesser HD, Fujimoto M (1968) J. Chem. Phys. 48: 4605
150. Fischer H, Hefter H (1968) Z. Naturforsch. A 23: 1763
151. Cater MK, Vincow G (1967) J. Chem. Phys. 47: 292
152. Vincow G, Morrell ML, Volland MV, Dauben HJ, Hunter FR (1965) J. Am. Chem. Soc. 87: 3527
153. Sutcliffe BT, Gaze C (1978) Mol. Phys. 35: 525
154. Thuomas KA, Eriksson A, Lund A (1980) J. Magn. Reson. 37: 223
155. Fessenden RW, Ogawa S (1964) J. Am. Chem. Soc. 86: 3591
156. Lawler RG, Bolton JR, Fraenkel GK, Brown TH (1964) J. Am. Chem. Soc. 86: 520
157. Lawler RG, Fraenkel GK (1968) J. Chem. Phys. 49: 1126
158. Volland WV, Vincow G (1968) J. Chem. Phys. 48: 5589
159. Carrington A, Longuet-Higgins HC, Moss RE, Todd PF (1965) Mol. Phys. 9: 187
160. Longuet-Higgins HC, Öpik U, Pryce MHL, Sack A (1958) Proc. Roy. Soc. London A 244: 1
161. Kuchitsu K, Oyanagi K (1977) Faraday Discuss. Chemical Society London 62: 20
162. Lide DR (1975) In: Buckingham AD (ed) MTP International Review of Science. Physical Chemistry Series Two. Butterworts, London p 1 (Molecular Structure and Properties, vol 2)
163. Herschbach DR, Laurie VW (1962) J. Chem. Phys. 37: 1668
163a. Nemes L (1984) In: Durig JR (ed) Vibrational Spectra and Structure, Elsevier, Amsterdam 13: 161
164. Kuchitsu K (1967) Bull. Chem. Soc. Japan 40: 498
165. Morino Y, Kuchitsu K, Oka T (1962) J. Chem. Phys. 36: 1108

166. Sykora S, Vogt J, Bösiger H, Diehl P (1979) J. Magn. Reson. 36: 53
167. Novick SE, Davies P, Harris SJ, Klemperer W (1973) J. Chem. Phys. 59: 2273
168. Hutson JM, Howard BJ (1981) J. Chem. Phys. 74: 6520
169. Ruoff RS, Emilsson T, Chuang C, Klots TD, Gutowsky HS (1987) Chem. Phys. Lett. 138: 553
170. Nelson DD, Klemperer W, Fraser GT, Lovas FJ, Suenram RD (1987) J. Chem. Phys. 87: 6364
171. Mills IM (1984) J. Phys. Chem. 88: 532
172. Barton AE, Howard BJ (1982) Faraday Discuss. Chemical Society London 73: 45
173. Dyke TR, Howard BJ, Klemperer W (1972) J. Chem. Phys. 56: 2442
174. Born M, Huang K (1954) Dynamical theory of crystal lattices, Oxford University Press, London
175. Smith JG (1978) Mol. Phys. 35: 461
176. Lacy M (1982) Mol. Phys. 45: 253
177. Whiffen DH (1980) Mol. Phys. 39: 391
178. Fessenden RW (1967) J. Phys. Chem. 71: 74
179. Evans DF (1960) Chem. Ind. 1961
180. Hindermann DK, Cornwell CD (1968) J. Chem. Phys. 48: 4148
181. Chesnut DB (1986) Chem. Phys. 110: 415
182. Chesnut DB, Foley CK (1986) J. Chem. Phys. 84: 852
183. Weller T, Meiler W, Köhler HJ, Lischka H, Höller R (1983) Chem. Phys. Lett. 98: 541
184. Schindler M, Kutzelnigg W (1983) Mol. Phys. 48: 781
185. Chesnut DB, Foley CK (1986) J. Chem. Phys. 85: 2814
186. Schindler M, Kutzelnigg W (1982) J. Chem. Phys. 76: 1919
187. Imachi M, Tanaka T, Hirota E (1976) J. Mol. Spectrosc. 63: 265
188. Kukolich SG, Nelson AC (1972) J. Chem. Phys. 57: 869
189. Helminger P, Beeson EL, Gordy W (1971) Phys. Rev. A 3: 122
190. Morino Y, Hirose C (1967) J. Mol. Spectrosc. 24: 204
191. Williams JR, Kukolich SG (1979) J. Mol. Spectrosc. 74: 242
192. O'Reilly DE, Peterson EM, Kadaba P (1970) J. Chem. Phys. 52: 6444
193. Kuczkowski RL (1973) J. Mol. Spectrosc. 45: 261
194. Arimondo E, Glorieux P, Oka T (1978) Phys. Rev. A17: 1375
195. Dössel KF, Sutter DH (1979) Z. Naturforsch. A 34: 469
196. DeLucia FC, Gordy W (1969) Phys. Rev. 187: 58
197. Kukolich SG, Reuben DJ, Wang JHS, Williams JR (1973) J. Chem. Phys. 58: 3155
198. Jones H, Takami M, Sheridan J (1978) Z. Naturforsch. A33: 156
199. Jones H, Sheridan J, Stiefvater OL (1977) Z. Naturforsch. A32: 866
200. Raynes WT, Lazzeretti P, Zanasi R (1988) Mol. Phys. 64: 1061
201. Zeroka D (1973) J. Chem. Phys. 59: 3835
202. Ishiguro E, Koide S (1953) Phys. Rev. 94: 350
203. Snijders JG, Van der Meer W, Baerends EJ, DeLange CA (1983) J. Chem. Phys. 79: 2970
204. Huber H (1985) J. Mol. Struct. THEOCHEM 121: 207
205. Rosenberg BJ, Ermler WC, Shavitt I (1976) J. Chem. Phys. 65: 4072
206. Ermler WC (1970) M. S. Thesis, Ohio State University
207. Engström S, Wennerstrom H (1978) Mol. Phys. 36: 773
208. Bacskay GB, Gready JE (1988) J. Chem. Phys. 88: 2526
209. Docken KK, Hinze J (1972) J. Chem. Phys. 57: 4928, 4936
210. Huo WM (1965) J. Chem. Phys. 43: 624
211. Green S (1973) J. Chem. Phys. 58: 3117
212. Sundholm D, Pyykkö P, Laaksonen L, Sadlej AJ (1986) Chem. Phys. 101: 219
213. Cummins PL, Bacskay GB, Hush NS, Ahlrichs R (1987) J. Chem. Phys. 86: 6908
214. Kowalewski J, Roos B, Siegbahn P, Vestin R (1974) Chem. Phys. 3: 70
215. Schulman JM, Lee WS (1980) J. Chem. Phys. 73: 1350
216. Lazzeretti P, Zanasi R, Raynes WT (1989) Mol. Phys. 66: 831
217. Raynes WT, Lazzeretti P, Zanasi R, Fowler PW (1985) J. Chem. Soc. Chem. Commun. 1985: 1538
218. Raynes WT, Lazzeretti P, Zanasi R (1986) Chem. Phys. Lett. 132: 173
219. Kowalewski J, Roos B (1975) Chem. Phys. 11: 123
220. Keil F, Ahlrichs R (1979) J. Chem. Phys. 71: 2671
221. Lazzeretti P, Rossi E, Taddei F, Zanasi R (1982) J. Chem. Phys. 77: 408

222. Rychlewski J, Raynes WT (1980) Mol. Phys. 41: 843
223. Bishop DM, Cheung LM (1979) J. Phys. B 12: 3135
224. Hunt JL, Poll JD, Wolniewicz L (1984) Can. J. Phys. 62: 1719
225. Rychlewski J (1980) Mol. Phys. 41: 833
226. Sunil KK, Jordan KD (1988) Chem. Phys. Lett. 145: 377
227. Oddershede J, Svendsen EN (1982) Chem. Phys. 64: 359
228. Temkin A (1978) Phys. Rev. A17: 1232
229. Lazzeretti P, Rossi E, Zanasi R (1981) J. Phys. Chem. B14: L269
230. Rosmus P, Werner HJ (1985) In: Jørgensen P, Simons J (eds) Geometrical derivatives of energy surfaces and molecular properties. Reidel, Dordrecht, p 265
231. Wolniewicz L (1976) Can. J. Phys. 54: 672
232. Lazzeretti P, Zanasi R, Fowler PW (1988) J. Chem. Phys. 88: 272
233. Meyer W, Rosmus P (1975) J. Chem. Phys. 63: 2356
234. Werner HJ (1981) Mol. Phys. 44: 111
235. Werner HJ, Rosmus P (1980) J. Chem. Phys. 73: 2319
236. Rosmus P, Werner HJ, Grimm M (1982) Chem. Phys. Lett. 92: 250
237. Werner HJ, Meyer W (1981) J. Chem. Phys. 74: 5802
238. Langhoff S, Bauschlicher CW, Partridge H (1983) Chem. Phys. Lett. 102: 292
239. Werner HJ, Rosmus P, Reinsch EA (1983) J. Chem. Phys. 79: 905
240. Amos RD (1978) Mol. Phys. 35: 1765
241. Raynes WT, Stanney G (1974) J. Magn. Reson. 14: 378
242. Cummins PL, Backsay GB, Hush NS (1987) J. Chem. Phys. 87: 416
243. Ishiguro E (1958) Phys. Rev. 111: 203
244. Raynes WT, Riley JP (1974) Mol. Phys. 27: 337
245. Chan SI, Ikenberry D, Das TP (1964) Phys. Rev. A 135: 960
246. Bishop DM, Lam B (1988) Chem. Phys. Lett. 143: 515
247. Elliott DS, Ward JF (1984) Mol. Phys. 51: 45
248. Bishop DM (1981) Mol. Phys. 42: 1219
249. Hindermann DK, Williams LL (1969) J. Chem. Phys. 50: 2839
250. Wasylishen RE (1982) Can. J. Chem. 60: 2194
251. Fabricant B, Muenter JS (1977) J. Chem. Phys. 66: 5274
252. Schrötter HW, Klöckner HW (1979) In: Weber A (ed) Raman Spectroscopy of gases and liquids. Springer, Berlin Heidelberg New York, p 123
253. Spelling RI, Meredith RE, Smith FG (1972) J. Chem. Phys. 57: 5119
254. Chackerian C, Farrenq R, Guelachvili G, Rossetti C, Urban W (1984) Can. J. Phys. 62: 1579
255. Person WB, Zerbi G (eds) (1982) Vibrational intensities in infrared and Raman spectroscopy, Elsevier, Amsterdam, p 96

11 Appendix

A guide to the literature is given in Tables A1 to A3.2. No attempt at completeness has been made. References to ab initio calculations of various property derivatives are given in Table A1 and rovibrational calculations (at various levels) using theoretical derivatives are given in Table A2, and inversion of data to obtain empirical estimates of derivatives are given in Table A3.

Table A1. Ab initio theoretical calculations of property surfaces or their derivatives

σ:	Review [18] H_2^+ [19] H_2 [44, 181, 201, 202] LiH [44, 93] HF [113, 44, 94]
	HCl [181] CO [95, 182–184] N_2 [96, 182, 184] CN^- [182] LiF, P_2 [113]
	CH_4 [70, 181, 185] H_2O [71, 181, 185, 186] SiH_4, PH_3, NH_3 [113, 181, 185]
	PH_4^+, NH_4^+, BH_4^-, BH_3, AlH_3, AlH_4^-, H_2S [181]
	HCCH, H_2CC_2, CH_3CH_3, CH_3F, HCN [182] SiF_4, PF_3, NF_3 [113]
C:	LiH [93] HF [94] CO [95] N_2 [96] H_2O [23]
a:	$\dot{C}H_3$ [88]
q:	Review [56] at H in H_2, CH_4 [203]
	at H in H_2, LiH, CH_4, NH_3, H_2O, HF, NaH, SiH_4, PH_3, H_2S, HCl [204]
	H_2O [205, 206] HCl, HCN [207]
	at H in HF_2^- [208] OH^-, SH^-, HF, HCl, $HO^{\cdot}$, $HS^{\cdot}$ [49]
	LiH [209] CO, BF [210] ClF [211] NO^+, N_2 [212]
	at N and H in N_2, NO, NO^+, CN, CN^-, HCN, HNC, NH_3 [213]
J:	Review [83] HD [45, 214, 215] CH_4 [216–218] NH_3 [219] PH_3 [220]
	HF [141] CO, N_2 [14] PH_2^- [221]
χ:	H_2 [222] LiH [93] H_2O [71, 206] CH_4 [70] HF [94] CO [95] N_2 [96]
α:	H_2^+ [223] H_2 [224, 225] LiH [93] CH_4 [200] CO [226, 227]
$\alpha(\omega)$:	N_2, CO, HCl, CL_2 [227] N_2 [26, 228] CO_2 [26]
γ, B, C:	HCl, HBr [229]
μ:	Review [230] HD [231] LiH [232, 233] CH_4, NH_3 [232] CO [95, 226, 234]
	HF [94, 232, 233, 235] HCl [233, 235] HBr [235] HI [17] BF [236] LiF [237]
	OF [238] BeH, BH, CH, NH, OH, NaH, MgH, AlH, SiH, PH, SH [233]
	H_2O [71, 205, 206, 232] OH, OH^-, OH^+ [239]
	H_2^+ [26, 223] H_2 [224] LiH, CH_4, NH_3, HF [232] N_2 [26] H_2O [23, 232]
Ω:	HF [240] H_2O [206]
g_J or μ_J:	H_2 [222] HF [94] CO [95] N_2 [96] H_2O [23]

Table A2.1. Rovibrational averaging of properties in diatomic molecules using ab initio theoretical derivatives

Property	Using 1st and 2nd deriv. only	3rd and higher
σ	H_2, LiH, HF, CO, N_2 [18]	H_2 [44, 92] LiH, HF [44]
	HF [180] $N_2(v = 0)$ [96]	CO [241]
C	HF [94, 180] LiH [93] N_2 [96]	
q	CO, OH^-, HCl, LiCl, FCl [242]	
	N_2, NO^+, NO, CN, CN^- [213]	
J	HD [45, 214, 215, 243, 244] HF [141]	
	CO, N_2 [14]	
χ	H_2 [245] HF$(v = 0)$ [94] N_2 [96]	CO [95]
α	LiH [93] HF [94]	
β, γ, B, C	H_2^+, H_2 [11, 246] HF(β) [11]	
μ	LiH, BeH, BH, CH, NH, OH, HF,	HF, HCl, HBr $v = 0–8$ [235]
	NaH, ... , HCl [233]	OH, OH^-, OH^+ $v = 0–5$ [239]
	HI$(v = 0)$ [17]	BF, BF^+ $v = 0–6$ [236]
		OF $v = 0–6$ [238]
g_J or μ_J	H_2 [245] HF [94] N_2 [96]	

Table A2.2. Rovibrational averaging of properties in polyatomic molecules using ab initio theoretical derivatives

1st and 2nd derivs. neglect 2nd order L tensor elements		1st deriv. only include 2nd order L tensor elements	1st and 2nd derivs. and 2nd order L elements	3rd and higher order
σ		H in CH$_4$ [103]	C in CH$_4$ [69]	H$_2$O [58]
C				H$_2$O [23]
a	$\dot{C}H_3$ [90]			
	pi model C$_6$H$_6^-$ [157]			
	naphthalene$^-$ [119]			
J		J(HH) in CH$_4$ [217]		^{2}J(HH) in NH$_3$
		J(CH) in CH$_4$ [82]		inversion mode [219]
χ			CH$_4$ [69]	H$_2$O [58]
α			CH$_4$ [200]	
β, γ	1st deriv. only			
	γ^vCH$_n$F$_{4-n}$, SF$_6$ [247]			
	β^v CHCl$_3$, CHF$_3$ [248]			
μ				H$_2$O [58, 97]
θ, g_J				H$_2$O [23, 97]

Table A3.1. Inversion of experimental data to obtain property derivatives of diatomic molecules

Property	1st deriv only	1st and 2nd derivs.	Complete surface
σ	H$_2$ [53] F$_2$ [249]		
	ClF, F$_2$ [68] CN$^-$ [98, 250]		
	CO [74, 111] N$_2$ [74, 112]		
C	H$_2$ [52, 53] F$_2$ [249]		
q	Review [56] HBr [47] ClF [251]	HCl [46]	I$_2$ [20]
a	$\dot{C}H_3$ [123]		
α	Review [252]		
μ		HF [253] HCl [46]	HF [15]
			HCl, HBr, HI [16]
			CO [254]

Table A3.2. Inversion of experimental data to obtain property derivatives of polyatomic molecules[a]

Property	1st deriv. only estimate $\langle \Delta r \rangle$	1st deriv. only neglect 2nd order L tensor	1st deriv. incl. 2nd order L tensor
σ	O in H_2O, H_3O^+ [101]	CF_4, BF_3, SiF_4 [41]	NF_3, PF_3, PCl_3, PBr_3
	Sn in SnH_3^- [117]		NH_3, PH_3, CO_2 [18]
	Sn in SnH_4, $SnH_{3-n}D_n^+$ [102]		CO_2 [129] CH_4 [110]
	cen. nucl. in BH_4^- CO_2 CS_2		SF_6 [124] COF_2 [128]
	NH_4^+ NO_3^- PO_4^{3-} MnO_4^-		$CF_{4-n}H_n$ [125]
	MoO_4^- TcO_4^- $PtCl_6^{2-}$		H_2Se, H_2O [109]
	$PtBr_6^{2-}$ [98]		$CF_3Cl \ldots CFCl_3$ [127]
	C in CO_2 [111]		MF_6 [80]
	N in NO_2^- NO_3^- [18]		$M(XY)_6$ [122, 131]
	P in PH_3, PH_2^- [117]		
J	SnH_4 [102] SnH_3^- [117]		$J(CH)$ in CH_4 [82]
	PH_2^- [117]		

[a] See also [56] for approximate inversion of eqQ of ^{75}As in AsF_3. See [255] for reviews of inversion of intensities to obtain derivatives of electric dipole moments and of electric dipole polarizabilities in polyatomics

Properties of Molecules in Excited States

M. Klessinger and T. Pötter

Organisch-Chemisches Institut der Westf. Wilhelms-Universität Orléansring 23,
D-4400 Münster, FRG

Quantum chemical calculations for organic molecules in excited states are considerably more complex than for ground state species, due to the fact that much fewer empirical data about excited state geometries are available and that geometry optimization on the CI level, which is a prerequisite for all calculations of excited state properties, is by no means simple. In this paper we discuss
− qualitative models, which may be helpful in locating minima on excited state potential surfaces,
− methods and results for calculations of excited state properties at various levels of sophistication; particular emphasis is laid on sudden polarization phenomena, *cis-trans* isomers of acetylene in its excited states, the pyramidalization of the carbonyl group, excited states of aromatic compounds (TICT states) and
− theoretical treatments of radiative and non-radiative deactivation of excited states.

1 Introduction

A molecule in an excited state is a different species from the same molecule in the ground state. It possesses additional energy and has a different structure, different spectra and different reactivity. A prerequisite for the understanding of any excited state process is the knowledge of the excited states themselves, i.e. their structure, their spin state, their energy, their change in stability as a function of nuclear displacements, plus the knowledge of their lifetimes [1]. Although the detailed study of excited states by fast experimental techniques is a rapidly growing field [2], quantum chemical calculations still represent the most widely applicable tool for an investigation of excited states. But whereas nowadays calculations at almost all levels of sophistication [3–7] can be performed routinely for ground state species, so that hundreds of calculations have been published in the last decade or so, comparatively few investigations of excited states [8–18] have been described. There are essentially three reasons for this:

i) Much less empirical data is available for excited states than for ground states of organic molecules. Whereas it is no problem to define standard geometries to be used as initial data for ground state calculations, very few general data for excited state geometries are known. In fact, very often the structure is the most interesting property of an excited species [19] and is frequently only established after searching large areas of the excited state potential energy surface.

ii) Molecules in excited states are in general not closed-shell species. SCF approaches are, therefore, in many cases not even suitable as first approximations, so that even in the first step of the excited state calculation, the geometry optimization, configuration interaction has to be taken into account [20]. But this is by no means an easy task.

iii) And finally, in order to decide whether a given excited state may be of practical interest, one has to know its lifetime; only if it is sufficiently long, may it be possible to detect the excited species either directly by spectroscopic techniques or indirectly from the reaction products. Thus, once the existence of a stable minimum on an excited state potential energy surface is established, the next problem is to discuss the photophysics of that species, which again can be rather involved.

In this paper we will try to shed some light on these three points. We will first discuss some qualitative models which may be helpful in locating minima or excited state hypersurfaces. The next section will be devoted to methods and results of excited state calculations, with particular emphasis on the geometry of the excited state species. And finally, the present status of the theoretical treatment of the photophysics of excited states is briefly discussed.

2 Qualitative Models for Predicting Minima and Barriers on Excited State Potential Hypersurfaces

In the Born Oppenheimer approximation the total wave function, Ψ^T of a vibronic level is written as

$$\Psi^T_{j,v}(\mathbf{q}, \mathbf{Q}) \sim \Psi^Q_j(\mathbf{q}) \, X^j_v(\mathbf{Q}) \,, \tag{1}$$

where j denotes the electronic state and v the vibrational level. The electronic wave function Ψ_j^Q for an excited state with coordinates $q_1, q_2, \ldots q_n \equiv q$ is obtained by solving the Schrödinger equation at the nuclear conformation Q, $E_j(Q)$ as a function of all geometrical variables $Q_1, Q_2, \ldots Q_F \equiv Q$ defines an F-dimensional hypersurface which serves as potential in the nuclear Schrödinger equation to get the nuclear wave functions $X_v^j(Q)$. As the Schrödinger equation has an infinite number of solutions for each Q, an infinite number of Born-Oppenheimer hypersurfaces may in principle be obtained, out of which the lowest ones of each multiplicity are of interest.

We shall use the nomenclature of photochemistry where spin orbit coupling is neglected in the first approximation, so that singlet and triplet states can be distinguished. At each geometry the Born-Oppenheimer states of one particular multiplicity are numbered consecutively with increasing energy. Surfaces of different multiplicity can cross, such crossings have no influence on the designation of the surfaces by $S_0, S_1, S_2, \ldots$ or $T_1, T_2, T_3, \ldots$, but if states of the same multiplicity cross, they change their labels.

2.1 Spectroscopic and Biradicaloid Minima

Minima on the excited state hypersurface $E_j(Q)$ which correspond to stable structures are, in general, not easy to predict, i.e. it is much more difficult to estimate, even qualitatively, the position of such minima, for instance on the S_1 and T_1 surfaces, than to locate the minima on S_0, which correspond to stable ground state structures. But there are two types of geometries at which minima in S_1 and T_1 can be expected, either those close to the ground state equilibrium geometry or biradicaloid geometries.

Minima of the first type are called spectroscopic minima [21], because in general they can be reached from the ground state by spectroscopic excitations, even if the transition probability might be limited by the Franck-Condon principle. From these spectroscopic minima the molecule can return to the ground state either by radiative processes such as fluorescence (F) or phosphorescence (P), or by non-radiative processes such as internal conversion (IC) or intersystem crossing (ISC). Even if the differences between the equilibrium geometries of the ground state and the excited states are small, they may be significant as for instance in formaldehyde which is planar in the ground state but pyramidal in the excited state [22, 23].

Minima of the second type are called biradicaloid minima [21]. A biradicaloid is a species which in the simple MO scheme contains two electrons in two nearly degenerate non-bonding orbitals. A prerequisite for the existence of two such orbitals is a certain geometry which is therefore called biradicaloid geometry. For the discussion of potential energy hypersurfaces it is important that closed-shell molecules may turn into biradicaloids at such geometries. These geometries may be reached e.g. (a) by bond stretching, (b) by twisting of a double bond, (c) by changing the bond angles at a triple bond, or (d) by pericyclic processes which correspond to ground state forbidden reactions and are topologically equivalent to antiaromatic [4n] Hückel or [4n + 2] Möbius annulenes.

2.1.1 Perfect Biradicaloids

In contrast to closed-shell species, biradicals cannot be described by a single configuration built up from MOs, not even as a first approximation. If in the simplest possible model only the two (nearly) degenerate orbitals φ and φ' are considered, the following configurations have to be taken into account [24]:

$$^1B = {}^1(\varphi\varphi') = [|\varphi\bar{\varphi}'| + |\varphi'\bar{\varphi}|]/\sqrt{2} \qquad \text{(a)}$$

$$^3B = {}^3(\varphi\varphi') = [|\varphi\bar{\varphi}'| - |\varphi'\bar{\varphi}|]/\sqrt{2} \qquad \text{(b)}$$

$$Z_1 = {}^1(\varphi'^2) = |\varphi'\bar{\varphi}'| \qquad \text{(c)} \qquad\qquad (2)$$

und

$$Z_2 = {}^1(\varphi^2) = |\varphi\bar{\varphi}| \qquad \text{(d)}$$

If φ und φ' are localized AOs, $\varphi = \chi_a$ and $\varphi' = \chi_b$, than these configurations are VB structures which may be termed formally covalent ($^{1,3}B$) or hole-pair structures (Z_1, Z_2) respectively. If on the other hand φ and φ' are delocalized MOs which in case of χ_a and χ_b being equivalent by symmetry can be written as

$$\phi_1 = (\chi_a + \chi_b)/\sqrt{2} \quad \text{and} \quad \phi_2 = (\chi_a - \chi_b)/\sqrt{2}, \qquad (3)$$

$^1B = {}^1(\phi_1\phi_2)$ corresponds to the VB structure $(\chi_a^2 - \chi_b^2)$, whereas $Z_1 - Z_2 = (\phi_1^2 - \phi_2^2)$ is equivalent to the formally covalent VB structure $^1(\chi_a\chi_b)$.

A perfect biradical is a biradical for which χ_a and χ_b are equivalent by symmetry and orthogonal. In the simple HMO approximation the four configurations given by Eq. (2) with $\varphi = \chi_a$ and $\varphi' = \chi_b$ are degenerate. Taking into account electron interaction the energy changes in such a way that with a corresponding choice of E_0 [24]

$$E = E_0 \pm K'_{ab} \pm K_{ab} = E_0 \pm K_{12} \pm K_{12}, \qquad (4)$$

where $2K'_{ab} = (J_{aa} + J_{bb})/2 - J_{ab}$ and $2K'_{12} = (J_{11} + J_{22})/2 - J_{12}$ with $J_{\mu\nu} = (mm \mid \nu\nu)$ being the usual Coulomb integral.

If $K_{\mu\nu} = (\mu\nu \mid \mu\nu)$ is the usual exchange integral, than

$$K'_{ab} = K_{12} \quad \text{and} \quad K'_{12} = K_{ab}, \qquad (5)$$

as may be seen from Eq. (3). Finally, as the overlap density for orbitals localized in different parts of the molecules is smaller than the overlap density of delocalized orbitals,

$$K'_{ab} \geqq K_{ab} \quad \text{and} \quad K'_{12} \leqq K_{12}. \qquad (6)$$

As K_{ab} and K'_{ab} are not negative it follows that $T (= {}^3B)$ is the most stable of the four states of Eq. (2), whereas $(\chi_a^2 + \chi_b^2) = Z_1 + Z_1$ is always of highest energy and therefore describes S_2. The energetical order of ${}^1(\chi_a\chi_b)$ or $(\phi_1^2 - \phi_2)$ and $(\chi_a^2 + \chi_b^2)$ or ${}^1(\phi_1\phi_2)$ on the other hand depends on the value of K_{ab} and K'_{ab}.

If K_{ab} and K'_{ab} are both not zero and not equivalent $(K'_{ab} \neq K_{ab})$ than the energies of the four states are different. Two special cases may be of interest: If $K_{ab} = 0$ then T and S_0 are degenerate and so are S_1 and S_2. As $K_{ab} = 0$ can only be true if χ_a and χ_b are infinitly apart, as for a pair of separate radicals, this is called a pair radical. Finally if $K'_{ab} = K_{ab}$, S_0 and S_1 are degenerate and the biradical is called axial.

2.1.2. Biradicaloids

Biradicals which do not fulfill the conditions of perfect biradicals are called biradicaloids [24]. In general they are non-symmetric. But again, two special cases of symmetric biradicaloids can be distinguished [24]: If the localized orbital χ_a and χ_b are equivalent by symmetry, but an interaction exists between χ_a and χ_b, the biradical is called homosymmetric. If on the other hand χ_a and χ_b are not equivalent but differ e.g. in electronegativity, and do not interact, the biradical is called heterosymmetric.

As a consequence of the deviation from perfect behavior configuration interaction has to be invoked even on the VB level. In the case of homosymmetric biradicaloids this leads to an interaction of ${}^1(\chi_a\chi_b)$ and $(\chi_a^2 + \chi_b^2)$ which stabilizes S_0 and destabilizes S_2 with increasing interaction. The energy gap ΔE between S_0 and S_1 increases with increasing interaction for instance in twisted ethylene with the angle of twist decreasing from 90° (perfect biradical) to values smaller than 90° (homosymmetric biradicaloid).

For heterosymmetric biradicaloids on the other hand only a mixing of the hole-pair structures $(\chi_a^2 + \chi_b^2)$ and $(\chi_a^2 - \chi_b^2)$ occurs which leads to a splitting of the corresponding states. As a consequence, if S_1 and S_2 are nearly degenerate and K is small, a small difference in electronegativity Δ of the orbitals χ_a and χ_b is sufficient for the mixing of the hole-pair structures to produce a strong polarisation into χ_a^2 and χ_b^2. This leads to the effect known as sudden polarization [25, 26].

The fact that S_1 and not S_0 of a perfect biradical is stabilized if the energies of χ_a and χ_b differ leads eventually to a crossing of the two states. S_0 and S_1 are degenerate, if

$$\Delta = \Delta_0 = 2\sqrt{K'_{ab}(K'_{ab} - K_{ab})}. \tag{7}$$

This leads to a further distinction between weakly heterosymmetric biradicals for which $\Delta < \Delta_0$ and S_0 is given by ${}^1(\chi_a\chi_b)$, whereas S_1 is described by a mixture of (χ_a^2) and (χ_b^2) with (χ_b^2) dominating. In the case of strongly heterosymmetric biradicaloids with $\Delta > \Delta_0$ the S_0 state is described by the hole-pair structures and may be below T. Large Δ values are given if ${}^1(\chi_a\chi_b)$ has a formal charge and (χ_b^2) not. This is the case for aminoborane, but is also observed in molecules where a donor is connected to an acceptor by a single bond. In this case the S_1 state is called a TICT state (twisted internal charge transfer) [27] and is associated with an exceptionally strong solvent dependence of the emission frequency [26]. In the case of a critically heterosymmetric biradicaloid with $\Delta = \Delta_0$, the S_0 and S_1 states are degenerate according to the simple model. This situation occurs most easily if neither $^1(\chi_a\chi_b)$ nor (χ_b^2) show a formal charge separation, i.e. for biradicaloids in which excitation induces just a charge shift, as for instance for $CH_2 = N^{\oplus}H_2$ [28].

This brief outline show clearly that the results of this simple model for biradicals and biradicaloids may be very helpful in predicting and understanding excited state structures. Some of the examples discussed in Sect. 3 will make this even more evident.

2.2 Correlation Diagrams

Very valuable hints as to where minima and barriers on the potential hypersurfaces of excited states are to be expected may be obtained from correlation diagrams. There are many different ways of constructing such diagrams [29]. Particularly useful in the present context are diagrams based on natural orbital correlations [30]. They are obtained by correlating those orbitals which are localized in the same region of the molecule and which show the same sign relations between the dominating LCAO coefficients. Symmetry is only invoked in converting the configuration correlation diagrams obtained from these natural correlations into state correlation diagrams.

As an example Fig. 1 shows the correlation diagram of the α cleavage reaction of formaldehyde. From this diagram it is predicted that the $^{1,3}(\pi, \pi^*)$ excited states of formaldehyde will be stable with respect to dissociation, i.e. that a spectrocopic minimum can be expected on the corresponding potential hypersurfaces. This is not trivial as may be seen below. Furthermore, it is quite easy to foresee that formaldehyde will tend to pyramidalize in the excited states, in particular in the (n, π^*) states, as the electron in the π^* MO will increase the population of the p-AO at the carbonyl

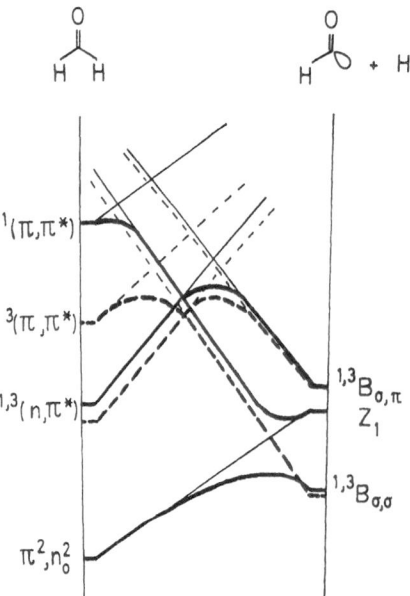

Fig. 1. State correlation diagram for the α cleavage reaction of formaldehyde, based on natural orbital correlations

carbon, so that the decrease of the p, p overlap with increasing pyramidalization will be compensated for by the increasing s character of the carbon AO; if occupied by two electron as in carbanions, this AO can be described as being sp^3 hybridized.

For larger molecules the existence of spectroscopic minima is rather the rule than an exception. Thus, in many aromatic hydrocarbons and other compounds it is not S_1 but S_2 which is dominated by the HOMO-LUMO excited configuration. This crossing in the configuration correlation diagram induces a barrier in the S_1 state which creates a stable spectroscopic minimum. This barrier is high enough to persist even in the case of dewarnaphthalene (1), where the photochemical conversion into naphthalene (2) is very exothermic with the correlation line between the two S_1 state going steeply downhill [30].

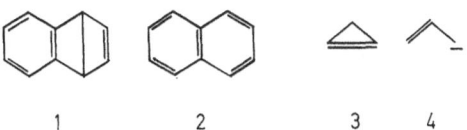

| 1 | 2 | 3 | 4 |

There are also molecules which in some of their excited states do not possess spectroscopic minima and therefore dissociate immediately on excitation. A hint to such a behavior can also be obtained from correlation diagrams. An interesting example is cyclopropene (3) which is stable in the T_1 state, whereas the S_1 state turns into vinylmethylene (4). This can be rationalized from the correlation diagram in Fig. 2 which is based on natural orbital correlations and configuration energies obtained from MNDOC-CI calculations.

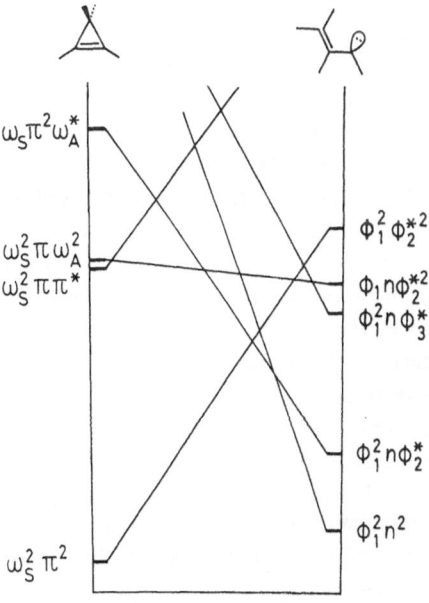

Fig. 2. Singlet configuration correlation diagram for the ring opening of cyclopropene, based on natural orbital correlations (ω_s, ω_a denote the Walsh-orbitals of cyclopropene, ϕ_i the π MOS of carbene)

3 Quantitative Determination of Excited State Geometries, Vibrational Frequencies and Charge Distributions

3.1 Experimental Results

Most experimental data of excited state properties are obtained by spectroscopic techniques, as for instance energies of excited state minima from fluorescence and phosphorescence spectra [31, 32], excited state lifetimes from an analysis of excited state kinetics [30], and dipole moments and charge distributions from solvent shifts [33]. If the absorption spectra are sufficiently structured and highly resolved, excited state geometries may be extracted from vibrational intensities, even for molecules of low symmetry [34–38]. Other methods for the determination of excited state geometries include rotational spectra [39–41], resonance Raman spectra [42, 43] and phosphorescence microwave double resonance spectra [44]. In spite of this diversity of methods reliable data are scarce, as may be seen from the review of Gustav and Sühnel [45]. Particularly thorough studies have been accomplished for formaldehyde, for which the geometries and vibrational frequencies of three excited states are known [22, 23, 46, 47]. References to experimental data of some other molecules will be given in Sect. 3.3.

3.2 Methods of Computation

As ground state SCF orbitals yield only a very poor description of excited states, configuration interaction (CI) has to be invoked in order to calculate excited state properties. The simplest approach is to use only configurations which are obtained from the closed-shell ground configuration by single excitations; this scheme (SCI) is commonly used in PPP type [48] or CNDO/S type calculations [8]. As in many cases (cf. Sect. 3.3.1) this method is rather poor, the inclusion of single as well as double excitations (SDCI) is to be preferred. Another way to obtain better excited state wave functions is based on the use of improved virtual orbitals [49].

As the composition of electronic states varies along the dissociation path and as minima may be due to avoided crossings (cf. Sect. 2.1) theoretical methods based on a truncated CI expansion with a single reference configuration do in general not satisfy the requirement of a balanced description of different geometries. Thus multi-reference calculations are required, the most widely used scheme in ab initio work being the MRD-CI method of Buenker and Peyerimhoff [50], which apart from the multi-reference CI including double excitations involves an energy extrapolation to approximate the contribution from configurations omitted.

Relatively few multi-configuration SCF methods (MCSCF) have been applied to excited states [51]. Other excited state calculations involve manybody perturbation theory (MBPT) [6] or coupled electron pair methods (CEPA) [52]. In all ab initio calculations the overall accuracy of excited state results depends not only on the CI, but also on the atomic orbital (AO) basis set employed, which in general must be considerably larger for excited states than for ground states [1].

Semiempirical methods other than PPP and CNDO/S have become available only recently. The main reason being that correlation effects, which are supposed to be taken care of by the parametrization of the method and must not be taken into account a second time by the configuration interaction treatment, which as mentioned above is essential to describe excited states. The only semiempirical method which was parametrized such that it allows an explicit treatment of correlation effects is the MNDOC method of Thiel [53, 54]. This method has been used to calculate excited states on the basis of small CI + perturbational treatment.

In this review we quite often use MNDOC-CI results to illustrate specific excited state phenomena. The MNDOC-CI method is based on a careful selection of configurations which are singly or doubly excited with respect to one or more reference configurations. The CI space is truncated to some hundreds of configurations by means of a judicious choice of the criteria for selecting configurations, of which the excitation indices for single and double excitations with respect to appropriate reference configurations were shown to be particularly important in order to achieve comparable accuracy for all states considered [15, 16, 17].

Among the molecular properties geometries are of fundamental significance and are therefore of primary interest also in excited state calculations [19]. But as CI is of outstanding importance for the description of excited states, it is, except for some special cases, unavoidable to determine the excited state geometries at the CI level of computation. Furthermore, as much less empirical data is available for excited state geometries, it is necessary to verify for each stationary point found on an excited state potential energy hypersurface, whether it is a minimum, a transition state or some other type of stationary point. This requires the vibrational frequencies to be calculated by diagonalization of the Hessian matrix which is the matrix of the second derivatives of the energy [55].

Of all methods available for function minimization, gradient algorithms are the most widely used in quantum chemistry for the optimization geometries [6, 7]. On the ab initio level of theory continuing progress has been made in recent years in the use of analytic first and second derivatives of the potential energy surface for correlated wave functions of the MCSCF type [6], for various orders of Møller-Plesset perturbation theory [6] and even to some extend of CI wave function [56, 57]. In these cases where the first derivatives of the energy can be computed efficiently, such algorithms have become preferred over those which employ function values only, or the direct use of the Newton-Raphson relaxation method which requires the more expensive calculation of the second derivatives. The gradient algorithms fall into two classes which may be described as conjugative gradient and variable metric or quasi-Newton methods, respectively. The former methods like those of Fletcher-Reeves [58] or Polak-Ribiere [59] do not build up a Hessian matrix and are preferred when not enough storage is available. The latter methods, on the other hand, start from an approximate (inverse) Hessian matrix and generally improve during the course of optimization using the geometry and gradient information generated by a sequence of Newton-like relaxation steps. In the well-known Murtagh-Sargent [60], Davidon-Fletcher-Powell [61, 62] and Broyden-Fletcher-Goldfarb-Shanno [63–66] algorithms, the geometry and gradients from the current and previous step are used to update the Hessian matrix.

For semiempirical SCF methods geometry optimization by means of gradient techniques [67] as well as the determination of the Hessian matrix by finited differences of gradients [68, 69], can be performed routinely [5]. Geometry optimization on the CI-level based on finite differences and determination of the Hessian matrix by means of finite second differences have been achieved within the MNDOC-CI method, which therefore is particularly well suited for computing properties of excited states [20, 70, 71].

3.3 Some Examples

In this section we will give some examples of properties, especially geometries of organic molecules in their excited states and of some of the problems which may be encountered when calculating these properties. Although we will try to cover most of the recent work, we do not intend to give a complete review of all theoretical work done on excited states of organic molecules. General recent reviews include those of Davidson and McMurchi (1982) [51], of Bruna and Peyerimhoff (1987) [1] and of Gustav and Colditz (1988) [72]; references of papers confined to specific aspects will be given in the following subsections.

3.3.1 Ethylene and Polyenes

Ethylene is the simplest unsaturated hydrocarbon, therefore many studies have been devoted to this molecule, either to explain the observed spectrum, the assignment of which to computed excitation energies being still not quite perfect [51, 73], or to study the mechanism of *cis-trans* isomerization [24] and finally in order to investigate the sudden polarization phenomenon [74–76].

Simple molecular orbital assignments and correlation diagrams suggest that the lowest excited valence states should be twisted by 90°, the triplet being a π, π^* state whereas the lowest singlet minimum occurs at a biradicaloid geometry due to the avoided crossing of the ground and the doubly excited $(\pi^*)^2$ configurations [29]. This simple picture is however obscured by the existence of low-lying Rydberg states [77, 78].

From the arguments in Sect. 2 concerning biradicaloid minima it follows that the excited valence states of ethylene can be described correctly in the framework of SDCI only if three reference configurations are used. At 90° twist the symmetry becomes D_{2d}; at the two-configuration level, the $^1B_2(\pi, \pi^*)$ state is above the $^1A_1(\pi^*)^2$ state, but when sufficient CI is included, the $^1A_1$ is lowest in energy [79] being 10 kJ/mol below the $^1B_2$ state.

Structural parameters of ethylene in ground and excited states are given in Table 1. Apart from the 90° twist, the main difference in geometry between ground and excited states is apart from the 90° twist a lengthening of the CC bond by 15 [80] to 17 pm [79] in the $^3A_2$ state and by 9 pm in both the $^1A_1$ and the $^1B_2$ states [79]. Semiempirical MNDOC-CI calculations, although not able to reproduce Rydberg states, are quite suited to describe the properties of ethylene in its lowest excited states including the vibrational frequencies, except for the CC bond length [20]. The failure to reproduce the lengthening of that bond seems to be a common feature to most semiempirical methods like MNDOC [54] and SINDO1 [81].

Table 1. Ethylene, geometries of ground and excited states (distances in pm, angles in degrees). MNDOC-CI results and experimental (in parentheses) or ab initio data [in brackets]

State	Sym.[a]	R_{CC}	R_{CH}	∢CCH	∢H'CCH	∢H''CCH
S_0	D_{2h}	133	109	123°	0°	180°
		(134)[b]	(109)[b]	(121°)[b]	(0°)[b]	(180°)[b]
T_1	S_4	143	109	121°	90°	255°
		[147][c]	[108][c]	[121°][c]	[90°][c]	[270°][c]
		[149][d]				
S_1	D_{2d}	133	110	126°	90°	270°
		[140][d]				
S_2	D_{2d}	133	110	126°	90°	270°
		[140][d]				

[a] Symmetry was determined after full optimization, whereas in Ref. c and Ref. d D_{2d} symmetry was assumed for all excited states.
[b] Ref. 53.
[c] Ref. 80.
[d] Ref. 79

The vibrational motion which is of particular interest is the twist mode, the frequency of which has been calculated to be 903 cm^{-1} (S_0), 593 cm^{-1} (T_1), 2336 cm^{-1} (S_1) and 872 cm^{-1} (S_2). This reflects well the steep torsional potential of the S_1 state due to its large contribution from high-lying doubly excited configurations [20].

As is apparent from the results in Sect. 2, slight distortions of the symmetry may lead to large charge polarizations. This phenomenon called „sudden polarization" by Salem and Bruckmann [74–76] has been investigated in great detail not only for ethylene [79, 82–85], but also for larger polyenes [86–91]. The role of non-adiabatic coupling in ethylene and derivatives was studied [92] and dipole moments of valence orbital states of linear polyenes were determined [93]. Experimental evidence of sudden polarization in acyclic alkenes has been obtained from the effects of substituents and solvent polarity on [1, 3] sigmatrope shifts [94], for a related result see [95].

As a typical example of sudden polarization Fig. 3 shows the charge translocation Δp between the two allyl fragments in twisted s-cis, s-trans-hexatrien as a function of the angle of twist ϑ. The MNDOC-CI results [70] are compared to ab initio results [96]. At 90° twist the agreement is perfect, but for other values of the MNDOC-CI method with 90 selected configurations yield slightly larger charge translocations than the STO-3G calculations based on a 3×3 CI. As is seen from this calculations the charge separation develops only at angles very close to 90° twist and is as large as 0.8 at the orthogonal geometry.

The excited states of other unsaturated hydrocarbons have been discussed including styrene [97, 98], stilbene [99–101, 109], substituted stilbene [102, 103] and cyclopropene [104]. As is to be expected from the correlation diagram of cyclopropene in Fig. 2, only the triplet state is stable, the lowest excited singlet state is calculated to undergo a ringopening process to form vinylmethylene.

Excited states of longer polyenes have been investigated thoroughly by theory as well as by experiment [105]. Thus, it is well established that the significantly shifted emission of longer polyenes derives from a totally symmetric excited state not observed in

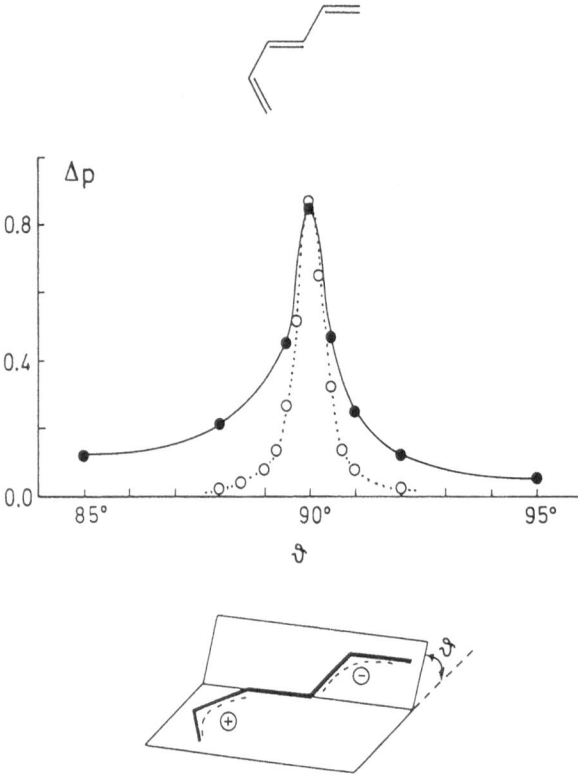

Fig. 3. Charge transloca-
tion Δp between the allyl
fragments of twisted s-cis,
s-trans-hexatriene as function of the twist angle ϑ
$(-\bullet-, $ MNDOC-CI
results [70]; ...O..., STO-3G
results [96]

absorption. From ab initio data it can be concluded, that in hexatriene this ^1A_g state
is already below the ^1B_u state [106]; for additional results see [107–110]. It is also
concluded with reasonable certainty that polyenes do not twist on excitation [105].

3.3.2 Acetylene

Acetylene is another small molecule, the excited states of which are of considerable
interest in so far as they may be stabilized by bending either cis or trans with only
a small energy difference between these conformations. The bent species are again
biradicaloids which has to be taken into consideration when chosing the CI although
the splitting (more than 40 kJ/mol) of the biradicaloid states is larger than for instance
in the case of ethylene.

In agreement with ab initio results [111, 112] MNDOC-CI calculations yield, for the
lowest singlet as well as for the lowest two triplet states, two minima corresponding
to a cis (C_{2v}) or trans (C_{2h}) arrangement, as is illustrated in Fig. 4 showing the
potential hypersurface of the three lowest excited states for cis-trans isomerization
by inversion as well as by rotation around the triple bond. As can be seen, the barriers
vary asccording to state and type of the process but are particularly low in the T_1 state.

As seen from Table 2 no trans geometry (C_{2h}) could be located for the S_2 state by
MNDOC-CI calculations [20], in contrast to ab initio results. The CC bond is

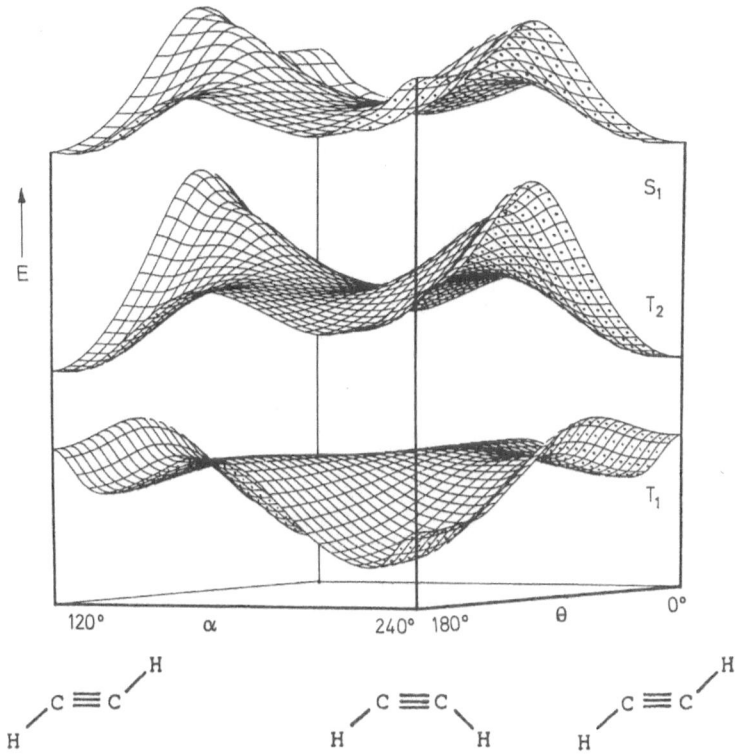

Fig. 4. Calculated potential hypersurfaces of $\pi \rightarrow \pi^*$ excited states T_1, T_2 and S_1 for the *cis-trans* isomerization in acetylene as a function of the bond angle α and the dihedral angle θ

considerably longer in the excited states than in the ground state, the lengthening being 12 pm if the conformation is *cis* and 15–18 pm if it is *trans*. From Fig. 4 it is seen that the order of the states is

$$^3B_2 < {}^3B_u < {}^3A_u < {}^3A_2 < {}^1A_u < {}^1A_2$$

where A_2 and B_2 refer to C_{2v} or *cis* geometry and A_u and B_u to C_{2h} or *trans*. Vibrational frequencies have been calculated for all minima [20, 111]. The experimentally observed frequency changes for an excitation in the *trans* S_1 state are $\Delta v = +437 \text{ cm}^{-1}$, -594 cm^{-1} and -354 cm^{-1} for the symmetrical CH bend, the CC stretch and the CH stretch mode respectively, the calculated differences being $\Delta v = +482 \text{ cm}^{-1}$, -566 cm^{-1} and -248 cm^{-1}. Similarly, for an excitation into the *cis* S_1 state $\Delta v = -540 \text{ cm}^{-1}$ is calculated for the CC stretch vibration, the observed frequency difference being $\Delta v = -508 \text{ cm}^{-1}$. Calculations show that the CC stretch vibration decreases in first and second $^{1,3}\pi, \pi^*$ excited states by -250 cm^{-1} for the *cis* conformations and $400–500 \text{ cm}^{-1}$ for the *trans* conformations [20, 111].

Table 2. Acetylene, geometries of ground and excited states (distances in pm, angles in degrees). MNDOC-CI results and experimental[a] (in parentheses) or ab initio[b] data [in brackets]

State	Sym.	R_{CC}	R_{CH}	$\angle CCH$
S_0	D_{2h}	119	106	180°
		(120	(106)	(180°)
T_1	C_{2v}	131	108	138°
		[133]	[109]	[129°]
	C_{2h}	133	107	142°
		[134]	[108]	[132°]
T_2	C_{2h}	137	108	135°
		[138]	[109]	[121°]
	C_{2v}	132	108	144°
		[135]	[109]	[132°]
S_1	C_{2h}	136	108	137°
		[137]	[110]	[122°]
	C_{2v}	131	109	148°
		[134]	[110]	[133°]
S_2	C_{2v}	130	110	156°
		[133]	[109]	[145°]
	C_{2h}			
		[134]	[108]	[146°]

[a] Ref. 53. [b] Ref. 111

3.3.3 Carbonyl Compounds

Formaldehyde is the simplest compound to show n, π^* and π, π^* excited states; therefore it has been studied quite often [22, 23, 46, 47], the most detailed experimental and ab initio investigation (formaldehyde, acetaldehyde, acetone, cyclopentanone and cyclobutanone) of carbonyl compounds being that of Baba et al. (1985) [113–115]. Other carbonyl compounds which have been studied include ketene [116], acrolein [117], β-hydroxy acrolein [118] as well as formamide, formic acid and formyl fluoride [119].

Some results for formaldehyde, acetaldehyde and acetone as well as for formamide and acetamide are collected in Table 3. From these data it is seen, that the most important geometry changes on excitation are the lengthening of the CO bond which is markedly larger in the π, π^* than in the n, π^* excited states, the pyramidalization of the carbonyl carbon, and a rotation of the methyl groups, which has also been observed in ab initio calculations [113, 120, 121]. In the ground state the methyl groups in acetaldehyde, acetone and acetamide are oriented in such a way that one hydrogen is eclipsed with the C=O bond, while in excited states the staggered conformations correspond to the minima [20]. If the dihedral angle between two planes containing the carbon atom and two of its three substituents is called ω_i with $\omega = 180°$ and $\omega = 120°$ for a planar sp^2 hybridized and a tetrahedral sp^3 hybridized carbon respectively, the degree of pyramidalization may be defined as

$$Py = \left[1 - \left(360 - \sum_{i=1}^{3} \omega_i\right)\middle/(360 - 540)\right] \cdot 100\% .$$

Table 3. Important geometry parameters of some carbonyl compounds, changing on excitation. CO bond length (R_{CO} in pm), pyramidalization (Py in %) and the rotation of the methyl group (φ in degrees). MNDOC-CI results, experimental (in parentheses) or ab initio data [in brackets]

Molecule	State	Excitation		R_{CO}	Py	φ[a]
H_2CO	S_0	–		121 (120)[b]	0	–
	T_1	n,π^*		128 (131)[c]	60	–
	S_1	n,π^*		129 (133)[c]	54	–
	T_2	π,π^*		137 [143][d]	56	–
CH_3	S_0	–		122 (122)[e]	0	0
	T_1	n,π^*		129 [139][f]	38	47 [45][g]
	S_1	n,π^*		130 (132)[h]	18	49 [48][i]
	T_2	π,π^*		140	45	54
$(CH_3)_2CO$	S_0	–		122 (122)[e]	0	0
	T_1	n,π^*		131	0	60 (60)[j]
	S_1	n,π^*		131	0	60
	T_2	π,π^*		141	28	87
H_2NCHO	S_0	–		122 (122)[k]	0	–
	T_1	n,π^*		131	69	–
	S_1	n,π^*	(t)[a]	133	60	–
			(c)	132	48	–
	T_2	π,π^*		134	60	–
CH_3CONH_2	S_0	–		123 (122)[l]	0	–
	T_1	n,π^*		132	53	38
	S_1	n,π^*	(t)	134	47	35
			(c)	134	45	45
	T_2	π,π^*	(t)	138	53	48
			(c)	137	45	48

[a] Pyramidalization both on C and N leads to transoid (t) and cisoid (c) geometries. [b] Ref. 22. [c] Ref. 46. [d] Ref. 47. [e] Ref. 124. [g] Ref. 120. [h] Ref. 125. [i] Ref. 113. [j] Ref. 121. [k] Ref. 126. [l] Ref. 127

The values given in Table 3 indicate that the pyramidalization of carbonyl compounds in excited states decreases with increasing degree of methyl substitution, and increases with amino substitution [20].

3.3.4 TICT States

Twisted internal charge transfer (TICT) has been proposed [128, 129] to be the reason for the dual fluorescence observed for molecules like *p-N,N*-dimethylamino-benzonitril which consists of a donor (Me$_2$N-) and an acceptor (C$_6$H$_4$CN) linked by a single bond. The simplest system of that kind is aminoborane H$_2$N-BH$_2$, and ab initio calculations [130] show that it behaves exactly as is to be expected for a strongly heterosymmetric biradicaloid (cf. Sect. 2.1.2). Thus, the wave function of S$_0$ is almost exactly the hole-pair structure (χ_b^2), whereas S$_1$ corresponds to the formally covalent $^1(\chi_a\chi_b)$ structure and is thus of biradicaloid zwitterionic character with the donor positively and the acceptor negatively charged.

Due to the large spatial separation of χ_a and χ_b and the large difference in geometry between S$_0$ and S$_1$ these states are normally populated by IC from other states and observed by emission only. As the energy of such dipolar states depends sensitively on the polarity of the environment, they can be readily detected by characteristic solvent shifts of the fluoresence [26, 131].

This description is inadequate for most compounds in which TICT emission has been observed in that the donor and the acceptor are both complicated structures with low-lying locally excited states, such as substituted benzene rings. These additional locally excited states favor planar geometries, and since the charge-separated state decreases in energy upon twisting form planarity, crossings or avoided crossings result. Consequently, the S_1 state can exhibit a barrier separating two minima, one close to the planar geometry, and the other at the twisted conformation.

In Fig. 5 it is shown how all these expectations show up in the results of MNDOC-CI calculations [70]. The energies E and the charge translocation Δp between the amino group (donor) and the cyano-phenyl group (acceptor) for the lowest two excited

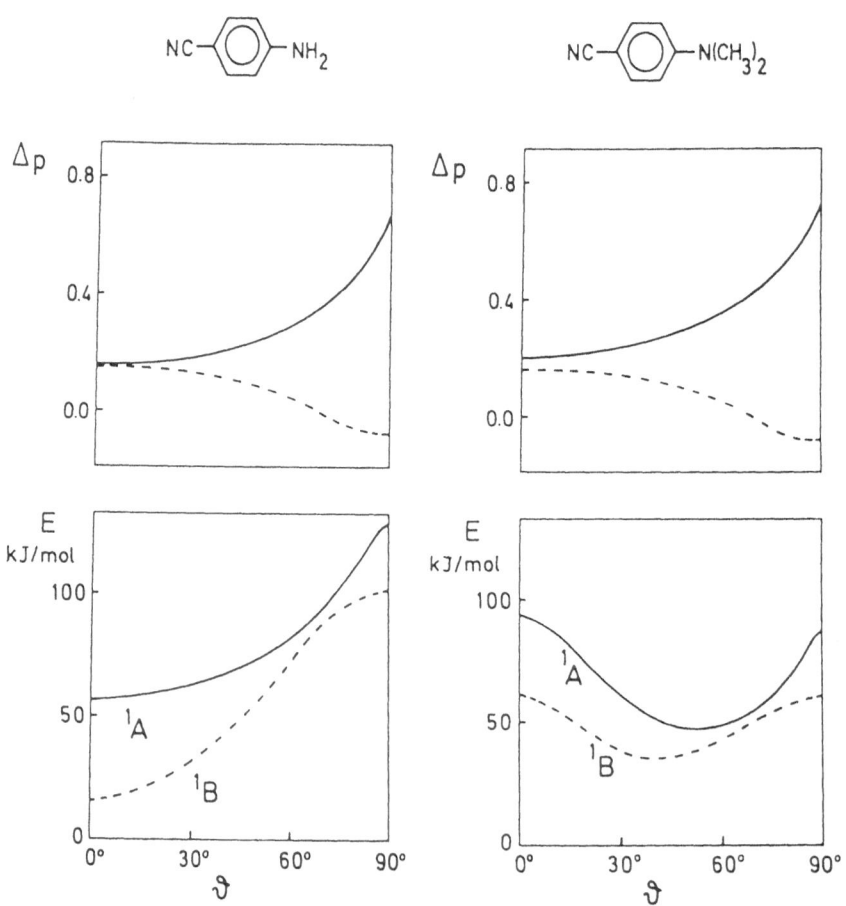

Fig. 5. Energies E (lower part) and charge separations Δp (upper part) of the first two excited states (———, ^{1}A; – – –, ^{1}B) of p-aminobenzonitrile and p-N,N-dimethylaminobenzonitrile as a function of the angle ϑ for rotation of the amino group [70]. The origin of the energy scale is orbitrary

states of p-aminobenzonitrile and p-N,N-dimethylaminobenzonitrile are shown as a function of the angle ϑ for rotation of the amino group which is assumed to be planar; the states of both molecules can therefore be classified according to C_2 symmetry. The ^1B state corresponds to a locally excited state of the substituted benzene, whereas the ^1A state, for which the calculated charge translocation corresponds nearly to $\Delta p = 0.8$ at $\vartheta = 90°$ – Alkyl substitution leads to minima at nonplanar geometries for both states. Polar solvents will stabilize the charge separated state until its minimum is below that of the locally excited state, and dual fluorescence can be observed [26, 131].

TICT states are still being investigated intensively [132–145], practical applications may be seen in molecular switching and in the development of new laser dyes [26].

4 Radiative and Non-radiative Deactivation of Excited States

Next to the geometries the most interesting properties of excited states may be their lifetimes and their decay mechanisms. Whereas stable ground states have in general an infinite lifetime, some of the excited states might possibly be so short lived that there is no way of detecting them either directly by spectroscopic techniques or indirectly by studying the reactions products. In this case it is important to know what deactivation channel exist, i.e. which other states are produced by the decay of a particular state. The lowest excited singlet state S_1 can in general decay only back to the ground state S_0 either by fluorescence (F) or by internal conversion (IC), or turn into the triplet manifold by intersystem crossing (ISC). All theses processes are competitive, their rates determine the efficiency of each of them. In many cases the chemistry of a molecule in its S_1 state is quite different from the chemistry of its T_1 state. Therefore, if the efficiency of the ISC process is known and understood, the photochemistry of the molecule may be influenced to yield either one or the other product.

The theoretical treatment of the rates of all radiative and non-radiative processes requires the knowledge of the fully optimized nuclear geometries of all states under consideration as well as of all normal vibrations for each of these states. In the previous section it was shown how these quantities may be obtained from quantum chemical calculations at various levels of approximation. This section gives a brief outline of the methods necessary in order to get the required rate constants from these data.

4.1 Radiative Transitions

The rate constant of fluorescence k_F from an excited state $\Psi_{av'}$ into the state Ψ_{0v} may be written as

$$k_F = \frac{64\pi^4}{3h} \sum_v \tilde{v}^3_{av' \to 0v} \, \bar{M}^2_{av' \to 0v} \, , \tag{8}$$

where $\tilde{v}$ is the wave number of the transition and $\bar{\mathbf{M}}_{av \to 0v}$ the vibronic transition moment [146], which within the Born-Oppenheimer approximation may be written as

$$\bar{\mathbf{M}}_{av \to 0v} = \langle \chi_{av'}(\mathbf{Q}) | \, \mathbf{M}_{a \to 0}(\mathbf{Q}) \, | \chi_{0v}(\mathbf{Q}) \rangle \tag{9}$$

with

$$\mathbf{M}_{a \to 0}(\mathbf{Q}) = \langle \Psi_a^Q(\mathbf{q}) | \, \mu(\mathbf{r}) \, | \Psi_0^Q(\mathbf{q}) \rangle . \tag{10}$$

Here, the electronic transition moment $\mathbf{M}_{a \to 0}(\mathbf{Q})$ is seen to depend on the normal coordinates through the parametrical dependence of the many-electron wavefunction $\Psi_j^Q(\mathbf{q})$ on these coordinates [see eq. (1)]. If this dependence is neglected, the Condon approximation [147]

$$\bar{\mathbf{M}}_{av' \to 0v} = \mathbf{M}_{a \to 0}(\mathbf{Q}_0) \, \langle \chi_{av'}(\mathbf{Q}) \, | \, \chi_{0v}(\mathbf{Q}) \rangle \tag{11}$$

results, according to which the static electronic transition moment is modulated by Franck-Condon overlap integrals.

In order to include for instance vibronic coupling the $\mathbf{Q}$-dependence of the electronic wave function has to be taken into account. An expansion in powers of $\mathbf{Q}$ up to the linear term yields the Herzberg-Teller approximation [148]

$$\begin{aligned} \bar{\mathbf{M}}_{av' \to 0v} = \mathbf{M}_{a \to 0}^0 \, \langle \chi_{av'} \, | \, \chi_{0v} \rangle + \sum_k [(\partial \mathbf{M}_{a \to 0} / \partial Q_k) \\ + \sum_{b \neq a} \langle \Psi_a^0 | \, A_k' \, | \Psi_b^0 \rangle \, \mathbf{M}_{b \to 0}^0 / (E_a^0 - E_b^0) \\ + \sum_{b \neq 0} \mathbf{M}_{a \to b}^0 \, \langle \Psi_b | \, A_k' \, | \Psi_0 \rangle / (E_0 - E_b)] \, \langle \chi_{av'} | \, Q_k \, | \chi_{0v} \rangle , \end{aligned} \tag{12}$$

where the superscript 0 refers to the equilibrium geometry $\mathbf{Q}_0$.

The first term is the Condon term of Eq. (11) and the expression within brackets corresponds to the variation of the vibronic transition moment $\bar{\mathbf{M}}_{av' \to 0v}$ with respect to the normal coordinate Q_k at Q_0, modified by the transition moment $\langle \chi_{av'} | \, Q_k \, | \chi_{0v} \rangle$. It involves the vibronic coupling elements $A_k' = \partial A / \partial Q_k$ which are just derivatives of the configuration interaction matrix $\mathbf{A}$. The first term in brackets is due to orbital following of the AOs with the vibrating atoms and may become important for strongly allowed electronic transitions.

Detailed expressions for all the quantities occuring in Eq. (12) including those needed for the evaluation of the integrals $\langle \chi_{av'} | \chi_{0v} \rangle$ and $\langle \chi_{av'} | Q_k | \chi_{0v} \rangle$ over multimode vibrational functions are given in a recent paper by Gustav et al. [149], and examples have been provided by the same group. Thus based on the π approximation and on geometries optimized by a combined force-field + PPP method [150] the phosphorescence of naphthalene and d_8-naphthalene as well as the vibronic features of the absorption and fluorescence spectra of perylene (**5**) could be explained on the basis of the Condon approximation [151, 152], Biphenylene (**6**) on the other hand provides an example for which the Herzberg-Teller coupling given by Eq. (2) is necessary in order to explain the spectral features [153].

5   6

The results suggest that the method developed on the basis of the π approximation may be extended to treat all kinds of molecules by evaluating the geometries and normal modes of vibration by the methods discussed in Sect. 3. A recent example for the application of ab initio methods is the analysis of the vibrational structure of the lowest singlet-triplet transition in ethylene [80].

4.2 Non-radiative Processes

Starting point for a theoretical treatment of intramolecular radiationless deactivation of vibronic states is the Fermi golden rule [154], which may be written as

$$k_{nr} = \frac{2\pi}{\hbar} \sum_{v'} |\langle \chi_{bv'}| \langle \Psi_b| I(Q) |\Psi_a\rangle |\chi_{a0}\rangle|^2 \, \delta(E_{a0} - E_{bv'}). \tag{13}$$

In the case of internal conversion $I(Q) = T_n(Q)$ is the operator of the nuclear kinetic energy, whereas in the case of intersystem crossing the spin-orbit coupling operator $H_{SO}(r, Q)$ is to be included as well, and the non-radiative transition moments are evaluated between the vibrational ground level 0 of the initial electronic state Ψ_a and all the vibrational levels v' of the final state Ψ_b for the interval described by the δ function. Invoking again the Herzberg-Teller expansion up to linear terms the vibronic transition moment may be written in the case of internal conversion as

$$V_{a0 \to bv'} = \hbar^2 \sum_k \frac{\langle \Psi_a^0| A_k' |\Psi_b^0\rangle}{E_b^0 - E_a^0} \langle \chi_{av'}| \partial/\partial Q_k |\chi_{a0}\rangle . \tag{14}$$

The first term in this equation, the electronic factor, may be evaluated along the same lines as for radiative transition moments. The integral over the multimode vibrational function may again be broken down using the non-interacting oscillator approach [149]. One of the major problems is the summation in Eq. (14) which even for medium size molecules extends over an extremely large number of states. Thus for benzene at $\Delta E = 40000$ cm^{-1} the density of states is already 5×10^{15} states/cm^{-1}. Although quite a number of counting algorithms are available [149], detailed calculations on larger molecules will still have to show which one is best suited.

If spin orbital coupling is taken into account in order to treat intersystem crossing processes, matrix elements of the spin-orbit coupling operator

$$H_{so} = \frac{l^2}{2m_e^2 c^2} \sum_j \sum_\mu \frac{Z_\mu}{|r_\mu|^3} l_j \cdot s_j \tag{15}$$

with $l_j = r_j \times p_j$ being this orbital angular momentum operator, s_j the spin operator and the summation running over all electrons j and all nuclei μ are to be evaluated. Several schemes have been designed [155], and work on the MNDOC level which incorporates the NDDO approximation in determining the corresponding AO integrals is in progress [156].

5 Conclusions

As shown in the present paper quantum chemical calculations of excited state properties have made some remarkable progress during the last few years. Excited state geometries are available on various levels of sophistication and semiempirical methods have been developed which in the very near future will make excited state calculations to be performed nearly as easily as ground state calculations. But as due to the CI these calculations will in general be much more time-consuming than the corresponding ground state calculations, as much information as possible should be gained from qualitative models in order to get some minimal estimates for the equilibrium geometry and in particular, in order to see whether due to some avoided crossings or other reasons the use of more than one reference configuration is required in order to get a proper wave function even if full CI can not be taken into account. Thus, although newly developed programs will make excited state calculations possible for all kinds of molecules, these calculations will always be of higher complexity then a ground state calculation and will therefore not easily become routine work.

This is even more important for the treatment of transition probabilities and lifetimes. The theoretical apparatus for evaluating rates of radiative processes is available and has been tested for large hydrocarbons within the π approximation, and calculations of this kind will become available in the near future for organic molecules of different types and sizes. But due to the complexity of the methods it will take some time before they become very popular. As far as non-radiative transitions and intersystem crossings are concerned, some work has still to be done. Judging from what is under way, it will not take long before these calculations are possible, too.

Thus we may conclude by stating that theoretical calculations on excited state properties are well within the scope of present day quantum chemical program systems as far as geometries, energies and wave functions are concerned. From the wave functions the density matrices may be obtained from which most properties like charge distributions, dipole moments and nuclear magnetic coupling constants, to mention just a few, may be calculated in the same way as for ground states. As the estimation of rate constants of radiative and non-radiative processes is also well within reach, excited state chemistry will soon be open to quantum chemical studies in the same way as ground state chemistry has been in the last decades. Considering the fact that due to experimental difficulties much less information is available for the excited state than for the corresponding ground state properties and taking into account the great diversity of excited states, a very wide field for future work is opening up.

6 References

1. Bruna PJ, Peyerimhoff SD (1987) In: Lawley KP (ed) Ab initio methods in quantum chemistry-I. Wiley, New York
2. Zewail AH (1986) Springer Ser Chem Phys 46: 356
3. Hehre WJ, Radon L, Schleyer v. PR, Pople JA (1986) Ab initio molecular orbital theory. Wiley, New York
4. Fogarsi G, Pulay P (1984) Ann Rev Phys Chem 35: 191
5. Dewar MJS, Zoebisch EG, Healy EF, Stewart JJP (1985) J Am Chem Soc 107: 3902
6. Pulay P (1987) In: Lawley KP (ed) Ab initio methods in quantum chemistry-II. Wiley, New York
7. Schlegel HB (1987) In: Lawley KP (ed) Ab initio methods in quantum chemistry-I. Wiley, New York
8. Del Bene J, Jaffé HH (1968) J Chem Phys 48: 1807, 4050; 49: 1221
9. Ellis RL, Jaffé HH (1970) In: Segal CA (ed) Modern theoretical chemistry. Plenum, New York, vol 8 chap 2
10. Dick B, Nickel B (1983) Chem Phys 78: 1
11. Buss S, Jug K (1987) J Am Chem Soc 109: 1044
12. Jug K, Iffert R, Müller-Remmers PL (1988) J Am Chem Soc 110: 2049
13. Sumathi K, Chandra AK (1987) J Photochem Photobiol A40: 265; (1988) 43: 313
14. Chandra AK (1988) J Mol Struct (Theochem) 181: 255
15. Reinsch M, Höweler U, Klessinger M (1987) Angew Chem 99: 250
16. Reinsch M, Höweler U, Klessinger M (1988) Mol Struct (Theochem) 167: 301
17. Reinsch M, Klessinger M (1990) J Phys Org Chem 3: 81
18. Dewar MJS, Fox MA, Campbell KA, Chen C-C, Friedheim JE, Holloway MK, Kein, SC, Liescheski PB, Pakiari AM, Tien T-P, Zoebisch E (1984) J Comput Chem 5: 480
19. Lee HY, Jaffé HH (1985) J Mol Struct (Theochem) 123: 301
20. Klessinger M, Pötter T, van Wüllen C (1991) Theoret Chim Acta (submitted)
21. Michl J (1974) Fortschr Chem Forschg 46: 1
22. Job VA, Sethuraman V, Innes KK (1969) J Mol Spectrosc 30: 365
23. Kirchoff WH, Lovas FJ, Johnson DR (1972) J Phys Chem Ref Data 1: 1011
24. Bonačić-Koutecký V, Koutecký j, Michl J (1987) Angew Chem 99: 216
25. Bonačić-Koutecký V, Bruckmann P, Hiberty P, Koutecký J, Forestier C, Salem L (1975) Angew Chem 87: 599
26. Rettig W (1986) Angew Chem 98: 969
27. Grabowski ZR, Dobkowski J (1983) Pure Appl Chem 55: 245
28. Bonačić-Koutecký V, Köhler J, Michl J (1984) Chem Phys Letters 104: 440
29. Klessinger M, Michl J (1990) Lichtabsorption und Photochemie organischer Moleküle. VCH, Weinheim
30. Devaquet A, Sevin A, Bigot B (1978) J Am Chem Soc 100: 2009
31. Herzberg G (1950) Molecular spectra and molecular structure. Van Nostrand, Princeton
32. v Bünau G, Wolff Th (1987) Photochemie. VCH, Weinheim
33. Lippert E (1957) Z Elektrochem 61: 962

34. Metz F, Robey MJ, Schlag EW, Dörr F (1977) Chem Phys Lett 51: 8
35. Sharp TE, Rosenstock HM (1964) J Chem Phys 41: 3453
36. Lucas NJD (1973) J Physics B 6: 155
37. Doktorov EV, Malkin IA, Manko VI (1977) J Mol Spectrosc. 56: 1
38. Faulkner TR, Richardson FS (1979) J Chem Phys 70: 1201
39. Innes KK (1975) in: Lim EC (ed) Excited states. Academic, New York, vol 2, p 1
40. Hollas JM (1973) Mol, Spectrosc 1: 62
41. Ross IG (1971) Adv Chem Phys 20: 341
42. Schrader B (1973) Angew Chem 85: 925
43. Sugawara Y, Hamaguchi H, Harada I, Shimanouchi T (1972) Chem Phys Lett 52: 323
44. El-Sayed MA (1974) in: Lim EC (ed) Excited states. Academic, New York, vol 1
45. Gustav K, Sühnel J (1980) Z Chem 20: 283
46. Jones VT, Coon JB (1960) J Mol Spectrosc 31: 137
47. Saxe P, Yamaguchi Y, Schaefer III HF (1982) 99: 250
48. Parr RG (1963) Quantum theory of molecular electronic structure. Benjamin, New York
49. Hunt WJ, Goddard WA (1969) Chem Phys Lett 3: 414, 418
50. Buenker RJ, Peyerimhoff SD, Butscher W (1978) Mol Phys 35: 771
51. Davidson ER, McMurchie LE (1982) In: Lim EC (ed) Excited states. Academic, New York, vol 6
52. Manz U, Rosmus P, Werner H-J, Botschwina P (1988) Chem Phys 112: 387
53. Thiel WJ (1981) J Am Chem Soc 103: 1415, 1420
54. Schweig A, Thiel W (1981) J Am Chem Soc 103: 1425
55. McIver Jr JW, Kormonicki A (1972) J Am Chem Soc 94: 2625
56. Lee TJ, Handy NC, Rice JE, Scheiner AC, Schaefer III HF (1986) J Chem Phys 85: 3930
57. Helgakar T, Jørgensen P (1989) Theoret Chim Acta 75: 111
58. Fletcher R, Reeves CM (1964) Comput J 7: 149
59. Polak E (1971) Computational methods in optimization. Academic, New York
60. Murtagh BA, Sargent RWH (1972) Comput J 13: 185
61. Fletcher R, Powell MJD (1963) Comput J 6: 163
62. Davidon W. Argonne National Lab Report ANL-5990
63. Broyden JC (1970) J Inst Math 6: 76
64. Fletcher R (1970) Comput J 13: 317
65. Goldfarb D (1970) Math Compur 24: 23
66. Shanno DF (1970) Math Comput 24: 647
67. McIver Jr JW, Kormonicki A (1971) Chem Phys Lett 10: 303
68. Dewar MJS, Ford GP (1979) J Am Chem Soc 99: 1685
69. Dewar MJS, Ford GP, McKee ML, Rzepa HS, Thiel W, Yamaguchi Y (1978) J Mol Struct 43: 135
70. Klessinger M (1989) J Mol Struct (Theochem) 202: 129
71. Pötter T, Klessinger M (1991) J Comp Chem (in press)
72. Gustav K, Colditz R (1988) Z Chem 28: 309
73. Mullien RS (1979) J Chem Phys 71: 556
74. Salem L, Bruckmann P (1975) Nature 258: 526
75. Bruckmann P, Salem L (1976) J Am Chem Soc 98: 5037
76. Salem L (1979) Acc Chem Res 12: 87
77. McMurchie LE, Davidson ER (1977) J Chem Phys 66: 2659
78. Buenker RJ, Peyerimhoff SD, Shin SK (1978) J Chem Phys 69: 3882
79. Brooks BR, Schaefer HF (1979) J Am Chem Soc 101: 307
80. Siebrand W, Zerbetto F, Zgierski MZ (1989) J Chem Phys 91: 5926
81. Mishra PC, Jug K (1982) Theoret Chim Acta 103: 1425
82. Bonačić-Koutecký V, Buenker RJ, Peyerimhoff SD (179) J Am Chem Soc 101: 5917
83. Buenker RJ, Bonačić-Koutecký V, Pogliani V (1980) J Chem Phys 73: 1836
84. Petsalakis ID, Theodorakopoulos G, Nicolaides CA, Buenker RJ, Peyerimhoff SD (1984) J Chem Phys 108: 597
85. Trinquier G, Malrieu JP (1980) 73: 1836
86. Malrieu JR, Trinquier G (1979) J Am Chem Soc 54: 59
87. Karafiloglou P, Hiberty PC (1980) Chem Phys Lett 70: 180

88. Bonačić-Koutecký V, Persio M, Döhnert D, Sevin A (1982) J Am Chem Soc 104: 6900
89. Karafiloglou P, Evleth EM (1983) J Mol Struct (Theochem) 35: 161
90. Orlandi G, Hennekor WH, Siebrand W, Zgierski MZ (1982) Chem Phys Lett 92: 19
91. Bonačić-Koutecký V, Gizek J, Doehnert D, Koutecký J (1978) J Chem Phys 69: 1168
92. Persico M (1980) J Am Chem Soc 102: 7839
93. Dinur U (1982) Chem Phys Lett 93: 253
94. Peijnenburg WJGM, Buck HM (1988) Tetrahedron 44: 4927
95. Tezuka T, Kikuchi O, Hank KN, Paddon-Row MN, Santiago M (1981) J Am Chem Soc 103: 1367
96. Bonačić-Koutecký V, Bruckmann P, Hiberty P, Koutecký J, Leforestier C, Salem L (1975) Angew Chem 87: 571
97. Nebot-Gil I, Malrieu JP (1981) Chem Phys Lett 84: 571
98. Orlandi G, Palmierei P, Poggi G (1981) J Chem Soc Faraday Trans-II 77: 71
99. Negri F, Orlandi G, Zerbetto FJ (1989) Phys Chem 93: 945
100. Hohlneicher G, Dick B (1984) J Photochem 27: 215
101. Troe J, Weitzel K-M (1988) J Chem Phys 88: 7030
102. Saltiel J, Charlton JL (1980) In: de Mayo (ed) Rearrangements in ground and excited states. Academic, New York, vol 3
103. Kikuchi D, Yoshida H (1985) Bull Chem Soc Jpn 58: 131
104. Yoshimine M, Pacansky J, Honjou N (1989) J Am Chem Soc 111: 2785, 4198
105. Hudson BS, Kohler BE, Schulten K (1982) In: Lim EC (ed) Excited states. Academic, London, vol 6
106. Nacimento MAC, Goddard WA (1979) Chem Phys 36: 147
107. Cave RJ, Davidson ER (1987) J Phys Chem 91: 4481
108. Hemley RJ, Lasaga AC, Vaida V, Karplus M (1988) J Phys Chem 92: 945
109. Negri F, Orlandi G, Brouwer AM, Langkilde FW, Wilbrandt R (1989) J Chem Phys 90: 5944
110. Meerman-Van Benthem CM, Jacobs HJC, Mulder JJC (1978) Nouv J Chim 2: 123
111. Lischka H, Karpfen A (1986) Chem Phys 102: 77
112. Perič M, Peyerimhoff SD, Buenker R (1987) J Mol Phys 62: 1339
113. Baba M, Hanazaki I, Nagshima U (1985) J Chem Phys 93: 425
114. Baba M, Hanazaki I, Nagshima U (1985) J Chem Phys 82: 3938
115. Baba M, Nagahima U, Hanazaki I (1985) J Chem Phys 83: 3514
116. Yoshimine M (1989) J Chem Phys 90: 378
117. Valenta K, Grein F (1982) Can J Chem 60: 601
118. Ha TK, Graf F (1980) Chem Phys Lett 72: 358
119. Ha TK, Keller L (1975) J Mol Struct 27: 225
120. Altman JA, Doust TAM, Osborne AD (1980) Chem Phys Lett 69: 595
121. Peterson MR, DeMare GR, Csizmadia I, Strausz OP (1981) J Mol Struct 86: 131
122. Hubbard LM, Bocian DF, Barge RR (1981) J Am Chem Soc 103: 3313
123. Crighton JS, Bell S (1985) J Mol Spectrosc 112. 285
124. Yadav JS, Goddard JD (1986) J Chem Phys 84: 2682
125 Hubbard LM, Bocian DF, Barge RR (1981) J Am Chem Soc 103: 3313
126. Wiberg KB, Laidig KE (1987) J Am Chem Soc 109: 5935
127. Popellier P, Lenstra ATH, van Alsenoy L, Geise HJ (1989) J Am Chem Soc 111: 5658
128. Grabowski ZR, Rotkiewicz K, Rubaszewska W, Kirkor-Kaminska E (1978) Acta Phys Pol A A56: 767
129. Grabowski ZR, Rotkiewicz R, Siemiarczuk A (1979) J Lumin 18–19: 420
130. Bonačić-Koutecký V, Michl J (1985) J Am Chem Soc 107: 1765
131. Lippert E, Rettig W, Bonačić-Koutecký V, Heisel F, Micke JA (1987) Adv Chem Phys 68: 1
132. Launay JP, Sowinska M, Leydier L, Gourdon A, Amouyal E, Boillot ML, Heisel F, Micke JA (1989) Chem Phys Lett 160: 89
133. Nag A, Kundu T, Bhattacharyya K (1989) Chem Phys Lett 160: 257
134. Cazeau-Dubroca C, Peirigua A, Ben Brahim M, Nouchi G, Cazeau P (1989) Chem Phys Lett 157: 393
135. Honjez BJ, Yazdi PT, Fox MA, Johnston KP (1989) J Am Chem Soc 111: 1915
136. Gormin D, Kasha M (1988) Chem Phys Lett 153: 574

137. Gormin D, Kasha M (1988) Chem Phys Lett 146: 121
138. Hara K, Suzuki H, Rettig W (1988) 145: 269
139. Gilabert E, Lapouyade R, Rulliere C (1988) 145: 262
140. Hayashi R, Tazuke S, Frank CW (1987) Chem Phys Lett 135: 123
141. Rulliere C, Grabowski ZR, Dobkowski J (1987) Chem Phys Lett 137: 408
142. Heisel F, Miehe JA (1986) Chem Phys Lett 128: 323
143. Rotkiewicz K (1986) Spectrochim Acta 42A: 575
144. Rettig W, Wermuth G (1985) J Photochem 28: 351
145. Cazeau-Dubroca C, Peirigua A, Lyazidi SA, Nouchi G, Cazeau P, Lapouyade R (1986) Chem Phys Lett 124: 110
146. Birks JB (1970) Photophysics of aromatic molecules. Wiley-Interscience, London
147. Condon EU (1928) Phys Rev 32: 858
148. Herzberg G, Teller E (1947) Rev Mod Phys 13: 75
149. Gustav K, Storch M, Jung Ch (1989) Acta Phys Pol A76: 883
150. Warshel A, Karplus M (1972) J Am Chem Soc 94: 5612
151. Gustav K, Storch M (1986) Mh F Chemie 117: 1007
152. Gustav K, Seydenschwanz C (1986) J Prakt Chem 328
153. Gustav K, Storch M (in preparation)
154. Robinson GW, (1974) in: Lim EC (ed) Excited states. Academic, London, New York, vol 1
155. Kotzian M, Roesch M, Zerner MC (1989) Chem Phys Lett 160: 168
156. Klessinger M, Zerner MC (in preparation)

Semiclassical Interpretation of Intramolecular Interactions

J. Tomasi[1], G. Alagona[2], R. Bonaccorsi[2], C. Ghio[2], and R. Cammi[3]

[1] Dipartimento di Chimica e Chimica Industriale, Università di Pisa, Via Risorgimento 35, 56126 Pisa, Italy
[2] Istituto di Chimica Quantistica ed Energetica Molecolare del C.N.R., Via Risorgimento 35, 56126 Pisa, Italy
[3] Istituto di Chimica Fisica, Università di Parma, Viale delle Scienze 1, 43100 Parma, Italy

1 Introduction.
The Theoretical Approach to the Study
of Intramolecular Interactions

Within the range of energies belonging to the domain covered by chemistry, there are interaction phenomena involving material systems which may vary enormously in size and internal composition. In many cases the basic unit in which the phenomenon occurs is very large and composite, in other cases it is very simple, indeed atomic, in nature. For chemists, however, the center of attention remains the molecule, even if the bewildering variety of structures and properties encountered poses a continuous challenge to anyone wrestling with a chemical problem, whatever his principal interest or goal may be.

Theoreticians, as well as other categories of chemists, have adopted the molecule as the starting unit for their investigations. The effort of interpreting known facts, as well as any attempt to predict something new (new phenomena, new properties, new compounds) must start with a deeper look inside the molecular edifice. The molecule is first dissected, by mental or computer experiments in the domain of theoretical chemistry, by other means in the domain of experimental chemistry, and the mutual interplay of the resulting portions may then be analyzed in order to understand the properties of the molecule and its interactions with its surroundings.

When the existence of the molecule was finally accepted, at the beginning of this century, as a palpable reality rather than a mere heuristic model [1], partitioning was done first in terms of groups of atoms, i.e. chemical functions. Subsequent advances in our knowledge and understanding led to the introduction of new models — the bond, the shared couple, the lone pair — and these were eventually linked to an even sounder theoretical foundation, quantum mechanics [2]. Since that period many years ago, no further resolutive changes have occurred in the basic principles of the molecular sciences, and the impressive, still on-going, progress in our understanding and in our exploitation of the molecular world is based on continual refinements of a rationale which maintains a flavor of the old concepts and methods, first expressed at the beginning of this century.

No more "revolutions", therefore, have been happened in the recent past and are to be expected, probably because they are not necessary in this range of energies. The frontier is instead represented by the complexity and variety of the phenomena under study and by the ingenuity needed to find the appropriate combination of simpler effects able to explain, or to predict, the phenomenon of interest. In such endeavours the methodology must continually be revised, however, at the basic level as well as at more advanced stages of the elaboration. The development of new interpretative models may be regarded as a challenge to old interpretations, while the revision of the methods means a greater precision in the interpretation and an enlarged field of application — extending to ever more complex molecular systems where as yet unidentified or only partially understood effects await the scrutiny of the molecular scientist.

There is a wide selection of molecular subunits available, and an even wider choice of interpretative approaches which may be used to exploit the characteristics of these subunits. In practice every researcher involved in a deep examination of the behaviour of molecules has elaborated his own version of these methodological tools and it should be possible to write the history of theoretical chemistry, following the evolution of the models for intra- and intermolecular interactions as a guide-line.

If we limit ourselves in this complex network to the models most directly related to a quantal description of the material system, it is easy to individuate three main lines of investigation, characterized by their choice of the basic subunit to be studied: the atom, the molecular orbital, or different partitions into fragments of the whole molecular charge distribution.

Each procedure presents certain advantages and disadvantages.

The atoms constitute a very reasonable choice. They may be studied independently, and the energy needed to assemble them in a molecule represent a modest portion of their total energy [3-5]. On the other hand, precisely that modest amount of energy corresponding to the building up of molecular edifices is the source of the effects under examination. Therefore one is obliged to introduce intermediate steps in the build-up process, such as for instance the "valence prepared atomic states", which have no correspondence with experiment and which are not easy to define accurately on a theoretical basis. If we include the valence bond theory [6, 7] in this line of approach to the molecular problem, however — and there are several reasons for doing so — the atomic approach reveals itself to be very fruitful, indeed. In the past a good deal of the rationalization of chemical properties has been done using the tools provided by this theory. Some books which rely heavily on VB concepts have been fundamental to the process of making the quantum mechanical description of chemical systems acceptable to chemists [8, 9]. Moreover, the VB theory, in a renewed version [10], seems to offer a promising alternative way of performing ab initio calculations on molecular systems.

The molecular orbital and similar one-electron subunits derive from a mathematical approximation used in the calculation of the wave function [11]. Their direct relationship with experimental quantities is modest, however, and valid only as a first approximation. Because of their definition, canonical MOs do not fit the underlying basic model which, as mentioned earlier, considers subunits confined in specific regions of the space as the ultimate components of the analysis. The researcher is thus obliged to introduce additional manipulations of the canonical MOs, or to shift his attention to the MOs of fragments of the molecule under examination. In the end he finds himself resorting to material models in which part of the original interactions are missing. In addition, the MOs do not account for electron correlation, and similar quantities (the natural orbitals [12], for example) are less transparent and less effective for the analysis of intramolecular interactions (see, however, Weinhold et al. [13]). Although thus far we have only listed the principal disadvantages of the approach based on MO-like one-electron subunits, the fundamental contribution made by MO theories to our understanding of the properties of molecular structures must be emphasized. Indeed, as every chemist is well aware, without MOs our knowledge of the molecular world would be lamentably poor [14-17].

The partition of the charge distribution into separate contributions derives from yet another approximation introduced in the quantum molecular approach. The charge distribution or, equivalently, the diagonal elements of the first order density matrix [18], carries less information than that given by the molecular wave function. This reduction in information makes impossible the direct use of some of the quantum mechanical methods amply exploited by the two approaches mentioned before, such as various applications of the perturbation theory. At the same time, it makes more transparent the quantum description of the system.

The adoption of the charge distribution as a starting point is not sufficient, however, to define the method. One also has to specify the criteria adopted in its partition. One criterion used in the past which still has its adherents is essentially based on information theory (the Daudel's loge concept [19–22]), but other approaches based on the use of classical concepts, especially electrostatic considerations, today seem more promising. The charge density approach has, in fact, the merit of constituting a natural bridge between quantum-mechanical and classical concepts and models. The semiclassical partition of the density function may be carried out in different ways, either by using different electrostatic quantities (such as the forces between the particles, or the reduction of charges to atomic components) or by utilizing the concept of chemical bond etc., as will be discussed below. The most elaborate and detailed methods belonging to this category require a considerable computational effort which reduces at present its field of application in the extensive sense, i.e. in the size of the molecular systems which may be analyzed in such a way, as well as in the intensive sense, i.e. in the possibility of using this description also for the examination of subtler properties of the system. We shall come back to these problems later on, introducing distinctions between those properties which are related to the molecular observables and those related to interactions with other physical systems. What is important to remark here is that the amount of computer time or the storage capacity of the computer required to apply the method should not be taken as the ultimate parameter for judging a method; the evolution of the hard and soft components of computers may overcome these difficulties if the method is worthy of the effort.

From this general and rather schematic overview of the main lines of approach it should be clear that "admixtures" of the different techniques are possible. The methods presently in use correspond in fact to "admixtures" in this very sense. From the partition of the density function it is possible to recover, indirectly, the atoms of which the molecule is composed. The MOs may then be transformed into localized orbitals and suitably collected, to give alternative descriptions of those portions of the molecular density, which are related to specific functional groups. The electron distribution of the atoms in their valence states may be decomposed into hybrids, which can be recomposed to give localized descriptions of components of the charge distribution, etc.

We do not intend to develop any further this particular analysis, since it would quickly become cumbersome and confusing, given the broad basis which has served as our point of departure. Suffice it to have stressed that the notable variety of approaches does not diverge, but rather converges into a few main lines, with frequent, and often fruitful, "admixtures".

It is our intention to examine in more detail the title subject, namely the semiclassical approach to the interpretation of intra-molecular interactions, paying special attention to the approach which we have elaborated during the course of our investigations. Many analogous procedures have been developed by others and we will discuss some of them at appropriate points in the text. We consider ourselves unable to write an account which would be both complete and readable, however, of all the available material pertinent to this field, and there surely exist other methods bearing interesting analogies with those considered here which have escaped our attention. As said before, the field under discussion here constitutes the very center of theoretical elaborations in chemistry, and the pertinent literature is so vast, having spread over the years into many different specialized journals, that useful connections among different interpretative models independently elaborated may easily be lost. The reader is thus warned that the following discussion has a partisan character, and that a satisfactory appreciation of the state of the art in this topical area will be reached only by comparing and integrating different partisan expositions, some of which are collected in the present set of volumes.

2 Why Semiclassical Approximations?

To answer the question posed in the title of this section, let us start with some general methodological considerations.

Models in chemistry, as well as in the other scientific domains, may be profitably divided into three categories: they can be iconic, analogic or symbolic [23].[1] The models considered in the introduction belong to the second category, where emphasis is placed on the functional aspects of the model and close connections with the form and the functional aspects of the chemical system being modeled are maintained. In these analogic models it is convenient to introduce a partitioning which brings out different aspects of the problem: the material composition of the model, the physical aspects of the model, and the mathematical definition of the model.

These three components, in short the material, the physical, and the mathematical models, may be defined and elaborated separately, though there are mutual limitations and interferences which we shall analyze in the following paragraphs.

The material model defines the portion of matter explicitly considered in the analogical description. In our case the material models are, prima facie, isolated molecules, and, in a more detailed analysis, the molecular subunits mentioned in the introduction.

The physical model regards the definition of the physical interactions taking place inside the material model, as well as that portion of the physical interactions with the exterior, which are retained in our interpretative model. An interpretative model is always reductive in nature, and consequently the physical model is also affected by this need of reduction.

The mathematical model defines all the aspects of the interpretative model which are subject to mathematical elaboration. Practical reasons prevent the consideration of this aspect of the model separately from the preceding ones. The selection of overly complex material models makes the use of sophisticated mathematical models

---

[1] We notice that a similar analysis of models has been carried out by Trindle [24]. Some of the remarks on the status of models in theoretical chemistry presented in the following discussion have also been briefly but lucidly expressed by Dewar [25].

impossible. Other examples of mutual relationships among the three aspects of the models may easily be found by the reader.

The definition of the molecular subunit is related to the material as well as to the mathematical aspects of the model. In other cases, not explicitly mentioned in the introduction, the physical model also has a role to play. We are here considering models based on a quantum mechanical description which neglects the small non-electrostatic terms of the Hamiltonian. Thus, we shall put aside in our discussion models which incorporate the small terms of the Hamiltonian, as well as those which do not consider Hamiltonians of every sort, since they are intrinsically intuitive in nature.

The description of the interaction among molecular subunits is mainly derived from the physical model and consequently, of course, from the mathematical model. With appropriate definitions of the molecular subunits it is possible to introduce a partition of the whole quantum mechanical interaction, defined above, into separate terms which may be considered or discarded in the physical model. Where the definition of the material subunits permits us to do so, we can introduce a separation into terms of classical and non-classical origin, and yet further partitions within these two terms.

We may now tentatively answer the question formulated in the title. Semiclassical approximations consider molecular subunits, which are defined within an appropriate quantum mechanical framework, and which interact via classical interaction terms.

Classical interactions are more easily interpreted, analyzed and modeled than quantum mechanical ones. Models based on classical interactions, if their validity and formal limits are clearly demonstrated, may be readily extended and applied to large molecular systems.

The "if" inserted in the preceding sentence makes it clear that what we are dealing with now is a hypothesis.

In our opinion, the explicit use of hypothetical statements which may be subjected to empirical scrutiny is an important point in the elaboration of models, whatever their nature or scope may be. A clear initial statement, including the identification of its supposed limits of validity, makes very much easier the analysis of the necessary experiments (of a numerical nature, in our case) performed on the basis of the initial assumption. Things in practice are not so clearly cut as in the models (or dreams) of the methodologists, however. The falsificability concept is seldom applicable in its original technical definition [26, 27], but often it can be used to delineate the limits of an assumption and the degree of confidence which we may place in it.

It is extremely advisable to use models which permit the definition of working hypotheses at different levels, in ascending or descending order of complexity and "realism". Only an examination of the results obtained with bracketing approximations, i.e. with approximations both more and less accurate than those under examination, can give the necessary confidence levels in the interpretative model selected. Otherwise, the researcher is obliged to rely on his own ingenuity to elucidate the key points of the phenomenon under examination, drawing his proof from the methodologically questionable procedure of discovering similarities between his model and the experimentally ascertained facts.

We have demonstrated in other papers [23, 28], with examples taken from a cognate field (i.e., the interactions between two molecules), how the use of a methodological strategy based on a sequence of provisional working hypotheses may lead to a clearer understanding of the validity, and the limits, of a given model. We shall not repeat that analysis here in a revised version for intramolecular interactions but we shall keep these methodological principles in mind in the following discussion. In working out a model every step and every assertion must correspond to a specific hypothesis, which may be rejected or reformulated by adding or deleting some features.

The semiclassical approach, of which we have given a definition above, must thus be considered a working hypothesis. A systematic and exhaustive examination of its limits should disclose those of its aspects which are intrinsically quantum in nature, and thus not amenable to a classical approximation. The final analysis might possibly tell us whether or not the semiclassical approach can be applied to the interpretation of intramolecular interactions. Molecules exist in such a wide variety of structures, however, that a definitive answer appears quite improbable. We must be content if our analyses lead to the definition of sufficiently large classes of compounds and phenomena in which the approach may be used.

To reach this goal it is necessary to have at our disposal methods of sufficient flexibility. A few examples will clarify this point. In the scientific literature there are many statements saying that a specific effect or a specific property cannot be reduced to its electrostatic contribution. In some cases the statement is, or may be, correct, but in other cases the statement is supported only by a poor description of the electrostatic effect, with hidden limitative hypotheses. Thus, for instance, in some cases the hidden hypothesis may be that Mulliken charges are a faithful and complete source of electric fields, in other cases that the mutual polarization of the molecular subunits is negligible, etc. These electrostatic versions of the semiclassical approach are generally not worked out satisfactorily, and the conclusions are often misleading. On the other hand, a satisfactory picture reached at a given level of approximation must be tested for its stability with respect to less restrictive hypotheses, or less restrictive mathematical formulations of the working hypothesis. Very persuasive models, elaborated in the past on the basis of semiempirical or minimal basis set results, have shown inconsistencies when checked against results of better quality (including, for example, the extension of the basis set with polarization or diffuse functions).

3 A Proposed Outline for a Research Program

We may now draw up the general outlines of a research program, based on the considerations put forward in the preceding Section and addressed to the verification of the semiclassical approximation. Many different programs respecting the same methodological indications and directed towards the same goal are conceivable. Here we shall disclose a "rearranged" version of the program which we have been following; the "rearrangement" is due to the fact that in our work semiclassical approximations to intramolecular interactions have been merely a side product of investigations primarily addressed to the study of intermolecular interactions.

The program may be summarized in the following points:

1) Definition of the basic subunits, with characterization and investigation of their intrinsic properties (degree of invariance and transferability of a given subunit to different molecules).

2) Analysis of the description of the molecular properties in terms of these subunits (additive and non-additive effects). The analysis is limited at this stage to an examination of the effects of chemical substitutions in molecular frameworks at a fixed geometry.

3) Interpretation of the effects considered in the preceding point in terms of semi-classical concepts.

4) Search for simplified expressions of the molecular subunits and for a semiclassical description of their mutual interactions.

5) Analysis of the description of molecular properties when coupled with geometry deformation effects. The geometry deformation may be internally or externally driven.

6) Interpretation of the effects considered in point (5) and search for approximate semiclassical descriptions.

7) Analysis of the description of the molecular properties when the molecule undergoes a change in electronic state. Similar to point (5), but now the representative point is displaced in the space of the geometrical configurations to another potential energy surface.

8) Interpretation and search for approximate descriptions of the effects considered in the preceding point.

9) Analysis of the effects produced on the molecule by external fields, especially those related to the presence of a solvent,

10) Interpretation in semiclassical terms of the effects considered in point (9) and search for approximate expressions.

Many other points might be added to this program. For instance, dynamic effects are not explicitly considered, although there are various indications, including results reached by other groups, which suggest that an extension of this kind in the program could be fruitful. We have limited our summary to those points for which we have some results worthy of attention; some of them will be treated in more detail, others more concisely in the discussion which follows, where we attempt to balance the large quantity of material available against the need for a clear and limpid exposition. As we have already said, the vast scope of the subject makes an exhaustive review of the literature almost impossible; therefore commentary and comparisons with similar approaches will be limited to what is strictly necessary.

4 Definition of the Basic Subunits

When this research project was started, around 1960, there was not a large choice of basic subunits derived from molecular wave functions available for study and manipulation. Since one of the first MO SCF ab initio programs for polynuclear molecules was then in the course of completion in our laboratory, we decided to try to exploit the invariance of the first order density matrix with respect to unitary transformations of the canonical MOs. The original program, based on the direct derivation of the **R** matrix according to McWeeny [29], was therefore supplemented with subroutines transforming the canonical MOs into LOs, by maximizing the sum of the squares of the distances between the orbital centroids. We were not satisfied with the procedure suggested at that time by Boys [30] based on the product of these distances, nor by ad hoc procedures derived from the projection of the MOs on pre-selected combinations of atomic hybrids. In the 1960s, the definition of the localization transformation was satisfactorily dealt with by several authors. Edminston and Ruedenberg [31] proposed a localization procedure based on a different criterion, giving at the same time a clear overview of the problem. The approach based on external criteria was satisfactorily dealt with by Magnasco and Perico [32]. In 1966, Boys proposed a revised version [33] of his localization procedure, which was completely equivalent to the one we had been using in the immediately preceding years. Boys' remarkable paper is still of great interest today for some of its not yet fully explored suggestions regarding both quantum chemistry in general and the problem which we are discussing here, namely the description of submolecular units (cf. the interesting remarks made by Dewar on this subject [34]).

Returning to the problem of localized orbitals, our story does not come to a close at the end of the sixties. The spectrum of possible definitions of LOs has been yet further enlarged and there are still contributions being made in this field [35–53]. The LOs have a wide range of applications in quantum chemistry, but here we are mainly interested to the use of LOs for a specific purpose and we refer the reader to some reviews for a more thorough presentation and analysis [54–56].

An extensive literature centering on the description of molecular structures in terms of LOs has grown up in the last twenty years. Even a summary of these descriptions is not possible here. It should be sufficient for our purposes to recall that for "normal" compounds the LO description corresponds to chemical intuition. Even in "difficult" cases, e.g. for molecules which cannot be described in terms of a single canonical structure, the LOs give a description which contains the features of the bonding properties established by other means. It will also be sufficient to recall that several concepts, such as the bent bond, the three-center bond, etc., have been accepted in the vocabulary of chemistry via analyses done in terms of localized orbitals. The assertion that the LO description of molecular bonding nowadays represents *the* classical description could be defended without excessive efforts at argumentation.

We have up to now being discussing the LOs in fairly broad terms, without intro-ducing distinctions among the different localization procedures, because these orbitals are not very sensitive to the localization method adopted. The shape and overall characteristics of the LOs are also, in general, almost independent of the basis set. These statements derive from the experience which we have gained over the years, looking at thousands of different LOs obtained by different authors using different methods and basis sets. There are, of course, points of disagreement, and in their papers various authors may stress the differences and merits of their own localization methods. This is understandable, but general statements should help to smooth out the small differences which may emerge in specific cases or for the defense of a personal approach.

The LO description is not completely amenable to a classical approach, however, for two reasons: the presence of delocalized tails and the not negligible spatial overlap of LOs belonging to the same set of atoms.

In the following analysis of the characteristics of partial invariance and transferability in LOs (the chemical groups have these characteristics) we must consider with particular attention the two aforementioned points.

4.1 Transferability of the LOs: The Tails

It is easy to partition the LOs obtained with atom-centered basis sets into a main component and a residuum, called the tail. If we take two simple cases of very common occurrence, namely LOs centered on a unique atom (a lone pair or an inner shell orbital) and LOs centered on two atoms (a single bond or a component of a multiple bond), formally we may write:

$$\lambda_A = c_A \bar{\lambda}_A + c_T \bar{\lambda}_T \tag{1}$$
$$\lambda_{AB} = c_{AB} \bar{\lambda}_{AB} + c_T \bar{\lambda}_T . \tag{2}$$

$\bar{\lambda}_i$ can be operationally defined as the projection of λ_i in the functional subspace $\chi_A \oplus \chi_B$ which is spanned by the functions belonging to atoms A and B: this definition may be easily extended to cases in which out-of-center basis functions are employed. In practice, this decomposition can be performed with elementary manipulations of the corresponding rows of the coefficient matrix C describing the LOs in the molecular basis set χ_M. The tails ensure orthogonality between the LOs, but they cannot be reduced to a mere consequence of the orthogonality constraints. This point, which will be numerically demonstrated in the following section, has been generally accepted either on the basis of similar numerical proofs (see e.g. Newton et al. [57]) or using other arguments (e.g. [58–60]).

4.1.1 The Extent of the Tails

The tails are a disturbing feature, and their role and their weight in the description of the binding must be carefully examined before either LOs (λ) or LOs without tails ($\bar{\lambda}$) can be accepted as a starting point to set up the molecular subunit sought.

The relative weight of the tails is easily derived from the coefficients c_A (or c_{AB}) and c_T in Eqs. (1) and (2), or from similar numerical indexes. We have extensively adopted the examination of the mean square deviation between λ_i and its main component $\bar{\lambda}_i$:

$$\int (\lambda_i - \bar{\lambda}_i)^2 \, d\tau = 2(1 - \langle \lambda | \bar{\lambda} \rangle) . \tag{3}$$

This index rarely reaches 0.1 and for "normal" compounds is between 0.01 and 0.03. Some statistics drawn from our files, regarding 1,530 LOs obtained by employing different basis sets in the various columns (from single zeta and double zeta quality Slater type orbitals to Gaussian basis sets of different complexities) are shown in

Table 1. Magnitude of the "tails" in molecular localized orbitals, as measured by the mean square deviation between the LO λ_i and its main component $\bar{\lambda}_i$

	A	B	C	D	E	F	G	H	I
$\langle MSD \rangle$	0.038	0.027	0.021	0.022	0.020	0.021	0.025	0.028	0.025
St. Dev.	0.050	0.012	0.011	0.012	0.010	0.011	0.012	0.044	0.010
St. Er.	0.006	0.001	0.001	0.001	0.000	0.001	0.002	0.006	0.008
Cases	109	101	456	222	208	148	108	98	80

A: Compounds containing cumulated double bonds
B: Three membered unsubstituted ring molecules
C: Compounds containing the $RR'C=O$ group (aldehydes, ketones, esters, acids)
D: Compounds containing the $R-O-X$ group (ethers, alcohols, etc.)
E: Compounds containing the $RR'R''N$ group (amines, imines, etc.)
F: Compounds containing the $RR'C=O$ and $RR'R''N$ groups (amides, etc.)
G: Compounds containing the $RN=NR'$ group
H: Heterocyclic compounds (5 and 6 membered rings, fused rings)
I: Complex metal beryllo- and borohydrides

Table 1. This statistical examination can be easily supplemented by values drawn from other sources, one of the most recent additions being by Murgich et al. [61].

These statistics include "normal" as well as "difficult" cases, and also take into account transition state geometries, but they are limited to compounds composed of atoms from the first two rows (with a few exceptions, such as K, Ca and Br) and they exclude the inner shell orbitals which in general have small tails. Our sampling of conjugated compounds is limited to aromatic and to heteroaromatic compounds formed at the most by two condensed rings. A detailed analysis of the LOs in compounds containing several condensed aromatic rings was carried out several years ago by England et al. [62]. The fact is that the sample covered by our statistics is lamentably poor, in view of the extreme variety of structures present in chemical compounds. In particular, the field of inorganic compounds is practically unexplored territory, with the exception of some complex metal berillo- and borohydrides, included in our statistics, and the case of the complex borohydrides, which has been thoroughly studied by Lipscomb and coworkers [63].

A more detailed analysis of the homogeneous classes of compounds bears out the feasibility of using the LOs as models of "true" localized substructures: the reader is referred to the examples provided in a less recent overview regarding a few classes of "difficult" cases [64].

4.1.2 The Role of the Tails

An analysis of the role of the tails is more complex than the mere evaluation of their weights. In fact, the contribution of the tails is different in principle for each observable. We shall consider in the following discussion some first order one-electron observables and the energy.

The analysis of the role of the tails may be done in two ways. 1) By examining the transferability properties of the contributions of specific λ_is to selected observables and by comparing these values with those obtained from the main portions $\bar{\lambda}_i$. If

$\hat{Q}$ is an observable, Q_i and $\bar{Q}_i$ will be the values corresponding to the functions λ_i and $\bar{\lambda}_i$, respectively. 2) By comparing the expectation values $\langle Q \rangle$ obtained from the original wave function with the values arising from the assembly of $\bar{\lambda}_i$s, $\langle \bar{Q} \rangle$. A final step would involve comparison with a model of localized subunits without tails, $\bar{\lambda}_i^0$ (or related quantities), and with the corresponding expectation values $\langle \bar{Q}^0 \rangle$. The two approaches are complementary.

The tails are strictly specific to the molecular framework corresponding to the molecular wavefunction from which λ_i has been extracted, while the main portion, $\bar{\lambda}_i$, should be less dependent on the molecular environment. There will be of course changes in the actual $\bar{\lambda}_i$s with respect to the ideal prototype $\bar{\lambda}_i^0$, due to interactions with the molecular remainder, but these should be second order effects if chemical experience is indeed applicable to models constructed on the basis of LOs.

We shall select as our one-electron observables for this analysis the kinetic energy T, the dipole moment μ, and the molecular electrostatic potential V, because they sample different characteristics of the LOs and their main components.

4.1.2.1 The Kinetic Energy

The orbital kinetic energies T_i are quantities indirectly related to the shape of the localized orbital. The actual values depend on the nature of the atoms involved in the definition of i, on the nature of λ_i (inner shell, lone pair, bond) and of course on the basis set employed in the calculations. The tails should have, in principle, a noticeable influence on the T_i values, because they increase the volume available for the electron couple. We must confess that some of the files containing our most recent work on T_i and $\bar{T}_i$ values have been lost, so we are obliged to rely on older results, partially published in preceding reports. Table 2, derived from Ref. 61 with some additions, reports the mean values and standard deviations for 155 T_i values grouped in broad classes. The specimens represented here are cumulenes, molecules with a cyclic strain, and eventually compounds not amenable to a canonical structure; "normal" compounds have been discarded. The calculations refer to Slater-type

Table 2. Mean value and standard deviation of the kinetic energy T_i for some classes of localized orbitals λ_i (values in hartrees: 1 hartree = 2626.1 kJ/mol)

λ_i	$\langle T_i \rangle$	St. Dev.	Cases
N lone pairs	1.756	0.079	34
O lone pairs	2.289	0.125	16
CH bonds	0.910	0.016	40
NH bonds	1.265	0.056	26
CO bonds[a]	1.626	0.023	8
CN bonds[a]	1.361	0.028	31
NN bonds[a]	1.625	0.032	8
CC bonds[a]	1.128	0.022	20

[a] Under this heading we have collected single and multiple bonds

orbitals (best atom zetas). Values obtained using GTOs of single and double zeta quality (data no longer available) were of similar quality.

In this sampling the $\bar{T}_i$ values are lower than the corresponding T_i ones by 0.2 hartrees on the average, with the standard deviations within each class being decidedly lower than those reported for the whole set. The correlation between the T_i and $\bar{T}_i$ values for the specimens in Table 2 is r = 0.94. A more detailed examination shows that the largest deviations in the T_i and $\bar{T}_i$ values are to be found in the nitrogen lone pairs of the linear molecules (sp hybridization). In these LOs the $\bar{T}_i$ remains remarkably constant, but the tails differ in extent, with "volumes" which depend on the chemical nature of the remainder.

The correlation between the T_i and $\bar{T}_i$ values in the "normal" molecules has a higher r value (0.97–0.98).

4.1.2.2 Electric Multipole Moments

For the comparison of dipole moments it is convenient to define electrically neutral subunits by performing a suitable partition of the nuclear charges. Every partition of nuclear charges is perforce arbitrary, but for neutral closed shell molecules the simplest and most natural partition consists in assigning a charge of $+2e$ to an inner shell orbital, the same charge to a lone pair, and $+1e$ charges to nuclei A and B for the $\bar{\lambda}_{AB}$ orbitals. This simple recipe may be recast into a more general definition [65], where the $n_c(A)$ inner shell orbitals of atom A each receive a $+2e$ charge, and the remaining $Z_A - 2n_c(A)$ charges are then distributed over the n_b bonds and the n_l lone pairs, each of the former receiving a nuclear charge equal to:

$$(Z_A - 2n_c(A))/(n_b(A) + 2n_l(A)) \tag{4}$$

and each of the latter twice this charge. This definition resolves some of the ambiguities which may be met with in compounds with cumulated bonds. Our experience with charged molecules has in practice been limited to MH^+ systems, where M is a neutral closed shell molecule, and to anions M^-, these also being closed shells. We have found it useful to confine the extra nuclear charge or the missing charge to the inner shell subunit of the relevant atom A. In such cases the inner shell group acts essentially as a monopole.

For brevity's sake, we shall consider here only the modulus of μ_i and of $\bar{\mu}_i$. In doing so we lose interesting information, for example the direction of the 1_A group moment and the magnitude of the component perpendicular to the internuclear axis in banana or bent bond orbitals. These features are important in subsequent steps of the modelling procedure, but they are too complex to be examined here. It must suffice to say that such detailed analyses do in fact confirm the general trend exposed here.

We report in Table 3 the mean values and standard deviations of μ_i and $\bar{\mu}_i$ for the set of examples already considered in Table 2. The dipole moment is particularly sensitive to tails located far away from the main component and the comparison between $\langle \mu \rangle$ and the value obtained from the antisymmetric product of the symmetrically orthogonalized $\bar{\lambda}_i$s may be taken as a test of the extent to which the tails are merely the consequences of orthogonality constraints. Such tests have already been

Table 3. Mean value and standard deviation of group dipole moments μ_i for some classes of neutral subunits i (values in Debyes: 1 Debye = 0.3937 a. u. = 3.3356×10^{-30} Cm)

i	$\langle \mu_i \rangle$	St. Dev.	Cases
N lone pairs	3.23	0.20	34
O lone pairs	2.81	0.16	16
CH bonds	1.75	0.13	40
NH bonds	9.86	0.19	26
CO bonds[a]	1.81	0.13	8
CN bonds[a]	1.79	0.20	31
NN bonds[a]	0.50	0.17	8
CC bonds[a]	0.32	0.15	20

[a] Under this heading we have collected single and multiple bonds

employed several times (see, e.g. Newton et al. [57], Bonaccorsi et al. [64]) and it has been shown that tails cannot be just adscribed to the use of orthogonality constraints in the definition of the LOs.

As reported in Table 3, the standard deviations of the $\bar{\mu}_i$ values for a given i are smaller by almost an order of magnitude than those of the μ_i values. It is worth noting that in Tables 2 and 3 we used very broad categories to separate the groups; for example, no distinction was made between the sp, sp^2, and sp^3 hybridized lone pairs.

The regression coefficient for the correlation between the μ_i and $\bar{\mu}_i$ values of the cases considered in Table 3 is r = 0.96. Similar correlations for "normal" molecules, performed with basis sets varying from STO-3G to 6-31G** produce r values on the order of 0.97–0.99. In addition, the standard deviations for the $\bar{\mu}_i$ values of "normal" molecules are decidedly lower (by a factor 1/2) than those reported in Table 3.

Our analyses of quadrupole and octupole electric moments have been conducted in much less detail; therefore the reader is referred to previous, albeit somewhat limited, analyses performed over the separate components of such tensors [66]. For single tensor components the deviations, with respect to the mean value computed over a sufficiently representative set of examples, are less than 5%. The quadrupole components are employed to define the "size" and "shape" of localized orbitals [67–69]: interesting results in this specific context have been obtained for the characterization of LOs in excited states [70–72].

4.1.2.3 The Molecular Electrostatic Potential

We have devoted much attention in the past to the analysis of the molecular electrostatic potential V(M, r) and to its dissection into a series of LO group contributions. The conclusions reached at the end of our review papers [63] and [69] have been amply confirmed by more recent results. V(M, r) is an observable which depends on the location r at which the electrostatic potential is sampled; the change in the absolute value and sign of V(M) as we pass from one region to another in the outer molecular space gives, as is well known, valuable information on the reactivity of the different

Table 4. Localized orbital contributions to the electrostatic potential (kcal/mol) in five selected points around methylformate (see Scheme 1) computed using the 4-31G basis set (1 kcal/mol = 4.185 kJ/mol)

λ	1	2	3	4	5
$lp(O_2)$	10.1	−45.7	−3.4	11.0	25.6
$lp(O_2)$	2.2	−44.4	17.1	−11.5	8.6
$\beta(CO)\ y > 0$	22.0	− 0.2	23.4	− 0.4	−18.9
$\beta(CO)\ y < 0$	22.0	− 0.2	23.4	− 0.4	61.0
$b(C_1H)$	− 2.3	12.3	5.2	6.1	18.5
$b(C_1O_1)$	− 0.6	8.6	17.4	− 7.3	−52.9
$lp(O_1)$	−61.8	6.3	−3.8	8.2	−67.2
$lp(O_1)$	−62.4	6.3	−3.8	8.2	69.7
$b(O_1C_2)$	− 5.0	− 6.5	−8.3	36.6	−26.3
CH_3	22.5	2.0	6.9	−26.6	21.9
Total[a]	−53.2	−61.2	73.9	23.9	39.8

[a] Including the contributions from the $1s$ orbitals

regions of M. In addition, the V_i values are long-range in character, and consequently the value of the observable at a given **r** derives from the composition of many terms, of different sign and magnitude. The breakdown given in Table 4 will help to clarify this point. The values refer to some selected points around the methylformate molecule (see Scheme 1).

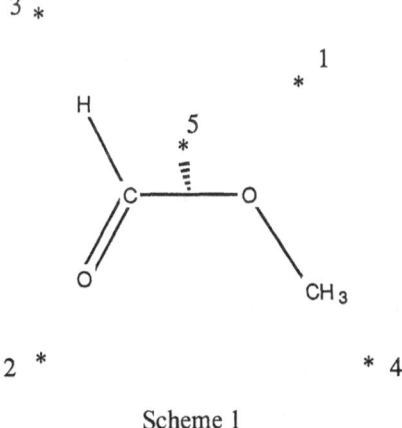

Scheme 1

The submolecular units defining the observable V(M, r) are the same neutral subunits which we used to describe the dipole moment. A detailed analysis of how the final value of V(M, r) is built up is not necessary here. It should suffice to note that various well known characteristics of V(M), such as the negative regions near the ethereal and carbonylic oxygens (points 1 and 2, respectively) or the positive regions near the H atoms (points 3 and 4) derive from a complex series of contributions of opposite sign.

It is thus evident that the conservation and transferability of the V_i and $\bar{V}_i$ components should be analyzed on the basis of a large number of points, located near the main component i as well as far away from it. We may dispense with this rather cumbersome task by remarking that at large distances the behavior of V_i and $\bar{V}_i$ can be expressed in terms of a multipole expansion:

$$V_i(r) = \frac{\mu_i r}{r^3} + \frac{r Q_i r}{r^5} + \dots . \tag{5}$$

The analysis of $V_i(r)$ and $\bar{V}_i(r)$ is thus related to the analysis of the multipolar expansion terms presented in the preceding section. One may conclude that the differences in electrostatic potential seen in for group i, generated by the presence or absence of tails in the charge distribution, are quite modest at large distances. At short distances, where the multipolar expansion is no longer of use, one must resort once again to numerical computations. A number of analyses performed in our laboratory are summarized in reviews on the electrostatic potential [66, 73] and in two papers dedicated to the same subject [74, 75].

A decomposition of $\langle \bar{V}(M) \rangle$ is reported in Table 5: it parallels the decomposition of $\langle V(M) \rangle$ given in Table 4. We purposely selected points very close to the molecular subunits (a distance of 1.2 Å from the nearest atom). This choice of distances emphasizes, of course, the difference between the V_i and $\bar{V}_i$ values. However, the more distant groups are fairly well represented by the description without tails; compare, for example, the b(C−H) contributions to points 2, 4, 5 in the two tables, or the β(C=O) contributions to points 1, 3, 4.

4.1.2.4 Total Energy

The SCF expression for the electronic energy can easily be decomposed into its LO contributions. For closed shell molecules one has:

$$E = \sum_i E_i = \sum_i \left[2 \left(T_i + \sum_\alpha^{nuc} Z_\alpha V_i^\alpha \right) + \sum_j (2J_{ij} - K_{ij}) \right], \tag{6}$$

Here the indexes i and j refer to the LOs and the usual shortened expressions for one- and two-electron integrals have been employed.

Each E_i term may be divided into two parts, one related to the intrinsic energy of the i group, and one related to the interaction of this electron couple with the other components of the molecule:

$$E = E_i^0 + U_i \tag{7}$$

where

$$E_i^0 = 2(T_i + \sum Z_\delta V_i^\delta) + J_{ii} \tag{8}$$

and

$$U_i = 2 \sum_{\alpha \neq \delta} Z_\alpha V_i^\alpha + \sum_i \sum_{j \neq i} (2J_{ij} - K_{ij}). \tag{9}$$

Table 5. Localized orbital without tails contributions to the electrostatic potential (kcal/mol) in five selected points around methylformate (see Scheme 1) computed using the 4-31G basis set

λ	1	2	3	4	5
$lp(O_2)$	14.9	−58.0	−2.7	12.9	32.9
$lp(O_2)$	4.5	−58.2	25.8	−12.7	9.8
$\beta(CO)$ y > 0	21.4	3.1	22.5	− 0.4	−27.7
$\beta(CO)$ y < 0	21.4	3.1	22.5	− 0.4	61.0
$b(C_1H)$	− 7.5	12.7	21.5	5.9	11.4
$b(C_1O_1)$	7.0	6.4	19.3	− 7.1	−56.1
$lp(O_1)$	−77.3	10.7	−1.4	7.6	−72.4
$lp(O_1)$	−77.9	10.7	−1.4	7.6	83.3
$b(O_1C_2)$	0.4	− 8.3	−5.2	38.6	−28.9
CH_3	16.7	1.1	5.6	−29.4	15.4
Total[a]	−76.3	−76.2	106.5	22.6	28.4

[a] Including the contributions from the $1s$ orbitals

The δ nuclei in Eq. (8) belong to the main portion of λ_i.

While E_i, for a fixed i, may show variations within a set of different molecules, E_i^0 remains fairly constant. We report in Table 6 the results of the above manipulation for the same set of wave functions already employed in Tables 2 and 3. In this table we give the mean values of E_i, their standard deviations, and the regression coefficient for the correlation between E_i and U_i:

$$E_i = \langle E_i^0 \rangle_{mean} + U_i . \tag{10}$$

Note that once again the entries have been assembled in groups which differ in chemical nature; a division into more homogeneous subgroups, with distinctions being made, for example, between the C–C, C=C and C≡C groups, would have merely emphasized the degree of conservation of E_i^0 with respect to the different molecular fields, measured here by U_i which lies in the range of −4 to +6 hartrees. We report in Table 7, as an example of a more homogeneous specimen, the case of

Table 6. Mean value of E_i^0, standard deviation and regression coefficient for the relation $E_i = \langle E_i^0 \rangle + U_i$ (values in hartrees: 1 hartree = 2626.1 kJ/mol)

λ_i	$\langle E_i^0 \rangle$	St. Dev.	Regr. coeff.	Cases
N lone pairs	− 9.01	0.225	0.980	34
O lone pairs	−11.93	0.370	0.827	16
CH bonds	− 7.19	0.097	0.994	40
NH bonds	− 9.46	0.163	0.982	26
CO bonds[a]	−16.92	0.420	0.948	8
CN bonds[a]	−14.04	0.285	0.978	31
NN bonds[a]	−15.28	0.310	0.969	8
CC bonds[a]	−12.18	0.372	0.998	20

[a] Under this heading we have collected single and multiple bonds

Table 7. Mean value of E_i^0, standard deviation and regression coefficient for the relation $E_i = \langle E_i^0 \rangle_{mean} + U_i$. The values refer to a set of alkanes (linear, branched, cyclic; values in hartrees: 1 hartree = 2626.1 kJ/mol)

λ_i	$\langle E_i^0 \rangle$	St. Dev.	Regr. coeff.	Cases
CH bonds	-7.17	0.062	0.998	230
CC bonds	-11.92	0.090	0.999	118

E_{CC} for saturated hydrocarbons of various types (linear, branched, cyclic) computed with the 3-21G basis set.

The changes in E_i^0, though remarkably small, are another index of the incomplete transferability of these intrinsic energies of the localized orbitals. The transferability is limited in part by the tails and in part by the modification of the main portion of the LO induced by intergroup interactions. The E_i^0 values, computed using an equation similar to Eq. (8) but based instead on the $\bar{\lambda}_i$ orbitals, seem to indicate that the tails are only partially responsible for the spread. In fact, the regression coefficient for the correlation between the E_i^0 and $\bar{E}_i^0$ values is 0.995 (mean value for all the categories in Table 6) while the standard deviation is reduced by a factor of less than 1/2, so apparently there is another factor influencing the transferability.

Additional information on the energy of the LOs may be obtained by looking at the decomposition of the total molecular energy:

$$W = E^{el} + E^{nuc} . \tag{11}$$

If we adopt the definition of neutral fragments that makes use of the partitioning of nuclear charges given in Eq. (4), we can write:

$$W = \sum_i W_i = \sum_i (E_i + N_i) \tag{12}$$

where E_i is given by Eq. (6) and N_i is defined by:

$$E^{nuc} = \sum_i N_i = \sum_i \left(N_{ii} + 2 \sum_{j \neq i} N_{ij} \right) \tag{13}$$

N_{ii} is present only when i is a bond, and it corresponds to the repulsion between the two unitary charges placed on the bonded atoms. N_{ij} corresponds to 1/4 of the repulsion energy between the positive charges of i and j. The nuclear attraction operators are partitioned in a similar way

$$\sum_\alpha^{nuc} Z_\alpha / r_\alpha = \sum_i 2 V_i^i . \tag{14}$$

The definition of the V^i operators then becomes straightforward. W (Eq. 12) may thus be rewritten:

$$W = \sum_i W_i = \sum_i (W_i^0 + Y_i) \tag{15}$$

Table 8. Mean value of W_i^0, standard deviation and regression coefficient for the relation $W_i = \langle W_i^0 \rangle + Y_i$ (values in hartrees: 1 hartree = 2626.1 kJ/mol)

λ_i	$\langle W_i^0 \rangle$	St. Dev.	Regr. coeff.	Cases
N lone pairs	0.52	0.042	0.982	34
O lone pairs	1.18	0.015	0.998	16
CH bonds	−0.14	0.040	0.995	40
NH bonds	0.424	0.120	0.999	26
CO bonds[a]	1.53	0.060	0.998	8
CN bonds[a]	1.11	0.068	0.997	31
NN bonds[a]	1.37	0.053	0.995	8
CC bonds[a]	0.769	0.034	0.992	20

[a] Under this heading we have collected single and multiple bonds

with

$$W_i^0 = 2(T_i + V_i^i) + J_{ii} + N_{ii} \tag{16}$$

and

$$Y_i = \sum_{j \neq i} Y_{ij} = \sum_{j \neq i} 2\left(V_j^i + V_i^j + J_{ij} - \frac{1}{2} K_{ij} + N_{ij}\right). \tag{17}$$

Comparing Tables 8 and 6, one can see that the "intrinsic" term W_i^0 has a higher degree of conservation than the E_i^0 terms. The mean values are, of course, quite different. The Y_i values lie, for all the cases shown in Table 8, within a range of -2 to -9 hartrees. Table 9 reports analogous results for the molecules already considered in Table 7.

The $\overline{W}_i^0$ terms correlate well with their corresponding W_i^0 values: the regression coefficient is greater than 0.997 (mean values for all the headings of Table 8). The higher degree of conservation seen in the W_i^0 terms compared to the E_i^0 terms is due to the different partitionings employed in the two definitions. E_i^0 contains, in fact, all the attractive energies between the two electrons in i plus the nuclear charges on the pertinent atoms, while W_i^0 contains only a portion of these attractive energies, counterbalanced by repulsive energies deriving from the same amount of nuclear charges. In conclusion, the partition given by Eqs. 12–15, suggests a higher transferability degree of the molecular subunits, than the partition given by Eqs. 6–9. This

Table 9. Mean value of W_i^0, standard deviation and regression coefficient for the relation $W_i = \langle \overline{W}_i^0 \rangle_{mean} + U_i$. The values refer to the same set of alkanes as that considered in Table 7 (values in hartrees: 1 hartree = 2626.1 kJ/mol)

λ_i	$\langle E_i^0 \rangle$	St. Dev.	Regr. coeff.	Cases
CH bonds	−0.12	0.026	0.997	230
CC bonds	0.75	0.028	0.998	118

shows how hasty conclusions drawn from the results of a single analytical tool, no matter apparently reasonable, may be actually misleading.

An analysis of the Y_i terms (Eq. 17) reveals other interesting features. All the terms in Y_{ij} except one, K_{ij}, are classical in origin. Formally, there is a partial compensation among the terms of opposite sign: $J_{ij} + N_{ij}$ and $V_j^i + V_i^j$. Conversely, K_{ij} has no counterparts of opposite sign, but it is short-range in character. Although the aforementioned compensations are only partial, it is fairly easy to divide the Y_{ij} terms into classes: one class in which the dominant term is the non-classical K_{ij} (i.e., a portion of those cases in which i and j have one atom in common), and other classes which are dominated by specific combinations of the classical terms. We cannot go into any greater detail here, but such further analyses can shed much light on the subgroup description of the total energy.

In the hope of achieving a more precise description of the properties and characteristics of molecular subunits using the different approaches mentioned in the introduction, many researchers have been investigating the conservation and stability properties of some of the energetic contributions. In the early stages of computational quantum chemistry these properties were exploited by including in the calculations experimental data on the fragment energy; only the mutual interaction energies were computed explicitly, as in the atoms-in-molecules method of Moffit [3] (see also Ref. 5) and in the molecules-in-molecules method of Longuet-Higgins and Murrell [76].

The theoretical bases for a partitioning of the ab initio energy were first put forward by McWeeny several years ago [29], and they were further extended and developed later on. Here we shall consider this approach in one particular context, linked to our investigations, for so many studies have been carried out on the energy contribution in molecular subunits that a mere enumeration of the methods and papers in this area would be exceedingly long. We shall limit ourselves to a concise summary, directing the interested reader to the literature which we consider most significant in this field.

A revised version of the above mentioned molecules-in-molecules method, based on a localization of the MOs coupled with projections over appropriate functional subspaces, has been proposed by von Niessen [77]. Practical computational methods have been developed by Klessinger with his fragments-in-molecules method [78–81]; the last paper is a very readable review on the general topic which we are considering here. A similar computational scheme was devised by our group several years ago [82].

A computational method based on the conservation of the elements of the Fock matrix, when expressed in terms of LOs, has been set up by Leroy and Peeters [83] and exploited in a number of papers [84–87]. The transferability of the Fock matrix elements is the distinguishing feature of the SAMO program [88], which was first proposed by Eilers and Withman [89], and later expanded by O'Leary and Duke. Ref. 55 is a review of the subject up to 1975; it discusses SAMO and other similar approaches not mentioned in these brief comments. Further extensions and applications of this approach are given in Refs. 90–96. The last reference considers the formal aspects of the transfer of the Fock matrix elements (on this question see also Refs. 97, 98). The Deplus et al. approach [99] contains some points of similarity with the SAMO method. The transferability of ab initio atomic potentials in molecular calculations has been exploited by Durand et al. [100–102].

Several methods have been proposed to exploit the conservation of some of the energy terms in conformational problems (internal rotations, inversion at an atomic center, etc.). We shall come back to this aspect later on in our exposition, merely pausing here to note the SIBFA procedure developed by the Pullman group [103–104]. By coupling the energies of the fragments to a partition of the interaction energy they arrived at an efficient method for computing molecular conformation energies.

The Hungarian school has also done much work in this field. A clear overview is provided in Ref. 105 and more recent contributions may be found in Refs. 106–107.

This short survey is far from complete however; Ravimohan and Gopinatham [108] single out 15 different methods of partitioning the molecular energy, none of which have been reported here, and the actual list is still longer. Nor have we attempted to distinguish between papers describing methods which have found real practical applications, and papers of a critical nature, or those describing methods which have not proved useful in their original version. We think that there is something to be learned from all of the above-quoted material (as well as from those papers which have not been mentioned for lack of space), even when the research does not conclude with a workable proposal.

One problem to which much space has been dedicated concerns the orthogonality between constituent groups. A complete separation between groups inside a molecule is not possible, of course, but there are methods which make this problem less disturbing. To close this section we would mention the paper by Palke [109] which, in a different approach to the orthogonality problem, computes the Coulomb and exchange components of the Fock matrix, without invoking wave functions that would violate the Pauli principle: to do so he exploits the non-diagonal Lagrangian multipliers.

4.2 Spatial Overlap of LOs

We remarked at the beginning of this chapter that two features of the LOs do not fit the "naive" description of chemical groups: the presence of tails, and their spatial overlap. Thus far we have analyzed in detail the question of the tails; we will now dedicate a few comments to the second point.

The LOs in a given molecule are orthogonal, but there is a spatial overlap which is not negligible, especially when the couple $\lambda_i - \lambda_j$ has an atom in common. This feature has sometimes been invoked against the eligibility of LOs to describe chemical substructures (see e.g. Ref. 110). Canonical MOs also have non-negligible spatial overlaps, and this is a feature shared by all the one-electron quantities directly derived from the molecular wave function (e.g. natural orbitals or VB structures). In our opinion the presence of spatial overlaps does not ipso facto preclude a "naive" description of the groups. However, the subject deserves further comment for another reason.

There exists an alternative way of introducing a partition of the first order density function, still relying on semiclassical electrostatic assumptions.

Pioneered by Berlin [111] and later refined by other researchers (especially the Bader's group [112]) this approach utilizes the direction of the electrostatic force, acting on different volume elements of the electronic charge distribution, to partition the real space occupied by the molecule into exclusive domains with sharply defined boundaries.

Later, Bader and his coworkers thoroughly reformulated the approach, adding new ideas and new features drawn from quantum mechanics, classical concepts and topology. It is not our intention to analyze Bader's model and the different phases of its evolution here, for they have been well documented in a set of review papers [113–117]. It is sufficient for our discussion to state that this approach is based on a partitioning of the molecular space into mutually exclusive regions with sharp boundaries, that are defined as the surface of zero flux in the gradient vector of the charge density:

$$\nabla \varrho(\mathbf{r}) \, n(\mathbf{r}) = 0 \, . \tag{18}$$

For these models of submolecular units the spatial overlap is zero, which is in agreement with the "naive" description of groups. Against such models quite another objection has been raised, however, namely that they lack the "fuzziness" inherent to quantum microscopic objects, the "fuzziness" present in the LO description of subunits.

In our opinion non-negligible spatial overlaps, or lack of fuzziness, are characteristics which may be accepted in a semiclassical molecular model. A model must fulfill certain prerequisites; it must be internally consistent and it should not contradict any fundamental physical laws (for more details, see Ref. 23). The phenomenon of spatial overlap is not inconsistent with any of the laws of physics (take, for example, the overlap present in assemblies of condensed molecules) and the lack of fuzziness is acceptable as a useful approximation (consider, for instance, the clamped nuclei approximation in quantum molecular descriptions).

We would like to point out that the LO description and the description with "virial fragments" [118] are both acceptable, and that they represent alternative model descriptions of the same object, the molecule. Competition between alternative models is welcome in scientific research. The possible superiority of one model over another must be determined firstly on the basis of the quality of the results. This guideline is not always sufficient, however — there are a lot of models which appear to give good results. Additional criteria concern the generality and flexibility of the method, on the one hand, and its simplicity on the other. The Bader approach has demonstrated good qualities of generality and flexibility in numerous applications and extensions carried out not only by Bader himself, but by Wiberg, Streitwieser, Cremer, and many others as well (the Refs. 119–123 are simply indicative). The term "simplicity" may be applied on two different levels — the conceptual, and the computational. Both the semiclassical and the Bader approach satisfy the criterion of conceptual simplicity. And now it may be asserted that, while the semiclassical approach poses no computational problems, the elaborate calculations required by the Bader approach (integrations over complicated surfaces, etc.) now also appear to be quite within the realm of feasibility [124].

5 The Behaviour of Basic Subunits in the Presence of Both Molecular and External Fields. Elaboration of Semiclassical Models

5.1 Polarization Effects on the SCF Description of Chemical Groups

The LOs are, as we have seen in the preceding chapter, relatively stable entities. Therefore, we can use the main portion of the localized orbital, i.e. $\bar{\lambda}_i$, as a starting point in the reconstruction of the electronic component of our basic subunit. Each electronic distribution

$$\bar{\varrho}_i(\mathbf{r}) = -2\bar{\lambda}_i(\mathbf{r})\,\bar{\lambda}_i(\mathbf{r}) \tag{19}$$

can be assigned an appropriate portion of the nuclear charges according to Eq. (4), thus giving rise to a composite electronic and nuclear distribution which we will designate by the symbol $\bar{\gamma}_i$. We can call this partition $(+2e, -2e)$ for neutral systems. A suitable subset of such basic entities may be collected in a single group.

We have compared the group descriptions obtained by this method, with those generated by an alternative procedure in which a suitable subset of LOs is selected at the very beginning (consider, as an example, a CH_3 or an $HN-CH=O$ group). To do this, it is necessary to change definitions (1)—(2) of the main portion and of the tails for these orbitals, and to use a projection over the basis subset belonging to the group itself (this will be called a $(+ne, -ne)$ description for neutral groups). The results have been encouraging, and from a practical point of view the two methods seem to be interchangeable, e.g. their respective descriptions of a CH_3 or $HN-CH=O$ group are almost equivalent. For more complex systems the situation may be different; we have spent considerable effort in an attempt to harmonize a $(+2e, -2e)$ description of some nucleic acid base derivatives with descriptions using the pyrimidinic or purinic rings as subunits. The attempt was successful, but regretfully has not been documented by a formal publication [125]. For benzene derivatives we considered the whole ring to be an appropriate subunit [126], but we haven't yet pursued investigations of this kind in the field of the aromatic compounds.

The total charge distribution (electrons and nuclei) of each molecule M in a given geometry, indicated by $\Gamma(M)$, may be taken to consist of the sum of the charge distributions of the molecular subunits g $[(+2e, -2e)$ or larger groups] placed at appropriate locations in the space.

A description of the total charge distribution may be realized at different levels of approximation. An exact expression, completely equivalent to the one directly obtainable from the solution of the Hartree-Fock equation, merely requires the localization of the canonical orbitals and the partition of the nuclear charges introduced in the preceding chapter:

$$\Gamma(M) = \sum_g \gamma(g, M) \qquad g \in M . \tag{20}$$

In that chapter we also established the premises for the approximation of (20):

$$\bar{\Gamma}(M) = \sum_g \bar{\gamma}(g, M) \qquad g \in M \tag{21}$$

which we have introduced here. The difference between Eqs. (20) and (21) stays in the use of the $\bar{\gamma}(g, M)$ models expressed in terms of its own LOs without tails, as in Eq. (19).

The approximation given by Eq. (21) may be used for interpretative purposes, but in practice one must first obtain a correct ab initio description at the molecular level, to be followed by other computations. In fact, to use this approximation in the interpretation of data or, what is more important, in laying the bases for further approximations, it is necessary to proceed with the basic strategy outlined in Sect. 3.

Thus we have explicitly formulated a working hypothesis which states that for selected classes of compounds the mutual interactions may be reduced to classical polarization interactions, and that the changes in the group descriptions may be reduced to polarization effects. For commodity's sake, we shall denote this hypothesis as the primitive classical substitution effect (PCSE) model. This description is used very widely in interpretative chemistry, but generally in association with other mechanisms based on very different concepts (e.g. induction versus resonance effects). The qualitative use of models containing opposite factors, which have not been subjected to an independent and careful evaluation, is rather risky, and any conclusions should be checked by other means. The evaluation of polarization effects is easy, once the molecular charge distribution and its partition into group contributions (Eq. 20) is known.

We define here the field of the molecular remainder acting on group g, $\mathbf{F}(M/g)$, as the sum of all the fields of the $\gamma(g', M)$ distributions ($g' \neq g, g' \in M$). For simplicity's sake, the components of the vector $\mathbf{F}(M/g)$ are given here in terms of the derivatives of the electrostatic potentials $V(g', M)$ related to the charge distributions $\gamma(g', M)$:

$$F_\alpha(M/g) = \sum_{g'} \frac{d}{d\alpha} V(g', M) \tag{22}$$

with

$$V(g', M) = \int \frac{\gamma(g', M)}{|\mathbf{r} - \mathbf{r}'|} d\mathbf{r}' \tag{23}$$

We may check the limits of validity of the PCSE model by applying the same strategy used in other phases of this investigation. After selecting one or more series of chemical compounds, each bearing different substituents attached to a common framework, one derives the group partitioning of $\Gamma(M)$ (Eq. 20), and computes the remainder field $\mathbf{F}(M/g)$ (Eq. 22) for all the groups of the set. To demonstrate the effects of substitutions within the selected set of molecules it is opportune to compare the properties of group g for each molecule to those of a parent compound within the set. The techniques for these comparisons are similar to those presented for the evaluation of the transferability properties of LOs discussed in the preceding Section.

The analyses we have performed regard LOs with tails (λ_i) and without tails ($\bar{\lambda}_i$) and cover a fairly large number of observables and numerical indexes. Changes in the properties of each $\gamma(g, M)$ and $\bar{\gamma}(g, M)$ distribution with respect to that found in the parent compounds, are correlated with changes in the molecular remainder field. To do this we designate another field, defined as the difference between the remainder field of molecule M and the remainder field of the parent compound P of the series. We may call it the "substituent field on g":

$$\Delta F_s(M/g) = F(M/g) - F(P/g) \, . \tag{24}$$

The s here stands for substitution, and it means that a given group A in the parent molecule (say a C—H group) is replaced by a new group B (for example, a C—CHO group). The definition of the substituent field is justified by the exact partition of the molecular charge distribution into additive contributions (orbitals λ_i with tails) and by the additivity of the electric fields for a given charge distribution.

The change in the electric field experienced by the target group g may be ascribed to two factors: the substitution of one group with another (A replaced by B), which we refer to as the "direct" term, and the modifications in the other groups induced by the A → B substitution, called "indirect" terms. The first term may be likened to a "through space" effect, and the second to a "through bond" effect, although both of them are actually related to the same physical effect.

The definition of the $F(M/g)$ and $\Delta F_s(M/g)$ fields does not settle all the problems which must be solved before we can begin our analysis, however. We must verify whether this scheme works for substitutions adjacent to the target group as well (remember the question of the spatial overlap of LOs, discussed in Sect. 4.2). We also have to check if the occurrence of substituent tails in the region covered by the main portion of the target group $\gamma(g, M)$ invalidates our hypothesis of classical interactions between distinct charge distributions. Finally, we have to evaluate the fluctuations of $\Delta F_s(M/g)$ in the region of interest, and define a numerical index (a scalar, or a vector quantity) at an appropriate point inside the region covered by the main portion of $\gamma(g, M)$. We have verified all these points including the problem of the numerical index, which was resolved by observing that in the main portion of $\gamma(g, M)$, $\Delta F_s(M/g)$ remains almost constant so that one may selected as the index the components of $\Delta F_s(M/g)$ evaluated at a generic point near the charge center of $\gamma(g, M)$.

This rather elaborate analysis has been performed thus far only on a few classes of compounds: $R_2C=CR_2$, ethylenes with one or two substituents; $R_2C=CR-CN$, acrylonitriles with one substituent; $R_2N-CR=O$, amides with three substituents; $OHCR-CR'=CR'-CRHO$, two symmetric substituents, using maleic anhydride as the parent molecule; $R_3C-O-CR_3$, ethers (two substituents); $R-O-C=O$, esters; diesters; $R-O-CH_2-O-R$, acetals (two substituents); $R_2C=NR$, imines (two substituents); and $OHCR-N=N-CRHO$ (one symmetric substituent). The analyses for some of these classes of compounds have not yet been completed.

Furthermore, the results of these analyses have been thoroughly documented only for few classes [127]. We report here as an example the decomposition of the substitution field into direct and indirect components for a set of substituted acrylonitriles (Fig. 1). The results obtained with a deletion of the tails are practically indistinguishable from those shown in the figure.

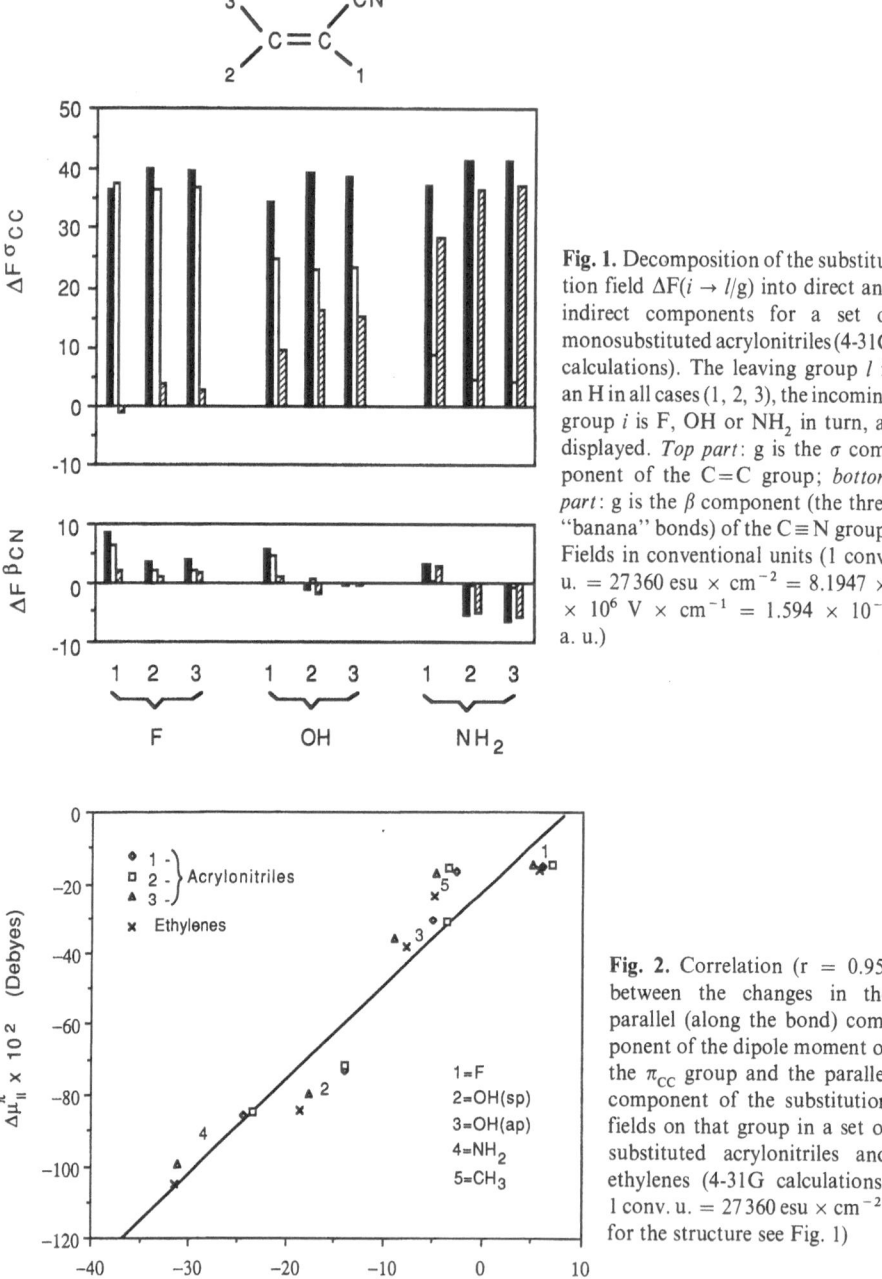

Fig. 1. Decomposition of the substitution field $\Delta F(i \rightarrow l/g)$ into direct and indirect components for a set of monosubstituted acrylonitriles (4-31G calculations). The leaving group l is an H in all cases (1, 2, 3), the incoming group i is F, OH or NH_2 in turn, as displayed. *Top part*: g is the σ component of the C=C group; *bottom part*: g is the β component (the three "banana" bonds) of the $C \equiv N$ group. Fields in conventional units (1 conv. u. = 27360 esu $\times$ cm^{-2} = 8.1947 $\times$ $\times$ 10^6 V $\times$ cm^{-1} = 1.594 $\times$ 10^{-3} a. u.)

Fig. 2. Correlation (r = 0.95) between the changes in the parallel (along the bond) component of the dipole moment of the π_{CC} group and the parallel component of the substitution fields on that group in a set of substituted acrylonitriles and ethylenes (4-31G calculations; 1 conv. u. = 27360 esu $\times$ cm^{-2}; for the structure see Fig. 1)

The numerical index given by $\Delta F_s(M/g)$ correlates well with the changes in the properties of g. We report in Figs. 2 and 3 its correlation with the changes in the local component of the dipole moment μ for the σ and π components of the C=C group in a set of substituted ethylenes and acrylonitriles. Figures 4 and 5 report similar

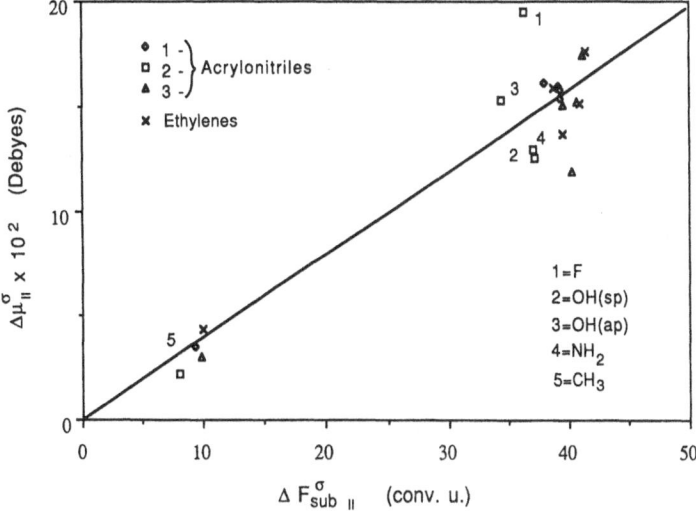

Fig. 3. Correlation (r = 0.94) between the changes in the parallel (along the bond) component of the dipole moment of the σ_{CC} group and the parallel component of the substitution fields on that group in a set of substituted acrylonitriles and ethylenes (4-31G calculations; 1 conv. u. = 27 360 esu $\times$ $\times$ cm^{-2}; for the structure see Fig. 1)

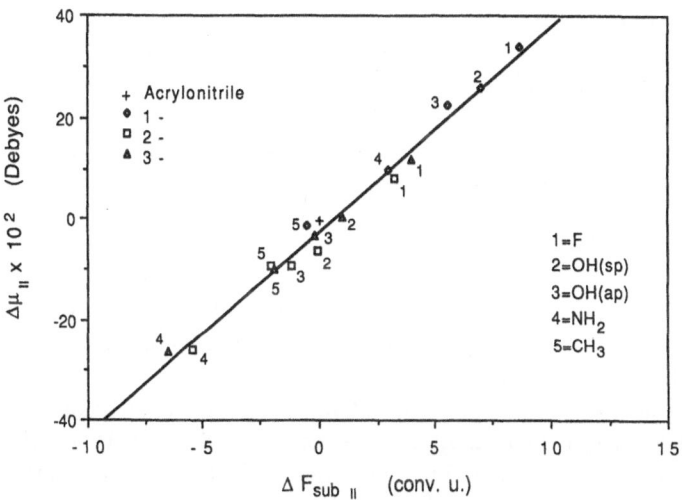

Fig. 4. Correlation (r = 0.99) between the changes in the parallel (along the bond) component of the dipole moment of the C≡N group and the parallel component of the substitution fields on that group in a set of substituted acrylonitriles (4-31G calculations; 1 conv. u. = 27 360 esu $\times$ cm^{-2}; for the structure see Fig. 1)

correlations for the banana bonds in C≡N and for the N lone pair group in the same set of substituted acrylonitriles.

One further step in the analysis of ab initio results is to evaluate the connection between the classical fields, $\mathbf{F}(M/g)$ and $\Delta\mathbf{F}_s(M/g)$, for which we have defined an appropriate molecular index, and the effects of chemical substitution on the group g

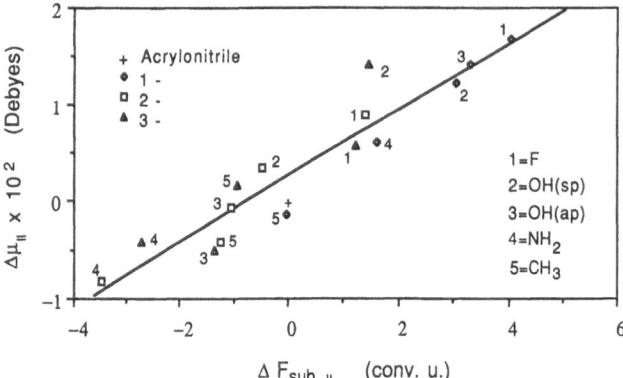

Fig. 5. Correlation (r = 0.94) between the changes in the parallel (along the bond) component of the dipole moment of the N lone pair and the parallel component of the substitution fields on that group in a set of substituted acrylonitriles (4-31G calculations; 1 conv. u. = 27360 esu × cm^{-2}; for the structure see Fig. 1)

for which we have elaborated the analytical tools above. In order to do so, it is convenient to use the (+2e, −2e) description of groups and to examine a gradient of smaller and smaller subunits of the partitioned molecule. The electronic part of the (+2e, −2e) model is composed (see Eq. 19) of $\bar{\lambda}_i$ orbitals which can be further decomposed into generalized hybrids. For commodity's sake we make use here of terminology already adopted in preceding papers: a $\bar{\lambda}_i$ describing a σ bond between atoms A and B is called b_{AB}, β_{AB} is the $\bar{\lambda}_i$ describing a banana bond, while the orbitals describing inner shells and lone pairs are called i_A and l_A, respectively. The terminology may be extended to other cases, but in our analyses we at most make use of bonds defined on three atoms, t_{AHB}.

In our analysis we examine the response of the components of each $\bar{\lambda}_i$ to various electric fields — primarily the substitution field $\Delta \mathbf{F}_s(M/g)$, but also static fields of external origin such as constant fields, fields generated by a small number of external point charges, q_{ext}, placed at different positions, and solvent reaction fields, with solvents of different dielectric constants.

The analysis, performed on basis sets of different sizes, produces almost constant results. The effect of $\mathbf{F}$ on $\bar{\varrho}_i$ is practically linear when the strength is that of the $\Delta \mathbf{F}_s(M/g)$ fields (substitution of neutral groups). The external fields may be, of course, of arbitrary strength, but as a general rule the strength of the field produced on the nearest group g by an Li$^+$ cation at the equilibrium distance is of the same order of magnitude as that of the substitution field. The solvent reaction field reaches analogous values, although only for the solvents with the highest dielectric constants. The only exception is the field produced by a proton at the equilibrium distance: a small deviation from linearity is found for the group closer to the proton.

In summary, for all the cases of interest the classical response of the electronic group distribution to the fields with a molecular origin is linear. We have mentioned above the 2 electron — 3 atom subunit t_{AHB}. Our experience is limited to various sets of complex metal-beryllo- and metal-boro-hydrides [128–130]. We have not yet published our semiclassical analysis of the binding in these representative classes of

Table 10. Values of $\Delta\mu_{CH}$ (along the bond) and its components deriving from variations (a) in the coefficients and (b) in the shape of the hybrids for a set of substituted ethylenes

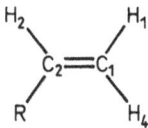

Molecule	Group[a]	Total value	(a)	(b)
Ethylene	CH	0.	0.	0.
F-Ethylene	C_1H_4	0.0682	0.0859	0.0177
	C_1H_1	0.0316	0.0297	0.0019
	C_2H_2	0.1795	0.1182	0.0040
OH(sp)-Ethylene	C_1H_4	−0.0483	−0.0443	−0.0040
	C_1H_1	0.0186	0.0277	−0.0109
	C_2H_2	0.1873	0.1924	−0.0051
OH(ap)-Ethylene	C_1H_4	0.0463	0.0623	−0.0160
	C_1H_1	−0.0174	−0.0224	0.0050
	C_2H_2	0.0967	0.0597	0.0370
CH_3-Ethylene	C_1H_4	−0.0189	−0.0136	−0.0053
	C_1H_1	−0.0211	−0.0227	0.0016
	C_2H_2	0.0189	0.0301	−0.0112
NH_2-Ethylene	C_1H_4	−0.0640	−0.0642	0.0002
	C_1H_1	−0.0484	−0.0464	−0.0020
	C_2H_2	0.1087	0.1032	0.0055

[a] The atom numbering is given according to the scheme

inorganic compounds, but we may anticipate the conclusion here and say that the behaviour of the 2-electron subunits is subject to the same rules which hold for organic compounds.

The λ_is which we have examined in terms of classical fields may be rewritten in terms of generalized hybrid orbitals:

$$b_{AB} = c_A h(b)_A + c_B h(b)_B$$
$$\beta_{AB} = c_A h(\beta)_A + c_B h(\beta)_B \tag{25}$$
$$t_{AHB} = c_A h(t)_A + c_H h(t)_H + c_B h(t)_B$$

etc.

The classical fields affect the hybrids in se, as well as the coefficients:

$$b_{ab} \rightarrow c_A^* h^*(b)_A + c_B^* h^*(b)_B \tag{26}$$

etc. It is useful to consider fields which have a generic orientation with respect to the bond, separating them into parallel, $F_{||}$, and perpendicular, $F_\perp$, components. For the parallel component the polarization changes are accounted for by changes in the coefficients and hybrids, with ratios on the order of 90/10 (or even higher):

an example is reported in Table 10. For the perpendicular components the results are less clear cut, as they depend on the nature of the group as well as on the characteristics of the basis set. For constant external fields exactly perpendicular to the bond, changes occur only in the hybrid if the group is placed in a symmetrical molecular environment (e.g. the β_{CC} group in RHC=CHR compounds, with $\mathbf{F}$ perpendicular to the double bond), but in other cases the ratio introduced above may reach values as high as 80/20.

Our analyses have shown that the parallel polarizability of the bond is decidedly higher than its perpendicular polarizability. On the basis of these results we have drawn a tentative practical conclusion. In order to describe in a first approximation the polarization effects in the molecule it is sufficient to use the index related to the parallel components of F for changes in the numerical coefficient of Eq. (26). This procedure allows us to recover, in the cases we have examined, almost 80% of the polarization effects on the electronic distribution. The lone pairs, l_A, however, have a small parallel polarization; when necessary one can retain the effects related to the perpendicular polarization, which induce a change in the orientation of l_A with respect to the molecular skeleton. The data available for groups of the t_{AHB} type are not sufficient to formulate hypotheses on simplifications of the polarization effects. We shall discuss in a following section how one may translate these indications into actual computational procedures.

Before closing this section we feel it opportune to mention that there exists an ample literature on the subject of the effects of electrostatic fields on the properties of molecular subunits. The descriptions which they offer clearly reflect the original orientations of the different studies, however.

Fields in crystalline environments have been the subject of numerous articles: indeed some of the methods introduced here and in the following section may find their counterpart in methods developed for molecular studies of crystals. The effects of the electrostatic fields deriving from interactions of molecules with charged species have been amply examined (with contributions coming from our group as well). Solvent effects are often reduced, in the pertinent literature, to solvent reaction field effects of semiclassical origin, and the analysis of a variety of chemicophysical processes in solution exploits concepts similar to those presented here. External fields are often employed to derive polarizability data, generally dissected into local contributions. Surface effects of different origin are sometimes reduced to electrostatic effects and analyzed accordingly. Spectroscopic studies in a large variety of experimental situations make use of the analysis of electrostatic fields of different origins. This enumeration could be continued, but it is clear that we are once again facing a familiar dilemma — a forced choice between giving a very condensed summary of the above mentioned topics, accompanied by a limited (and barely representative) number of references, or simply dropping the subject. Above, we have tried to offer a compromise in the form of a brief enumeration of subjects related to the theme treated here.

5.2 Introduction and Use of the "Prototypes"

5.2.1 Definition of the Prototypes

We may pass now to the second step of our research based on the PCSE model.

A simpler description of the total molecular electronic distribution given in Eqs. (20)—(21) is the following:

$$\Gamma^0(M) = \sum_g \gamma(g, 0).$$ (27)

In previous papers we have called the $\gamma(g, 0)$ models of the group description "directly transferable localized orbitals" (TLO); it is convenient to denote them here as "prototypes".

The prototypes have no direct relationship with the molecule M. They derive from the calculation of the $\gamma(g, P)$ functions, performed over a set $\{P\}$ of simpler parent molecules, and averaged for the same g over different Ps. The material presented in Sect. 4 gives the reader the tools needed to judge the soundness of this operation.

The $\gamma(g, 0)$ functions are collected in a library, and placed in spatial locations which permit one to define $\Gamma^0(M)$ without other manipulations (i.e., without mutual ortho-gonalization). Equation (27) may be a useful approximation in many cases: in the following discussion we shall give some examples. Calculations are in fact more economical using Eq. (27): in this approximation the calculation of the wave function of M, which may be a large molecule, is no longer required, and in addition there is a considerable saving of computer time in the calculations of the observables. The prototypes collected in (27) are expressed in a given basis set; collectively they span the same basis set $\chi_M = \chi_A \oplus \chi_B \oplus \chi_C \oplus \ldots$ as the molecule M. The matrices $\mathbf{Q}$ containing the integrals related to the observable Q, expressed in the χ_M basis set, may be reduced to a limited number of blocks, discarding blocks not covered by one of the $\gamma(g, 0)$ distributions. For example in a triatomic molecule, A—B—C, the non-diagonal block containing the integrals between the χ_A and χ_C functions may be dropped from the calculations. With a modest programming effort one may save a considerable amount of time, especially when the one-electron observable has to be computed many times, as is the case for molecular electrostatic potential maps. The time gained becomes more important as the size of M increase and obviously depends on the connectivity of the atoms inside the molecule.

We must now develop our analysis in two opposite directions. First, we shall make Eq. (27) more realistic, by taking into account the polarization effects produced by the molecular remainder (see Sect. 5.1), while still maintaining the low level of computational effort mentioned above. Second, we shall further simplify the $\gamma(g, 0)$ descriptions, retaining their ability to reproduce polarization effects while reducing the computational effort required.

We shall start with the first objective.

5.2.2 Polarization of the Prototypes

The description in terms of prototypes does not take into account the mutual inter-actions inside M. We have to resort to definitions of the molecular remainder fields and the substitution fields, akin to expressions (22) and (24) but given now in terms of $\gamma(g, 0)$ distributions.

$$F_\alpha(M/g, 0) = \sum_{g'} \frac{d}{d\alpha} V(g', 0) \tag{28}$$

$$\Delta F_s(M/g, 0) = F(M/g, 0) - F(P/g, 0) . \tag{29}$$

The description which emerges from this model is much more easily verified than the one based on SCF results. A direct comparison in the physical space between the fields defined according to Eqs. (22) and (24), and those defined in terms of Eqs. (28) and (29) shows that the latter give a sufficiently accurate description of the former.

This finding indicates that polarization effects, though important in the description of some features of the electronic distribution, are less important than the primary effects arising from the presence of groups g' (or their substitution) and their mutual positioning.

The definition of the set of groups g present in the molecule is given by our initial definition of the material model (see Sect. 2). Their mutual orientation and position is assumed to be known, i.e. is assumed to correspond to the experimental or quantum mechanically determined geometry of M and from this point of view the geometry may be considered to be a non-classical, quantal element of the model. Actually, there are a number of models in the literature which furnish predictions for these parameters, or for a part of them. Some of the empirical models, such as the VSEPR [131] are in widespread use; even if they have sometimes been criticized, some positive confirmations of their validity have appeared with respect to ab initio calculations [132–138].

Even more common is the use of models addressed to the derivation of preferred molecular conformations. Often these are not pure semiclassical models, since they also include features of different nature (e.g. dispersion contributions, etc.). The well known molecular mechanical models may be considered as further examples from this category, because many of their parameters derive from empirical values governed by semiclassical effects, and because electrostatic (i.e. semiclassical) interactions are often directly introduced into the model itself. The literature on this subject is so vast that we shall refrain from citing the pertinent papers: it would be of little use to give a long list of references without comments, and at the same time it would be unfair to select just a few as examples. The subject is well known and has been the object of numerous reviews.

Coming back to our theme, we would like to note that the semi-classical approach discussed here can provide valuable information on both molecular conformations and internal geometry deformations, as will be shown later. For the moment we shall ignore this additional feature contenting ourselves to consider fixed molecular geometries.

The good correspondence between the molecular fields described by Eqs. (22)–(28) and Eqs. (24)–(29) permits us to use the relationship between the fields and the polarization effects on the $(+2e, -2e)$ groups as a practical computational method. When the $F(M/g, 0)$ field is computed at the appropriate place, a modification of $\gamma(g, 0)$ which includes its polarization effects may be obtained:

$$\gamma(g, 0) \xrightarrow{F(M/g, 0)} \gamma^*(g, M, 0) . \tag{30}$$

This equation makes use of both the correlation equation between applied fields, and modifications in the coefficients and hybrid shape of $\gamma(g, 0)$. In accordance with the remarks made earlier, we generally apply this modification only to the coefficients of the bonds. The modified group descriptions obtained in this way are collected in a new first order density matrix (compare with definitions (27) and (21)):

$$\Gamma^*(M) = \sum_g \gamma^*(g, 0) . \tag{31}$$

Some results are displayed here. Figure 6 reports the correlation between the SCF values (full molecule calculation) and the $\gamma(g, 0)$ and $\gamma^*(g, M, 0)$ values calculated for the $>C=O$ component of μ in several compounds. An analogous comparison is made in Table 11 for the electric field gradient, eq, on the N nucleus of some compounds containing the $-CN$ group.

All the results presented in the above examples are based on expressions (28) and (31) of the first order density matrix. The expectation values of the one-electron observables were calculated simply by summing the independent g contributions, discarding the overlap between the γ's belonging to different groups. It is possible, of course, to check this approximation which, at first sight, could be considered too drastic. Comparisons between values obtained with expressions (28) and (31) and the corresponding values obtained after a previous orthogonalization of the fragments have been reported in several of our preceding papers (see, e.g., [72, 139]). The orthogonalization procedure also requires the evaluation of the complete matrices $\mathbf{Q}$ of the various observables on the whole χ_M space, and in this way one of the numerical advantages of the method is lost. In numerous checks we have never found a higher correlation with the SCF values after orthogonalization. Similar tests based on a density matrix without tails (definition (21)) and the introduction of a reorthogonaliza-

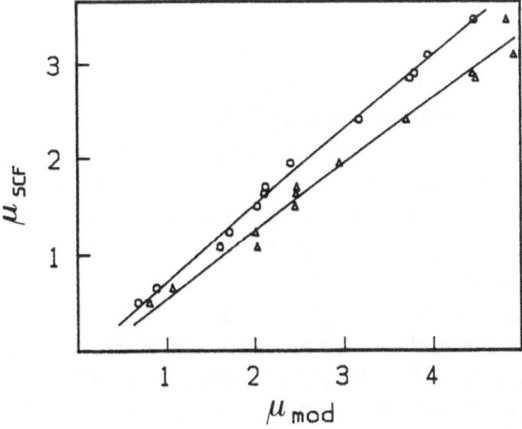

Fig. 6a, b. Comparison of the dipole moments (debyes obtained from SCF calculations and from two different models for a set of carbonylic compounds (STO-3G calculations). The models are: **(a)** rigidly transferred group contributions ($\triangle$); **(b)** the same group contributions but polarized by the field of the molecular remainder ($\bigcirc$)

Table 11. Comparison of eq_{mol} (^{14}N) values (electric field gradient at the nucleus) obtained with accurate ab initio SCF wavefunctions and with approximations (27) and (30)
(values in a.u.: 1 a.u. $= 2.3497 \times 10^{30}$ MHz m^{-2})

Molecule	ab initio	from γ^* (g/0)	from γ(g/0)
$H-C\equiv N$	-1.1573	-1.1402	-1.1161
$F-C\equiv N$	-0.7114	-0.7082	-0.7396
$Cl-C\equiv N$	-0.9417	-0.9388	-0.9611
$N\equiv C-C\equiv N$	-1.1908	-1.1890	-1.2104
$H-C\equiv C-C\equiv N$	-1.0940	-1.0890	-1.1080
CN^-	-1.0800	-1.0038	-0.9890
OCN^-	-0.1819	-0.2011	-0.2418
SCN^-	-0.5552	-0.5450	-0.6123

tion have produced analogous results. We consider these numerical checks to be further evidence that the tails in LOs are not a consequence of the orthogonality constraints; on the contrary, electronic fragments forced into mutual orthogonality show a worse fit than their unconstrainted counterparts. We will not present any graphic or numerical documentation for these statements due to lack of space, and in the following discussion we shall continue to use non-orthogonal fragments for our evaluation of one-electron properties.

Models similar to ours have been developed by Náray-Szabó and coworkers [107, 140–143]. The dissection of the molecule into subunits based on localized orbitals, their expression in terms of prototypes written as combination of atomic hybrids (like Eq. 25) and the polarization of prototypes accounted for by changes in the numerical coefficients, are all features common to both approaches. The Hungarian group has developed several computer programs for distribution, and has demonstrated the viability of their approach by applying the methods to a considerable number of chemical processes, especially in the area of molecular biology.

Coming back to our exposition, one could ask if the considerations and the procedures outlined above hold also for the energy of the system. In the H-F approximation the energy may be formally written in terms of one-electron density matrices:

$$E_{tot}(M) = 1/2 \ tr\{H\Gamma(r'_1|r_1)\ \Gamma(r'_2|r_2) - \Gamma(r'_1|r_2)\ \Gamma(r'_2|r_1)\} \tag{32}$$

on which manipulations similar to those already performed on the diagonal element could be introduced. We shall not follow this suggestion here, however.

Calculations of E using our definitions of γ(g/M), γ(g/0) and γ^*(g/0) (Eqs. 21, 27, 31) are routinely carried out in our laboratory as supplementary tests of the quality of the various approximations. These routine calculations are performed using energy expectation values built up in terms of the various γ(g) s, subjected to a previous orthogonalization. These calculations have nothing to do with the semiclassical approximation, and are of little practical use, because to calculate E in this way one must compute all the two-electron integrals over the χ_M basis. These tests give positive answers within the limited scope for which they were intended: the deviation from the correct SCF energy is fairly constant for the γ(g, M) descriptions ($\Delta E\% =$ $= 0.060$–0.070); it is a little higher for the γ(g, 0) descriptions ($\Delta E\% = 0.080$–0.100)

and it reaches values comparable to the initial ones when the polarization of the molecular remainder is added to the $\gamma^*(g, 0)$ description ($\Delta E\% = 0.060\text{--}0.075$).

This test is a particularly valuable tool for examining the description of the molecule in term of prototypes when the molecule is subjected to geometrical deformation. The test is in fact very sensitive to inaccurate descriptions of the molecular subunits. We shall discuss geometry deformations in a later chapter.

Using the strategy disclosed in the preceding pages for one-electron observables, we can now examine energy values obtained with the release of the strong orthogonality constraints.

Many attempts have been made by quantum chemists to incorporate non-orthogonal subunits in their calculations of the energy of the molecule. Classical and spin-coupled VB [10] theories use as their starting points non-orthogonal subunits. Within the SCF scheme as well various attempts and proposals for the use of non-orthogonal subunits have been made. The overlap between subunits is never discarded, but is usually introduced somewhat later in the various methods. A comparison of the different approaches, in terms of how they treat the overlap and exchange problems and to what point they separate classical and non-classical effects in the calculation of E, cannot be given here, indeed it is a topic to which an entire review could be dedicated. We will limit ourselves to a couple of references [144, 145] which should be sufficient to indicate the problems inherent in this approach; references 39–60 quoted above also deal in part with this topic.

Returning to our semiclassical model, it is evident that this procedure cannot be directly applied to the calculation of the energy, even in the quasi-independent particle approach, because the $\Gamma(r'_1 | r_2)$ distributions of Eq. (32) are not of classical origin.

A reduction in the number and in the importance of the non-classical (i.e., exchange) energy terms may be attempted, however. In a paper already cited above, Edminston and Ruedenberg [31] remarked that their localization procedure reduces to a minimum the exchange contributions to the energy. Taking into account the similarity of the results which may be obtained using different localization procedures, this observa-

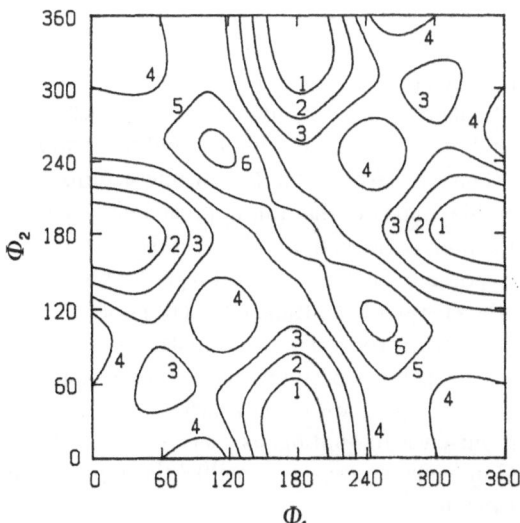

Fig. 7. Conformational energy map (kcal/mol) obtained from SCF/3-21G calculations for CHO—CH$_2$—CHO (malonaldehyde). The torsion angles ϕ_1 and ϕ_2 refer to rotations of the aldehydic groups

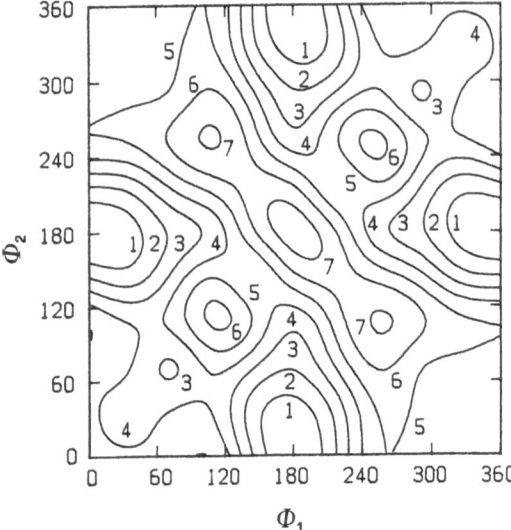

Fig. 8. Conformational energy map (kcal/mol) obtained from $\gamma^*(g, 0)$/3-21G distributions (see text) for $CHO-CH_2-$ $-CHO$ (malonaldehyde). Torsion angles ϕ_1 and ϕ_2 as in Fig. 7

tion can be extended to the other LO descriptions. A considerable portion of the exchange energy in an LO description is related to the tails, as a few numerical tests which we have performed in the past indicate. The partition of the molecular energy given by Eqs. (15)–(17) may suggest a partial deletion of the non-classical components, K_{ij}, for those groups i and j which have a relatively large spatial separation.

We compare in Figs. 7 and 8 two conformational energy maps, the first one derived from normal ab initio SCF calculations and the second derived from calculations using the $\gamma^*(g, 0)$ distributions while eliminating of all the K_{ij} contributions between groups at least one bond apart. This case has been drawn from a limited set of fully worked examples, in which some relatively good results are mixed with a few failures. The failures are mainly due to changes occurring in the geometry which is sensitive to changes in the hybridization of some of the atoms, such as for instance the passage from tetrahedral to planar conformations at a given nitrogen atom.

Our checks have not been accompanied by a parallel search for optimized algorithms capable of exploiting in a general and coherent way this hypothesized reduction of the non-classical elements of E. It would seem to be much simpler, and more appropriate, to restrict the hypothesis to certain subclasses of molecular interactions. For example, the excellent work of Mehler [144] indicates that this approach may be used in the study of weak non-covalent interactions between molecules. Another subclass of molecular interactions to which this hypothesis might be applied regards substituent effects on the barrier height (i.e., on the energy of the transition state) once the TS geometry and the group description are known for a parent example.

Yet another subclass for which this approach has been proven useful concerns the analysis of internal rotation barriers. Here again we cannot survey the extensive literature on the subject and will limit ourselves to quoting, as a significant example, a paper by the Magnasco group [146] in which concepts not unlike those presented here have been exploited in different manner.

5.2.3 Simplified Descriptions of the Prototypes

At the beginning of Sect. 5.2 we suggested two methods for the further elaboration of prototype descriptions. In Sect. 5.2.2 we followed one line, attempting to make the descriptions more realistic by adding molecular polarization effects. Actually we interrupted our elaboration of the model of these non-linear effects at the end of the first step (mutual polarization of prototypes), without considering any of the higher order effects (mutual polarization of polarized prototypes). We felt encouraged to do so by the relatively good results generated and by the linearity of the values found for the target group with respect to fields of internal and external origin within a reasonable range of force strengths.

We may pass now to the second point. If it is sufficient to use the molecular remainder and the molecular substitution fields to procure, via the $\gamma^*(g', 0)$ functions (Eq. 30), more accurate descriptions of the molecular charge distribution, is it nevertheless necessary to use all the information encoded in the $\gamma(g', 0)$ functions to compute these fields? In fact we can try to derive simplified descriptions of the $\gamma(g', 0)$s for the purposes of computing these fields.

This working hypothesis meets the criterion of flexibility discussed at the end of Sect. 2. Indeed we can use different approximations for the $\gamma(g', 0)$, and we may introduce a hierarchy of approximations as a function of the distance from the target group. This problem is closely related to the question of finding approximate expressions for the molecular electrostatic potential and for the description of charge distributions in terms of atomic components (atomic charges). Most of the abundant literature in this field concentrates on the description of the molecular electrostatic potentials (MEP) in the external regions of the molecule. These methods will be reviewed in another chapter of the present set of books.

The sources of some of the group fields $\mathbf{F}(g')$ are very close to the main portion of the target group for which they have to serve as the index. For adjacent $(+2e, -2e)$ groups there is a spatial overlap between the source of the field and the target distribution. These characteristics make the use of the conventional expansion techniques of classical electrostatics extremely suspect, because the conditions for convergency and asymptotic stability of the expansion are not satisfied.

We do not consider it necessary to formulate a working hypothesis for the simplified expression of the field generated by all the groups g, as an intermediate approach appears to be more practical. Let us define for each target group g a surrounding space, called S_g. All the groups g' lying within this space are treated as sources of the $\mathbf{F}(g')$ in the original LCAO description, while for the groups outside of S_g we may search for simplified expressions. The surrounding space is loosely defined, and its definition could be given in terms of the bond connectivity of the target group, or in terms of a radius. In the first case we may define S_g as the space containing either the first neighbours alone, or the first and second neighbours together, and we can make use of chemical connectivity graphs. In the second case we have to rely on an examination of the actual molecular geometry; thus we lose the facilities offered by the connectivity graphs but at the same time we gain flexibility in that some of the groups added to S_g, though not directly linked, may be near the target in some specific conformations of the molecule.

For the groups outside of S_g we chose first to explore the behaviour of multicenter expansions truncated to the first term, i.e. to a collection of point charges. Starting with a definition of S_g corresponding to the second neighbours (or, in parallel, to a radius of about 4.0Å from the center of the target group), we have found that the expansion which we elaborated several years ago for the analogous evaluation of the MEP in the outer regions of the molecule still gives satisfactory results.

These expansions are composed of the charge on the pertinent nuclei or nucleus (i.e. the $+2e$ part of the group), plus either two negative charges symmetrically placed with respect to the electronic charge center on the median of the bond axis (for b_{AB} and β_{AB} groups) or a -2 charge alone (for lone pairs). The relevant numerical procedures are reported in Ref. 75. The results, as anticipated, practically coincide with those obtained using the LCAO description of these groups outside the S_g region.

Our analysis has thus far been limited to "simple" molecules, covering only a portion of the list of groups given in Sect. 3. More "difficult" cases such as, for example, condensed aromatic systems have not yet been considered. It would be interesting to refine the elaboration of our hypothesis within such a context. In these cases, groups at larger distances could be adequately represented by a simple monopole, bringing us back to the use of opportune atomic charges (not necessarily the Mulliken charges). In recent years a large number of papers have been published which suggest various ways of deriving the atomic charges best able to reproduce the MEP (see, e.g. Refs. 147–150). A reduction of the radius of S_g to include the first neighbours only, probably requires a finer description of the nearer groups. We have not yet considered these points, however.

Since we have now arrived at the end of this section it may be appropriate to introduce a critical remark. We have not at any point made explicit use of the concept of group polarizability. Much data has been accumulated over the years regarding the segmentation of molecular polarizabilities into bond or atom components. The "classical" review on molecular polarizability [151] gives space to this topic, but does not cover the more recent developments. A critical examination of the studies of Applequist, Birge, Boyd, Dewar, Oxtoby, Miller, Sundberg, and Thole, to name only a few [152–163] probably would have suggested to us other working hypotheses or other solutions for the implementation of semiclassical methods. Polarizabilities are in fact easily included in a semiclassical formulation, and in accordance with the general considerations on modelling set forth in Sect. 2, we ought to have added the concept of polarizability when we formulated and tested our working hypotheses. There are no justifications for this omission, which we mention here just to emphasize that even in our work not all the general methodological premises set forth in Sect. 2 have been fully satisfied.

6 Changes in the Molecular Geometry

6.1 Geometry Deformation Effects

This section, and the ones to follow, will be shorter than the preceding ones for two reasons. There will be no need to reiterate earlier analyses, such as for example our discussion concerning the transferability of the LOs subjected to geometry deforma-

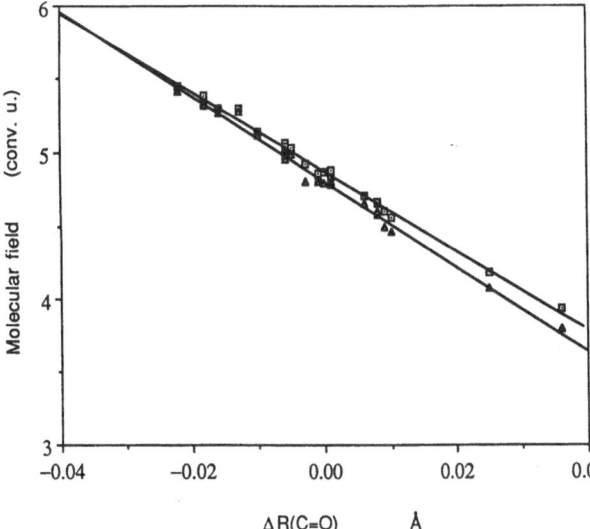

Fig. 9. Changes in the equilibrium distance of the carbonyl group with respect to the formaldehyde one versus the parallel component of the field of the molecular remainder at the midpoint of the C=O bond in a series of related aldehydes, computed using (□) SCF charge distributions and (△) charge distributions in terms of prototypes $\gamma(g, 0)$

tions. Furthermore, we have a lower quantity of results worked out, not sufficient in fact to corroborate our conclusions with a statistical elaboration of the data.

Group interactions in molecules produce changes in the bond lengths and angles. According to the strategy we have selected, we must formulate specific hypotheses regarding the influence of the substitution field for each chemical group, or better for each collection of chemical groups under examination. There exists, in fact, the possibility that a semiclassical hypothesis, successfully tested for some specific groups, will fail when applied to a collection of these groups, because other factors (such as non-classical effects, or classical effects of a higher order) make their influence felt in this specific collection of groups. Our work thus far has been limited to relatively simple chemical systems, such as those described in the following examples.

Figure 9 reports the changes in the equilibrium distance of the carbonyl group with respect to a reference compound found in a representative set of aldehydes. These changes are related to the parallel component of the molecular remainder field $F(M/C=O)$ computed at the midpoint of the C=O bond. The correlation is good and is representative of the correlations which have been found in similar analyses performed on the bond lengths and bond angles of the most important simple chemical groups. The correlation between the SCF $\Delta R(C=O)$ values and the molecular remainder field $F(M/C=O, 0)$, reported in the same figure, is of analogous quality. It thus seems that the description of the charge distribution of M in terms of prototypes $\gamma(g, 0)$ may be used for predictive as well as for interpretative purposes.

Predictions may be of some use but clearly interpretations are of greater interest. The literature is rich of interpretations of internal geometry changes arising from substitution or conformation effects. As we have stressed in our introduction, a wide variety of interpretative methods exist, which use different definitions of the molecular subunits and which place more or less emphasis on the classical versus the quantal elements of the analysis. We consider our approach detailed enough to illustrate both the classical elements of the effect, and then, by a comparison using accurate calculations, the role played by the true quantal effects.

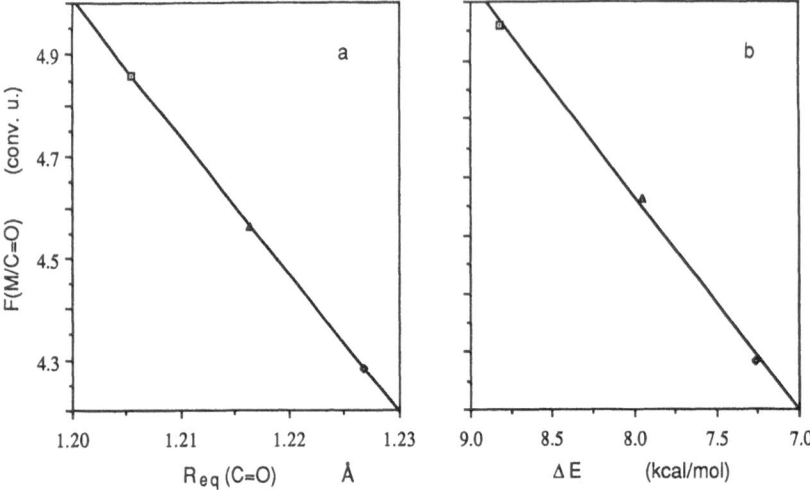

Fig. 10 a, b. Comparison of the C=O equilibrium bond length (**a**) and of the energy of elongation of the C=O bond by 0.1 Å (**b**) with the field of the molecular remainder for formaldehyde (□), formamide (△) and urea (◇)

The analysis may be extended to other observables as well, and the effects of chemical substitutions can be profitably supplemented by a consideration of the other factors or fields, which may influence the molecular geometry. Among these additional factors we can adduce thermal excitation (i.e., molecular vibrations), and external fields, either created by incoming reagents (with a net charge or bearing a permanent dipole), or arising from a generic source (especially linear fields) or from solvent effects. In all these cases nuclear geometry changes are connected with electronic distribution changes. Some of these topics will be considered in later sections. We will limit ourselves here to giving a few examples.

In Fig. 10 the field $\mathbf{F}(M/C=O)$ is compared once again with $R_{eq}(C=O)$ and also with the energy of elongation, by a fixed amount, of the C=O bond. The molecular remainder field linearly correlates with the energy changes as well as with the geometry changes. When there is an elongation of a bond (for example, during a vibration), the point where the molecular fields are probed is slightly shifted. The difference between the two values of the fields correlates with the changes in energy (Fig. 11) as well as with changes in the dipole moment and in other properties.

A description of small changes in the length of the prototypes for the single as well as the double bonds, (i.e. the $(+2e, -2e)$ and $(+4e, -4e)$ models), seems to be easily arrived at. We have achieved good results simply by keeping the LCAO description of the group (see, e.g. Eq. 25) obtained with a given value of the internuclear distance, and subjecting it to a normalization procedure with the new internuclear distance.

The semiclassical interpretation of the effects of external charges on the molecular properties, including the geometry, may be better appreciated if one looks at specific problems, like intermolecular interactions (a subject which will not be treated in this review), or solvation effects, a topic which will be examined in a separate section. Actually, an investigation of the effects of separate fields represents the first step in

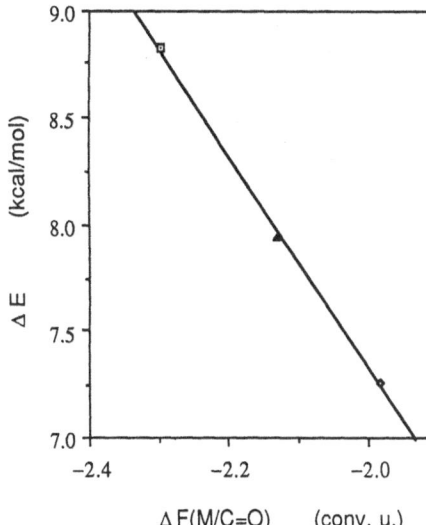

Fig. 11. Comparison of the energy of elongation of $R(C=O)$ by 0.1 Å with the difference between $F(M/C=O)$ measured at $R_{eq} + 0.1$ Å and R_{eq}. Codes as in Fig. 10

the analysis of a physical model which corresponds more closely to the situation actually encountered in experiments — i.e. the interaction between fields of different origin. In this approach, it is advisable to also consider fields with a simpler mathematical expression, such as constant electric fields with variable strength and direction.

6.2 Effects of External Fields on the Geometry of Molecular Complexes

In the course of some preliminary investigations on the aforementioned subject, we obtained some results which should be discussed here because they relate to the question of the limits of the semiclassical model. The investigation concerned the effect of a constant field $\mathbf{F}_{ext}$ on the structure and properties of a dimeric system, linked by noncovalent interactions. A considerable amount of work has been done on the semiclassical interpretation of noncovalent intermolecular interactions, with relatively good results. Although the subject is not pertinent to a review centering on intramolecular effects, we may on this occasion consider the dimer as a supermolecule. The analysis of the interaction energy for a dimer:

$$\Delta E_{AB}(R) = E_{AB}(R) - (E_A^0 + E_B^0) \tag{33}$$

is usually carried out using the well known Kitaura and Morokuma method [164]. This decomposition may be viewed as a specialized version of the energy decomposition formulas given in Sect. 4, in which the two monomers are the two groups composing the AB system. Some of the components of the interaction energy:

$$\Delta E_{AB}(R) = E_{ES}(R) + E_{PL}(R) + E_{EX}(R) + E_{CT}(R) + E_{MIX}(R) \tag{34}$$

are clearly semiclassical in nature, such as E_{ES}, the coulombic contribution, and E_{PL}, the polarization contribution. Another term E_{EX}, the exchange-repulsion contribu-

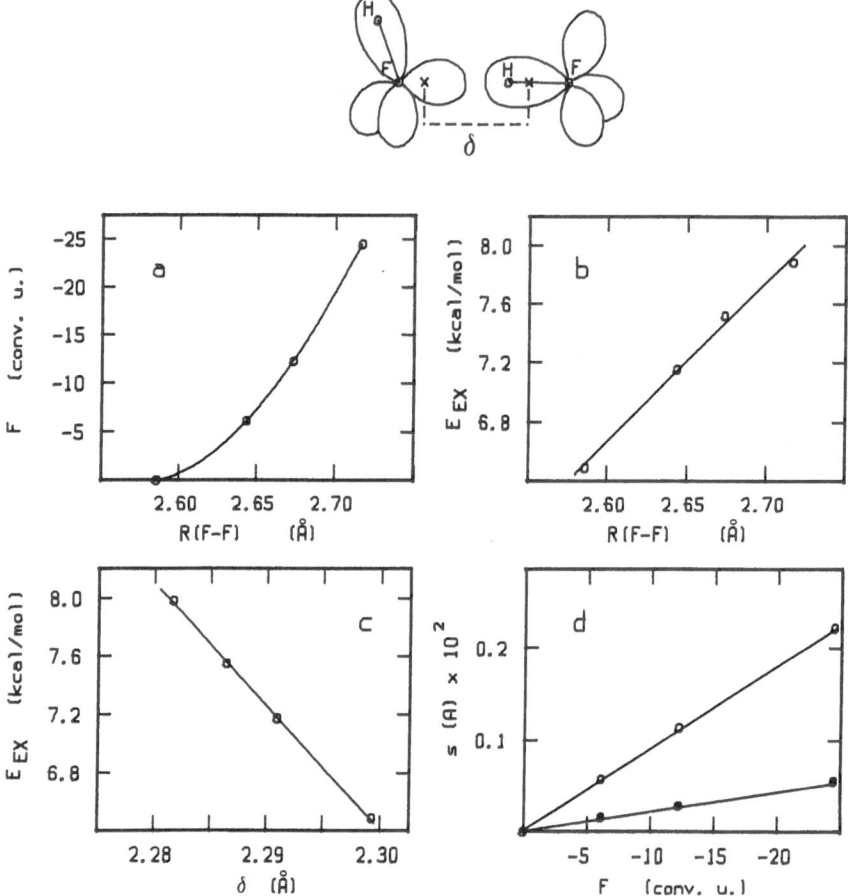

Fig. 12a–d. Effects determined by the application of constant external fields of different intensity to the HF dimer (its arrangement is displayed at the top): **(a)** changes in the equilibrium distance R(F—F) in function of the external fields applied; **(b)** relationship between the exchange component E_{EX} of the interaction energy and the equilibrium distance R(F—F); **(c)** relationship between the exchange component E_{EX} of the interaction energy and the distance δ between the $\gamma(lp_F)$ and $\gamma(b_{HF})$ charge centers; **(d)** shift of the $\gamma(lp_F)$ (●) and $\gamma(b_{HF})$ (○) charge centers in function of the external fields applied

tion, has a non-classical component which is related to the exchange of electrons between A and B.

By applying a constant field to the dimer $HF_a \cdot HF_b$ we have found, inter alia, a non-negligible dependence of the R(F — F) distance upon the strength of the applied field, with a parabolic correlation. The change in the equilibrium value of R(F — F) also correlates with the E_{EX} value. Therefore, this effect on the internal geometry of the dimer is related to a non-classical energy term. Actually, the semiclassical model may also be of some use in this case. The external field produces changes of different magnitudes on the lone pairs of the monomer A involved in the hydrogen bond, and on the H—F bond of monomer B. There is a linear correlation between the strength of $\mathbf{F}_{ext}$ and the parameters related to the shape of the two (+2e, −2e) groups. The effect of $\mathbf{F}_{ext}$ on $E_{EX}(R_{eq})$ may thus be reduced to a classical description, namely the polarization of the charge clouds which change their overlap, and thus their repulsion, according to the strength and direction of $\mathbf{F}_{ext}$. Some numerical results in support of this analysis are reported in Fig. 12.

6.3 Changes of Geometry of Chemical Reactions

In the preceding sections we have considered molecular systems placed in the concave region of the potential energy hypersurface (PES) corresponding to an equilibrium structure. The geometry deformation which makes it possible for the system to reach a separate minimum, i.e. the changes which give rise to a chemical reaction are, of course, of primary importance in chemistry.

For reasons of economy we have limited our attention (as have many other researchers) to a few selected points on the PES, namely to those points which correspond to the transition state (TS), supplemented by a selection of points along the intrinsic reaction coordinate (IRC) [165].

An examination of the behaviour of LOs during the chemical reaction has frequently been used to shed more light on the reaction mechanism. In bimolecular encounters it has been used to ascertain the distance at which deformations of the electronic distribution of the attacked group make themselves evident (see e.g. Refs. 166–168). The evolution in position of the LO charge centroids along the IRC has been developed as an interpretative tool by the Leroy group and has been employed successfully for many reactions (a clear overview is given in Ref. 169, with more recent additions in Refs. 170–173). These analyses have been often accompanied by a closer scrutiny of the submolecular units, especially the spatial distribution of the electron charge (extension of the tails, polarization, second moment ellipsoids of the LOs [173], etc.).

In our preliminary analysis of the characteristics of chemical groups expressed in terms of LOs with and without tails we have also considered some examples of chemical reactions. The main results, with the analyses performed on the equilibrium geometries, have been collected and documented in Sect. 4.

Our experience is limited to a few classes of chemical reactions, and what will be said here refers only to the cases we have studied: as we have repeated many times, hurried generalizations are not advisable. In all the cases examined we have found a constant trend which may be summarized in a few words.

Changes in the characteristics of chemical groups along the IRC are relatively modest and amenable to a classical interpretation until very near the TS. The differential field acting on a target group, g, is of the same order of magnitude as the substitution fields. Near the TS there is an abrupt change involving only a limited number of groups, in effect those implicated in the breaking and formation of covalent bonds.

The noticeable change in shape of these LOs is not accompanied by significant changes in the weight of the LO tails. An illustration of the deformations occurring in the LOs during the reaction act is given in Fig. 13; other pictorial examples may be found in Refs. 168, 174. The deformed LOs, like the normal LOs, are modified by chemical substitutions in their electronic part. Unfortunately, we cannot supply a large statistical basis to corroborate these statements, as was done in Sect. 4 for molecules at the equilibrium geometry.

Instead, we will examine here in more detail some results for the reaction:

$$H_2C=O + LiBH_4 \rightarrow H_3C-OLi + BH_3$$

in which at the TS there is a transfer of an H atom from the BH_4 to the CH_2 group, as depicted in Scheme 2:

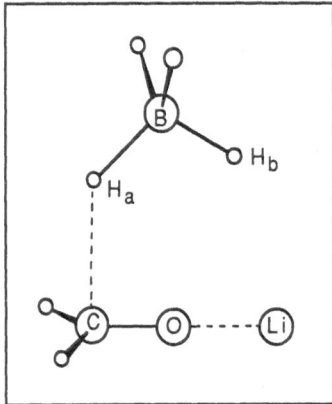

Scheme 2

The LO localization degree (according to Eq. 3) in the reactant, TS and product regions are reported in Table 12. The weight of the tails is not significantly different from that reported in Sect. 4.

The effect of the substituents on the TS geometry is modest, and the changes are well correlated with classical forces expressed in terms of group contributions: a formal derivation is given as an appendix in Ref. 162.

Similar analyses hold for the other reactions we have examined.

Table 12. Localization degree of formaldehyde localized orbitals in the reactants, initial complex, transition state, and products

Reactants		Initial complex		TS		Products	
b(CH)	0.025	b(CH)	0.024	b(CH)	0.023	b(CH)	0.018
b(CH)	0.025	b(CH)	0.024	b(CH)	0.023	b(CH)	0.018
lp(O)	0.032	lp(O)	0.038	β(OLi)	0.034	β(OLi)	0.017
lp(O)	0.032	lp(O)	0.038	β(OLi)	0.034	β(OLi)	0.017
β(CO)	0.002	β(CO)	0.010	β(CO)	0.028	β(OLi)	0.017
β(CO)	0.002	β(CO)	0.010	β(CO)	0.014	β(CO)	0.031
b(BH$_a$)	0.012	b(BH$_a$)	0.012	b(BH$_a$)	0.056	b(CH$_a$)	0.018
b(BH$_b$)	0.034	b(BH$_b$)	0.032	b(BH$_b$)	0.040	b(BH)	0.005
b(BH)	0.034	b(BH)	0.030	b(BH)	0.010	b(BH)	0.005
b(BH)	0.034	b(BH)	0.030	b(BH)	0.010	b(BH)	0.005

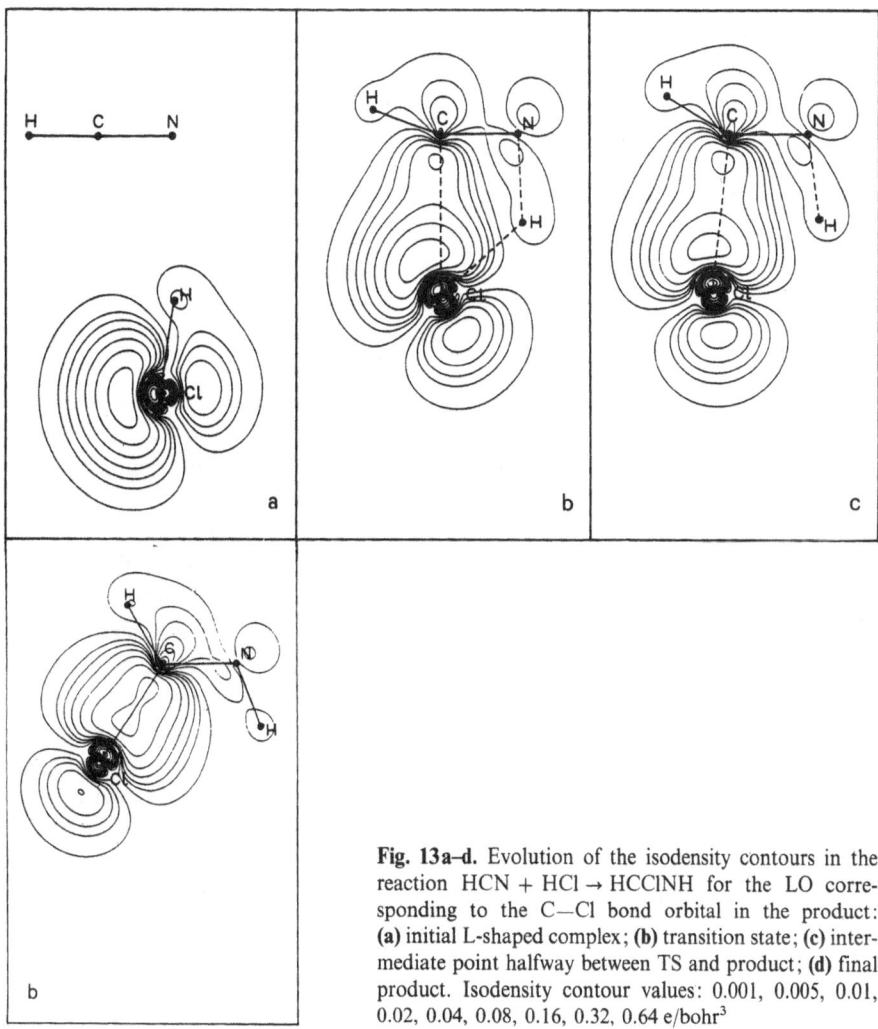

Fig. 13a–d. Evolution of the isodensity contours in the reaction HCN + HCl → HCClNH for the LO corresponding to the C–Cl bond orbital in the product: **(a)** initial L-shaped complex; **(b)** transition state; **(c)** intermediate point halfway between TS and product; **(d)** final product. Isodensity contour values: 0.001, 0.005, 0.01, 0.02, 0.04, 0.08, 0.16, 0.32, 0.64 e/bohr³

In summary, we can make the following observations based on our investigations. For a given class of reactions only a small portion of the molecular system is directly involved in the chemical reaction (see Scheme 3). The remaining part of the system is only slightly affected. Substituents S or S' in the remainder have an influence on the reaction which may be reduced to a classical component.

Scheme 3

The active part of the system may be divided into a small number of classes, according to the number and topology of the bonds involved in the reaction. The early stage of the reaction, as well as the later one, is principally governed by semiclassical interactions, and the semiclassical approximation may give some information about the electronic structure, geometry and energy of the TS when some non-active groups are changed.

These general indications must be carefully translated into well defined working hypotheses, and the hypotheses must be scrupulously checked step by step. This work remains to be done; in particular, we have not attempted as yet to establish prototypes for the active groups in the TS.

6.4 On the Inclusion of Correlation Effects in the Semiclassical Description

We shall pass in this following section to the consideration of more complex physical and material models (molecules in solution). The mathematical model also will become more complex, and so it will be expedient to keep the description of the molecular system at a relatively low level of sophistication. Thus in the following discussion we will retain the description at the ab initio full electron SCF level.

For the isolated molecules (or supermolecules) which we have considered thus far, this type of simplification is not justifiable. Post H—F methods are available and they are being increasingly used. One could then ask why we have limited our analyses to the SCF level, and whether the basic concepts of the semiclassical approach are compatible with molecular descriptions which include the electron correlation.

There are no objections, in principle, to the application of the semiclassical approach to submolecular units modified by the electron correlation. The question must be reformulated in a different way: are there correlation methods available which preserve the identity of molecular subunits of a certain kind, i.e. those that constitute the ideal basic building blocks of chemistry (see Sect. 1)?

Leaving aside "brute force" calculations, all the effective computational schemes for correlation are based on some physical model of the molecular subunit. The heuristic role played by the concept of "geminal" [17] must not be forgotten. The use of LOs as the starting point for CI calculations has been examined by a number of authors (for a review, see Ref. 175, to which Refs. 174, 176–185 may be added as partial list of further contributions) on the whole with positive results. It appears that the scheme which we have adopted at the SCF level can be extended to post H—F descriptions.

Efforts concentrating on the implementation of concurrent computational procedures have shifted the emphasis of the approach toward the evaluation of intrageminalic contributions, which actually represent a sizable portion of the whole correlation energy. The situation, however, is not qualitatively different from the one existing at the H—F level: a complete decoupling of the quantal interaction terms among subunits is not possible at both the correlated and the H—F levels. The descriptions at the two levels of the theory are also comparable quantitatively, the correspondence being even better in the case of correlation couplings. A recently published partitioning of the correlation energy contribution according to the "neighbourhood order" shows that the restriction of intra-group contributions to the first neighbour LO electron couples gives 97–98 % of the total correlation energy in hydrocarbons [186].

Some newly developed approaches may also be exploited, or adapted, to the ends considered in this section [186–190]. It is evident, however, that the procedure eventually chosen must generate as one of its outputs the charge distribution of the subunit. In many correlation methods the attention is focused on the energy, while for the utilizations we are talking about here the emphasis is on the charge distribution.

In conclusion, our opinion is relatively optimistic. The extension of the semiclassical approach to correlated descriptions of the electronic distribution seems to be possible, and the additional difficulties encountered in the implementation of the method do not appear to be insurmountable. Unfortunately, we have no direct experience in this field as yet.

7 Solvent Effects

Environmental effects, and solvent effects in particular, are good candidates for a semiclassical description. Specific solvent effects, like the formation of hydrogen bonds between the solute and solvent, are well described by the semiclassical model (see, e.g. Refs. 191–194). Non-specific solvent effects, a theme more appropriate to the subject of this review, are often treated with semiclassical approximations.

7.1 The Semiclassical Continuum Description of Solvent Effects

Models which introduce a continuum distribution of the solvent are classical or semiclassical in nature. The most commonly used approximations consider the electrostatic terms alone (i.e., the continuous distribution is reduced to a continuous dielectric), but the approach pioneered by Linder [195] and afterwards elaborated by Rivail [196] and by Olivares del Valle [197] indicate that the dispersion contribution to the solvation energy may also be recast in semiclassical terms.

We shall make use of a computational model developed in our laboratory [198–200] and applied to a number of chemical problems. In the version of the model considered here only the electrostatic contributions are retained.

The starting point is a pseudo-Hartree-Fock equation in which the Hamiltonian of the solute in vacuo is coupled to a solute-solvent interaction operator:

$$H_M(\text{in sol}) = H_M^0(\text{in vacuo}) + V_\sigma(M, \text{solvent}) \tag{35}$$

namely:

$$H_M \Psi' = E' \Psi'. \tag{36}$$

$V_\sigma(M)$ depends on several factors: the dielectric constant of the surrounding medium $\varepsilon(r)$, generally assumed to be constant and equal to the solvent bulk value at a given temperature; the size and shape of the cavity cut in the dielectric medium and containing M (a realistic cavity, defined in terms of intersecting van der Waals spheres); and the electrostatic potential of M, V'(M), computed using the solvent polarized wave function. V_σ is in fact defined in terms of an apparent charge distribution $\sigma(M)$ spread on the surface of the cavity containing M:

$$\sigma(M, s) = -[(\varepsilon - 1)/4\pi\varepsilon] [\partial(V'(M) + V(\sigma, M))/\partial n]_{s-} .$$

The partial derivative of Eq. (37) is computed at the points s defining the cavity surface, along the outer normal **n** to the cavity. Equation (37) is an implicit equation; in fact from Eq. (37) one may derive:

$$V(\sigma, M, \mathbf{r}) = \int (\sigma(M, s)/|\mathbf{r} - \mathbf{s}|) \, ds . \tag{38}$$

V'(M) in Eq. (37) is defined in terms of $\Gamma'(M)$, i.e. the first order density function of M, whose electronic part derives from Ψ' of Eq. (36). It is thus evident that (36) is also an implicit equation, its solution depending, via Eqs. (37) and (38), on the solution of Ψ' itself.

The energy of the system (actually a free energy, G) is related to the solution of Eq. (36), modified by an additive factor that depends once again on the solute charge distribution $\Gamma'(M)$ and on $V(\sigma, M)$:

$$G_{SCF}(M, sol) = E'_{tot}(M, sol) - 1/2 \int \Gamma'(M) \, V(\sigma, M) \, dr . \tag{39}$$

The electrostatic contribution to the free energy of solution is given by:

$$\Delta G^{sol}_{el}(M) = G_{SCF}(M, sol) - E^0_{tot}(M, vac) . \tag{40}$$

This computational procedure introduces, via $V(\sigma, M)$ a realistic and detailed description of the solvent reaction field $F(\sigma, M)$. The derivation of $F(\sigma, M)$ from $V(\sigma, M)$ is straightforward and is made even easier by the decomposition of $V(\sigma, M)$ in the local point charges which we use in the actual calculations [198, 201]. The use of cavities carefully modeled on the shape of the molecule, the iterative evaluation of V_σ (generally in the literature the reaction potential, or field, is limited to a first order estimate), and the direct evaluation of the potential without truncations in an expansion set all make the evaluation of this field quite realistic, and as accurate as our evaluations of the other fields considered in this paper.

Other people has been engaged in similar elaborations of more refined and powerful solvation models based on a continuous distribution of the bulk solvent. In a recent review [202] we suggested that there are generally three stages in the elaboration of such models: the first stage corresponds to the use of simple classical descriptions of the solute; the second to the simultaneous definitions of the quantum mechanical structure of the solvent, and of the (semi)classical reaction potential (this is the stage which we are discussing here); the third to the elaboration of more complex mathematical and physical models addressed to problems of greater complexity, such as those actually occurring in the experimental laboratory or in natural systems. The

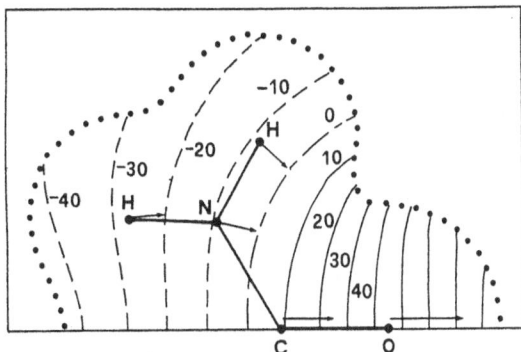

Fig. 14. Solvent ($\varepsilon = 78.5$) reaction potential $V(\sigma, M)$ for urea in the molecular plane. The isopotential lines are spaced by 10 kcal/mol. The strength and direction of the solvent reaction field can be fairly well evaluated by considering the number of isopotential lines and the direction of their normal vector. The fields at the C and O nuclei are 12.37 and 16.77 conv. u., respectively

very thorough review by Tapia [203] reflects the state of the art for the second step; it would be premature to attempt a review of the third stage at present, even though we have already obtained interesting results.

7.2 On the Use of the Solvent Reaction Potential and Field in the Semiclassical Interpretation of Solvent Effects

We shall first examine the characteristics of $V(\sigma, M)$ and $F(\sigma, M)$, considered as the sources of changes in both the total charge distribution of M and in the related observables. The solvent reaction field may be coupled to fields of different origin (such as the chemical substitution field, the geometry deformation field, the chemical interaction field, and the electronic excitation field, all of which we have either introduced in preceding sections, or which we shall introduce later on) in order to obtain models applicable to a wider range of phenomena at the molecular level.

We show in Fig. 14 a graphic representation of the solvent reaction potential $V(\sigma, M)$ for urea $(H_2N)_2C=O$ in water. In our opinion, the reaction potential is more significant, as an iconic model, than the corresponding field. The arrows on the atoms indicate the direction and strength of the reaction field on the atoms. The symmetry of the molecule makes a verbal analysis easy. The reaction field shifts the electrons of the carbonyl group towards the O end. In fact, the apparent charge

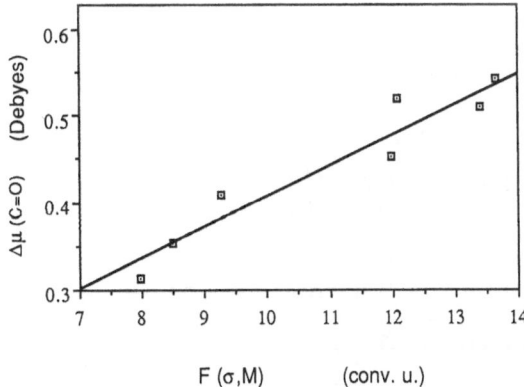

Fig. 15. Changes in the carbonyl dipole moments produced by the solvent reaction field as measured on the C and O nuclei in a set of carbonylic compounds

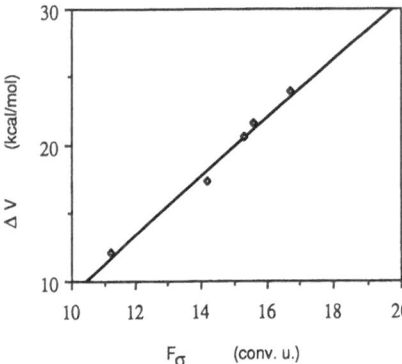

Fig. 16. Changes in the electrostatic potential V_{min} in the minimum along the CO bond produced by the solvent reaction field as measured in the same region in a set of carbonylic compounds

distribution σ is positive in the vicinity of the O atom: this gives rise to forces, $F(\sigma) Q(C)$ and $F(\sigma) Q(O)$, which push the electrons around the C and O atoms towards the periphery. The force acting on the C atom is smaller than that acting on the O atom and the net solvent effect gives more weight to the polar $C^+ - O^-$ structure in solution than in vacuo. The effect is primarily supported by the β_{CO} groups, since the lone pairs have, in this particular case as well as in general, a low polarizability. This effect influences all the local quantities (local dipole moments, nuclear charges, valence and bond indexes, molecular electrostatic potential), with correlations with respect to the value of $F(\sigma, M)$, computed either at the middle of the double bond or on the two atoms, similar to those found for substitution fields. The example of Fig. 15 refers to the correlation between $\Delta\mu(C=O)$ and $F(M/C=O)$.

As further example in Fig. 16 we have taken, for a specific set of compounds, the changes produced by the solvent in the values of the electrostatic potential at the bottom of the well in the proximity of the $C=O$ group, and compared them with the solvent electric field computed at the bottom of the potential well. The data of this figure are also reported in Table 13, which gives more details concerning the influence of the solvent field on the shape of the electrostatic potential in this region of the molecular space. All the aforementioned results refer to 4-31G calculations.

Table 13. Comparison of the molecular electrostatic potentials V (kcal/mol) along the CO direction at their minimum in vacuo and in solution, with respect to the solvent reaction field[a]

Molecule	in vacuo		in solution		
	R_{min}	V_{min}	R_{min}	V_{min}	F_σ
HCHO	1.22	−57.9	1.16	− 75.2	14.2
NH_2CHO	1.16	−79.3	1.16	−100.8	15.6
CH_3NHCHO	1.16	−82.1	1.16	−102.7	15.3
$(NH_2)_2CO$	1.15	−83.0	1.14	−106.9	16.7
CH_3OCHO	1.20	−61.2	1.17	− 73.3	11.2

[a] Electric field intensities in conventional units, at the separation R_{min} (Å) between the carbonyl oxygen and V_{min}

Table 14. Localized orbital contributions to the electrostatic potential (kcal/mol) in five selected points around methylformate (see Scheme 1) in water solution ($\varepsilon = 78.5$) computed using the 4-31G basis set

λ	1	2	3	4	5
$lp(O_2)$	10.4	−47.2	− 3.1	11.0	26.2
$lp(O_2)$	2.6	−47.3	17.8	−11.5	9.3
$\beta(CO)$ y > 0	23.2	− 3.4	24.8	− 0.4	−15.7
$\beta(CO)$ y < 0	23.2	− 3.4	24.8	− 0.4	62.6
$b(C_1H)$	− 2.6	11.7	9.9	5.8	17.0
$b(C_1O_1)$	− 0.7	8.6	17.5	− 7.3	−52.9
$lp(O_1)$	−62.1	6.3	− 4.1	8.4	−67.5
$lp(O_1)$	−63.0	6.4	− 4.1	8.4	69.3
$b(O_1C_2)$	− 6.3	− 6.5	− 8.6	37.3	−27.1
CH_3	21.6	1.6	6.5	−25.4	20.8
Total[a]	−53.6	−73.1	81.5	25.9	42.0
V_σ	20.8	46.3	−19.3	−12.3	− 2.1

[a] Including the contributions from the $1s$ orbitals

Let us come back again to Fig. 14. The effect of the reaction field on the N—H of the urea groups obviously depends on the orientation of the field with respect to the group. The apparent charge distribution σ is negative in the vicinity of an X—H group and pushes electrons from the molecular periphery towards the centre. This local effect is superseded, however, by the global shape of $F(\sigma, M)$ which in the case of urea is dominated by a dipolar component. Thus, the effect is visible for the N—H bonds which are parallel to the C—O axis (changes in the atomic population and in the local stretching force constant, for example) and are practically zero for the other two N—H groups.

We reported in Table 4 (Sect. 4) a decomposition of the molecular electrostatic potential for some selected positions near the methylformate molecule. This same decomposition is reconsidered in Table 14, now using solvent-polarized LOs. The data is supplemented by values for $V(\sigma)$ computed at the same positions. The solvent effects of the $V_i(r)$ values may be understood in qualitative terms which are very similar to those used above to explain the case of urea. Note how the ratios $V(M)/V(\sigma, M)$ differ in the set of points we have selected.

As a further element needed to understand the entity of solvent effects, we report in Figs. 17–19 a comparison of the effects of external fields acting on the group dipole moments of another chemical system, the $H_2C{=}NH$ molecule. Figures 17 and 18 show the effects arising from a positive point charge placed at varying distances along the N lone pair axis, while Fig. 19 illustrates the effects of solvents with increasing dielectric constants. The upper limit for the field strength, which in Figs. 13 and 14 corresponded to the field produced by a unitary charge placed at the equilibrium distance found for a Li^+ cation, in Fig. 19 corresponds to the field produced by a solvent with $\varepsilon = 100$. The different signs of the slopes in these figures are a result of the characteristics of the different fields, but what is important to remark here is that the absolute values of the fields all lie in the same range, and that there is a coherent trend in the slopes for all the groups. The N lone pair is the least polarizable group in the molecule.

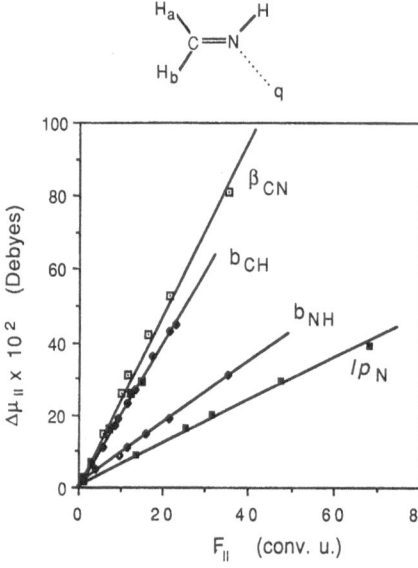

Fig. 17. Changes in the parallel component of the group dipole moments of $H_2C=NH$ produced by an external field arising from a positive point charge placed at different sites along the N lone pair axis (SCF/4-31G calculations). The fields are measured at the group charge centers

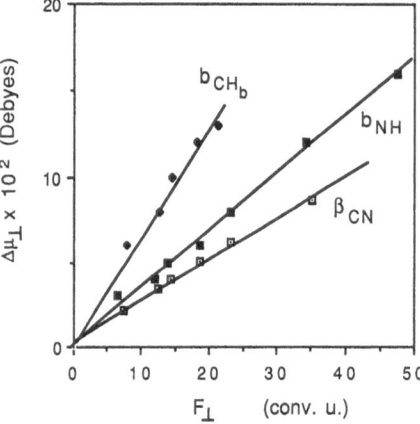

Fig. 18. As in Fig. 17, but referred now to the perpendicular components of $\Delta\mu$ and F

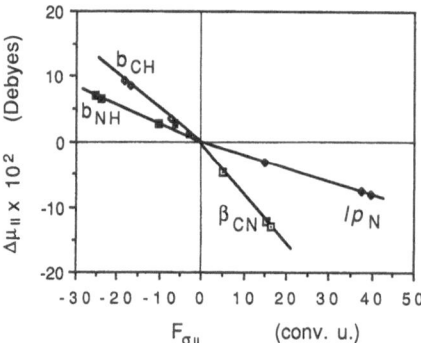

Fig. 19. Changes in the parallel component of the group dipole moments of $H_2C=NH$ produced by solvent reaction fields of increasing magnitude ($\varepsilon = 1$ to $\varepsilon = 80$) (SCF/4-31G calculations)

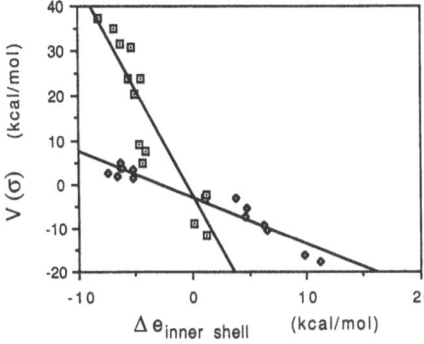

Fig. 20. Changes in the orbital energies of the inner shells of heteroatoms ($\Diamond$, r = 0.96) and of C atoms ($\Box$, r = 0.92) produced by the solvent reaction potential at the atom site in various conformers of N-methylformamide and methylformate

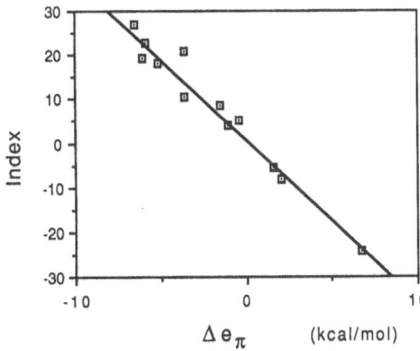

Fig. 21. Correlation (r = 0.98) between the changes in the orbital energies of the π orbitals, produced by the solvent reaction potential at the atom site, and a suitable index (see text)

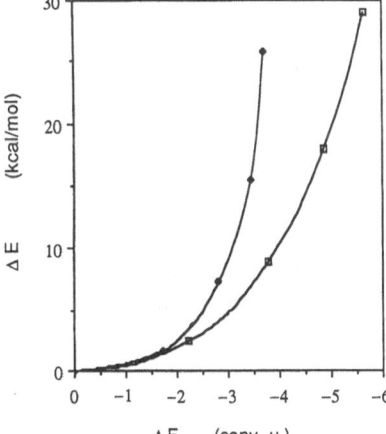

Fig. 22. Trend of the energy increase for successive constant increments of 0.05 Å in the C=O bond length for formaldehyde with respect to the difference in the fields acting on the C=O group in vacuo ($\Box$) and in solution ($\Diamond$). At R_{eq} = 1.206 Å, the field is 7.0 conv. u. in vacuo and 20.5 conv. u. in solution

The strength of the substitution field is also correlated with the strength of $F(\sigma, M)$: in Table 15 we compare the strength of the substituent field of some mono-substituted acrylonitriles with the $F(\sigma, M)$ of their parent compound. Other semiclassical correlations with the solvent field are of great potential interest. Among these we could mention the effects of solvent fields on the orbital energies, on the local force constants (stretching or bending), and on the potential energy surface in general.

Table 15. Electric fields (conv. u.) produced by different substituents as compared to the solvent ($\varepsilon = 78.5$) reaction field on a few selected bonds in acrylonitrile, computed using the 4-31G basis set (1 conv. u. $= 27\,360$ esu $\times$ cm^{-2})

	b(CH$_1$)	b(CH$_2$)	b(CH$_3$)	π(CC)	β(CN)	lp(N)
1-F	—	-6.00	2.20	-7.00	8.70	4.04
2-F	22.23	—	1.85	5.10	3.25	1.40
3-F	26.60	-0.40	—	5.80	3.97	1.24
1-NH$_2$	—	-15.63	-13.40	23.50	3.03	1.59
2-NH$_2$	8.12	—	-14.30	-31.20	-5.43	-3.48
3-NH$_2$	12.65	-11.10	—	-31.45	-6.48	-2.73
F$_\sigma$	5.15	3.40	3.09	4.60	11.68	13.27

As examples of the first item, the orbital energies, we can report some correlations between $V(\sigma, M)$ and the changes in the orbital energies Δe_i found for selected geometrical conformations of N-methylformamide and methylformate. Figure 20 refers to the inner shell (carbon atoms and hetero atoms), and Fig. 21 refers to the π orbitals. In the latter case the semiclassical quantity is a weighted average of the reaction potentials on the relevant atoms [204].

As an example of the second item, the local force constants, we report in Fig. 22 the correlation between the changes in energy, ΔE, arising from the elongation of the $C=O$ bond length, and the changes in the field experienced by the group, in vacuo and in solution. In solution the field also includes the contribution coming from the solvent reaction field.

The third item, the effects on the potential energy surface of M, can be divided into many separate sub-items. Changes in the equilibrium geometry for classical molecules are generally modest in magnitude; they correlate well, in the examples we have worked out, with the prediction given by $\mathbf{F}(\sigma, M)$. Changes in the relative stability of different local minima are also explained considering the shape and strength of $\mathbf{F}(\sigma, M)$ [204–207]. Here again the rotational barriers offer a promising field of investigation. The rotational barrier of a simple group, such as CH$_3$ in asymmetric environments, is greatly influenced by the solvent. We have studied the rotations in some selected geometries of N-methylformamide and methylformate and in Fig. 23 we report the correlation between the difference in the barrier heigth, $\Delta\Delta G$, of these rotations and a composite index which takes into account the changes in $\mathbf{F}(\sigma, M)$ experienced by the 3 C—H groups during the rotation. This is a first attempt which surely will be improved.

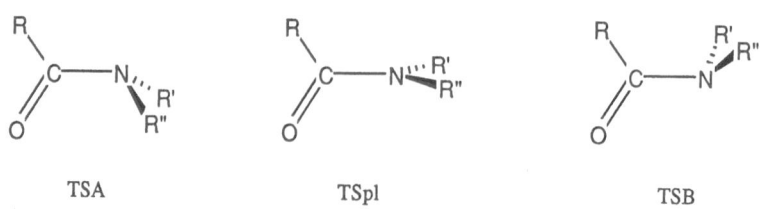

TSA   TSpl   TSB

Scheme 4

The N-methylformamide molecule can be used to illustrate other types of solvent effects which occur in the proximity of transition states. The rotation around the $N—C_\alpha$ axis in amides presents an energy barrier. The almost planar conformation of the $—NR_2$ group in equilibrium geometries is lost in the proximity of this barrier, and in principle the pyramidalization at the N atom could be directed in two opposite directions (see Scheme 4).

Actual calculations at the SCF level (confirmed by introducing electron correlation effects) indicate that in vacuo only the channel through TSA is open [204, 208]. This effect is confirmed by a semiclassical analysis of the forces acting on the groups. In solution ($\varepsilon = 78.5$) both channels TSA and TSB are open, and have barriers of comparable height. This effect is solely due to the influence of $F(\sigma, M)$ which acts in favor of the second channel.

A schematic energy drawing which compares the relative energies of TSA, TSpl and TSB, in vacuo and in solution, is given in Fig. 24. More details will be provided in a forthcoming subsection, where simplified expressions of the reaction field will be considered.

In closing this subsection, we would like to note that the exposition presented here was rather more impressionistic and less detailed than those of preceding sections. One reason for this is the lower abundance of fully worked examples in this relatively new field, but another reason was our desire to avoid overwhelming the reader with excessive numerical documentation. It must suffice to remark that every result presented in the preceding pages, despite its relatively sparse commentary and limited mathematical documentation, represents the fruit of repeated tests on many aspects of the mathematical and physical model (basis set effects, the transformation and truncation of molecular orbitals, the definition of the indexes, as in the in vacuo studies, etc.), supplemented by various new features introduced by the solvent model: the shape and dimension of the cavity, number of iteration cycles, etc.

The literature documenting the use of the solvent reaction field concept in the interpretation of physico-chemical phenomena at the molecular or submolecular level is quite extensive. The classical stage of the continuum theory (the first step, as defined at the end of Sect. 7.1) has been, and still continues to be, the standard model in chemistry for explaining solvent effects. In this vast literature we would like to cite the meticulous, on-going work of Ehrenson, who in many papers (see, e.g. Refs.

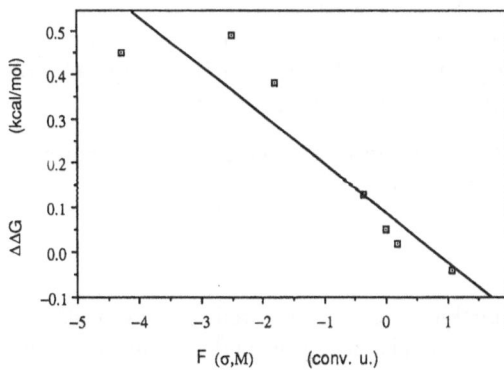

Fig. 23. Correlation ($r = 0.92$) between the changes in the ΔG produced by a rotation of the methyl group and the solvent reaction field acting on the group itself

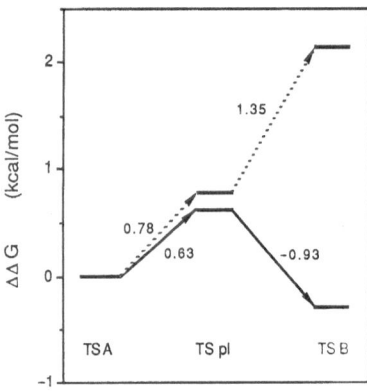

Fig. 24. $\Delta\Delta G$ in vacuo (dashed line) and in water (solid line) for the various transition states of N-methylformamide with respect to TSA taken as zero

209–212) has made valuable contributions to the semiclassical interpretation of these effects, including considerable improvements to the original classical methodology.

7.3 The Use of Prototypes in the Description of Solvent Effects

In Sect. 7.2 we have considered the semiclassical interpretation of solvent effects on molecular properties. In the present section we shall try to introduce the simplified expressions discussed in Sects. 5 and 6, in the description of environmental effects.

Equations (36)–(38) make it clear that the solute one-electron density function has an important role to play in assessing the solvent effects on the electronic distribution and on the energy of the solute itself.

We may therefore explore the effect of introducing into $\Gamma(M)$ the simplifications discussed above.

For the sake of completeness we shall also introduce in this sequence of approximations some levels in which the partition of $\Gamma(M)$ into group contributions is not exploited. The main quantities of interest will be the reaction potential $V(\sigma, M)$ (and the related field) and the solvation energy ΔG^{sol}.

— *Approximation 1.* The full ab initio description of the system.

In general we have employed the theory at the SCF level (RHF, UHF, and EHP for excited states) with a few exceptions [205]. The inclusion of correlation effects has been attempted thus far in only a few cases, and we feel it would be premature to take such effects into consideration here. We shall use $\Gamma^{(1)}(M)$ or $\Gamma^{sol}(M)$ to indicate the electrostatic charge distribution obtained at the end of the SCF iterative procedure outlined in Eqs. (36)–(38). The free energy of M, and the solvation free energy (electrostatic components) are given by expressions (39) and (40), respectively.

— *Approximation 2.* The level of the theory in this case is the same as in Approx. 1, but the solvation free energy is expressed in a simpler form:

$$\Delta G_2^{sol} = 1/2 \int \Gamma^{(1)}(M) \, V(\sigma, M) \, dr \ . \tag{41}$$

The expression may be justified by invoking the virial theorem and by neglecting differences between the mean electronic energies in vacuo and in solution. When applied to conformational problems the local effects on conformational surfaces are in general less than 0.2 kcal/mol (0.8 kJ/mol).

— *Approximation 3*. The solute charge distribution is described here in terms of the SCF wave function in vacuo, $\Gamma^{vac}(M)$. This approximation corresponds to the neglecting of the polarization of M induced by the reaction field. Polarization effects are important for the evaluation of the solvent effects on various molecular observables, including the energy. We have seen in Sect. 7.2 that solvent fields and internal substitution fields are of the same order of magnitude, and that in some cases they add to each other, whereas in other cases there is a partial compensation.

The effect of this approximation on G depends on the size of the molecule and on the basis set. To give a rough idea of the numerical values, we may quote the case of the simplest aminoacids, in which the polarization contribution to G may reach 8–10 kcal/mol when evaluated with a good basis set. The differential effects on the conformational surfaces of the same molecules are less than 0.3–0.4 kcal/mol.

The solvation free energy thus assumes the simple form:

$$\Delta G_3^{sol} = 1/2 \int \Gamma^{vac}(M) \, V(\sigma^0, M) \, d\mathbf{r} \tag{42}$$

and the iterative solution of Eq. (38) is no longer necessary. Only the solution of the Schroedinger equation in vacuo, accompanied by the calculations related to Eq. (36) are needed.

Thus far we have introduced levels of approximations which make use of a quantum mechanical description of the solute charge distribution. In the following steps we shall introduce semiclassical descriptions of the solute.

— *Approximation 4*. The prototype functions $\gamma(g, 0)$ are here introduced and subjected to modifications due the molecular remainder field, and to the solvent field. Formally, we can write (compare with expression (30)):

$$\gamma(g, 0) \xrightarrow{\mathbf{F}(M/g, 0) + \mathbf{F}(\sigma, M, 0)} \gamma^*(g, M, \sigma) . \tag{43}$$

The resulting sum of $\gamma^*(g, M, \sigma)$ group distributions, i.e.:

$$\Gamma_4^*(M, \sigma) = \sum_g \gamma^*(g, M, \sigma) \tag{44}$$

is used in Eq. (36). This equation, as previously remarked, is an implicit equation, to be solved iteratively. In actual calculations we have interrupted the iterative cycle at the first step (convergency in SCF calculations is generally achieved in 3 steps). The resulting reaction potential will be called:

$$V_4(\sigma, M) = \int (\Gamma_4^*(M, \sigma)/|\mathbf{r} - \mathbf{r}'|) \, d\mathbf{r}' . \tag{45}$$

As was discussed in Sect. 5.2.2, it is possible to arrive at a rough evaluation of E, and then of G, by using a description of the charge distribution in non-orthogonal fragments, which proved to be of some use in conformational problems. We have not exploited this possibility, since we prefer to use an expression for ΔG^{sol} similar to that given in Approx. 2:

$$\Delta G_4^{sol} = 1/2 \int \Gamma_4^*(M, \sigma) \, V_4(\sigma, M) \, d\mathbf{r} \, . \tag{46}$$

By rearranging Eq. (40), the energy of the system may be written as:

$$G(M) = E_{tot}^0(M, vac) + \Delta G_{el}^{sol}(M) \, . \tag{47}$$

For calculations on large molecules, the ab initio evaluation of $E_{tot}^0(M)$ may be replaced by faster methods (e.g. MNDO or AM1 calculations) without detracting much from the quality of the results. Calculations now under completion seem to indicate that the coupling of empirical methods (MM2 [213], AMBER [214]) to compute E_{tot}^0, with ΔG^{sol} values computed with the aid of Approx. 4 (and of Approximations 5 and 6) give fairly good results.

— *Approximation 5.* The prototype functions $\gamma(g, 0)$ are subjected to solely those modifications coming from the molecular remainder:

$$\gamma(g, 0) \xrightarrow{\ F(M/g,\ 0)\ } \gamma^*(g, M, 0) \, . \tag{48}$$

Solvent polarization effects are neglected. This approximation is the counterpart of Approx. 3, using transferable groups instead of the in vacuo SCF wave function.

The resulting sum of group distributions $\Gamma^*(M)$, defined in Eq. (31), is used in Eq. (36), and gives rise to a reaction potential called:

$$V_5(\sigma, M) = \int (\Gamma^*(M, 0)/|\mathbf{r} - \mathbf{r}'|) \, d\mathbf{r}' \, . \tag{49}$$

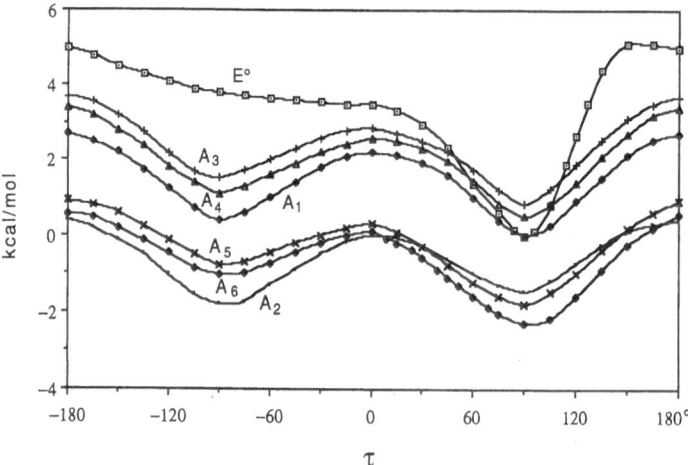

Fig. 25. A comparison of the conformational curves (referring to the rotation of the OH group in HOCH$_2$NHNO) obtained using the approximations described in the text. The E^0 curve has been arbitrarily shifted in order to make it coincident with the curve for A_1 at $\tau(HOCN) = 90°$. The E^0 and $A_1 = \Delta G_{el}$ curves have been obtained with full 3-21G calculations, whereas the others refer to models making use of the same basis set

The subsequent steps are similar to those already described for Approx. 4.

— *Approximation 6.* The prototype functions $\gamma(g, 0)$ here are employed without modifications. The resulting charge distribution is applied in Eq. (36) to give:

$$V_6(\sigma, M) = \int (\Gamma(M, 0)/|\mathbf{r} - \mathbf{r}'|) \, d\mathbf{r}' .$$

(50)

The following steps are similar to those described for Approx. 4.

To this sequence of six approximation levels we could add further steps, introducing simplifications in the descriptions of the $\gamma(g, 0)$ functions, or in a part of them. We have pointed out before that the LCAO description of $\gamma(g, 0)$ may be replaced by simpler descriptions. In the case of solvent effects the description of groups far from the cavity boundary is less critical. For large solutes, like enzymes, a significant portion of the system can, in our opinion, be roughly represented, or even reduced to a continuous dielectric.

A parallel approach to simplifying the description of solvent effects may be derived from an examination of the coupling effects involved in the determination of $V(\sigma)$. The electrostatic effects which give rise to the reaction potential, and subsequently to the solvation energy, are not linear. Preliminary attempts to decouple the various contributions have been documented on an earlier occasion [72], and we do not consider it appropriate to introduce here a topic for which the numerical results are still too scarse. We can say, however, that a decoupling for another solvation effect, that due to dispersion forces, has already led to the proposition of less costly computational procedures [200, 215].

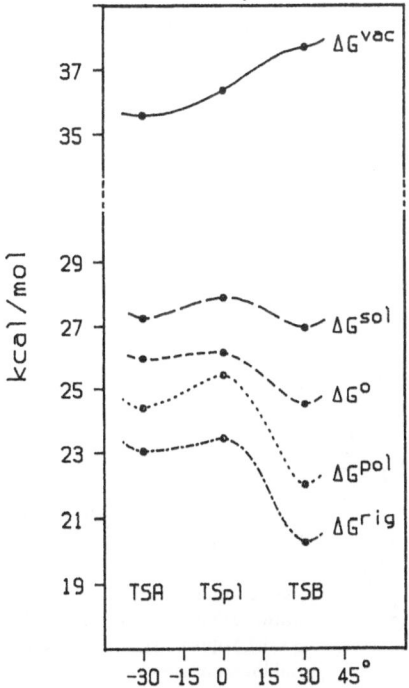

Fig. 26. Trend of the potential energy curves corresponding to the different pyramidalizations at the N atom for *N*-methylformamide, with respect to the ΔG^{sol} of the *trans* conformer taken as zero: ΔG^{vac} (*solid line*); ΔG^{sol} (*long dashed line*); ΔG^0 (*dashed line*, Eq. 41); ΔG^{pol} (*dotted line*, Eq. 50); ΔG^{rig} ($- \cdot - \cdot -$, Eq. 49)

It must be stressed that the long sequence of approximations outlined in the preceding pages is addressed not only to the search for more effective computational procedures, but, more importantly, is being used to elaborate tools able to shed more light on the very nature of environmental processes.

After this long formal description, we can now present some numerical results. Figure 25 reports the unscaled values of a conformational curve for $HOCH_2NHNO$, obtained with the sequence of approximations 1–6. In the same graph we have also added the in vacuo energy profile (E_{tot}^0), scaled to have the same value as Approx. 1 at the minimum. This curve has been added here to show that solvation effects may also produce qualitative changes in the conformational energy profile. An accurate evaluation of G in solution requires, as said before, the evaluation of other contributions [202, 204, 216].

In Sect. 7.2 we noted that in polar solvents there are two pathways, and two TSs, for the rotation about the amidic bond $OHC-N<$. We report in Fig. 26 a description of the energy of the three conformations TSA, TSpl and TSB, depicted in Scheme 4, obtained with Approximations 1, 2, 5 and 6. It is interesting to note that even the crudest approximation (Approx. 6) is able to reproduce this feature of the PES in solution. This finding strongly suggests that the appearance of the second TS in amides is mainly due to direct solvent effects, and not to secondary effects caused by polarization. The polarization inside the molecule, as well as the solvent polarization, partially smooth this direct effect.

Our interpretation of solvent effects is essentially based on the examination of the $V(\sigma, M)$ and $F(\sigma, M)$ maps. An example of $V(\sigma, M)$ obtained with Approx. 1 has been already reported in Fig. 14. We can now compare (Figs. 27, 28 and 29) the more complex maps obtained for methylformate in its trans conformation with Approximations 1, 5 and 6. Analogous good results have been found for the *cis* conformation and for distorted geometries. Comparable examples may be found in Ref. 217.

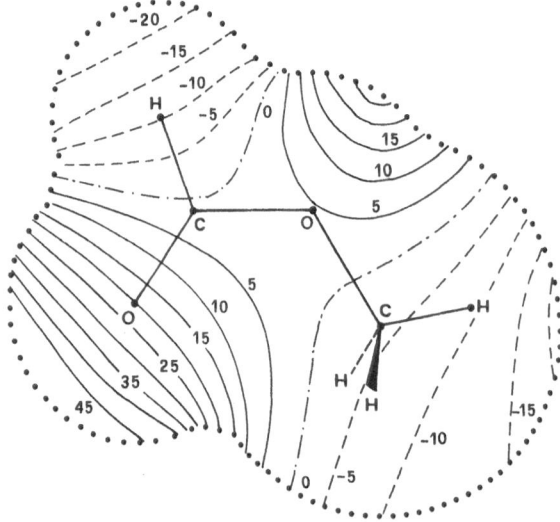

Fig. 27. Solvent ($\varepsilon = 78.5$) reaction potential $V(\sigma, M)$ for *trans* methylformate in the molecular plane, obtained from a full ab initio calculation at the 4-31G level (Approx. 1). The isopotential lines are spaced by 5 kcal/mol

8 Electronic Excitation Effects

8.1 Semiclassical Methods in the Study of Vertical Excitations

When the electronic excitation involves a valence state for a well localized chromophore, the passage from the ground state (GS) to the excited state (EX) may be considered as a special kind of chemical substitution. Our experience is limited to the lowest singlet and triplet $(n \to \pi^*)$, $(\pi \to \pi^*)$, $(n \to \sigma^*)$ and $(\pi \to \sigma^*)$ excitations in simple chromophores like $X=Y$, $(X, Y = C, N, O)$, and aromatic rings (benzene derivatives and heterocycles).

We shall consider as an example here the $^1(n \to \pi^*)$ excitation in $C=X$ chromophores. The new group present in EX differs from the corresponding group in GS because of the presence of an electron in the π^* orbital of the fragment and because of the lack of one electron in the lone pairs of the group. Formula (4), which gives the partition of nuclear charges necessary to derive the neutral subunits, may be easily rearranged to cope with these electronic distributions.

Excitation processes may be described in terms of canonical orbitals, which can ultimately produce a localized description of the process [72], but it is advisable to use LOs from the very beginning [218]. Our calculations have, in general, been carried out using simple methods to describe the excitation process — in particular, the Electron Hole Potential (EHP) of Iwata and Morokuma [219] which optimizes by using a variational method, both the description of the hole in the space of the occupied orbitals, and the description of the electron in the space of the virtual SCF orbitals. This description is better than the one given by the complete account of single excitation configurations (CSECI), and for simple chromophores it is not too far from those generated by much more sophisticated methods [220].

The EHP procedure induces changes in all the occupied MOs and therefore our prototypes $\gamma(g, 0)$ must be changed accordingly. Actually, in the cases we have

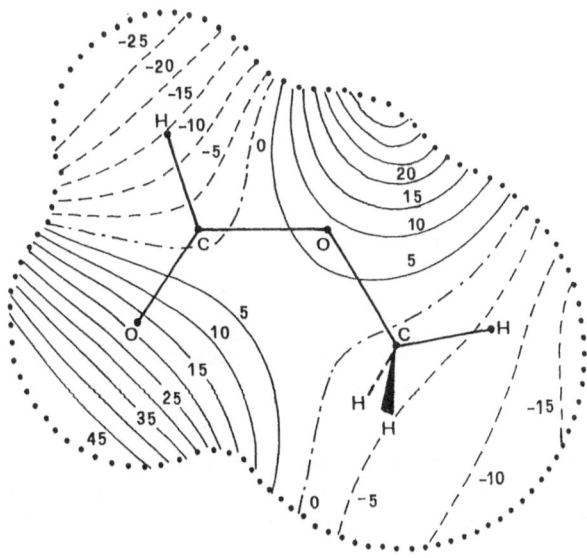

Fig. 28. Solvent $(\varepsilon = 78.5)$ reaction potential $V(\sigma, M)$ for *trans* methylformate in the molecular plane, obtained from a group description modified by the molecular field (Eq. 49), at the 4-31G level (Approx. 5). The isopotential lines are spaced by 5 kcal/mol

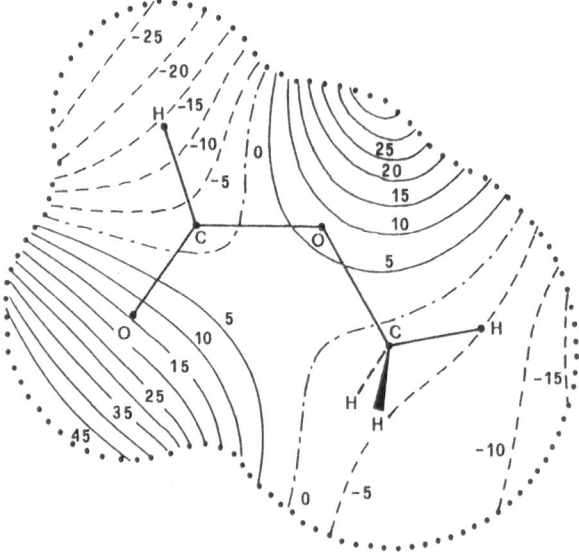

Fig. 29. Solvent ($\varepsilon = 78.5$) reaction potential $V(\sigma, M)$ for *trans* methylformate in the molecular plane, obtained from a given general library group description (Eq. 50), at the 4-31G level (Approx. 6). The isopotential lines are spaced by 5 kcal/mol

examined the changes required have been quite modest. So it seems that the same set of prototypes, $\gamma(g, 0)$, may be used in the description of groups of M(EX) not directly involved in the description [72].

We have built up a small library (with only a limited number of cases thus far) of $\gamma(g^{ex}, 0)$ groups which may be used to describe the (1e, −1e) electron function in

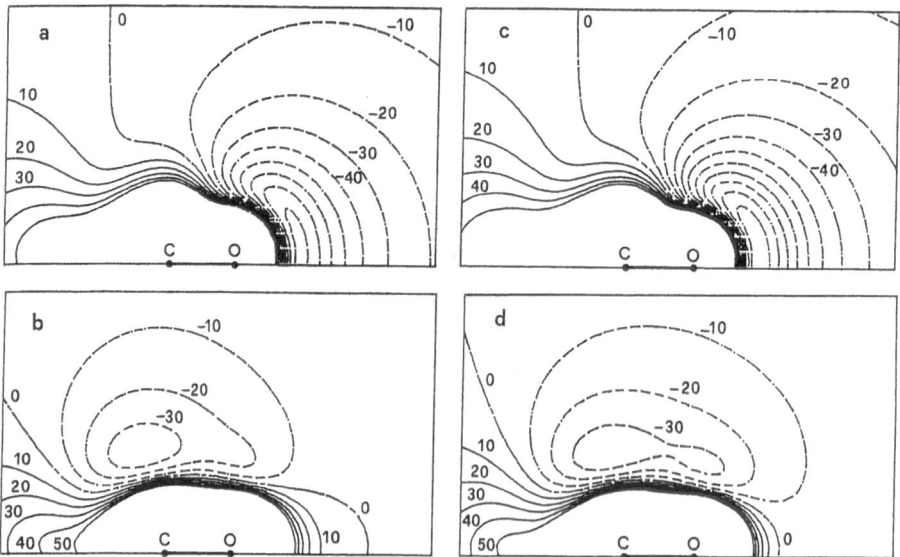

Fig. 30a–d. Comparison of the electrostatic potential (4-31G) along the C=O group in the plane perpendicular to the molecular one for urea: **(a)** ground state (GS) in vacuo; **(b)** GS in solution ($\varepsilon = 78.5$); **(c)** $^1(n \rightarrow \pi^*)$ excited state (EX) in vacuo; **(d)** $^1(n \rightarrow \pi^*)$ EX in solution ($\varepsilon = 78.5$). The isopotential lines are spaced by 10 kcal/mol

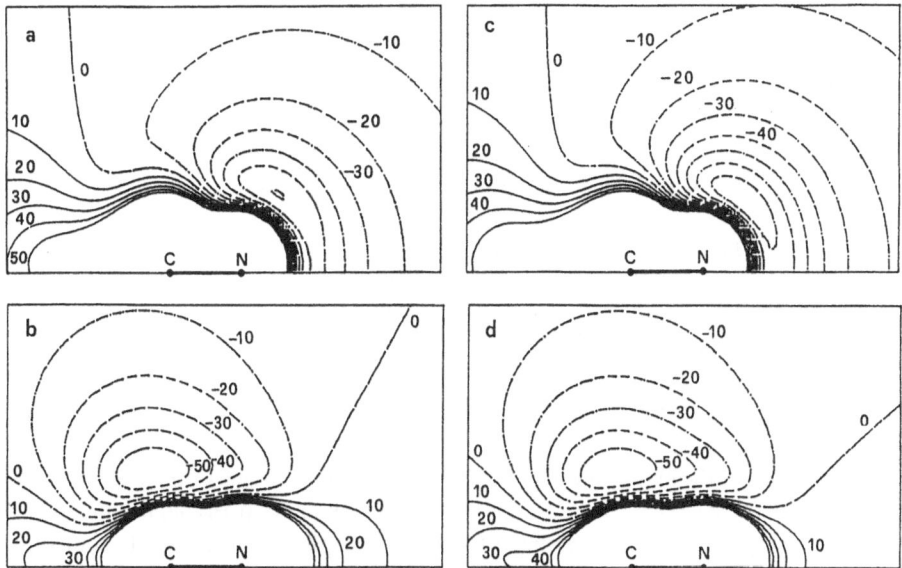

Fig. 31 a–d. Comparison of the electrostatic potential (4-31G) along the C=NH group in the plane perpendicular to the molecular one for guanidine. Same remarks as in Fig. 30

the EHP method. The tails are larger than for normal closed shell groups, but still of a reasonable extent. We are not yet in a position to give, however, a meaningful statistical elaboration of these submolecular units. The hole left in the GS closed shell distribution is described by replacing the $(+ne, -ne)$ description of the pertinent group by a $(+(n-1)e, -(n-1)e)$ description without other changes. This rather crude definition (corresponding to a working hypothesis of presumed low accuracy) is addressed not to the evaluation of the vertical excitation energy, but rather to an appraisal of the changes taking place in the molecular field.

The formulation of this preliminary working hypothesis is justified by an examination of the electrostatic potential and the field generated by the EX wave functions in molecules with simple chromophores.

The molecular electrostatic potential is more suited than the corresponding charge distribution to be represented iconically. Therefore, we report in Figs. 30 and 31

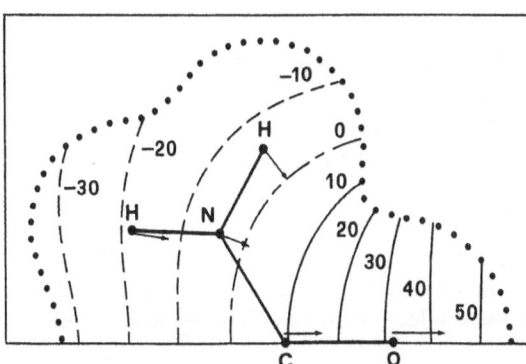

Fig. 32. Solvent ($\varepsilon = 78.5$) reaction potential $V(\sigma, M)$ for the $^1(n \rightarrow \pi^*)$ state of urea in the molecular plane. The isopotential lines are spaced by 10 kcal/mol. The strength and direction of the solvent reaction field can be fairly well evaluated by considering the number of isopotential lines and the direction of their normal vector. The fields at the C and O nuclei are 8.16 and 10.32 conv. u., respectively

some examples of V(M) maps for the GS and the EX. These maps are accompanied by similar plots referring to the same systems in solution. We shall consider this second part of the figures in the following paragraphs.

A simple visual comparison of the maps for GS and EX demonstrates that our simple picture should be adequate to reproduce the essential features of the electronic rearrangements which follow electronic excitation. Similar comparisons dealing with different EX states may be found in Refs. 70–72.

We have thus far examined the vertical excitation GS $\rightarrow$ EX. Electronic excitations are often accompanied by a change in the equilibrium geometry, however, and we have not yet examined this point in detail, even though it is of paramount importance for assessing the photochemical behavior of a molecule.

In the $>C=O$ chromophores, to which we have devoted our utmost attention, the $^{1,3}(n \rightarrow \pi^*)$ excitation is accompanied by a lengthening of the $C=O$ distance and by a pyramidalization at the carbon atom.

A dissection of the classical forces acting on the various atoms of the EX molecule, on the geometry of the GS, or on a slightly distorted geometry (to avoid the metastable condition produced by the presence of a symmetry plane), gives results similar to those found in comparable analyses of the reaction TS, when chemical substitutions are introduced. Once again our numerical data are not sufficient for a statistical analysis, but we have not yet found any counterexample which invalidates our hypothesis that the semiclassical methods applied to GS systems can be reformulated and used for molecules in their excited states.

8.2 The Combination of Fields of Different Origin; Electronic Excitation and Solvation

One of the final goals of our research is the evaluation and interpretation of a combination of different fields acting on the molecule at the same time.

In this final subsection of our review the combination of environmental effects with electronic excitations will be briefly considered.

At the beginning of our investigation of this subject we examined the influence of the physical composition of the model on the energetics of the excitation process, by comparing several cases: a) the supermolecule model, $M + nH_2O$; b) the mixed model, $M + nH_2O + $ continuum; and c) the simple solvation model, $M + $ continuum [218, 221, 222]. With the exception of those cases in which the solvent takes an active part in the process (e.g., solvolysis induced by electron excitation) we reached the conclusion that the simplest model, (c), shows performances comparable to the mixed model, (b), and superior to that of the supermolecule one (a). A comparison of the electrostatic potential maps for the GS and the EX in vacuo suggests that the significant differences seen between them should also affect the values of the apparent charge distribution σ on the cavity surface, and the resulting $V(\sigma, M)$ potential for the two states.

Actually the description of solvent effects on electronic transitions is relatively complex, because during the electronic excitation the positional, orientational and electronic polarization of the medium occur on very different timescales. Some suggestions regarding how to treat these effects with time-independent techniques are given in Ref. 223.

For very short timescales corresponding to the excitation act, orientational changes in the solvent molecules will not be effective. The surface polarization charge must therefore be divided into components — the orientational component, which depends on the charge distribution (and the electrostatic potential) of the GS, and the inductive component which depends on the electrostatic potential of the EX:

$$\sigma_{tot}(\text{vert. exc. GS} \rightarrow \text{EX}) = \sigma_{or}(V(GS)) + \sigma_{ind}(V(EX)). \tag{51}$$

In a solvent with a high dielectric constant, σ_{or} is much more important than σ_{ind}; therefore the shape assumed by $V(\sigma, M)$ immediately after the electronic excitation is not too different from that of the molecule at equilibrium in the GS. Figure 32 shows the map of $V(\sigma, EX)$ for urea obtained under these assumptions. The insert d in Figs. 30 and 31 report the electrostatic potential maps of urea and guanidine computed with this form of the physical model.

The results are different if one examines the EX state after a time sufficient enough to allow for solvent reorientation. In this case, discarding for the sake of simplicity all changes in the internal geometry of M (relaxation after the vertical excitation), the apparent charge distribution is a function of $V(EX)$ only:

$$\sigma_{tot}(EX) = \sigma_{or}(V(EX)) + \sigma_{ind}(V(EX)). \tag{52}$$

The shape of $V(\sigma)$ is now noticeably different. The effects of the solvation field on the molecular properties of M(EX), in equilibrium with the solvent, are different as well.

This qualitative analysis could be extended to the electronic processes involved in the passage to a lower excited state, or to the GS, for which other definitions of σ may be devised [202, 223, 224]. This being said, it is sufficient to reiterate that the extension and verification of the limits of the semiclassical interpretation of intramolecular effects in excited states is still in its preliminary stages.

In the introduction to this review we stressed the enormous variety of chemical structures which are of actual or potential interest in chemistry. Here at the end, we are compelled to consider the even larger variety of physical and chemical interactions in which these structures may be involved. The efforts of different research groups active in the field, whose numbers far exceed what we have been able to document here, seem lamentably restricted when compared to the number and variety of problems which still await theoretical and experimental investigation.

9 References

1. Perrin J (1913) Les Atomes, Alcan, Paris
2. Heitler W, London F (1927) Z. Physik. 44: 455
3. Moffit W (1951) Proc. Roy. Soc. (London) A210: 224, 245
4. Arai T (1960) Rev. Mod. Phys. 32: 370
5. Maksić ZB, Eckert-Maksić M, Rupnik K (1984) Croat. Chem. Acta 57: 1295
6. Pauling L (1931) J. Am. Chem. Soc. 53: 1367, 3225
7. Slater JC (1931) Phys. Rev. 37: 481
8. Pauling L (1960) The nature of the chemical bond, Cornell University Press
9. Wheland GW (1955) Resonance in organic chemistry, Wiley, New York
10. Cooper DL, Gerratt J, Raimondi M (1986) Adv. Chem. Phys. 69: 318

11. Mulliken RS (1932) Phys. Rev. 41:49; (1949) J. Chim. Phys. 46: 497, 675
12. Davidson ER (1972) Rev. Mod. Phys. 44: 451
13a. Foster JP, Weinhold F (1980) J. Am. Chem. Soc. 102: 7211
13b. Carpenter JE, Weinhold F (1988) J. Am. Chem. Soc. 110: 368
14. Fukui K, Fujimoto H (1968) Bull. Chem. Soc. Jpn. 41: 1989
15. Jorgensen WL, Salem L (1970) The organic chemist's book of orbitals, Wiley-Interscience, New York
16. Dewar MJS, Dougherty RC (1975) The PMO theory of organic chemistry, Plenum Press, New York
17. Epiotis ND, Cherry WR, Shaik S, Yates L, Bernardi F (1977) Top. Curr. Chem. 70: 1
18. McWeeny R (1960) Rev. Mod. Phys. 32: 335
19. Daudel R (1953) C. R. Acad. Sci. 237: 601
20. Aslangul C, Constanciel R, Daudel R, Kottis P (1972) Adv. Quant. Chem. 6: 93
21. Ludeña EV (1975) Int. J. Quantum Chem. 9: 1069
22. Leroy G, Peeters D, Tihange M (1985) J. Mol. Struct. (Theochem) 123: 243
23. Tomasi J (1988) J. Mol. Struct. (Theochem) 179: 273
24. Trindle C (1984) Croat. Chem. Acta 57: 1231
25. Dewar MJS (1984) J. Am. Chem. Soc. 106: 669
26. Popper K (1959) The logic of scientific discovery, Basic Book, New York
27. Lakatos I (1970) In: Lakatos I, Musgrave A (eds) Criticism and growth of knowledge, Univ. Press, Cambridge
28. Alagona G, Bonaccorsi R, Ghio C, Tomasi J (1986) J. Mol. Struct. (Theochem) 135: 39
29. McWeeny R (1957) Proc. Roy. Soc. (London) Ser. A235: 496
30. Foster JM, Boys SF (1960) Rev. Mod. Phys. 32: 300
31. Edmiston C, Ruedenberg K (1963) Rev. Mod. Phys. 35: 457
32. Magnasco V, Perico A (1967) J. Chem. Phys. 47: 971
33. Boys SF (1966) In: Löwdin PO (ed) Quantum theory of atoms, molecules and the solid state, Academic, New York, p 253
34. Dewar MJS (1985) J. Phys. Chem. 89: 2145
35. Adams WH (1961) J. Chem. Phys. 34: 89, (1962) 37: 2099, (1965) 42: 4030
36. Gilbert TL (1964) In: Löwdin PO, Pullman B (eds) Molecular orbitals in chemistry, physics and biology, Academic, New York, p 405
37. Gilbert TL (1972) Phys. Rev. A6: 580, (1974) J. Chem. Phys. 60: 3835
38. Anderson PW (1968) Phys. Rev. Letters 21: 13, (1969) Phys. Rev. 181: 25
39. Schlosser H (1972) J. Chem. Phys. 57: 4332, 4342, (1974) 61: 2814
40. Matsuoka O (1977) J. Chem. Phys. 66: 1245
41. Peters D (1963) J. Chem. Soc. 2003, 2015, 4017
42. Daudey JP (1974) Chem. Phys. Letters 24: 574
43. Levy M, Nee TS, Parr RG (1975) J. Chem. Phys. 63: 316
44. von Niessen W (1971) J. Chem. Phys. 55: 1948, (1972) 56: 4290
45. Payne PW (1977) J. Am. Chem. Soc. 99: 3787
46. Sundberg KR, Bicerano J, Lipscomb WN (1979) J. Chem. Phys. 71: 1515
47. Stoll H, Wagenblast G, Preuss H (1978) J. Am. Chem. Soc. 100: 7742, (1978) Theor. Chim. Acta 49: 67, (1980) 57: 169
48. Böhm MC (1981) Theor. Chim. Acta 59: 609
49a. Aufderheide KH (1980) J. Chem. Phys. 73: 1777
49b. Aufderheide KH, Chung Phillips A (1982) J. Chem. Phys. 76: 1897
50. Leonard JM, Luken WL (1982) Theor. Chim. Acta 62: 107
51. Luken WL, Culberson JC (1984) Theor. Chim. Acta 66: 279
52. Reed AE, Weinhold F (1985) J. Chem. Phys. 83: 1736
53. Rajzmann M, Brenier B, Purcell KF (1987) Theor. Chim. Acta 72: 13
54. Weinstein H, Pauncz R, Cohen M (1971) Adv. Atom. Mol. Phys. 7: 97
55. O'Leary B, Duke BJ, Eilers JE (1977) Adv. Quant. Chem. 9: 1
56. Chalvet O, Daudel R, Diner S, Malrieu JP (eds) (1975) Localization and delocalization in quantum chemistry, vol I, Reidel, Dordrecht
57. Newton MD, Switkes E, Lipscomb WN (1970) J. Chem. Phys. 53: 2645
58. Levy M (1976) J. Chem. Phys. 65: 2473

59. Surján PR, Mayer I (1981) Theor. Chim. Acta 59: 603
60. Surján PR, Mayer I, Kertész M (1982) J. Chem. Phys. 77: 2454; Mayer I, Surján PR (1984) J. Chem. Phys. 80: 5649
61. Aray Y, Gomperts R, Saavedra J, Urdaneta C, Murgich J (1988) Theor. Chim. Acta 73: 279
62. England W, Salmon LS, Ruedemberg K (1971) Top. Curr. Chem. 23: 31
63. Lipscomb WN (1973) Acc. Chem. Res. 8: 257
64. Bonaccorsi R, Ghio C, Scrocco E, Tomasi J (1980), Israel J. Chem. 19: 109
65. Daudey JP, Malrieu JP, Rojas O (1975) In: Chalvet O et al. (eds) Localization and delocalization in quantum chemistry, vol. I, Reidel, Dordrecht, p 155
66. Scrocco E, Tomasi J (1973) Top. Curr. Chem. 42: 95
67. Csizmadia IG (1975) In: Chalvet O et al. (eds) Localization and delocalization in quantum chemistry, vol I, Reidel, Dordrecht, p 349
68. Kapuy E, Kozmutza C, Stephens ME (1976) Theor. Chim. Acta 43: 175
69. Stephens ME, Kapuy E, Kozmutza C (1977) Theor. Chim. Acta 45: 111
70. Cimiraglia R, Tomasi J (1977) J. Am. Chem. Soc. 99: 1135
71. Tomasi J (1980) In: Daudel R, Pullman A, Salem L, Veillard A (eds) Quantum theory of chemical reactions, vol 1, Reidel, Dordrecht, p 191
72. Alagona G, Bonaccorsi R, Ghio C, Montagnani R, Tomasi J (1988) Pure Appl. Chem. 60: 231
73. Scrocco E, Tomasi J (1978) Adv. Quant. Chem. 11: 115
74. Bonaccorsi R, Scrocco E, Tomasi J (1976) J. Am. Chem. Soc. 98: 4049
75. Bonaccorsi R, Scrocco E, Tomasi J (1977) J. Am. Chem. Soc. 99: 4546
76. Longuet-Higgins HC, Murrell JN (1955) Proc. Phys. Soc. London A68: 601
77. von Niessen W (1973) Theor. Chim. Acta 31: 111, (1973) 31: 297, (1974) 32: 13, (1974) 33: 7
78. Klessinger M (1978) Theor. Chim. Acta 49: 77
79. Klessinger M (1983) Int. J. Quantum Chem. 23: 535
80. Klessinger M (1983) Croat. Chim. Acta 56: 397
81. Klessinger M (1984) Croat. Chim. Acta 57: 887
82. Scrocco E, Tomasi J (1964) In: Löwdin PO (ed) Molecular orbitals in chemistry, physics and biology, Academic, New York, p 263
83. Degand P, Leroy G, Peeters D (1973) Theor. Chim. Acta 30: 243
84. Leroy G, Peeters D (1975) In: Chalvet O et al. (eds) Localization and delocalization in quantum chemistry, vol I, Reidel, Dordrecht, p 207
85. Leroy G, Peeters D (1975) Theor. Chim. Acta 36: 11
86. Clarisse F, Leroy G, Peeters D (1976) Bull. Chem. Soc. Belg. 85: 375
87. Delhalle J, André JM, Delhalle S, Pivont-Malherbe C, Clarisse F, Leroy G, Peeters D (1977) Theor. Chim. Acta 43: 215
88. O'Leary B, Duke BJ, Eilers JE (1975) Program SAMO Q.C.P.E. n.263
89. Eilers JE, Whitman DR (1973) J. Am. Chem. Soc. 95: 2067
90. Duke BJ, Eilers DR, Eilers JE, Kang S, Liberles A, O'Leary B (1975) Int. J. Quant. Chem. Q.B.S. 2: 155
91. Collins MPS, Duke BJ, Eilers JE, O'Leary B (1976) Int. J. Quant. Chem. 10: 629
92. Duke BJ, Collins MPS (1978) Chem. Phys. Letters 54: 304
93. Duke BJ, O'Leary B (1980) Chem. Phys. Letters 69: 517
94. Duke BJ, Collins MPS (1981) J. Chem. Phys. 74: 4746
95. Liebmann SP, Duke BJ, Collins MPS, Hirst DM (1979) Mol. Phys. 37: 579
96. Duke BJ, O'Leary B (1983) Theor. Chim. Acta 62: 223
97. Gregson K, Hall GG (1969) Mol. Phys. 17: 49
98. Kollmar H (1978) Theor. Chim. Acta 50: 235
99. Deplus A, Leroy G, Peeters D (1974) Theor. Chim. Acta 36: 109
100. Nicolas G, Durand P (1979) J. Chem. Phys. 70: 2020
101. André JM, Burke LA, Delhalle J, Nicolas G, Durand P (1979) Int. J. Quantum Chem. Symp. 13: 283
102. Brédas JL, Chance RR, Silbey R, Nicolas G, Durand P (1981) J. Chem. Phys. 75: 255
103. Gresh N, Claverie P, Pullman A (1984) Theor. Chim. Acta 66: 1
104. Gresh N, Pullman A, Claverie P (1985) Theor. Chim. Acta 67: 11
105. Surján PR (1984) Croat. Chim. Acta 57: 833

106. Poirier RA, Surján PR (1987) J. Comp. Chem. 8: 436
107. Náray-Szabó G, Kramer G, Nagy P, Kugler S (1987) J. Comp. Chem. 8: 555
108. Ravimohan C, Gopinathan MS (1985) Theor. Chim. Acta 67: 199
109. Palke WE (1979) J. Chem. Phys. 71: 4664
110. Daudel R, Stephens ME, Kapuy E, Kozmutza C (1976) Chem. Phys. Letters 40: 194
111. Berlin T (1951) J. Chem. Phys. 19: 208
112. Bader RFW, Henneker WH, Cade PE (1976) J. Chem. Phys. 46: 3341
113. Bader RFW (1975) Acc. Chem. Res. 8: 34
114. Bader RFW, Nguyen-Dang TT (1981) Adv. Quant. Chem. 14: 63
115. Bader RFW, Nguyen-Dang TT, Tal Y (1981) Rep. Progr. Phys. 44: 893
116. Bader RFW (1985) Acc. Chem. Res. 18: 9
117. Bader RFW (1988) Pure Appl. Chem. 60: 145
118. Srebrenik S, Bader RFW (1975) J. Chem. Phys. 63: 3945
119. Stutchbury NCJ, Cooper DL (1983) J. Chem. Phys. 79: 4967
120. Cremer D, Gauss J (1986) J. Am. Chem. Soc. 108: 7467
121. Streitwieser A, Rajca A, McDowell RS, Glaser R (1987) J. Am. Chem. Soc. 109: 4184
122. Richtie JP, Bachrach SM (1987) J. Am. Chem. Soc. 109: 5909
123. Gatti C, Fantucci P, Pacchioni G (1988) Theor. Chim. Acta 72: 433
124. Glaser R (1989) J. Comp. Chem. 10: 118
125. Bonaccorsi R, Scrocco E, Tomasi J (1974) Communication to the V CQTEL Congress, Morelia (Mexico)
126. Agresti A, Bonaccorsi R, Tomasi J (1979) Theor. Chim. Acta 53: 215
127. Ghio C, Scrocco E, Tomasi J (1978) Theor. Chim. Acta 50: 117, (1980) 56: 61, (1980) 56: 75
128. Cimiraglia R, Persico M, Tomasi J, Charkin OP (1984) J. Comp. Chem. 5: 263
129a. Charkin OP, Bonaccorsi R, Tomasi J, Zjubin AS, Gorbik AA (1987) Zhurn. Neorg. Khim. 32: 2644
129b. Charkin OP, Bonaccorsi R, Tomasi J, Zjubin AS, Musaev DG (1987) Zhurn. Neorg. Khim. 32: 2907
130. Bonaccorsi R, Charkin OP, Tomasi J Inorg. Chem. (in press)
131. Gillespie RJ, Nyholm RS (1957) Quart. Rev. (London) 11: 339
132. Kepert DL (1979) Progr. Inorg. Chem. 25: 41
133. Naleway CA, Schwarz ME (1975) J. Am. Chem. Soc. 95: 8235
134. Palke WE, Kirtman B (1978) J. Am. Chem. Soc. 100: 5717
135. Hall MB (1978) J. Am. Chem. Soc. 100: 6333
136. Schniedekamp A, Cruikshank DWJ, Skaarup S, Pulay P, Hargittay I, Boggs JE (1979) J. Am. Chem. Soc. 101: 2002
137. Bartell LS, Barshad YZ (1984) J. Am. Chem. Soc. 106: 7700
138. Røeggen I (1986) J. Chem. Phys. 85: 969; Røeggen I, Nilssen EW (1987) 86: 2869
139. Bonaccorsi R, Ghio C, Tomasi J (1984) Int. J. Quant. Chem. 26: 637
140. Náray-Szábó G (1979) Int. J. Quant. Chem. 16: 265
141. Náray-Szábó G, Grofcsick A, Kósa K, Kubinyi M, Martin A (1981) J. Comp. Chem. 2: 58
142. Náray-Szábó G, Surján PR (1983) Chem. Phys. Letters 96: 499
143. Náray-Szábó G (1984) Croat Chim. Acta 57: 901
144. Mehler EL (1977) J. Chem. Phys. 67: 2728, (1981) 74: 6298
145. Smits GF, Altona C (1984) Theor. Chim. Acta 67: 461
146. Musso GF, Magnasco V (1984) Mol. Phys. 53: 615
147. Cox SR, Williams DE (1981) J. Comp. Chem. 2: 304
148. Singh UC, Kollman PA (1984) J. Comp. Chem. 5: 129
149. Zakrzewska K, Pullman A (1985) J. Comp. Chem. 6: 265
150. Chirlian LE, Miller-Francl M (1987) J. Comp. Chem. 8: 894
151. Le Fèvre RJW (1965) Adv. Phys. Org. Chem. 3: 1
152. Applequist J (1977) Acc. Chem. Res. 10: 79
153. Applequist J, Sundberg KR, Olson ML, Weiss LC (1979) J. Chem. Phys. 70: 1240
154. Applequist J (1979) J. Chem. Phys. 71: 4324, (1985) 83: 809
155. Applequist J, Felder CE (1981) J. Chem. Phys. 75: 1863
156. Birge RR (1980) J. Chem. Phys. 72: 5312
157. Birge RR, Schick GA, Bocian DF (1983) J. Chem. Phys. 79: 2257

158. Boyd RH, Kesner L (1980) J. Chem. Phys. 72: 2179
159. Dewar MJS, Stewart JJP (1984) Chem. Phys. Letters 111: 416
160. Oxtoby DW (1980) J. Chem. Phys. 72: 5171
161. Miller CK, Orr BJ, Ward JF (1977) J. Chem. Phys. 67: 2109
162. Sundberg KR (1977) J. Chem. Phys. 66: 1475, (1978) 68: 5271
163. Thole BT (1981) Chem. Phys. 59: 341
164. Kitaura K, Morokuma K (1976) Int. J. Quantum Chem. 10: 325
165. Fukui K (1970) J. Phys. Chem. 74: 4161
166. Dixon DA, Lipscomb WN (1973) J. Am. Chem. Soc. 95: 2853
167. Nagase S, Takatsuka K, Fueno T (1976) J. Am. Chem. Soc. 98: 3838
168. Bonaccorsi R, Palla P, Tomasi J (1982) J. Mol. Struct. (Theochem) 87: 181
169. Leroy G, Sana M, Burke LA, Nguyen M-T (1979) In: Daudel R, Pullman A, Salem L, Veillard A (eds) Quantum theory of chemical reactions, vol 1, Reidel, Dordrecht, p 91
170. Nguyen M-T, Sana M, Leroy G, Dignam KJ, Hegarty AF (1980) J. Am. Chem. Soc. 102: 573
171. Ha TK, Nguyen M-T, Hendricks M, Vanquickenborne LG (1983) Chem. Phys. Lett. 96: 267
172. Nguyen M-T, Hegarty AF (1983) J. Am. Chem. Soc. 105: 3811
173. Nguyen M-T (1983) J. Mol. Struct. (Theochem) 105: 343
174. Alagona G, Tomasi J (1983) J. Mol. Struct. (Theochem) 91: 263
175. Kutzelnigg W (1975) In: Chalvet O, Daudel R, Diner S, Malrieu JP (eds) Localization and delocalization in quantum chemistry, vol I, Reidel, Dordrecht, p 143
176. Mehler EL (1974) Theor. Chim. Acta 35: 17
177. Levy M (1974) J. Chem. Phys. 61: 1857
178. Klein DJ (1976) J. Chem. Phys. 64: 4868
179. Prime S, Robb MA (1976) Theor. Chim. Acta 42: 181
180. Kollmar H (1980) Theor. Chim. Acta 58: 19
181. Dykstra CE, Chiles RA, Garrett MD (1981) J. Comp. Chem. 2: 266
182. Kapuy E, Csépes Z, Kozmutza C (1983) Int. J. Quantum Chem. 23: 981
183. McLean AD, Ellinger Y (1985) Chem. Phys. 94: 25
184. Pauzat F, Chekir S, Ellinger Y (1986) J. Chem. Phys. 85: 2861
185. Pipek J, Ladik J (1986) Chem. Phys. 102: 445
186. Kapuy E, Bartha F, Bogár F, Kozmutza C (1987) Theor. Chim. Acta 72: 337
187. Cullen JM, Lipscomb WN, Zerner MC (1985) J. Chem. Phys. 83: 5182
188. Kirtman B, Dykstra CE (1986) J. Chem. Phys. 85: 2791
189. Pulay P, Saebø S (1986) Theor. Chim. Acta 69: 357
190. Saebø S, Pulay P (1988) J. Chem. Phys. 88: 1884
191. Tomasi J (1982) In: Ratajczak H, Orville-Thomas WJ (eds) Molecular interactions, vol 3, Wiley, Chichester, p 119
192. Alagona G, Ghio C, Cammi R, Tomasi J (1988) In: Maruani J (ed) Molecules in physics, chemistry and biology, vol 2, Reidel, Dordrecht, p 507
193. Alagona G, Ghio C, Tomasi J (1989) J. Phys. Chem. 93: 5401
194. Alagona G, Ghio C, Latajka Z, Tomasi J (1990) J. Phys. Chem. 94: 2267
195. Linder B (1967) Adv. Chem. Phys. 12: 225
196. Rinaldi D, Costa Cabral BJ, Rivail JL (1986) Chem. Phys. Letters 125: 495
197. Aguilar M, Olivares del Valle FJ Chem. Phys. (to be published)
198. Miertuš S, Scrocco E, Tomasi J (1981) Chem. Phys. 55: 117
199. Bonaccorsi R, Cimiraglia R, Tomasi J (1983) J. Comp. Chem. 4: 567
200. Floris F, Tomasi J (1989) J. Comp. Chem. 10: 616
201. Pascual-Ahuir JL, Silla E, Tomasi J, Bonaccorsi R (1987) J. Comp. Chem. 8: 778
202. Tomasi J, Alagona G, Bonaccorsi R, Ghio C (1987) In: Maksić Z (ed) Modelling of structures and properties of molecules, Ellis Horwood, Chichester, p 330
203. Tapia O (1982) In: Ratajczak H, Orville-Thomas WJ (eds) Molecular interactions, vol 3, Wiley, Chichester, p 47
204. Alagona G, Ghio C, Igual I, Tomasi J (1990) J. Mol. Struct. (Theochem) 204: 253
205. Persico M, Tomasi J (1984) Croat. Chim. Acta 57: 1395
206. Alagona G, Bonaccorsi R, Ghio C, Tomasi J (1986) J. Mol. Struct. (Theochem) 137: 263
207. Montagnani R, Tomasi J Int. J. Quantum Chem. (in press)

208. Wiberg KB, Laidig KE (1987) J. Am. Chem. Soc. 109: 5935
209. Ehrenson S (1976) J. Am. Chem. Soc. 98: 7510, (1981) 103: 6036, (1984) 104: 4793
210. Ehrenson S (1981) J. Comp. Chem. 2: 41, (1989) 10: 77
211. Ehrenson S (1977) J. Phys. Chem. 81: 1520
212. Brunschwig BS, Ehrenson S, Sutin N (1985) J. Phys. Chem. 90: 3657, (1987) 91: 4714
213. Sprague JT, Tai JC, Yuh Y, Allinger NL (1987) J. Comp. Chem. 8: 581
214a. Weiner PK, Kollman PA (1981) J. Comp. Chem. 2: 287
214b. Weiner SJ, Kollman PA, Case DA, Singh UC, Ghio C, Alagona G, Profeta S, Weiner P (1984) J. Am. Chem. Soc. 106: 765
215. Floris F, Pascual-Ahuir JL, Tomasi J (to be published)
216. Ventura ON, Lledós A, Bonaccorsi R, Bertrán J, Tomasi J (1987) Theor. Chim. Acta 72: 175
217. Alagona G, Ghio C, Igual J, Tomasi J (1989) J. Am. Chem. Soc. 111: 3417
218. Bonaccorsi R, Ghio C, Tomasi J (1982) In: Carbó R (ed) Current aspects of quantum chemistry Elsevier, Amsterdam, p 389
219. Morokuma K, Iwata S (1972) Chem. Phys. Letters 16: 192
220. Daudel R, Le Rouzo H, Cimiraglia R, Tomasi J (1978) Int. J. Quantum Chem. 13: 537
221. Cimiraglia R, Miertuš S, Tomasi J (1981) Chem. Phys. Letters 80: 286
222. Bonaccorsi R, Ghio C (1981) IV Amer. Conf. on Theoret. Chemistry, Boulder (USA) (unpublished)
223. Bonaccorsi R, Cimiraglia R, Tomasi J (1983) Chem. Phys. Letters 99: 77
224. Bonaccorsi R, Cimiraglia R, Tomasi J (1984) J. Mol. Struct. (Theochem) 107: 197

The Analysis of Potential Energy Surfaces in Terms of the Diabatic Surface Model

Fernando Bernardi

Dipartimento di Chimica' G. Ciamician', Universita' di Bologna, 40126 Bologna, Italy

Massimo Olivucci and Michael A. Robb

Department of Chemistry, King's College London, Strand, London WC2R 2LS, U.K.

In this paper, we summarize the salient features of the Diabatic Surface model, which represents a theoretical instrument for analysing a Potential Energy Surface (PES). In particular this model provides information about the global properties of a PES and gives a rationalization of the main features of a critical point (e.g. intermediates, transition states, etc.), such as its origin, existence and index.

The Diabatic Surface model has the important feature of being both quantitative and qualitative: on the one hand this method reproduces the numerically computed quantities exactly, on the other, it provides a basis for discussing the experimental facts in a qualitative manner.

The Diabatic Surface model is based upon Valence Bond theory and in the present paper we illustrate the formalism for a four electrons/four orbitals problem. For illustrative purposes the Diabatic Surface model is applied to the analysis of the PES associated with the cycloaddition of two ethylenes.

1 Introduction

In this paper, we shall review the salient features of the model [1–9] we have recently developed for the analysis of Potential Energy Surfaces (PES), which we shall refer to as the Diabatic Surface model. In this model, a given reaction surface is analyzed in terms of two component surfaces, one associated with the bonding situation of the reactants (reactant diabatic surface) and the other with the bonding situation of the products (product diabatic surface). The Diabatic Surface model provides information about the global properties of a PES and gives a rationalization of the main features of a critical point (e.g. minimum, transition state, etc.) such as its origin, existence and index in simple concepts that stem from Valence Bond (VB) theory.

The origin of a critical point can be understood in terms of the behaviour of the two diabatic surfaces along the reaction coordinate. In practice, two situations arise: (i) the critical point arises from the crossing of the two diabatic surfaces (a situation where the two diabatic components have a similar weight in the adiabatic wavefunction) or (ii) the critical point lies in a region of the reaction surface dominated by one of the two diabatic surfaces (a situation where one of the two diabatic components has a largely dominant weight in the adiabatic wavefunction). We denote the first type of critical point as a critical point of electronic origin and the second type as a critical point of conformational origin.

In contrast, our discussion about the existence and index of critical points will be formulated in terms of the behaviour of the coulomb and exchange energies that are similar to the concepts encountered in Heitler-London VB theory. The existence of a critical point is related to the conditions that make the first derivatives of the total energy equal to zero with respect to all the geometrical coordinates while the index of a critical point represents the number of directions with negative curvature which is usually expressed by the number of negative eigenvalues of the Hessian.

While the main features of the critical points of conformational origin are well understood since conformational effects have already been extensively investigated at a theoretical level [10], the main features of the critical points of electronic origin are less well understood. Our goal is a quantitative model (that can also be applied qualitatively) for critical points of electronic origin. This model is defined by the transformation of MC-SCF or CI wavefunctions to Valence Bond Space (6).

2 Potential Energy Surfaces

PES represent the most useful conceptual framework for the discussion of chemical reactions. The concept of PES arises from the Born-Oppenheimer approximation [11], (also known as the adiabatic approximation) which allows us to separate the nuclear and electronic motions and therefore to separate the Schrodinger equation into an electronic part and into a nuclear part.

Since a detailed treatment of the Born-Oppenheimer approximation can be found in any textbook of quantum chemistry [12–14], we will be content with a brief summary of the essential formalism. The physical behaviour of a molecular system containing N nuclei and M electrons in a stationary state is described by the Schrodinger equation:

$$H_T \psi(x, z) = E_T \psi(x, z), \tag{1}$$

where E_T denotes the total energy, x the set of nuclear coordinates and z the set of electron coordinates. The full Hamiltonian operator H_T has the form:

$$H_T = Tx + Tz + U(x, z) \tag{2}$$

and includes:

i) the kinetic energy operator of the nuclei

$$Tx = -\frac{1}{2}\sum_i \frac{1}{m_i} \nabla_i^2, \tag{3}$$

m_i being the corresponding nuclear masses;

ii) the kinetic energy operator of the electrons

$$Tz = -\frac{1}{2}\sum_K \nabla_K^2 ; \tag{4}$$

iii) the classical potential energy $U(x, z)$ of the system of nuclei and electrons.

The Born-Oppenheimer approximation consists of neglecting the kinetic energy operator Tx in the full Hamiltonian [2], which means solving the wave equation [1] at fixed nuclear coordinates by representing the wavefunction for a given electronic state by the product

$$\psi_{s,v}(x, z) = \varphi_{s,v}(x) \, \Phi_v(x, z), \tag{5}$$

where v denotes the set of electronic quantum numbers and s the set of the nuclear quantum numbers. In this way we obtain two separate wave equations for any electronic state v.

The first one is the electronic Schrodinger equation which can be written as follows:

$$H\Phi_v(x, z) = E_v(x) \, \Phi_v(x, z), \tag{6}$$

where the electronic Hamiltonian H has the form:

$$H = Tz + U(x, z) \tag{7}$$

and therefore includes the kinetic energy operator Tz of electrons and the total potential energy of the system. This equation describes the motion of M electrons in the field of N nuclei at a fixed position. The quantities $\Phi_v(x, z)$ and E_v represent the wavefunction and the energy of an electronic state of the system defined by the electronic quantum number v and correspond to an eigenvector and an eigenvalue of H. In this equation both Φ_v and E_v depend parametrically on the nuclear coordinates x.

The nuclear part of the Schrodinger equation can be written as:

$$H_x \varphi_{s,v}(x) = \varepsilon_{s,v} \cdot \varphi_{s,v}(x),$$ (8)

where the Hamiltonian operator

$$H_x = Tx + E_v(x)$$ (9)

is a sum of the kinetic energy operator Tx of nuclei and the potential energy function $E_v(x)$. This equation describes the motion of the nuclei for a given electronic state defined by v. In this equation the PES $E_v(x)$ represents the potential energy for the nuclei and its value, for the electronic state defined by v, is univocally determined by specifying the nuclear configuration.

If N nuclei are present, 3N-6 coordinates are needed in a polyatomic molecule to define the nuclear configuration (3N-5 in a linear molecule). Therefore for a system of N nuclei the PES has 3N-5 dimensions, (3N-4 dimensions if the molecular system is linear throughout the reaction), since to the dimensions for each nuclear coordinate must be added the dimension that measures the energy. Thus even for a triatomic molecule it is impossible to plot the full PES since we need a four-dimensional diagram. The PES is also denoted as adiabatic surface since its existence is based upon the adiabatic approximation. Since the adiabatic surface is a function of the electronic quantum numbers v, there is an adiabatic surface for every value of v, i.e. in addition to the PES of the ground state one can define a PES for every excited state.

3 Decomposition of PES

The PES contains a large amount of information about chemical reactivity. Therefore it is of great importance to construct suitable theoretical models which allow us to rationalize and predict the main features of PES.

A general procedure to implement this objective involves the decomposition of the PES in component surfaces which behave in a very simple (possibly monotonically) way and are identified with a clear chemical model. The well known models for discussing reactivity, such as the Woodward-Hoffman approach [15], the MO-VB model [16] and the curve crossing model [17], represent specific applications of this general decomposition technique.

In this manuscript, we describe in detail the decomposition method we have recently developed, i.e. the Diabatic Surface model [1–9], which has the important feature of being both quantitative and qualitative. On the one hand this method reproduces the numerically computed quantities exactly, on the other it provides a basis for discussing the experimental facts in a qualitative manner.

Within the Diabatic Surface model a given adiabatic surface can be analyzed in terms of two component surfaces, one associated with the bonding situation of the reactants (reactant diabatic surface) and the other with the bonding situation of the products (product diabatic surface). The Diabatic Surface model is based upon VB theory, (rather than Molecular Orbital (MO) theory) since the processes of bond making/breaking can be more easily formulated in this method.

4 The Diabatic Surface Model for a Four Electrons/ Four Orbitals Problem

In this section, we illustrate the development of the theory with respect to the following specific example, i.e. a reaction involving four electrons and four orbitals where bonds 1–3 and 2–4 are broken and the new bonds 1–2 and 3–4 are formed:

$$
\begin{array}{cc}
\begin{array}{cc} 1 & 2 \\ | & | \\ 3 & 4 \end{array} & \begin{array}{c} 1-2 \\ \\ 3-4 \end{array} \\
\text{Reactants} & \text{Product}
\end{array}
$$

where the sequence numbers 1–4 refer to singly occupied orbitals centered on each of the four sites in the reaction. The theoretical formalism presented in this paper refers only to this type of problem, which however is of significant chemical importance since it can be used for describing relevant chemical reactions such as all types of cycloaddition reactions, electrocyclic reactions, sigmatropic shifts and various types of ionic reactions.

At a quantum mechanical level the decomposition of the PES discussed in the present section corresponds to a decomposition of the electronic wavefunction in terms of two components:

$$\Phi = N \cdot (\Phi_R + C \cdot \Phi_P) \tag{10}$$

where Φ_R and Φ_P represent the diabatic functions and describe the bonding situation of the reactants and products respectively. C is the ratio C_P/C_R between the coefficients of the two diabatic components and N the normalization factor. The adiabatic wavefunctions are found as the solutions of a two level secular equation:

$$\begin{pmatrix} E_R - E & H_{RP} - E \cdot S \\ H_{RP} - E \cdot S & E_P - E \end{pmatrix} \cdot \begin{pmatrix} C_R \\ C_P \end{pmatrix} = 0, \tag{11}$$

where:

$$E_R = \langle \Phi_R | H | \Phi_R \rangle, \tag{12}$$

$$H_{RP} = \langle \Phi_R | H | \Phi_P \rangle, \tag{13}$$

$$S = \langle \Phi_R | \Phi_P \rangle. \tag{14}$$

The quantities E_R and E_P are the diabatic energies of the reactants and product configurations respectively, H_{RP} the resonance energy that results from the interaction between the reactant and product diabatic functions and S their overlap. The precise form of the expressions for E_R, E_P, H_{RP} and S depends upon the specific choice of the diabatic functions Φ_R and Φ_P.

We must now make a choice of diabatic functions. The Rumer bond eigenfunctions [13, 18] are the obvious choice. We use $|R\rangle$ and $|P\rangle$ to denote the Rumer functions associated with the reactants and products. Thus in $|R\rangle$ orbitals 1–3 and 2–4 are singlet spin coupled and in $|P\rangle$ orbitals 1–2 and 3–4 are singlet spin coupled. The wavefunction which describes the adiabatic surface can be expressed in terms of $|R\rangle$ and $|P\rangle$ as:

$$\Phi = N(|R\rangle + C \cdot |P\rangle), \tag{15}$$

where the coefficient C varies from 0 to ∞ as the reaction proceeds along the reaction coordinate. In particular C assumes the value 0 for the reactants, the value 1 when $E_R = E_P$ and tends to ∞ at the products.

From the solution of the two level secular equation we can obtain the energy of the adiabatic surface E as

$$E = \frac{(E_R + E_P - 2 \cdot S \cdot H_{RP})}{2(1 - S^2)}$$
$$- \sqrt{\frac{(E_P - E_R)^2}{4(1 - S^2)} + \frac{(H_{RP} - S \cdot E_R)(H_{RP} - S \cdot E_P)}{(1 - S^2)^2}}, \tag{16}$$

where S, the overlap between the two bond eigenfunctions, is a negative constant $(S = -1/2)$.

The quantities E_R, E_P and H_{RP} expressed in terms of the coulomb (Q) and exchange (K_{ij}) integrals assume the following expressions:

$$E_R = Q - \tfrac{1}{2}(K_X + K_P) + K_R, \tag{17}$$

$$E_P = Q - \tfrac{1}{2}(K_X + K_R) + K_P, \tag{18}$$

$$H_{RP} = -\tfrac{1}{2}Q + K_X - \tfrac{1}{2}(K_R + K_P), \tag{19}$$

where $K_R = K_{13} + K_{24}$ represents the exchange energy of the reactants, $K_P = K_{12} + K_{34}$ represents the exchange energy of the product and $K_X = K_{14} + K_{23}$ represents the non-bonded exchange energy. It should be noted that according to expression [19], H_{RP} is always a positive quantity $(H_{RP} > 0)$.

Insertion of expressions [17], [18], [19] in Eq. [16] leads to the following expression for the total energy E:

$$E = Q - T, \tag{20}$$

$$T = \sqrt{RK^2 + H_R \cdot H_P}, \tag{21}$$

where $RK = K_P - K_R$, $H_R = K_X - K_P$, $H_P = K_X - K_R$.

Thus the total energy E can be expressed in terms of the coulomb integral Q and of the exchange energy T. In turn, T can be expressed in terms of the three exchange terms RK, H_R and H_P. Since usually along the reaction coordinate the dominant term in E_R and E_P is K_P (which is a negative quantity), E_R is repulsive and E_P is attractive. Consequently there is usually a crossing bnetween the two diabatic surfaces which occurs when $E_R = E_P$. At the crossing the total energy E becomes:

$$E = (2/3) \cdot (E_R - H_{RP}) = Q - H_R . \tag{22}$$

Furthermore at the crossing the following expressions hold:

$$K_P = K_R , \tag{23}$$

$$E_R = Q - (1/2) \cdot H_R , \tag{24}$$

$$E_P = Q - (1/2) \cdot H_R , \tag{25}$$

$$H_{RP} = -(1/2) \cdot Q + H_R . \tag{26}$$

Therefore at the crossing the exchange contribution of the two diabatic surfaces is exactly the same and the crossing occurs at the midpoint between their values of the coulomb integral Q and of the adiabatic energy E. Figure 1 illustrates the behaviour of the various quantities along the reaction coordinate and their relationship. It can be seen that the reactant diabatic surface E_R tends to coincide with the adiabatic surface in the region of the reactants and the product diabatic surface E_P tends to coincide with the adiabatic curve in the product region. Furthermore the crossing between the two diabatic curves is almost coincident with the maximum of the adiabatic curve.

While the Rumer spin coupling is appropriate in the region of the transition state, with this spin coupling, the product diabatic surface at the reactant geometry and the reactant diabatic surface at the product geometry do not represent pure spectrocopic spin states (see Fig. 1).

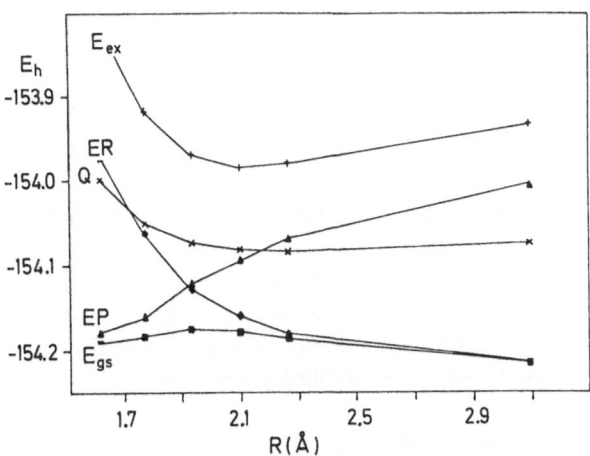

Fig. 1. Diabatic decomposition along the reaction coordinate in a non-orthogonal basis for the ethylene + ethylene anti attack at the STO-3G level

This limitation can be easily overcome by using an orthogonal basis, where the reactant and product Rumer diabatic surfaces have been appropriately orthogonalized. Using $|R\rangle$ and $|P\rangle$ to denote two non-orthogonal Rumer eigenfunctions (with overlap $S = -1/2$) we now introduce $|R^{\perp}\rangle$ and $|P^{\perp}\rangle$ to denote the corresponding orthogonalized functions. The wavefunction which describes the adiabatic surface can be expressed as

$$\Phi = N \cdot (|R^{\perp}\rangle + C^{\perp} \cdot |P^{\perp}\rangle) \tag{27}$$

The energy of the adiabatic surface can now be expressed in the following way

$$E = \frac{(\alpha_R + \alpha_P)}{2} - \sqrt{\frac{(\alpha_R - \alpha_P)^2}{2} + \beta^2} \tag{28}$$

where α_R and α_P denote the diabatic energies in the orthogonal basis and β the resonance energy. These quantities expressed in terms of the coulomb (Q) and exchange (k_{ij}) integrals become:

$$\alpha_R = Q + K \tag{29}$$

$$\alpha_P = Q - K \tag{30}$$

$$K = a \cdot K_X + b \cdot K_R + c \cdot K_P \tag{31}$$

$$\beta = ((c - b) \cdot K_X + (a - c) \cdot K_R + (b - a) \cdot K_P)/\sqrt{3} \tag{32}$$

Here the quantities K_R, K_P and K_X represent the reactant, product and nonbonded exchange as previously defined. The coefficients a, b and c depend upon the spin coupling. It must be emphasized that the simplicity of the expressions for α_R, α_P, K and β depends upon the use of an orthogonal set of spin eigenfunctions. Insertion of the expressions [29] and [30] in Eq. (27) leads to the following equation for the total energy E:

$$E = Q - T, \tag{33}$$

$$T = \sqrt{K^2 + \beta^2}. \tag{34}$$

It is remarkable that in the orthogonal case, the total energy E can be expressed in terms of the coulomb integral Q and of the total exchange energy T, which assumes a much simpler form in terms of only two exchange terms K and β. The exchange term K represents the exchange contribution associated with the diabatic surfaces, while the exchange term β represents the resonance energy. Further T can always be written in a way independent of any chosen spin coupling scheme as

$$T = ((K_P - K_X) \cdot (K_R - K_X) + (K_P - K_R)^2)^{1/2}. \tag{35}$$

Of particular interest is the form that these expressions assume at the crossing between the two diabatic surfaces. The condition $E_R = E_p$ implies that $K = 0$ so that at the crossing we obtain the important result that

$$\alpha_R = \alpha_p = Q . \tag{36}$$

In addition the following expressions hold:

$$E = Q - \beta , \tag{37}$$

$$K = (\sqrt{3}/2) \cdot (K_R - K_p) = 0 , \tag{38}$$

$$K_R = K_p , \tag{39}$$

$$\beta = K_p - K_X . \tag{40}$$

The comparison with the correspondling expressions obtained in the non orthogonal model shows that the use of an orthogonal model leads to a significant simplification as illustrated in Fig. 2. In particular in the orthogonal model the

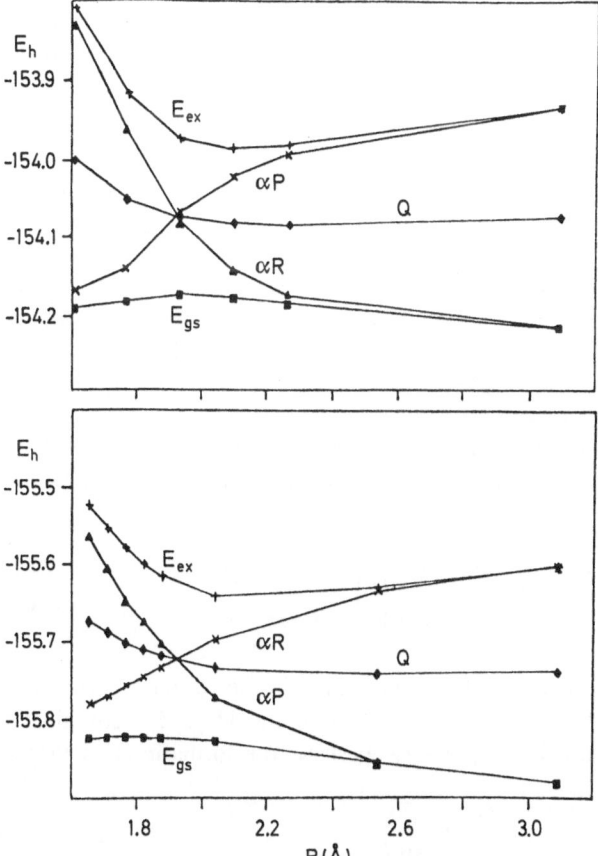

Fig. 2. Diabatic decomposition along the reaction coordinate in a basis for the ethylene + ethylene anti attack at the STO-3G and 4-31G levels

coulomb integral corresponds to the average of the energy values of the two diabatic surfaces. Furthermore the interaction energy T coincides at the reactant geometry with K (the exchange term associated with the diabatic surfaces) and at the crossing with β, the resonance stabilization.

As previously pointed out, in the orthogonal Diabatic Surface model, the reactant and product Rumer diabatic functions are orthogonalized. The orthogonalization procedure chosen here consists of determining the two orthogonal functions $|R^{\perp}\rangle$ and $|P^{\perp}\rangle$, starting from $|R\rangle$ and $|P\rangle$, in such a way that at every point of the surface the coefficient $C^{\perp}$ in Eq. (27) is equal to the coefficient C in Eq. (15). This orthogonalization scheme corresponds to Schmidt orthogonalization schemes at the reactant and product geometries and to a symmetric orthogonalization scheme at the crossing and has the additional feature of being continuously defined in all regions of the potential energy surface.

5 Rigorous Implementation of the Diabatic Surface Model via Effective Hamiltonians

As previously pointed out, in the Diabatic Surface model the PES is described as a solution of a VB problem in the space of the active atomic orbitals (i.e. those involved in the bond making/breaking process). In general the VB problem is formulated in the space of the possible linearly independent spin projections of n active singly occupied orbitals and n active electrons (in this case n is equal to 4). In general the reactants correspond to one of the possible independent spin projections while the product is associated with another one.

Our purpose in this section is to describe a procedure for the implementation of the Orthogonal Diabatic Surface model in a rigorous way so that the energy obtained reproduces the results of a CAS-SCF computation exactly. This objective is obtained through the computation of an effective Hamiltonian. The field of effective Hamiltonian theory has recently been the subject of an extensive review (19). Thus here we briefly summarize only the essential ideas.

We begin by considering the CI eigenvalue problem in a partitioned form defined by a reference space V spanned by functions $\{\Lambda_R\}$ and a secondary space Z spanned by functions $\{\varphi_S\}$:

$$\begin{pmatrix} H_{VV} & H_{VZ} \\ H_{ZV} & H_{ZZ} \end{pmatrix} \begin{pmatrix} A & D \\ B & F \end{pmatrix} = E_{\lambda} \begin{pmatrix} A & D \\ B & F \end{pmatrix}, \tag{41}$$

where H_{VV} is the submatrix of the full CI Hamiltonian in the reference space, H_{ZZ} is the corresponding submatrix in the secondary space and H_{VZ}, H_{ZV} contain the matrix elements connecting the two subspaces. One seeks a transformation U such that

$$U^{-1} \begin{pmatrix} H_{VV} & H_{VZ} \\ H_{ZV} & H_{ZZ} \end{pmatrix} U = \begin{pmatrix} H_{eff} & 0 \\ 0 & W \end{pmatrix}, \tag{42}$$

where the diagonalization of

$$H_{eff}A = E_\lambda A \tag{43}$$

must reproduce n eigenvalues exactly where n is the dimension of the reference space V. In the case that all the eigenvalues of the full CI Hamiltonian are known, the transformation U can be easily obtained without recourse to perturbation expansions. If we define

$$\mathbb{C} = \mathbb{B}\mathbb{A}^{-1}, \tag{44}$$

$$\mathbb{S} = (1 + \mathbb{C})^t (1 + \mathbb{C}) \tag{45}$$

then it can be shown that

$$H_{eff} = \mathbb{S}^{-1} \cdot (1 - \mathbb{X}) H (1 + \mathbb{X}), \tag{46}$$

where

$$\mathbb{X} = \begin{pmatrix} 0 & \mathbb{C}^t \\ \mathbb{C} & 0 \end{pmatrix}. \tag{47}$$

The effective Hamiltonian defined by Eq. (46) is the Bloch Hamiltonian and is not Hermitian. The Hermitian Hamiltonian is given as

$$H'_{eff} = \mathbb{S}^{-1/2}(1 - \mathbb{X}) H (1 + \mathbb{X}) \mathbb{S}^{-1/2} \tag{48}$$

which is usually referred to as the Des Cloizeau Hamiltonian. These two effective Hamiltonians have the same eigenvalues. For interpretative purposes we find more convenient to use H'_{eff}, which can be interpreted to be the symmetric form of an Effective Hamiltonian

$$H_{eff}^{NO} = (1 - \mathbb{X}) H (1 + \mathbb{X}) \tag{49}$$

defined in a non-orthogonal basis of configurations of the form

$$\Lambda'_R = \Lambda_R + \sum_s C_{SR}\Phi_S \tag{50}$$

with metric S.

The preceding discussion is completely general and can be applied to any problem that can be expressed as a CI problem. We now limit our discussion to a particular type of CI space and reference space/secondary space partition that is commonly used in MC-SCF computations, defined as follows:

i) A complete CI corresponding to all possible arrangements of the active electrons in the active orbitals, i.e. a Complete Active Space (CAS).

ii) The number of active electrons is equal to the number of active orbitals.

iii) The reference space consists of those configurations where each active orbital is singly occupied and the secondary space is spanned by the remainder of the configurations that may have doubly occupied orbitals.

In CAS-SCF the energy is invariant to transformation of the active orbitals among themselves. Thus the active orbitals are rather arbitrary. Then we can choose a particular set of orbitals localized onto the atomic sites, with which we can make the following identifications:

iv) The reference space configurations $\{\Lambda_R\}$ correspond to neutral valence only VB structures.

v) The secondary space configurations $\{\Phi_S\}$ correspond to ionic configurations where the orbitals may have double occupancy.

Thus the CI expansion corresponds to a VB expansion in the active orbitals where the active orbitals are orthogonal. Now H_{eff}^{NO} as defined by Eq. [49] can be interpreted as a VB Hamiltonian where the configurations Λ_R as defined by Eq. [50], that form the basis on which this Hamiltonian acts, can be interpreted as if they had been built from non-orthogonal orbitals. Further the overlap matrix $\mathbf{S}$ as defined by Eq. [45] represents the overlap matrix between non-orthogonal configurations built from non-orthogonal orbitals. Thus we can make the following identifications

$$(\mathbf{S})_{RS} = \langle \Lambda_R' \mid \Lambda_S' \rangle, \tag{51}$$

$$[(1 - \mathbf{X}) H (1 + \mathbf{X})]_{RS} = \langle \Lambda_R' \mid H \mid \Lambda_S' \rangle. \tag{52}$$

With the preceding definitions of the reference space we can also identify H_{eff}' as a type of Heisenberg Hamiltonian, i.e. a matrix representation of the Hamiltonian in the space of all neutral only VB determinants that can be built from n orbitals and n electrons:

$$H_S = \mathbf{S}^{-1/2} (1 - \mathbf{X}) H (1 + \mathbf{X}) \mathbf{S}^{-1/2}. \tag{53}$$

The Heisenberg Hamiltonian has the property that it can be expressed exactly in terms of one and two center coulomb and exchange integrals. Thus the matrix elements of H_{eff}' will have the form:

$$(H_{eff}')_{KK} = Q + f_1(K_{ij}), \tag{54}$$

$$(H_{eff}')_{KL} = f_2(K_{ij}), \tag{55}$$

where Q and K_{ij} are intended to have the same physical interpretation as in the Heitler-London treatment of the H_2 molecule and f_1 and f_2 are functions that depend upon the details of the spin coupling. On the other hand the basic expressions of the Q and K_{ij} integrals are defined as if the orbitals were not orthogonal, according to Eq. [50]. In particular K_{ij} is assumed to have the form:

$$K_{ij} = [ij \mid ij] + 2[i \mid h \mid j] \cdot S_{ij} \tag{56}$$

with [ij | ij] being the usual exchange integral, [i| h |j] the kinetic and nuclear attraction integral and S_{ij} the orbital overlap.

Therefore, via effective Hamiltonian theory, we have provided a theoretical justification to the following empirical rules for the construction of the Heisenberg Hamiltonian:

1) Evaluate the matrix elements using the rules for determinants built from orthogonalized AO's

2) Replace (implicitly) the $K_{ij} = [ij | ij]$ by $K_{ij} = [ij | ij] + 2 \cdot S_{ij} \cdot [i| h |j]$.

On the other hand, the important point is that the Heisenberg Hamiltonian constructed from the effective Hamiltonian reproduces the eigenvalues of the CI Hamiltonian exactly. Thus the subsequent projection is exact as well. Therefore the above derivation illustrates a complete procedure for the rigorous computation of the Diabatic Surfaces model as summarized in the flow diagram of Fig. 3.

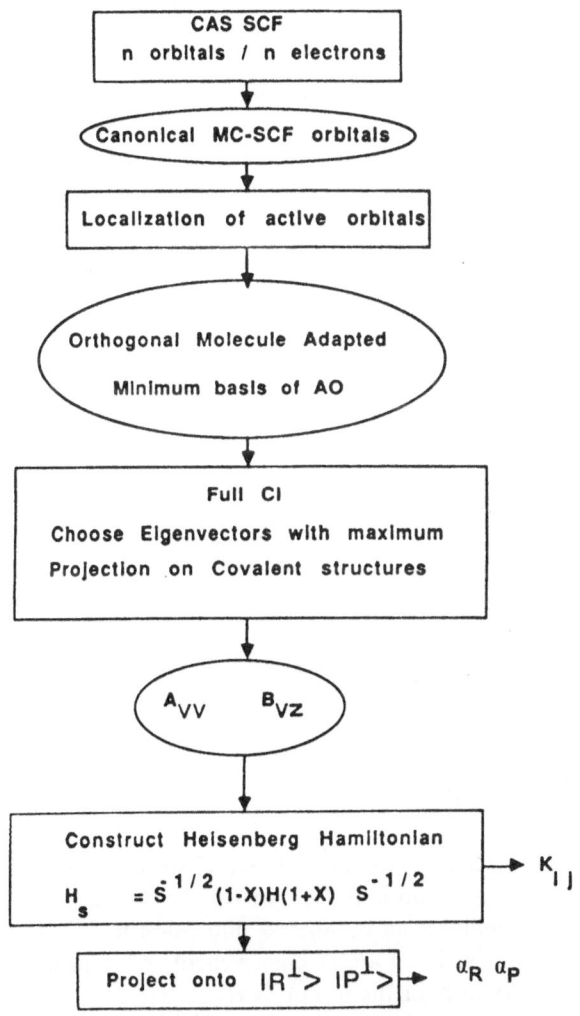

Fig. 3. Flow Diagram for the rigorous computation of the Diabatic Surface model

6 Illustrative Applications

In the Sect. 4, we compared two VB formalisms to describe the diabatic surfaces for a four electron/four orbital problem, one involving non-orthogonal Rumer-type diabatic functions, the other orthogonalized Rumer-type diabatic functions. We showed that the orthogonal formalism presents various advantages over the non-orthogonal one, concerning in particular a more clear separation of the effects of the coulomb and exchange contributions and a decomposition involving at the same time the ground and related excited state. Therefore in this section we discuss the use of the orthogonal VB formalism to analyze a PES. This formalism can be used either at a qualitative or at a quantitative level. In the quantitative application, the various coulomb and exchange integrals Q and K_{ij} which appear in the expressions for α_R, α_P, K and T, are computed using the effective Hamiltonian procedure previously described. However the VB formalism can also be used at a qualitative level: in this case the behaviour of the coulomb integral Q is discussed in terms of simple steric effects, while that of the exchange integrals K_{ij} in terms of the expressions [56] and therefore in terms of the overlap integrals S_{ij}.

We now discuss two different types of application of our model, a Global Analysis which provides information about the regions at high and low energy of the surface, and a Local Analysis which provides information about the topological features of a critical point, such as its origin, existence and index, and about the energy features such as the activation barriers.

6.1 Global Analysis

Within a Global Analysis (9), a PES is decomposed into various regions by "ridges" of zero bond echange energy (Eq. [38]), zero resonance (Eq. [40]) and zero total exchange (Eq. [35]) as defined below:

(a) Diabatic Intersection

$$K = 0 \quad \text{when} \quad K_R = K_P,$$ (57)

(b) Zero Resonance

$$\beta = 0 \quad \text{when} \quad K_X = K_P \quad \text{or} \quad K_X = K_R,$$ (58)

(c) Conical Intersection (both a and b hold)

$$T = 0 \quad \text{when} \quad K_X = K_P = K_R.$$ (59)

These relations are satisfied only in certain points of the surface. If n is the dimension of the surface, the sets of points where conditions (a) and (b) are satisfied have dimension $n - 1$ and those where condition (c) is satisfied dimension $n - 2$. The points where conditions (a) and (b) are satisfied are at higher energy since in these points the contribution of either the term K or the term β is zero.

The "point" where condition (c) is satisfied corresponds to a point of real crossing between the ground and excited surfaces and is a conical intersection.

Figure 4 shows the results of a Global analysis for the case of an hypothetical two-dimensional surface. In this case the sets of points where $K = 0$ or $\beta = 0$ have dimension 1 and represent curves, while the sets where $T = 0$ have dimension 0 and represent just points on the surface. The "ridge" defined by Eq. 57 (condition (a)) divides the domain of the reactants and products. In contrast the "ridge" of zero resonance defined by Eq. 58 (condition (b)) separates different reaction paths. These high energy curves divide the surface into six regions at low energy, where are located the minima which correspond to reactants, product and intermediates. It is easy to recognize three different reaction paths between the regions 1 and 4,2 and 5,3 and 6 separated by the curves where $\beta = 0$. The curve $K = 0$, which separates for every reaction path the two regions at lowest energy represents the region of the transition states.

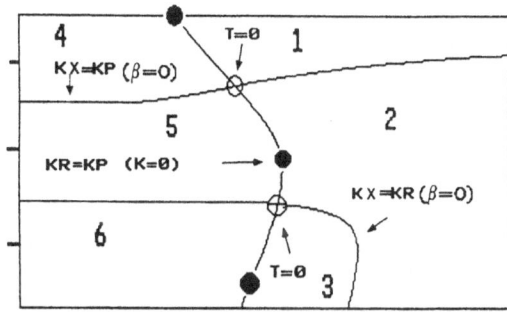

Fig. 4. Global analysis for an hypothetical two-dimensional surface

The curves of Fig. 4 can be interpreted in chemical terms on the basis of the quantities α_R, α_P, K and β. Infact, the curve where $K = 0$ coincides with the curve of intersection between the two diabatic surfaces α_R and α_P and therefore represents the set of points where $T = \beta$, i.e. where the interaction energy arises only from the resonance energy. On the other hand, the curves where $\beta = 0$ represent the sets of points where the electronic structure is rigorously described by only one bond eigenfunction and in these cases the interaction energy arises only by the term K. In turn, the points where these two types of curves intersect represent points where $T = 0$ and therefore in these points $E = Q$.

6.2 Local Analysis

In this section we discuss how to apply this model to rationalize or predict the main features of a critical point such as its origin, existence and index (20). As previously pointed out, information about the origin can be obtained from an analysis along the reaction coordinate, while information about the existence and index can be obtained by extending the analysis along the other relevant coordinates.

6.2.1 Analysis Along the Reaction Coordinate

This type of analysis provides information about the geometry of the critical point, i.e. the position along the reaction coordinate where the crossing occurs. Such information are derived from the analysis of the behaviour of the two diabatic surfaces, that can be discussed in terms of the expressions [29] and [30], and therefore in terms of coulomb (Q) and exchange (K_{ij}) integrals. A typical behaviour of Q along the reaction coordinate is illustrated in Fig. 2: it can be seen that Q remains almost constant along the reaction coordinate and increases rapidly only for values of the reaction coordinate in the region of the products. Consequently the behaviour of the two diabatic surfaces is dominated by the exchange term K, which has a destabilizing contribution in the reactant diabatic surface and a stabilizing contribution in the product diabatic surface. It follows that the two diabatic surfaces cross at a certain point of the reaction coordinate and such a crossing is related to a critical point of the adiabatic surface. This analysis can be used to compare reaction paths belonging either to the same PES or to different PES. In the comparison of paths belonging to the same PES the coulomb term Q has a similar behaviour, at least for paths with similar steric effects. Thus the different behaviour of the two diabatic surfaces along reaction paths belonging to the same PES can be discussed simply in terms of the exchange contribution K. On the other hand in the comparison of reaction paths belonging to different PES also the coulomb term Q can play an important role.

This analysis also provides information about the magnitude of the activation barrier ΔE, which is the difference between the total energy values computed at the geometry of the crossing (Ec) and of the reactants (Er). We can, infact, decompose ΔE in the following way, making use of Eqs. [33] and [34]:

$$\Delta E = Ec - \bar{E}r, \tag{60}$$

$$= (Qc - Tc) - (Qr - Tr), \tag{61}$$

$$= \Delta Q + Tr - Tc, \tag{62}$$

$$= \Delta Q + |Kr| - |\beta c|, \tag{63}$$

where the subscript c denotes quantities computed at the geometry of the crossing and r quantities computed at the geometry of the reactants. Thus the energy barrier is given as a sum of a coulomb contribution ΔQ and an exchange contribution ($|Kr| - |\beta c|$). For some types of reaction surfaces (i.e. some types of cycloaddition reactions), the coulomb contribution ΔQ is negligible since Q is almost constant, except for reaction paths characterized by large steric effects. Furthermore for a given reaction surface the exchange term of the reactants Kr is constant. Thus in these cases the relative magnitudes of the various barriers are controlled by βc, the resonance energy computed at the geometry of the crossing, which is given by Eq. [40]. On the other hand, in the comparison of the energy barriers associated with reaction paths belonging to different potential energy surfaces, all these terms can be important.

6.2.2 Analysis Along the Other Relevant Coordinates

In a one-dimensional analysis along the reaction coordinate, the crossing between the two diabatic curves is usually associated with a maximum on the adiabatic curve which is assumed to be a transition structure. However a transition structure is a maximum along the reaction coordinate and a minimum along the other coordinates. In other words it is a saddle point with index equal to 1. Therefore a transition structure can not be identified in terms of a two-dimensional analysis along the reaction coordinate, since at this level it is not possible to distinguish between a saddle point with index equal to 1 or larger than 1.

To obtain information about the real nature of the critical point we have to analyze also the behaviour of the total energy along the other relevant coordinates in correspondence of the critical point. A natural choice of coordinates to study the index of critical points is the ensemble of normal coordinates. Let us denote with n the number of such coordinates. Obviously, also the reaction coordinate belongs to this ensemble and therefore the analysis has to be performed along the remaining $n - 1$ coordinates, which can be denoted with q_i ($i = 1, 2 \dots, n - 1$). At the critical point the following expressions hold:

$$\frac{\partial E}{\partial q_i} = \frac{\partial Q}{\partial q_i} - \frac{\partial T}{\partial q_i} = 0 \quad \text{for all values of} \quad i = 1, 2, \dots n - 1, \tag{64}$$

$$\frac{\partial^2 E}{\partial^2 q_i} = \frac{\partial^2 Q}{\partial^2 q_i} - \frac{\partial^2 T}{\partial^2 q_i} \lessgtr 0 \quad \text{for all values of} \quad i = 1, 2 \dots n - 1, \tag{65}$$

where we have used expression [33] for the total energy E.

The analysis of the first derivative expression (see Eq. [64]) provides additional information about the existence of a critical point. In particular we can distinguish two cases:

i)
$$\frac{\partial Q}{\partial q_i} = \frac{\partial T}{\partial q_i} = 0 \quad \text{for all values of i}. \tag{66}$$

In this case a critical point for the adiabatic surface is also a critical point of the Q and T surfaces

ii)
$$\frac{\partial Q}{\partial q_i} = \frac{\partial T}{\partial q_i} \neq 0 \quad \text{for at least one value of i} \ (i = 1, 2 \dots n - 1). \tag{67}$$

In this case a critical point of the adiabatic surface is not a critical point of the Q and T surfaces. In these cases the critical point on the adiabatic surface arises when the slope of the Q surface is equal to the slope of the T surface.

The analysis of the second derivative expression (see Eq. [65]) provides information about the curvature of the surface at the critical point along the various coordinates

q_i. On this basis the curvature along a coordinate q_i can be discussed in terms of the curvature of the Q and T components and is determined by that component that has the largest curvature.

7 An Illustrative Example: The Ethylene + Ethylene Cycloaddition Reaction

For illustrative purposes we discuss here the application of the Diabatic Surface model to the analysis of the PES associated with the cycloaddition reaction of two ethylenes (8). We present here a quantitative analysis where the basic quantities Q, K_R, K_P and K_X are obtained via the effective Hamiltonian treatment from MC-SCF computations. A quantitative analysis is just a Local Analysis with the objective to obtain information about the origin of the various critical points and to rationalize the relative energy and geometry of the critical points.

The critical points found in the MC-SCF study of the ethylene + ethylene PES have the structures shown in Fig. 5. In particular the supra-supra, the *syn* and

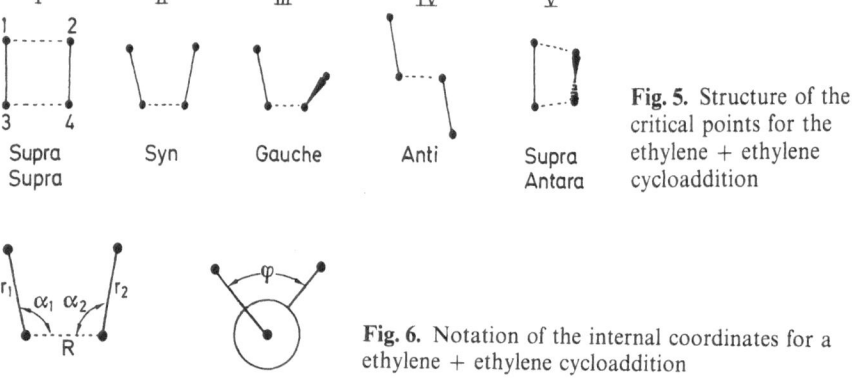

Fig. 5. Structure of the critical points for the ethylene + ethylene cycloaddition

Fig. 6. Notation of the internal coordinates for a ethylene + ethylene cycloaddition

the supra-antara critical points are second order saddle points (i.e. critical points with index 2) while the *gauche* and *anti* critical points are transition states (i.e. critical points with index 1). The notation used for the internal coordinates is shown in Fig. 6.

This analysis can be significantly simplified using the following expressions for the total exchange energy T and the bond exchange energy K:

$$T = (\sqrt{3}/2) \cdot |K_R - K_X| \quad \text{when} \quad |K_R| > |K_P|, \tag{68}$$

$$T = (\sqrt{3}/2) \cdot |K_P - K_X| \quad \text{when} \quad |K_R| < |K_P|, \tag{69}$$

$$T = |K_P - K_X| \quad \text{or} \quad |K_R - K_X| \quad \text{when} \quad |K_P| = |K_R|, \tag{70}$$

$$K = (K_R - K_P). \tag{71}$$

These expressions have been tested numerically for various cycloaddition reactions and have been found to reproduce accurately the values of T and K. We have also found numerically that at the various critical points of the ethylene + ethylene PES the following relation holds:

$$|K_P| > |K_R|. \tag{72}$$

Consequently in this PES the total exchange energy T is described by Eq. [69].

7.1 Analysis Along the Reaction Coordinate

Let us consider first the analysis along the reaction coordinate which is dominated by the interfragment distance R (for the notation of the internal coordinates see Fig. 6). In this case the analysis of the behaviour of the two diabatic surfaces can be discussed in terms of Q and K (Eqs. [29] and [30]). Since Q is almost constant along R (except for values of R in the region of the product where it begins to increase very rapidly), the behaviour of the two diabatic surfaces along R is determined by K. The behaviour of the parameter K can be discussed using Eq. [71] and therefore on the basis of the reactant and product exchange contributions K_R and K_P. Both these terms are negative quantities. At the reactant geometry K_P is zero while K_R has its largest value in absolute magnitude. With the decrease of R, K_P increases in absolute magnitude because the interfragment overlap between the carbon atoms increases, while K_R decreases in absolute magnitude because the intrafragment overlap between the carbon centers decreases. Thus as K increases along R the reactant diabatic curve increases and the product diabatic curve decreases so that at the value of the reaction coordinate where $K_R = K_P$, the diabatic curves cross. This behaviour is illustrated in detail in Fig. 2 for the anti attack and a similar behaviour has been found for all the other types of attack, i.e. supra-supra, *syn, gauche* and supra-antara (see Fig. 7). Therefore in each case the maximum along the reaction coordinate found for each type of attack originates from the crossing of the two diabatic curves.

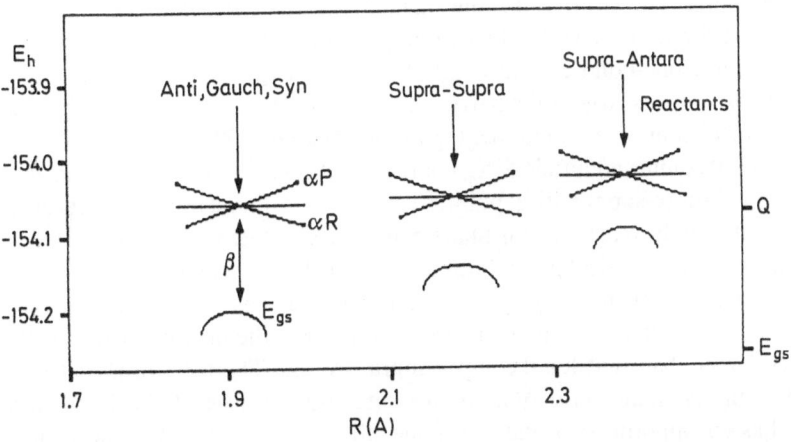

Fig. 7. Analysis along the reaction coordinate for the **anti, gauche, syn,** supra-supra and supra-antara approaches for the ethylene + ethylene cycloaddition reaction

These results are not significantly affected by the computational level. The behaviour of the diabatic surfaces computed at the STO-3G (22) and 4-31G (23) levels is very similar, as shown in Fig. 2 for the anti attack. The only major difference is found in the behaviour of the coulomb integral Q which shows a slight minimum in the transition state region at the STO-3G level, while in this region it is always repulsive at the 4-31G level. This different behaviour is responsible of the fact that the fragmentation barrier of the *anti* intermediate (i.e. the difference between the energies of the *anti* intermediate and *anti* transition state) decreases significantly going from the STO-3G to the 4-31G level, as shown by the behaviour of the ground state adiabatic curve in Fig. 2.

The position of the crossing along the reaction coordinate is determined mainly by the behaviour of K_P: at a given value of R, the value of K_P for the supra-supra rectangular attack is twice as large in absolute magnitude as the K_P values for the *syn*, *gauche* and *anti* attacks, where only one bond is forming. Consequently the reactant diabatic curve is more repulsive and the product diabatic curve more attractive for the supra-supra rectangular attack. Thus one has an early crossing along the supra-supra path, and more advanced crossings (smaller R) for the other three types of attack.

Similarly for the supra-antara attack, one has the formation of two bonds and the crossing occurs early, even earlier than that associated with the supra-supra rectangular attack. This trend is determined by the behaviour of K_R whose absolute value decreases more rapidly than in the other cases. Along this path one ethylenic fragment is more distorted and this effect makes the reactant diabatic more repulsive and the product diabatic more attractive, with a crossing occurring at a larger value of R.

The relative magnitude of the energy barriers ΔE associated with the various reaction paths can be rationalized by the values of the coulomb (ΔQ) and exchange (Tr − Tc) terms, which are listed in Table 1. We shall discuss first the situation of the supra-supra, *syn*, *gauche* and *anti* paths. As expected, ΔQ is small and the barrier along these paths arises almost completely from the exchange term. Since the interaction energy Tr is independent of the reaction path, the trend of the various energy barriers is determined by Tc, the interaction energy computed at the geometry of the critical point. The larger is Tc, the smaller is the energy barrier.

From Table 1 the value of Tc for the two-bond supra-supra critical point is almost one half that for a one-bond critical point. Thus the energy barrier of a two-bond process is almost twice as large as the barrier of a one-bond process. This trend is determined by the behaviour of the component terms of Tc (see Eq. [69]), mainly by K_P. One can see that the absolute magnitude of K_p is larger for the one-bond critical points than for the two-bond supra-supra critical point, because the one-bond critical points occur at a shorter value of the reaction coordinate. Since the absolute value of K_x is slightly larger for the two-bond critical point, both terms contribute to determine the trend of the energy barriers, with K_P playing the dominant role. The comparison with the results obtained for the supra-antara process shows that the energy barrier in this case is even larger than that for the supra-supra process. The term responsible for this trend is the coulomb term ΔQ. In this case ΔQ, because of the large steric repulsion, has an opposite sign and is significantly larger than in the supra-supra critical point (see Table 1). The exchange contribution on the other hand is very similar to that found for the supra-supra barrier.

Table 1. Values of the total energy (E), of the coulomb integral (Q), of the interaction energy (T) and of the exchange contributions (KX, KR, KP) computed at the STO-3G level for the various critical points (see Fig. 5) and reactants

Structures	E^a	Q^a	KX^a	KR^a	KP^a	T^a	ΔQ^b	$(T_R - T_C)^b$	ΔE^b
Ethylene + Ethylene									
I	−154.1273	−154.0728	−.0323	−.0859	−.0877	.0545	− 8.9	75.6	66.7
II	−154.1661	−154.0694	−.0223	−.1182	−.1198	.0967	− 6.8	49.1	42.3
III	−154.1730	−154.0722	−.0203	−.1227	−.1195	.1008	− 8.5	46.5	38.0
IV	−154.1751	−154.0733	−.0187	−.1229	−.1181	.1019	− 9.2	45.9	36.7
V	−154.0930	−154.0325	−.0014	−.0582	−.0650	.0605	+16.4	71.8	88.2
Reactants	−154.2336	−154.0586	.0000	−.1750	.0000	.1750	—	—	0.0

Table 2. Values of the total energy (E), of the coulomb integral (Q), of the interaction energy (T) and of the exchange contributions (KX, KR, KP) computed at the 4-31G level for the critical points and reactants of the ethylene + ethylene reaction

Structures	E^a	Q^a	K_x^a	K_R^a	K_P^a	T^a	ΔQ^b	$(T_R - T_C)^b$	ΔE^b
I	−155.7829	−155.7368	−.0417	−.0850	−.0900	.0461	− 4.0	81.2	77.2
II	−155.8096	−155.7025	−.0404	−.1217	−.1617	.1071	+17.6	42.9	60.4
III	−155.8203	−155.7009	−.0366	−.1307	−.1709	.1194	+18.6	35.2	53.7
IV	−155.8240	−155.7021	−.0333	−.1313	−.1698	.1219	+17.8	33.6	51.4
V	−155.7656	−155.6973	−.0024	−.0644	−.0757	.0683	+20.8	67.3	88.0
Reactants	−155.9059	−155.7305	.0000	−.1755	.0000	.1755	—	—	—

a Values in a.u.
b Values in Kcal/mol

Also in this case the results are not significantly affected by the computational level. As shown in Table 2 the trend of the various energy barriers at the 4–31G level is very similar to that found at the STO-3G level, with again the exchange term largely dominant. The only significant difference is found in the coulomb term for the diradicaloid critical points, that becomes significantly larger with also a change in sign: however these differences are not large enough to modify the trend of the -barriers.

7.2 Analysis Along the Other Coordinates

For illustrative purposes we discuss here only the analysis along the motion described by the angle α (see Fig. 6), which varies from a value of 90 degrees for structure I (the supra-supra second order saddle point) to a value of about 117 degrees for structure II (the syn second order saddle point). The behaviour of Q, −T and E along this coordinate is shown in Fig. 8 and can be rationalized in very simple terms.

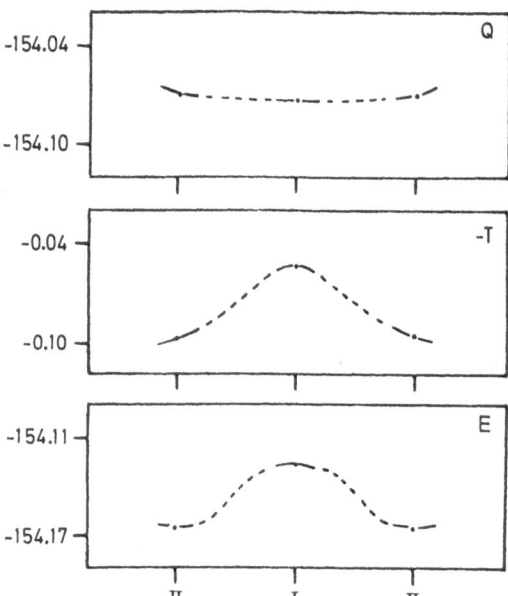

Fig. 8. Behaviour of Q, −T and E along the coordinate α

In fact Q is almost constant along this transformation and begins to increase only for α > 117 degrees when the steric repulsions between the adjacent methylene groups begin to be significant. On the other hand −T decreases in the I → II transformation following the trend of the product exchange K_P, which is larger, in absolute magnitude, for a situation involving a shorter and a longer C–C bond rather than two equal bonds of intermediate length.

The behaviour of the total energy is dominated by T for values of α < 117 degrees and by Q for values of α > 117 degrees. Consequently, along this mode structure I is a maximum and structure II is a minimum. Also the conditions of existence of the two critical points are different. The critical point I arises from a situation where condition [66] is satisfied, i.e. a situation where the critical point of E is also a critical point of the Q and $-T$ curves, while the critical point II arises from a situation where condition [67] is satisfied, i.e. a situation where there are not critical points in the Q and T curves, but they have the same slope. Thus the critical point II arises because we pass from a situation dominated by T to a situation deminated by Q. Furthermore this analysis shows also that condition [66] can be satisfied only when the coordinate under examination is a non-totally symmetric mode for the critical point and this situation occurs here only for structure I.

As already pointed out the index of the two critical points along this mode is different and arises from two different possibilities. In the case of structure I, the curvatures of Q and $-T$ are opposite, with $-T$ being the dominant factor (the curvature is negative giving rise to a second order saddle point). In the case of structure II the curvature (positive) of Q and $-T$ is the same, giving rise to a first order saddle point.

To obtain information about the index of the various critical points we have to perform this analysis for each critical point along all relevant coordinates. As previously pointed out, a proper choice of coordinates to study the nature of critical points is the ensemble of normal coordinates, which are obtained from analytical computation of the complete hessian matrix at the MC-SCF level. The analysis is not performed along all the normal coordinates, but only along those chemically relevant, which are a small subset of coordinates that describe the motions of the atoms directly involved in the bond(s) making/breaking process.

8 References

1. Bernardi F, Robb MA, Mol. Phys. 1983, 48: 1345–1355
2. Bernardi F, Robb MA, J. Am. Chem. Soc. 1984, 106: 54–58
3. Bernardi F, Paleolog SAHD, McDouall JJW, Robb MA, J. Mol. Struct. Theochem, 1986, 138: 23–38
4. Bernardi F, Olivucci M, Robb MA, Tonachini G, J. Am. Chem. Soc. 1986, 108: 1408–1415
5. Bernardi F, Robb MA, Adv. Chem. Phys. 1987, 67: 155–248
6. Bernardi F, Olivucci M, McDouall JJW, Robb MA, J. Chem. Phys. 1988, 89: 6365–6375
7. Bernardi F, Robb MA (1989) In: Bertran J, Csismadia IG (eds.) New theoretical concepts for understanding organic reactions, NATO ASI Series
8. Bernardi F, Olivucci M, Robb MA, Research Chem. Inter., 1989, 12: 217–249
9. Olivucci M, Thesis for "Dottorato di ricerca in Scienze Chimiche", Bologna (Italy), 1988
10. Epiotis ND, Cherry WR, Shaik S, Yates R, Bernardi F, Top. Curr. Chem. 1977, 70
11. Born M, Oppenheimer JR, Ann. Physik 1927, 84: 457
12. Pilar LF (1968), Elementary quantum chemistry, MacGraw-Hill: New York
13. McWeeny R, Sutcliffe B (1969), Methods of molecular quantum mechanics; Academic, New York
14. Daudel R, Leroy G, Peeters D, Sana M, Quantum Chemistry; Wiley: New York, 1983
15. Woodward RB, Hoffmann R, Angew. Chem. Int. Ed. Engl. 1969, 8: 781–853
16. (a) Epiotis ND, Lect. Notes Chem., 1982, 29. (b) Epiotis ND, Lect. Notes Chem., 1983, 34
17. (a) Pross A, Shaik SS, Acc. Chem. Res. 1983, 16: 363. (b) Shaik SS, Prog. Phys. Org. Chem. 1985, 15: 197. (c) Pross A, Adv. Phys. Org. Chem. 1985, 21: 93

18. Eyring H, Walter J, Kimball G, Quantum Chemistry; Wiley: New York, 1944
19. Durand P, Malrieu JP, Adv. Chem. Phys. 1987, 67: 321−412
20. Mezey PG, Potential Energy Hypersurfaces; Elsevier: Amsterdam, 1987
21. Bernardi F, Bottoni A, Robb MA, Schlegel HB, Tonachini G, J. Am. Chem. Soc. 1985, 107: 2260−2264
22. Hebre WJ, Stewart RF, Pople JA, J. Chem. Phys. 1969, 51: 2657−2664
23. Ditchfield R, Hebre WJ, Pople, JA, J. Chem. Phys. 1971, 54: 724−728

Theoretical Models of Chemical Bonding

Ed. **Z. B. Maksić**

Part 1

Atomic Hypothesis and the Concept of Molecular Structure

1990. XXVIII, 324 pp. 40 figs. 51 tabs. Hardcover DM 350,– ISBN 3-540-51578-X

Subscription price (valid only for subscribers to the complete work):
Hardcover DM 280,–

Contents: *B. T. Sutcliffe:* The Concept of Molecular Structure. – *O. E. Polansky:* Topology and Properties of Molecules. – *J. P. Dahl:* Symmetry in Molecules. – *L. D. Barron:* Chirality of Molecular Structures – Basic Principles and their Consequences. – *J. E. Boggs:* Interplay of Experiment and Theory in Determining Molecular Geometries. A. The Experiments. – *J. E. Boggs:* Interplay of Experiment and Theory in Determining Molecular Geometries. B. Theoretical Methods. – *A. Y. Meyer:* Molecular Mechanics alias Mass Points and Elastic Springs Model of Molecules. – *K. B. Wiberg:* Atoms in Molecular Environments. – *Z. B. Maksić:* The Modelling of Molecules as Collections of Modified Atoms.

Part 2

The Concept of the Chemical Bond

1990. X, 643 pp. 181 figs. 88 tabs. Hardcover DM 450,– ISBN 3-540-51553-4

Subscription price (valid only for subscribers to the complete work):
Hardcover DM 280,–

Contents: *W. Kutzelnigg:* The Physical Origin of the Chemical Bond. – *R. G. Pearson:* Absolute Electronegativity and Absolute Hardness. – *K. Jug, M. S. Gopinathan:* Valence in Molecular Orbital Theory. – *J.-P. Malrieu:* The Magnetic Description of Conjugated Hydrocarbons. – *Z. B. Maksić:* Directional Properties of Covalent Bonding in Molecules. – *P. R. Surján:* The Two-Electron Bond as a Molecular Building Block. – *C. Edmiston:* Interpretation of Molecular Behaviour by Localized Molecular Orbitals (LMOs). – *W. L. Luken:* Properties of the Fermi Hole and Electronic Localization. – *P. J. Kuntz:* The Diatomics-in-Molecules Method and the Chemical Bond. – *P. Fulde:* Calculation of Electron Correlations by Using Local Operators. – *M. Grodzicki:* The Concept of the Chemical Bond in Solids. – *E. Kraka, D. Cremer:* Chemical Implication of Local Features of the Electron Density Distribution. – *A. A. Low,* Deformation Densities and Chemical Bonding in Transition Metal Complexes. – *W. H. E. Schwarz:* Fundamentals of Relativistic Effects in Chemistry.

Springer-Verlag
Berlin
Heidelberg
New York
London
Paris
Tokyo
Hong Kong
Barcelona
Budapest

Theoretical Models of Chemical Bonding

Ed. Z. B. Maksić

Part 3

Molecular Spectroscopy, Electronic Structure and Intramolecular Interactions

1991. Approx. 660 pp. 172 figs. 126 tabs. Hardcover DM 350.- ISBN 3-540-52252-2

Subscription price (valid only for subscribers to the complete work): Hardcover DM 280,-

Contents: *J. E. Boggs:* Nuclear Vibrations and Force Constants. - *H. P. Figeys, P. Geerlings:* Some Aspects of the Quantum-Chemical Interpretation of Integrated Intensities of Infrared Absorption Bands. - *S. P. McGlynn, K. Wittel, L. Klasinc:* The Orbital Concept as a Foundation for Photoelectron Spectroscopy. - *E. Honegger, E. Heilbronner:* The Equivalent Bond Orbital Model and the Interpretation of PE Spectra. - *M. Eckert-Maksić:* Through-space and Through-bond Interactions as Mirrored in Photoelectron Spectra. - *K. Ohno, Y. Harada:* Penning Ionization - The Outer Shape of Molecules. - *K. Jug, Z. B. Maksić:* The Meaning and Distribution of Atomic Charges in Molecules. - *Z. B. Maksić:* Electron Spectroscopy for Chemical Analysis (ESCA) - Basic Features and Their Model Description. - *K. T. Leung:* Experimental Momentum-Space Chemistry by (e, 2e) Spectroscopy. - *J. Kowalewski, A. Laaksonen:* Theoretical Parameters on NMR Spectroscopy. - *D. Feller, E. R. Davidson:* Theoretical Approaches to ESR Spectroscopy. - *C. J. Jameson:* Rovibrational Averaging of Molecular Electronic Properties. - *M. Klessinger, T. Pötter:* Properties of Molecules in Excited States. - *J. Tomasi, G. Alagona, R. Bonaccorsi, C. Ghio, R. Cammi:* Semiclassical Interpretation of Intramolecular Interactions. - *F. Bernardi, M. Olivucci, M. A. Robb:* The Analysis of Potential Energy Surfaces in Terms of the Diabatic Surface Model.

Part 4

Theoretical Treatment of Large Molecules and Their Interactions

1991. Approx. 470 pp. 104 figs. 52 tabs. Hardcover DM 350,- ISBN 3-540-52253-0

Subscription price (valid only for subscribers to the complete work): Hardcover DM 280,-

Contents: *J. G. Angyán, G. Náray-Szabó:* Chemical Fragmentation Approach to the Quantum Chemical Description of Extended Systems. - *Ch. L. Brooks:* Semiclassical Methods for Large Molecules of Biological Importance. - *G. M. Maggiora, J. D. Petke, R. E. Christoffersen:* Electronic Excited States of Biomolecular Systems: Ab Initio FSGO-based Quantum Mechanical Methods with Applications to Photosynthetic and Related Systems. - *A. J. Stone:* Classical Electrostatics in Molecular Interactions. - *V. Magnasco, R. McWeeny:* Weak Interactions Between Molecules and Their Physical Interpretation. - *S. Scheiner:* An Initio Studies of Hydrogen Bonding. - *J. Tomasi, R. Bonaccorsi, R. Cammi:* The Extramolecular Electrostatic Potential. An Indicator of the Chemical Reactivity. - *S. Shaik, P. C. Hiberty:* Curve Crossing Diagrams as General Models for Chemical Reactivity and Structure. - *R. A. van Santen, E. J. Baerends:* Orbᵇ ᵃ ᵇ and Chemical Reactivity of Metal Particles and Metal Surfaces. - *A. van der Avoird:* Intermolecular Forces and the Properties of Molecular Solids. - *O. Tapia:* Theoretical Evaluation of Solvent Effects.

Springer-Verlag
Berlin
Heidelberg
New York
London
Paris
Tokyo
Hong Kong
Barcelona
Budapest